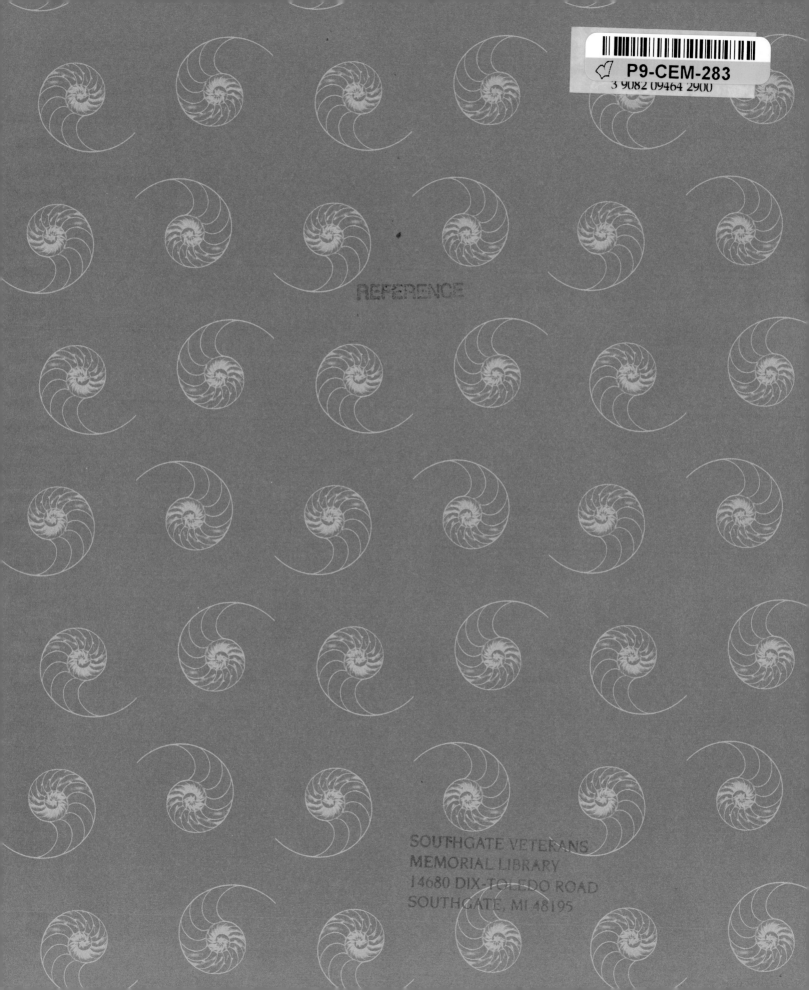

THE ENCYCLOPEDIA OF
ANIMALS
A Complete Visual Guide

THE ENCYCLOPEDIA OF
ANIMALS
A Complete Visual Guide

University of California Press
Berkeley   Los Angeles

University of California Press
Berkeley and Los Angeles, California

Published by arrangement with
Weldon Owen Pty Ltd
59 Victoria Street, McMahons Point
Sydney NSW 2060, Australia
Copyright © 2004 Weldon Owen Inc.

Chief Executive Officer John Owen
President Terry Newell
Publisher Sheena Coupe
Creative Director Sue Burk
Vice President International Sales Stuart Laurence
Administrator International Sales Kristine Ravn

Project Editors Stephanie Goodwin, Angela Handley
Designers Clare Forte, Hilda Mendham, Heather Menzies,
Helen Perks, Sue Rawkins, Jacqueline Richards, Karen Robertson
Jacket Design John Bull
Picture Research Annette Crueger
Copy Editors Janine Flew, Lynn Humphries
Editorial Administrator Jessica Cox

Text Jenni Bruce, Karen McGhee, Luba Vangelova, Richard Vogt

Species Gallery Illustrations MagicGroup s.r.o. (Czech Republic) —
www.magicgroup.cz
Pavel Dvorský, Eva Göndörová, Petr Hloušek, Pavla Hochmanová,
Jan Hošek, Jaromír a Libuše Knotkovi, Milada Kudrnová, Petr Liška,
Jan Maget, Vlasta Matoušová, Jiří Moravec, Pavel Procházka, Petr Rob,
Přemysl Vranovský, Lenka Vybíralová
Feature Illustrations Guy Troughton
Maps Andrew Davies Creative Communication and Map Illustrations
Information Graphics Andrew Davies Creative Communication

Editorial Coordinator Jennifer Losco
Production Manager Caroline Webber
Production Coordinator James Blackman

Cataloguing-in-Publication data is on file with the Library of Congress

ISBN 0-520-24406-0

13 12 11 10 09 08 07 06 05 04
10 9 8 7 6 5 4 3 2 1

Color reproduction by Chroma Graphics (Overseas) Pte Ltd
Printed by Kyodo Printing Co. (S'pore) Pte Ltd
Printed in Singapore

A Weldon Owen Production

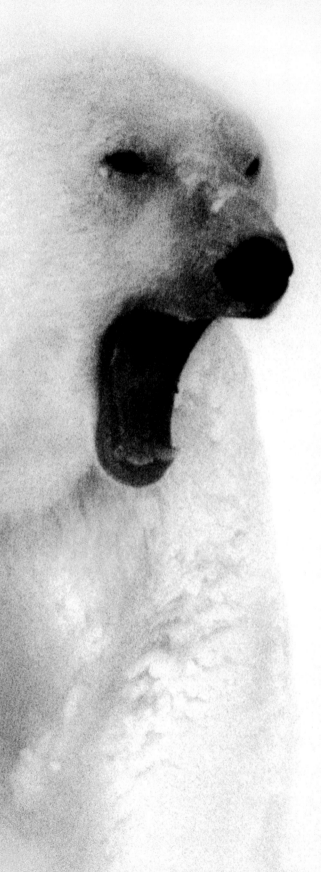

CONSULTANTS

Dr Fred Cooke
President-Elect
American Ornithologists' Union
Norfolk, UK

Dr Hugh Dingle
Professor Emeritus
University of California
Davis, USA

Dr Stephen Hutchinson
Visiting Senior Fellow
Southampton Oceanography Centre
Southampton, UK

Dr George McKay
Consultant in Conservation Biology
Sydney, Australia

Dr Richard Schodde
Fellow
Australian National Wildlife Collection
CSIRO
Canberra, Australia

Dr Noel Tait
Consultant in Invertebrate Biology
Sydney, Australia

Dr Richard Vogt
Curator of Herpetology and Professor
National Institute for Amazon Research
Manaus, Amazonas, Brazil

CONTENTS

FOREWORD

Like any excellent natural history reference, *The Encyclopedia of Animals* will eventually be revised for two reasons, one of them admirable and the other profoundly sad. On the positive side, zoologists are still finding undescribed species, as well as discovering many new and surprising things about animals. Twenty-five years ago we knew of fewer than three thousand species of frogs and now the count stands at almost five thousand; only a decade ago parental care was regarded as exceedingly rare among snakes, but recently field biologists have determined that most female pitvipers attend their young for about a week after birth. Better ways to classify organisms are working their way into popular writing too, so we now correctly discuss crocodilians as more closely related to birds than they are to lizards—and thereby emphasize that nest building and vocal communication by alligators, eagles, and extinct dinosaurs likely reflect their shared heritage as archosaurs.

The Encyclopedia of Animals impressively conveys the real texture of biodiversity, that marvelously great variation among organisms themselves. Written by a team of experts, this book's coverage emphasizes mammals and birds, the two most popular groups of vertebrates, but it also treats other kinds of animals in timely detail. Here readers can push beyond familiar facts and trendy pandas to become acquainted with big-brained elephant fishes, gastric brooding frogs, the bizarre solenodon, and other improbable creatures; here they will meet an oft-slighted but fascinating group of amphibians, the limbless caecilians, and read about diverse, previously obscure invertebrates. Throughout the text beautiful artwork accurately depicts hundreds of species in life, while numerous outstanding photographs portray their special relationships with particular habitats.

The other, disturbing reason for one day updating this wonderful volume will be that the status of some species will change from threatened to extinct, and many others will become endangered. Environmental conservation is among our most important and formidable responsibilities, one that must start with education. The more we understand how richly diverse and how inspiringly adapted organisms are for their environments, the more we will care about them. Thus, to whatever extent humanity increasingly values animals in practical and esthetic terms, we will more readily make difficult personal and political choices in their favor. I hope that *The Encyclopedia of Animals* will encourage you to get out in nature, to appreciate animals all the more for what you have learned, and then to do what you can to conserve them.

HARRY W. GREENE
Professor of Ecology and Evolutionary Biology at Cornell University

HOW TO USE THIS BOOK

The first section of this book provides an overview of animal life, animals through the ages, how animals are classified, their biology and behavior, habitats and adaptations, and animals in danger. Then follow six taxonomic sections, namely Mammals, Birds, Reptiles, Amphibians, Fishes, and Invertebrates. Each of the sections comprises an introductory feature discussing the group in general, then is broken down into chapters devoted to particular subgroups. The sections do not necessarily provide a similar amount of coverage; for example, there is a discrepancy between Mammals and Invertebrates, the latter profiling larger taxonomic groups to give sufficient coverage to Invertebrates as a whole. The book concludes with a comprehensive glossary and index.

HABITAT ICONS

The 19 habitat icons below indicate at a glance the various habitats in which a species or group can be found. It should be noted that the icons are used in the same order throughout the book, rather than in their order of significance. A more detailed profile of each habitat can be found on pages 40–53.

- Tropical rain forest
- Tropical monsoon forest
- Temperate forest
- Coniferous forest
- Moorlands and heath
- Open habitat, including savanna, grassland, fields, pampas, and steppes
- Desert and semidesert
- Mountains and highlands
- Tundra
- Polar regions
- Seas and oceans
- Coral reefs
- Mangrove swamps
- Coastal areas, including beaches, oceanic cliffs, sand dunes, intertidal rock pools, and/or coastal waters (as applicable to group)
- Rivers and streams, including river and stream banks
- Wetlands, including swamps, marshes, fens, floodplains, deltas, and bogs
- Lakes and ponds
- Urban areas
- Parasitic

Section and chapter
This indicates the group of animals under discussion.

Classification box
This indicates the taxonomic groups to which the animals belong in the animal kingdom.

Detailed diagram
Where appropriate, diagrams are included to illustrate points about anatomy or adaptation.

SEALS AND SEA LIONS

CLASS	Mammalia
ORDER	Carnivora
FAMILIES	3
GENERA	21
SPECIES	36

With flexible, torpedo-shaped bodies, limbs modified to become flippers, and insulating layers of blubber and hair, seals, sea lions, and walruses are superbly adapted to a life in water. They have not, however, completely severed their link with land and must return to shore to breed. Collectively known as pinnipeds, these marine mammals were once placed in their own order, but are now considered to be part of Carnivora. Most feed on fish, squid, and crustaceans, but some also eat penguins and carrion and may attack the pups of other seal species. They can dive to great depths in search of prey, with the elephant seal able to stay submerged for up to 2 hours at a time.

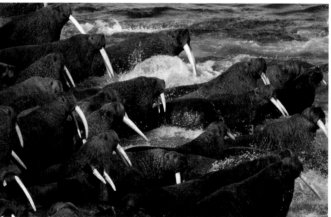

Group global distribution
A map shows the worldwide distribution of the group being profiled, followed by text that discusses the distribution of particular groups in more detail.

Cold-water creatures Although monk seals are found in warmer waters, most seals, sea lions, and walruses are restricted to the colder, highly productive seas of the world's polar and temperate regions. The fossil record shows that the three families all originated in the North Pacific. They are now most abundant in the North Pacific, North Atlantic, and Southern oceans.

Communal living Most pinnipeds are gregarious animals and tend to live in large colonies. Walrus herds can number in the hundreds or even thousands and may be single sex or mixed, with both body and tusk size determining rank.

THREE GROUPS
There are three pinniped families. The Phocidae are known as the true seals. They swim mainly with strokes of their hind flippers, which cannot bend forward to act as feet, making their movement on land particularly ungainly. Although their hearing, especially under water, is good, true seals lack external ears.

Sea lions and fur seals belong to the family Otariidae. These "eared seals" have small external ears. They rely mostly on their front flippers for swimming, and can bend their hind flippers forward when on land, allowing them to walk "four-footed" and sit in a semi-upright position.

The third family, Odobenidae, contains a single species, the walrus, instantly recognizable by the long canine teeth that form tusks on both sexes. Like true seals, walruses use their hind flippers for swimming and lack external ears. Like eared seals, however, walruses can bend their hind flippers forward.

BRINGING UP BABY
All pinnipeds return to land or ice to give birth and mate. Mating takes place just days after the usually single pup is born, but the fertilized egg does not become implanted in the uterus for months. This delayed implantation allows birthing, nursing, and mating to occur in one season so that the animals live on land, where they are most vulnerable, only once a year. Pups are dependent for varying lengths of time. Harp seals (right), for example, are nursed for only 12 days or so, while walruses stay with their mother for 2 years.

Insulating layers Pinnipeds have a thick layer of blubber that provides insulation, buoyancy, and a fat store. For further protection, all but the walrus have hairy bodies, and fur seals have dense secondary hairs that form a waterproof barrier.

guard hair
secondary hair
blubber
sebaceous gland

CONSERVATION WATCH

The commercial sealing operations that began in the 16th century had a devastating effect on pinniped populations. Of the 36 species of pinnipeds, 36% are listed on the IUCN Red List, as follows:

- 2 Extinct
- 1 Critically endangered
- 2 Endangered
- 7 Vulnerable
- 1 Near threatened

Feature box
This support text describes in detail a facet of the species' behavior or biology, and is accompanied by relevant photographs, illustrations, or diagrams.

Conservation watch box
This provides information about the status of a particular species or group of animals, according to the IUCN Red List of Threatened Species. These boxes may also outline factors that threaten the animal's survival.

CONSERVATION INFORMATION

Within the fact files, each profiled species is allocated a conservation status, using IUCN and other conservation categories, as follows:

✝ Indicates that a species is listed under the following categories:

Extinct (IUCN) It is beyond reasonable doubt that the last individual of a given species has died.

Extinct in the wild (IUCN) Only known to survive in captivity or as a naturalized population outside its former range.

⚡ Indicates that a species is listed under the following categories:

Critically endangered (IUCN) Facing a very high and immediate risk of extinction in the wild.

Endangered (IUCN) Facing a very high risk of extinction in the wild in the near future.

The following categories are also used:
Vulnerable (IUCN) Facing a high risk of extinction in the wild in the foreseeable future.

Near threatened (IUCN) Likely to qualify for one of the above categories in the near future.

Conservation dependent (IUCN) Dependent upon species- or habitat-specific

conservation programs to keep it out of one of the above threatened categories.

Data deficient (IUCN) Inadequate information available to make an assessment of its risk.

Not known Not evaluated or little studied.

Common Widespread and abundant.

Locally common Widespread and abundant within its range.

Uncommon Occurs widely in low numbers in preferred habitat(s).

Rare Occurs in only some of preferred habitat or in small restricted areas.

FACT FILE STATISTICS

Important or interesting facts about that species or group, using the following icons and information. All measurements given are maximums.

Length
- 🐾 **Mammals:** head and body
- ⚓ **Birds:** tip of bill to tip of tail
- 🦎 **Reptiles:** snakes and lizards: snout to vent; other reptiles: head and body including tail
- 🐢 **Turtles:** length of carapace
- 🐸 **Amphibians:** head and body, including tail
- 🐟 **Fishes:** head and body, including tail

Height
- 🐘 **Mammals:** shoulder height
- 🐦 **Birds:** head and body height

Tail: Mammals
- 🐁 Length of tail

Wingspan: Birds
- ➤ From tip of one wing to tip of the other

Weight/Mass
- ⚖ Body weight

Social Unit: Mammal
- 🔹 Solitary
- 🔹🔹 Pair
- 🔹🔹🔹 Small to large group
- 🔹🔹 Varies between the above

Plumage: Birds
- ∥ Sexes alike
- ∥ Sexes differ

Reproduction: Birds and Reptiles
- ● Number of eggs

Migration: Birds
- ↻ Migrant
- ↶ Partial migrant
- ⊘ Sedentary
- ∼ Nomadic

Habit: Reptiles and Amphibians
- ◔ Terrestrial
- ◑ Aquatic
- ◓ Burrowing
- ◉ Arboreal
- ◐ Varies between the above

Breeding season: Amphibians
- ⇌ When breeding occurs, e.g. spring

Breeding: Reptiles and Fishes
- ✎ Viviparous (producing live young)
- ○ Oviparous (producing eggs that develop outside the maternal body)
- ◎ Ovoviviparous (producing eggs that develop within the maternal body)

Sex: Reptiles and Fishes
- ♀♂ Reptiles: indicates whether a species is temperature sex determined (TSD) or genetically sex determined (GSD); Fishes: indicates separate male and female, hermaphrodite, or sequential hermaphrodite

Number of genera and species: Birds, Reptiles, and Invertebrates
The number of genera and species in the relevant taxonomic group

---

SEALS AND SEA LIONS **MAMMALS** 143

New Zealand fur seal
Arctocephalus forsteri

Length of male up to 7¼ ft (2.2 m), female up to 5½ ft (1.7 m)

South American sea lion
Otaria byronia

Males can be up to three times larger than females

Pronounced mane on male

South African fur seal
Arctocephalus pusillus

Length of male up to 7 ft (2.1 m), female up to 5 ft (1.5 m)

Northern fur seal
Callorhinus ursinus

Male has massive, maned neck

Length of male up to 8 ft (2.5 m), female up to 6 ft (1.8 m)

Californian sea lion
Zalophus californianus

Short stubble on black flippers

Largest of the eared seals

Steller's sea lion
Eumetopias jubatus

FACT FILE

New Zealand fur seal In late spring, male fur seals establish territories on rocky shorelines, where they are joined by females for breeding. After the pups are born, the females visit the ocean to forage but the males stay put until the end of the breeding season.
- ⚖ Male up to 795 lb (360 kg), female up to 245 lb (110 kg)
- 🔹🔹🔹 Harem
- ⚡ Common

S.W. Australia to New Zealand

Northern fur seal These seals migrate south in winter, returning north to breed in spring. Some travel more than 6,000 miles (10,000 km) per year.
- ⚖ Male up to 605 lb (275 kg), female up to 110 lb (50 kg)
- 🔹🔹🔹 Harem
- ⚡ Vulnerable

North Pacific, Bering Sea

Californian sea lion The pinniped most often used in animal acts, this vocal, gregarious species sticks close to shore and frequently hauls itself onto land or structures such as jetties and piers.
- ⚖ Male up to 880 lb (400 kg), female up to 265 lb (120 kg)
- 🔹🔹🔹 Harem
- ⚡ Common, increasing

Coastal W. North America

TERRITORIAL MALES
Eared seals tend to be highly social and will gather in large numbers during the breeding season. Males defend their patch of shore and harem of females against other males, using aggressive postures and barks before resorting to actual fighting.

Fact file
This profiles one or more of the illustrated species or groups, with information about its size, appearance, habitat, range, reproduction, migratory habits, behavior, calls, or regional variations.

Distribution map
This shows the species' or group's range (and former range, where appropriate). If distributed throughout the world, a world map is shown; if regional distribution, a map of that area only is given.

Habitat icons
The icons indicate the various habitats in which the profiled animal(s) can be found, for example coral reefs or rain forest. The full list can be found on the opposite page.

Panel feature box
Illustrations and text provide information about an interesting aspect of a species or group of species.

Conservation symbols
A red cross above the species' name indicates that it is extinct or extinct in the wild; a red flash indicates that it is critically endangered or endangered according to the IUCN Red List.

Name labels
Labels indicate the common and scientific names, as well as the species' family, order, or class, where appropriate.

Labels
The labels highlight distinguishing aspects or characteristics of the species, such as color variations, behavior, habitat, size, and anatomical features.

ANIMALS

THE ANIMAL KINGDOM

Animalia is one of five kingdoms into which biologists usually divide the living world. Monera includes bacteria and blue-green algae; Protista contains mostly large, unicellular organisms such as amoeba and paramecium, which were once considered animals but are now assigned to a kingdom of their own; molds, mildews, and mushrooms fall within the kingdom Fungi; and, as its name suggests, Plantae contains the plants. With more than 1 million of the 1.75 million living species currently described on Earth, kingdom Animalia is by far the largest and most diverse. It contains all the organisms most of us easily recognize as animals, as well as some species whose status would confuse many non-scientists. The overwhelming majority of animals are invertebrates—animals without a backbone. Among those, the insects dominate in terms of both individual numbers and species diversity. It is, however, the vertebrates with which most of us are familiar—fishes, amphibians, lizards, birds, and mammals—the group to which our own species is most closely affiliated in the evolutionary sense.

Long-distance travelers
A delicate, frail appearance can often belie great endurance. Moths and dragonflies migrate, but the star travelers in the insect world are butterflies. The American painted lady (*Vanessa virginiensis*) (right) heads northward in spring from Mexico and the southern United States, traveling more than 1,500 miles (2,400 km).

DEFINING FEATURES

The kingdom Animalia is also often referred to as Metazoa, a term that indicates the multicellular nature of all its members. As in plants, the tissues of animals are always made up of eukaryotic cells. In animals, however, these lack cell walls. Instead, animal cells are held in place by an extracellular matrix that inevitably contains collagen and provides a mostly flexible framework within which the cells are organized.

Another key feature of all animals is that they are heterotrophs. Unlike plants, which are termed autotrophs, they cannot produce their own food and so, instead, must consume other organisms, either directly or indirectly, to get their nourishment. This has helped drive the extraordinary biodiversity seen among animals today and a vast range of ways to track, capture and consume food has evolved within the group.

Food requirements have also had an impact on animal body plans; most have some sort of centralized digestive system into which they take food and break it down.

The necessity for seeking out, or being near, food sources has also led to capabilities for movement. And, although, it is not always clear-cut, mobility separates most plants from most animals. Questions of mobility have caused confusion in the past among scientists about the status of some groups, in particular the sponges, the animals that have the longest evolutionary history. These are the only living animals that lack organization beyond the cell level

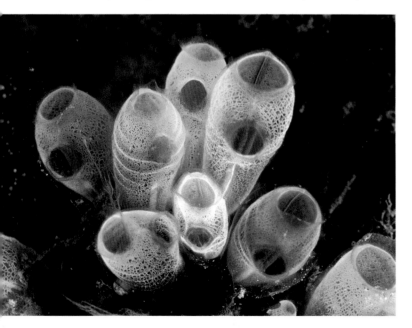

Glued to the spot Some animals, such as sea squirts (above), are not capable of moving around. The tadpole larva of sea squirts does not feed; its sole purpose is to find a suitable place to settle. Once found, the larva uses the adhesive papilllae on the head to secure itself to a hard surface. It remains there from that point on, evolving into its simple, filter-feeding adult form.

Food for warmth The brown bear (*Ursus arctos*), one of the largest ursids, feeds on tubers, berries, fishes, and carrion. Before winter, the northern temperate species stores up body fat and retreats into a den. The bears go into a winter dormancy, a state distinguished from true hibernation because their body temperature does not drop, and live entirely off their body fat.

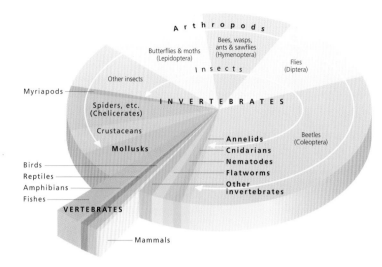

Arthropods

Butterflies & moths
(Lepidoptera)

Bees, wasps,
ants & sawflies
(Hymenoptera)

Flies
(Diptera)

Insects

Other insects

Myriapods

INVERTEBRATES

Spiders, etc.
(Chelicerates)

Crustaceans

Mollusks

Annelids

Cnidarians

Nematodes

Flatworms

**Other
invertebrates**

Beetles
(Coleoptera)

Birds

Reptiles

Amphibians

Fishes

VERTEBRATES

Mammals

Counting the species

Scientists have described in the region of 1.75 million species of life-forms on Earth, but that is thought to be only a small proportion of the true figure. Estimates place this at between 5 and 100 million species. Even the figures for the number of known species are hard to pin down because new species are being discovered all the time. Vertebrates (see chart at left) are the best-described group but make up only about 5 percent of animal species. There are about 1 million known species of insects but the true number could be more in the region of 30 million.

so that they have no organs or tissues, a unique feature that has added to past confusions about their status. Their cells are, however, capable of small movements, most have a free-swimming larval stage, and today there is no doubt they are animals and not plants.

The necessity and ability for independent movement has also helped impel in most animals the development of a nervous system with associated sensory equipment to coordinate and guide movements. Most also have body tissues, such as muscle, that facilitate movement.

Sexual reproduction is another unifying feature. Almost all animal species create offspring at some stage in their life-cycle by sexual reproduction; some have the capabilities to reproduce by asexual methods.

Aerial agility The range of the golden eagle (*Aquila chrysaetos*) extends over mountains and lowlands across the Northern Hemisphere. Most of its mammalian prey is taken on the ground from a low flight, but it is so agile and fast in the air that it will also take birds in flight. Some golden eagle pairs hunt together.

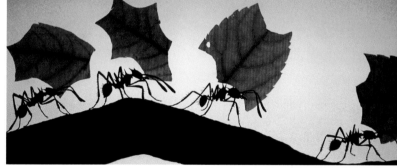

Teamwork Ants are members of a highly social family of insects. The collaborative efforts of leaf-cutter ants (*Atta cephalotes*) ensure there is a constant source of food. The ants bite off pieces of leaves, carry them underground, and then live on the fungi that subsequently forms on the plant matter and breaks it down.

Building blocks All life on Earth is made of cells. The first were prokaryotes—simple assemblages of genetic material within a cell wall. This structure is seen in bacteria. The development from prokaryotes of the mostly larger, far more complex, eukaryotic cells underpinned the evolution of all plants and animals. The genetic material of the eukaryotes is contained within a membrane-bound nucleus and separate organelles that perform specific metabolic functions.

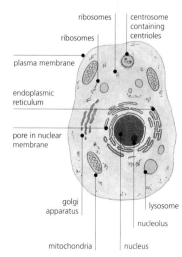

ribosomes

centrosome
containing
centrioles

ribosomes

plasma membrane

endoplasmic
reticulum

pore in nuclear
membrane

golgi
apparatus

lysosome

nucleolus

mitochondria

nucleus

CLASSIFYING ANIMALS

Humans have a natural tendency to sort and organize. And it seems that, ever since the days of the ancient Greek philosopher Aristotle, we have been attempting to do just that with the many different living organisms with which we share the planet. Biologists have so far discovered, described, and assigned names to about 1.7 million species of plant, animal and microorganism that presently exist on Earth—a mere fraction of what the total number is thought to be. They have also named many species that lived in the past but have since gone extinct. Modern classification attempts, in part, to provide order and structure to the huge amount of information that has been gathered on different organisms. This is done by providing a unique name for each of these organisms and sorting them into hierarchies of increasingly exclusive groups, or "taxa," based on their evolutionary relationships. This allows for individual organisms to be unequivocally recognized while, at the same time, being associated with other organisms sharing a common ancestry. All classifications evolve and change with the gathering of more knowledge and new discoveries. In recent years, the capabilities to investigate and compare the DNA of organisms through genetic techniques have forced scientists to rethink the classification of many animals.

SCIENTIFIC SORTING

Systematics is the name given to the science of discovering the diversity and evolutionary relationships of organisms. Aspects of systematics involved with naming and classifying organisms form the sub-discipline of taxonomy. The basic taxon, or group, for classification is the species. Ideally, all higher or more inclusive taxa comprise an ancestral species and all its descendants. Determination of this relationship is based on all members of a group sharing one or more derived features (evolutionary novelties).

NAMING RIGHTS

Animals are known by different vernacular or common names that vary from language to language, country to country, and often even within the same country. And so, to avoid confusion, scientists use Latinized names for groups of organisms. This provides universality and stability and avoids the necessity of translating names into many different languages. In this way, no matter what one's native tongue, the Latinized scientific name is immediately associated with the same group of organisms.

The basic category in all classifications is the species. Species are populations of organisms that share one or more similarities not found in related organisms. Species also form closed genetic systems. This means that individuals can only reproduce sexually with another individual of the same species, although closely related species may occasionally hybridize.

Close relations Scientists have been witnessing and documenting behavioral and physiological similarities between humans and chimpanzees (*Pan troglodytes*) since before Darwin. In recent years, they have been able to compare the DNA of the two. Recent revelations that both share more than 98 percent of their genes have reinforced the view that they should be classified within similar categories.

Linnaean classification Each grouping in this system of nested categories contains organisms with progressively similar characteristics. The domestic cat, for example, belongs to kingdom Animalia; phylum Chordata (animals with a centralized nerve chord); subphylum Vertebrata (the chord is within a bony vertebral column connected to the head); and class Mammalia (warm-blooded vertebrates with hair, milk glands, and a four-chambered heart). Categories continue through subclass Eutheria (placental mammals that bear live young); order Carnivora (with specialized teeth for eating meat); family Felicidae (all cats); and genus *Felis* (all small cats). Finally, no other organism shares the species scientific name, *Felis catus*.

SPECIES: *catus* – domestic cat

GENUS: *Felis* – domestic cat, sand cat, jungle cat, black-footed cat, Chinese desert cat

FAMILY: Felidae – domestic cat, lion, tiger, leopard, panther, puma, lynx, *Smilodon*

ORDER: Carnivora – domestic cat, seal, wolf, dog, bear, thylacine

CLASS: Mammalia – domestic cat, human, lemur, dolphin, platypus, woolly mammoth

PHYLUM: Chordata – domestic cat, fish, salamander, dinosaur, albatross

KINGDOM: Animalia – domestic cat, stick insect, sea urchin, sponge

Common confusion This South American insect (left) is known as both the pitbull katydid and the flat-headed katydid. Adding to the confusion, katydids are also called long-horned grasshoppers and bush crickets in North America, and the term tizi is sometimes used in Europe. Use of the taxonomic name assigned to this species, *Lirometopum coronatum*, avoids such issues of ambiguity.

New discoveries New invertebrates are still being discovered regularly, but it is rare to encounter previously undescribed mammal species. So it was unexpected when, late last century, the forests of Southeast Asia yielded three new species of small deer known as muntjacs.

organisms, both living and dead, with which the named organism shares its closest evolutionary relationships.

The second name is the specific name. Only one organism within any genus will ever be assigned this name. In this way, each organism receives its own unique two-part name. Scientists throughout the world know that when they use this name they are invariably talking about the same organism.

The generic name always begins with a capital letter, while the specific name is always written in the lower case. Both names are always printed in italics.

In instances when populations of a species are separated geographically and have relatively consistent differences, they may be recognized as a subspecies and assigned a third name. Like the species name, this is written in the lower case and also printed in italics.

AN ENDURING SYSTEM
In the early 18th century, the Swedish naturalist Carl Linnaeus developed a scheme for naming, ranking, and classifying different organisms according to the presence or absence of observable similarities. His earliest version of this system was published in 1735 under the name *Systema Naturae*. Linnaeus went on to revise and refine it many times in his lifetime, and it has been developed and improved many times since. It still provides, however, foundations for the approach to classification now used by biologists throughout the world. As a result, Linnaeus is often referred to as the "father of taxonomy."

The principal feature of the Linnaean system, and one which continues to apply today, is that it assigns a unique two-part name to each and every different organism. In this so-called binomial system, the first part of the name indicates the genus to which the organism belongs. Known as the generic name, this indicates other

Looks can deceive
Evolutionary relationships are not always obvious from superficial anatomical features. Consider, for example, the African and Asian family of mammals known as hyraxes (right). All living species are roughly rabbit-sized and all look very much like rodents. But these animals are actually ungulates, or hoofed animals, and their closest living relatives are elephants and ocean dwelling manatees. Most species in one hyrax genus lead a solitary existence in the trees, making them the only living hoofed animals to have an arboreal lifestyle.

Leaping alike All frogs and toads have one adaptation that sets them apart from other amphibians: the ankle bones are greatly elongated to form an extra segment in the hind leg, providing greater leverage for leaping. This is a distinguishing condition of the order Anura, within which all frogs and toads are placed. Frogs and toads also possess another adaptation to leaping; they have a short vertebral column with no more than 10 free vertebrae followed by a bony rod (the coccyx, representing fused tail vertebra).

MAMMALS

The 26 orders of mammals are divided into three major groups based on the structure of their reproductive tracts. The most primitive are the egg-laying mammals with a single order, the monotremes. Marsupials, which give birth to young in a very early stage of development, are now considered to consist of seven orders. The other 18 orders are made up of placental mammals. The most recent evidence from DNA sequence analysis has revealed that whales and even-toed ungulates are more closely related to each other than to any other group. DNA also indicates that there have been three major radiations of placental mammals: in Africa, South America, and the Northern Hemisphere.

Class Mammalia

EGG-LAYING MAMMALS
Order Monotremata
Monotremes

MARSUPIALS
Order Didelphimorphia
American opossums

Order Paucituberculata
Shrew opossums

Order Microbiotheria
Monito del monte

Order Dasyuromorphia
Quolls, dunnarts, marsupial mice, numbat, and allies

Order Peramelemorphia
Bandicoots

Order Notoryctemorphia
Marsupial moles

Order Diprotodontia
Possums, kangaroos, koalas, wombats, and allies

PLACENTAL MAMMALS
Order Xenarthra
Sloths, anteaters, and armadillos

Order Pholidota
Pangolins

Order Insectivora
Insectivores

Order Dermoptera
Flying lemurs

Order Scandentia
Tree shrews

Order Chiroptera
Bats

Order Primates
Primates

Suborder Strepsirhini
Prosimians

Suborder Haplorhini
Monkeys and apes

Order Carnivora
Carnivores

Family Ursidae, page 130

Family Canidae
Dogs and foxes

Family Ursidae
Bear and pandas

Family Mustelidae
Mustelids

SEALS AND SEA LIONS
Family Phocidae
True seals

Family Otariidae
Sea lions and fur seals

Family Odobenidae
Walrus

Family Procyonidae
Raccoons

Family Hyaenidae
Hyenas and aardwolf

CIVETS AND MONGOOSES
Family Viverridae
Civets, genets, and linsangs

Family Herpestidae
Mongooses

Family Felidae
Cats

Order Proboscidea
Elephants

Order Sirenia
Dugong and manatees

Family Trichechidae
Manatees

Family Dugongidae
Dugong

Order Perissodactyla
Odd-toed ungulates

Family Equidae
Horses, zebras, and asses

Family Tapiridae
Tapirs

Family Rhinocerotidae
Rhinoceroses

Order Hyracoidea
Hyraxes

Order Tubulidentata
Aardvark

Order Artiodactyla
Even-toed ungulates

Family Cervidae, page 190

Family Bovidae
Cattle, antelopes, and sheep

DEER
Family Cervidae
Deer

Family Tragulidae
Chevrotains

Family Moschidae
Musk deer

Family Antilocapridae
Pronghorn

Family Giraffidae
Giraffe and okapi

Family Camelidae
Camels and llamas

Family Suidae
Pigs

Family Tayassuidae
Peccaries

Family Hippopotamidae
Hippopotamuses

Order Cetacea
Cetaceans

Suborder Odontoceti
Toothed whales

Suborder Mysticeti
Baleen whales

Order Rodentia
Rodents

Suborder Sciurognathi
Squirrel-like rodents, mouse-like rodents, and gundis

Suborder Hystricognathi
Cavy-like rodents

Order Lagomorpha
Hares, rabbits, and pikas

Order Macroscelidea
Elephant shrews

BIRDS

Since Charles Darwin, birds have been classified according to perceptions of their natural relationships. This approach clusters similar-looking, interbreeding populations in species—related or sister species in genera; sister genera in families; and sister families in orders—representing the "tree" of bird evolution.

We now identify species as similar-looking populations of organisms that can interbreed freely. Genera, families, and orders are based on structural similarities in limbs, skeletons, and feathers. Thus birds of the parrot family have a foot with two toes forward and two back, and a hooked bill with vertically twisted palate bones.

The accumulated data led the American Alexander Wetmore to devise a classification of the orders and families of birds in the 1930s. This epic work became the standard for bird classifications in the 20th century. Since then, DNA and other molecular studies have shown that many structural traits used for classifying birds are unreliable due to convergent evolution, especially within the families of passerines, or songbirds, which involve over half the world's bird species. In particular, they found that the wrens, flycatchers, robins, and warblers of Australasia were unrelated to Eurasian look-alikes. Modern research indicates that these Australasian groups are the old ancestral lineages of the world's songbirds, from Gondwana.

The bird classification used here takes account of these changes. At species, generic, and family levels, it is based on the most up-to-date and authoritative world checklist current, the Howard and Moore *Complete Checklist of Birds of the World*, published in 2003. That checklist did not consider orders; but here we have adapted its families to the familiar Wetmore arrangement.

Class Aves

Order Tinamiformes
Tinamous

Order Struthioniformes
Ostrich

Order Rheiformes
Rheas

Order Casuariiformes
Cassowaries and emus

Order Apterygiformes
Kiwis

Order Galliformes
Gamebirds

Order Anseriformes
Waterfowl

Order Sphenisciformes
Penguins

Order Gaviiformes
Divers

Order Podicipediformes
Grebes

Order Procellariiformes
Albatrosses and petrels

Order Phoenicopteriformes
Flamingos

Order Ciconiiformes
Herons and allies

Order Pelecaniformes
Pelicans and allies

Order Falconiformes
Birds of prey

Order Gruiformes
Cranes and allies

Order Charadriiformes
Waders and shorebirds

Order Pteroclidiformes
Sandgrouse

Order Columbiformes
Pigeons

Order Psittaciformes
Parrots

Order Cuculiformes
Cuckoos and turacos

Order Apterygiformes, page 251

Order Pelecaniformes, page 271

Order Strigiformes
Owls

Order Caprimulgiformes
Nightjars and allies

Order Apodiformes
Hummingbirds and swifts

Order Coliiformes
Mousebirds

Order Trogoniformes
Trogons

Order Coraciiformes
Kingfishers and allies

Order Piciformes
Woodpeckers and allies

Order Passeriformes
Passerines

Order Passeriformes, page 324

REPTILES

Reptilia—living reptiles—traditionally includes turtles, crocodilians, tuatara, and squamates (lizards, snakes, and worm lizards). Crocodilians are most closely related to birds, but as birds are treated separately this section is assumed to cover non-avian reptiles. The position of turtles is in controversy; some studies show that the lineage of turtles is so distant from the squamates that they belong in a separate class. Lizards lost their limbs to become snakes or limbless lizards or amphisbaenians, so the formation of suborders within Squamata is controversial. Only the status of the tuatara as an ancient order by itself within the reptiles is agreed upon. The traditional separation of groups has been used, but the reader should be aware of the artificial nature of these groupings. The classification and names used for families, genera, and species generally follows the EMBL Reptile Data Base (http://www.embl-heidelberg.de/~uetz/LivingReptiles.html).

Class Reptilia

Order Testudines
Tortoises and turtles

Order Crocodilia
Crocodilians

Order Rhyncocephalia
Tuatara

Order Squamata
Lizards and snakes

Suborder Amphisbaenia
Worm lizards

Order Testudines, page 358

AMPHIBIANS

Lissamphibia, the living amphibians, includes three orders: the frogs and toads (Anura), the salamanders, newts, and sirens (Urodela), and the caecilians (Gymnophiona). Amphibians are all derived from the same common ancestor (monophyletic). The main common feature all amphibians have is smooth skin without scales. The use of families, genera, species, and common names follows Amphibian Species of the World: an on-line reference (http://research.amnh.org/herpetology/amphibia/index.html). Other common names were sourced from regional guides or compendiums.

Class Amphibia

Order Caudata
Salamanders and newts

Order Gymnophiona
Caecilians

Order Anura
Frogs and toads

Order Anura, page 428

FISHES

Any study of the fishes shows that they are an immensely diverse array of animals, differing greatly in the range of habitats they occupy and their body forms and adaptations. As a consequence, most biologists regard the term "fishes," as a convenient name, rather than a closely defined taxonomic entity, that describes aquatic vertebrates such as hagfishes, lampreys, sharks, rays, lungfishes, sturgeons, gars, and the advanced ray-finned fishes. There are a number of classification schemes for the fishes but one of the most widely accepted recent ones recognizes five classes of living species and three classes that are now extinct. The five classes, whose classification is detailed below, are hagfishes, lampreys, cartilaginous fishes, lobe-finned fishes, and ray-finned fishes. These are grouped into two superclasses: jawless fishes and jawed fishes. The three extinct classes are the pteraspidomorphs—jawless armored fishes, the jawed placoderms that were encased in bony plates, and the acanthodians, small true bony fishes with two long dorsal spines.

JAWLESS FISHES
Superclass Agnatha
Lampreys and hagfishes

JAWED FISHES
Superclass Gnathostomata
(includes all the groups below)

Superclass Agnatha, page 453

   CARTILAGINOUS FISHES "Chondrichthyes"
   Class Chondrichthyes
   Sharks, rays, and allies

      Subclass Elasmobranchii
      Sharks
      Rays and allies

      Subclass Holocephali
      Chimaeras

   BONY FISHES "Osteichthyes"

   Class Sarcopterygii
   Lungfishes and allies

   Class Actinopterygii

      Subclass Chondrostei
      Bichirs and allies

Subclass Elasmobranchii, page 462

      Subclass Neopterygii

         Primitive Neopterygii
         (gars and bowfin)

         Division Teleostei

            Subdivision Osteoglossomorpha
            Bonytongues and allies

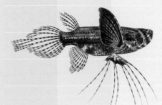

            Subdivision Elopomorpha
            Eels and allies

            Subdivision Clupeomorpha
            Sardines and allies

            Subdivision Euteleostei
            (includes all the groups below)

Subdivision Osteoglossomorpha, page 472

               Superorder Ostariophysi
               Catfish and allies

               Superorder Protacanthopterygii
               Salmons and allies

               Superorder Stenopterygii
               Dragonfishes and allies

               Superorder Cyclosquamata
               Lizardfishes and allies

               Superorder Scopelomorpha
               Lanternfishes

               Superorder Polymixiomorpha
               Beardfishes

               Superorder Lampridiomorpha
               Opahs and allies

               Superorder Paracanthopterygii
               Cod, anglerfishes, and allies

               Superorder Acanthopterygii
               Spiny-rayed fishes

Superorder Acanthopterygii, page 498

INVERTEBRATES

Over 95 percent of all animals are invertebrates. They are characterized by a structure that they all lack: a backbone or vertebral column. Invertebrates are divided into about 30 phyla, each displaying a distinct body form. Their evolutionary relationships can be inferred from their anatomy, their early development, and more recently from molecular analyses, particularly DNA, the genetic code. Features that define phyla include the organization of the body from a loose association of cells (Porifera), through tissue formation (Cnidaria) to the development of organs (Platyhelminthes). The acquisition of a fluid-filled body cavity was a defining point in animal evolution that allowed animals, such as Nematoda, Annelida, and many other phyla of worms, to move about by an hydraulic system driven by fluid pressure. The origin and form of these body cavities characterize different phyla. While these phyla are soft-bodied, others are protected and supported by various types of skeletons, such as shells in Mollusca and a jointed exoskeleton in Arthropoda. The division of the body into segments allowed for specialization of parts of the body. In arthropods, this has led to the development of segmental appendages that carry out specific functions, such as sensory perception, feeding and locomotion. Details of early embryonic development divide many advanced phyla into two lineages, one leading through the Echinodermata to the Chordata, the phylum to which vertebrates belong, the other containing the bulk of animal phyla. While molecular analyses have confirmed many of our ideas about the course of evolution based on anatomy and development, there are a number of instances where they are at variance. Hence, our classificatory system is undergoing revision. Furthermore, the continual identification of new species of invertebrates indicates that we are nowhere near their full inventory, and certainly far from understanding their vital roles in the sustainability of ecosystems.

Phylum Cnidaria, page 520

Phylum Chordata
Invertebrate Chordates

      Subphylum Urochordata
      Sea squirts

      Subphylum Cephalochordata
      Lancelets

Phylum Porifera
Sponges

Phylum Cnidaria
Cnidarians (sea anemones, corals, jellyfishes, etc.)

Phylum Platyhelminthes
Flatworms

Phylum Nematoda
Roundworms

Phylum Mollusca
Mollusks (bivalves, snails, squids, etc.)

Phylum Annelida
Segmented worms

Phylum Arthropoda
Arthropods

Phylum Mollusca, page 525

      Subphylum Chelicerata
      Chelicerates

         Class Arachnida
         Arachnids

         Class Merostomata
         Horseshoe crabs

         Class Pycnogonida
         Sea spiders

      Subphylum Myriapoda
      Myriapods (centipedes, etc.)

Class Arachnida, page 536

Subphylum Crustacea
Crustaceans

Subphylum Hexapoda
Hexapods

Class Insecta
Insects

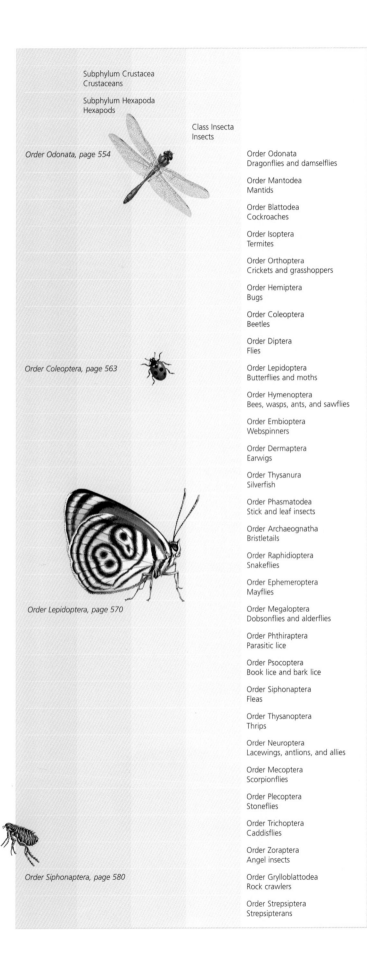

Order Odonata, page 554

Order Coleoptera, page 563

Order Lepidoptera, page 570

Order Siphonaptera, page 580

Order Odonata
Dragonflies and damselflies

Order Mantodea
Mantids

Order Blattodea
Cockroaches

Order Isoptera
Termites

Order Orthoptera
Crickets and grasshoppers

Order Hemiptera
Bugs

Order Coleoptera
Beetles

Order Diptera
Flies

Order Lepidoptera
Butterflies and moths

Order Hymenoptera
Bees, wasps, ants, and sawflies

Order Embioptera
Webspinners

Order Dermaptera
Earwigs

Order Thysanura
Silverfish

Order Phasmatodea
Stick and leaf insects

Order Archaeognatha
Bristletails

Order Raphidioptera
Snakeflies

Order Ephemeroptera
Mayflies

Order Megaloptera
Dobsonflies and alderflies

Order Phthiraptera
Parasitic lice

Order Psocoptera
Book lice and bark lice

Order Siphonaptera
Fleas

Order Thysanoptera
Thrips

Order Neuroptera
Lacewings, antlions, and allies

Order Mecoptera
Scorpionflies

Order Plecoptera
Stoneflies

Order Trichoptera
Caddisflies

Order Zoraptera
Angel insects

Order Grylloblattodea
Rock crawlers

Order Strepsiptera
Strepsipterans

Class Collembola
Springtails

Class Protura
Proturans

Class Diplura
Diplurans

Phylum Echinodermata
Echinoderms (sea stars, sea urchins, sea cucumbers, etc.)

Phylum Nemertea
Ribbon worms

Phylum Entoprocta
Goblet worms

Phylum Tardigrada
Water bears

Phylum Ctenophora
Comb jellies

Phylum Rotifera
Rotifers (wheel animals)

Phylum Hemichordata
Hemichordates (acorn worms)

Phylum Chaetognatha
Arrow worms

Phylum Gastrotricha
Gastrotrichs

Phylum Kinorhyncha
Spiny-crown worms

Phylum Phoronida
Horseshoe worms

Phylum Onychophora
Velvet worms

Phylum Brachiopoda
Brachiopods (lamp shells)

Phylum Bryozoa
Bryozoans (lace animals)

Phylum Sipuncula
Peanut worms

Phylum Echiura
Spoon worms

Phylum Loricifera
Brushheads

Phylum Priapulida
Phallus worms

Phylum Nematomorpha
Horsehair worms

Phylum Acanthocephala
Spiny-headed worms

Phylum Pogonophora
Beard worms

Phylum Gnathostomulida
Sand worms

Phylum Cycliophora
Cycliophorans

Phylum Placozoa
Placozoans

Phylum Orthonectida
Orthonectids

Phylum Rhombozoa
Rhombozoans

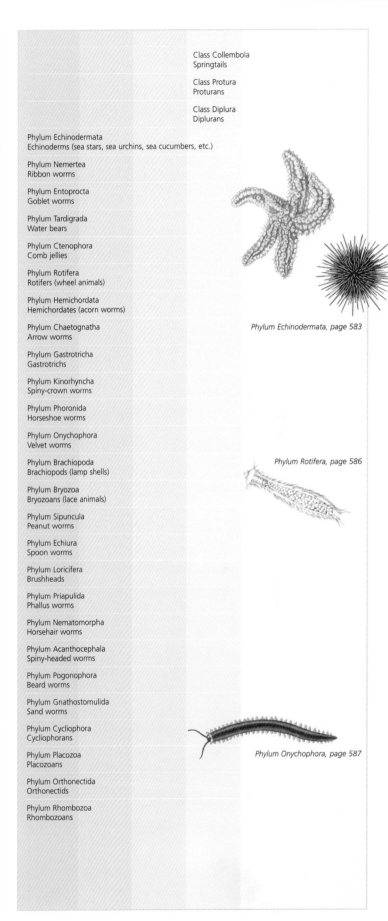

Phylum Echinodermata, page 583

Phylum Rotifera, page 586

Phylum Onychophora, page 587

EVOLUTION

Questions of how and where life began have fascinated and confounded people for millennia, giving rise to creation myths and legends in all human cultures. Scientifically, the answers continue to be debated and are unlikely ever to be known for certain. There are, however, two things upon which scientists tend to agree: that all life on Earth shares common ancestral origins; and that the crucial events that led to life's earliest precursors probably occurred more than 4 billion years ago. At that time, Earth would have been just hundreds of millions of years old and mostly ocean. It would have been a fiercely inhospitable place with high ultraviolet light levels, violent electrical storms, intense volcanic activity, and an atmosphere containing very little oxygen but dominated by methane, hydrogen, and ammonia. However unlikely it may seem, it was from this hostile cradle that life first emerged. And slowly, over the ensuing billions of years, it was the appearance of life itself, and all the processes that this requires, that helped to slowly transform the planet.

IN THE BEGINNING

The most popular scientific theory on the origin of life is that it began with simple organic molecules, upon the likes of which all life is built, arising spontaneously from inorganic compounds contained in Earth's early environment. How this might have occurred was first suggested by a famous experiment carried out in the early 1950s by biochemists Stanley Miller and Harold Urey at the University of Chicago. Within a laboratory simulation of Earth 4 billion years ago, they produced organic building blocks common to all life, including amino acids and nucleotides.

Aggregations of such molecules are thought to have led to the first cells. These may have been similar to the types of non-air-breathing bacteria that cause fermentation today. The oldest evidence of such cells, found in ancient rocks in Greenland, indicate such organisms would have formed before 3.5 billion years ago. The next crucial step required cells to harness the energy of the Sun.

Evidence of past life The fossil record provides consistent support for the theory of evolution as an explanation for the ascent of life on Earth. Most people think of fossils as bones, or the shapes of bones, preserved in rock, soil, or sediment. However, a fossil is any physical evidence of plants or animals from the ancient past. These prehistoric artefacts of life have been recovered from all continents.

Living fossils At 3.5 billion years of age, stromatolites are the world's oldest known fossils. These ancient reef-building structures were formed mainly by photosynthesizing colonial cyanobacteria, also known as blue-green algae, believed to have been Earth's dominant life form for about 2 billion years. Stromatolites were built vertically in layers as mats of the bacteria trapped minerals and sediment grains. New bacterial colonies grew up toward the light upon each successive layer. This process continues in Shark Bay in Western Australia, one of only two locations in the world where living stromatolites still occur.

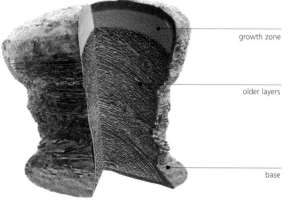

growth zone

older layers

base

REVOLUTIONARY THEORY

Eventually, some of those early cells developed the ability to photosynthesize, using energy from sunlight to produce their own food in the form of simple sugars. Importantly, that process released oxygen as a by-product. As oxygen gradually accumulated in the atmosphere, another type of cell, the eukaryotic cell, appeared. Evidence of metabolic activity by such cells found recently in rocks in northwestern Australia indicates these had arisen by 2.7 billion years ago. It is this type of cell that ultimately gave rise to all multicellular life, both plant and animal.

Until late last century, the oldest known fossil proof of animal life dated to almost 600 million years ago. However, in 1998 researchers announced the discovery of what is believed to be evidence of animals that existed much earlier—trails known as "trace fossils" thought to be created about 1 billion years ago by worm-like creatures in bottom sediments of an ancient sea that once covered India.

Regardless of when life of any type first appeared, most scientists agree that its diversification into the billions of different forms that have since appeared is a product of evolution. The core concept of evolution—that similar species share similar ancestries—harks back to ancient Greek philosophers. But the formal theory of evolution, as we now know it, was set down in 1859 by English naturalist Charles Darwin in his book *On the Origin of Species by Means of Natural Selection*. The theory is now seen as a crucial development in science and remains pivotal to virtually all biological research. It has been refined and developed since it was first published, but its basic tenets, which propose that different species appear in response to changing conditions, still hold.

Ever since Earth formed some 4.5 billion years ago, its environments have been changing. According to evolutionary theory, those organisms that successfully adapt, persist. Those that cannot, perish. This process later became known as "survival of the fittest." Darwin's theory posits that organisms that are better adapted to their surroundings are more likely to persist over time than those that are not; as a result, they have greater opportunities to produce descendants. These, in turn, are likely to carry and pass on the same traits that helped their parents survive.

Overwhelming evidence for evolution, and the mechanisms of natural selection that drive it, has come from the fossil record. The theory is also rigorously supported by comparisons between species, both surviving and extinct, in the areas of anatomy, embryological development, and, more recently, gene biochemistry.

Brave new work Darwin's theory of evolution was condemned by many church followers in the deeply religious England of the mid-1800s. Darwin was ridiculed in caricatures and cartoons (below), and became embroiled in controversy because his theory challenged the mainstream thinking of the day on the genesis of Earth and its life-forms. The public was particularly outraged by what Darwin's work insinuated about human origins, although this topic was specifically avoided in the original publication of his theory.

CONTINENTAL DRIFT

Until the early 1900s, it was widely believed that the continents held fixed positions. But by the 1960s, extensive geological, palaeontological, and biological evidence suggested ancient connections between continents and indicated that these land masses were still drifting inches (cm) apart annually. It is thought that about 300 million years ago, all the present-day continents were joined in the one large supercontinent known as Pangea. This huge land mass began separating about 200 million years ago to create two great continents. Biological similarities between past species preserved in the fossil record as well as between present-day species support evidence that Australia, India, Africa, South America, and Antarctica were once part of a southern supercontinent known as Gondwana. Similarly, Asia, North America, and Europe were once joined in a northern land mass, known as Laurasia.

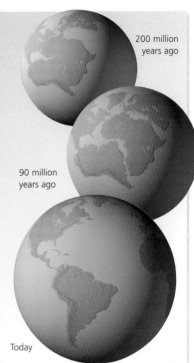

200 million years ago

90 million years ago

Today

Cephalopod evolution Nautiluses are the most ancient of the surviving cephalopods. Their ancestry can be traced back more than 500 million years to the Cambrian. The group was once considerably more diverse than it is today and included giant, fiercely predatory, straight-shelled species that may have filled ecological roles later taken over by sharks.

Natural selection in action In Britain in the early 1800s, the peppered moth (*Biston betularia*) was typically light with dark spots, perfectly camouflaged against the bark of birch trees. As industrial pollution slowly blackened the trees' bark, a darker form of the moth, better camouflaged from avian predators, began predominating with successive generations.

LIFE IN PERSPECTIVE

Earth's history is documented in vast intervals of time, and repeatedly punctuated by accounts of organisms that have evolved and gone extinct. It is a record that, if anything, reveals the insignificance of our species compared with what has preceded us and what will inevitably follow.

The history of modern humans (the species *Homo sapiens*) spans just over 100,000 years—a mere speck on a time-line covering more than 4 billion years.

Evidence of significant events in Earth's history comes largely from geology and palaeontology, and inevitably involves the sophisticated dating and interpretation of rocks and fossils. Most animal fossils have been recorded from the last 600 million years, since the beginning of the Cambrian period. Many of Earth's geological periods that have occurred since then are characterized by the animals that dominated them. The Jurassic, for example, is often referred to as "the age of dinosaurs" because these creatures reigned supreme during that time, in terms of both total numbers and species diversity. That gave way to the "age of mammals;" however, as the geological record repeatedly reveals, dominance does not always ensure survival.

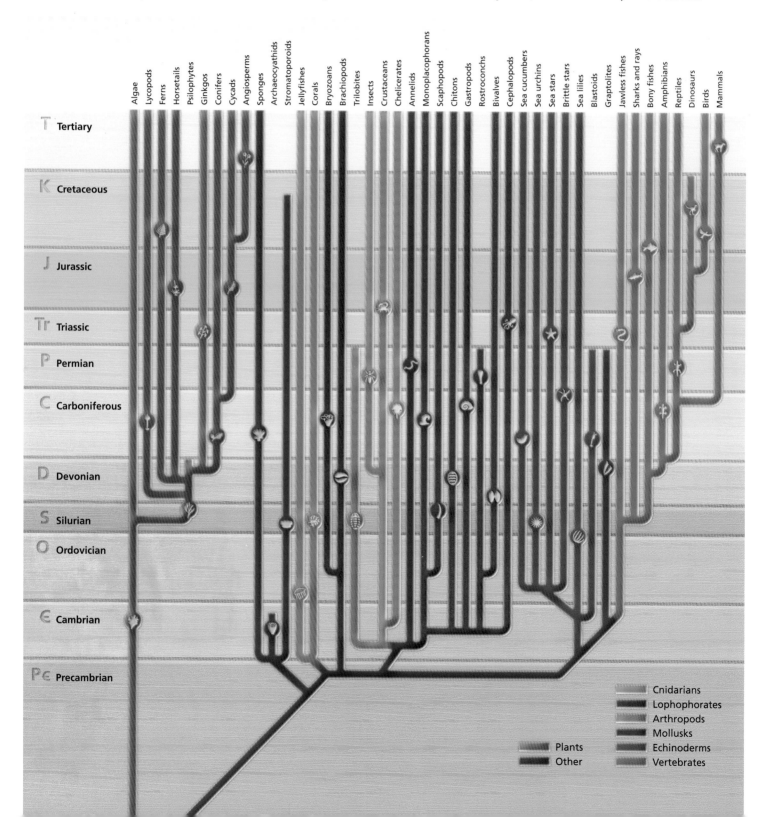

SEXUAL EVOLUTION

Without sex, evolution would not exist, because the opportunities for change and adaptation that under-pin natural selection require the genetic variation that is provided by sexual reproduction.

The first form of reproduction was undoubtedly asexual: the simple splitting of cells into daughter cells, as is still seen in today's single-celled organisms. Many "primitive" multi-cellular organisms such as worms and sponges can reproduce asexually through "budding," or regeneration from body cells or tissues. Slightly more advanced multicellular animals use the asexual process of partheno-genesis to produce offspring from unfertilized eggs. Asexual reproduction allows for rapid increase in numbers, but individuals are almost always genetically identical.

Sexual reproduction probably arose some 1.5 billion years ago and became the dominant, and usually only, form of reproduction in higher animals. Sexually produced organisms carry half the genes of each parent, usually in a unique combination. Thus, within any given species, a range of genes will be available for adaptations to a changing environment. When genes are advantageous, the species persists and/or eventually gives rise to a new species. Other genes may be benign or irrelevant; disadvantageous genes can cause a species to die out.

Budding "babies" The phylum Cnidaria, which includes corals, jellyfishes, hydroids, and sea anemones, is one of the few phyla in which asexual reproduction occurs commonly. Budding among species in the genus *Obelia* (above left) results in genetically identical individuals. In colonial species such as corals, individuals remain attached, creating branching tree-like organisms joined with a common flesh.

Reproductive success Each generation of a species represents a slightly new range of genetic combinations. Thus, any group or species with rapid reproductive rates—such as many of the insects—will be better equipped than less prolific breeders to deal with changing environ-ments. This partly explains why insects have representatives in virtually every available terrestrial habitat.

Populate or perish Sexual reproduction works better for some species than others. Those with low reproductive rates, such as the giant Galápagos tortoise (*Geochelone elephantopus*) (below), survive well in stable environments. But this species is not coping with recent rapid change (such as introduced competitors and predators) on the Galápagos archipelago, and its future is now threatened.

WHEN SPECIES DIE

Today, the extinction of a species invariably evokes widespread anger and sadness; rightly so, most people would argue, because extinctions documented in very modern times (that is, the past few hundred years) have invariably occurred as a direct result of human activity.

But extinction is not necessarily a dirty word. It is in fact a natural part—indeed, a consequence—of evolution. More than 99 percent of the species that have ever evolved on Earth—estimated to number at least 2 billion—have gone extinct. Those that survive today—perhaps as many as 30 million, although less than 2 million have so far been identified and described—represent a tiny fraction of the species that have gone before them.

The evidence, particularly from the fossil record, suggests that most species exist for between 1 million and 5 million years before dying out. It is highly likely, however, that many species live for considerably shorter periods of time. There are also many that manage to survive longer. Such species that are still alive today are often referred to as "living fossils." Examples include the coelacanths, two ancient fish species believed to have appeared more than 65 million years ago. Remarkably, one of these remained undiscovered until the late 1990s.

EXTINCTION EN MASSE

Under "normal" circumstances, scientists believe that between one and 10 species would be dying out annually due to the natural forces of evolution. This is known as the background extinction rate. At various times during Earth's long geological history, however, that rate has been considerably higher, indicating a period of "mass extinction."

The fossil record suggests that at least five mass extinctions have occurred to date, the last—and one of the biggest—at the end of the Cretaceous period, about 65 million years ago. More than three-quarters of all species alive at the time, including the last of the dinosaurs, disappeared forever. Likely reasons for mass extinctions have included large-scale climate change, huge meteors hitting Earth, and intense geological activity.

These extinctions occurred over many thousands or, in some cases, millions of years. There is widespread concern that we are currently experiencing a sixth mass extinction, proceeding at a faster and more intense rate than any before, caused by human activity.

Convergent evolution Sometimes animals evolve under such similar conditions that they develop comparable anatomical modifications, making them seem more closely related than they really are. For example, penguins (left) and auks were once mistakenly thought to share recent ancestral origins. Both live in cool oceanic environments and are flightless, with flipper-like wings well adapted for underwater swimming. Penguins, however, are confined to the Southern Hemisphere and are most closely allied to petrels and albatrosses. Auks are strictly Northern Hemisphere relatives of puffins and gulls. The warm waters of the tropics are thought to have kept the two groups apart for many millennia.

Resisting desiccation The development of the amniotic egg occurred first in the reptiles and was critical in the colonization of land by vertebrates and their pervasive penetration of terrestrial environments. A series of fluid-filled sacs cushion the developing embryo within a calcareous shell that is permeable to gases, such as oxygen and carbon dioxide, but which restricts water loss.

Distant relations Echinoderms such as crinoids (left) and starfishes (lower left) are thought to be more closely related to humans than their appearance might suggest. This group of exclusively marine organisms is considered an evolutionary offshoot of the line from which rose the chordates, a group which contains all the vertebrates. The association is suggested by similarities in early embryonic development.

Amber The bodies of insects and spiders easily become distorted and crushed, so they are very poorly represented in the traditional fossil record. Their features survive much better when preserved in fossil tree resin known as amber (below). Other small animals such as centipedes and lizards, as well as plants, have been found preserved in amber, the most significant deposits of which are found in sands lining the shores of the Baltic Sea.

Dinosaur demise Climate change following a collision between Earth and a vast meteor remains the most widely supported theory on why the dinosaurs, among them *Triceratops* (below), went extinct. Evidence of dinosaurs disappears from the fossil record at about the same time that geological proof of such an impact appears.

EARLIEST ANIMALS

Although the fossil record indicates that multicellular animals first appeared on Earth much earlier, the oldest "complete" fossils come from about 600 million years ago, toward the end of the Precambrian. They belong to a group of soft-bodied marine organisms known as the Ediacaran fauna, evidence of which was found in 565-million-year-old rocks in the Ediacaran Hills, in South Australia's Flinders Ranges, in the 1940s. Since then, similar fossils have been found on all continents except Antarctica.

Ediacaran fauna included many bizarre animals unlike anything known from other times. Other forms were reminiscent of worms, jellyfishes, arthropod-like creatures, and organisms that looked like large sea-pens, similar to those seen within the phylum Cnidaria.

The relationship between the Ediacaran animals and those that have appeared on Earth since is unclear. Some scientists believe they were predecessors of groups that came later. Others consider them an evolutionary dead-end: a natural history "experiment" that ended without leading onto anything else.

CAMBRIAN EXPLOSION

Before the discovery of the Ediacaran fauna and the recognition of their significance, the oldest known complete fossils of multicellular animal life came from the Cambrian. This period is still known and celebrated for the explosion of life, exclusively marine, that suddenly appears in the fossil record in rocks from this time: an abrupt and huge rise in biodiversity, the scale of which has not been known since. Ancestors of almost all major living animal phyla appear in the fossil record for the first time during the Cambrian, mostly within a period spanning just 40 million years. Of particular significance is the earliest evidence of hard shells and exoskeletons, the evolution of which may have been partly driven by a rise in predatory species.

Long-since-extinct trilobites, early forms of arthropod with unusually well-developed eyes, proliferated. So, too, did sponges. Other distinctive Cambrian fauna included large, burrowing, wormlike predators known as echiurans. Echinoderms were also plentiful and represented by several large classes. However, it is believed that only one from that time, Class Crinoidea, still survives today.

Both gastropod and cephalopod mollusks are represented in fossils from the Cambrian, but these groups did not reach their heyday until later. The earliest chordates, ancient ancestors of our own species and all other vertebrates, also appeared during this period.

In addition, there were also many strange life forms that do not appear to be directly allied to any known group. One of the most famous was *Wiwaxia*. A small bottom-dweller, up to 2 inches (5 cm) long, *Wiwaxia* had a body covered by plate-like armor and rows of spines. Beneath this, it is believed to have been slug-like, but its relationship to any living animal group remains unclear.

Wiwaxia is known from the most famous and extensive Cambrian fossil deposits, at a site known as the Burgess Shale, located within British Colombia's Yoho National Park in the Canadian Rocky Mountains. The site was first identified in 1909, and was given World Heritage status in the early 1980s. More than 60,000 fossils have been recovered from these deposits. Some of them are so exquisitely well preserved that it is possible to determine their stomach contents and details of muscles and internal organs.

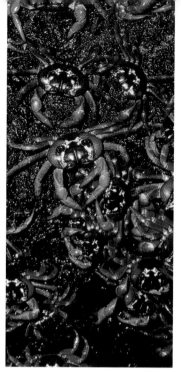

Mostly terrestrial Behavior can provide clues to the origins and evolutionary affiliations of living species. Adult red crabs (*Gecarcoidea natalis*) live almost exclusively in burrows on the floors of rain forests on Christmas Island, in the Indian Ocean. Each year, however, they migrate en masse (left) to cliffs where females cast fertilized eggs into the ocean, their ancestral home. Larvae develop there for up to 4 weeks before returning to land as juveniles.

Avian enigma The evolutionary lineage of South America's hoatzin (*Opisthocomus hoazin*) baffled scientists for more than 200 years. Unlike any other bird, the hoatzin ferments plant matter in a foregut, as cows do. Wing claws, used by young to climb trees, are reminiscent of those of the famed prehistoric *Archaeopteryx*. Its morphology suggests an ancestry tied to pheasants and chickens. Recent genetic investigations, however, finally confirmed it to be most closely related to cuckoos.

BEYOND THE CAMBRIAN

The fossil record provides evidence of two Cambrian creatures that could have been, or were probably very much like, the direct ancestor of the vertebrates. Small fish-shaped filter-feeders similar to modern-day lancelets, *Pikaia* (from the Burgess Shale deposits, in Canada) and *Cathaymyrus* (found in southern China) are the oldest known fossil chordates. The first vertebrates are thought to have arisen from such creatures just after the close of the Cambrian, about 480 million years ago. They were jawless fishes, known as Agnathans, with mainly benthic lifestyles. They were, in many cases, protected from predators by heavy armor-like plates.

The first cartilaginous fishes, sharks, had appeared by the late Ordovician. The origins of the bony fishes—now the most diverse group of vertebrates—show a little later in the fossil record, around the late Silurian, just over 400 million years ago. By the Devonian, the fishes proliferated to such an extent that this period is often referred to as "the Age of Fishes."

Just before this—toward the end of the Silurian, about 420 million years ago, and shortly after vascular plants emerged on land—the first animals began moving out of the oceans and into terrestrial habitats. These were arthropods, whose hard external skeleton (which may have evolved initially as protection against predators) proved valuable in restricting water loss. Arachnids and centipedes would have been among the earliest terrestrial forms of this group, which went on to become one of the planet's most successful.

LAND VERTEBRATES

The move by vertebrates onto land was preceded by the evolution of lobe-finned bony fishes (the Sarcopterygii) in the early Devonian. By the middle of that period, a group of sarcopterygiians known as rhipidistians are thought to have begun using their fleshy fins to clamber around the bottom of shallow, warm, oxygen-depleted waters and probably even on land for brief periods. These fishes shared features in common with both modern lungfishes (among the few surviving sarcopterygiians) and amphibians. And it is likely they, or fishes very similar to them, gave rise to the first amphibians about 380 million years ago, near the end of the Devonian.

It is thought the first reptiles—probably small insectivores—evolved from amphibian ancestors very quickly, in evolutionary terms, within the next 30 million years. At that time, Earth was warm and humid. As the climate began to get considerably drier—about 280 million years ago, at the start of the Permian—reptiles began to diversify. By 275 million years ago, mammal-like reptiles had appeared. They became the dominant land vertebrates for a short period, but most disappeared by around 230 million years ago.

However, around 220 million years ago, one of the remaining groups of these reptiles, the cynodonts, gave rise to the first mammals. At about the same time, amphibians similar in form to modern frogs and salamanders also emerged. Crocodiles and turtles first appeared around this time too, as did the dinosaurs.

Indricotherium This type of giant hornless rhinoceros, which became extinct about 25 million years ago, may have been the largest mammal ever to walk the Earth. Adults attained heights of about 18 feet (5.5 m) at the shoulder and weights in excess of 15 tons (13.6 tonnes).

Land-loving fish Watching mudskippers "climbing" trees on mudflats (left), it is not hard to imagine how the early colonization of land may have proceeded. However, the skeletal structure of mudskippers' fins and the associated musculature differ significantly from those of the fishes that ultimately gave rise to land vertebrates.

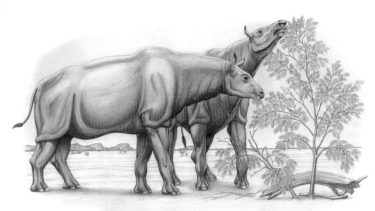

Adaptive radiation When a species arrives or appears in a new habitat to which it is well suited but where it faces little competition, it may undergo a rapid type of evolution known as adaptive radiation. This occurred in the lemurs (below, right, and below right), a group of primitive primates known only from the Indian Ocean island of Madagascar. The ancestral lemur is thought to have left its biological competitors on the African mainland about 55 million years ago and to have arrived on the island on a natural raft. Finding no rivals with similar requirements, it rapidly and repeatedly evolved into a unique group that is now represented by more than 30 surviving species, each of which occupies its own slightly different ecological niche, and some of which have adaptations to deal with preferred prey.

Living fossils Two species of tuatara, *Sphenodon punctatus* and *S. guntheri*, are the only survivors of a large group of reptiles that was widespread around the world during the Triassic, more than 225 million years ago. Most became extinct about 60 million years ago. Fossil evidence shows the relict tuatara species, which are now confined to small islands off New Zealand, have changed little since their early evolution. Current populations stand at some 400 *S. guntheri* and more than 60,000 *S. punctatus*.

Ancient aeronauts Odonata, to which dragonflies and damselflies belong, is one of the oldest surviving insect orders. Some of the earliest dragonflies had wingspans of up to 3 feet (1 m). Flapping flight was unlikely. Instead, it is believed they were gliders with limited capability for aerial maneuvering. Wings emerged very early in insect evolution and are believed to have developed from gill-like organs.

DINOSAURS RULE

The earliest dinosaurs were small bipedal carnivores. The group diversified quickly, and by the end of the Triassic, 206 million years ago, many herbivorous forms had appeared. Some had already begun to approach the massive sizes for which these animals are renowned. The radiation of dinosaurs coincided with the appearance and global spread of coniferous trees, which are thought to have been the main food source for the largest of the dinosaurs.

For 150 million years, dinosaurs dominated the planet, the last disappearing 65 million years ago as part of the mass extinction at the end of the Cretaceous. Around the middle of this extraordinary reign, the first birds are thought to have arisen from small, bipedal, carnivorous dinosaurs called theropods. It is believed that feathers first evolved not for flight, but warmth.

MAMMALS RISE

Mammals first appeared about the same time as the earliest dinosaurs, but early mammalian radiation is thought to have been inhibited by the rise of the latter. As a result, the mammals diversified and evolved relatively modestly at first, with most species remaining nocturnal and small—about the size of today's shrews and mice—for millions of years. Once competition from, and predation by, the dinosaurs ended, mammals flourished and diversified.

One important and largely tree-dwelling genus that appeared just before the Cretaceous mass extinction, and survived it, was *Purgatorius*. It is thought that within this group was the ancestor that ultimately gave rise to modern primates, including our own species, which probably first appeared just over 100,000 years ago.

BIOLOGY AND BEHAVIOR

All animals share basic survival requirements such as the need to respire, eat, and find shelter, and they meet these by using a combination of biological and behavioral responses. Such strategies are frequently consistent between different groups across the animal kingdom. This enables identification of biological and behavioral blueprints—patterns that are often constrained by the universal laws of physics—that are revisited repeatedly within and between different groups of animals. For example, all animals take in oxygen via a thin, moist surface. This occurs over the entire outer body in many worms; across thin gill filament folds in most fishes; and in lungs in all of the land vertebrates.

Food chains There is far more diversity at the bottom of a food chain than at the top. Solar energy drives the food chain. In marine ecosystems, the bulk of this energy is taken up by phytoplankton. In turn, this plant material is grazed by zooplankton (left) and larger filter feeders such as the giant Japanese spider crab (*Macrocheira kaempferi*) (below). In their turn, zooplankton are grazed by larger animals.

SIZE MATTERS

There are good reasons why huge insects and other arthropods are the stuff of science fiction. One is the fact that they have a hard external skeleton. To grow, arthropods must molt: the exoskeleton is shed to allow expansion of the body within before a new exoskeleton is formed. This leaves soft tissues unsupported against gravity during each of the growth periods.

The bigger an animal, the greater the gravitational force exerted upon it. Gravity would collapse oversized insects into a heap during molting. As a result, the largest arthropods inhabit aquatic environments, where they are supported by water.

Gravity also helps to define the upper size limits of mammals, so it is no coincidence that the biggest mammal and largest animal ever, the blue whale (*Balaenoptera musculus*), is oceanic. It attains lengths of more than 110 feet (33.5 m) and weights of almost 209 tons (190 tonnes) because water helps to support its massive bulk.

In a large mammal, the ratio between body surface area and volume is low. Because small animals have higher surface-area-to-volume ratios, they lose heat faster. Larger mammals, consequently, tend to be found in colder climes while the smaller species proliferate closer to the tropics. For similar reasons, species of polar animals such as seals, penguins, and the musk ox (*Ovibus moschatus*) usually have compact extremities—including legs, ears, and snouts—to keep their surface-area-to-volume ratio as low as possible.

On a small scale Smaller mammals need to consume more food in proportion to their overall body size and often have higher metabolic rates and shorter lengths of life than larger animals. The large tree shrew (*Tupaia tana*) averages 4 ounces (100 g) in weight. Highly active, it lives between 2 and 3 years. Scent marking their territory and their young is a vital part of the daily routine of all shrews.

Gain without pain Although it may look vicious, a confrontation between animals is often little more than sound and fury. Gaining dominance without either party sustaining severe injury is important to the survival of the species. In an encounter between two male gray wolves (*Canis lupus lupus*) (right), the resolution comes with the vanquished signaling submission by lying supine with his stomach exposed.

Battle of the giants Mating among seals is controlled by a small group of dominant males, with a single male controlling up to 50 females on a small stretch of beach. Dominant males often inflate their noses and produce a noise that sounds like a drum to warn lesser males away. Conflict of a more physical nature occurs between males such as the northern elephant seals (*Mirounga angustirostris*) (right) when a territory is seriously threatened or one male seeks to displace an established male.

SEX WARS

All species are innately impelled to pass on genes to future generations. Most achieve this transfer through sexual reproduction, which involves the meeting of sperm and eggs.

Hermaphroditism is a reasonably widespread phenomenon among animals, but more often sex cells are contained within separate males and females, and many biological and behavioral adaptations have evolved to ensure that reproduction occurs among the fittest individuals. These animals are the best adapted to the environment and, therefore, the ones most likely to contain the genes that ensure the survival of their species.

In many instances, males prove their reproductive worth by battling one another and this is one reason males are frequently larger than the females of their species. This has also driven the evolution of male combat equipment such as horns and antlers in species as diverse as rhinoceros beetles and deer. Behavioral displays are often important, particularly among mammals, to avoid or reduce the severity of aggressive encounters so that the combatants are not left dead or disabled from battle.

FANFARE AND DEMONSTRATION

In some species, the males secure reproductive rights by proving their genetic fitness directly to prospective mates, and have evolved physical and behavioral displays specifically for this purpose. This underlies the spectacular colors, adornments, and elaborate courtship displays that are common among the males of many bird and some fish species.

There are many other strategies that males deploy in order to attract females. Male bower birds construct impressive arches of grass decorated with collections of small offerings aimed at influencing the decision of prospective mates. In a similar way,

male stickleback fishes build nests to win female attention.

Some males demonstrate their genetic fitness to mates by providing gifts—usually offerings of food. This tactic is common among insects and some spiders. With the latter, the offering is usually the male himself. It is thought that consuming the gift distracts the female spider from mating with other males, while also providing nutrients that will give her offspring a good start at life.

Among birds, courtship feeding is used by the males of some species to show their paternal capabilities for finding food for offspring.

Armed and dangerous The rhinoceros beetle (*Xylotrupes australicus*) uses its horns in its battle for territory and, ultimately, a mate. The horns are very strong, enabling the beetle to forage through heavy leaf litter on the forest floor or dig a burrow.

Show time Usually it is the male bird that plays the lead role in courtship by showing off his plumage to duller-colored females. Possibly the most stunning fanning display is that of the male Indian peafowl (*Pavo cristatus*). The radial alignment of feathers in the semi-circular fan requires the root of each one to be pointed with an acute degree of precision. Remarkably, they are put into position by muscles in the male's tail. Not only can he deploy the feathers, but he can make them vibrate and hum.

Tooth and claw A lioness (*Panthera leo*) brings down a zebra (*Equus burchelli*), risking lethal injury from the equid's flying hooves as she does so. Despite the danger, zebras are a mainstay in the diet of lions on the East African savanna.

Tree dweller The maned three-toed sloth (*Bradypus torquatus*) is both a slow eater and a slow digester of food. As much as a month may pass before foodstuffs finally pass from the multichambered stomach to the small intestine.

YOU ARE WHAT YOU EAT

Diet influences metabolism and activity levels and can in some cases provide protection against predators. For example, by eating certain ants, South American poison-arrow frogs (*Dendrobates* sp.) produce a toxin in their skins to which only one snake species is known to be immune.

Diets that are of low nutritional value typically produce slow-moving, lethargic animals. Tropical rain-forest-dwelling three-toed sloths consume mainly plant matter with a high level of cellulose that is difficult for them to digest. A slow metabolism, low body temperature, and the habit of sleeping for up to 18 hours a day help them to survive on this low-energy diet. In Australia, the koala displays similar physiological and behavioral adaptations to a low-energy diet that is comprised of large amounts of leaves from a few species of eucalypts.

EAT OR BE EATEN

The most readily recognizable physical and behavioral features of animal species often relate directly to the need to locate and consume food as efficiently as possible. Most carnivorous mammals, such as the big cats of the African savanna, have large, sharp teeth designed to rip and tear tough flesh, lightning-fast reflexes, and capabilities for rapid acceleration. Herding behavior provides the grazing herbivores on which they prey, such as zebras and gazelles, with a safety-in-numbers response. Also, these herbivores have evolved hooves that can deliver a harmful blow, and keen senses of sight and smell.

Many of the most acute feeding adaptations are seen among those animals with specialist diets. The long tongues and curved bills of honeyeaters, for example, are the ideal design for the extraction of high-energy nectar from flowers. And the South American leaf-cutter ant supplements its largely herbivorous diet with fungus that groups of the insects farm on leaves cut and composted specifically for that purpose.

Colorful Lesser flamingos (*Phoenicopterus minor*) get their pink hue from carotenoid proteins in their diet of plant and animal microplankton.

Rain-forest food chain As with other food chains, in a South American rain forest (right), there is far more diversity and a greater number of species at the bottom end of it than there is at the top. The most stable forest environment is found at ground level. Here, wind seldom penetrates and temperature, humidity, and ambient light levels rarely change. Along the chain, predator and prey wage a constant battle for biological superiority, pushing each other into more and more evolutionary developments.

Silent killer Like other owl species, the great gray owl (*Strix nebulosa*) swoops unheard on its prey. Its outer flight feathers have a jagged edge, which slows the flow of air over their wings, minimizing the noise made as they flap. Most flying birds' flight feathers have smooth edges.

Boom or bust Locusts are usually solitary insects. But when food is abundant, they may band together in huge swarms, some containing several billion insects, and then take to the air en masse. The scourge of farmers, locusts wreak havoc, destroying vast fields of grain in a few hours.

Silky anteater
Arboreal anteaters are threatened by birds of prey.

Kinkajou
Although this animal is a carnivore, it mainly eats fruit. Occasionally it will feed on birds and small mammals.

Toco toucans
The toucan feeds on fruit and nuts, which it picks up in the tip of its bill and moves into position with its tongue. The toucan then throws its head back and tosses the food down its throat.

Northern tamandua
A form of anteater, the tamandua catches ants with its long, sticky tongue. This animal is strictly nocturnal.

Jaguar
Top of the food chain, and small in numbers, this big cat eats many forest canopy species, including monkeys and birds.

Hoatzin
This bird has a very large crop, which is a highly specialized adaptation for grinding food—the leaves, buds, fruit, and flowers of only a few tree and shrub species.

BREAKING IT DOWN

Whether they do so directly or in-directly, all animals ultimately derive nutrition from plants. Herbivores are not, however, the main consumers of vegetation. Rather, a vast amount is eaten by the decomposers, which include burrowing earthworms and insect larvae.

There are also many animals such as flies and beetles that play a role in the decomposition of animal flesh in land habitats. And, in aquatic environments, crustaceans such as crabs, scavenging fishes, filter-feeding zooplankton and bottom-dwelling anemones, starfishes, and other invertebrates, all contribute directly to breaking down organic matter.

The role of decomposers is critical to the survival of all living things, because their actions release minerals and nutrients and make them available to be recycled in food webs.

WHEN IS ENOUGH ENOUGH?

In the case of many herbivores, the simple answer to this question is never. When compared with flesh, vegetation is low in nutrients such as protein. It is also hard for most animals to digest because they do not produce the enzymes necessary to break down the tough cellulose that commonly strengthens plant-cell walls. Herbivores have evolved ways to overcome these challenges. They spend the majority of their time eating, and have long, complex intestines in which symbiotic bacteria break down cellulose.

Nevertheless, even the most efficient herbivores swallow large quantities of vegetation that pass through them undigested. An adult African elephant, for example, is unable to digest more than half of the several hundred kilograms of vegetation that it consumes daily.

Because flesh is higher in nutrients than plant matter and easier to digest, carnivores have relatively short intestines and feed less frequently. The white shark (*Carcharodon carcharias*), a fierce oceanic predator, is one of many carnivores believed to need only one large meal every few days.

Bird-eating spider
As an adult, this spider is large enough to devour small birds. It also eats lizards, beetles, and small snakes.

Vegetation
The forest floor has a slow-growing woody shrub layer and a herbaceous ground layer.

Bushmaster snake
A classic ambush predator, this snake selects a suitable ambush site beside a mammal trail and lies in wait.

IN ATTACK AND DEFENSE

The eternal conflict between prey and predator has underpinned the evolution of key adaptations in many animals. Camouflage is a common strategy; it helps to hide predators while hunting, and is also used by prey seeking concealment.

Spines keep birds from eating some caterpillars, just as they deter dingoes from eating echidnas. Prey animals that are poisonous are often brightly colored to warn of their toxic flesh and unpalatability. To deter predators, some animals that are not toxic even mimic the markings of those that are.

Predators, too, may use mimicry. Anglerfishes dangle worm-like baits in front of their mouths. Some snakes use the tips of their tails in a similar way to attract prey.

GETTING AROUND

Every animal species is capable of movement during some part of its life. Exactly how depends on where it lives and its reasons for moving. Almost all fishes, for example, have the ability to swim, but torpedo-shaped pelagic species that chase their food are invariably faster and more agile than those that sit and wait for prey in ocean sediments.

Marine mammals, of course, are all proficient swimmers. But a surprising number of terrestrial mammals can also swim well when they need to. More often, however, they move by walking, running, crawling, or jumping. A few tree-dwellers, such as possums and squirrels, can glide. However, bats are the only mammals that can fly as proficiently as most birds.

Some animals change their mode of movement at different stages of their life-cycle. Many tadpoles, for example, lead free-swimming lives until metamorphosing into hopping adult frogs or toads.

Quick change Chameleons are famed for their highly developed camouflage, which, combined with their leaf-like shape, is especially useful when slowly stalking prey. Chameleons may also change color in threat displays (such as from one male to another) or when they are disturbed.

Cheetah hunting Cheetah generally hunt in open terrain, choosing a victim then pursuing it to the exclusion of all others, even those nearer or more available. Chases seldom last more than 20 seconds; such a burst of speed requires tremendous amounts of energy and builds up massive amounts of heat, so that the cat soon finds itself overheated and out of breath.

Playing possum When faced with danger, opossums (right) commonly become comatose and appear dead. The opossum will lie on its side, become stiff, and drool. Its eyes will glaze over and its tongue will loll out of its mouth. This response, which is entirely involuntary and may last as long as 4 hours, helps the opossum survive an attack, as many predators will give up if it seems that the opossum is already dead.

Nudibranch feeding on coral Many nudibranchs (far right) are brightly colored to warn potential predators that they are unpalatable or toxic. Others are subtly colored for camouflage. Many nudibranchs feed on cnidarians such as anemones, which contain stinging cells that the nudibranch adopts into its own body to form part of a defensive mechanism against any predator that attacks it.

On the hop Macropods such as the red kangaroo (*Macropus rufus*) have large, powerful hindlimbs and long hindfeet that enable them to hop fast, using their long and rather inflexible tail for balance. This is a very energy-efficient form of locomotion; a hopping kangaroo uses less energy than a four-legged animal of the same size moving at the same speed. However, this specialization has meant that kangaroos and wallabies are unable to walk. When moving slowly, they raise their hindquarters on a tripod formed by the forelimbs and the tail, then swing both hindlegs forward at the same time.

From tree to tree Flying squirrels glide rather than fly, using a membrane down each side of the body (the patagium) as a parachute and the tail as a rudder. Direction is controlled by the legs, tail, and the stiffness of the paragium. Before landing, the squirrel brakes by flexing its body and tail upward.

IN THE FAST LANE

The ability to move fast is a boon for predators. Much of the physiology of the cheetah (*Acinonyx jubatus*) is built for speed—from its streamlined body; long, powerful legs; large nostrils, lungs, heart, and liver; to its long rudder-like tail and feet designed to improve traction. Clocked at top speeds of about 70 mph (112 km/h), no other land animal is faster. Its favored prey, gazelles (*Gazella* sp), are swift, but cannot outrun cheetahs. They often, however, outmaneuver them.

The fastest swimmers are sailfish (*Istiophorus* sp.), pelagic carnivores that prey on other fishes and cephalopods. They have been recorded at speeds of about 68 mph (110 km/h).

The peregrine falcon (*Falco peregrinus*) can achieve a similar pace when in horizontal flight. However, it has been shown to travel at up to 273 mph (440 km/h) when plummeting to swoop on prey.

Limbs are not always necessary for rapid movement. The venomous black mamba (*Dendroaspis polylepis*), an East African snake, can slither, with its head and front body raised off the ground, to speeds of about 7 mph (11 km/h).

Swimming machines Barracuda such as the blackfin (*Sphyraena genie*) are designed for maximum hydrodynamic efficiency. The body is streamlined, and has grooves and depressions into which the fins can be tucked away to reduce turbulence. While barracuda do not hunt cooperatively, they will frequently attack schools of prey en masse, breaking up the school structure so that individual fishes are easily captured.

FINDING THE WAY

Scientists have only recently begun to understand the navigational tools that animals use to guide their movements. Bees use polarized light to navigate according to the sun's position during daily foraging forays. Birds, too, probably use the sun as a guide on long migrations, but the Earth's magnetic field is also thought to play a role. The latter is thought also to be important to migrating whales; they probably also recognize visual landmarks close to coastlines. Animals as diverse as salamanders and rodents use scent trails for short distances. Scent is probably also an important aid to marine turtles looking for nesting beaches.

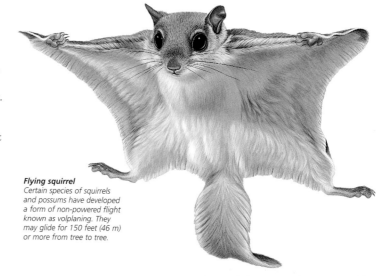

Flying squirrel
Certain species of squirrels and possums have developed a form of non-powered flight known as volplaning. They may glide for 150 feet (46 m) or more from tree to tree.

LIVING TOGETHER

Animals can gain many advantages by living with other members of the same species. There are many tasks in which a group approach can make life considerably easier, such as watching out for and fighting off predators, and finding, hunting for, or gathering food.

Many birds live in simple family groups with a monogamous adult pair raising chicks from year to year. Single-parent families are also common among birds as well as mammals. In these situations it is usually the female, rather than the male, that remains in close association with the offspring until they become mature enough to fend for themselves.

Extended families are formed when different clutches or litters of offspring stay on with their parents after reaching maturity, sometimes even helping them to raise the next generation. The most extreme examples are among the social insects, such as bees and termites, which live in large groups with complex divisions of labor. The naked mole rat (*Heterocephalus glaber*) is the only mammal known to have a similar social structure.

EXTERNAL RELATIONS

Many animals form close relationships with other species, with which they have often evolved in tandem. An association that benefits both parties is called mutualism. For example, oxpeckers (*Buphagus* sp.) obtain nutrition by picking parasites from the skin of grazing herbivores such as zebra (*Equus* sp.). In return they act as early warning signals, flying up into the air and screeching at the approach of a predator.

In situations of commensalism, only one species benefits, while the other is unaffected. Remoras gain protection and food scraps by living in association with large fishes, which neither gain nor lose anything from the relationship.

In parasitic relationships, one party, the host, is deleteriously affected, sometimes even killed.

Group living Mandrills (*Mandrillus sphinx*) live in troops of a few to several dozen animals. The young learn about selecting and preparing food by observing their mothers. In this way, group traditions in food preference develop. Adults will also prevent juveniles from eating unfamiliar food. Like other social primates, mandrills use various forms of touch, including grooming, to cement social bonds.

Cubs beware Lions (*Panthera leo*) live in prides of five to 15 adult females and their offspring, and one to six adult males. Newly installed dominant males usually kill all cubs sired by previous males. The absence of cubs prompts the females to come into season and to mate with the new male. Infanticide thus improves the chances of survival for the new male's cubs.

Mutual benefit Some animals that would normally be potential prey for another species gain its protection by offering a useful service. Although all moray eels are carnivorous, they do not eat cleaner shrimp, instead relying on the shrimp to remove parasites (left). In return, the shrimp gets both food and protection, as the eel's presence deters other predators.

At home in the sea The sea otter (*Enhydra lutis*) is the only marine mammal to use tools, hammering mollusks with a stone to break the shell and extract the flesh. It dives for its food, feeling with agile forefeet for mollusks and crustaceans such as clams, sea urchins, abalone, crabs, and mussels. Most dives last a little more than 1 minute, although some otters can stay down for up to 5 minutes. The otter stores its prey in loose skin folds at its armpits, then surfaces to eat, or to breathe and dive again. A social animal, it lives in sex-segregated groups, known as "rafts," of a few to a couple of hundred animals.

LEARNING POWER

All animals are born with innate capabilities, responses, and adaptations, which are known collectively as instinct. Many, however, also have some capacity to learn. Even some invertebrates are capable of the simplest form of learning, known as habituation. This involves the reduction in a reflex as a response to repeated but inconsequential stimulation. For example, the marine snail (*Aplysia californica*) will retract its siphon (which is used to expel waste matter) in response to touch. When repeated touching has no negative effect, reflex retraction of the siphon ceases.

Through conditioning, an animal learns to anticipate a consequence in response to a stimulus. A bird, for example, will recognize and avoid individuals of a particular caterpillar species if it has already tried to eat one and found it distasteful.

Animals living within well-defined social structures often have strong capabilities for trial-and-error learning. This occurs when an animal learns, often through repeated mistakes, the most advantageous way to perform a particular behavior. Many young animals learn survival skills in this way through play. They also learn vital lessons by watching and imitating a parent or other elder.

CULTURAL RELATIONS

Through imitation, some animal populations transfer learned skills between individuals and even generations so that they develop a form of culture. Animals in which this has been recognized include primates, cetaceans, and elephants. A famous example, documented in the 1950s, occurred within a macaque troop living on a small island off Japan. Scientists fed the monkeys sweet potatoes. One monkey washed its potato in water to remove the dirt. Within a short time, all the monkeys were doing this, and it became an entrenched behavior among that troop.

There are also many examples of chimpanzees developing tools, such as sticks to extract termites from mounds, and then passing on the skill to peers and offspring.

Interlopers Cuckoos are famed for their habit of laying eggs in the nests of other species, such as a sedge warbler (below). However, only about 50 of the 130-odd species are truly parasitic. In some parasitic species, the cuckoo's single young pushes other eggs or chicks out of the nest; in others, the young cuckoos out-compete the host's hatchlings for food and they die. Cuckoos' eggs often mimic those of the host, and different females within one species of cuckoo may each be adapted to parasitize one species of host.

HAVING AN IMPACT

Many animals have a profound influence—be it beneficial or deleterious—on the environments in which they live. For example, earthworms' actions break down vegetative material, thus improving soil fertility. Swarming locusts, however, can reduce or destroy an area's productivity by stripping vegetation before moving on.

Elephants can be similarly destructive in their pursuit for food, often pushing over entire trees to reach the most succulent leaves.

Beavers can create an even more significant effect through their dam-building, by changing the nature of the water flow in streams and rivers.

UNDER CONSTRUCTION

Although beavers are the best-known builders in the animal world, many other species have impressive construction skills. Among the most accomplished avian nest-builders are African weaver birds, the males of which carefully interlace and knot grass stems and leaves into huge structures with several entries.

Many insects construct shelters and traps, but the supreme insect builders are termites. The mounds of some species can be 30 feet (9 m) tall and house millions of insects.

CALLING CARDS

By emitting chemical signals known as pheromones, animals can leave behind or transmit messages to other members of their species. Recipients use their sense of smell, and sometimes also taste, to detect these signals.

Pheromones often produce a behavioral response among recipients and can sometimes even have a physiological reaction. For example, a pheromone produced by the queen in honey-bee hives suppresses female workers' reproductive systems and stimulates drones to mate with the queen only. Chemicals in the urine of queens in naked mole rat colonies are thought to perform a similar function.

RECEIVING AND SENDING MESSAGES

All of the five basic senses that humans have—sight, touch, taste, smell, and sound—also function throughout the animal kingdom, although the organs of reception often vary. Antennae in many ants, for example, respond to touch in

the same way the legs of spiders detect vibrations. And, like the antennae of many moths, they can also detect chemical messages in a sense akin to smell.

The range of perception of sense organs also varies. Elephants and cetaceans, for example, can perceive and communicate by deep rumbling infrasound undetectable to the human ear. Certain animals, such as canids and sharks, are able to follow scent trails so faint that they are undetectable to humans. Likewise, the visual capacity of animals that rely on sight can vary significantly between different animal groups.

Animals also have senses for which humans have no equivalent. Fishes, for example, have lateral lines, systems of jelly-filled canals under the skin that are connected to the outside by open pores. These perceive tiny pressure changes in the surrounding water, which indicate the presence of an object or another animal.

ECHOING

Echolocation has evolved independently several times in the animal kingdom. Bats, toothed whales, porpoises, and some birds use this elaborate form of sonar to navigate, to locate and capture prey, and for social interactions. Even some mammals, such as shrews and tenrecs, use a basic form of echolocation to find their way in lightless conditions.

An animal using echolocation will emit sounds, such as clicks in dolphins and squeaks in bats, that are beyond the range of human hearing. It then interprets the angle and quality of the echoes that bounce off solid objects nearby. The frequencies used range from about 1,000 hertz in birds; 30,000 to 120,000 hertz in bats; and at least 200,000 hertz in whales.

Seeing in the dark To be able to chase and capture prey in the dark, a bat such as this greater horseshoe bat (*Rhinolophus ferrumequinum*) uses echolocation. This enables it to simultaneously recognize all other objects in its vicinity—the ground, trees, bushes, rocks, other bats, and owls. The bat changes the frequency and rapidity of the sound pulses to enable it to discern different types of information about its prey, such as how far away it is and what speed it is travelling at.

Warning signal Rattlesnakes, such as the northern Pacific rattlesnake (*Crotalus viridis oreganus*), have the tip of the tail modified into a warning device, the rattle, which is formed of specially shaped dry scales. When the tail is vibrated, the segments move against each other to create a buzz.

Attracting and repelling Like most moths, the African moon moth (*Argema mimosae*) releases pheromones as signals to other moths. In general, female moths release pheromones to attract males from a distance, while males do so only when close to a female, to sexually stimulate her. As with most insects, scent detection is performed by the antennae. Some moths also emit ultrasounds, which interfere with the echolocation of predatory bats.

House proud Males of most bowerbird species build and decorate elaborate structures that are neither nest nor shelter, but exist only to impress females. The type of bower and its decorative scheme are species specific, with most species having distinct color preferences. The western bowerbird (*Chlamydera guttata*) (left) builds two parallel walls of sticks; others build platforms or mats. But the greatest avian architect is the Vogelkop bowerbird (*Amblyornis inornatus*), which builds a maypole-like structure up to 5 feet (1.5 m) high. Males devote most of their time to building and decorating their bowers.

Good and bad impacts When food is scarce, African elephants (*Loxodonta africana*) can devastate large areas by pushing over trees to reach the topmost twigs. However, they play a pivotal role in their ecosystem in various other ways: by dispersing seeds and distributing nutrients through their dung; by providing water by digging waterholes that other species use; and by enlarging existing waterholes when bathing and wallowing. They make paths, usually leading to waterholes, which act as firebreaks. Also, they provide protection for other species, as their height enables them to see far and warn of danger.

HABITATS AND ADAPTATIONS

Habitats are the locations or surroundings in which living organisms survive. They are the providers of food, shelter, and other fundamental requirements necessary for plants and animals. Different habitats are most often characterized by a combination of their climate and geography but sometimes also by the communities of species that dominate them. Coral reefs, for example, occur in shallow and mostly tropical oceanic waters in which the coral polyps have laid down their calcium-carbonate skeletons. Different forest types, however, are defined by combinations of characteristics such as their latitude, total rainfall, whether the rainfall is received consistently throughout the year or in a marked season, and the dominant vegetation.

BETWEEN WORLDS

While most species exist within one habitat, there are many that exploit or regularly move between more than one. Emperor penguins huddle together on ice in Antarctic breeding colonies during the bitter polar winters but feed on fishes in open oceanic waters long distances away. Migrating species, too, often move between habitats to exploit or avoid different or changing conditions. Within habitats, animals function in ecological niches. This defines the roles that animals play and includes what they eat, what eats them, where they find shelter, where they find food, and how they respond to and interact with other organisms. Some species are adapted to a very specific niche with a narrow range of parameters. Koalas, for example, survive only where their food trees—specific eucalypt species—grow.

Other animals such as some cockroach pest species have more generalist adaptations and are able to operate within wide niches or across several niches.

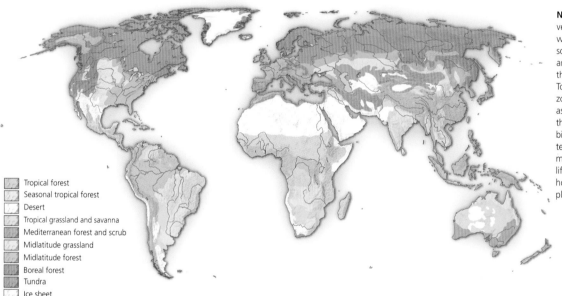

- Tropical forest
- Seasonal tropical forest
- Desert
- Tropical grassland and savanna
- Mediterranean forest and scrub
- Midlatitude grassland
- Midlatitude forest
- Boreal forest
- Tundra
- Ice sheet
- Mountain vegetation

Natural vegetation This map shows the vegetation that would occur naturally without human interference. Climate and soil determine the vegetation of an area, and the vegetation, in turn, determines the extent and diversity of animal life. Together, the plants and animals of each zone form an ecological community known as a biome. The rainfall and warmth of the tropics have given rise to the greatest biodiversity in the world. In addition to the terrestrial zones indicated, freshwater and marine biomes contain abundant, diverse life-forms. The oceans, for example, are home to life-forms ranging from minuscule plankton to giant whales.

Tree dweller The koala (*Phascolarctos cinereus*) builds neither nest nor den. For this marsupial, eucalypts provide both a home and a food source. The species of eucalypts that the koala eats have a high fiber and cellulose content as well as high concentrations of toxic compounds. The koala must chew on large quantities of leaves to extract sufficient nutrients.

Hibernating hedgehog Hibernation is dictated by environmental conditions and is not a species trait. When conditions are unfavorable, hedgehogs undergo a period of dormancy. Their body temperature drops to a level close to that of the surrounding air. Tropical hedgehogs don't hibernate but, if they are artificially exposed to low food levels or low ambient temperatures, they will exhibit the ability to do so.

DEMARCATION ISSUES

The line where one habitat begins and another ends is not necessarily abrupt or clearly defined. Many forest habitats, for example, give way gradually at the outer margins to open habitats such as grasslands. In some cases, however, it is very clear where a habitat begins and ends. Caves are habitats that are frequently well delineated by their geology. Life within is defined by factors such as low light levels and the availability of food. Conditions at a cave's entrance are often very different to those further into the cave. Similarly, the deep-sea hydro-thermal vent habitats discovered in recent years occur only within very well-defined and highly productive oceanic areas.

From high to low Altitude as well as latitude play a part in delineating habitats. The snow-covered peak of Mt. Kilimanjaro contrasts with the semiarid veldt below.

Carlsbad Caverns National Park This seemingly inhospitable, vast cave complex in New Mexico is a habitat for bats during the summer breeding season. They hang from the ceilings in their thousands.

ENVIRONMENT CHANGE

Habitats are not fixed or static, and it is natural that they change over time. Many are shaped by sudden, dramatic geological events such as earthquakes and volcanoes, while some shift more slowly in response to a persistent natural force such as erosion. Changing weather patterns often create seasonal, recurring, and reversible change. In addition, some habitats change in response to the animals and plants that are living and evolving within them.

However, many changes being witnessed within Earth's habitats are not the result of natural forces or events. Rather, it is human activity that now drives the most significant transformations. Most habitats, for example, are shrinking in area. Major exceptions include the ever-expanding desert, semi-desert, and urban habitats. It is also significant that the diversity of life within most habitats is seen to be on the decline.

The more complex and species-rich habitats, such as the rain forests and coral reefs, were once thought to be the most resilient to change. In recent times, however, they have exhibited signs of being among the most fragile.

RAIN FORESTS

Rain forest is a moist and exceptionally biodiverse habitat dominated by broad-leaved trees with shallow root systems. Annual precipitation exceeds 40 inches (1,000 mm) and falls consistently throughout the year. Tropical rain forest occurs in more than 80 countries (largely those in South and Central America, West Africa and Southeast Asia), stretching around the equator between the tropics of Capricorn and Cancer.

It covers just 7 percent of Earth's land surface but supports at least half of all the plant and animal species. Temperate rain forest occurs in cooler climes, particularly along North America's Pacific coast and in pockets of southern Australia, New Zealand, Norway, Japan, and the United Kingdom.

Egg strategy The red-eyed tree frog (*Agalychnis callidryas*) lays its eggs on leaves overhanging water. As soon as the tadpoles hatch, they wriggle down into the water.

Under cover The ocelot (*Leopardus pardalis*) is dependent on the dense cover afforded by parts of the rain forest for its protection. Its spotted coat also acts as camouflage. Ocelots came under the protection of the International Convention of Trade in Endangered Species in 1975.

The Nouragues Reserve The canopy of this rain forest in French Guiana, northern South America, is home to numerous bird species such as macaws and toucans. Well hidden from view, in the lower regions of the forest, are jaguars and other big cats.

LAYERED LIFE

Rain-forest life takes place within vertical layers. The highest treetops reach up to 130 feet (40 m), creating an emergent sunlit, windy stratum patrolled by large predatory eagles.

Most tree crowns, however, stop just below this height to form a leafy canopy rich in insect life and so dense it is almost impenetrable to sunlight. The birds and mammals that abound here rarely descend to the forest floor. Communication by sound is common and this stratum rings constantly with a cacophony of howls, whistles, and screeches.

Low-light adapted shrubs and small trees frame an understory in which big cats, such as the jaguar in South America and the clouded leopard in Southeast Asia, hunt.

The forest floor is home to an array of insects as well as rodents and large mammals such as tapirs, elephants, and gorillas.

Exceptionally high moisture levels mean that frogs and other animals normally associated with aquatic environments are found in greater abundance in rain forests than in other terrestrial habitats.

TROPICAL MONSOONAL FORESTS

Tropical monsoonal forest, known also as tropical deciduous forest, grows in areas of high annual precipitation that falls mostly during a distinct wet, or monsoon, season. This habitat experiences a distinct dry season during which no, or scant, rain falls and most trees and shrubs lose their leaves. Monsoon forest trees do not grow as tall as those in rain forests, and the understory is more developed because it receives higher levels of light. There is a distinct seasonality to life in this habitat because most plants flower and fruit at the onset of the wet season. Monsoon forest is most common in Southeast Asia.

Formidable lizards The Komodo dragon (*Varanus komodoensis*) is found on a few tropical islands in central Indonesia. These fierce animals feed primarily on carrion but also prey on large mammals such as deer and water buffalo.

Enveloped city Almost hidden in the encroaching tropical forest is the city of Mrauk-U, Myanmar (Burma). Depletion of forests in some areas to service the needs of the human population is causing loss of animal and plant habitat.

CONIFEROUS FORESTS

This habitat is dominated by evergreen coniferous trees such as firs and pines, which have mostly thin, needle-like leaves and produce seeds in cones. They usually occur in areas with long, very cold winters and moist, short summers. Animal life here often avoids harsh winters by hibernation or migration. Huge areas of land in the Northern Hemisphere are covered with coniferous forest, known as taiga or boreal forest, that stretches right around the northern end of the planet, between about 50 and 60 degrees north. Coniferous forest is also a feature of mountainous areas throughout the rest of the world.

Gypsy moths A naturalist brought the gypsy moth (*Lymantria dispar*) to the United States from Europe in 1869. Some escaped and the species quickly became a serious pest. The caterpillars consume huge amounts of leaf material in spring. Pine and spruce trees are most affected.

Giant among trees On California's wet north coast, Redwood National Park is a vast area with stands of Earth's tallest tree, the redwood. These slow-growing giants live up to 2,000 years, forming a complex ecosystem in which hundreds of different plants and animals live. Logging poses a continued threat to old-growth forests.

TEMPERATE FORESTS

Temperate forests occur around the world at latitudes between 25 and 50 degrees and are usually dominated by broad-leaved trees. In areas with cold winters beset by frost, as in eastern North America, eastern Asia, and Western Europe, these forests are usually deciduous. The four seasons are well defined, with fall marked by a spectacular color change in leaves before they drop in their entirety. Animals in these forests typically cope with the harsh winters by hibernating or migrating. In temperate forests in South America, southern China, New Zealand, eastern Australia, Japan, South America, and Korea, where the winters are much milder and most often frost-free, the dominant trees are evergreen.

Grand Teton National Park In fall, this vast area turns golden. The Snake River meanders through the park. Bald eagles and ospreys are common along the shores. Beavers, weasels, and coyotes frequent the marshes, meadows, and woods.

Lookout duty The raccoon (*Procyon lotor*) is a solitary mammal that spends its days sleeping in hollow trees, or a burrow in winter. The forest floor is rich in organic matter that feeds the animal population.

Marsupial downunder The Leadbeater's possum (*Gymnobelideus leadbeateri*) feeds on plant exudates and insects. Extensive logging in areas of Australia's old-growth forest that it inhabits has put its survival at risk. It is now classified as endangered.

SEASONAL LIFE

A distinct seasonality marks the life of animals that dwell in temperate deciduous forests. As spring nears, the weather warms and food begins to become plentiful. Hibernating residents such as the hedgehogs and bears awaken. Great numbers of species return from over-wintering in warmer climes. Many migratory birds, in particular, arrive to feast on the booming insect populations so typical of these forests in spring. Also, courtship and reproduction proceed at a frenzied pace so that offspring can be born and have a solid start to life while conditions are at their most favorable.

Seasonal lines aren't always so well defined in evergreen temperate forests. Life doesn't batten down the hatches to survive through the cold months in the same extreme way, and the peak period of growth and reproduction tends to start earlier, in the late winter. These forests also tend to have more well-developed understorys of shrubs and the smaller trees so that ground-dwelling species are more plentiful than they are in their deciduous counterparts.

An ancient oakwood forest, England
Oaks are hardy and long-lived shade trees. They bear catkins (male flowers) and spikes (female flowers) on the same tree.

MOORLANDS AND HEATHS

This harsh habitat features dwarfed bushes and grasses and little other vegetative cover, so most animal life tends to live close to or under the ground. Although often associated with the United Kingdom, moorlands occur worldwide in exposed or high open areas, along coastlines and on mountains. The Ethiopian highlands, for example, are dominated over large sections by moorlands inhabited by the critically endangered Ethiopian wolf (*Canis simensis*). In Europe, the moorlands are often associated with heath in a habitat that was originally created, up to 5,000 years ago, when woodland was cleared by early farmers. This now supports endangered species such as the natterjack toad (*Bufo calamita*).

Torres del Paine National Park This park in Patagonia, Chile (far right), comprises steppes, heath, forest, and desert—and is home to numerous species of mammals, including the guanaco (*Lama guanicoe*). These, the larger of the two wild camelid species, thrive in the rough terrain.

Drawn to honey Honeyeaters are largely restricted to the Australasian region. These birds have a uniquely structured tongue. It has a deep cleft and is fringed at the tip so that it forms four parallel brushes. This is an adaptation to nectar feeding. Here, a yellow-faced honeyeater (*Lichenostomus chrysops*) is feeding on banksia nectar.

OPEN HABITATS

Low annual rainfall of 20–30 inches (500–750 mm) limits tree growth in these places, which often represent transition zones between forests and deserts. Sprawling flatlands dominated by grasses offer few hiding places for large animals. Nevertheless, this habitat type is often dominated by big herbivores. Grazing together exploits a safety-in-numbers strategy that is intended to confuse and overwhelm potential predators. Huge herds of bison and pronghorns once roamed North America's prairies. Africa's savanna is still renowned for massive, albeit declining, mobs of zebras, antelope, and other large mammals that are hunted by big cats and wild dogs. Other open habitats include the Ukrainian steppes, the South American pampas, and southern Africa's veldt.

High rise Termite mounds are conspicuous on the dry, eucalyptus-dotted grasslands of north-eastern Australia. Some bird species make their nests in them. The mounds are ventilated by tunnels to keep temperature and humidity in the optimum range.

Serengeti elephant herd Serengeti is the Masai word for endless plain. Tanzania's oldest and most famous park consists of flat, treeless plains stretching as far as the eye can see. An astounding assemblage of animals grazes the vast area.

DESERTS AND SEMIDESERTS

These habitats, which cover about one-third of the planet's land surface, are characterized by a lack of moisture. This is caused by high evaporation or low precipitation rates, or a combination of both. The deserts average less than 10 inches (250 mm) of rain per year, while semiarid habitats receive 10-20 inches (250-500 mm) annually. Precipitation is typically sporadic and unpredictable. Most of the world's driest areas lie between 15 and 40 degrees north and south of the equator. Searing daytime temperatures are common but do not define these habitats. They frequently experience freezing overnight temperatures. Much of the Antarctic, which is bitterly cold year-round, fits the definition of desert.

A sideways move In order to move over surfaces such as loose sand and soft mud, snakes adopt a locomotion method known as sidewinding. The snake uses a point of contact with the ground as purchase, then lifts its trunk clear of the ground to secure another point of contact. Shown here is a Peringuey's desert adder (*Bitis peringueyi*) sidewinding its way across a dune in the Namib Desert in Africa.

HIDDEN LIFE

These habitats invariably support more life-forms than most people appreciate. Extreme temperatures and dryness are the major concerns for desert-dwelling animals, and most manage to cope by using a combination of physiological and behavioral adaptations.

The smaller animals are typically nocturnal, hunting and foraging under cover of darkness to avoid the daytime higher temperatures and evaporation rates. These animals include a range of arthropods such as spiders, centipedes, and scorpions as well as gerbils, ground squirrels, kangaroo rats, and other mammals. Underground burrows are the most common source of shelter during the day, due not only to the lack of vegetative cover but also to the milder conditions that usually prevail below ground.

Heat absorption in large animals, including camels, goats, donkeys, emus, and ostriches, is limited by thick coats of fur or feathers. These limit heat loss as well, however, and so these animals also have sparsely covered regions of the body for the purpose of emitting excess body heat.

Fog catcher Fog rolling in from the coast is the only regular moisture in the Namib Desert. To utilize this resource, the darkling beetle (*Onymacris baccus*) points its rear into the wind and patiently waits while fog condenses on its back and water droplets trickle along dorsal grooves to its mouth.

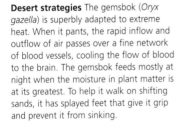

Desert strategies The gemsbok (*Oryx gazella*) is superbly adapted to extreme heat. When it pants, the rapid inflow and outflow of air passes over a fine network of blood vessels, cooling the flow of blood to the brain. The gemsbok feeds mostly at night when the moisture in plant matter is at its greatest. To help it walk on shifting sands, it has splayed feet that give it grip and prevent it from sinking.

Ostriches on a salt plain The powerful legs and large toes of the ostrich enable it to run swiftly, compensating for its inability to fly. Ostriches eat fruits, seeds, succulent leaves, grasses, and the shoots of shrubs. They cover great distances of inhospitable terrain such as the Etosha Pan salt plains of Namibia, in search of food and water.

MOUNTAINS AND HIGHLANDS

Mountains are like terrestrial islands. The process of evolution often occurs in isolation, so that the animals on one mountain can be unique from those of other mountains—even those located nearby. Life in the mountains, however, always displays a layered pattern to its structure, mainly because the climatic conditions alter with altitude. Air gets colder, for example, with increasing elevation, so that even tropical mountains such as Mt. Kenya and Mt. Kilimanjaro, both in Africa, can be permanently capped with snow. Agility is a feature of large mountain animals such as the chamois in Europe and the snow leopard in Asia. The smaller animals often shelter from environmental extremes in burrows and hibernate in winter.

Scavenger of the sky The king vulture (*Sarcoramphus papa*) rises on thermals, and soars great distances. It is unable to smell carrion and often works with turkey vultures, relying on them to detect a carcass. Once one is discovered, the king vulture descends, rips open the carcass, eats its fill, and then lets other birds feed.

Fog in the Pyrenees The thick fogs, high peaks, and winter snows of the Pyrenees create a daunting barrier between France and Spain. The areas of high pasture in the Pyrenees were once a habitat for raising livestock. Now, the economy of European high mountains is increasingly focused on tourism and recreation.

TUNDRA

True tundra is one of the harshest and least biologically diverse habitats: a vast, treeless, Northern Hemisphere world north of 55 degrees latitude covering one-fifth of Earth's land surface. Very little snow or rain falls and nutrients are scarce. During the long, bitterly cold winters, unbroken darkness endures for months and the ground remains frozen to a depth of a least 10 inches (25 cm). Summers are exceptionally short. The Sun shines 24 hours a day, the upper permafrost melts, and shallow, ephemeral lakes form in the wet ground. Migratory insects, birds, and mammals move in to take advantage of this very limited growing season.

Musk oxen, Alaska In the bitter cold, fur and feathers are important survival tools. The musk ox (*Ovibus moschatus*) has guard hair 24–36 inches (60–90 cm) long, which is the longest of any animal. Ptarmigan fly into soft banks of snow to sleep and are among the most heavily feathered of birds.

Arctic tundra in fall Some Arctic animals such as the snowy owl (*Nyctea scandiaca*) and the snowshoe hare (*Lepus americanus*) turn from brown in the warmer months to white in winter. The ability to change color allows them to blend in with the season's colors and go unnoticed by predators.

POLAR REGIONS

Antarctica is the coldest, windiest, driest continent and the Arctic is almost as inhospitable. Despite this, a surprising level of animal life is able to survive at the poles. The Arctic Ocean, for example, supports some of the world's most productive fisheries, while in Antarctica, Wilson's storm-petrel (*Oceanites oceanicus*) is a bird that breeds in its millions. Extreme cold is the main driver of animal adaptations. Thick layers of fat and fur or feathers help prevent heat loss in mammals and birds. Often the animals migrate seasonally to avoid the harshest conditions. Many fishes and invertebrates are protected by a natural "antifreeze" in their body fluids.

A parade of penguins From a young age, the thickset Adélie penguin (*Pygoscelis adeliae*) grows dense, short, fur-like feathers that provide protection from icy conditions in the Antarctic.

Cold comfort Polar bears (*Ursus maritimus*) have a mechanism limiting heat loss. Warm blood from the heart transfers heat to cool blood returning from the skin.

Beneath the ice Seafloor-dwelling species such as sea anemones and sponges are able to survive and flourish in the relatively tranquil waters where the spread of polar ice does not extend.

LAKES AND PONDS

Some 5 million lakes and many more ponds are scattered across the continents. The older they are, the more biodiverse they tend to be, and the more endemic species they contain. Most pond and lake fauna falls into three distinct categories. Zooplankton live mainly in the surface waters and are tiny animals such as crustaceans and rotifers. Nekton includes fishes and some large invertebrates that consume zooplankton and live throughout the water column. They also feed on benthic animals, mostly invertebrate bottom-dwellers such as insects, crustaceans, and mollusks. Most ponds and lakes also have animals living on and in surrounding vegetation and in bank sediments.

The Pantanal This area is a haven for one of South America's greatest concentrations of wildlife—home to more than 600 bird species. In the wet season, it becomes the world's largest freshwater wetland.

At rest before migration The snow goose (*Anser caerulescens*) breeds in the Arctic regions of North America on the tundra and winters on the Pacific coast in salt marshes and marshy coastal bays.

SEAS AND OCEANS

Covering almost three-quarters of Earth's surface and with an average depth of 10,500 feet (3,000 m), these vast marine environments comprise more area than any other habitat. Most animals are pelagic (living within the water column) or benthic (bottom-dwellers). Pelagic fauna ranges from the tiny planktonic creatures drifting at the mercy of the currents to strong swimmers such as marine mammals, fishes, and squid. Sunlight has a direct impact on life in the surface layers of open water and in shallow benthic communities. And its absence, along with increasing pressure, has been a key force behind many adaptations of animals living in the depths of the oceans.

Breaching dolphins The spinner dolphin (*Stenella longirostris*) is famous for its fantastic spinning leaps. Its main threat has come from tuna fisheries, which have caused hundreds of thousands of deaths.

Kelp forest A dependence upon sunlight for photosynthesis restricts kelp forests to clear, shallow water. Various invertebrates attach to drift kelp, including nudibranchs and barnacles.

MANGROVE SWAMPS

Mangrove swamps stabilize shorelines in wave-protected river deltas, estuaries, and coasts in the tropics and subtropics. Almost three-fourths of tropical coastlines are narrowly fringed with this valuable habitat. It is dominated by salt-tolerant, shallow-rooted trees and shrubs. The tangled, extensive roots of these plants stabilize sediment, capture vegetation, and create an environment in which decomposing organisms ranging from bacteria to worms can flourish. They also provide substrate and shelter for a wide range of larger invertebrate fauna from mollusks to crustaceans. Although rarely entered by large vertebrates, mangrove swamps play a critical role as nurseries for a wide range of fishes at the larval and juvenile stages.

Barnacle feast Aerial roots that carry air down to the bottom roots of mangrove trees form vertical microhabitats with distinctive zoned communities of animals. The position of this mangrove crab (*Scylla serrata*) on the root is determined by the animal's tolerance of tidal exposure.

Coastal mangroves Cordgrasses and mangrove plants have root systems that exclude most of the salt in the water they take in. Any that does make its way into the plant sap is excreted through pores on the surface of the grass blade or mangrove leaf.

CORAL REEFS

The foundations of coral reefs are formed largely by the hard external skeletons of small invertebrate animals called polyps, close relatives of sea anemones and jellyfishes. Not all coral polyps are capable of forming reefs. Those that are capable are classified as hard corals and live in interconnected colonies in which individuals extrude and sit within hard, protective, cup-shaped deposits of calcium carbonate. As these animals die, they leave behind their skeletons and successive generations of polyps build on top of them. Other invertebrate animals that produce calcium carbonate, including sponges and mollusks, as well as certain algae, also contribute to coral-reef formation.

Life on the reef Coral reefs support an abundant, colorful, and diverse group of animals that live on, burrow into, and even eat, the hard coral. Inhabitants face strong competition for food, space, and mates, and the constant threat of predators.

 Deep-water coral reefs

 Warm ocean

 Warm-water coral reefs

RESTRICTIVE NEEDS

Limestone coral polyp skeletons provide a substrate upon which other sedentary invertebrates and algae settle and grow, and within which a huge diversity of animal life lives, feeds, and shelters.

Most of the world's coral reefs are located between the tropics of Capricorn and Cancer. They are

Fragile habitat Every year, the world's reefs are damaged by oil pollution, sewage, sedimentation, over-fishing, and the effects of global warming. Cold-water corals are also under threat from trawling.

best developed in clear waters with temperatures of 70–85°F (21–29°C) and at depths of less than 33 feet (10 m). This is due for the most part to the temperature and light requirements of unicellular algae, known as zooxanthellae, which live in a symbiotic relationship within coral polyp tissue. The hard, reef-building corals cannot live without zooxanthellae. They produce the essential nutrients for the polyps, which in turn provide food and shelter for the algae.

The most common coral reefs are the fringing reefs that run parallel to the coast and are separated from shore waters by shallow lagoons. The largest are barrier reefs which, similarly, run parallel to coastlines but also run beside wider and deeper lagoons.

Great Barrier Reef, Australia This reef is the biggest of all coral structures, and the largest biological entity on Earth. Warm-water coral reefs rival tropical rain forests in the range of flora and fauna they support.

COASTAL AREAS

Beaches, oceanic cliffs, sand dunes, intertidal rock pools, and shallow coastal waters are all included in this habitat. Life in these coastal areas can be exposed to fierce winds, salt-laced sea spray, crashing waves, and tidal influences. Invertebrates are the dominant animals, from predatory polychaetes that burrow and hunt on moist beach sands to well-camouflaged, hard-shelled gastropod mollusks clinging with amazing tenacity to rocks constantly battered by waves. Marine birds breed on cliff faces and patrol beaches and rocky shorelines for food. And in some protected coastal areas, marine mammals such as seals haul themselves up to mate and raise their young. Marine turtles, too, find their nesting sites on certain select beaches.

Dugong grazing Dugongs (*Dugong dugon*) are found in the coastal waters of the Pacific and Indian oceans. Their few defenses, coupled with low reproduction rates, have resulted in their classification as vulnerable.

Rocky shores The Twelve Apostles in Victoria, Australia, is a marine national park and sanctuary. Underwater arches, canyons, caves, and walls attract schools of reef fishes and breeding colonies of birds.

RIVERS AND STREAMS

Animal life in these habitats is strongly influenced by water-quality characteristics such as dissolved nutrient levels, oxygen content, and turbidity as well as the pace of water flow. The more consistent these are, in both space and time, the more diverse life tends to be. Some rivers and streams experience major seasonal fluctuations, their waters swelling and racing with summer rains or spring thaws, and animals often have life-cycles that accommodate such extremes. Life also tends to be best adapted to either benthic or open-water environments, as it does in other aquatic habitats. River and stream banks and vegetation provide additional niches.

Leaping food Grizzly bears (*Ursus arctos horribilis*) emerge hungry from their winter hibernation. They forage on lush grass beside streams until the salmon runs begin each summer. Streams can be thick with salmon, the fishes jammed side by side, tail to head. Bear hierarchy along the riverbank ensures that the biggest bears secure the best fishing spots.

Hippopotamuses in a water hole The hippopotamus (*Hippopotamus amphibius*) has its ears, nose, and eyes on the top of its head so it can remain submerged and still know what is going on around it. These animals move onto land at sunset.

WETLANDS

Wetlands are often vital to the health of other habitats because they soak up and slow down rising floodwaters and filter out and capture the excess nutrients, sediments, and pollutants before they can reach bodies of open water. It is for the latter reason that they are sometimes described as Earth's kidneys. Water-soaked and characterized by sodden soil, wetlands encompass all the fresh and brackish water environments that lie between dry land and open water. Different types include swamps, marshes, fens, floodplains, deltas, and bogs. Many are permanent while some form seasonally. They are found worldwide along rivers, streams, and lakes and in flat areas inundated by melting snow or rising water tables.

CRITICAL REASSESSMENT

Up until the latter decades of the past century, wetlands were widely held to be unproductive wastelands worthy of little other than in-filling. As a result, many have been destroyed or damaged. These days, about 6 percent of the world's land surface is thought to be still covered by these habitats.

It is now understood and widely appreciated that as well as helping to protect and purify other habitats, wetlands support an extensive variety of plant and animal species, most of which have both terrestrial and aquatic adaptations. Permanent wetland residents include fishes; amphibians such as frogs and toads;

turtles, crocodiles, snakes, and lizards; and, a range of invertebrates, including insects and crustaceans.

But wetlands are also often crucial, transient places of residence for migrating animals, particularly birds. Some exploit wetlands as stopovers for feeding and resting while en route to other locations. But many, such as herons, egrets, pelicans, and other waterbirds, also use them as breeding grounds.

Horses in the Camargue A breed of wild, small, white horses is found only on the Camargue—a wetland and salt marsh area in southeastern France. The hardy animals are able to weather the bleak, cold winters and very hot summers of the region.

Alligator resting The American alligator (*Alligator mississippiensis*) resides in the swamps of the southeastern United States. When hunting, it uses its heavy head to smash through densely vegetated areas, and then grabs its victim in its blunt, broad jaws.

Preening spoonbills Roseate spoonbills (*Platalea ajaja*) inhabit the coastal swamps of the Americas. To feed, a bird sweeps its elongated, partly open bill from side to side in the water. When prey is found, the bill snaps shut over the food in an instant.

URBAN AREAS

In urban habitats, humans and their activities are dominating forces. Upper levels of food chains, for example, are controlled by people and their domesticated animals, such as dogs and cats. Such factors combine to ensure that the animal life of urban areas is comprised of a small number of species with many individuals. Animals with opportunistic or wide-ranging diets and shelter requirements abound, including sparrows, pigeons, rats, cockroaches, and other insects. Those animals with boom-or-bust population cycles tend to do particularly well. However, species with naturally low reproductive rates rarely survive. Nor do those with narrow habitat requirements or those sensitive to air and water pollution, such as many frogs.

Swallow feeds its young Swallow nests are made of layers of grasses and small mud or cow-dung pellets. To protect them from disintegrating in the rain, swallows choose a site such as rafters in barns, or under bridges.

High-rise living Largely a cliff-nesting species, the Peregrine falcon (*Falco peregrinus*) has adapted to nesting on the ledges of large buildings in North American and European cities. This fierce, fast predator can kill in mid air.

PARASITIC LIFESTYLES

Some animals, mostly small invertebrates, exist for all or much of their lives inside other organisms. To survive, these parasites need to be adapted to the internal chemical and physical environments of their plant or animal hosts. They also need to resist or respond to the defences targeted at them by the immune systems of the organisms in which they live. Animals that lead parasitic lifestyles require a way of passing between hosts. For some this means surviving the rigors of the external environment with encysted or resilient dormant stages. Others use the bodies of other organisms as intermediate hosts to avoid external habitats.

Unwelcome guest Nematodes, or round worms (left) of the *Trichostrongylidae* family are a common parasite of ruminants. Lodged in the small intestine, they weaken their host and arrest its development.

Overwhelmed Parasitic wasp insects (left) often lay their eggs in paralyzed but still-living caterpillars so that the young have a source of fresh food. Sometimes, more than one parasite lays in a single caterpillar.

Grasshopper residents (above). The egg larvae of parasitic mites usually attach at the base of an adult grasshopper's wings and feed on the host until adulthood. It is thought that they cause the host no harm.

ANIMALS IN DANGER

Biodiversity is the total of all animals, plants, and microorganisms living on Earth, and the genetic information that they carry. The term encompasses the way in which species interact with one another as well as with the ecosystems in which they live. Research has shown that preserving biodiversity is essential to habitat health and stability, which, in turn, can have profound quality-of-life and economic implications for human populations. Another argument for maintaining as many of Earth's species as possible is that their genes often provide a source of new substances for industrial or medical applications. Further, because biodiversity has been involved in the development of all human cultures, there are spiritual arguments in support of species conservation.

Amazon clearance The rain forest is not inexhaustible, yet logging continues (left). Though the lushness of the forest suggests that it is growing on rich soil, the reality is that regular heavy rain leaches most of the nutrients from it: nutrients within the ecosystem are fixed in the living wildlife. When organisms die, they break down and the minerals are taken up by the forest. If large areas are cleared, the soil is depleted in a few years and the trees cannot regenerate.

Modern medicine In the search for new pharmaceuticals, the skin secretions of toads and frogs such as Australia's green and golden bell frog (*Litoria aurea*) are receiving attention. Some frog skins are known to have antibiotic properties, and research suggests that these host defense compounds have great potential in the world of medicine. Hence, the need to protect frogs such as the vulnerable green and golden bell frog gains added impetus.

MAJOR THREATS

As a result of the expanding human population, Earth is experiencing high levels of biodiversity loss with extinction rates estimated to be at least a thousand times higher than they would have been before our own species first evolved. Climate change threatens animals and plants, as does the pollution created by the sheer magnitude of people and the technologies employed in daily life. Further hazards are habitat loss and degradation due to inappropriate land use and land clearing, and overexploitation through poaching and legally sanctioned hunting and commercial harvesting. Competition from species of flora and fauna that have been introduced accidentally or intentionally has also left an indelible mark.

According to a 2004 report by an international team of experts published in the scientific journal *Nature*, climate change alone shows the potential to cause staggering ecological damage in the near future. The authors of the report, which is entitled "Extinction Risk from Climate Change," warn that among the effects of global warming could be the extinction of well over one million species of mammals, birds, reptiles, frogs, invertebrates, and plants by 2050.

The Tsingy de Bemaraha Strict Nature Reserve, Madagascar The Indian Ocean island of Madagascar comprises just 1.9 percent of African land area, but has more orchids than the entire African mainland and is home to about 25 percent of African plants, all of the continent's lemurs, and most of its reptiles and amphibians. The Tsingy de Bemaraha Strict Nature Reserve (right) includes habitat for rare and endangered lemurs and birds, but many of Madagascar's other areas of biological wealth are at risk from slash-and-burn agriculture, timber exploitation, uncontrolled ranching of livestock, and hunting. Erosion is causing devastating problems and the rivers often run blood-red with soil.

Biodiversity hotspots
1 California Floristic Province
2 Caribbean
3 Mesoamerica
4 Chocó-Darién–Western Ecuador
5 Tropical Andes
6 Brazilian Cerrado
7 Atlantic Forest
8 Central Chile
9 Mediterranean Basin
10 Caucasus
11 Guinean Forest of West Africa
12 Eastern Arc Mountains and Coastal Forests
13 Madagascar and Indian Ocean Islands
14 Succulent Karoo
15 Cape Floristic Region
16 Western Ghats and Sri Lanka
17 Mountains of Southwest China
18 Indo-Burma
19 Philippines
20 Sundaland
21 Wallacea
22 Southwest Australia
23 New Zealand
24 New Caledonia
25 Polynesia and Micronesia

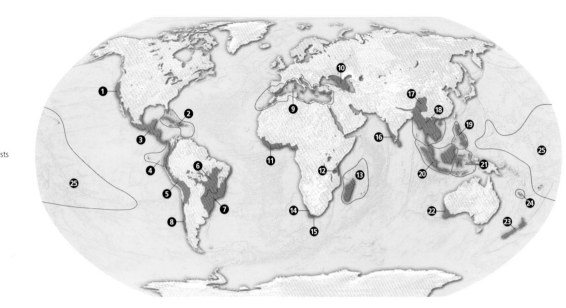

Targeted strategies for survival The hotspot concept identified the richest and most threatened areas of biodiversity, where conservation resources could be used to maximum benefit. The biological basis for hotspot designation is plant diversity. To qualify, a region must support 1,500 endemic plant species, 0.5 percent of the global total. Existing primary vegetation is the basis for assessing human impact. A region must have lost more than 70 percent of its original habitat. The 25 hotspots shown on the map have all been badly affected by human activities. Combined, they contain 44 percent of all plant species and 35 percent of all terrestrial vertebrates, in only 1.4 percent of Earth's land area.

Long gone The loss of the dodo (*Raphus cucullatus*) is a classic example of human disregard for wildlife. Discovered in 1507 on islands in the Indian Ocean, the dodo weighed about 50 pounds (23 kg) and had useless stubs for wings. With no means of escape from seafarers who killed it for sport and for food, the bird was extinct by 1680.

Stop the traffic Trafficking in wild animals is a multimillion-dollar business. It feeds the demand for exotic creatures as pets, for food, and as material for decorative objects. The majority dies in transit. Below, river terrapins (*Batagur baska*) are examined after authorities intercepted the traffic of 1,500 live turtles bound for China in 2002.

Agile rock climber Chamois (*Rupicapra rupicapra*) are fleet-footed inhabitants of the Pyrenees and other mountain regions in Europe. Although not threatened as a species, the numbers of some subspecies are declining due to excessive hunting and competition from livestock. In some areas, chamois species are being introduced or reintroduced with good results.

Do not disturb Overfishing of Tasmania's giant freshwater lobster (*Astacopsis gouldi*) has reduced breeding-age animals to a dangerous low. Males and females of this species take 9 and 14 years respectively to reach sexual maturity. Here (above right), a specimen is being tagged and measured before being returned to the wild.

Red for danger The bald uakari (*Cacajao calvus*) lives in Brazil and Peru in swampy, flooded forests. It and its subspecies are classified as near threatened by the IUCN due to mining and logging. These activities have also provided access for hunters who kill uakaris for food or sell them as pets.

THE SIXTH EXTINCTION

Scientists believe that, under natural extinction rates, the planet should be losing an average of one species every 4 years. However, estimates put the loss at about 17,000 species each year, the vast majority vanishing before their existence has even been scientifically documented. For many esteemed ecological commentators, this loss constitutes the sixth mass extinction to have taken place in the history of life on Earth. The last one occurred 65 million years ago when more than three-fourths of all species that were alive at the time, including the vast array of dinosaurs, disappeared for ever.

Scientists agree that all other mass extinctions can be linked to some form of natural phenomenon. The blame for the current crisis, however, rests firmly, they contend, on global human activities. It is possible—some scientists believe inevitable—that the same factors driving the current mass extinction could ultimately bring our own species to the brink. Such a bleak prognosis is alleviated by the news that many experts believe we have the means, and just enough time, to significantly slow, and possibly even halt, the current extinction drama that is unfolding. What is even more vital, though, is the will to do it.

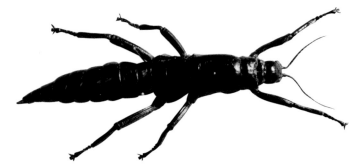

Creature of myth and mystery Snow leopards (*Uncia uncia*) are elusive inhabitants of high mountain areas of central Asia. Our knowledge of these animals has increased as a result of a program of radio collaring and tracking. Also, farmers are being taught ways to protect livestock without resorting to killing the endangered snow leopard.

On the brink Australia's Lord Howe Island phasmid (*Dryococelus australis*), thought to have become extinct in the early 1920s, was rediscovered in 2001. Only three specimens of this large, flightless insect (above) were found on the island, but it is hoped that a species recovery program will result in its reintroduction to its former range.

LOSING ANIMALS

The World Conservation Union estimates that one in four species of mammal and one in eight species of bird are now at risk of extinction, as are 25 percent of reptiles, 20 percent of amphibians, and 30 percent of fishes. Countries with the highest number of known threatened bird and mammal species include China, Peru, Brazil, and India. Percentages of known threatened species are smaller for insects and other groups of invertebrates, but that is probably because they contain many more species that remain undescribed.

Amphibians are known to be particularly sensitive to environmental degradation. Hence, recent declines in their numbers are widely regarded as a significant sign that the planet's capabilities to maintain current levels of biodiversity are being dismantled.

Among the many rescue attempts now being staged around the world are the last-ditch efforts to save the human race's closest relatives—all four species of great apes—from extinction. The United Nations has predicted that gorillas, orangutans, chimpanzees, and bonobos (pygmy chimpanzees) have just a few decades left before they become extinct in the wild.

Food for thought The range of the giant panda (*Ailuropoda melanoleuca*) is steadily shrinking as land clearance affects its habitat. The bamboo on which it feeds is subject to large-scale, periodic die-back. In the past, pandas could migrate to other areas to feed, but habitat fragmentation is making this difficult. Stringent punishment imposed by Chinese authorities has seen a substantial decline in poaching.

Long-distance traveler The wandering albatross (*Diomedea exulans*) (left) has a vast ocean range and is known to be able to cover up to 9,000 miles (15,000 km) in a single trip. Classified as vulnerable, the albatross is at risk from long-line tuna fisheries, as well as predators such as cats and dogs.

Deadly oil Classified as vulnerable, the jackass penguin (*Spheniscus demersus*) is a flightless bird endemic to an area off the coast of southern Africa. Oil resulting from illegal discharges from tankers, harvesting of penguin eggs for human consumption, and a reduction in food supply because of commercial fishing are all taking their toll on these birds. On the positive side, the South African National Foundation for the Conservation of Coastal Birds has a high rate of success in saving oiled seabirds.

OCEANS ON THE BRINK

Pollution and overfishing have been of concern in marine environments for a long time. Land clearing, too, is having a deleterious impact because it leads to sediment run-off which interferes directly with life in coastal areas and, ultimately, elsewhere in deeper-water environments. Now, global warming is also seen as a major threat to marine biodiversity.

Inundation in low-lying mainland coastal areas is likely to affect many species dependent upon intertidal environments. Island shores will suffer similar fates and there is a likelihood that some islands will disappear completely beneath the waves, along with their often unique flora and fauna.

Of particular concern are recent scientific predictions that the largest coral system in the world, the Great Barrier Reef in Australia, is likely to lose as much as 95 percent of its coral by 2050 as a result of rising sea levels and ocean temperatures. It is possible that the system could collapse completely by 2100.

Scientists have also discovered that phytoplankton, upon which virtually all marine food webs ultimately depend, are declining at alarming rates, a trend observed most markedly in the waters of the North Pacific.

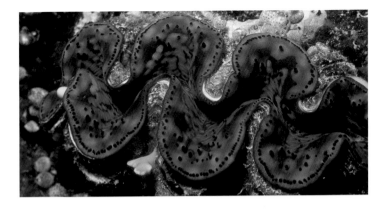

The threat of poaching The giant clam (*Tridacna gigas*) lives only in the tropical waters of the Indo-Pacific. These clams are endangered because their flesh (above) is heavily exploited for human consumption and their shells (right) sold to tourists.

Global warming Current data indicates that the average global rate of sea level rise is 8 inches (20 cm) per century. This may not seem very much, but for low-lying islands and coral atolls in the South Pacific, this rise will prove disastrous. Many will be submerged by the end of the century.

Shark sale The fins of many species of sharks (below right) are among the world's most sought-after fishery products. Shark fin soup has been a delicacy for centuries among the Chinese. Excessive harvesting of sharks is putting many species at risk.

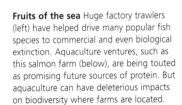

Fruits of the sea Huge factory trawlers (left) have helped drive many popular fish species to commercial and even biological extinction. Aquaculture ventures, such as this salmon farm (below), are being touted as promising future sources of protein. But aquaculture can have deleterious impacts on biodiversity where farms are located.

FINDING ANSWERS

Delegates to the 2002 World Summit on Sustainable Development held in Johannesburg, South Africa, agreed to strive for a significant reduction in the current rate of biodiversity loss by 2010. While this agreement was obviously a step in the right direction, the important issue is how this reduction is going to be achieved. At the global level, there is now a range of international protocols and treaties in place that provide for the increased protection and reduced exploitation of species, particularly those that are classified as threatened or vulnerable. These include the Convention on the Conservation of Migratory Species of Wild Animals, the Convention on International Trade in Endangered Species, and the Convention on Biological Diversity.

Within many countries, in particular the more affluent nations, legislation carrying potent penalties is becoming increasingly important in deterring industrial polluters, poachers, and others threatening the future of endangered species and habitats. Laws that govern the appropriate use of terrestrial and marine habitats and resources are also becoming critically important.

It is widely agreed, however, that education and changes at grass-roots level—in communities, schools, and backyards—will ultimately ensure the future of Earth's biodiversity.

Creating a safe haven National parks, wildlife reserves, and other protected areas are the custodians of nature, critical to the survival of many thousands of species of wildlife. In Yellowstone National Park (left), Wyoming, USA, conservation-dependent bison (*Bison bison*) roam in herds. However, when the park opened in 1872, there was no money to manage it and poaching went unchecked, reducing the last bison herd in the United States to 22 animals.

Extinct in the wild The rapid cultivation of the steppes of southern Russia and the Ukraine in the 19th century contributed to the disappearance of wild horses. The only subspecies still extant, Przewalskii's horse (*Equus ferus przewalskii*), survives only in captive breeding programs. While there are plans to return these animals to the wild, the success of this venture will depend on efforts to preserve their natural habitat.

Mistaken identity Although gentle unless provoked, the gray nurse shark (*Carcharias taurus*) is at risk from hunting because of its fierce look. Coupled with a very low reproduction rate, it is classified vulnerable. This species adapts readily to captivity, so is often seen in aquariums and marine parks.

Modern-day Noah's arks Captive breeding programs for disappearing animal species are already underway in zoos around the world, with the ultimate intention of rebuilding wild populations that are either threatened or already extinct. As a safeguard, however, scientists at several locations internationally have begun collecting and storing frozen tissue samples of endangered animals for which emerging cloning and genetic engineering techniques could prove to be a last resort. Work has already progressed toward cloning the Sumatran tiger (*Panthera tigris sumatrae*) and China's giant panda (*Ailuropoda melanoleuca*).

MAMMALS

MAMMALS

CLASS	Mammalia
ORDERS	26
FAMILIES	137
GENERA	1,142
SPECIES	4,785

The astonishing diversity of the class Mammalia ranges from tiny field mice no bigger than a thimble, to the massive blue whale, which weighs 1,750 times more than a man. Their flexibility, adaptability, and intelligence have allowed mammals to occupy all continents and almost all habitats, exploiting niches on land, below ground, in trees, in the air, and in fresh and salt water. Despite their diverse body forms, mammals share some unique characteristics. Their defining feature is that the lower jaw bone, the dentary, attaches directly to the skull. Mammals are endothermic ("warm-blooded"), nurse their young with milk produced by mammary glands, and usually have hair on the body.

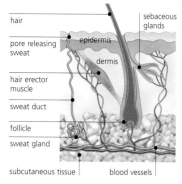

In the skin Mammalian skin is made up of two layers: the epidermis, an outer layer of dead cells, and the dermis, which contains blood vessels, nerve-endings, and glands. Muscles can raise or lower hairs to trap or release an insulating layer of air.

In water Streamlined for sleek movement through water, cetaceans such as dolphins have lost their body hair and rely instead on a layer of blubber as insulation.

ORIGINS AND ANATOMY

Mammals evolved from a group of mammal-like reptiles that had a powerful bite and complex teeth. The first true mammals, with the distinctive jaw hinge between the dentary and the skull, emerged some 195 million years ago. These morganucodontids were nocturnal, shrew-like creatures, only about an inch (2.5 cm) long, that fed on insects. For the next 130 million years, Earth was dominated by dinosaurs, and mammals remained small. About 65 million years ago, however, a cataclysmic change in global climate led to the extinction of 70 percent of all animal species, including the dinosaurs. Mammals survived and began filling the newly empty niches, the start of a process that led to the remarkable variety of species that we see today.

The key to the survival of mammals during a period of severe climate change may have been the ability to regulate their internal body temperature through such methods as adjusting their metabolic rate or blood flow, and shivering, sweating, or panting. Endothermy allowed mammals to remain active regardless of extreme external temperatures and thus to colonize a great range of habitats.

One feature unique to mammals is the presence of hair on the body. Many mammals have a double coat of soft underfur and coarser guard hairs, which insulates against heat and cold. Specialized hairs form whiskers, or vibrissae, which are extremely sensitive to touch. Coat patterns can provide camouflage or communicate information, while spines and quills are modified hairs that create a defensive shield.

The skin of mammals contains a number of different types of glands. In females, mammary glands secrete milk for nourishing their young. Sebaceous glands produce an oily lubricant to protect and waterproof the fur. Sweat glands are important in temperature control, allowing many mammals to release sweat and cool down through its evaporation. Scent glands produce complex odors that convey a mammal's status, sexual condition, and other information. In some species, such as skunks, scent is also used in defense.

The brains of mammals are large (relative to body size) and complex. The sense of smell is important in communication. Color vision has arisen independently in a number of mammals, and binocular vision allows primates to judge distances. All mammals have three bones in the middle ear, and most have an external ear to collect sound.

Fur and whiskers Almost all mammals, from Australia's tree-dwelling brush-tailed possums (right) to the aquatic common seal (far right), rely on a fur coat for insulation. Made of cells strengthened by keratin, hairs can also take the form of vibrissae, touch-sensitive whiskers that provide important tactile information.

Megaherbivore The largest of all land mammals, the African elephant supports its great bulk on a herbivorous diet, collecting grasses, leaves, branches, flowers, and fruit with its dextrous trunk. As it must consume great quantities, it devotes three-quarters of its time to the search for food.

On the wing The ability to fly, along with a sophisticated echolocation system, has allowed bats such as the little brown bat (*Myotis lucifugus*) to exploit the abundant food source of nocturnal flying insects.

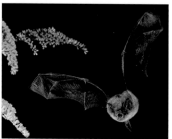

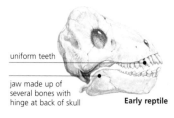

uniform teeth

jaw made up of several bones with hinge at back of skull

Early reptile

zygomatic arch

single jawbone, the dentary, with hinge farther forward

Early mammal

large zygomatic arch

specialized teeth

Modern mammal

Jaws and teeth Mammals are the only animals that chew their food. They differ from their reptilian ancestors by having a single jawbone, powerful jaw muscles in a zygomatic arch, and complex teeth.

Captured prey Meat provides a high-energy, easily digestible meal, but requires great effort and intelligence to catch. The snowshoe hare makes up between 35 and 97 percent of the Canadian lynx's diet.

DIVERSE LIFESTYLES

As mammals moved into different habitats, they adapted their anatomy, locomotion, diet, and social habits to suit their particular niche. Most terrestrial mammals walk on four feet, but some move on two. Some, such as bears, plant the entire sole on the ground, while others walk on just the digits (cats) or tips of the toes (deer). Many shelter below ground, with moles and some other efficient diggers spending almost their entire lives in their burrows. Trees are home for many primates, rodents, and other mammals. In some, grasping hands and feet provide a secure grip on branches and a prehensile tail acts as a fifth limb. A number of species have a gliding membrane that allows them to glide from tree to tree. Bats have developed true flight, with flapping wings allowing them to

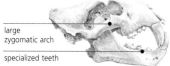

Serious play For many young mammals, such as these wrestling brown bear cubs, play is an important way to learn the hunting and social skills they will need later in life.

traverse great distances through the air. Many mammals are good swimmers, but aquatic mammals such as whales, dolphins, seals, and manatees have fully adapted to a life spent in water. Their shape has become streamlined and their limbs have been modified into flippers.

To fuel the constant regulation of their body temperature, mammals must consume a rich or plentiful diet. The structure of the jaw bones and muscles provides mammals with a powerful bite and the ability to cut and chew their food. Teeth are highly specialized and reflect the diet of the animal. The cheek teeth of carnivores, for example, are sharp for cutting flesh and bones; those of herbivores are broad for grinding plant matter; and omnivores have multicusped cheek teeth that enable them to chew both animals and

plants. While many mammals are opportunistic feeders and will take advantage of any meal they come across, others are specialists. Some carnivores concentrate on insects and other invertebrates, or on small or large vertebrates. Herbivores may focus on fruit, leaves, or grasses. Plant-eaters tend to have a complex digestive system and depend on microorganisms in the gut to break down plant cellulose.

The adaptability of mammals is reflected in the diversity of their social organization. Adult mammals may be solitary, live in pairs or small family groups, or form harems, herds, or colonies. These groupings may be flexible and temporary, or highly stable and permanent. In all mammals, the fact that young are nourished by their mother's milk develops some kind of social bond.

Leaf-eating monkey The douc langur lives in the rain-forest canopy, gripping branches with its grasping hands and feet and leaping from tree to tree. Its potbelly contains a complex digestive system to break down large quantities of leaves.

Frenzied feeder To fuel a fast metabolism, the Eurasian common shrew must eat every few hours and consumes up to 90 percent of its weight in invertebrates each day. It shelters in a burrow, often taking over one abandoned by another species.

MAMMAL REPRODUCTION

In all mammals, the female's eggs are fertilized internally by the male's sperm. Monotremes lay the eggs after several days and incubate them externally until they hatch. Marsupials and placental mammals both give birth to live young. The young of marsupials have short gestations and are poorly developed at birth, but latch onto a teat and are nourished by the mother's milk. Placental mammals grow for longer inside the uterus, nourished by a placenta. At birth, they tend to be in a more advanced state of development.

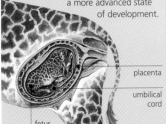

placenta

umbilical cord

fetus

This may be rudimentary—the tree shrew, for example, visits its young only every couple of days for a brief nursing session, and some rodents are weaned after a few weeks. Many young mammals, however, remain dependent on the mother for many months or even years, as is the case with primates, elephants, and whales.

Mammals communicate with one another through scent, touch, sound, posture, and gesture. Both competition and cooperation are highly developed in many mammal species. Individuals of the same species will compete for territory, food resources, the right to mate, and social dominance. Social species may rely on fellow group members to help raise the young, to hunt cooperatively or share information about food sources, to warn of danger, and to deter predators.

Herd mentality Large grazing mammals often form herds, which provide many eyes and ears to detect danger, and lessen the chance that any one individual will be caught. Male Uganda kobs defend a mating territory known as a lek, which is visited by female herds (right) for breeding.

Along for the ride Although marsupials are born in an almost embryonic state, by the time they detach from the teat they are as developed as newborn placental mammals. A young koala leaves its mother's pouch at 7 months, but continues to ride on her back for another 5 months.

MONOTREMES

CLASS	Mammalia
ORDER	Monotremata
FAMILIES	2
GENERA	3
SPECIES	3

Like other mammals, monotremes are covered in fur, lactate to feed their young, and have a four-chambered heart, a single bone in the lower jaw, and three bones in the middle ear. They are unusual, however, because they lay eggs rather than give birth to live young, and also have some anatomical similarities to reptiles, such as extra bones in their shoulders. The two families in this order are Tachyglossidae, which includes two species of echidna, and Ornithorhynchidae, which contains just the platypus. With its duck-like bill, webbed feet, furred body, and beaver-like tail, the platypus has fascinated scientists since the first specimen was sent to Britain in 1799.

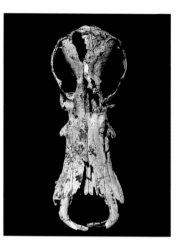

Fossil platypus This 15-million-year-old platypus skull is among the handful of monotreme fossils so far discovered. The fossil record suggests that monotremes originated at least 110 million years ago, when Australia was still part of Gondwana.

EGG-LAYING MAMMALS

Monotreme eggs are soft-shelled and hatch after about 10 days. Once hatched, the young rely on their mother's milk for several months, a dependence typical of mammals.

After mating in spring, female platypuses lay up to three eggs in a bankside nesting burrow, and curl up to incubate them between the tail and body. The hatched young stay in the burrow for 3–4 months, feeding on milk that oozes from two nipple-like patches onto the mother's fur. After they emerge from the burrow, the young are gradually weaned and eventually disperse to a largely solitary life. Wild platypuses have been known to live for at least 15 years.

During their winter mating season, several male short-nosed echidnas may follow a female for up to 14 days, conducting digging and pushing contests until one wins the right to mate. The female then lays a single egg into her pouch. After hatching, the young stays in the pouch until it develops spines and moves to a burrow. The mother lacks teats but her mammary glands open into patches of skin in her pouch. She may continue to feed the young for up to 7 months. A short-nosed echidna has survived for 49 years in captivity, but 16 years is the oldest recorded age in the wild.

Although little is known of the long-nosed echidna's breeding cycle, it is thought to resemble that of the short-nosed echidna.

relatively large nostrils

long, sticky tongue

Sticky feeder The short-nosed echidna's narrow snout and long tongue can search for ants and termites in narrow cavities.

Underwater forager When diving underwater, the platypus closes the groove that contains both its eyes and ears. It relies instead on its soft bill, which not only has sensitive touch but can also detect electrical signals coming from its prey of bottom-dwelling invertebrates.

Spiny defense To protect itself from predators such as dingoes, the short-nosed echidna can speedily burrow vertically down into the earth until only the tips of its spines are visible. If threatened on hard ground, the echidna curls into a spiky ball.

Fat stored in broad tail

Partially webbed hindfeet used as rudders

Platypus
Ornithorhynchus anatinus

Webs of forefeet can be folded back to free claws for walking or digging

Each spine is an individual hair anchored in muscle

Pliable, sensitive bill helps platypus find food and navigate

Long snout can be used as a snorkel when crossing water

Short-nosed echidna
Tachyglossus aculeatus

Mouth at end of snout

Long-nosed echidna
Zaglossus bruijni

Walks with a rolling gait

FACT FILE

Platypus Perhaps one of the most unusual of all animals, this amphibious mammal has a pliable duck-like bill, thick fur, and webbed feet. It lives in riverbank burrows and feeds on insect larvae and other invertebrates.

🐾 Up to 16 in (40 cm)
🐾 Up to 6 in (15 cm)
⚖ Up to 5½ lb (2.4 kg)
Solitary
Locally common

E. Australia, Tasmania, Kangaroo I., King I.

Short-nosed echidna Its stout body covered in long spines and shorter fur, this echidna walks with a rolling gait. It survives in a wide range of habitats, from semiarid to alpine, and feeds mainly on ants and termites.

🐾 Up to 14 in (35 cm)
🐾 Up to 4 in (10 cm)
⚖ Up to 15½ lb (7 kg)
Solitary
Locally common

Australia, Tasmania, New Guinea

Long-nosed echidna This echidna has more hair and fewer spines than the short-nosed species. Tiny spines on its tongue help capture the earthworms that make up the bulk of its diet.

🐾 Up to 31½ in (80 cm)
🐾 None
⚖ Up to 22 lb (10 kg)
Solitary
Endangered

New Guinea

POISONOUS SPUR

The male platypus can use the spur on its rear ankle to inflict a paralyzing sting, making it one of the world's few venomous mammals. This feature may help it fight other males for territory and mates during the breeding season.

venom gland

venom duct

spur

CONSERVATION WATCH

Long-nosed echidna Found only in the alpine forests and meadows of New Guinea, the current population of approximately 300,000 long-nosed echidnas is hunted by humans for food. It is also threatened by habitat loss, as more and more farms are created through land-clearing.

MARSUPIALS

CLASS	Mammalia
ORDERS	7
FAMILIES	19
GENERA	83
SPECIES	295

Commonly known as the pouched mammals, marsupials are virtually embryonic when born and must immediately drag themselves to their mother's teats, which are usually enclosed in a pouch of some kind. They latch firmly onto a teat for some weeks or months and come off only after reaching a level of development comparable to that of newborn mammals that have been nourished by a placenta in the womb. Most marsupials differ from placental mammals in other ways as well, with more incisors in each jaw, an opposable toe on the hindfoot, a relatively smaller brain, and a slightly lower body temperature and metabolism.

Marsupial success While some have flourished in the Americas, marsupials are most diverse in Australia and New Guinea, where there were no placental mammals. They have been introduced in New Zealand, Hawaii, and Britain.

Sibling rivalry The common opossum may give birth to more than 50 young at a time, but only those that attach to one of her 13 nipples will survive. Once the young have developed further, but are still quite helpless, the mother will leave them in the nest while she forages for food.

FINDING A NICHE

Once considered a single order, marsupials are much more diverse than any order of placental mammals and are now classified into seven orders. Of these, Didelphimorphia (containing American opossums), Paucituberculata (shrew opossums), and Microbiotheria (monito del monte) are found in the Americas, while the Australia–New Guinea region hosts Dasyuromorphia (quolls, dunnarts, marsupial mice, numbat), Peramelemorphia (bandicoots), Notoryctemorphia (marsupial moles), and Diprodontia (possums, kangaroos, koala, and wombats).

The fossil record suggests that marsupials and placental mammals diverged more than 100 million years ago. In North America and Europe, marsupials died out as placental mammals diversified. South America, on the other hand, was isolated from North America for about 60 million years, and marsupials evolved to fill a variety of ecological niches. When the Americas joined up, 2–5 million years ago, northern carnivores such as the jaguar soon displaced South America's large carnivorous marsupials, but small omnivorous marsupials persisted, with some, including the common opossum, recolonizing North America. It was in the region of Australia and New Guinea, however, that marsupials remained free of competition for the longest period and attained their greatest diversity.

Life in the pouch Wallabies, kangaroos, and other large marsupials give birth to a single young and carry it in a capacious, forward-opening pouch. Even once fully weaned, the young continues to use the pouch for transport and sleeping.

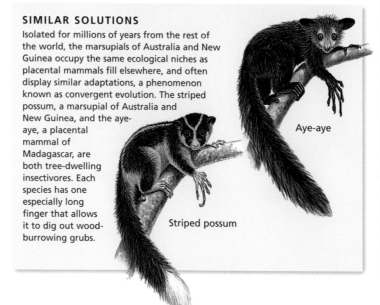

SIMILAR SOLUTIONS

Isolated for millions of years from the rest of the world, the marsupials of Australia and New Guinea occupy the same ecological niches as placental mammals fill elsewhere, and often display similar adaptations, a phenomenon known as convergent evolution. The striped possum, a marsupial of Australia and New Guinea, and the aye-aye, a placental mammal of Madagascar, are both tree-dwelling insectivores. Each species has one especially long finger that allows it to dig out wood-burrowing grubs.

Aye-aye

Striped possum

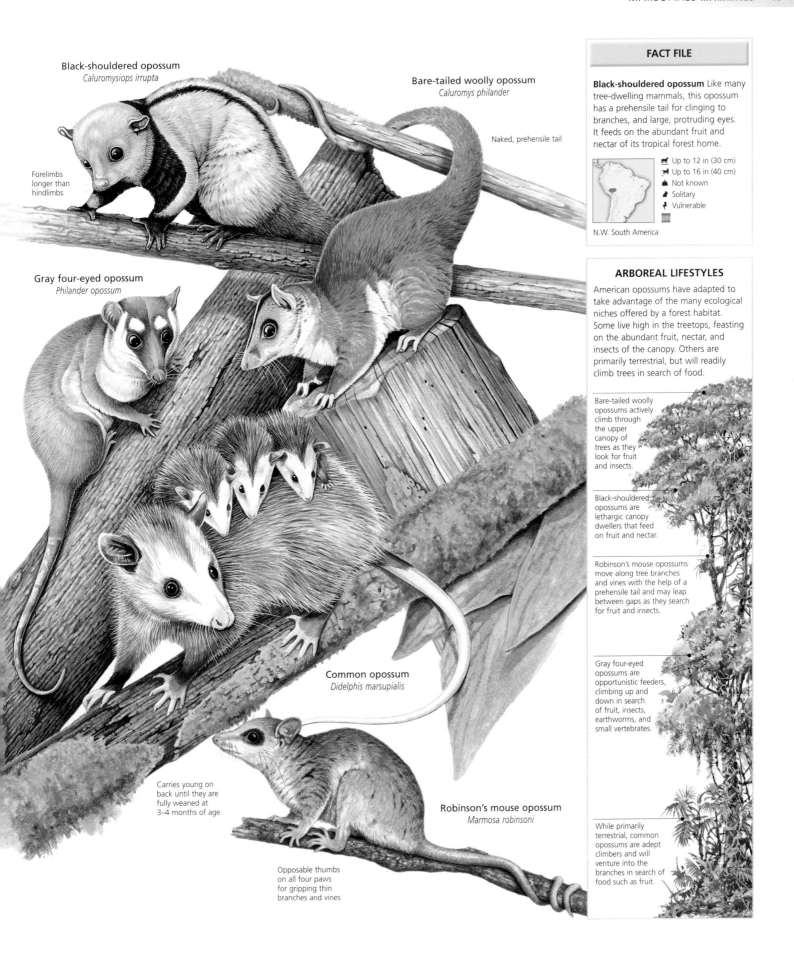

Black-shouldered opossum
Caluromysiops irrupta

Forelimbs longer than hindlimbs

Bare-tailed woolly opossum
Caluromys philander

Naked, prehensile tail

Gray four-eyed opossum
Philander opossum

Common opossum
Didelphis marsupialis

Carries young on back until they are fully weaned at 3–4 months of age

Robinson's mouse opossum
Marmosa robinsoni

Opposable thumbs on all four paws for gripping thin branches and vines

FACT FILE

Black-shouldered opossum Like many tree-dwelling mammals, this opossum has a prehensile tail for clinging to branches, and large, protruding eyes. It feeds on the abundant fruit and nectar of its tropical forest home.

Up to 12 in (30 cm)
Up to 16 in (40 cm)
Not known
Solitary
Vulnerable

N.W. South America

ARBOREAL LIFESTYLES

American opossums have adapted to take advantage of the many ecological niches offered by a forest habitat. Some live high in the treetops, feasting on the abundant fruit, nectar, and insects of the canopy. Others are primarily terrestrial, but will readily climb trees in search of food.

Bare-tailed woolly opossums actively climb through the upper canopy of trees as they look for fruit and insects.

Black-shouldered opossums are lethargic canopy dwellers that feed on fruit and nectar.

Robinson's mouse opossums move along tree branches and vines with the help of a prehensile tail and may leap between gaps as they search for fruit and insects.

Gray four-eyed opossums are opportunistic feeders, climbing up and down in search of fruit, insects, earthworms, and small vertebrates.

While primarily terrestrial, common opossums are adept climbers and will venture into the branches in search of food such as fruit.

IN THE SWIM

The only truly aquatic marsupial, the yapok of Central and South America is superbly adapted to a life in water. It has long webbed toes on its hindfeet, water-repellent fur, and a pouch that closes during dives. It hunts its aquatic prey of fish, frogs, crustaceans, and insects mostly at night, and rests in a riverbank den by day.

Strong stroke
The yapok uses its webbed hindfeet to move through the water, leaving its forefeet free to sift through the stream bottom for food.

MARSUPIAL ORIGINS

DNA studies have confirmed that the monito del monte of Argentina and Chile is the only living member of the Microbiotheriidae, a family of South American marsupials more closely related to Australian marsupials than to the other marsupials of South America. Together with fossil marsupials found in the Antarctic Peninsula, the monito del monte studies support the theory that marsupials spread from South America to Australia via Antarctica between 100 and 65 million years ago, when these continents formed a single landmass known as Gondwana.

North America

South America   Antarctica   Australia

Separate populations
Marsupials radiated from South America to Antarctica and Australia. They flourished in Australia, where there was little competition, but vanished from Antarctica, which split away and moved south. When North and South America joined up again, northern carnivores displaced South America's large marsupials.

Little monkey
The monito del monte ("little monkey of the mountain") lives in the dense, humid forests of highland Chile and Argentina.

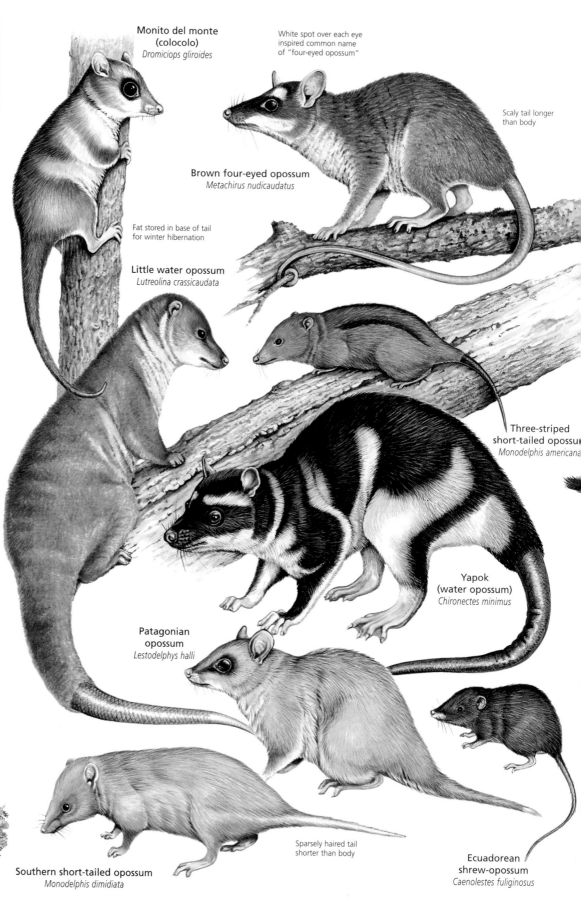

Monito del monte
(colocolo)
Dromiciops gliroides

White spot over each eye inspired common name of "four-eyed opossum"

Brown four-eyed opossum
Metachirus nudicaudatus

Scaly tail longer than body

Fat stored in base of tail for winter hibernation

Little water opossum
Lutreolina crassicaudata

Three-striped short-tailed opossum
Monodelphis americana

Yapok
(water opossum)
Chironectes minimus

Patagonian opossum
Lestodelphys halli

Southern short-tailed opossum
Monodelphis dimidiata

Sparsely haired tail shorter than body

Ecuadorean shrew-opossum
Caenolestes fuliginosus

Thylacine (Tasmanian tiger)
Thylacinus cynocephalus

13–19 dark vertical stripes along back

Stiff tail thick at base but tapers to a point

Padded, five-toed paws

Massive head with strong jaw and heavy, bone-crushing molar teeth

Tasmanian devil
Sarcophilus harrisii

Numbat
Myrmecobius fasciatus

Sticky cylindrical tongue can extend up to 4 inches (10 cm) from small mouth

Vertical white bands along back

Long, bushy tail

Lacks external ears; tiny, non-functional eyes are hidden in fur; horny shield on nose

Southern marsupial mole
Notoryctes typhlops

Spade-like claws for digging through sandy soil

Silky, iridescent fur stained pinkish or reddish by iron-rich soil

---

FACT FILE

Tasmanian devil About the size of a terrier and now the largest carnivorous marsupial, the Tasmanian devil will hunt live prey such as possums and wallabies, but prefers to scavenge. It screeches or barks when threatened.

🐃 Up to 25½ in (65 cm)
🐁 Up to 10 in (26 cm)
⚖ Up to 20 lb (9 kg)
🐾 Variable
🕯 Uncommon
🌡

Tasmania

Numbat The only marsupial that is fully active by day, the numbat spends most of its time looking for termites, which make up almost its entire diet. It digs termites from loose earth with its front claws and captures them with its long, sticky tongue.

🐃 Up to 11 in (27.5 cm)
🐁 Up to 8½ in (21 cm)
⚖ Up to 24½ oz (700 g)
🐾 Solitary
🕯 Vulnerable
🌡

S.W. Australia
● Former range

Southern marsupial mole Below the desert sands of Australia, the marsupial mole searches for burrowing insects and small reptiles. Instead of building a tunnel, it "swims" through the ground, allowing the sand to collapse behind it.

🐃 Up to 6½ in (16 cm)
🐁 Up to 1 in (2.5 cm)
⚖ Up to 2½ oz (70 g)
🐾 Solitary
🕯 Endangered
🌡

C. Australia

EXTINCT TIGER

Thylacine The largest carnivorous marsupial to survive into historical times, the thylacine resembled a wolf, but had a distinctive striped coat and a long, stiff tail. It mainly hunted birds, wallabies, and smaller mammals. Although competition with the dingo led to its disappearance from mainland Australia 3,000 years ago, the thylacine was widespread on Tasmania until Europeans arrived. A reputation for killing sheep soon saw it hunted to extinction, with the last confirmed sightings occurring in the 1930s.

🐃 Up to 51 in (130 cm)
🐁 Up to 27 in (68 cm)
⚖ Up to 77 lb (35 kg)
🐾 Solitary, small group
✝ Extinct
🌡

Tasmania (until 1930s)
● Range prior to extinction

FACT FILE

Kowari In Central Australia's stony deserts and dry grasslands, this little marsupial carnivore preys on insects and small birds, reptiles, and mammals. To survive in the arid climate, the kowari shelters in burrows and obtains all its moisture from its food, eliminating the need to drink water.

🐾	Up to 7 in (18 cm)
🐾	Up to 5½ in (14 cm)
⚖	Up to 5 oz (140 g)
♦	Solitary
⚑	Vulnerable

C. Australia

ANNUAL DIE-OFF

One of the most unusual life-cycles occurs in all *Antechinus* and two *Phascogale* species, which breed just once annually. At the same time every year, following an exceptionally intense 2-week mating season, all the males within a population die off. Their vigorous mating efforts induce very high stress levels, which allow them to forgo food in favor of reproduction but also make them vulnerable to diseases such as stomach and intestinal ulcers. Females may live for a second year, but usually produce only one or two litters in their lifetime.

Sole parent *The female dusky antechinus (Antechinus swainsonii) raises her litter after all the males in her population have died off.*

SLOWING DOWN

Dunnarts and other small insectivorous marsupials sometimes enter periods of torpor, during which the metabolism decreases and the heart and respiratory system slows down. By conserving energy, this state reduces the need to eat, a great advantage during winter months when food can be scarce. A torpid period may last from a few hours up to several days.

A winter's tail *In winter, the fat-tailed dunnart may enter a torpid state and live off the fat that has been stored in its tail during more plentiful times.*

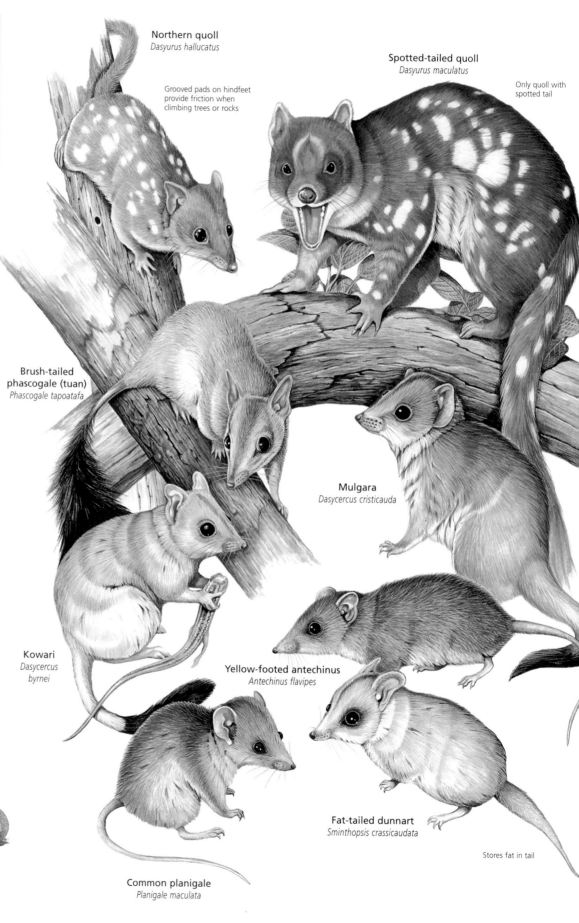

Northern quoll
Dasyurus hallucatus

Grooved pads on hindfeet
provide friction when
climbing trees or rocks

Spotted-tailed quoll
Dasyurus maculatus

Only quoll with
spotted tail

Brush-tailed
phascogale (tuan)
Phascogale tapoatafa

Mulgara
Dasycercus cristicauda

Kowari
*Dasycercus
byrnei*

Yellow-footed antechinus
Antechinus flavipes

Fat-tailed dunnart
Sminthopsis crassicaudata

Stores fat in tail

Common planigale
Planigale maculata

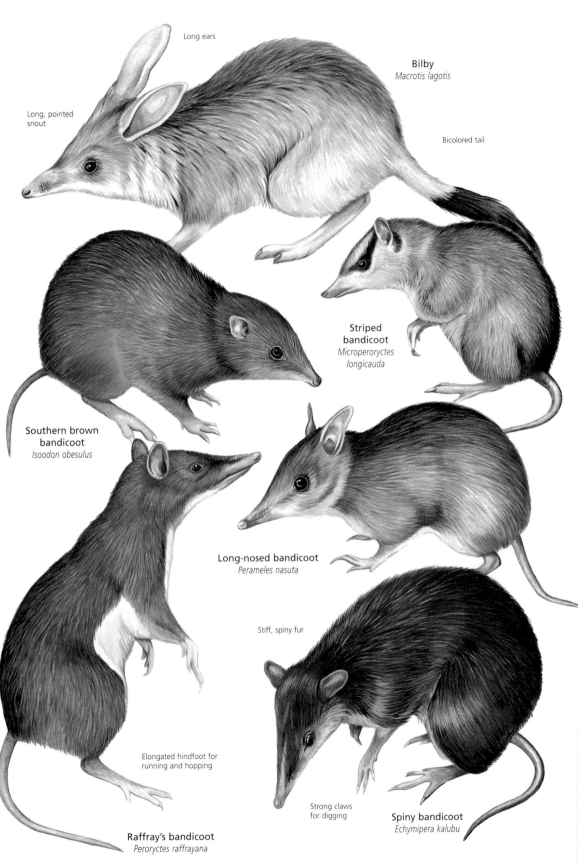

Long ears

Long, pointed snout

Bilby
Macrotis lagotis

Bicolored tail

Southern brown bandicoot
Isoodon obesulus

Striped bandicoot
Microperoryctes longicauda

Long-nosed bandicoot
Perameles nasuta

Elongated hindfoot for running and hopping

Stiff, spiny fur

Strong claws for digging

Spiny bandicoot
Echymipera kalubu

Raffray's bandicoot
Peroryctes raffrayana

BANDICOOTS

The relationship of the omnivorous bandicoots to other marsupials is unclear. They have an arrangement of teeth similar to that of carnivorous marsupials, but they also have fused toes on the hindfoot, a feature known as syndactyly that is shared with herbivores such as kangaroos and wombats. Bandicoots are divided into two families: the mainly Australian Peramelidae (comprising the *Perameles* and *Isoodon* species) and the mainly New Guinean Peroryctidae (including the spiny bandicoot).

Rapid reproduction
After a short gestation period lasting 12 days or so, bandicoot young develop quickly and may reach sexual maturity in about 90 days.

Spotted cuscus This rain-forest marsupial spends most of its time in trees, sleeping by day in a small bed of leaves and feeding at night on a diet of fruit, flowers, and leaves. While males are white with gray spots, females tend to be uniformly gray and unspotted. Intensively hunted in New Guinea, the spotted cuscus is also threatened by habitat loss from logging and farming.

- Up to 23 in (58 cm)
- Up to 17½ in (45 cm)
- Up to 11 lb (4.9 kg)
- Solitary
- Vulnerable

N. Australia, New Guinea, some islands

MOUNTAIN MEALS

The only Australian marsupial to live above the snowline, the mountain pygmy-possum makes the most of its habitat's resources by eating a wide range of food types, but specializing according to the season. During the warmer months, it feeds mostly on Bogong moths, which migrate annually to the Australian Alps, and also eats small amounts of other insects. As the moths become less abundant around January, it switches to seeds and berries, storing caches of these for the cold winter months ahead.

- Up to 4½ in (12 cm)
- Up to 6 in (15 cm)
- Up to 3 oz (80 g)
- Solitary
- Endangered

Australian Alps

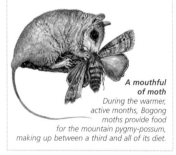

A mouthful of moth
During the warmer, active months, Bogong moths provide food for the mountain pygmy-possum, making up between a third and all of its diet.

Mountain pygmy-possum With a total population of fewer than 2,000 adults, this marsupial species is classified as endangered. Once believed to be extinct, the mountain pygmy-possum has a very restricted range in the mountains of eastern Australia. Much of its habitat has been destroyed or fragmented by the building of roads, dams, and ski resorts, and by recent wildfires.

Bobuk
(mountain brushtail possum)
Trichosurus caninus

Gray cuscus
Phalanger orientalis

Spotted cuscus
Spilocuscus maculatus

Brushtail possum
Trichosurus vulpecula

Feathertail glider
Acrobates pygmaeus

Gliding membrane extends from wrist to knee

Feather-like arrangement of fur on tail unique among mammals

Scaly-tailed possum
Wyulda squamicaudata

Mountain pygmy-possum
Burramys parvus

Prehensile tail covered in thick scales

Rock ringtail possum
Petropseudes dahli

Green ringtail possum
Pseudochirops archeri

**Lemuroid
ringtail possum**
*Hemibelideus
lemuroides*

**Common
ringtail possum**
*Pseudocheirus
peregrinus*

Two opposable
digits on each
front paw

Herbert River ringtail possum
Pseudochirulus herbertensis

Highly arboreal, rarely
descends to ground

Tail tightly curled
when not in use

FACT FILE

Rock ringtail possum By day, this
possum stays cool and safe in rock
crevices, emerging at night to feed
in trees. Males and females share care
of the young equally, a level of male
participation that is rare in mammals
and unknown in other marsupials.

- Up to 15½ in (39 cm)
- Up to 10½ in (27 cm)
- Up to 4½ lb (2 kg)
- Pair
- Locally common

N. Australia

Green ringtail possum Bands
of black, yellow, and white on its
hairs combine to give this possum
a distinctive lime-green appearance,
helping it to remain hidden from
predators in the trees of its rain-forest
home. It feeds primarily on leaves.

- Up to 15 in (38 cm)
- Up to 15 in (38 cm)
- Up to 3 lb (1.3 kg)
- Solitary
- Near threatened

N.E. Australia

Common ringtail possum Most of
this nocturnal animal's diet is made up
of eucalyptus leaves, but it also feeds
on fruit, flowers, and nectar, even
eating rosebuds in urban areas. Small
family groups live in dreys—nests of
bark, twigs, and ferns built in the fork
of a tree or in dense shrubs.

- Up to 15 in (38 cm)
- Up to 15 in (38 cm)
- Up to 2 lb (1 kg)
- Solitary
- Locally common

E. Australia

A POISONOUS DIET

The common ringtail primarily eats
eucalyptus leaves, which are usually
toxic and lacking in nutritional value.
A specialized digestive system featuring
an enlarged cecum (a pouch in the
large intestine) detoxifies the leaves
and then releases soft fecal pellets that
the possum can eat. Any undigested
material is then expelled as hard pellets.
To cope with this low-energy diet, the
ringtail has a slow metabolism.

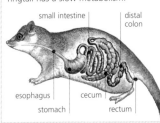

small intestine

distal
colon

esophagus

cecum

stomach

rectum

FACT FILE

Leadbeater's possum This possum has exploited a niche created by the wildfire ecology of its highland forest home. When fires sweep through an area, they may kill some old trees and clear the way for new growth of wattle trees. Family groups may share nests of shredded bark in the hollow center of large trees, feeding on the insects that breed in the bark.

🐾 Up to 6½ in (17 cm)
🐾 Up to 7 in (18 cm)
⚖ Up to 5½ oz (160 g)
🐾 Pair, family group
⚡ Endangered

S.E. Australia

Possum pickings
Leadbeater's possums feed not only on the insects that breed in the bark of wattle trees but also on the sap of the surrounding wattles.

SAP SUCKERS

A membrane extending from the wrists to the ankles allows gliders to travel substantial distances through the air from one feeding tree to another. Once landed, they cut notches in the bark with their teeth and lap up the sap and gum. The yellow-bellied glider targets a number of eucalypt species, while the sugar glider prefers wattles and *Eucalyptus resinifera* trees.

Sticky feeders
Yellow-bellied gliders will vigorously defend their sap-feeding sites.

⚡ CONSERVATION WATCH

Specialized nests The Leadbeater's possum relies on a very specialized habitat, nesting in hollow old-growth trees that can take up to 150 years to become suitable. Presumed extinct after a wildfire in 1939 that burned nearly 70 percent of its range, Leadbeater's possums now number about 5,000, but are endangered by timber logging. Even with conservation measures, there may not be enough remaining nesting sites to sustain this species.

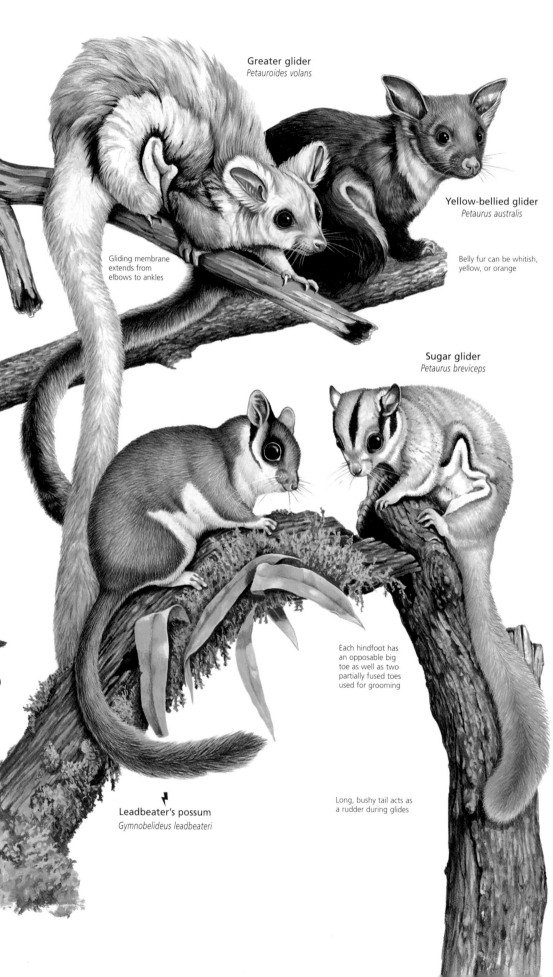

Greater glider
Petauroides volans

Yellow-bellied glider
Petaurus australis

Belly fur can be whitish, yellow, or orange

Gliding membrane extends from elbows to ankles

Sugar glider
Petaurus breviceps

Each hindfoot has an opposable big toe as well as two partially fused toes used for grooming

Long, bushy tail acts as a rudder during glides

⚡ Leadbeater's possum
Gymnobelideus leadbeateri

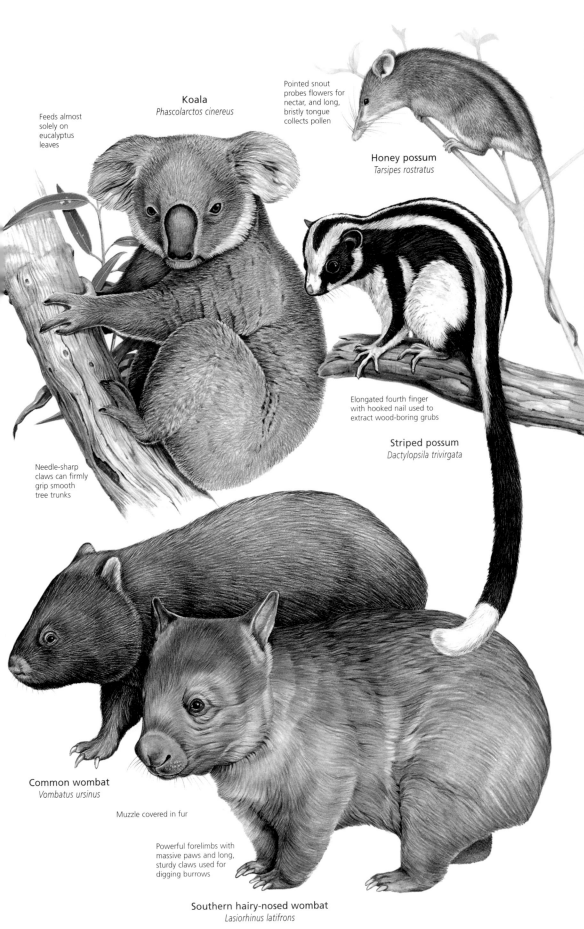

Koala
Phascolarctos cinereus

Feeds almost
solely on
eucalyptus
leaves

Pointed snout
probes flowers for
nectar, and long,
bristly tongue
collects pollen

Honey possum
Tarsipes rostratus

Elongated fourth finger
with hooked nail used to
extract wood-boring grubs

Striped possum
Dactylopsila trivirgata

Needle-sharp
claws can firmly
grip smooth
tree trunks

Common wombat
Vombatus ursinus

Muzzle covered in fur

Powerful forelimbs with
massive paws and long,
sturdy claws used for
digging burrows

Southern hairy-nosed wombat
Lasiorhinus latifrons

FACT FILE

Goodfellow's tree kangaroo Limbs of equal length and sharp claws help this marsupial climb through rain-forest trees, where it shelters in small groups and feeds on leaves and fruit.

New Guinea

- Up to 25 in (63 cm)
- Up to 30 in (76 cm)
- Up to 18½ lb (8.5 kg)
- Solitary
- Endangered

Northern nail-tailed wallaby Named after the horny spur on its tail, this animal spends the day in a shallow nest beneath a bush. At night, it feeds on the roots of the grasses that grow in its savanna or open woodland home.

N. Australia

- Up to 27½ in (70 cm)
- Up to 29 in (74 cm)
- Up to 20 lb (9 kg)
- Solitary
- Locally common

Brush-tailed bettong Adapted to a range of habitats, from temperate forests to arid grasslands, the brush-tailed bettong does not consume green plants or water. Its main food is fungi, which it digs from the ground.

S.W. Australia
● Former range

- Up to 15 in (38 cm)
- Up to 14 in (35 cm)
- Up to 3½ lb (1.6 kg)
- Solitary
- Conserv. dependent

Red-legged pademelon The only ground-dwelling wallaby to live in wet tropical forests, this nocturnal creature searches the dense understory for food such as leaves, fruit, bark, and cicadas.

E. Australia, New Guinea

- Up to 21 in (54 cm)
- Up to 18½ in (47 cm)
- Up to 14 lb (6.5 kg)
- Solitary
- Locally common

⚡ CONSERVATION WATCH

Brush-tailed bettong Although it once ranged over 60 percent of Australia, the brush-tailed bettong is now restricted to a handful of small areas, a victim of predation and competition from introduced species, and habitat loss from farming. In recent years, captive-breeding programs and fox controls have enabled the reintroduction of this marsupial in South Australia.

Joey carried in pouch until it is a subadult

Bennett's tree kangaroo
Dendrolagus bennettianus

Goodfellow's tree kangaroo
Dendrolagus goodfellowi

Four limbs of roughly equal length

Northern nail-tailed wallaby
Onychogalea unguifera

Long-nosed potoroo
Potorous tridactylus

Brush-tailed bettong
Bettongia penicillata

Red-legged pademelon
Thylogale stigmatica

Musky rat-kangaroo
Hypsiprymnodon moschatus

Rufous bettong
Aepyprymnus rufescens

MAKING MARSUPIALS

The unique nature of marsupial reproduction begins with the anatomy of the parents. On the outside, the female system seems simpler than in placental mammals, with a single opening, called a cloaca, for the digestive and reproductive tracts. Inside, however, there is a double reproductive tract involving two uteri, each with its own vagina. Many male marsupials have a forked penis, which directs semen into both vaginas. A pregnant female develops a third vagina as a birth canal. After a short gestation, ranging from 12 days in some bandicoot species to 38 days in the eastern gray kangaroo, almost embryonic young are born that crawl to a teat, which is usually protected by a pouch. Once the young are fully formed, they leave the teat, but are weaned gradually, usually taking several months to become fully independent.

Doubling up Featuring two uteri and two vaginas, the internal anatomy of female marsupials differs markedly from that of placental mammals. Once a female marsupial becomes pregnant, she develops a third canal for the birth of her young.

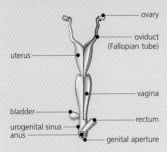

ovary
oviduct (Fallopian tube)
uterus
vagina
bladder
rectum
urogenital sinus
anus
genital aperture

Female placental mammal
This system has a single uterus and a single vagina, and separate openings for the reproductive and digestive tracts.

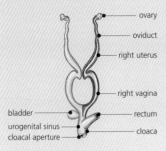

ovary
oviduct
right uterus
right vagina
bladder
rectum
urogenital sinus
cloaca
cloacal aperture

Non-pregnant female marsupial
Two uteruses lead to two vaginas. Both vaginas and the rectum lead to a single opening, which is known as the cloaca.

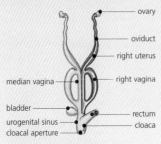

ovary
oviduct
right uterus
median vagina
right vagina
bladder
rectum
urogenital sinus
cloaca
cloacal aperture

Pregnant female marsupial
A median birth canal develops during pregnancy. In most marsupial species, this disappears after the birth, but it remains as a permanent structure in kangaroos and the honey possum.

Raising a family Most kangaroos and wallabies give birth to a single young, but mate within a day or two of its birth. When a female is suckling a joey that can move in and out of the pouch, she usually also has a pouch embryo—a younger offspring attached to a teat inside the pouch. In addition, she is likely to be carrying a blastocyst, a fertilized egg that stays quiescent until the pouch embryo is ready to detach from the teat.

Pouch time A marsupial is tiny when born—a red kangaroo's newborn is 0.003 percent of the mother's weight, while a human baby is about 5 percent of its mother's weight. By the end of weaning, however, the ratio of a marsupial offspring's weight to the mother's is comparable to that of placental mammals.

1. On their way
Newborn eastern quolls (Dasyurus viverrinus) crawl through their mother's belly hairs to find the teats in her pouch. At this stage, their eyes, ears, and hindlimbs are embryonic, but their nostrils, mouth, and forelimbs are all large and functioning. While as many as 30 quolls may be born, only the 6 or so that attach to the mother's teats will survive.

2. Latched on
The young quolls remain firmly attached to their mother's teats during 8 weeks of passive development, when they are known as pouch embryos. To avoid the danger of choking, a large glottis shuts off a baby's mouth from its air passage.

3. Into the world
Once fully formed, the young quolls detach from the teats and leave the pouch, but remain with the mother for some months, clinging to her back while she is foraging, sharing a den for sleeping, and feeding on her milk.

FACT FILE

Parma wallaby The smallest wallaby, this animal was believed to be extinct for decades until its rediscovery in 1965 on New Zealand's Kawau Island, where it had been introduced long before. Other surviving populations were later found in Eastern Australia.

🐃 Up to 21 in (53 cm)
🦘 Up to 21¼ in (54 cm)
⚖ Up to 13 lb (6 kg)
♟ Solitary
⚡ Near threatened

E. Australia

Red kangaroo The largest of the marsupials, the red kangaroo usually hops slowly, but can reach speeds of 35–45 miles per hour (55–70 km/h). Males of this species have a reddish coat, but females are bluish gray.

🐃 Up to 55 in (140 cm)
🦘 Up to 39 in (99 cm)
⚖ Up to 187 lb (85 kg)
♟♟♟ Herd
⚡ Common

Australia

MOB MEMBERSHIP

Large kangaroos often congregate in groups, or mobs, of 50 or more, a strategy that helps deter predators such as dingoes. While males are not territorial, their access to mates depends on their position in the mob's hierarchy, which tends to be based on size. A dominant male Eastern gray kangaroo may father up to 30 young in a season, but most males never get the chance to mate.

Kickboxing
Male kangaroos sometimes fight to establish dominance, delivering kicks with their powerful back legs.

⚡ CONSERVATION WATCH

Of the 295 species of marsupials, 56% are listed on the IUCN Red List, as follows:

10	Extinct
5	Critically endangered
27	Endangered
47	Vulnerable
45	Near threatened
32	Data deficient

Yellow-footed rock wallaby
Petrogale xanthopus

Striking coat pattern distinguished by rich red to yellow limbs and striped tail

Nabarlek
Petrogale concinna

Brush-tailed rock wallaby
Petrogale penicillata

Hindfeet have rough soles for gripping rocks

Parma wallaby
Macropus parma

Spectacled hare-wallaby
Lagorchestes conspicillatus

Eastern gray kangaroo
Macropus giganteus

Red kangaroo
Macropus rufus

Quokka
Setonix brachyurus

Hops on fourth and fifth toes of hindfeet

XENARTHRANS

CLASS Mammalia
ORDER Xenarthra
FAMILIES 4
GENERA 13
SPECIES 29

Some of the world's most bizarre animals—anteaters, sloths, and armadillos—make up Xenarthra, an ancient order that was once much more diverse and included ground sloths larger than elephants, and armored mammals bigger than polar bears. Originating in and confined to the Americas, xenarthrans all have extra joints, known as xenarthrales, in the lower spine, which limit twisting and turning but strengthen the lower back and hips, a particular advantage for the burrowing armadillos. The brains of xenarthrans are small and their teeth are rudimentary or, in the case of anteaters, absent. A slow metabolism has allowed these species to take advantage of narrow niches.

Clingy child Following a gestation period of a year, female sloths give birth to a single young and nurse it for roughly a month. The young then stays with the mother for several months, clinging to her thick fur with its curved claws.

SLOW AND STEADY

The low metabolic rate and body temperature of anteaters and sloths have enabled them to become highly specialized feeders, exploiting food sources that are abundant but low in energy content. Armadillos, on the other hand, have a varied diet, but live in deep underground tunnels, where their slow metabolism helps them to avoid overheating.

Ranging from the terrestrial giant anteater to the tree-dwelling silky anteater and tamanduas, anteaters use their keen sense of smell to detect ants and termites. From an especially long, tubular snout, an even longer tongue darts out, covered in tiny spines and sticky saliva to capture the prey.

Remarkably sluggish, sloths spend almost all their waking hours feeding on forest leaves, consuming such quantities that a full stomach can account for nearly a third of their body weight. Inside a sloth's multichambered stomach, toxins in the leaves are neutralized and the leaves slowly break down, taking a month or more to be fully digested.

With a carapace of bony horn-covered plates, armadillos are not only shielded from predators but are also at ease foraging in dry, thorny vegetation. While much of their diet is made up of invertebrates, they may also consume fruit and small reptiles. They use their powerful limbs and sharp claws to excavate up to 20 burrows in their home range and to dig for prey.

Toilet stop About once a week, a sloth leaves the trees to defecate on the ground. Unable to support its body weight, the animal use its long forelimbs to drag itself along. The toilet site is carefully selected, suggesting that the sloth may be fertilizing favorite feeding trees.

BUILT-IN BLANKET

The giant anteater spends up to 15 hours a day resting. Its digs a shallow depression in the ground and lies down, curling its bushy, fan-like tail around itself. As well as providing warmth, this arrangement helps to disguise the anteater when it is at its most vulnerable.

Digging deep An anteater's sharp front talons can rip into concrete-like termite mounds, allowing its long, sticky tongue to collect the insects. An attack causes little permanent damage to a nest, however, because it lasts just a few minutes and only a small number of termites are eaten. The mound is repaired by the surviving termites.

FACT FILE

Maned three-toed sloth Apart from a black mane on the shoulders, this animal's coarse, shaggy fur is grizzled tan, often tinged green by algae. The green helps to camouflage the slow-moving sloth in its treetop home.

- Up to 19½ in (50 cm)
- Up to 2 in (5 cm)
- Up to 9½ lb (4.2 kg)
- Solitary
- Endangered

N.E. South America

Silky anteater This little tree-dwelling anteater clings to the branches with its long claws and prehensile tail, feasting on arboreal ants and termites. A single young is raised by both parents.

- Up to 8½ in (21 cm)
- Up to 9 in (23 cm)
- Up to 9½ oz (275 g)
- Solitary
- Uncommon

Central America and N. South America

IDENTICAL QUADRUPLETS

The only xenarthran in the United States, the nine-banded armadillo has rapidly expanded its range in the past 150 years. Along with the other species in the genus *Dasypus,* it is unique among vertebrates because the female produces a single fertilized egg that divides into a number of genetically identical embryos.

Family likeness
The nine-banded armadillo usually gives birth to four identical pups.

- Up to 22½ in (57 cm)
- Up to 17½ in (45 cm)
- Up to 13 lb (6 kg)
- Solitary
- Common

North America & South America

⚡ CONSERVATION WATCH

Vulnerable sloths So specialized that they have few natural competitors or predators, sloths have been remarkably successful in Central and South America. Their future, however, depends on the survival of their rain-forest habitat, which has been disappearing at an alarming rate. Already, the maned three-toed sloth is endangered, restricted to small areas of Brazil's coastal forest.

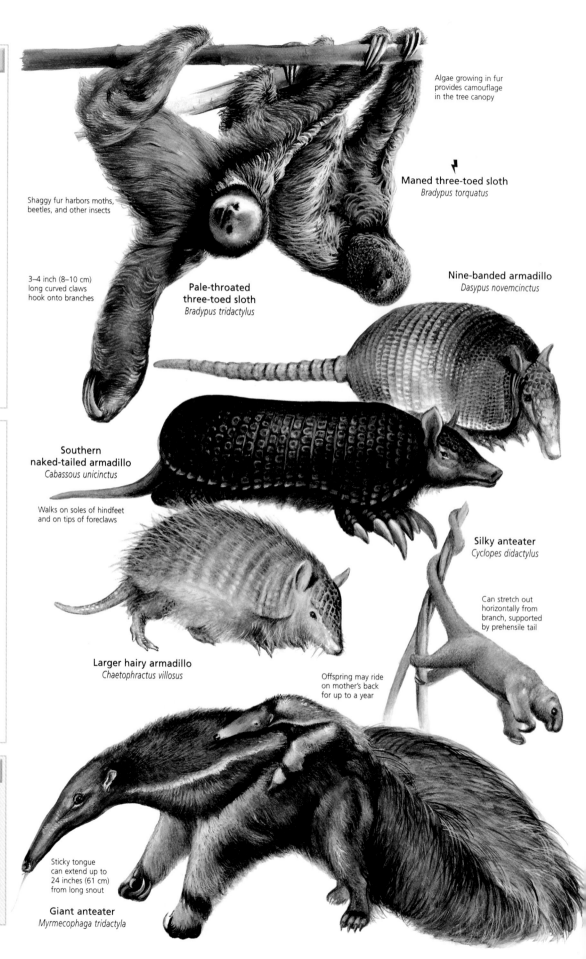

Algae growing in fur provides camouflage in the tree canopy

Maned three-toed sloth
Bradypus torquatus

Shaggy fur harbors moths, beetles, and other insects

3–4 inch (8–10 cm) long curved claws hook onto branches

Pale-throated three-toed sloth
Bradypus tridactylus

Nine-banded armadillo
Dasypus novemcinctus

Southern naked-tailed armadillo
Cabassous unicinctus

Walks on soles of hindfeet and on tips of foreclaws

Silky anteater
Cyclopes didactylus

Can stretch out horizontally from branch, supported by prehensile tail

Larger hairy armadillo
Chaetophractus villosus

Offspring may ride on mother's back for up to a year

Sticky tongue can extend up to 24 inches (61 cm) from long snout

Giant anteater
Myrmecophaga tridactyla

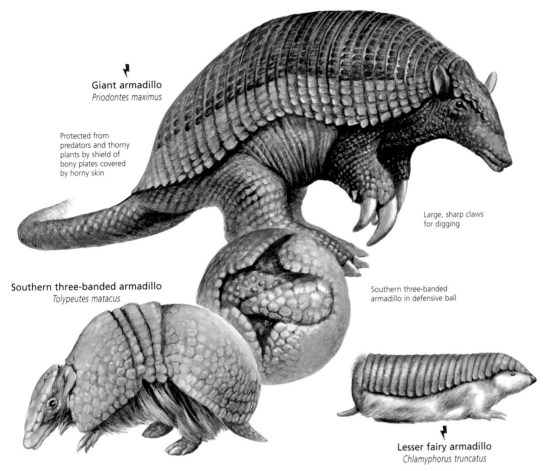

Giant armadillo
Priodontes maximus

Protected from predators and thorny plants by shield of bony plates covered by horny skin

Large, sharp claws for digging

Southern three-banded armadillo
Tolypeutes matacus

Southern three-banded armadillo in defensive ball

Lesser fairy armadillo
Chlamyphorus truncatus

FACT FILE

Giant armadillo About the size of a German shepherd dog, the giant armadillo has huge front claws that it uses for building its own burrows and for digging into termite mounds. While it prefers termites, it will also eat other insects and worms, snakes, and carrion.

- Up to 39½ in (100 cm)
- Up to 21½ in (55 cm)
- Up to 66 lb (30 kg)
- Solitary
- Endangered

N. South America

Southern three-banded armadillo Unlike other armadillos, which will speedily burrow if threatened, this species can roll itself into a tight ball, so that it is entirely protected by armor. It may leave a slight opening that can snap shut on the probing paw of a predator. This defense strategy is ineffective against humans, however, who hunt this armadillo for food.

- Up to 10½ in (27 cm)
- Up to 3 in (8 cm)
- Up to 2½ lb (1.2 kg)
- Solitary
- Near threatened

C. South America

PANGOLINS

CLASS	Mammalia
ORDER	Pholidota
FAMILY	Manidae
GENUS	*Manis*
SPECIES	7

A covering of horny body scales growing from a thick underlying skin distinguishes pangolins from all other mammals. An extraordinary tongue, longer than the animal's head and body, is coiled in the animal's mouth when at rest, but can be extended and flicked into ant nests and termite mounds. Pangolins lack teeth, relying instead on powerful muscles and small pebbles in their stomachs to grind up their food. Terrestrial species, such as the giant ground pangolin, dig out underground burrows for shelter during the day. The long-tailed pangolin and other tree-dwelling species have a prehensile tail for help with climbing, and curl up into balls in tree hollows when resting.

Asia and Africa Pangolins are found in much of Southeast Asia and subtropical Africa. While the Asian species have external ears and grow hair at the base of their scales, the African species lack external ears and have no scales on the tail's underside. Like anteaters in the Americas and echidnas in Australia and New Guinea, pangolins are specialized to feed on ants and termites.

Overlapping scales cover entire body except for belly, inner limbs, and underside of tail

Scales are shed and replaced throughout life

Giant ground pangolin
Manis gigantea

CONSERVATION WATCH

Hunted scales Pangolins have been relentlessly hunted, prized as meat in Africa and for the medicinal value of their scales in Asia. Destruction of their rain-forest habitat threatens these specialized creatures. The Cape pangolin (*Manis temminckii*) of Africa, as well as all of the three Asian species—the Indian pangolin (*M. crassicaudata*), Malayan pangolin (*M. javanica*), and Chinese pangolin (*M. pentadactyla*)—are listed on the IUCN Red List as near threatened.

INSECTIVORES

CLASS	Mammalia
ORDER	Insectivora
FAMILIES	7
GENERA	68
SPECIES	428

Small, quick creatures with long, narrow snouts, shrews, moles, hedgehogs, and other insectivores make up a diverse order whose classification is much debated. While they share primitive features such as a small, smooth brain, simple bones in the ear, and rudimentary teeth, many also display specializations such as burrowing adaptations, defensive spines, or poisonous saliva. Insectivores were named after their tendency to eat insects, but many will take advantage of any available food source and readily consume plants and other animals. Usually shy and nocturnal, they rely on acute senses of smell and touch rather than vision and have very small or even minute eyes.

Worldwide spread While three insectivore families—hedgehogs and moonrats; moles and desmans; and shrews—are found in much of the world, solenodons, tenrecs, and otter shrews are highly localized.

All in a row To avoid getting lost, young of the white-toothed shrew (*Crocidura russula*) "caravan," form a chain behind the mother by tenaciously gripping the rear end of the animal in front.

Fast food Because of their extremely fast metabolisms, shrews must eat vast amounts for their size and usually live where food is abundant. The diet of the Eurasian water shrew (*Neomys fodiens*) features aquatic invertebrates, fish, and frogs.

CONVERGING SPECIES

The order Insectivora contains numerous examples of convergent evolution, with animals in similar habitats displaying similar behaviors or physical adaptations, even though they are not closely related.

Some insectivores have exploited an aquatic niche. The desmans of Europe and the web-footed tenrec (*Limnogale mergulus*) of Madagascar evolved in total isolation from each other, but they share a dense, waterproof coat, a streamlined body, partially webbed feet, a long tail that acts as a rudder, and specialized mechanisms for breathing and detecting underwater prey.

European moles and African golden moles are very distantly related, with true moles evolving from a shrew-like animal, and golden moles more closely related to tenrecs. Nevertheless, they look very much alike and both display adaptations to a burrowing lifestyle. Their bodies are compact and cylindrical, their limbs are short and powerful, with large digging claws on the forefeet, and their eyes are minute and hidden by fur or skin.

The hedgehogs of Europe and the tenrecs of Africa employ similar means of self-defense. Both have a thick coat of spines and will curl up when threatened, becoming a spiky ball to discourage predators.

The solenodons of Cuba and Hispaniola and the tenrecs of Africa appear to have developed echolocation, a method of locating prey by bouncing sounds off the surrounding environment.

Opportunity knocks Insectivores are often opportunistic feeders, prepared to eat a wide variety of prey and plants. The Western European hedgehog mostly eats invertebrates such as earthworms, slugs, beetles, and grasshoppers, but will also devour vertebrate carrion and young birds.

Flat-footed walkers
Almost all insectivores have a plantigrade gait, keeping heels, soles, and toes on the ground when walking.

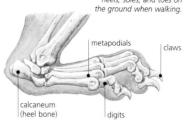

metapodials

claws

calcaneum
(heel bone)

digits

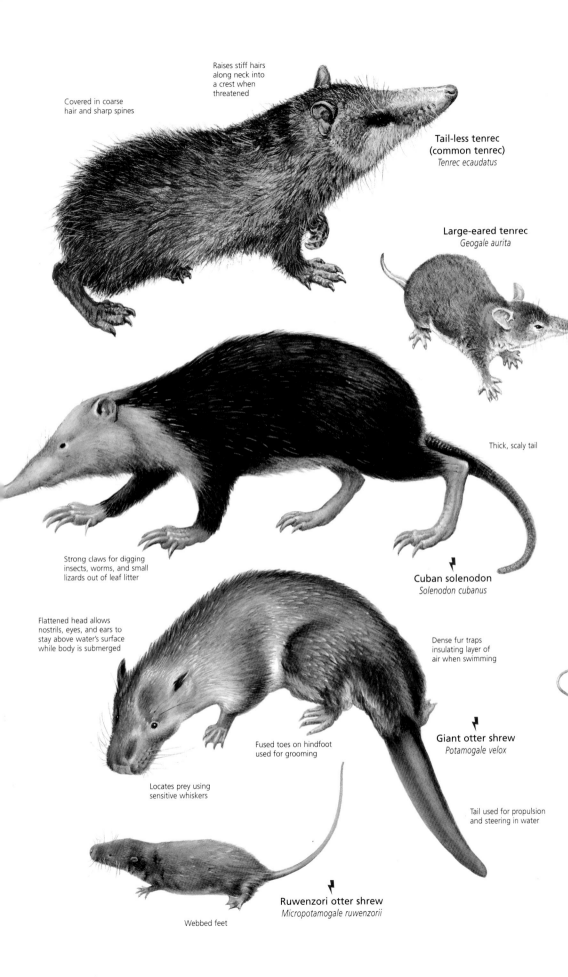

Raises stiff hairs along neck into a crest when threatened

Covered in coarse hair and sharp spines

Tail-less tenrec (common tenrec)
Tenrec ecaudatus

Large-eared tenrec
Geogale aurita

Thick, scaly tail

Strong claws for digging insects, worms, and small lizards out of leaf litter

Cuban solenodon
Solenodon cubanus

Flattened head allows nostrils, eyes, and ears to stay above water's surface while body is submerged

Dense fur traps insulating layer of air when swimming

Fused toes on hindfoot used for grooming

Giant otter shrew
Potamogale velox

Locates prey using sensitive whiskers

Tail used for propulsion and steering in water

Ruwenzori otter shrew
Micropotamogale ruwenzorii

Webbed feet

FACT FILE

Cuban solenodon Like *Solenodon paradoxus* of Hispaniola, the Cuban solenodon releases venomous saliva through a groove in one of its lower incisors. It may use the venom to paralyze larger prey such as frogs. With only a handful of individuals seen in the wild since the 1980s, the future of this species is uncertain.

	Up to 15 in (39 cm)
	Up to 9½ in (24 cm)
	Up to 2 lb (1 kg)
	Solitary
	Endangered

E. Cuba

Giant otter shrew Appearing in African folklore as part fish, part mammal, the giant otter shrew swims powerfully through the water with thrusts of its laterally flattened tail. It hunts at night, searching out prey such as crabs, frogs, fish, and insects by scent and touch. After foraging, it returns to its burrow through an underwater entrance.

	Up to 14 in (35 cm)
	Up to 11½ in (29 cm)
	Up to 14 oz (400 g)
	Solitary
	Endangered

C. Africa

TENRECS OF MADAGASCAR

Among the first mammals to arrive on the island of Madagascar after its separation from mainland Africa 150 million years ago, tenrecs had few competitors. They diversified to fill a variety of ecological niches, with species adapting to life in water, on land, in trees, and underground.

Bony tail *The lesser long-tailed shrew tenrec (*Microgale longicaudata*) has 47 vertebrae in its tail.*

CONSERVATION WATCH

Declining solenodons Both species of solenodon are listed as endangered, with populations taking a dramatic dive since Europeans colonized the West Indies. Because they had few natural predators on their island homes, solenodons never developed effective defenses and made easy prey for introduced species such as mongooses and domestic dogs and cats. Clearing for agriculture has also taken its toll, destroying the forest habitat of these insectivores.

Moonrat The smell of rotting onions, stale sweat, or ammonia may indicate the presence of a moonrat, which produces a strong scent from two glands near the anus. The animal uses this powerful odor to mark its territory. A solitary creature, the moonrat will emit threatening hiss-puffs and low roars if it encounters another member of its species. It rests by day in hollow logs or crevices, and hunts its prey of insects, earthworms, crustaceans, mollusks, frogs, and fish at night.

	Up to 18 in (46 cm)
	Up to 12 in (30 cm)
	Up to 4½ lb (2 kg)
	Solitary
	Locally common

Malay Peninsula, Sumatra, Borneo

Lesser moonrat Spending most of its time on the floor of humid mountain forests, this animal moves in short leaps and sometimes climbs through bushes. It often nests under rocks.

	Up to 5½ in (14 cm)
	Up to 1 in (3 cm)
	Up to 3 oz (80 g)
	Solitary
	Locally common

Indochina, Malaysia, Indonesia

Hottentot golden mole A horny pad on the nose and four clawed toes on each forepaw help this golden mole build extensive tunnel networks beneath the surface. Like the unrelated Australian marsupial mole, it is sightless, with fur covering its residual eyes.

	Up to 5½ in (14 cm)
	None
	Up to 3½ oz (100 g)
	Solitary
	Locally common

S. South Africa

CONSERVATION WATCH

Population pressure Among the 1,000 rarest animals in the world, the giant golden mole *Chrysospalax trevelyani* survives in just a few small areas in South Africa's East Cape region. Already endangered, the species is under ever greater pressure as the human population increases. It has been wrongly blamed for crop damage that was actually caused by mole-rats and other rodents. Predation by domestic dogs has had a serious impact, as has the fragmentation of the mole's forest habitat, with people collecting firewood, cutting down trees for building, and introducing livestock.

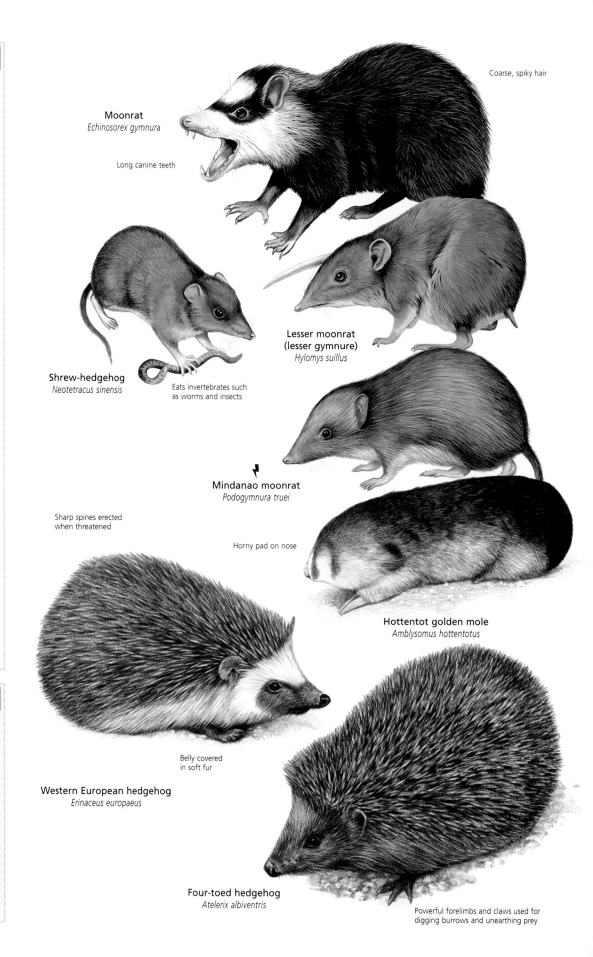

Coarse, spiky hair

Moonrat
Echinosorex gymnura

Long canine teeth

**Lesser moonrat
(lesser gymnure)**
Hylomys suillus

Shrew-hedgehog
Neotetracus sinensis

Eats invertebrates such as worms and insects

Mindanao moonrat
Podogymnura truei

Sharp spines erected when threatened

Horny pad on nose

Hottentot golden mole
Amblysomus hottentotus

Belly covered in soft fur

Western European hedgehog
Erinaceus europaeus

Four-toed hedgehog
Atelerix albiventris

Powerful forelimbs and claws used for digging burrows and unearthing prey

House shrew
Suncus murinus

Good sense of smell
and sharp hearing

Piebald shrew
Diplomesodon pulchellum

Small eyes with
poor vision

Feet fringed with hairs to
help shrew run across sand

Vocalizations include
screams and twitters

Eurasian common shrew
Sorex araneus

Northern
short-tailed shrew
Blarina brevicauda

Himalayan
water shrew
Chimarrogale himalayica

Webbed feet
for swimming

Elegant water shrew
Nectogale elegans

HEDGEHOG HIBERNATION

To cope with the lack of food during cold winters, hedgehogs will go into hibernation, lowering their body temperature and heart and respiration rates to minimize their energy needs. Their metabolic rate can become 100 times slower and they may even stop breathing for up to 2 hours. Hibernating hedgehogs live off the extra body fat accumulated in late summer, but also rouse themselves for a day or so every couple of weeks to forage and defecate.

Getting settled *While the Western European hedgehog makes a hasty nest in summer, it takes more care in winter, choosing a well-insulated site.*

FACT FILE

Russian desman Once found across Europe, this gregarious species is now restricted to a handful of river basins in Russia. Its soft coat made it a target of trappers, but it is now legally protected.

🐾 Up to 8½ in (22 cm)
🐾 Up to 8½ in (22 cm)
⚖️ Up to 8 oz (220 g)
🐾 Solitary
⚡ Vulnerable

E. Europe

American shrew mole About the size of a shrew and without the large forelimbs of other moles, this animal is North America's smallest mole. It is also the only mole able to climb bushes.

🐾 Up to 3 in (8 cm)
🐾 Up to 1½ in (4 cm)
⚖️ Up to ½ oz (11 g)
🐾 Solitary
⚡ Locally common

W. North America

SENSITIVE STARS

An arrangement of fleshy tentacles around its nose helps the star-nosed mole sense small fishes, leeches, snails, and other aquatic prey. When this agile swimmer leaves the water, it usually retreats to its network of tunnels.

🐾 Up to 5 in (13 cm)
🐾 Up to 3 in (8 cm)
⚖️ Up to 3 oz (85 g)
🐾 Solitary
⚡ Locally common

E. North America

Perceptive rays Each of the star-nosed mole's 22 rays bears thousands of sensory organs.

⚡ CONSERVATION WATCH

Of the 428 species of insectivores, 40% are listed on the IUCN Red List, as follows:

5	Extinct
36	Critically endangered
45	Endangered
69	Vulnerable
5	Near threatened
9	Data deficient

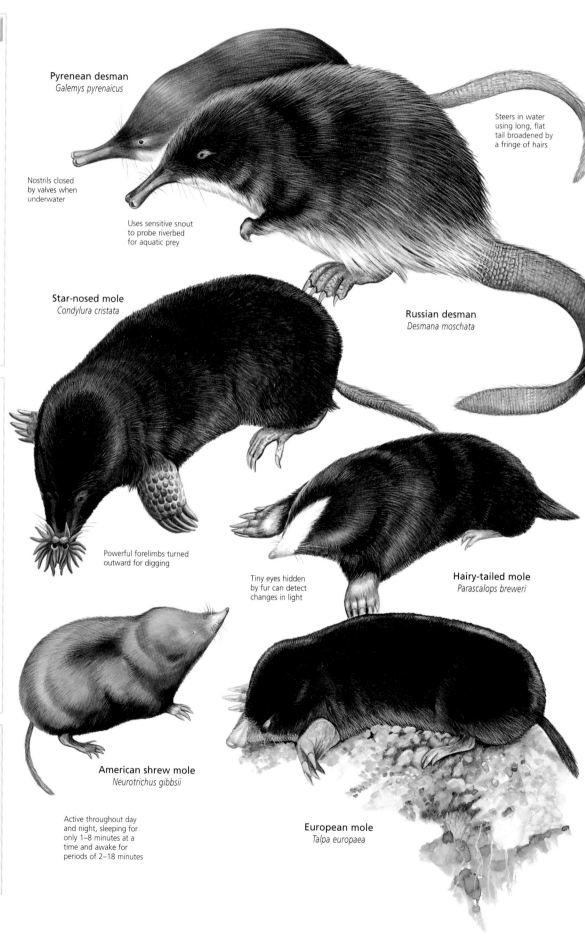

Pyrenean desman
Galemys pyrenaicus

Steers in water using long, flat tail broadened by a fringe of hairs

Nostrils closed by valves when underwater

Uses sensitive snout to probe riverbed for aquatic prey

Star-nosed mole
Condylura cristata

Russian desman
Desmana moschata

Powerful forelimbs turned outward for digging

Tiny eyes hidden by fur can detect changes in light

Hairy-tailed mole
Parascalops breweri

American shrew mole
Neurotrichus gibbsii

Active throughout day and night, sleeping for only 1–8 minutes at a time and awake for periods of 2–18 minutes

European mole
Talpa europaea

LIFE BELOW THE SURFACE

Often the only sign of moles in an area is the presence of molehills, small mounds of dirt created when a mole digs a vertical shaft to the surface. Spending almost their entire lives underground, moles dwell in a network of tunnels, sleeping and raising young in a subterranean nest and foraging in the tunnels for earthworms, insect larvae, slugs, and other soil invertebrates. A mole can create up to 65 feet (20 m) of tunnels per day. It usually comes to the surface only to collect grasses and leaf litter to line a nest, or if a stronger animal has evicted it from its home and it needs to find a new territory.

Designed to dig Most moles have enormous, powerful forelimbs with outward-facing hands and spade-like claws. To dig a tunnel, a mole scoops soil sideways and backward, using the smaller hindlimbs to brace itself.

Raised in the dark Moles mate in the female's burrow during a frantic breeding season of just 24–48 hours. An average of three young are born a month later, and will be nursed in the nest for a further month. After initially exploring the tunnel system with the mother, young moles must soon leave and build their own tunnels elsewhere.

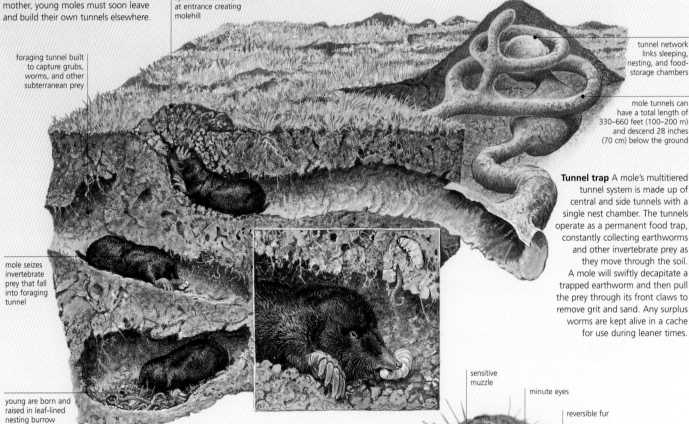

vertical shaft links tunnel system to surface, with dirt at entrance creating molehill

foraging tunnel built to capture grubs, worms, and other subterranean prey

tunnel network links sleeping, nesting, and food-storage chambers

mole tunnels can have a total length of 330–660 feet (100–200 m) and descend 28 inches (70 cm) below the ground

mole seizes invertebrate prey that fall into foraging tunnel

Tunnel trap A mole's multitiered tunnel system is made up of central and side tunnels with a single nest chamber. The tunnels operate as a permanent food trap, constantly collecting earthworms and other invertebrate prey as they move through the soil. A mole will swiftly decapitate a trapped earthworm and then pull the prey through its front claws to remove grit and sand. Any surplus worms are kept alive in a cache for use during leaner times.

young are born and raised in leaf-lined nesting burrow

sensitive muzzle

minute eyes

reversible fur

huge forelimbs

spade-like claws

Subterranean special A mole's anatomy is highly specialized for an underground life. Muscular shoulders and enlarged forelimbs make it a powerful digger. The dense fur can lie in any direction, allowing the mole to move forward or backward with ease. Virtually hidden by the fur, the tiny eyes do sense changes in light, but can barely see. Instead, the mole relies on touch, with a sensitive, mobile muzzle to seek out food.

FLYING LEMURS

CLASS	Mammalia
ORDER	Dermoptera
FAMILY	Cynocephalidae
GENUS	Cynocephalus
SPECIES	2

Also known as colugos, flying lemurs do not really fly and are not true lemurs. Instead, they glide through the air, and are placed in their own small order, Dermoptera ("skin wing"). These cat-sized animals are helpless on the ground and climb awkwardly in a lurching fashion, but travel easily between the tall trees of their rain-forest home. To minimize the risk of being picked off by a swift bird of prey during a glide, flying lemurs are nocturnal. They spend the day resting, either nestled in tree hollows or clinging to a trunk with their needle-sharp claws. A female flying lemur usually gives birth to a single, rather undeveloped young and then carries it on her belly until it is weaned. Her gliding membrane can be folded to create a snug pouch for the young, and becomes a hammock when she hangs upside down from a branch.

Narrow distribution The Malayan flying lemur is found in Malaysia, Thailand, and Indonesia, while the Philippine flying lemur lives only in the Philippines. Both species are hunted by local people for their fur and meat and are threatened by the destruction of their rain-forest habitat.

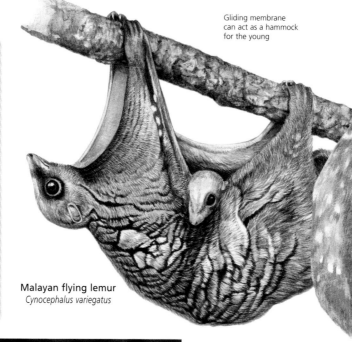

Gliding membrane can act as a hammock for the young

Malayan flying lemur
Cynocephalus variegatus

LEAF-EATER
A specialized stomach allows flying lemurs to digest the large amounts of leafy vegetation that make up the bulk of their diet. Other foods include buds, flowers, fruit, and possibly tree sap.

Ascends tree trunk in a series of hops, clinging to bark with sharp claws

Like a kite Stretching from the neck to the fingers, toes, and tail, the flying lemur's gliding membrane is the most extensive of any glider. A flying lemur can travel through the air for up to 450 feet (135 m), making an accurate landing with the help of its stereoscopic vision. Four other groups of forest mammals—flying squirrels (*Sciuridae*), scaly-tails (*Anomaluridae*), sugar gliders (*Petauridae*), and the greater glider (*Pseudocheiridae*)—independently evolved with the ability to glide.

Philippine flying lemur
Cynocephalus volans

TREE SHREWS

CLASS	Mammalia
ORDER	Scandentia
FAMILY	Tupaiidae
GENERA	5
SPECIES	19

In some Asian tropical forests, small, squirrel-like mammals known as tree shrews scurry along the ground and up trees, foraging for insects, worms, small vertebrates, and fruit. Their sharp claws and splayed toes keep a firm grip on bark and rock alike, while a long tail helps with balance. When eating, they hold food in their hands and may sit on their haunches, alert for predators such as birds of prey, snakes, and mongooses. An average of three young are born in a nest of leaves, which is often made by the father in a tree hollow. Maternal care tends to be limited, with some females visiting only once every 2 days. Considered a primitive form of placental mammal, tree shrews have no direct relationship to true shrews. While a few tree-shrew species spend almost all their time in trees, most are semi-terrestrial and some rarely venture into trees at all.

Divided family Tree shrews live in the tropical rain forests of south and Southeast Asia. Initially considered to be insectivores and then grouped with primates, they are now classified in their own order (Scandentia) in a single family (Tupaiidae), which is divided into two subfamilies. The subfamily Ptilocercinae contains just one species, the pen-tailed tree shrew, which is found on Borneo and the Malay Peninsula. The subfamily Tupaiinae contains the other 18 species of tree shrew. The majority of these make their home on the island of Borneo, while the remainder are distributed throughout eastern India and Southeast Asia.

Philippine tree shrew
Urogale everetti

Pen-tailed tree shrew
Ptilocercus lowii

Only fully nocturnal tree shrew

Scaly tail twitches continuously

Common tree shrew
Tupaia glis

Uses long snout to root through leaf litter on forest floor for insects and seeds

Large tree shrew
Tupaia tana

High fidelity During their 2- to 3-year life, common tree shrews form a permanent pair and display a high degree of fidelity to their mate. While each partner forages alone by day, the pair shares a territory and will defend it against other members of the species. To advertise ownership, they mark strategic sites and new objects with urine or feces or with a scent produced by glands on the chest and abdomen, rubbing their bodies along branches and other surfaces to deposit the scent. Tree shrews also scentmark their partners and young. If the scent is rubbed off, a female shrew will fail to recognize her offspring and may eat them.

⚡ **CONSERVATION WATCH**

Of the 19 species of tree shrews, 32% are listed on the IUCN Red List, as follows:

2 Endangered
4 Vulnerable

BATS

CLASS	Mammalia
ORDER	Chiroptera
FAMILY	18
GENERA	177
SPECIES	993

The only mammals with flapping wings and therefore the only ones capable of true flight, bats can travel through the air at speeds of up to 30 miles per hour (50 km/h). This ability has enabled them to cover great distances, exploiting food sources over a wide range and colonizing most parts of the globe, including far-flung islands such as New Zealand and Hawaii, where they are the only native land mammals. Almost 1,000 species of bat form the order Chiroptera, which is the second-largest mammal order. Chiroptera is split into two suborders: the Old World fruit bats, known as Megachiroptera, and the mostly insect-eating New World bats, known as Microchiroptera.

All over the world Bats make up nearly one-quarter of all mammal species. While most numerous in warmer regions, bat species are found worldwide except for the polar zones and a few isolated islands. Forest bats usually have relatively large, wide wings that offer maneuverability, while bats in open habitats tend to have small, narrow wings, which provide speed.

On the wing A bat has very long arms and highly modified hands, with all digits except the thumb greatly elongated to support the wing membrane, which is called the patagium. A double layer of skin, the patagium is both flexible and tough.

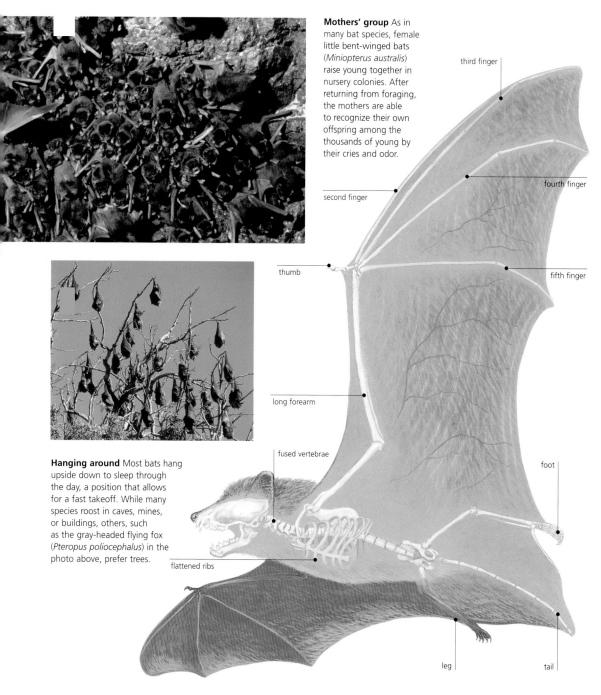

Mothers' group As in many bat species, female little bent-winged bats (*Miniopterus australis*) raise young together in nursery colonies. After returning from foraging, the mothers are able to recognize their own offspring among the thousands of young by their cries and odor.

Hanging around Most bats hang upside down to sleep through the day, a position that allows for a fast takeoff. While many species roost in caves, mines, or buildings, others, such as the gray-headed flying fox (*Pteropus poliocephalus*) in the photo above, prefer trees.

third finger

second finger

fourth finger

thumb

fifth finger

long forearm

fused vertebrae

foot

flattened ribs

leg

tail

FLYING FEEDERS

Bats are popularly portrayed as blood-sucking fiends, but only three species of bat drink blood and even these vampire bats display genuine altruism, sharing their food with hungry companions. Most bats are gregarious creatures, with some living in colonies of thousands or even millions of individuals.

More than 70 percent of bats consume insects that fly at night, a source of food exploited by few other animals. In addition to their flying ability, these hunters rely on echolocation, a means of locating both obstacles and prey by emitting high-pitched sounds and detecting the echoes. In many ecosystems, bats play a crucial role in keeping insect populations under control.

Most other bats are herbivores, using a keen sense of smell and effective night vision to locate fruit, flowers, nectar, and pollen. Such bats can be vital for pollination and seed dispersal, and some plants have adapted specifically to attract them, producing large fruits and strongly scented night-blooming flowers.

To minimize their consumption of energy, many bats regulate their body temperature, lowering it when roosting during the day. To cope with the scarcity of food in winter, some temperate species enter longer periods of hibernation, living off body fat deposited in autumn. Others migrate to warmer climes, with one species, the European noctule, flying up to 1,200 miles (2,000 km).

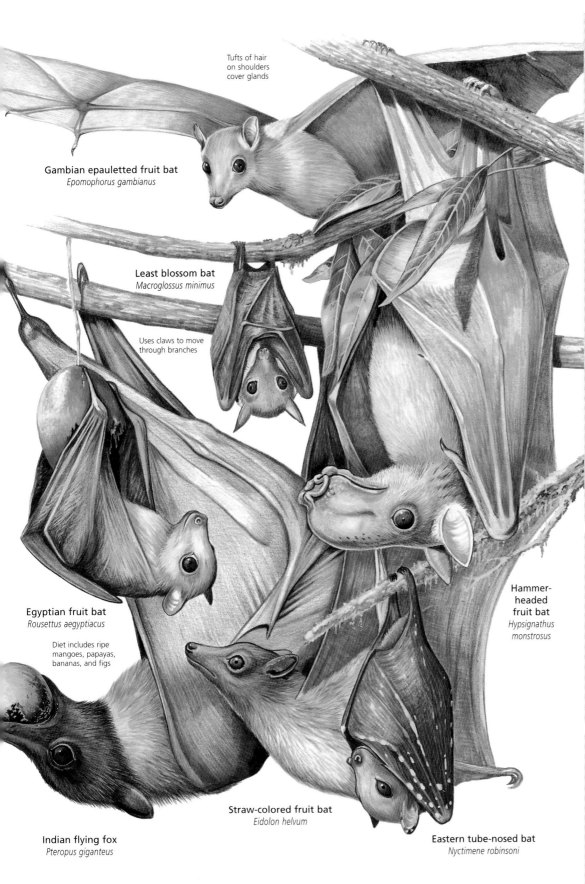

Tufts of hair
on shoulders
cover glands

Gambian epauletted fruit bat
Epomophorus gambianus

Least blossom bat
Macroglossus minimus

Uses claws to move
through branches

Egyptian fruit bat
Rousettus aegyptiacus

Diet includes ripe
mangoes, papayas,
bananas, and figs

Indian flying fox
Pteropus giganteus

Straw-colored fruit bat
Eidolon helvum

Hammer-
headed
fruit bat
*Hypsignathus
monstrosus*

Eastern tube-nosed bat
Nyctimene robinsoni

MAKING A TENT

Rather than hang in a cave or from a branch, the little Honduran tent bat is one of several fruit bats that create their own shelter. It curls a palm frond by chewing through the connection between the leaf's midrib and edges.

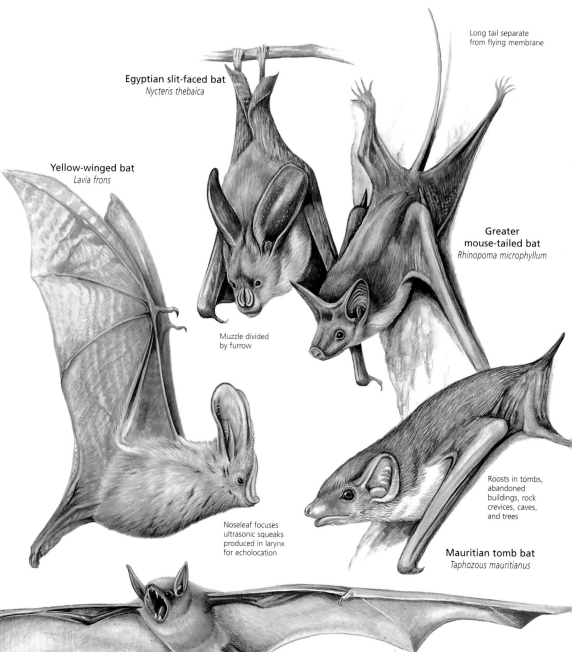

Long tail separate from flying membrane

Egyptian slit-faced bat
Nycteris thebaica

Yellow-winged bat
Lavia frons

Greater mouse-tailed bat
Rhinopoma microphyllum

Muzzle divided by furrow

Noseleaf focuses ultrasonic squeaks produced in larynx for echolocation

Roosts in tombs, abandoned buildings, rock crevices, caves, and trees

Mauritian tomb bat
Taphozous mauritianus

Cheek pouches store chewed fish so bat can continue fishing

Greater bulldog bat
Noctilio leporinus

Long hindlimbs with huge feet and strong claws for snatching fish from water

⚡ CONSERVATION WATCH

Of the 993 species of bats, 52% are listed on the IUCN Red List, as follows:

  12 Extinct
  29 Critically endangered
  37 Endangered
173 Vulnerable
209 Near threatened
  61 Data deficient

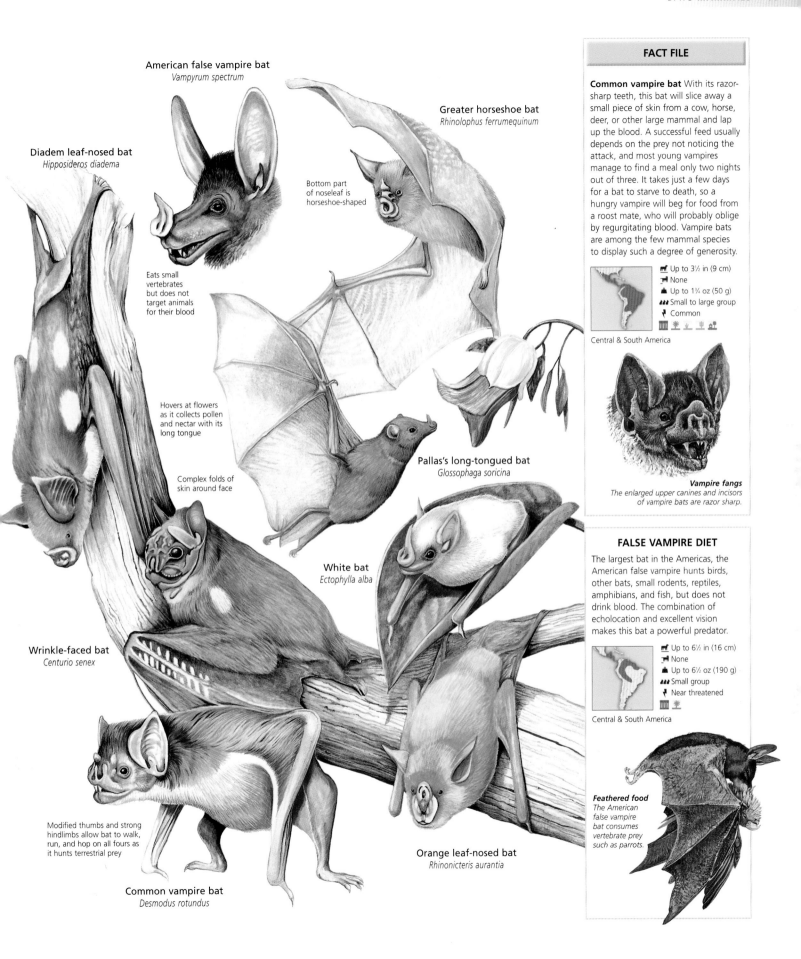

American false vampire bat
Vampyrum spectrum

Greater horseshoe bat
Rhinolophus ferrumequinum

Diadem leaf-nosed bat
Hipposideros diadema

Bottom part
of noseleaf is
horseshoe-shaped

Eats small
vertebrates
but does not
target animals
for their blood

Hovers at flowers
as it collects pollen
and nectar with its
long tongue

Pallas's long-tongued bat
Glossophaga soricina

Complex folds of
skin around face

White bat
Ectophylla alba

Wrinkle-faced bat
Centurio senex

Modified thumbs and strong
hindlimbs allow bat to walk,
run, and hop on all fours as
it hunts terrestrial prey

Orange leaf-nosed bat
Rhinonicteris aurantia

Common vampire bat
Desmodus rotundus

FACT FILE

Common vampire bat With its razor-
sharp teeth, this bat will slice away a
small piece of skin from a cow, horse,
deer, or other large mammal and lap
up the blood. A successful feed usually
depends on the prey not noticing the
attack, and most young vampires
manage to find a meal only two nights
out of three. It takes just a few days
for a bat to starve to death, so a
hungry vampire will beg for food from
a roost mate, who will probably oblige
by regurgitating blood. Vampire bats
are among the few mammal species
to display such a degree of generosity.

Up to 3½ in (9 cm)
None
Up to 1¾ oz (50 g)
Small to large group
Common

Central & South America

Vampire fangs
The enlarged upper canines and incisors
of vampire bats are razor sharp.

FALSE VAMPIRE DIET

The largest bat in the Americas, the
American false vampire hunts birds,
other bats, small rodents, reptiles,
amphibians, and fish, but does not
drink blood. The combination of
echolocation and excellent vision
makes this bat a powerful predator.

Up to 6½ in (16 cm)
None
Up to 6½ oz (190 g)
Small group
Near threatened

Central & South America

Feathered food
The American
false vampire
bat consumes
vertebrate prey
such as parrots.

FACT FILE

Western barbastelle This medium-sized bat roosts in caves, mines, cellars, hollow trees, or under loose bark. It emerges at dusk to search for moths. Although widespread, it appears to be rare almost everywhere in its range.

- Up to 2½ in (6 cm)
- Up to 1½ in (4 cm)
- Up to ¼ oz (10 g)
- Small group
- Vulnerable

W. Europe, Morocco, Canary Islands

Schreiber's bent-winged bat In some regions, this species migrates to warmer locations in winter. It roosts by day in caves or buildings, with the young placed in a communal group separate from the adults.

- Up to 2½ in (6 cm)
- Up to 2½ in (6 cm)
- Up to ½ oz (20 g)
- Large group
- Near threatened

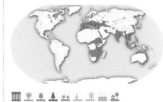

Europe, Africa, S. Asia, Australasia

Large mouse-eared bat Consuming up to half its body weight each night, this bat preys on beetles and moths. Groups of 10–100 individuals share roosts in caves and attics. Young are born in April to June and must build up fat to survive the winter hibernation.

- Up to 3 in (8 cm)
- Up to 2½ in (6 cm)
- Up to 1½ oz (45 g)
- Small to large group
- Near threatened

Europe & Israel

⚡ CONSERVATION WATCH

Vespertilionid bats This page shows just a few of the bats in the family Vespertilionidae, the largest, most widely distributed bat family, which includes a species that lives along the arctic treeline. Its members use almost every kind of roosting site, including buildings. Despite this resourcefulness, 2 species are extinct and many depend on conservation efforts, with 7 critically endangered, 20 endangered, and 52 vulnerable species. Even Britain's most abundant bats, the pipistrelles, have declined by more than 60 percent since 1986.

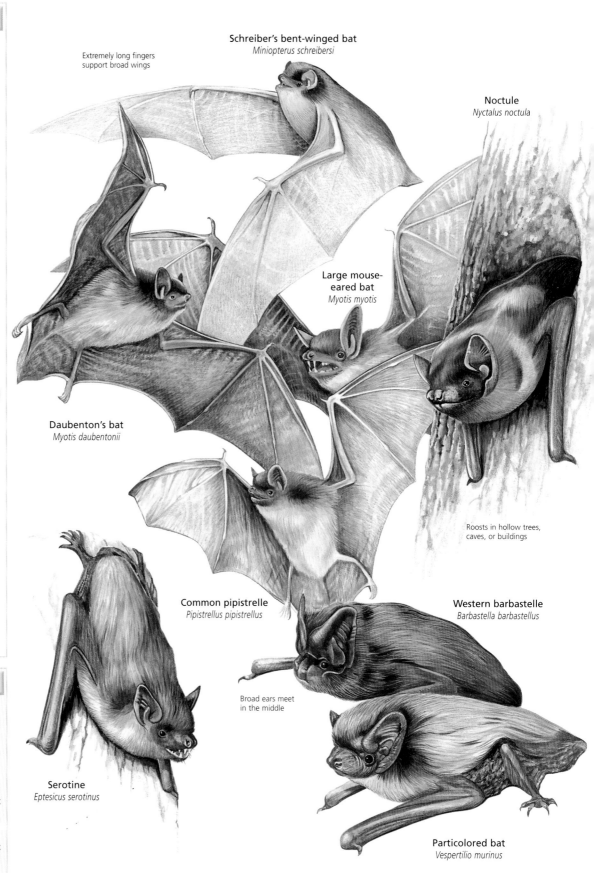

Schreiber's bent-winged bat
Miniopterus schreibersi

Extremely long fingers support broad wings

Noctule
Nyctalus noctula

Large mouse-eared bat
Myotis myotis

Daubenton's bat
Myotis daubentonii

Roosts in hollow trees, caves, or buildings

Common pipistrelle
Pipistrellus pipistrellus

Western barbastelle
Barbastella barbastellus

Broad ears meet in the middle

Serotine
Eptesicus serotinus

Particolored bat
Vespertilio murinus

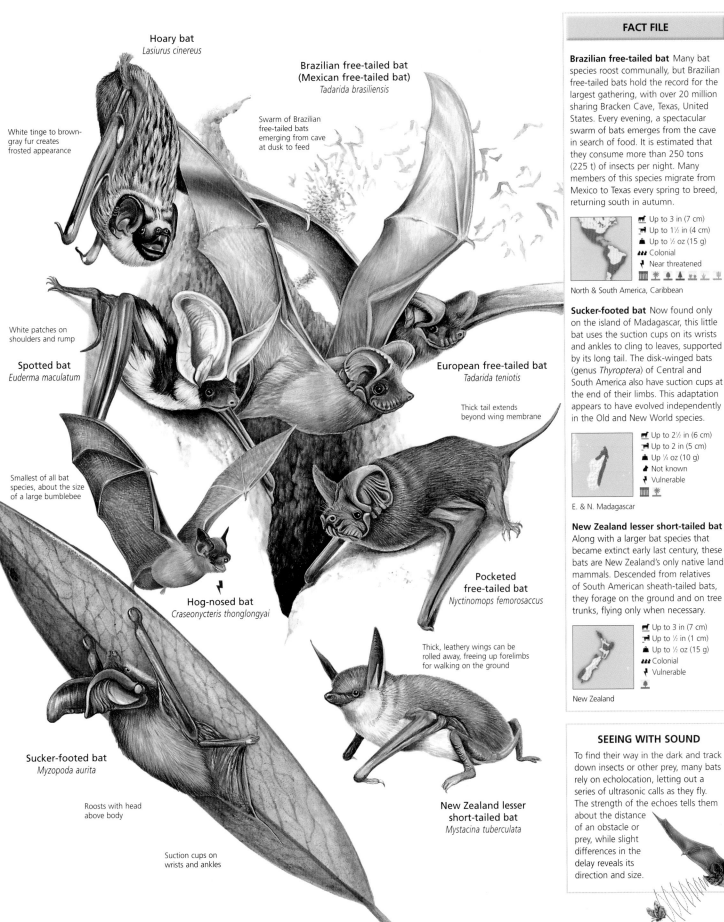

Hoary bat
Lasiurus cinereus

White tinge to brown-gray fur creates frosted appearance

**Brazilian free-tailed bat
(Mexican free-tailed bat)**
Tadarida brasiliensis

Swarm of Brazilian free-tailed bats emerging from cave at dusk to feed

White patches on shoulders and rump

Spotted bat
Euderma maculatum

European free-tailed bat
Tadarida teniotis

Thick tail extends beyond wing membrane

Smallest of all bat species, about the size of a large bumblebee

Hog-nosed bat
Craseonycteris thonglongyai

**Pocketed
free-tailed bat**
Nyctinomops femorosaccus

Thick, leathery wings can be rolled away, freeing up forelimbs for walking on the ground

Sucker-footed bat
Myzopoda aurita

Roosts with head above body

Suction cups on wrists and ankles

**New Zealand lesser
short-tailed bat**
Mystacina tuberculata

FACT FILE

Brazilian free-tailed bat Many bat species roost communally, but Brazilian free-tailed bats hold the record for the largest gathering, with over 20 million sharing Bracken Cave, Texas, United States. Every evening, a spectacular swarm of bats emerges from the cave in search of food. It is estimated that they consume more than 250 tons (225 t) of insects per night. Many members of this species migrate from Mexico to Texas every spring to breed, returning south in autumn.

- Up to 3 in (7 cm)
- Up to 1½ in (4 cm)
- Up to ½ oz (15 g)
- Colonial
- Near threatened

North & South America, Caribbean

Sucker-footed bat Now found only on the island of Madagascar, this little bat uses the suction cups on its wrists and ankles to cling to leaves, supported by its long tail. The disk-winged bats (genus *Thyroptera*) of Central and South America also have suction cups at the end of their limbs. This adaptation appears to have evolved independently in the Old and New World species.

- Up to 2½ in (6 cm)
- Up to 2 in (5 cm)
- Up to ¼ oz (10 g)
- Not known
- Vulnerable

E. & N. Madagascar

New Zealand lesser short-tailed bat Along with a larger bat species that became extinct early last century, these bats are New Zealand's only native land mammals. Descended from relatives of South American sheath-tailed bats, they forage on the ground and on tree trunks, flying only when necessary.

- Up to 3 in (7 cm)
- Up to ½ in (1 cm)
- Up to ½ oz (15 g)
- Colonial
- Vulnerable

New Zealand

SEEING WITH SOUND

To find their way in the dark and track down insects or other prey, many bats rely on echolocation, letting out a series of ultrasonic calls as they fly. The strength of the echoes tells them about the distance of an obstacle or prey, while slight differences in the delay reveals its direction and size.

PRIMATES

CLASS	Mammalia
ORDER	Primates
FAMILIES	13
GENERA	60
SPECIES	295

Charismatic and intelligent, lemurs, monkeys, apes, and their close relatives make up the order known as Primates. Early primates were tree-dwellers and developed adaptations for an arboreal lifestyle: forward-facing eyes with stereoscopic vision to help judge distances when traveling through the trees; dextrous hands and feet to firmly grasp branches; and long, flexible limbs to enhance agility when foraging. Most primates are still largely arboreal, but even those that have opted for life on the ground retain at least some of these adaptations. Perhaps the most fascinating aspect of this order is the complex social behavior displayed by many species.

Tropical distribution Although most primates live in the tropical rain forests between 25°N and 30°S, a handful of species are found farther afield, in North Africa, China, and Japan.

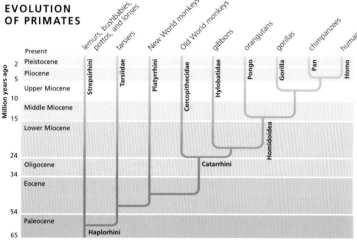

EVOLUTION OF PRIMATES

Strepsirhine primate Like other members of the suborder Strepsirhini, the Coquerel's dwarf lemur is distinguished by a moist, pointed snout and a keen sense of smell. Nocturnal and solitary, this small primate forages for flowers, fruit, and insects.

Meat eater For most primates, leaves, fruit, and insects make up the bulk of the diet, but baboons, chimpanzees, and humans also hunt large vertebrates. Here, an olive baboon (*Papio hamadryas anubis*) returns with its kill of a young gazelle.

Haplorhine primate Along with monkeys and humans, the gorillas and other apes make up the suborder Haplorhini. Gorillas are social animals, living in troops of one or two silverback males, a few younger males, and several females and young.

PRIMATE DIVERSITY

While some small primates are solitary foragers, relying on hiding and nocturnal habits to escape predators, many larger primates are active by day and form groups as protection. A group offers many pairs of eyes to watch for predators. Even when a predator does attack, there is a good chance that it will take another member of the group. Some primates will fight off an attack together—baboons have been known to kill an attacking leopard.

The size and organization of primate groups vary enormously. Some species live in monogamous pairs, while others form troops of several females and one or more males. Stable troops of 150 geladas sometimes combine into herds of 600 individuals. The most common structure is based on related females and their offspring, often headed by a single male. Group living leads to greater competition for food and mates, which is negotiated through complex hierarchies and alliances. These elaborate social networks rely on precise communication, with many species employing a range of nuanced visual and vocal signals. Relative to body size, primates have much larger brains than most other mammals, a feature that may be linked to their complex social lives.

The life of a primate moves slowly. Gestation periods are long; birth rates are low, with just one or two young in a litter; growth rates are slow, with long periods of dependency as infants and juveniles; and lifespans are long. This may be the cost of the large primate brain, which uses energy that would otherwise be available for growth and reproduction.

Primates range from the pygmy mouse lemur (*Microcebus myoxinus*), at 4 inches (10 cm) long and 1 ounce (30 g) in weight, to the gorilla, standing more than 5 feet (1.5 m) high and weighing 400 pounds

Temperate life Japanese macaques are among the few primates to live outside the tropics and subtropics. During the cold, snowy winters, their furry coats thicken and they feed on bark, buds, and stored food reserves, and warm up in thermal pools.

Vertical posture Chimpanzees and other apes sit and sometimes walk upright, supported by a shorter back, broader rib-cage, and stronger pelvis than those of monkeys and lemurs. The arms, often used in locomotion, are longer than the legs, and the wrists are highly flexible.

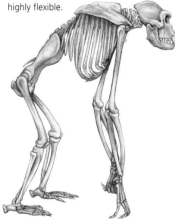

Swinging through the trees Gibbons use their extremely long arms to swing from tree to tree, a form of movement known as brachiation. They rely on stereoscopic vision to accurately judge the distance of their next handhold.

(180 kg). Many small primates feed primarily on insects, a quickly processed food source to fuel their fast metabolisms. Larger species need greater volumes of food and often concentrate on leaves, shoots, and fruit, which are slow to digest but are in abundant supply. The general reliance on fruit, shoots, leaves, and insects has probably helped restrict primates largely to the tropics, where these foods are available year round.

The order Primates is split into two suborders: Strepsirhini, made up of lemurs and their relatives, and Haplorhini, comprised of tarsiers, monkeys, apes, and humans. Tarsiers and strepsirhines share a number of primitive features and have been collectively referred to as prosimians.

Busy social lives The large brains of primates such as baboons may be needed to manage the complex social relationships associated with living in a hierarchical group. While group living leads to greater competition for food, it reduces the risk of being attacked by predators.

Primate survival About a third of all primate species are at risk of extinction, victims of habitat loss and hunting. Large species such as the orangutan are especially vulnerable as they are easy for hunters to find.

PROSIMIANS

CLASS	Mammalia
ORDER	Primates
FAMILIES	8
GENERA	22
SPECIES	63

The name *prosimians* means "before monkeys," a reference to the fact that these creatures retain many features of the early primates. Absent from the Americas, prosimians include the lemurs of Madagascar, the bushbabies and pottos of Africa, and the lorises of Asia—all members of the suborder Strepsirhini. These species all possess a moist, pointed snout and most have a light-reflecting disk in the eye, a long toilet claw, and compressed lower teeth forming a dental comb. Tarsiers, now classified in the suborder Haplorhini, are also often referred to as prosimians because their appearance and solitary, nocturnal behavior is like that of many strepsirhines.

Getting grubby The only strepsirhine to lack the dental comb and toilet claw, the aye-aye has huge, continually growing incisors for gnawing through bark, and a spindly middle finger for extracting grubs.

SPECIALIZED SENSES

Prosimians tend to be relatively small, nocturnal tree-dwellers who forage alone but may occasionally form limited groups. Most are largely insectivorous but will also consume fruit, leaves, flowers, nectar, and gum. They display two specializations for grooming: a long claw on the second toe of the foot, known as the toilet claw; and the dental comb, a compressed row of protruding lower teeth. The dental comb also seems to be used for scraping resin off trees.

The moist, dog-like snout shared by lemurs, lorises, and the other strepsirhines supplies them with a wealth of information through smell. Sight is also important, but they do not have the full color vision of monkeys and apes. Sophisticated color vision would be of limited use

Arboreal attitude Like other prosimians, the slender loris is superbly adapted for a life in trees. Its large, forward-facing eyes provide stereoscopic vision to accurately judge distances, while its dextrous hands and feet keep a firm grip on the branches.

to creatures that forage in the low-light conditions of night. Instead, most prosimians have a tapetum lucidum, a crystalline layer behind the retina that reflects light and produces the characteristic eyeshine of many nocturnal mammals.

While sound is important in prosimian communication, with many species employing alarm and territorial calls, scent plays a major role. Territories are often marked with urine, feces, or secretions from special scent glands, which can convey an individual's sex, identity, and breeding status.

Terrestrial troops Unusually among prosimians, ring-tailed lemurs are active during the day and spend most of their time on the ground. Their troops of between 3 and 20 animals are dominated by female lemurs, a matriarchal structure found in few other primates. Females give birth to one or two young, whose care is shared by the troop.

Broad-nosed gentle lemur
Hapalemur simus

Mongoose lemur
Eulemur mongoz

Ring-tailed lemur
Lemur catta

Large forward-facing eyes

Black lemur
Eulemur macaco

Brown lemur
Eulemur fulvus

Coquerel's dwarf lemur
Microcebus coquereli

Long tail can store fat reserves

Characteristic forked stripe on head

Fork-marked lemur
Phaner furcifer

LEMURS OF MADAGASCAR

For millions of years, lemurs have been isolated on the island of Madagascar, where they have adapted to a variety of forest niches and developed the remarkable diversity evident today. They now range from the size of a mouse to that of a medium domestic dog, but the recently extinct giant sloth lemur (*Archaeoindris fontoynontii*) was bigger than a male gorilla. Most lemurs are arboreal and nocturnal, but some live on the ground and are active by day. They can be solitary, live in pairs, or form larger permanent troops.

Habitat loss Like all Madagascar's lemurs, the red-bellied lemur (Eulemur rubriventer) is declining as its rain-forest habitat is destroyed.

HANDS AND FEET

The contrasting lifestyles of primates are evident in the shape of their hands and feet. The indri and the tarsiers cling vertically and leap from tree to tree, while the aye-aye climbs through the branches. The gorilla can climb but spends most of its time on the ground.

Clinging indri
The indri's stout thumb and big toe help it to cling firmly to tree trunks.

Digging aye-aye
Rather than cling, the aye-aye climbs by digging its long claws into the bark.

Tarsier friction
A tarsier's grip is strengthened by the friction of the disk-like pads on its fingers and toes.

Broad-handed gorilla
Largely terrestrial, a gorilla has broad hands and feet to help support its great weight.

 CONSERVATION WATCH

Of the 63 species of prosimians, 76% are listed on the IUCN Red List, as follows:

- 3 Critically endangered
- 8 Endangered
- 12 Vulnerable
- 17 Near threatened
- 8 Data deficient

Weasel sportive lemur
Lepilemur mustelinus

Red-tailed sportive lemur
Lepilemur ruficaudatus

Rests in vertical position and moves between trees in short leaps

Aye-aye
Daubentonia madagascariensis

Long, powerful hindlimbs power great leaps

Long middle finger for extracting grubs

Coat of a ruffed lemur can be black and red, or black and white

Black face, hands, feet, and tail regardless of coat color

Ruffed lemur
Varecia variegata

Woolly lemur
Avahi laniger

Thick, woolly coat

Large, black,
tufted ears

Indri
Indri indri

Furless black face

Only lemur with
a very short tail

Verreaux's sifaka
Propithecus verreauxi

Diadem sifaka
Propithecus diadema

Almost
completely
arboreal

LEAVING A TRACE

Lemurs scentmark their territories with secretions from glands on the head, hands, or rear. The indri's scent glands are in its cheeks, while the woolly lemur has them in its neck.

CLING AND LEAP

The indri, sifakas, and woolly lemurs are all vertical clingers and leapers. They remain upright as they travel from tree to tree, propelled for up to 30 feet (10 m) by their long, powerful legs. On the rare occasions that these lemurs descend to the ground, they jump about on their legs, holding their arms above the head for balance.

Sifaka on the move
On the ground, Verreaux's sifaka hops sideways, with arms raised. Small skin membranes at the base of its arms may help with gliding between trees.

FACT FILE

Lesser bushbaby This nocturnal animal prefers to feed on grasshoppers and other insect prey. When insects become rare during times of drought, it will feed solely on acacia gum, allowing it to survive in drier habitats.

🐀 Up to 8 in (20 cm)
🐁 Up to 12 in (30 cm)
⚖ Up to 10½ oz (300 g)
👣 Family band
🚩 Common

C. & S. Africa

Demidoff's galago By day, Demidoff's galago sleeps in an elaborate spherical nest of leaves. One of the smalllest primates, it has a fast metabolism and requires an energy-rich diet, 70 percent of which is made up of insects.

🐀 Up to 6 in (15 cm)
🐁 Up to 8½ in (21 cm)
⚖ Up to 4 oz (120 g)
👣 Family band
🚩 Locally common

C. Africa

CONTRASTING FAMILIES

Bushbabies or galagos belong to the family Galagonidae. Using their long hindlimbs for propulsion and long, bushy tails for balance, they leap from tree to tree with great agility (below). In contrast, the lorises, pottos, and angwantibos of the family Loridae creep slowly through the branches. As with other primates that move on all fours, their limbs are of roughly equal length and their tails are short.

Lesser bushbaby
Galago senegalensis

Large, bat-like ears help to detect insects at night

Eastern needle-clawed bushbaby
Euoticus inustus

Large hands and feet with nails forming claws

Demidoff's galago
Galagoides demidoff

Needle-like claws on digits to grip branches

Western needle-clawed bushbaby
Euoticus elegantulus

Allen's bushbaby
Galago alleni

Bushy tail longer than body acts as stabilizer during leaps

Eyes cannot move but head
can rotate almost a full circle

Western tarsier
Tarsius bancanus

Naked tail tipped
with tuft of hair

Slender loris
Loris tardigradus

Climbs on all fours
with slender limbs
of equal length

**Angwantibo
(golden potto)**
Arctocebus calabarensis

Potto
Perodicticus potto

Moves slowly through
branches on all fours

Spectral tarsier
Tarsius spectrum

Very long, skinny digits
for grasping branches

Slow loris
Nycticebus coucang

FACT FILE

Potto This animal attempts to avoid detection by moving very slowly and freezing in place for hours if necessary. Its hands and feet have areas for blood storage that allows them to maintain their hold without muscle fatigue. When confronted by a predator, the potto adopts a defensive posture but will bite if attacked.

🐾 Up to 17½ in (45 cm)
🐁 Up to 4 in (10 cm)
⚖ Up to 3½ lb (1.5 kg)
♦ Solitary
🏛 Locally common

C. Africa

Defense position
A threatened potto tucks its head down and presents its neck, which has a shield of spiny vertebrae covered by horny skin.

TARSIERS

Tarsiers are properly classified in the suborder Haplorhini with the monkeys and apes. They lack the wet snout shared by the strepsirhines, but do bear many superficial resemblances to this group. With their long legs, slender fingers and toes, large ears, and enormous eyes, they look much like a bushbaby with a thin tail. The only carnivorous primates, tarsiers can turn their head almost 360 degrees as they search for prey such as insects, lizards, snakes, birds, and bats.

Western tarsier Like other tarsiers, this nocturnal species lacks a light-reflecting disk to enhance its night vision, but it does have incredibly large eyes. Each eye is bigger than its brain.

🐾 Up to 6 in (15 cm)
🐁 Up to 10½ in (27 cm)
⚖ Up to 6 oz (165 g)
♦ Solitary
⚑ Data deficient

Sumatra, Borneo, Bangka, Belitung, Serasan

Spectral tarsier Pushing off with its very long legs, this small primate can leap up to 20 feet (6 m) between trees. On the ground, it moves about by hopping on its back legs.

🐾 Up to 6 in (15 cm)
🐁 Up to 10½ in (27 cm)
⚖ Up to 6 oz (165 g)
♦ Solitary
⚑ Near threatened

Sulawesi, Sangihe, Peleng, Salayar

MONKEYS

CLASS	Mammalia
ORDER	Primates
FAMILIES	3
GENERA	33
SPECIES	214

Geography has determined two separate lineages of monkeys: the New World monkeys of the Americas, classified within the suborder Haplorhini as platyrrhines; and the Old World monkeys of Africa and Asia, grouped with apes and humans as catarrhines. New and Old World monkeys are most easily distinguished by the shape of their noses and by the arrangement of their teeth. All New World monkeys live in trees and many have strong prehensile tails to grip branches. While most Old World monkeys are also arboreal, none has a prehensile tail and some species are semi-terrestrial. Some Old World monkeys have callous pads on their rumps, a feature found on no New World species.

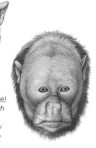

Old and new
Old World monkeys (above) have prominent noses with narrow, forward-facing nostrils. The noses of New World monkeys (right) are flattened with nostrils facing sideways.

Treetop life South America's gray woolly monkey (*Lagothrix cana*) is superbly adapted for an arboreal lifestyle. It swings through the trees, aided by muscular shoulders and hips, long, strong limbs, and grasping hands and feet. The prehensile tail acts as a fifth limb and features a bare gripping pad on the underside near the tip.

Pulling rank Many baboons live in large multi-male troops, and males will fight furiously to establish their dominance and gain access to the females. A troop's rank order, decided by both fighting and alliances, is not permanent, changing as the dominant male grows older or as males join or leave the troop.

SOCIAL SIMIANS

Monkeys tend to be medium-sized, ranging from the pygmy marmoset, with a length of 6 inches (15 cm) and a weight of 5 ounces (140 g), to the mandrill, measuring 30 inches (76 cm) long and weighing in at 55 pounds (25 kg). Most live in social groups, are active by day, and eat mainly fruit and leaves. All Old World monkeys and many New World monkeys have developed full color vision, which allows them to easily spot fruit among the foliage.

Like apes, monkeys differ from lemurs and other strepsirhines by having a dry, slightly hairy muzzle, a greater dependence on sight over smell, and a larger brain relative to body size. Not only are the brains of monkeys and apes larger, but the neocortex, the brain's outer sheath, is especially well developed. The neocortex is associated with creative thinking, an advantage when dealing with the machinations of group life. Monkeys are known to deliberately deceive other members of their group, raising false alarm calls, for example, in order to distract them from a source of food.

The social arrangements of monkeys display great variety: small family groups may contain just one monogamous breeding pair and their offspring; harems of several females may be dominated by a single adult male; and large troops may include several adult males and many females. While large groups often involve fierce competition to establish rank, they also feature extensive cooperation. Relationships among monkeys tend to be close and enduring, cemented by regular episodes of mutual grooming.

Golden lion tamarin
Leontopithecus rosalia

Striking reddish-gold
coat with long mane
framing black face

**Golden-headed
lion tamarin**
Leontopithecus chrysomelas

Pygmy marmoset
Callithrix pygmaea

Claws on all
digits except
big toe, which
has a flat nail

White ear tufts on
adults and juveniles,
absent from infants

**Cotton-top
tamarin**
Saguinus oedipus

Geoffroy's tamarin
Saguinus geoffroyi

Common marmoset
Callithrix jacchus

FACT FILE

Golden lion tamarin There are only
about 800 golden lion tamarins left
in the wild. Their striking appearance
made them popular pets and zoo
exhibits, with many falling victim to the
live-animal trade until it became illegal
in the 1970s. Deforestation continues
to have a serious impact.

🐾	Up to 11 in (28 cm)
🦴	Up to 16 in (40 cm)
⚖	Up to 1½ lb (650 g)
👪	Family band
⚑	Endangered

Coastal forest in Brazil

Pygmy marmoset The smallest of the
world's monkeys, this species gouges
out holes in trees to release its favorite
food of sap and gum. It runs on all
fours along branches, leaping from
tree to tree. Group members use high-
pitched trillings to communicate.

🐾	Up to 6 in (15 cm)
🦴	Up to 8½ in (22 cm)
⚖	Up to 5 oz (140 g)
👪	Family band
⚑	Locally common

W. Amazon basin

Cotton-top tamarin After spending
the night resting in the forks of their
sleeping tree, a group of 3–9 cotton-
top tamarins will move through
the trees of the canopy searching for
insects, fruit, and gum.

🐾	Up to 10 in (25 cm)
🦴	Up to 16 in (40 cm)
⚖	Up to 1 lb (500 g)
👪	Family band
⚑	Endangered

N. Colombia

COOPERATIVE BREEDING

Marmosets and tamarins usually live in
small groups of several unrelated adult
males and females. There tends to be
only one breeding female, who often
mates with more than one male and
gives birth to twins. Uniquely among
primates, the young are then cared
for by all members of
the group, including
the unrelated males.

Babysitting duties
A male Geoffroy's
tamarin will carry
the young of a
female in his
group even
though they
may not be
his own.

FACT FILE

Northern night monkey The world's only nocturnal monkeys, night monkeys use their acute sense of smell and large eyes to find insects, fruit, nectar, and leaves in low light. They live in monogamous pairs, with the male responsible for the bulk of infant care.

- 🐾 Up to 18½ in (47 cm)
- 🐾 Up to 16 in (41 cm)
- ⚖ Up to 2½ lb (1.2 kg)
- 👥 Pair
- 🏛 Common

S.W. Venezuela, N.W. Brazil

Dusky titi Titi monkeys live in closely bonded family groups, and a pair will often sit side by side with their tails entwined. They are active by day, when they consume large amounts of fruit.

- 🐾 Up to 14 in (36 cm)
- 🐾 Up to 18 in (46 cm)
- ⚖ Up to 3 lb (1.4 kg)
- 👥 Family band
- 🏛 Locally common

C. Amazon basin

White-faced saki The long hindlimbs of these active tree-dwellers enable them to leap up to 30 feet (10 m) between trees. At night, they sleep curled up on branches like cats.

- 🐾 Up to 19 in (48 cm)
- 🐾 Up to 17½ in (45 cm)
- ⚖ Up to 5½ lb (2.4 kg)
- 👥 Family band
- 🏛 Uncommon

Guianas, Venezuela, N. Brazil

Black-headed uakari This species lives in groups of up to 50 individuals, which include more than one adult male. Adult females and young engage in social grooming. For a tree-dweller, the uakari has an unusually short tail.

- 🐾 Up to 19½ in (50 cm)
- 🐾 Up to 8½ in (21 cm)
- ⚖ Up to 9 lb (4 kg)
- 👥 Large troop
- 🏛 Rare

Upper Amazon basin

⚡ CONSERVATION WATCH

Of the 214 species of monkeys, 56% are listed on the IUCN Red List, as follows:

- 14 Critically endangered
- 32 Endangered
- 32 Vulnerable
- 2 Conservation dependent
- 26 Near threatened
- 13 Data deficient

Northern night monkey
Aotus trivirgatus

Large eyes for better night vision

Dusky titi
Callicebus moloch

White-faced saki
Pithecia pithecia

White-nosed saki
Chiropotes albinasus

Bald uakari (red uakari)
Cacajao calvus

Black-headed uakari
Cacajao melanocephalus

Black howler
Alouatta caraya

Males are black, females are brown or olive, and infants are golden

Prehensile tail grips branches

Common squirrel monkey
Saimiri sciureus

Mantled howler
Alouatta palliata

Red howler
Alouatta seniculus

White-fronted capuchin
Cebus albifrons

Weeping capuchin
Cebus olivaceus

Brown capuchin
Cebus apella

White-throated capuchin
Cebus capucinus

FACT FILE

Common squirrel monkey These agile monkeys live in large troops of 50 or more. Unlike most other monkeys, they are seasonal breeders, mating promiscuously within their own group between September and November. Young are born between February and April, the wettest part of the year when food is most abundant.

- Up to 12½ in (32 cm)
- Up to 17 in (43 cm)
- Up to 3 lb (1.4 kg)
- Large troop, herd
- Common

Amazon basin, Guianas

Red howler The red howler is the largest of the nine howler monkey species. Howlers have a very deep jawbone to help chew the leaves that make up most of their diet, and an enormous gut to help digest them.

- Up to 27 in (69 cm)
- Up to 31 in (79 cm)
- Up to 24½ lb (11 kg)
- Family band
- Locally common

Venezuela to upper Amazon basin

Brown capuchin Troops of a dozen or so brown capuchins forage together by day, led by a dominant male who has first pick of the food. These monkeys sometimes use simple tools, cracking open a nut with a stone, for example.

- Up to 19 in (48 cm)
- Up to 19 in (48 cm)
- Up to 10 lb (4.5 kg)
- Small troop
- Locally common

N.E. South America

HOWLING

One of the loudest calls in the animal world is made by howler monkeys. At dawn, howler troops announce their presence with a deafening conversation of howls that resonate through the forest, traveling up to 3 miles (5 km) away. By helping troops avoid one another, the howls prevent territorial skirmishes that would waste time and energy that could be better spent eating or resting.

FACT FILE

Common woolly monkey This heavy monkey spends most of its time in trees, but often descends to the forest floor, where it can walk upright on its back legs. Large, multi-male groups of up to 70 individuals sleep together at night, but may split into smaller family groups to look for fruit and other food by day.

- Up to 23 in (58 cm)
- Up to 31½ in (80 cm)
- Up to 22 lb (10 kg)
- Variable
- Uncommon

Upper Amazon basin

Woolly spider monkey Also known as muriquis, woolly spider monkeys are found only in undisturbed high forest, 95 percent of which has been destroyed. Fewer than 500 individuals remain in the wild. Unusually for primates, males stay with their birth troop all their lives, while females must leave and join another troop when they reach adulthood.

- Up to 25 in (63 cm)
- Up to 31½ in (80 cm)
- Up to 33 lb (15 kg)
- Variable
- Endangered

S.E. Brazil

Black spider monkey Groups of about 20 black spider monkeys will cooperatively defend their territory or mob a predator, but they will split into subgroups of up to six for foraging.

- Up to 24½ in (62 cm)
- Up to 35½ in (90 cm)
- Up to 28½ lb (13 kg)
- Variable
- Locally common

N. of Amazon & E. of Rio Negro

TREETOP ANATOMY

Extremely agile climbers, spider monkeys have a slender body, long limbs, thumbless hands that act as simple hooks, and a flexible prehensile tail that functions as a fifth limb. They often scurry along branches on all fours, but will also swing swiftly through the trees, holding on with their hands and tail. A troop tends to travel in line, with the first member testing branches for the followers.

Thumbless hand acts as hook when swinging

Very long prehensile tail strong enough to support monkey's weight

Common woolly monkey
Lagothrix lagotricha

Woolly spider monkey
Brachyteles arachnoides

Coat can be reddish, dark to light brown, or dark to light gray

Long-haired spider monkey
Ateles belzebuth

Face color ranges from pink to black

Black spider monkey
Ateles paniscus

Black-handed spider monkey
Ateles geoffroyi

Douc langur
Pygathrix nemaeus

Hanuman langur
Semnopithecus entellus

Coat can be dark brown, fawn, or gray

Capped leaf monkey
Trachypithecus pileatus

Male's pendulous nose adds resonance to calls

Female has brown face, male has blue face

Chinese snub-nosed monkey
Rhinopithecus roxellana

Slight webbing between digits helps to make proboscis monkey an excellent swimmer

Proboscis monkey
Nasalis larvatus

FACT FILE

Proboscis monkey This monkey is named after the pendulous nose of the male. It lives in stable harems of one adult male and several females. Sexual contact is initiated by the female, who indicates her interest in a male by pursing her lips. If the male returns her gaze, she shakes her head. The male then mirrors her pouting expression and she presents her hindquarters to him. During copulation, both male and female proboscis monkeys maintain a pouting expression and the female keeps rapidly shaking her head.

🐂 Up to 30 in (76 cm)
🐃 Up to 30 in (76 cm)
🌢 Up to 51 lb (23 kg)
👥 Harem
⚡ Endangered
🏛

Lowland Borneo

INFANTICIDE

Although most thoroughly studied among Hanuman langurs, infanticide is practiced by many primates. It occurs when a new male becomes dominant in a troop, often after a bachelor band has invaded. The new male kills all the troop's unweaned infants, presumably because lactation prevents the mothers from conceiving. Although the mothers often try to defend their young, they are usually unsuccessful. With their infants dead, the females stop lactating and are able to conceive the offspring of the newly dominant male.

***Competition and cooperation** The intense rivalry that leads to infanticide among Hanuman langurs is in stark contrast to the high level of care and cooperation within an established troop.*

***Rival males** When young male Hanuman langurs are expelled from their birth troop, they form bachelor bands, which may invade a breeding troop, usurp the resident male's position, and kill the nursing infants.*

COLOBUS MONKEYS

Like the langurs of Asia, the colobus monkeys of Africa have a specialized stomach that has allowed them to exploit the most reliable food source in their forest habitat. Divided into a very large, chambered upper region and a lower acid region, the stomach can hold up to a quarter of the animal's weight in leaves, and contains a special bacteria to break down the vegetation and neutralize any toxins. Colobus monkeys demonstrate great agility as they gallop along branches and take flying leaps to neighboring trees, using their thumbless hands as hooks to secure themselves. Most colobus monkeys live in groups of 10 or so individuals, with a fixed core of related females. Females often "babysit" young that are not their own and may even suckle them.

Lift off Colobus monkeys can leap spectacularly from one tree to another, either to reach a new source of food or to escape a pursuing predator.

Mixed alliances
Colobus monkeys often form temporary or even stable associations with other monkey species. Red colobus monkeys and vervets, for example, may cooperate during the dangerous activity of drinking from a water hole, taking turns to keep an eye out for predators.

⚡ CONSERVATION WATCH

Disappearing red colobus All eight subspecies of red colobus are classified as endangered or critically endangered. Another subspecies, Miss Waldron's red colobus monkey (*Procolobus badius waldroni*), was declared extinct in 2000, the first documented extinction of a primate since 1900. Hunted for meat, red colobus monkeys make easy targets because they are brightly colored and vocal. They are also rapidly losing much of their forest habitat to logging, road building, and farming.

Olive colobus
Procolobus verus

Often forms permanent associations with Diana monkeys, which act as sentinels

Pied colobus
Colobus guereza

U-shaped white mantle on sides and back

Hook-like, thumbless hand allows quick movement through trees

Red colobus
Procolobus badius

Characteristic solemn expression

Black colobus
Colobus satanas

White tuft on end of tail

King colobus
Colobus polykomos

Rhesus macaque
Macaca mulatta

Cheek pouches
store food

Barbary ape
Macaca sylvanus

Mane of gray hair

Pig-tailed
macaque
Macaca nemestrina

Lion-tailed
macaque
Macaca silenus

Coat thickens
in winter

Short tail is
almost hairless

Bear macaque
(stump-tailed macaque)
Macaca arctoides

Japanese macaque
Macaca fuscata

FACT FILE

Rhesus macaque Up to 200 of these gregarious monkeys live together in a group. Adapted to a wide range of habitats, they vary their diet according to season and location, with some in urban areas raiding gardens and trash cans. The species has been extensively used in medical research.

Up to 25½ in (65 cm)
Up to 12 in (30 cm)
Up to 22 lb (10 kg)
Large band
Near threatened

Afghanistan & India to China

Barbary ape This species lives in multi-male troops of up to 40 animals who maintain a home range. Females mate with all males in the troop. Each male chooses a single infant to help raise, but this may not be his own offspring.

Up to 27½ in (70 cm)
None
Up to 22 lb (10 kg)
Variable
Vulnerable

N. Morocco, N. Algeria; introd. Gibraltar

SNOW MONKEYS

Japanese macaques live farther north than any other primate (apart from humans). During the cold, snowy winters, they live off tree buds and bark, as well as stores of fat. Troops of 20–30 Japanese macaques are headed by a dominant male. Their social lives tend to be harmonious, with much time spent grooming each other and sharing the care of the young.

Up to 23½ in (60 cm)
Up to 6 in (15 cm)
Up to 22 lb (10 kg)
Variable
Data deficient

Japan

Making snowballs
Entire troops of Japanese macaques make snowballs, fashioning a small ball in their hands then rolling it along the ground so it grows in size.

Male is grayish brown with shaggy mantle of silver hair on head

Hamadryas baboon (savanna baboon)
Papio hamadryas

Female is olive-brown

Mane of hair on head of male

Gelada
Theropithecus gelada

Bright red callous pads on rump

Heart-shaped patch of hairless skin on chest

Chacma baboon
Papio ursinus

Highly opposable thumbs provide dexterity to select the best grass blades, rhizomes, and seeds

FACT FILE

Hamadryas baboon At night, troops of up to 750 of these baboons sleep together on rocky outcrops, splitting into bands of 20–70 animals at dawn to forage for grass, fruit, leaves, flowers, and small vertebrates. These bands are made up of harem families—a single dominant male, with several females, and their juvenile offspring.

🐾 Up to 35½ in (90 cm)
🐾 Up to 27½ in (70 cm)
⚖ Up to 44 lb (20 kg)
👣 Family band, troop
⚡ Near threatened
❋ ☀ ⚘ ⚘ ⚘

Ethiopia, Somalia, Eritrea, Sudan, Arabian P.

Chacma baboon This species lives in a variety of habitats, including the most arid environments inhabited by any primate other than humans. One troop of chacma baboons in the Namib Desert was observed to survive without water for 116 days, obtaining all their moisture from figs. A more typical diet includes fruit, leaves, roots, and insects.

🐾 Up to 35½ in (90 cm)
🐾 Up to 29½ in (75 cm)
⚖ Up to 88 lb (40 kg)
👣 Herd, large troop
⚡ Least common
☀ ⚘ ⚘

S. Africa

On alert The dominant male in a troop is always on the alert for challenges from bachelor bands.

GRASS-EATING GELADAS

The only surviving species of a genus once widespread throughout Africa, geladas are now restricted to the highlands of northwest Ethiopia. Here, they sleep on rocky cliffs beyond the reach of most predators, and forage in nearby grasslands by day. They survive almost entirely on grass, a degree of specialization that makes the species especially vulnerable as the local human population burgeons and requires ever larger areas of pasture for grazing livestock.

ring of red beading on female indicates she is ready to mate

Bleeding heart Both male and female geladas have bald chests, which alter in color and appearance according to the female's sexual cycle. On other baboons, such sexual advertising is conveyed on the rump. By appearing on the chest, it allows geladas to remain in a heat-conserving squatting position.

Social creatures The basic unit of gelada society comprises one male, several females, and their offspring, but several families form foraging bands of around 70 animals. At times, a number of bands may congregate in vast herds of 600 or more individuals.

White-cheeked mangabey
Lophocebus albigena

Long, slender, prehensile tail suits arboreal lifestyle

Male is twice as large as female

Brightly colored rump with short tail

Drill
Mandrillus leucophaeus

Bright red and blue face on male, more subdued blue face on female and juvenile

Color of naked rump ranges from blue to purple

Mandrill
Mandrillus sphinx

Agile mangabey
Cercocebus galeritus

FACT FILE

Mandrill The largest of all monkeys, Africa's mandrills are instantly recognizable by their striking red and blue faces. During the day, they come down from their sleeping trees to search for fruit, seeds, insects, and small vertebrates on the rain-forest floor. Troops of up to 250 mandrills are made up of smaller multi-male groups, each with about 20 animals led by a dominant male who fathers most of the young.

Up to 30 in (76 cm)
Up to 3 in (7 cm)
Up to 55 lb (25 kg)
Family band, troop
Vulnerable

Equatorial W. Africa

Showing off
To threaten a rival or predator, a male mandrill will spread its arms wide and yawn, displaying its daunting teeth.

Attractive colors *In addition to their vibrant faces, male mandrills have a yellow beard, mauve rump, red penis, and lilac scrotum. The skin colors are most intense on the dominant male—they seem to be linked to testosterone levels and may be advertising his virility.*

CONSERVATION WATCH

Threatened drills and mandrills Both drills and mandrills are under threat as logging, farming, and human settlement continue to destroy much of their rain-forest habitat. Both species are also hunted for food, with their large groups and loud calls making them easy to find. Drills are now endangered, their numbers reduced by 80 percent over recent years. Mandrills, considered vulnerable, are expected to suffer a similar decline in the near future.

FACT FILE

Sykes' monkey Among this species, a single adult male dominates troops of 10–40 females and their offspring. The females help raise one another's young.

🐾 Up to 26½ in (67 cm)
🐒 Up to 33½ in (85 cm)
⚖ Up to 26½ lb (12 kg)
👣 Family band, variable
✹ Locally common

C., E. & S. Africa

Vervet monkey Although it prefers to live in woodland along rivers, this adaptable monkey is found in many habitats, including human settlements.

🐾 Up to 24½ in (62 cm)
🐒 Up to 28½ in (72 cm)
⚖ Up to 20 lb (9 kg)
👣 Herd, troop
✹ Declining

Sub-Saharan Africa

Diana monkey Spending their lives high in the trees, Diana monkeys live in single-male groups of 15 or more individuals. The young gain their arboreal skills through constant play.

🐾 Up to 23½ in (60 cm)
🐒 Up to 31½ in (80 cm)
⚖ Up to 16½ lb (7.5 kg)
👣 Family band
✹ Endangered

Coastal W. Africa

Mona monkey Like many other Old World monkeys, this small primate stores fruit and insects in its cheek pouches while it forages.

🐾 Up to 27½ in (70 cm)
🐒 Up to 27½ in (70 cm)
⚖ Up to 15½ lb (7 kg)
👣 Herd, troop
✹ Locally common

W. & C. Africa

MIXING SPECIES

Large groups of monkeys sometimes contain more than one species. In East Africa, for example, Sykes' and redtail monkeys form stable associations, traveling and feeding together. Such an arrangement reduces the risk of predation while avoiding the increased competition for food involved in a larger single-species group.

Sykes' monkey
(blue monkey)
Cercopithecus mitis

Coat can be blue, reddish brown, or grayish brown

Vervet monkey
Chlorocebus aethiops

Male has turquoise-blue scrotum

Diana monkey
Cercopithecus diana

White stripe across forehead thought to resemble shape of goddess Diana's bow, inspiring this monkey's common name

Long white tufts on ears

Will freeze in position if it senses danger

Mona monkey
Cercopithecus mona

Redtail monkey
Cercopithecus ascanius

Grasping hands for
gripping branches
and collecting fruit

Allen's swamp monkey
Allenopithecus nigroviridis

Lives in swamp forests
and forages on ground
or in shallow water

Chestnut fur
on underside of
tail gave rise to
common name

Webbing between digits
helps with swimming

Patas monkey
Erythrocebus patas

Slender legs of equal
length allow monkey
to run at speeds of up
to 35 mph (55 km/h)

Red-eared monkey
Cercopithecus erythrotis

MONKEY SIGNALS

Sometimes referred to as guenons,
the monkeys in the genus *Cercopithecus*
all use a range of signals to communicate
with other members of their species.
While many of these signals are vocal,
encompassing barks, grunts, screams,
booms, and chirps, others are tactile
or visual. Nose-to-nose contact is a
form of friendly greeting used by
many species. Tail position can indicate
an animal's level of confidence.
Staring, head-bobbing, and yawning
are often used as a threat display
to intimidate a potential opponent,
while a display of clenched teeth is a
fear grimace, employed as a gesture
of appeasement.

On the nose *A nose-to-nose greeting
between two redtail monkeys is often
followed by grooming or play.*

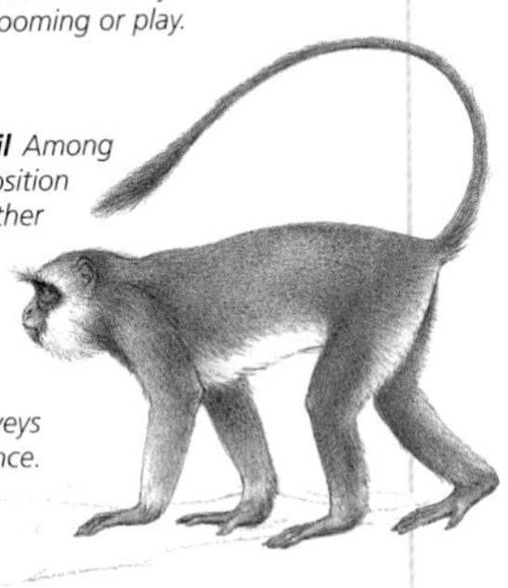

Confident tail *Among
vervets, tail position
indicates whether
the animal is
fearful or not.
When on all
fours, a tail
arched over
the body conveys
great confidence.*

Raising the alarm
*To warn the troop
of a predator,
a vervet uses
specific alarm calls:
a "chutter" sound
for snakes makes the
monkeys stand up and check the
grass; a double cough for eagles
prompts them to look up in the air and
take cover; and a barking call for leopards
sends them running to the trees.*

⚡ CONSERVATION WATCH

Patas vulnerability This terrestrial
monkey lives on the savannas of
central Africa. Already vulnerable to
rainfall fluctuations in this drought-
prone region, more and more patas
monkeys are being killed by hunters
for their meat and by farmers
because they feed on crops. Their
habitat is also becoming increasingly
fragmented by human activities.

APES

CLASS	Mammalia
ORDER	Primates
FAMILIES	2
GENERA	5
SPECIES	18

Like humans, apes are intelligent and quick to learn, form complex social groups, and spend years on the care of each young. They are divided into two families: the gibbons of Hylobatidae, and the great apes—orangutans, chimpanzees, gorillas—of Hominidae, which includes humans. Although apes and Old World monkeys have a similar nose shape and dental structure and are grouped together as catarrhine primates, they differ in many ways. Apes have skeletons suited to sitting or standing upright. They have no tail, their lowest vertebrae fused to form the coccyx instead. Their spines are shorter, their chests are barrel-shaped, and their shoulders and wrists are very mobile.

GIBBON SONGS

Gibbon pairs often begin the day with duets led by the female and punctuated by the male. While these songs may reinforce the pair bond, they are also territorial and act as spacing calls, helping the gibbons to avoid other pairs during their daytime foraging. Many gibbon species feature throat sacs that amplify their calls. Among the siamangs, the largest gibbons, both males and females possess enormous throat sacs. The sacs produce a booming sound as they are inflated, which is followed by a deafening bark or shriek.

CLEVER APES

Ape societies are organized in various ways. Monogamous gibbons live in pairs with their offspring, making groups of up to six animals. Orangutans with overlapping ranges form loose associations and meet occasionally, but the male tends to be a solitary forager, and the female usually lives just with her single young. Chimpanzee communities involve 40–80 individuals, but these are rarely together at once and tend to forage in smaller groups. Gorillas live in harems, with one dominant male, possibly one or two lower-ranking adult males, several females, and their young.

The gibbons and the great apes evolved into distinct families at least 20 million years ago. Chimpanzees are believed to be the closest relatives

Parenthood All infant apes stay dependent on their parents for a long time. Female gorillas usually give birth to a single young, who takes up to 3 years to be weaned. Because more than a third of all infant gorillas die before this, most females take 6–8 years to produce a surviving offspring.

of humans, sharing a common ancestor until about 6 million years ago. The great apes appear to work through problems much as humans do. Chimpanzees and orangutans are known to fashion tools in the wild, while in research centers all the great apes have been taught to use implements. By recognizing themselves in mirrors, great apes show a concept of self, and some have been taught to recognize and use symbols such as sign language.

Moving apes While all apes other than humans have arms longer than their legs, orangutans and gibbons are the only ones in which the arms are genuinely elongated compared to their trunk. The orangutan's head and body measure about 5 feet (1.5 m), while their arms have a spread of more than 7 feet (2.2 m). Gibbons move by brachiation, using their arms to swing from one branch to another. Orangutans do not brachiate, but clamber slowly through the trees using some combination of all four limbs. Chimpanzees spend up to three-quarters of their time on the ground, but will brachiate when in trees. Gorillas are largely terrestrial and rarely climb trees.

A flexible grip
The orangutan's strong hands and feet are hook-like, the thumb and big toe are short, and the other digits are long.

Hanging around
Although it does hang under branches, the orangutan uses all its limbs to move through the trees.

Energy conservation
Long arms allow the orangutan to reach fruit with minimum exertion.

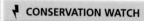

🐾 CONSERVATION WATCH

Extensive logging and clearing of tropical forest have placed most apes in jeopardy. The bushmeat trade has also taken a heavy toll. Of the 18 species of apes, 100% are listed on the IUCN Red List, as follows:

- 3 Critically endangered
- 7 Endangered
- 3 Vulnerable
- 4 Near threatened
- 1 Data deficient

Kloss's gibbon
Hylobates klossii

White face rings, hands, and feet with reddish or black coat

Hoolock
Hylobates hoolock

Lar gibbon
Hylobates lar

Throat sac bigger than head

Males are black; adult females are golden or buff, sometimes with black patches

Siamang
Hylobates syndactylus

Elongated arm with hook-like, grasping hand

Black gibbon
Hylobates concolor

FACT FILE

Orangutan Asia's only great apes, orangutans are also the world's largest arboreal mammals. Most hardly ever descend to the forest floor, moving through the forest by swinging a tree back and forth until they can grasp the next one. Every night, an orangutan constructs an elaborate nest in the crown of a tree and covers itself with vegetation to sleep.

- Up to 5 ft (1.5 m)
- None
- Up to 200 lb (90 kg)
- Solitary, pair
- Endangered

Borneo, Sumatra

Western gorilla The largest of all primates, male gorillas grow until they are about 12 years old, developing a saddle of silvery-white hair on the back, hence the term "silverback" for a mature male. Gorillas spend most of their time on the ground, covering substantial distances on all fours.

- Up to 6 ft (1.8 m)
- None
- Up to 400 lb (180 kg)
- Variable
- Endangered

C. Africa

TOOL TIME

The ingenuity and manual dexterity of chimpanzees are demonstrated by their tool use. They strip twigs and grass stems to make wands for probing ant and termite nests. Specially selected stones are used to open nuts and hard-shelled fruit. During displays of strength or hunting, some chimps use sticks and rocks as missiles. Tool use is a socially learned tradition and varies from one chimp population to another.

- Up to 3 ft (93 cm)
- None
- Up to 110 lb (50 kg)
- Large troop, herd
- Endangered

C. & W. Africa

Powerful grip

Flexible arms and legs can swing in most directions

Orangutan
Pongo pygmaeus

Male has large cheek pads and throat pouch with beard and mustache

Chimpanzee
Pan troglodytes

Arms longer than legs, with fingers longer than those of humans

Western gorilla
Gorilla gorilla

Walks on soles of feet and knuckles of hands

Bonobo
Pan paniscus

Slimmer body and more slender limbs than those of common chimpanzee

Extinction Miss Waldron's red colobus monkey (*Procolobus badius waldroni*), once found in Ghana and Côte d'Ivoire, was the only primate to be documented as extinct during the 20th century. The extinction appears to have been caused by hunting for the bushmeat trade, exacerbated by logging operations that gave access to formerly remote forests. Many living primate species continue to face similar threats. While the live-animal trade in endangered species has been outlawed, some primates are still captured and sold illegally as pets or for medical research, and many are killed as bushmeat.

CONSERVING PRIMATES

According to Conservation International, 195 primate species and subspecies—about a third of all primate types—are at risk of extinction within the next few decades. About half of all colobus monkeys and gibbons are threatened, while in the family Hominidae, humans are the only secure species, with all of the great apes considered endangered. The trade in live animals as pets or for biomedical research has contributed to the declining numbers, as has the hunting of primates as food or pests. The greatest threat, however, is from habitat destruction through logging, land clearing, and collection of wood for fuel. Because primates reproduce slowly, their populations take a long time to recover. Almost all are tropical animals and live in poorer countries, so conservation efforts are complicated by the pressing needs of the growing human population.

Habitat destruction When an area of rain forest is cut down, several primate species can be affected at once and the damage is permanent. Rain-forest soil is thin and low in nutrients, but it supports lush plant growth because the nutrients are efficiently recycled by the ecosystem. Once the tree cover is lost, the soil is washed away by rain and the area soon becomes barren.

Orangutan rescue When primates being kept illegally as pets are rescued, they do not have the skills to survive in the wild. At the Sepilok Orangutan Rehabilitation Centre, on the island of Borneo, rescued orangutans are trained to fend for themselves before being released back into the forest. More than 100 orangutans have joined Sepilok's wild population after rehabilitation.

3. Outward bound school
The Centre's staff gradually reduce the amount of food and emotional support they offer, encouraging the orangutans to fend for themselves.

Asian primates About 45 percent of endangered primates are found in Asia, particularly in Indonesia (35 endangered primates), China, India, and Vietnam (15 each). This photograph shows Delacour's langur (*Trachypithecus delacouri*), one of the Vietnamese species at risk.

1. Quarantine All new arrivals are kept in quarantine for 3–6 months to prevent them transmitting any diseases to the other orangutans at the Centre.

2. Nursery Wildlife rangers train young orangutans (up to 3 years of age) in basic survival skills, from climbing trees to building sleeping nests and finding fruit and other food in the forest.

4. Survival training When an orangutan starts showing signs of independence, even less food is offered. Eventually, most rescued animals join Sepilok's wild orangutan population.

CARNIVORES

CLASS	Mammalia
ORDER	Carnivora
FAMILIES	11
GENERA	131
SPECIES	278

From massive polar bears to little weasels, speedy cheetahs to lumbering elephant seals, pack-living wolves to solitary tigers, members of the order Carnivora display remarkable diversity. Although they are commonly referred to as carnivores, a term also used for any meat-eating animal, some of them eat meat only rarely, if at all. The one thing members of Carnivora have in common is a predatory ancestor with four carnassial teeth—scissor-like molars that can shear through meat. Most carnivores retain the carnassial teeth, which sets them apart from other meat-eating mammals. In mainly insectivorous or herbivorous carnivores, the carnassial teeth have been modified for grinding.

Sweet tooth Originally classified as a primate, the kinkajou is a nocturnal, tree-dwelling carnivore. A prehensile tail allows it to hang upside down as it feeds on fruit or uses its long, narrow tongue to gather nectar or honey. It sometimes supplements this sweet diet with insect prey.

Prowling for meat Like most animals that belong to the order Carnivora, the jaguar is a predator and usually eats meat it has caught itself. A solitary hunter, it relies on stealth and keen senses as it stalks prey. The large ears capture sound waves, while the large eyes provide excellent binocular vision by day and night.

Vegetarian carnivore While the giant panda may eat small mammals, fish, and insects if they are available, bamboo makes up more than 99 percent of its diet. A plentiful, year-round food source, bamboo is low in nutrition, so giant pandas must spend 10–12 hours a day feeding in order to satisfy their energy requirements.

PRIME HUNTERS

As the dominant land predators on all continents except Antarctica, carnivores are built for hunting. They rely on their acute senses of sight, hearing, and smell to help them detect prey. The complex ear region, which often features more than one inner chamber, increases sensitivity to the frequencies that are produced by their prey species.

Intelligence, agility, and speed help carnivores stalk, chase, and catch their prey efficiently. Even apparently cumbersome species such as bears are capable of impressive sprints, while the cheetah is the fastest land animal in the world. All carnivores have fused bones in the forefeet, a feature that helps absorb the shock of running. A reduced collarbone increases the mobility of the shoulder muscles, allowing a longer stride and greater speed.

To kill their prey, carnivores generally use their strong jaws and sharp teeth. Weasels smash the prey's skull by biting into the back of its head. Cats strike small prey at the neck to snap the spinal cord, while dogs dislocate the neck by violently shaking the animal in their jaws.

Adaptable animals Carnivores are found in almost all of Earth's habitats. Polar bears, arctic foxes, and these gray wolves survive in the icy landscape of the Arctic; otters and seals spend most of their time in the water; big cats prowl both jungles and savanna; and jackals live in deserts.

Cooperative hunting allows wolves, lions, and other pack animals to tackle much larger prey such as buffalo and wildebeest.

Almost all the larger meat-eating carnivores hunt vertebrates. Smaller carnivores mostly eat invertebrates, which are easier to catch but can rarely sustain a large animal. Various carnivores concentrate on termites, worms, fish, and crustaceans, and some are largely vegetarian, with a preference for berries, other fruit, nuts, seeds, nectar, or bamboo, but all tend to be opportunistic feeders and will take advantage of any easy meal that presents itself.

About 50 million years ago, the order Carnivora split into two lineages. The cat-like carnivores include civets (family Viverridae), cats (Felidae), hyenas (Hyaenidae), and mongooses (Herpestidae). Dog-like carnivores include dogs (Canidae), bears (Ursidae), raccoons (Procyonidae), weasels (Mustelidae), and seals (Otariidae and Phocidae). Until recent years, seals were often treated as a separate order known as Pinnipedia, but genetic studies indicate that they share a common ancestry with the other carnivores.

Group living For many smaller carnivores, living in a group helps to reduce the risk of predation. In suricate troops, which are made up of two or three family units, members take turns to watch for danger.

Designed to hunt The skeleton of a cat displays many of the anatomical features that have made carnivores such effective predators. A flexible spine, long limbs, fused wrist bones, and reduced collarbone all contribute to the cat's speed and agility.

INTRODUCING CARNIVORES

The exceptional hunting abilities of carnivores have inspired many misguided attempts to use them for pest control in regions where they are not native. The consequences of such introductions have usually been disastrous. The native fauna of New Zealand is under threat from stoats (right), which were introduced in the 1880s to take care of a rabbit problem. In the Caribbean and Hawaii, the small Indian mongooses (*Herpestes javanicus*) imported to control rodents and snakes ended up spreading rabies instead. On a number of the world's far-flung islands, feral cats, which were supposed to get rid of rats, preferred the easier prey of flightless birds and destroyed their populations.

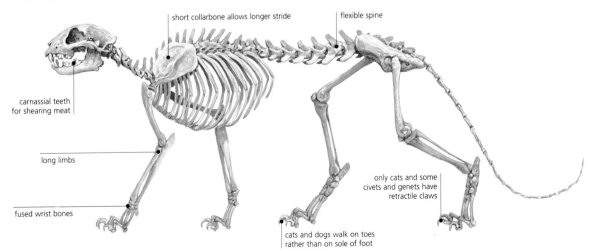

short collarbone allows longer stride

flexible spine

carnassial teeth for shearing meat

long limbs

fused wrist bones

only cats and some civets and genets have retractile claws

cats and dogs walk on toes rather than on sole of foot

THE DOG FAMILY

CLASS	Mammalia
ORDER	Carnivora
FAMILY	Canidae
GENERA	14
SPECIES	34

Perhaps no creatures have such an ambivalent relationship with humans as the dogs, wolves, coyotes, jackals, and foxes that make up the carnivore family Canidae. Domesticated at least 14,000 years ago, dogs were the first animals to enter a partnership with humans and have been used extensively for hunting, guarding, warmth, and companionship. At the same time, wild canids have been relentlessly persecuted, blamed for the loss of livestock and the spread of rabies and hunted for sport and fashion. While some species such as red foxes and coyotes have adapted and thrived in the midst of urban development, others such as red wolves are on the verge of extinction.

A wide distribution Originating in North America 34–55 million years ago, wild canids now occur on every continent except Antarctica, but are absent from some islands, including Madagascar, Hawaii, the Philippines, Borneo, and New Zealand. They were introduced to New Guinea and Australia in prehistoric times. The domestic dog is now found worldwide.

Cooperative hunters Gray wolves usually live in family units of 5–12 dogs dominated by an alpha pair. The pack cooperates to snare larger prey such as deer. They most often target a young, old, or weak herd member, pursuing it until it is exhausted.

GRASSLAND CARNIVORES

Most canids live in open grasslands, where they capture their prey either by sudden pouncing or by extended pursuit. With their slender build, muscular, deep-chested bodies, and long, sturdy legs, they are capable of great endurance. In addition to the fused foot bones common to other carnivores, canid forelimb bones are locked to prevent them rotating as they run. A pointed muzzle houses the large scent organs that allow canids to track prey over long distances, while large, erect ears contribute to their acute hearing.

Opportunistic and adaptable, canids prefer to eat freshly killed meat but will take advantage of whatever food is locally available and may eat fish, carrion, berries, and human garbage. Their social organization is similarly flexible, varying between and within species and often reflecting their diet. Smaller species such as jackals and foxes mostly eat small animals and often live alone or in pairs. Larger species such as gray wolves and African wild dogs usually live and hunt in hierarchical social groups, cooperating to bring down prey larger than themselves. Canids that hunt especially large animals tend to live in the largest groups, while in areas where smaller prey are abundant, some gray wolves choose to live in pairs.

Group living confers benefits other than hunting advantages, however, and some canids live communally but hunt alone. These animals work together to care for the pack's young and defend their territory from rival packs.

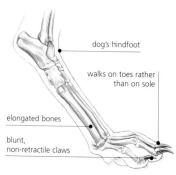

Walking on tiptoes Members of the dog family have elongated feet and move by digitigrade locomotion: walking on just the toes rather than on the entire sole of the foot. As the claws cannot be retracted, they become blunt from wear. In a dog's forefeet, fused bones at the wrists help to absorb the impact of running.

dog's hindfoot

walks on toes rather than on sole

elongated bones

blunt, non-retractile claws

FAMILY LIFE

Unusually among mammals, a lifelong monogamous pair bond forms the basis of the jackal social unit. Both parents take an active role in caring for the young, who are nursed for about 8 weeks and then fed regurgitated food for several weeks more. In many jackal families, one or two young stay with the parents for a year after reaching sexual maturity and help to raise the next litter.

Slender, pointed muzzle

Coyote
Canis latrans

Drooping, bushy tail with black tip

Reddish brown forelegs, feet, and muzzle

⚡ **Red wolf**
Canis rufus

Smaller than gray wolf, with longer ears and shorter fur

Size and fur color of gray wolf vary according to location

**Gray wolf
Alaskan form**
Canis lupus tundrarum

**Gray wolf
Scandinavian form**
Canis lupus lupus

COYOTE COMMUNICATION

While many coyotes are solitary, others live and hunt in pairs or larger packs. Like other canids, coyotes use scent marking, facial expressions, body and tail posture, and a range of squeaks, yelps, and howls to communicate.

friendly

submissive

playful

attacking

defending

⚡ CONSERVATION WATCH

Waning wolves While the versatile coyote's numbers and range have been expanding, wolves have been suffering. Once found throughout most of the Northern Hemisphere, gray wolves now survive in just a handful of restricted areas in North America and Eurasia. Red wolves are critically endangered, with only some 200 individuals remaining in the wild. Both wolf species are the target of reintroduction campaigns.

Side-striped jackal Rather than pursuing prey, this nocturnal jackal will quickly pounce on insects, mice, and birds, or scavenge the kills of other predators. Each family has a particular yipping call recognized only by its own members.

Up to 32 in (80 cm)
Up to 16 in (40 cm)
Up to 26½ lb (12 kg)
Pair
Uncommon

C. & S. Africa

Golden jackal The most widely distributed jackal, this species has lived on the edge of human settlements since ancient times, when it featured prominently in Egyptian mythology.

Up to 39½ in (100 cm)
Up to 12 in (30 cm)
Up to 33 lb (15 kg)
Pair
Common

N. Africa, S.E. Europe to Thailand, Sri Lanka

Black-backed jackal Near villages, this jackal is nocturnal, but elsewhere it may be active day or night. About half of its diet consists of insects, with small mammals and fruit making up the rest. Males and females form long-term pairs and share the care of their pups.

Up to 35½ in (90 cm)
Up to 16 in (40 cm)
Up to 26½ lb (12 kg)
Pair
Locally common

E. & S. Africa

Ethiopian wolf This species is found only in a dozen isolated pockets in Ethiopia. With fewer than 500 adults remaining in the wild, it is one of the world's most endangered mammals.

Up to 39½ in (100 cm)
Up to 12 in (30 cm)
Up to 42 lb (19 kg)
Solitary, small group
Critically endangered

Highlands of Ethiopia

Dingo Probably brought to Australia by Asian traders at least 3,500 years ago, the dingo became the dominant predator in many areas, cooperatively hunting large marsupials such as kangaroos and wallabies.

Up to 39½ in (100 cm)
Up to 14 in (36 cm)
Up to 53 lb (24 kg)
Solitary, small group
Locally common

Mainland Australia

Side-striped jackal
Canis adustus

Shorter legs and ears than other jackals, with white and black side stripe on drab coat

Color can vary from brown-tipped yellow in rainy season to pale gold in dry season

Golden jackal
Canis aureus

Black fur runs from back of neck to tail

Ethiopian wolf (Simien jackal)
Canis simensis

Long, pointed muzzle for catching small mammals

Black-backed jackal
Canis mesomelas

Dingo
Canis lupus dingo

Dhole
Cuon alpinus

**Arctic fox
winter coat**
Alopex lagopus

Paws covered
in dense fur
during winter

**Arctic fox
summer coat**
Alopex lagopus

Short, broad
muzzle suits
almost exclusive
diet of large
vertebrate prey

Teeth include up to eight extra
molars for grinding up termites,
dung beetles, and other insects

Bat-eared fox
Otocyon megalotis

**Cape wild dog
(African wild dog,
African hunting dog)**
Lycaon pictus

Each Cape wild dog has
a unique coat pattern

Raccoon dog
Nyctereutes procyonoides

Molts twice a year as
it grows a thick winter
coat followed by a
lighter summer coat

FACT FILE

Arctic fox For winter, this fox grows a white coat to match its snowy habitat and feeds on marine mammals, sea birds, fish, and crustaceans, as well as on stored food. As summer approaches, its coat turns brown or black to blend in with the tundra vegetation.

- Up to 27½ in (70 cm)
- Up to 16 in (40 cm)
- Up to 20 lb (9 kg)
- Solitary
- Common

N. North America, N. Eurasia, Iceland, Greenland

Cape wild dog This canid is entirely carnivorous. After a kill, the healthy adult dogs will regurgitate food for any young, sick, or wounded pack members.

- Up to 35½ in (90 cm)
- Up to 16 in (40 cm)
- Up to 79½ lb (36 kg)
- Solitary, small group
- Endangered

E. & S. Africa

UNUSUAL CANID

The raccoon dog is the only canid to enter a dormant state in winter and the only canid that does not bark. While most other canids live in open grasslands, the raccoon dog prefers the dense undergrowth of forests. Originally from East Asia, it was introduced to Europe by the fur trade.

Raccoon resemblance
The raccoon dog's black face mask, thick body, bushy tail, and ability to climb trees give it a great resemblance to the raccoon.

✦ CONSERVATION WATCH

Of the 34 species of canids, 53% are listed on the IUCN Red List, as follows:

1. Extinct
2. Critically endangered
1. Endangered
2. Vulnerable
1. Near threatened
2. Conservation dependent
9. Data deficient

FACT FILE

Kit fox Found in arid regions, the kit fox avoids the heat of the day by resting in its underground den, coming out at night to hunt rabbits and kangaroo rats. This fox relies on the moisture in its prey for fluids and therefore kills more animals than are needed to meet its energy requirements alone.

🐾 Up to 20½ in (52 cm)
🐾 Up to 12½ in (32 cm)
⬛ Up to 6 lb (2.7 kg)
♦ Solitary
⚘ Conserv. dependent

S.W. USA & N. Mexico

THE SUCCESSFUL RED FOX

One of the most widely distributed species in the world, the red fox's versatile feeding habits have allowed it to thrive in forest, prairie, farmland, and suburban areas. Hunted for sport and bred for fur, this canid has been blamed for killing poultry and spreading rabies. Where it has been introduced, as in Australia, the red fox poses a major threat to native fauna.

🐾 Up to 19½ in (50 cm)
🐾 Up to 13 in (33 cm)
⬛ Up to 13 lb (6 kg)
♦ Solitary, pair
⚘ Common

North America, Europe, N. & C. Asia, N. Africa, Arabia; introd. Australia

Fox feast
Red foxes will eat almost anything, from rodents and rabbits to fruit and garbage.

⚡ CONSERVATION WATCH

Swift reintroduction Once ranging throughout the North American prairies, the swift fox lost habitat to agriculture and urbanization and was also hunted and poisoned. By 1978, it had disappeared from Canada, but reintroduction programs have established small populations in Alberta and Saskatchewan and the species has moved from endangered to lower risk status.

Tibetan fox
Vulpes ferrilata

Broad ears listen for rustling rodents

Corsac fox
Vulpes corsac

Blanford's fox
Vulpes cana

Displays cat-like movements

Kit fox
Vulpes macrotis

Pale fox
Vulpes pallida

Red fox
North American form
Vulpes vulpes fulva

Short reddish summer coat replaced by longer, thicker gray coat for winter

Swift fox
Vulpes velox

Bengal fox
Vulpes bengalensis

Red fox
Central European form
Vulpes vulpes crucigera

Gray fox
Urocyon cinereoargenteus

Strong, hooked claws
used for climbing trees

Black-tipped tail

Mane of erect
black hairs

Crab-eating fox
(common zorro)
Cerdocyon thous

Maned wolf
Chrysocyon brachyurus

Argentine gray fox
Pseudalopex griseus

Long legs help maned
wolf see above tall
grass of pampas

Small-eared zorro
Atelocynus microtis

Culpeo fox
Pseudalopex culpaeus

Bush dog
Speothos venaticus

Webbed feet for
swimming after
aquatic prey

Coat colors most
vivid in northern
part of range

Pampas fox
Pseudalopex gymnocercus

FACT FILE

Crab-eating fox An opportunistic feeder, this fox often eats crabs and other crustaceans during the wet season, but shifts to insects in the dry season. Its diet also features fruit, turtle eggs, small mammals, birds, reptiles, amphibians, fish, and carrion.

Up to 30 in (76 cm)
Up to 13 in (33 cm)
Up to 17½ lb (7.9 kg)
Pair
Common

Colombia to Argentina except Amazon basin

Maned wolf This omnivorous canid eats large amounts of bananas, guavas, and other fruit as well as animals such as armadillos, rabbits, rodents, snails, and birds. It hunts at night in a fox-like manner, suddenly pouncing on its prey.

Up to 39½ in (100 cm)
Up to 16 in (40 cm)
Up to 53 lb (24 kg)
Solitary, pair
Near threatened

Brazil to Paraguay & Argentina

Argentine gray fox This nocturnal fox usually lives in small groups made up of a mating pair, their offspring, and sometimes a second adult female who helps to raise the young. The group maintains a year-round territory.

Up to 26 in (66 cm)
Up to 16½ in (42 cm)
Up to 12 lb (5.4 kg)
Solitary, pair
Locally common

Chile & Argentina

Bush dog With its squat stature and short face, the bush dog looks less like a dog than any other canid. Packs of up to 10 animals forage together in forest undergrowth, keeping in touch via high-pitched peeps and whines.

Up to 29½ in (75 cm)
Up to 5 in (13 cm)
Up to 15½ lb (7 kg)
Solitary to large group
Vulnerable

W. Panama to Paraguay & N. Argentina

Pampas fox When threatened by humans, the pampas fox freezes and may remain motionless if handled. Pairs live together only during the mating season and share the care of the young.

Up to 28½ in (72 cm)
Up to 15 in (38 cm)
Up to 17½ lb (7.9 kg)
Solitary, pair
Locally common

E. Bolivia & S. Brazil to N. Argentina

THE BEAR FAMILY

CLASS Mammalia
ORDER Carnivora
FAMILY Ursidae
GENERA 6
SPECIES 9

In spite of their fearsome reputation, bears tend to be the most herbivorous of the carnivores. Only one species, the polar bear, is primarily a meat-eater, while berries, nuts, and tubers make up the bulk of the American black bear's diet, the sloth bear feeds mostly on insects, and the giant panda eats almost nothing but bamboo. The first bear species emerged from the canid family 20–25 million years ago in Eurasia. About the size of a raccoon, these early creatures had a long tail and the shearing carnassial teeth common to most members of Carnivora. Over time, most bears became much bigger, their tails shortened, and their carnassials were flattened for grinding vegetable matter.

Northern diversity Most abundant in the Northern Hemisphere, bears live in Europe, Asia, and North and South America. The brown bear was also found in North Africa until the 1800s. Today, all but two bear species are in danger, victims of habitat loss and overhunting.

Fish feeder In the northwest coastal regions of North America, brown bears will wait at waterfalls to catch spawning salmon as they swim upstream. This annual glut of fish provides important protein before winter begins.

Full size Grizzly bears (*Ursus arctos horribilis*) will attempt to intimidate a rival or foe by rearing up on their hindfeet and even walking bipedally for a short distance. The upright stance increases their already impressive size. They may also growl and display their long canine teeth.

STRENGTH AND BULK

Although the red panda may weigh only about 6 pounds (3 kg), most bears classified in the Ursidae family are substantial animals, with the polar bear and brown bear vying for the title of the largest terrestrial carnivore. As they often spend more of their time foraging than hunting, bears are built for strength rather than speed, with a stocky, muscular body, thick legs, and a massive skull. An enlarged snout reflects their keen sense of smell. Vision and sound are less important, and their eyes and ears are relatively small.

Bears are found in the tropics, but it is in the cold northern lands that they are most numerous. Here, their great size allows them to put on fat deposits through spring and summer when food is abundant. When the weather cools, the bears retreat to a den or cave and enter a long sleep that can last half the year. During this dormant period, they live entirely on their body fat; they do not eat, urinate, or defecate,

and their heart and respiration rates drop. Unlike in true hibernation, however, their body temperature hardly falls. Remarkably, females give birth to young during the winter dormancy, maximizing the cubs' chance to grow and accumulate fat deposits before the next winter.

Most bears are solitary, although cubs often stay with the mother for 2–3 years. Rival males can be aggressive in the breeding season, with fights resulting in injury or even death.

DELIBERATE MOVEMENT

Bears usually move slowly and deliberately on all fours, but are capable of great bursts of speed when pursuing prey. They walk with a plantigrade gait, placing the entire sole on the ground. This not only supports their great bulk, but also allows them to stand up on their hindfeet. While mainly terrestrial, most bears are also agile climbers.

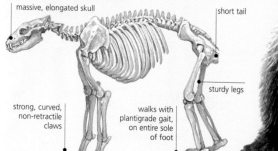

massive, elongated skull

short tail

strong, curved, non-retractile claws

walks with plantigrade gait, on entire sole of foot

sturdy legs

Himalayan brown bear
Ursus arctos isabellinus

Brown bear's coat can be brown, blonde, silver-tipped, or nearly black

Pronounced shoulder hump

American black bear
Ursus americanus

Black bear's coat can be black or brown

European brown bear
Ursus arctos arctos

Massive bulk helps polar bear withstand cold and store fat for times of scarcity

Largest species of bear

Polar bear
Ursus maritimus

Large, paddle-like paws for swimming

Largest of all brown bear subspecies

Males twice as large as females

Kodiak bear
Ursus arctos middendorffi

Asiatic black bear
Ursus thibetanus

FACT FILE

Brown bear This species once lived throughout Eurasia and North America and south to North Africa and Mexico. Its subspecies include the European brown bear (*Ursus arctos arctos*), the North American brown bear or grizzly (*U. a. horribilis*), the Kodiak bear (*U. a. middendorffi*), and the Himalayan brown bear (*U. a. isabellinus*).

- Up to 9 ft (2.8 m)
- Up to 8½ in (21 cm)
- Up to 1,320 lb (600 kg)
- Solitary
- Locally common

Pockets in N.W. North America, Wyoming, W. & N. Europe, Himalayas, Japan

American black bear Easily the most common bear, this species has adapted to a range of habitats and varies its omnivorous diet with the season.

- Up to 7 ft (2.1 m)
- Up to 7 in (18 cm)
- Up to 530 lb (240 kg)
- Solitary
- Locally common

Forested North America, nearby islands

Asiatic black bear This primarily herbivorous species readily climbs trees to gather fruit and nuts. To startle predators such as tigers, it will stand up and display its white chest mark.

- Up to 6 ft (1.9 m)
- Up to 4 in (10 cm)
- Up to 375 lb (170 kg)
- Solitary
- Vulnerable

Afghanistan & Pakistan to China, Korea & Japan

BEAR PAWS

The feet of bears reflect their varied habitats and diet. American black bears have hook-like claws for climbing trees and digging up roots. To grasp bamboo, giant pandas have an L-shaped pad, part of which covers a modified wrist bone that acts as a sixth digit.

American black bear

Giant panda

forefoot hindfoot forefoot hindfoot

FACT FILE

Giant panda This bear consumes up to 40 percent of its weight in bamboo per day, selecting shoots in spring, leaves in summer, and stems in winter. It does not hibernate during winter.

- Up to 5 ft (1.5 m)
- Up to 4 in (10 cm)
- Up to 355 lb (160 kg)
- Solitary
- Endangered

Isolated mountains in W. China

Red panda This species was once placed in the raccoon family because of a superficial resemblance, but genetic studies place it with the giant panda. It sleeps in trees by day and eats bamboo and fruit on the ground at night.

- Up to 25½ in (65 cm)
- Up to 19 in (48 cm)
- Up to 13 lb (6 kg)
- Pair
- Endangered

Nepal to Myanmar & W. China

Sloth bear After ripping open termite mounds and ant nests with its long, curved claws, this bear uses its long tongue to suck up the insects. It also climbs trees to extract honey from hives.

- Up to 6 ft (1.8 m)
- Up to 5 in (12 cm)
- Up to 320 lb (145 kg)
- Solitary
- Vulnerable

Sri Lanka, India, Nepal

BAMBOO BEAR

As China's human population has burgeoned, most of the giant panda's habitat has been destroyed. Only about 1,000 pandas remain in the wild, restricted to isolated mountain pockets of bamboo forest. Widely recognized as a symbol of conservation, the giant panda is still poached for its skin.

CONSERVATION WATCH

Of the nine species of bears, seven are listed on the IUCN Red List, as follows:

- 2 Endangered
- 3 Vulnerable
- 1 Conservation dependent
- 1 Data deficient

Giant panda
Ailuropoda melanoleuca

Distinctive black and white markings make the giant panda one of the most widely recognized of all animal species

Spectacled bear
Tremarctos ornatus

Only bear species in South America

Giant panda's front paws have "pseudo-thumb," an extra opposable digit for grasping bamboo

Shaggy coat may insulate from heat in tropical environment

Sloth bear
Melursus ursinus

Mobile snout with long tongue for capturing termites and ants

Long tongue licks up larvae, insects, and honey

Smallest bear apart from red panda

Sun bear
Helarctos malayanus

Long, curved claws help this highly arboreal bear to climb trees

Red panda
Ailurus fulgens

Coat made up of long, coarse hairs and dense undercoat to insulate against the cold weather of its high-altitude habitat

Only bear with a long tail

A YEAR AS A POLAR BEAR

Adapted to the harsh climate of the Arctic, the polar bear lives near the ice-covered water that contains its main prey, the ringed seal. Unlike other bears in cold environments, this semi-aquatic carnivore tends to remain active through winter, but can enter a dormant state and live off its fat deposits at any time of the year if food becomes scarce. Its solitary existence is interrupted only during the breeding season or when adult males fast together in groups known as sloths.

Cub protection Male polar bears sometimes kill cubs, presumably to free up the female for breeding. A mother often stands guard over her young and will attack a much larger male to protect them.

Seal hunter Polar bears will wait at a seal's breathing hole for hours, ready to pounce on the prey as it surfaces. They feed almost exclusively on seals and other marine mammals.

Expert swimmer Using its enormous forepaws as paddles, the polar bear can swim for several hours. On land, it is surprisingly agile and can reach speeds of 25 miles per hour (40 km/h). Dense fur and a layer of fat beneath the skin insulate against the cold.

April–July feeding Polar bears spend summer preying on the abundant and unwary pups of ringed seals. When the sea ice melts in late July, the bears come ashore and fast until it freezes once more.

April–May mating Female polar bears spend so long raising their young that they are available for mating only once every 3 years. This leads to intense competition among males for mates.

February–April emergence When the cubs are large enough to venture onto the sea ice, the mother leads them out of the den. Cubs stay with their mother for about 2½ years, learning the hunting skills that are crucial to their survival.

November–January birth While other polar bears remain active through winter, pregnant females retreat to a snow den, where they give birth to their young. Most litters contain two cubs, who are nursed until around late March.

MUSTELIDS

CLASS	Mammalia
ORDER	Carnivora
FAMILY	Mustelidae
GENERA	25
SPECIES	65

As a group, the weasels, otters, skunks, and badgers of the family Mustelidae are the most successful and diverse carnivores, with many more species than any other family. Mustelids are found in almost every type of habitat, including forests, deserts, tundra, and fresh and salt water, and may be arboreal, terrestrial, burrowing, semiaquatic, or fully aquatic. While a few species, such as sea otters and wolverines, can weigh more than 55 pounds (25 kg), most members of the family are medium-sized, with the smallest, the least weasel, weighing as little as 1 ounce (30 g). Highly carnivorous, mustelids are voracious hunters and will often tackle prey much larger than themselves.

Widespread family Absent only from Australia and Antarctica, mustelids are found throughout Europe, Asia, Africa, and the Americas. Despite their prevalence, only a small number have been well studied. Mustelids have been introduced in many places, either accidentally when they have escaped from fur farms or deliberately to control rodents and rabbits.

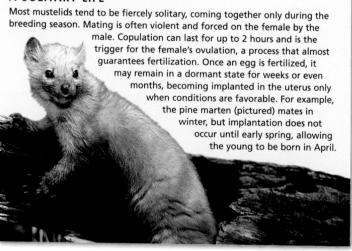

Scented defense Almost all mustelids possess scent glands near the anus that produce musk, a foul-smelling liquid used to mark territory. In skunks, this feature has developed into a defense system, with a spray of noxious musk deterring all but the most determined predators.

RELENTLESS HUNTERS

With an elongated body and short legs, mustelids can pursue rodents and rabbits down burrows. Weasels tend to be slender and lithe, with a flexible spine that allows them to scamper and leap. Badgers, on the other hand, have squat bodies and shuffle along with a rolling gait. Many mustelids are good swimmers and climbers and readily exploit aquatic and tree-dwelling prey as well as terrestrial species.

The mustelid head has a low, flat skull and short face with small ears and eyes. Smell is usually the most important of the senses and is used to locate and track prey and for communication among members of a species, with territories marked

Leaping weasel Agile and remarkably strong, weasels can run carrying up to half their weight in food. The smallest of all mustelids, the least weasel will relentlessly pursue its prey of mice and voles through thick grass or under snow.

by scent. Most mustelids have long, curved, non-retractile claws that they use for digging. Aquatic and semiaquatic species often have webbing between the toes to help with swimming.

Mustelids have a double coat, with a layer of soft, dense underfur interspersed with longer guard hairs. This warm, water-repellent coat enables mustelids to hunt in water and stay active throughout cold winters, but it has also made many species the target of the fur trade.

A SOLITARY LIFE

Most mustelids tend to be fiercely solitary, coming together only during the breeding season. Mating is often violent and forced on the female by the male. Copulation can last for up to 2 hours and is the trigger for the female's ovulation, a process that almost guarantees fertilization. Once an egg is fertilized, it may remain in a dormant state for weeks or even months, becoming implanted in the uterus only when conditions are favorable. For example, the pine marten (pictured) mates in winter, but implantation does not occur until early spring, allowing the young to be born in April.

Glands at base of tail
produce musk scent

European otter
Lutra lutra

Enters and leaves the
water at fixed locations
within its home territory

Smooth-coated otter
Lutra perspicillata

Neotropical otter
Lontra longicaudis

Fingers lack webbing
and claws but have
opposable thumb and
are highly dextrous
and sensitive

Cape clawless otter
Aonyx capensis

Sensitive whiskers
help locate prey

Giant otter
Pteronura brasiliensis

Spotted-necked otter
Lutra maculicollis

Sea otter
Enhydra lutris

Broad, flipper-like back
paws with webbing to
the tips of the toes

Uses stone as tool to
crack open sea urchin

Insulating layer of air
trapped between long guard
hairs and dense underfur

FACT FILE

Neotropical otter Much of this
solitary otter's day is spent diving for
fish. It eats small prey in the water but
takes larger prey to the shore.

	Up to 32 in (81 cm)
	Up to 22½ in (57 cm)
	Up to 33 lb (15 kg)
	Solitary
	Data deficient

Mexico to Uruguay

Giant otter Family groups of several
giant otters share a streamside burrow
and may forage together. Overhunting
has made this species the rarest otter.

	Up to 4 ft (1.2 m)
	Up to 27½ in (70 cm)
	Up to 75 lb (34 kg)
	Family band
	Endangered

S. Venezuela & Colombia to N. Argentina

Sea otter This often solitary species
can spend its entire life in the ocean,
sleeping on the surface and devoting
up to 5 hours a day to foraging, but
it may rest on shore in large sexually
segregated groups. The only non-
primate mammal known to employ
tools, it will use a stone to dislodge
abalone or crack open shellfish.

	Up to 4 ft (1.2 m)
	Up to 14 in (36 cm)
	Up to 99 lb (45 kg)
	Solitary, rests in group
	Endangered

North Pacific

AQUATIC PAWS

All otters have dextrous paws, but they
vary in the degree of webbing and the
presence or absence of claws. The
paws of most river otters (such as
those of the European otter, pictured)
are webbed for
swimming,
clawed for
digging, and
rounded for
walking on land.

⚡ CONSERVATION WATCH

Hunted otters By 1911, fur hunters
had depleted sea otter populations
to fewer than 2,000 individuals.
Legal protection and reintroduction
programs have helped increase their
numbers to about 150,000 animals,
but the species remains endangered,
threatened by poaching, marine
oil pollution, persecution by fishing
crews, and predation by orcas.

FACT FILE

Hog badger Although the hog badger sometimes falls prey to leopards and tigers, it will put up a fight. When threatened, it arches its back, bristles its hair, and growls. It may also emit a noxious fluid from its anal glands.

🐾 Up to 27½ in (70 cm)
🐾 Up to 6½ in (17 cm)
⚖ Up to 31 lb (14 kg)
🐾 Not known
🐾 Not known

N.E. India to N.E. China & S.E. Asia

Eurasian badger While most badgers tend to be solitary, this species lives in extended family groups. A powerful digger, it lives in permanent burrow complexes known as setts, which are passed down through the generations.

🐾 Up to 35½ in (90 cm)
🐾 Up to 8 in (20 cm)
⚖ Up to 35½ lb (16 kg)
🐾 Family band
🐾 Locally common

Britain & W. Europe to China, Korea & Japan

American badger This solitary species spends much of its time busily digging for rodents such as prairie dogs and ground squirrels. It will rapidly burrow its way out of danger if threatened by a predator on the surface.

🐾 Up to 28½ in (72 cm)
🐾 Up to 6 in (15 cm)
⚖ Up to 26½ lb (12 kg)
🐾 Solitary
🐾 Locally common

N. Canada to Mexico

CLIMBING BADGER

The smallest badger, the Chinese ferret badger forages at night for worms, insects, frogs, small rodents, and fruit. By day, it takes refuge in a burrow or rock crevice, or may use its long claws to climb a tree and then nest in the branches.

🐾 Up to 17 in (43 cm)
🐾 Up to 9 in (23 cm)
⚖ Up to 6½ lb (3 kg)
🐾 Not known
🐾 Locally common

E. India, S.E. Asia, S. China, Taiwan

Sunda stink badger
Mydaus javanensis

May squirt foul secretion from anal glands when threatened

Hog badger
Arctonyx collaris

Elongated snout with pig-like nostrils

Eurasian badger
Meles meles

Powerful forelimbs and claws for digging

Chinese ferret badger
Melogale moschata

Long, bushy tail

American badger
Taxidea taxus

Burmese ferret badger
Melogale personata

Hooded skunk
Mephitis macroura

Longer, softer fur
than striped skunk

Eastern hog-nosed skunk
Conepatus leuconotus

Striped skunk
Mephitis mephitis

Long claws on
forefeet for digging

Spotted skunk
Spilogale putorius

Striped hog-nosed skunk
Conepatus semistriatus

No white stripe down
center of face

Patagonian hog-nosed skunk
Conepatus humboldtii

Andean hog-nosed skunk
Conepatus chinga

Bare, projecting nose

FACT FILE

Striped skunk This nocturnal creature is an opportunistic feeder and tends to consume a wide range of foods, from small mammals, insects, and fish to fruit, nuts, grains, and grasses. During winter, it becomes inactive and rarely emerges from its den.

Up to 31½ in (80 cm)
Up to 15½ in (39 cm)
Up to 14 lb (6.5 kg)
Solitary
Common

N. Canada to Mexico

Spotted skunk The only skunk that can climb, this species also tends to be more alert and active than other skunks. During the mating season in March and April, males may succumb to "mating madness," spraying any large animal they come across.

Up to 13 in (33 cm)
Up to 11 in (28 cm)
Up to 2 lb (900 g)
Solitary
Common

USA (E. of Rocky Mts)

Andean hog-nosed skunk This species dens in rocky crevices, hollow logs, or burrows abandoned by other animals. It pounces on insects and will also hunt small vertebrates such as rodents, lizards, and snakes. It has some resistance to the venom of pit vipers.

Up to 13 in (33 cm)
Up to 8 in (20 cm)
Up to 6½ lb (3 kg)
Solitary
Common

South America

SKUNK DEFENSE

Before a skunk shoots out foul-smelling musk from its anal glands, it gives an enemy fair warning, raising its tail, stamping its feet, pretending to charge, or even doing a handstand. The skunk's distinctive black and white markings also serve to warn predators that it is a meal best avoided.

Danger display
When mortally threatened, a striped skunk will present both its head and rump to the enemy.

FACT FILE

American marten This agile mustelid speeds through the trees in pursuit of red squirrels and other prey. It will also hunt birds and insects and forage for fruit, nuts, and carrion. Although primarily a tree-dweller, it is also at ease on the ground or in water.

- Up to 17½ in (45 cm)
- Up to 9 in (23 cm)
- Up to 3 lb (1.3 kg)
- Solitary
- Uncommon

Alaska & Canada to N. California & Colorado

Fisher The largest marten, the fisher will tackle a porcupine. While other predators are put off by the porcupine's quills, the fisher stands at the perfect height to bite the unprotected face. Once repeated bites send the porcupine into shock, the fisher will flip it over and feed on its soft belly.

- Up to 31 in (79 cm)
- Up to 16 in (41 cm)
- Up to 12 lb (5.5 kg)
- Solitary
- Uncommon

Alaska & Canada to N. California

COVETED FUR

A creature of the dense taiga forest of northern Asia, the sable hunts and dens on the forest floor. The species was once found as far west as Scandinavia, but so many animals have been trapped for the fur trade that its range and numbers have been substantially reduced.

- Up to 22 in (56 cm)
- Up to 7½ in (19 cm)
- Up to 4 lb (1.8 kg)
- Solitary
- Uncommon

N. Asia

Valuable fur
The sable's long, silky winter coat has made its pelt one of the most prized by the fur trade.

American marten
Martes americana

Relatively large eyes and cat-like ears

Yellow-throated marten
Martes flavigula

Fisher
Martes pennanti

Long tail aids balance when climbing trees

Color varies from yellowish brown to dark brown

Japanese marten
Martes melampus

Sable
Martes zibellina

Beech marten
Martes foina

Pine marten
Martes martes

Soles of feet covered in fur in winter

Semiretractile claws used for climbing

Great variation in size, from 1–9 oz (35–250 g), depending on location

Coat becomes white in winter in northern part of range

Long-tailed weasel
Mustela frenata

Stoat (ermine) winter coat
Mustela erminea

Least weasel
Mustela nivalis

White winter coat helps stoat blend in with snow

Head paler than body

Malaysian weasel
Mustela nudipes

Stoat (ermine) summer coat
Mustela erminea

THE WOLVERINE

The largest terrestrial mustelid, the wolverine (*Gulo gulo*) lives in the northern coniferous forest and tundra of Eurasia and North America. It looks somewhat like a bear, but its behavior is definitely mustelid. Remarkably strong and ferocious for its size, it can take down reindeer and caribou, but will scavenge carrion if it is available. Any surplus food is stored in tunnels under the snow, and may be eaten up to 6 months later. The wolverine's massive head houses jaws and teeth that can crunch through large bones and frozen meat. Large feet and a plantigrade gait in which the entire sole is planted on the ground allow the wolverine to travel quickly over snow and outrun hoofed prey. This northern mustelid is also an agile climber and strong swimmer.

RELENTLESS CARNIVORES

Weasels will pursue prey underground or under snow and can carry up to half their own weight in meat as they run. While smaller weasels prefer mice and voles, and larger weasels favor rabbits, they all avail themselves of whatever animals they come across.

FACT FILE

Black-footed ferret While most mustelids are opportunistic feeders, the black-footed ferret preys almost solely on prairie dogs and uses their burrows for shelter.

- Up to 18 in (46 cm)
- Up to 5½ in (14 cm)
- Up to 2½ lb (1.1 kg)
- Solitary
- Extinct in the wild

S. Canada to N.W. Texas (until late 1980s); reintrod. Montana, Dakota & Wyoming
● Former range

BATTLE OF THE MINKS

Both the European and American minks are versatile feeders that hunt in or near water. The European mink has been in decline ever since the American mink escaped from European fur farms and became a direct competitor in the wild.

American mink
- Up to 19½ in (50 cm)
- Up to 8 in (20 cm)
- Up to 2 lb (900 g)
- Solitary
- Common

North America; introd. Europe, Siberia
● Introduced range

Mink variation
While most American minks are brown, about 10 percent have blue-gray fur.

European mink
- Up to 17 in (43 cm)
- Up to 7½ in (19 cm)
- Up to 26 oz (740 g)
- Solitary
- Endangered

France & Spain; Finland to Romania & Georgia
● Former range

⚡ CONSERVATION WATCH

Ferret decline In 1920, more than 500,000 black-footed ferrets lived on the North American plains but, as humans exterminated their prairie-dog prey, ferret numbers fell until they were thought extinct. A small population found in the 1980s led to a captive-breeding program, but the species remains North America's most endangered mammal.

Oily guard hairs make coat water-repellent

Siberian weasel
Mustela sibirica

American mink
Mustela vison

Partially webbed toes for swimming

Black face mask

✝ **Black-footed ferret**
Mustela nigripes

Polecat
Mustela putorius

Males can weigh twice as much as females

Steppe polecat
Mustela eversmannii

European mink
Mustela lutreola

Always has white patch on upper lip

Tough skin is so loose that ratel can swing around to attack a predator that has bitten down on its neck

Ratel (honey badger)
Mellivora capensis

Foul-smelling liquid secreted from anal glands

Patagonian weasel
Lyncodon patagonicus

Tayra
Eira barbara

Large hindfeet with long claws

Grison
Galictis vittata

Stands on hind legs to search for prey

Saharan striped weasel
Ictonyx libyca

Striped polecat (zorilla)
Ictonyx striatus

African weasel
Poecilogale albinucha

Marbled polecat
Vormela peregusna

FACT FILE

Ratel This mainly terrestrial mustelid will climb trees to reach honey. It has developed a symbiotic relationship with the honey guide bird, which will sing a distinctive song to lead it to a bee-hive. After opening the hive with its powerful claws, the ratel eats most of the honey, but leaves the wax and bee larvae for the bird. The ratel's omnivorous diet also includes insects and both large and small vertebrates.

Up to 30½ in (77 cm)
Up to 12 in (30 cm)
Up to 28½ lb (13 kg)
Solitary
Uncommon

W. Africa, sub-Saharan Africa, Arabia, Iraq, Turkmenistan, Pakistan, India

Shaggy attack
The ratel sometimes preys on poisonous snakes. Its long, thick coat and tough skin make it difficult for a snake to bite the flesh.

PLAYING DEAD

When threatened, a striped polecat will fluff up its long tail and growl or scream. If that does not work, it then squirts the attacker with foul-smelling fluid from its anal glands. As a final resort, the polecat will pretend to be dead. Despite these various defense strategies, striped polecats do fall victim occasionally to predators such as domestic dogs and wild cats. More commonly, however, they are killed by cars. Once one animal has been hit, its family members do not move from the scene and thus often meet the same fate.

A repulsive meal
Although feigning death makes the striped polecat easier to attack, it gives the predator a chance to taste the anal gland secretions on the fur. These are so unpleasant that the predator may well decide to abandon the meal.

SEALS AND SEA LIONS

CLASS	Mammalia
ORDER	Carnivora
FAMILIES	3
GENERA	21
SPECIES	36

With flexible, torpedo-shaped bodies, limbs modified to become flippers, and insulating layers of blubber and hair, seals, sea lions, and walruses are superbly adapted to a life in water. They have not, however, completely severed their link with land and must return to shore to breed. Collectively known as pinnipeds, these marine mammals were once placed in their own order, but are now considered to be part of Carnivora. Most feed on fish, squid, and crustaceans, but some also eat penguins and carrion and may attack the pups of other seal species. They can dive to great depths in search of prey, with the elephant seal able to stay submerged for up to 2 hours at a time.

Cold-water creatures Although monk seals are found in warmer waters, most seals, sea lions, and walruses are restricted to the colder, highly productive seas of the world's polar and temperate regions. The fossil record shows that the three families all originated in the North Pacific. They are now most abundant in the North Pacific, North Atlantic, and Southern oceans.

Communal living Most pinnipeds are gregarious animals and tend to live in large colonies. Walrus herds can number in the hundreds or even thousands and may be single sex or mixed, with both body and tusk size determining rank.

THREE GROUPS

There are three pinniped families. The Phocidae are known as the true seals. They swim mainly with strokes of their hind flippers, which cannot bend forward to act as feet, making their movement on land particularly ungainly. Although their hearing, especially under water, is good, true seals lack external ears.

Sea lions and fur seals belong to the family Otariidae. These "eared seals" have small external ears. They rely mostly on their front flippers for swimming, and can bend their hind flippers forward when on land, allowing them to walk "four-footed" and sit in a semi-upright position.

The third family, Odobenidae, contains a single species, the walrus, instantly recognizable by the long canine teeth that form tusks on both sexes. Like true seals, walruses use their hind flippers for swimming and lack external ears. Like eared seals, however, walruses can bend their hind flippers forward.

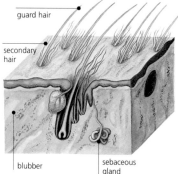

Insulating layers Pinnipeds have a thick layer of blubber that provides insulation, buoyancy, and a fat store. For further protection, all but the walrus have hairy bodies, and fur seals have dense secondary hairs that form a waterproof barrier.

BRINGING UP BABY

All pinnipeds return to land or ice to give birth and mate. Mating takes place just days after the usually single pup is born, but the fertilized egg does not become implanted in the uterus for months. This delayed implantation allows birthing, nursing, and mating to occur in one season so that the animals live on land, where they are most vulnerable, only once a year. Pups are dependent for varying lengths of time. Harp seals (right), for example, are nursed for only 12 days or so, while walruses stay with their mother for 2 years.

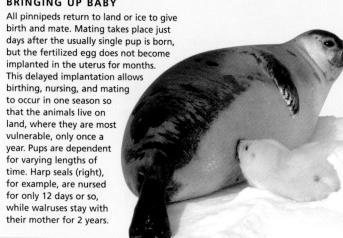

⚡ CONSERVATION WATCH

The commercial sealing operations that began in the 16th century had a devastating effect on pinniped populations. Of the 36 species of pinnipeds, 36% are listed on the IUCN Red List, as follows:

- 2 Extinct
- 1 Critically endangered
- 2 Endangered
- 7 Vulnerable
- 1 Near threatened

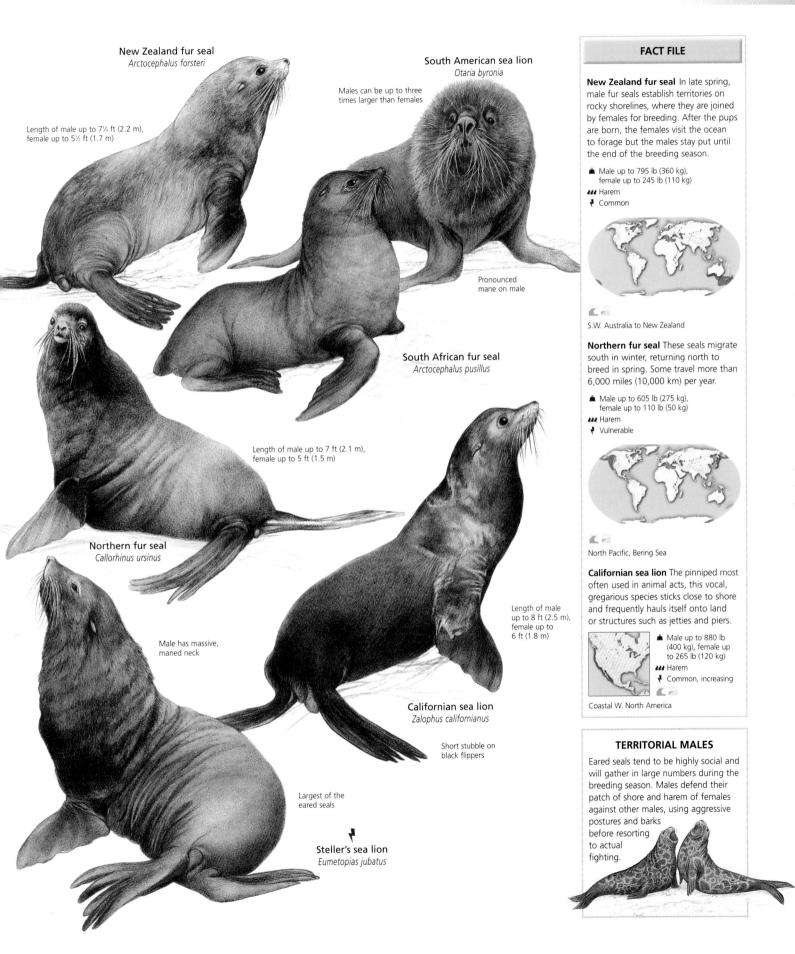

New Zealand fur seal
Arctocephalus forsteri

Length of male up to 7¼ ft (2.2 m),
female up to 5½ ft (1.7 m)

South American sea lion
Otaria byronia

Males can be up to three
times larger than females

Pronounced
mane on male

South African fur seal
Arctocephalus pusillus

Length of male up to 7 ft (2.1 m),
female up to 5 ft (1.5 m)

Northern fur seal
Callorhinus ursinus

Male has massive,
maned neck

Length of male
up to 8 ft (2.5 m),
female up to
6 ft (1.8 m)

Californian sea lion
Zalophus californianus

Short stubble on
black flippers

Largest of the
eared seals

Steller's sea lion
Eumetopias jubatus

FACT FILE

New Zealand fur seal In late spring,
male fur seals establish territories on
rocky shorelines, where they are joined
by females for breeding. After the pups
are born, the females visit the ocean
to forage but the males stay put until
the end of the breeding season.

⬛ Male up to 795 lb (360 kg),
female up to 245 lb (110 kg)
🐾 Harem
⚡ Common

S.W. Australia to New Zealand

Northern fur seal These seals migrate
south in winter, returning north to
breed in spring. Some travel more than
6,000 miles (10,000 km) per year.

⬛ Male up to 605 lb (275 kg),
female up to 110 lb (50 kg)
🐾 Harem
⚡ Vulnerable

North Pacific, Bering Sea

Californian sea lion The pinniped most
often used in animal acts, this vocal,
gregarious species sticks close to shore
and frequently hauls itself onto land
or structures such as jetties and piers.

⬛ Male up to 880 lb
(400 kg), female up
to 265 lb (120 kg)
🐾 Harem
⚡ Common, increasing

Coastal W. North America

TERRITORIAL MALES

Eared seals tend to be highly social and
will gather in large numbers during the
breeding season. Males defend their
patch of shore and harem of females
against other males, using aggressive
postures and barks
before resorting
to actual
fighting.

FACT FILE

Common seal The most widely distributed pinniped, the common seal tends to be solitary but will gather in groups on land. A Canadian subspecies, the Ungava seal, lives in fresh water.

🏔 Male up to 330 lb (150 kg), female up to 245 lb (110 kg)
🐾 Variable
🦷 Common

North Atlantic, North Pacific

Walrus Walruses rely on a mustache of sensitive whiskers to locate their main food of clams and mussels, which they dig out of the sand with their snout. The largest animal with the longest tusks usually dominates the herd.

🦷 Up to 11½ ft (3.5 m)
🏔 Up to 3,640 lb (1,650 kg)
🐾 Large herd
🦷 Locally common

Shallow Arctic seas

SNOW LAIRS

The ringed seal lives in waters that are covered in ice for at least part of the year. A pregnant female digs a cave into the snow above her breathing hole, which provides her pup with protection from both the extreme Arctic weather and hungry predators such as polar bears.

⚡ CONSERVATION WATCH

Harp seal slaughter In the late 1980s, a public outcry halted the clubbing of Canada's whitecoats, the very young pups of harp seals. Older pups have since been hunted throughout the Atlantic Ocean, however, often at levels that could precipitate a serious decline in overall numbers.

Bearded seal
Erignathus barbatus

Long, sensitive whiskers used to locate clams, snails, crabs, and shrimp

Baikal seal
Phoca sibirica

Common seal (harbor seal)
Phoca vitulina

Length of male up to 6¼ ft (1.9 m), female up to 5½ ft (1.7 m)

Ribbon seal
Phoca fasciata

Harp seal
Phoca groenlandica

White coats of pups shed by 3 weeks of age

Common name inspired by harp-shaped marking on back and sides of adult

Tusks on both males and females

Walrus
Odobenus rosmarus

Ringed seal
Phoca hispida

Spots surrounded by ring of lighter fur

Hooded seal
Cystophora cristata

Lining of left nostril inflated as part of mating display

Crabeater seal
Lobodon carcinophagus

Dark gray or brown back after January molt, becomes almost entirely blonde later in year

Gray seal
Halichoerus grypus

Swims with strokes of large flippers, while most other pinnipeds use tail

Weddell seal
Leptonychotes weddellii

Leopard seal
Hydrurga leptonyx

Can be colored brown, gray, or black

Mediterranean monk seal
Monachus monachus

Southern elephant seal
Mirounga leonina

Largest of all pinnipeds: length of male up to 20 ft (6 m), female up to 10 ft (3 m)

FACT FILE

Mediterranean monk seal This species was once common throughout Mediterranean coastal waters, but an ever-growing human presence has seen its numbers dwindle. Today, its main refuges are small, barren islands.

🐂	Up to 9 ft (2.8 m)
⚖	Up to 660 lb (300 kg)
♨	Harem
⚡	Critically endangered

Coastal W. Africa, Aegean Sea

Southern elephant seal Males of this species can be more than six times heavier than the females. Only the largest 10 percent of males ever have the chance to mate. To attract females, a male inflates his trunk-like nose.

⚖	Male up to 8,160 lb (3,700 kg), female up to 1,320 lb (600 kg)
♨	Harem
⚡	Locally common

Argentina, New Zealand, sub-Antarctic islands

KRILL EATERS

Rather than eating crabs as their name suggests, crabeater seals feed almost exclusively on the abundant supply of krill in Antarctic waters. They sieve the tiny crustaceans through their complex, knobbly cheek teeth.

INFLATED NOSE

When it is trying to attract a mate, or is threatened or excited, mature male hooded seals have the unusual ability to blow up their black hood, which is an extension of the nasal cavity. They can also make the lining of the left nostril inflate into a red bladder.

Blowing up *As part of its mating display, the male hooded seal may inflate either its red nostril bladder or its entire black hood.*

THE RACCOON FAMILY

CLASS	Mammalia
ORDER	Carnivora
FAMILY	Procyonidae
GENERA	6
SPECIES	19

Restricted to the New World, the family Procyonidae includes raccoons, coatis, kinkajous, ringtails, and olingos. All are medium-sized with a long body and tail, broad faces, and erect ears. Apart from the kinkajou, they also share mask-like markings on the face and alternating light and dark rings on the tail. Omnivorous feeding habits have enabled members of this family to thrive in habitats as varied as coniferous forest, tropical rain forest, wetlands, desert, farmland, and urban areas. Procyonids tend to be highly vocal, barking and squeaking to maintain complex social structures. Raccoons often sleep in communal dens, males may travel in groups, and females form "consortships" with one to four males. Male coatis are solitary, but bands of about 15 females groom one another, share the care of young, and fight off predators together. Kinkajous rest in groups of a female and her young plus two adult males.

Opportunistic omnivores Although their preferred habitat is woodland near water, raccoons have adapted to live alongside humans in rural and urban environments. They often visit North American backyards to raid trash cans for food scraps, and may den in old buildings, cellars, or attics.

FACT FILE

Kinkajou Originally classified as a lemur, the nocturnal kinkajou has a prehensile tail; large, forward-facing eyes; a primarily fruit-based diet; and a treetop lifestyle. It is nonetheless a carnivore, with genetic studies showing that it belongs in the Procyonidae family with raccoons and coatis.

- Up to 21½ in (55 cm)
- Up to 22½ in (57 cm)
- Up to 7 lb (3.2 kg)
- Solitary, pair
- Locally common

S. Mexico to Bolivia & Brazil

White-nosed coati Snuffling through the leaf litter on the forest floor by day, coatis use their flexible snouts and keen sense of smell to locate insect prey in crevices. They also consume large quantities of fruit and small vertebrates such as lizards and rodents. At night, they retreat to the treetops.

- Up to 27 in (69 cm)
- Up to 24½ in (62 cm)
- Up to 10 lb (4.5 kg)
- Family band
- Locally common

Arizona to Colombia & Ecuador

Raccoon These nocturnal masked creatures eat whatever is available, from small vertebrates, insects, and worms to fruit, nuts, and seeds. They prefer aquatic prey, such as fish, crustaceans, and snails, which they appear to wash with their dextrous hands.

- Up to 21½ in (55 cm)
- Up to 16 in (40 cm)
- Up to 35½ lb (16 kg)
- Solitary
- Common

North & Central America

Only New World carnivore with prehensile tail

Kinkajou
Potos flavus

Ringtail
Bassariscus astutus

14–16 alternating black and white rings on tail

Mobile tail used for balance

Raccoon
Procyon lotor

Forepaws are sensitive and dextrous

White-nosed coati
Nasua narica

⚑ CONSERVATION WATCH

Of the 19 species of procyonids, 63% are listed on the IUCN Red List, as follows:

1	Extinct
7	Endangered
3	Near threatened
1	Data deficient

HYENAS AND AARDWOLF

The four species in the family Hyaenidae—the aardwolf and the brown, striped, and spotted hyenas—look rather like dogs but are classified as cat-like carnivores because they are more closely related to the cats and civets. Between their long front legs and shorter back legs, the spine distinctively slopes downward to the tail. A relatively massive head with a broad muzzle houses powerful, bone-crunching jaws and teeth. Unlike most other mammals, hyenas can digest skin and bone. They often scavenge the kills of lions and other predators, but sometimes capture their own prey. Spotted hyenas are the most effective hunters and will cooperate to bring down large prey such as zebra and wildebeest. The aardwolf is primarily an insect-eater, using its long, sticky tongue and peg-like teeth to harvest up to 200,000 termites per night.

CLASS Mammalia
ORDER Carnivora
FAMILY Hyaenidae
GENERA 3
SPECIES 4

African distribution The range of the striped hyena extends into the Middle East and South Asia, but the other Hyaenidae members are restricted to Africa. Hyenas and aardwolf are most often found in grassland or savanna habitats, where they take refuge in caves, thick vegetation, or abandoned burrows. Although no hyena species is endangered, they are often loathed and widely persecuted, and the spotted hyena depends on conservation measures for its survival.

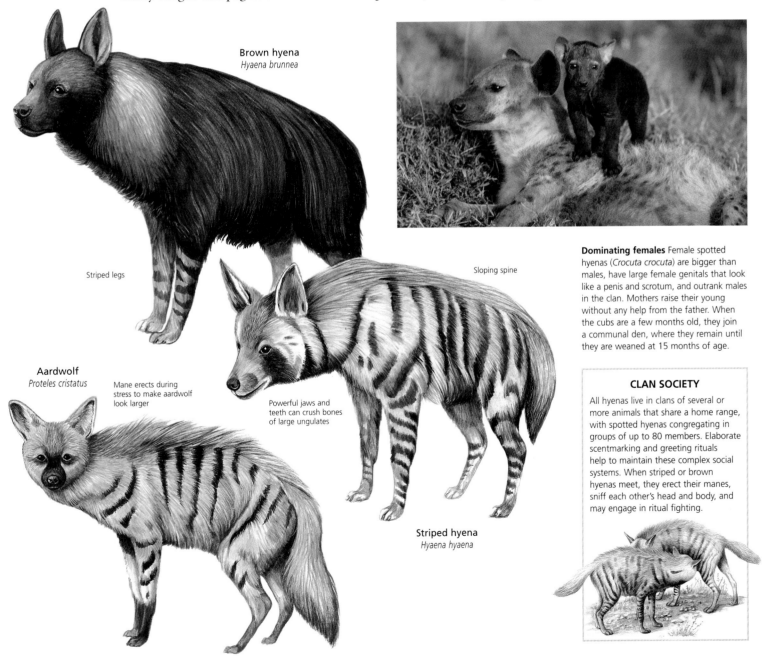

Brown hyena
Hyaena brunnea

Striped legs

Sloping spine

Dominating females Female spotted hyenas (*Crocuta crocuta*) are bigger than males, have large female genitals that look like a penis and scrotum, and outrank males in the clan. Mothers raise their young without any help from the father. When the cubs are a few months old, they join a communal den, where they remain until they are weaned at 15 months of age.

Aardwolf
Proteles cristatus

Mane erects during stress to make aardwolf look larger

Powerful jaws and teeth can crush bones of large ungulates

Striped hyena
Hyaena hyaena

CLAN SOCIETY

All hyenas live in clans of several or more animals that share a home range, with spotted hyenas congregating in groups of up to 80 members. Elaborate scentmarking and greeting rituals help to maintain these complex social systems. When striped or brown hyenas meet, they erect their manes, sniff each other's head and body, and may engage in ritual fighting.

CIVETS AND MONGOOSES

The carnivore family Viverridae includes civets, genets, and linsangs. It also once included mongooses, but these have now been placed in their own family Herpestidae. Related to cats and hyenas, viverrids and herpestids tend to be medium-sized with a long neck and head, a long, slender body, and short legs. Their skeletal structure and teeth closely resemble those of the earliest carnivores, but the inner-ear region is highly developed. Viverrids are generally nocturnal, arboreal forest-dwellers with long tails, retractile claws, and erect, pointed ears. Many have scent glands near the genitals that in some species produce civet oil, once widely used as a perfume base. Mongooses tend to be found in more open country. Usually terrestrial and often diurnal, they have shorter tails, non-retractile claws, and small, rounded ears.

CLASS	Mammalia
ORDER	Carnivora
FAMILIES	2
GENERA	38
SPECIES	75

Old World families The civets, genets, and linsangs of Viverridae and the mongooses of Herpestidae are native to much of the Old World, with a number of species found only on the island of Madagascar. Famed as ratters, some mongooses have also been introduced on many islands in the New World, often with disastrous consequences. In the West Indies and Hawaii, for example, the introduced small Indian mongoose (*Herpestes javanicus*) is now a pest that attacks poultry and native fauna.

CONSERVATION WATCH

While a number of viverrids and mongooses are so successful that they are considered pests in areas, others are suffering as their habitat is destroyed. Deforestation is a critical issue on Madagascar, which has four endangered species from these families. Of the 75 species in Viverridae and Herpestidae, 32% are listed on the IUCN Red List, as follows:

- 1 Critically endangered
- 8 Endangered
- 9 Vulnerable
- 6 Data deficient

Social life While all viverrids and many herpestids usually live alone or in pairs, some mongoose species are gregarious and live in colonies. Suricates live in troops of up to 30 animals, which cooperate to care for the young and watch for danger. Nonbreeding members will babysit, and sentinel duties are rotated.

Blotched genet (large-spotted genet)
Genetta tigrina

Coat is most striking, with stronger colors and more pronounced pattern, in wetter parts of range

Angolan genet
Genetta angolensis

Fishing genet
Osbornictis piscivora

Naked palms to locate fish hiding in crevices

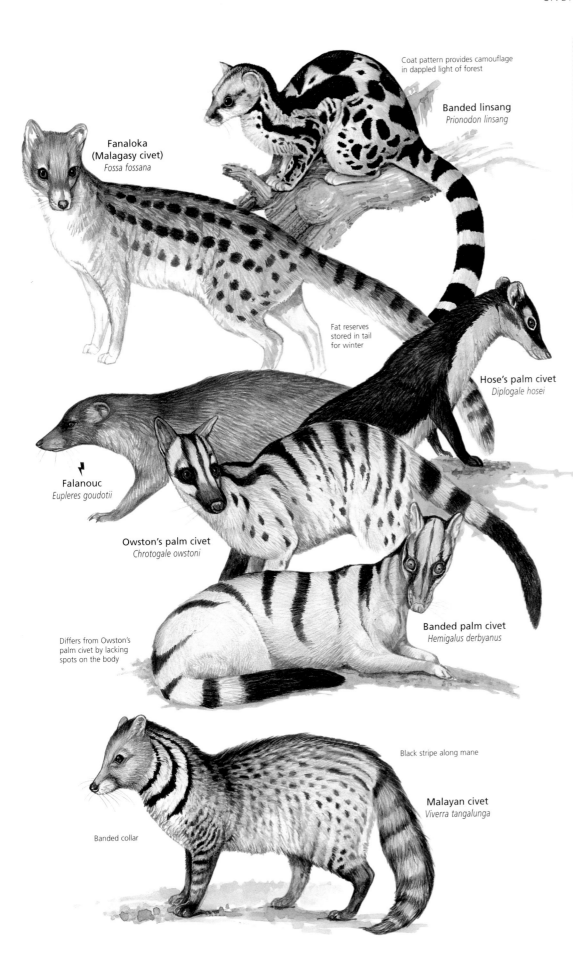

Coat pattern provides camouflage in dappled light of forest

Banded linsang
Prionodon linsang

Fanaloka
(Malagasy civet)
Fossa fossana

Fat reserves stored in tail for winter

Hose's palm civet
Diplogale hosei

Falanouc
Eupleres goudotii

Owston's palm civet
Chrotogale owstoni

Differs from Owston's palm civet by lacking spots on the body

Banded palm civet
Hemigalus derbyanus

Black stripe along mane

Malayan civet
Viverra tangalunga

Banded collar

THE FOSSA

The dominant predator on the island of Madagascar is the fossa (*Crytoprocta ferox*). This agile civet pursues lemurs through the trees, but will also hunt snakes, tenrecs, and guinea fowl on the ground. During the mating season, a female fossa waits high in a tree for males to congregate beneath her, then mates with a number of them for almost 3 hours.

Balancing act
The fossa's tail is about as long as its body and helps it to balance as it chases lemurs through the trees.

MASKED MAMMAL

The masked palm civet of China and Southeast Asia plays a key role in its ecosystem. Through an omnivorous diet, it keeps insect and small vertebrate populations under control and helps disperse fruit seeds. In turn, it is a prey species of tigers, hawks, and leopards. To discourage predation, this civet relies on the powerful odor produced by its anal glands. Its distinctive facial markings may serve to warn predators of this noxious scent.

Only viverrid with prehensile tail

Binturong
Arctictis binturong

Jerdon's palm civet
Paradoxurus jerdoni

Common palm civet (toddy cat)
Paradoxurus hermaphroditus

Brown palm civet
Macrogalidia musschenbroekii

African palm civet
Nandinia binotata

Masked palm civet
Paguma larvata

Three-striped palm civet
Arctogalidia trivirgata

Squirrels form part of civet's omnivorous diet, which also includes frogs, birds, insects, and fruit

Narrow-striped mongoose
Mungotictis decemlineata

Ring-tailed mongoose
Galidia elegans

If threatened, bristles fur and arches back to increase apparent size

Large gray mongoose
Herpestes ichneumon

White-tailed mongoose
Ichneumia albicauda

Indian gray mongoose
Herpestes edwardsii

Yellow mongoose
Cynictis penicillata

Yellow coat in south of range, grayish in north

Stands upright to survey terrain for danger

Suricate (meerkat)
Suricata suricatta

Banded mongoose
Mungos mungo

SNAKE ATTACK

Some mongoose species, including the Indian gray mongoose, are famed for their snake-killing ability. Although widely believed to be immune to the bites of cobras and other poisonous snakes, mongooses can tolerate only small amounts of the venom and must avoid being bitten. They rely on their speed and agility to conquer such dangerous prey, which they kill by sinking their sharp teeth into the neck and breaking the spine.

Standoff *A mongoose will attempt to exhaust its cobra prey by making a series of lunges, while skillfully dodging the venomous snake's strikes.*

CATS

CLASS Mammalia
ORDER Carnivora
FAMILY Felidae
GENERA 18
SPECIES 36

The ultimate hunters, cats eat very little other than meat, making them the most carnivorous of the carnivores. Their predatory expertise has placed them at the top of many food chains on all continents except Australia and Antarctica and in habitats ranging from deserts to Arctic regions. Within the family Felidae, there are considerable size differences but few variations in general form. All species have a strong, muscular body; a blunt face with large, forward-facing eyes; sharp teeth and claws; and acute senses and quick reflexes. They generally rely on stealth to capture prey, which they stalk or ambush. Although largely terrestrial, cats are agile climbers and often good swimmers.

Worldwide spread Wild felids are the dominant predators on most continents, and are absent only from Australasia, Madagascar, Greenland, and Antarctica. First domesticated in ancient Egypt thousands of years ago, the domestic cat has since spread everywhere except Antarctica, and feral populations have had a serious impact on native ecosystems.

Surprise attack Small cats such as bobcats, wildcats, and lynx tend to hunt smaller mammals such as rodents, lizards, and birds. They creep up, then suddenly pounce with a jack-in-the-box movement, killing the prey with a bite to the neck.

Roars and purrs A lion's roar usually advertises its control of territory. Only the big cats have a flexible-enough larynx to produce roars. All cats can purr, but big cats can purr only as they exhale, while small cats can produce a constant purr.

*Open wide
Because they rarely eat vegetation, cats lack grinding teeth. Instead, their powerful jaws house blade-like teeth, with sharp stabbing canines and shearing carnassials. A rough tongue covered with tiny projections is used to rasp meat off bones and to groom the fur.*

PRIME PREDATORS

Cats are split into three subfamilies: the big cats of Pantherinae include tigers, lions, leopards, and jaguars; the small cats of Felinae include pumas (which can be bigger than some "big cats"), lynx, bobcats, and ocelots; and cheetahs are on their own in Acinonychinae. The chief difference between the big and small cats lies in the flexibility of the larynx, which allows big cats to roar. Cheetahs are distinguished by their non-retractile claws and by their exceptional speed, which allows them to outrun fast prey such as gazelles over short distances.

The sense of smell is used for communication among cats, which scentmark their territory. When hunting, however, felids depend more on vision and sound. Forward-facing eyes provide binocular vision that helps with judging distances. A light-reflecting disk in the eye and rapidly adjusting irises enhance their night vision, which is up to six times sharper than that of humans. Large, mobile ears funnel sound to the sensitive inner ear, which can pick up the faint, high-pitched sounds of small prey such as mice.

The first felids began to evolve about 40 million years ago and led to today's species and to a branch that included the saber-toothed tiger, a massive creature with huge canines that disappeared only 10,000 years ago at the end of the last ice age. Cats were domesticated in the Middle East 7,000 years or so ago, a development that saw them spread to almost every part of the globe as companion animals.

⚡ CONSERVATION WATCH

Because they need a wide range to find an adequate supply of prey, felids are particularly affected by habitat loss. Hunting has also taken a heavy toll on most cat species. Of the 36 species of felids, 69% are listed on the IUCN Red List, as follows:

  1 Critically endangered
  4 Endangered
 12 Vulnerable
  8 Near threatened

Lion
Panthera leo

Male is 30–50 percent heavier than female

Some males have dark mane, but most have golden mane

Lionesses do most of the hunting, but the male will usually eat first

"King" cheetah

In some cheetahs, a recessive gene produces a blotchy coat pattern with stripes down the spine

Cheetah
Acinonyx jubatus

Cheetah cubs stay with mother until 13–20 months old

FACT FILE

Lion While most felids tend to be solitary, lions are famed for their close, enduring group relationships. Prides of 4–20 lionesses occupy a home range, cooperate to bring down large prey, and share care of the cubs. Males live alone or in coalitions with other males.

🐾 Up to 7½ ft (2.3 m)
🐾 Up to 39½ in (1 m)
🐾 Up to 495 lb (225 kg)
👪 Family band
🏃 Vulnerable
☀ ⬇ ⬆

Sub-Saharan Africa, India

Cheetah With a burst of speed, the cheetah attempts to capture hoofed mammals such as Thomson's gazelle and wildebeest calves. Only about half such pursuits end in success. In turn, cheetahs are sometimes killed by lions.

🐾 Up to 4½ ft (1.4 m)
🐾 Up to 31½ in (80 cm)
🐾 Up to 159 lb (72 kg)
🐾 Solitary
🏃 Vulnerable
☀ ⬇ ⬆

Sub-Saharan Africa

LION INFANTICIDE

The cubs in a pride of lions are usually sired by a coalition of two or three adult males. This resident coalition's tenure is usually short-lived, lasting only a few years before invading males move in. The intruders kill all the smaller cubs, which frees up the females to breed.

Protective mother
Lionesses carry their cubs in the mouth and will try to protect them from invading males.

FAST AS A FLASH

Reaching speeds of 60 miles per hour (95 km/h) in pursuit of prey, the cheetah is the fastest mammal on land. Such sprints can last for only 20–60 seconds before the cheetah overheats and must rest. Sometimes it is too out of breath to defend its kill from scavengers.

FACT FILE

Tiger The largest cats, these solitary hunters mainly rely on prey species much larger than themselves. They may travel 12 miles (20 km) in a day searching for food and need to kill a hoofed mammal every 3–5 days to sustain themselves. Although tiger habitats are diverse, they all feature some form of dense vegetation that creates dappled light patterns. A tiger's striking stripes actually help it to blend into this background, allowing it to creep up on prey without being seen. Such camouflage is essential because a tiger cannot outrun its large prey and relies instead on the element of surprise. Even so, only about 5 percent of attacks succeed.

- Up to 12 ft (3.6 m)
- Up to 39 ½ in (1 m)
- Up to 795 lb (360 kg)
- Solitary
- Endangered

India to E. Siberia
● Former range

Bengal tiger
Panthera tigris tigris

Stripe pattern unique to individual

Largest of all felids

Coat becomes lighter in winter

Siberian tiger (Amur tiger)
Panthera tigris altaica

VANISHING TIGERS

At the beginning of the 20th century, there were about 100,000 tigers roaming the tropical jungles, savanna, grasslands, mangrove swamps, deciduous woodlands, and snow-covered coniferous forests of Asia, from eastern Turkey to far eastern Russia. Today, perhaps fewer than 2,500 breeding adults remain in the wild. The Bali, Caspian, and Javan tigers—three of the eight tiger subspecies—are now extinct. Of the remaining subspecies, South China tigers exist only as a remnant population of 20–30 animals, and merely 500 or so Siberian or Amur tigers and 500 Sumatran tigers survive. The largest populations are of the Bengal tiger (shown at left) and the Indochinese tiger, but even these are in serious danger of extinction.

For years, tigers have been shot or poisoned because they are considered pests; because their skins as well as their body parts (used in traditional Asian medicine) fetch good prices; and because they are prized as trophies by sport hunters. At the same time, a continuing increase in local human populations has led to the degradation and fragmentation of much tiger habitat, as well as the decimation of hoofed mammals, the tiger's main prey, by subsistence hunters.

Although the tiger is now protected in most countries, illegal poaching persists. Conservation measures include preserving tiger habitat and reintroducing tigers in some areas.

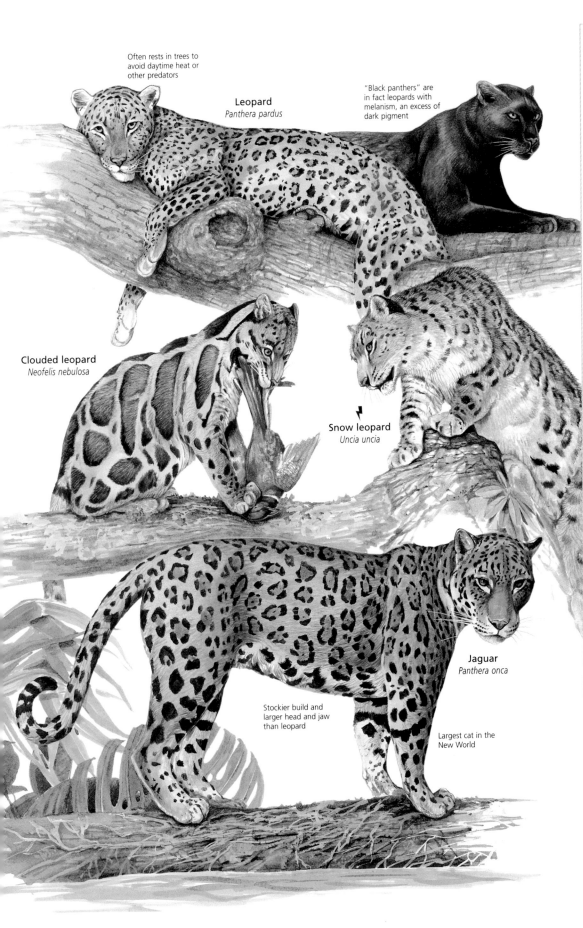

Often rests in trees to avoid daytime heat or other predators

Leopard
Panthera pardus

"Black panthers" are in fact leopards with melanism, an excess of dark pigment

Clouded leopard
Neofelis nebulosa

Snow leopard
Uncia uncia

Jaguar
Panthera onca

Stockier build and larger head and jaw than leopard

Largest cat in the New World

FACT FILE

Leopard The most widespread big cat, the leopard owes its success to a varied diet, which includes gazelles, jackals, baboons, storks, rodents, reptiles, and fish. An agile climber, it often drags its kill to the safety of high branches.

- Up to 7 ft (2.1 m)
- Up to 3½ ft (1.1 m)
- Up to 200 lb (90 kg)
- Solitary
- Locally common

Sub-Saharan & N. Africa; S., S.E. & C. Asia

Clouded leopard This largely arboreal cat will wait in a tree to ambush prey such as deer and pigs below. It will also seize primates and birds in the branches. The species is the smallest of the big cats and has only a soft roar.

- Up to 3½ ft (1.1 m)
- Up to 35½ in (90 cm)
- Up to 51 lb (23 kg)
- Solitary
- Vulnerable

Nepal to China, S.E. Asia

Jaguar This New World cat looks like the leopard, but fills a similar ecological niche to the tiger. It tends to live in dense vegetation near water and stalk large prey such as deer and peccaries. It also takes fish and other aquatic prey.

- Up to 6¼ ft (1.9 m)
- Up to 23½ in (60 cm)
- Up to 355 lb (160 kg)
- Solitary
- Near threatened

Mexico to Argentina

SNOWED IN

Adapted to high altitudes, the snow leopard lives in the remote mountains of Central Asia. It has a dense coat with very large, furry paws that act as snowshoes. Up to five cubs are born in a rocky den lined with the mother's fur.

FACT FILE

Bobcat Rare in some parts of its range but more common in others, the bobcat usually hunts small prey such as rabbits and rodents by night but will also eat carrion. It rests by day, often in a cave.

🐾 Up to 41 in (105 cm)
📏 Up to 8 in (20 cm)
⚖️ Up to 68½ lb (31 kg)
♟️ Solitary
❗ Locally common

Temperate North America to Mexico

Eurasian lynx Mostly found in remote forested areas frequented by its main prey of small deer, this species remains active throughout the winter months.

🐾 Up to 4½ ft (1.3 m)
📏 Up to 9½ in (24 cm)
⚖️ Up to 84 lb (38 kg)
♟️ Solitary
❗ Near threatened

France, Balkans, Iraq, Scandinavia to China

Puma This cat's once vast range is now largely restricted to remote mountains, where it hunts white-tailed deer, moose, and caribou. The puma hisses, growls, whistles, and purrs, but cannot roar.

🐾 Up to 5 ft (1.5 m)
📏 Up to 38 in (96 cm)
⚖️ Up to 265 lb (120 kg)
♟️ Solitary
❗ Near threatened

Canada to S. Argentina & Chile

Caracal The swiftest of the small cats, the caracal can leap 10 feet (3 m) to snatch birds from the air. It also pounces on rodents and antelopes.

🐾 Up to 36 in (92 cm)
📏 Up to 12¼ in (31 cm)
⚖️ Up to 42 lb (19 kg)
♟️ Solitary
❗ Uncommon

Africa, Middle East, India & N.W. Pakistan

JACKKNIFE CLAWS

Apart from the cheetah, all cats have retractile claws, which remain sharp because they are unsheathed only for capturing prey or climbing trees.

Bobcat
Lynx rufus

Canadian lynx
Lynx canadensis

Eurasian lynx
Lynx lynx

Coat can be mainly striped, mainly spotted, or plain

In winter, coat thickens and becomes paler, and furry feet help lynx to travel over soft snow

Largest of the small cats

Puma
(cougar, mountain lion)
Puma concolor

Tufts of black fur on long, slender ears

Caracal
Caracal caracal

Iberian lynx
Lynx pardinus

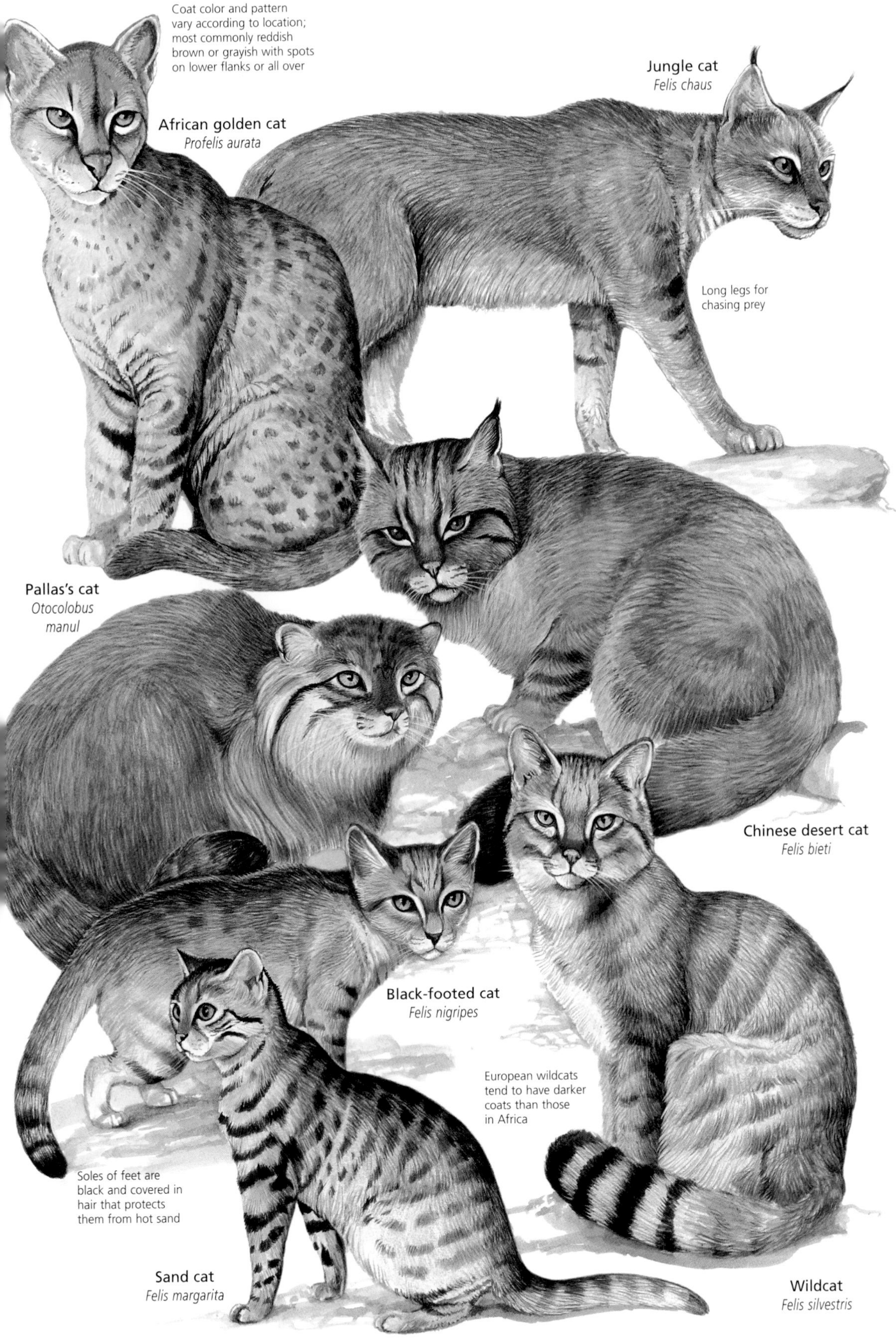

Coat color and pattern vary according to location; most commonly reddish brown or grayish with spots on lower flanks or all over
African golden cat
Profelis aurata
Jungle cat
Felis chaus
Long legs for chasing prey
Pallas's cat
Otocolobus manul
Chinese desert cat
Felis bieti
Black-footed cat
Felis nigripes
European wildcats tend to have darker coats than those in Africa
Soles of feet are black and covered in hair that protects them from hot sand
Sand cat
Felis margarita
Wildcat
Felis silvestris

FACT FILE

Sand cat This desert creature can survive in extremely arid conditions, gaining enough water from its prey of rodents, hares, birds, and reptiles so that it does not need to drink.

- Up to 21 in (54 cm)
- Up to 12¼ in (31 cm)
- Up to 7½ lb (3.5 kg)
- Solitary
- Near threatened

Sahara Desert (N. Africa)

Wildcat This species looks like a large domestic tabby cat, but with a broader head and darker fur. A solitary creature, it hunts mainly at night for rodents, birds, small reptiles, and insects.

- Up to 29½ in (75 cm)
- Up to 14 in (35 cm)
- Up to 17½ lb (8 kg)
- Solitary
- Locally common

Africa, Europe to W. China & N.W. India

FROM THE WILD

Cats were first domesticated in ancient Egypt when stores of grain attracted rats and mice to human settlements. The rats and mice drew wildcats, which were tolerated because they kept the rodents under control. The Romans spread domesticated cats throughout Europe. Today, there are more than 100 million domestic or feral cats in the United States alone.

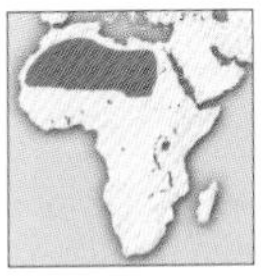

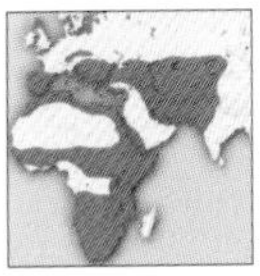

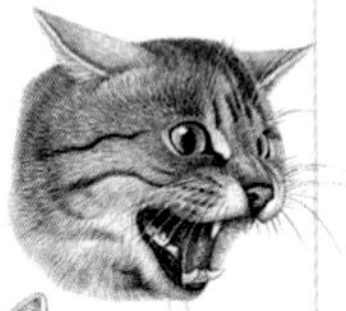

Wild feline *Although wildcats tend to be much more fierce than domestic cats, domestic cats retain the hunting instinct and readily revert to a feral state.*

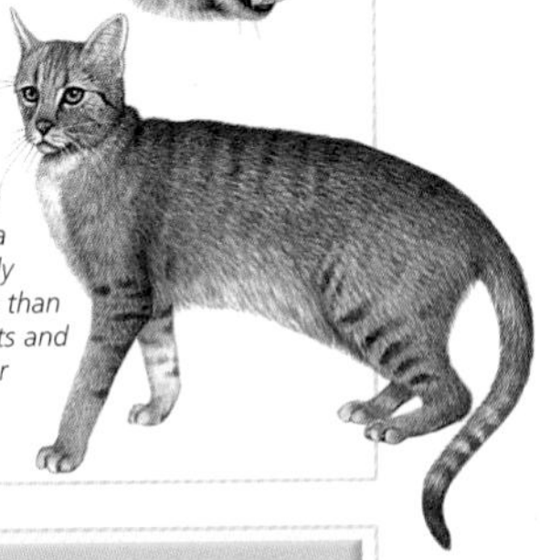

African wildcat *Wildcats in Africa live in less densely wooded habitats than European wildcats and tend to be lighter in coat color.*

CONSERVATION WATCH

Small cat status Most small cats suffered drastic losses during the 20th century. For some, such as the pumas of Florida, fragmentation of habitat has isolated populations and led to inbreeding. The fur trade also had a great impact, especially on the ocelot and Geoffroy's cat, but was gradually curtailed as consumers became aware of the problems.

FACT FILE

Marbled cat Resembling a smaller version of the clouded leopard, this highly arboreal species feeds mainly on birds. Hidden among the branches of dense tropical forests, it has rarely been seen and little is known of its behavior.

- Up to 21 in (53 cm)
- Up to 21½ in (55 cm)
- Up to 11 lb (5 kg)
- Solitary
- Vulnerable

Nepal, N.E. India, S.E. Asia

Asiatic golden cat This felid mostly preys on small mammals and birds, but will hunt in pairs to bring down larger quarry such as buffalo calves. Females give birth to one or two young in a den made in a hollow tree or among rocks. Unusually among cats, the male helps to raise the kittens.

- Up to 41 in (105 cm)
- Up to 22 in (56 cm)
- Up to 33 lb (15 kg)
- Solitary
- Vulnerable

Nepal to China, Indochina, Malaya, Sumatra

Leopard cat The swimming ability of the leopard cat may help to explain its presence on many islands in Asia. A subspecies, the Iriomote cat, is found only on the small Japanese islands of Iriomote and Ryukyu.

- Up to 42 in (107 cm)
- Up to 17½ in (44 cm)
- Up to 15½ lb (7 kg)
- Solitary
- Locally common

Pakistan & India to China, Korea & S.E. Asia

Bay cat This species is so rare that it was not photographed until 1998. It lives in jungles and among limestone outcrops near forests on the island of Borneo. The coat is most commonly chestnut red but can also be gray.

- Up to 26½ in (67 cm)
- Up to 15½ in (39 cm)
- Up to 9 lb (4 kg)
- Solitary
- Endangered

Borneo

Fishing cat This cat will tap the water surface to attract fish. It will also climb trees, then dive headfirst into water, grabbing aquatic prey in its mouth.

- Up to 34 in (86 cm)
- Up to 13 in (33 cm)
- Up to 31 lb (14 kg)
- Solitary
- Vulnerable

India, Nepal, Sri Lanka, S.E. Asia

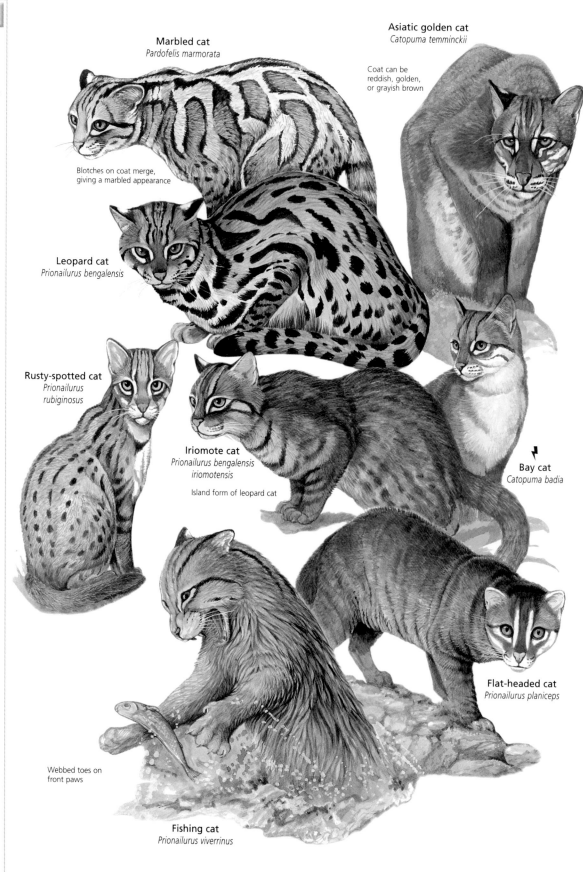

Marbled cat
Pardofelis marmorata

Blotches on coat merge, giving a marbled appearance

Asiatic golden cat
Catopuma temminckii

Coat can be reddish, golden, or grayish brown

Leopard cat
Prionailurus bengalensis

Rusty-spotted cat
Prionailurus rubiginosus

Iriomote cat
Prionailurus bengalensis iriomotensis

Island form of leopard cat

Bay cat
Catopuma badia

Flat-headed cat
Prionailurus planiceps

Webbed toes on front paws

Fishing cat
Prionailurus viverrinus

Elongated rather than rounded head

Coat can be chestnut or brownish gray

Jaguarundi
Herpailurus yaguarondi

Oncilla
Leopardus tigrinus

Ocelot
Leopardus pardalis

Margay
Leopardis wiedii

Kodkod
Oncifelis guigna

Pampas cat
Oncifelis colocolo

Geoffroy's cats have ocher coats in northern part of range, silvery gray coats in south

Geoffroy's cat
Oncifelis geoffroyi

Long, thick fur provides protection against exposed mountain environment

Andean cat
Oreailurus jacobita

FACT FILE

Jaguarundi This species is not closely related to the other South American felids and does not look like a typical cat. Sometimes called the weasel cat or otter cat, its long, slender body, short legs, and very long tail do resemble those of a mustelid. The jaguarundi is an agile climber and good swimmer.

Up to 25½ in (65 cm)
Up to 24 in (61 cm)
Up to 20 lb (9 kg)
Solitary
Uncommon

Arizona & Texas to S. Brazil & N. Argentina

Ocelot The wide variety of prey species taken by this opportunistic hunter has allowed it to live in a variety of densely covered habitats, from lush rain forests to semiarid brush. The ocelot's striking coat provides camouflage among thick vegetation, but has also made it a prime target for hunters. International trade in ocelot skins is now banned.

Up to 18½ in (47 cm)
Up to 16 in (41 cm)
Up to 26½ lb (12 kg)
Solitary
Locally common

S.E. Texas to N. Argentina

Geoffroy's cat This agile cat can walk upside down along a branch and hang from a tree by its feet. Its populations were decimated by the fur trade, with more than 340,000 skins exported from Argentina between 1976 and 1979.

Up to 26½ in (67 cm)
Up to 14½ in (37 cm)
Up to 13 lb (6 kg)
Solitary
Near threatened

S. Bolivia & Paraguay to Argentina & Chile

FELINE DEFENSE

Cats of the same species try to avoid conflict by marking their territory with scent, but hostile encounters can occur. Small cats are also at risk from larger cats. When threatened, cats will use a series of body and tail positions and facial expressions in an effort to avert what could be a costly attack.

Initial warning *A wide-eyed stare shows that the margay is prepared to defend itself.*

Last warning *With its ears folded back and mouth wide open to display its teeth, the margay gives the foe a last chance to retreat.*

HUNTING STRATEGIES

While some carnivores are chiefly foragers, surviving on carrion, easily captured invertebrate prey, or vegetation, most hunt vertebrates at least some of the time and many are capable of taking down an animal much larger than themselves. Those that target small mammals, birds, and reptiles tend to hunt alone, while larger prey often, but not always, encourages cooperative efforts. Group hunting is most common among dogs, lions, and spotted hyenas, while raccoons, civets, mongooses, mustelids, brown and striped hyenas, and most cats are more likely to be solitary predators. Dogs kill small prey by shaking it in their jaws to dislocate the neck, and defeat larger prey by disemboweling it or seizing the throat or nose. Almost all cats sink their canine teeth into the neck of their victim, while weasels, civets, mongooses, and jaguars are occipital crunchers, biting into the back of the head so the claws and teeth of the prey are kept out of the way. The larger carnivores usually need to make a substantial kill every few days. To find enough to eat, these predators often roam over extensive home ranges, which vary in size according to the abundance of prey. Lions and tigers, for example, have ranges of 8–200 square miles (20–500 sq. km) and may cover up to 12 miles (20 km) in a day.

Exception to the rule Most large carnivores need to eat vertebrates to support themselves. While insects make easy prey, they are small and can usually sustain carnivores no larger than a badger or aardwolf. The sloth bear, however, is at least five times larger than a badger but manages to survive mostly on a diet of invertebrates. Its muscular limbs and long, curved claws enable it to rip open termite mounds, releasing thousands of insects at a time.

Hyena variations Brown and striped hyenas depend largely on carrion, supplemented by invertebrates and small prey. Consequently, they tend to forage alone, as group foraging offers little advantage and would lead to aggressive competition. The larger spotted hyenas depend more on hunting and pursue larger prey, such as zebra, wildebeest, gemsbok, and impala. When the prey is substantially larger than themselves, hyenas rely on cooperation to bring it down. Even when the prey is a young, smaller animal and a single hyena could complete the task alone, the kill supplies enough food to feed several hyenas, so pack hunting avoids waste.

Family affair Spotted hyenas may live in clans of up to 80 animals, but their hunting packs are smaller and tend to be made up of closely related individuals. This ensures that a successful hunt benefits relatives rather than unrelated hyenas.

Hot pursuit Maintaining speeds of up to 35 miles per hour (60 km/h), hyenas will pursue their prey for up to 2 miles (3 km) until it tires and becomes vulnerable to attack. Only about a third of such pursuits end in success for the hyena pack.

The kill Hyenas tend to focus on young, weak, sick, or injured hoofed mammals, which they separate from the herd. When the target prey falters, the pack members will simultaneously bite the belly to disembowel it.

Moving in Even once a pack of spotted hyenas takes down an animal, they are not assured of a feast. Their main competitors are lions, which hunt the same prey and, attracted by the hyena's squabbling, may move in to steal the kill. While a pack of whooping hyenas advancing shoulder to shoulder may intimidate female or younger lions, they are no match for a male and usually surrender their meal.

Fast eaters If undisturbed, spotted hyenas will gorge themselves, eating up to a third of their weight in meat at a time. They often hide part of a carcass in muddy water for later consumption.

Making the most of a meal When hyenas lose their kill to lions, they wait for the lions to finish eating and then return for the leftovers. A massive skull with a short, powerful jaw and immensely strong teeth allows hyenas to cut through the tough hides of their large prey and crush the bones to get to the marrow. Their acidic digestive system extracts all possible nutrition from the bones.

Success and failure While predators are designed to outwit, outrun, or overpower their prey, the odds are not always in their favor. Most attempts at capture fail. The cheetah is capable of great bursts of speed in pursuit, but overheats in less than a minute and needs to rest. If the prey manages to stay ahead until this point, it will escape.

A broad hunter While some carnivores are highly specialized for capturing a particular kind of prey, others are generalists, pursuing any opportunity. The mink will hunt crustaceans and fish in water, take rabbits and birds on land, and chase small mammals down burrows. During times of scarcity, mink suffer from competition with more effective specialist predators, but can usually turn to an alternative source of food.

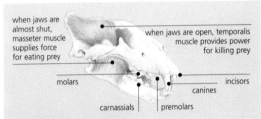

when jaws are almost shut, masseter muscle supplies force for eating prey

when jaws are open, temporalis muscle provides power for killing prey

molars

incisors

canines

carnassials

premolars

MEAT-EATING TEETH

Carnivore jaws and teeth are uniquely adapted to a diet of meat. Jaws tend to be extremely strong, able to suffocate or crush the bones of prey even when the mouth is wide open, and providing enough force to cut through flesh. Typically, carnivores have 44 teeth, with 3 incisors, 1 canine, 4 premolars, and 3 molars on each side of each jaw. The last premolar in the upper jaw and the first molar in the lower jaw form the carnassials, sharp-tipped cusps that act like scissors to shear through meat. In carnivores that primarily eat insects or plants, the equivalent teeth have flattened grinding surfaces, but their ancestors possessed carnassials. In some carnivores, particularly the cat family, the canines are especially large and are used for stabbing prey.

UNGULATES

CLASS	Mammalia
ORDERS	7
FAMILIES	28
GENERA	139
SPECIES	329

Some 65 million years ago, an order of hoofed mammals known as Condylarthra began to evolve into many diverse orders, of which seven remain. These surviving orders are all now referred to as ungulates, meaning "provided with hoofs." In fact, only two orders possess true hoofs: odd-toed ungulates (Perissodactyla) include horses, tapirs, and rhinoceroses, while even-toed ungulates (Artiodactyla) encompass pigs, hippopotamuses, camels, deer, cattle, sheep, and goats. The other five orders, each with its own specializations, are elephants (Proboscidea), aardvark (Tubulidentata), hyraxes (Hyracoidea), dugong and manatees (Sirenia), and whales and dolphins (Cetacea).

Head of the harem Like many other ungulates, zebras are territorial and live in harems of several mares dominated by a single stallion. The stallion will defend his harem from the attentions of other stallions by biting and kicking.

Mineral supplements Mountain goats, deer, and other ungulates may all converge on particular mineral-rich salt licks. It is thought that licking the rocks supplies them with nutrients that are lacking from their otherwise herbivorous diet.

HOOFS AND HERDS

Although less closely related to each other than to other members in the ungulate group, odd-toed and even-toed ungulates both stand on the tips of their toes, which are encased in hoofs. Together with their elongated metapodials (the hand and foot bones in humans), this unguligrade stance extends the leg, providing a longer stride and greater speed. As the dominant terrestrial herbivores, ungulates depend on their ability to run faster and farther than almost any large predator. They also have mobile ears, sharp, binocular vision, and a keen sense of smell to help them detect danger.

Another strategy for survival that is employed by many grassland ungulates is the formation of large herds. Living in a group increases the chance that a predator will be detected, and reduces any one individual's chance of being taken. Large herds are practical only on open plains, where many animals can remain in close contact. Many forest ungulates live in small family groups or are solitary.

Almost all ungulate species are herbivores, with teeth adapted for grinding. Their specialized digestive systems can break down cellulose, the usually indigestible component of plant cell walls. Food is fermented by microorganisms in the hindgut or in a special stomach chamber. Ruminants such as deer regurgitate the fermented food and chew it a second time, a practise known as "chewing the cud."

Elephants, hyraxes, and aardvarks lack true hoofs and do not have an unguligrade stance. While elephants have just the bones of their toes (encased in a fatty matrix) touching the ground, hyraxes and aardvarks walk on the entire foot.

Whales, dolphins, dugongs, and manatees have all evolved for life in the water, with a streamlined body and flippers. In fact, whales and dolphins were only recently classified as ungulates, when genetic studies showed that they are closely related to hippopotamuses. Some experts have even suggested placing them with even-toed ungulates in a single order called Certartiodactyla.

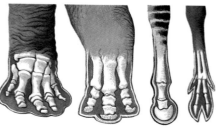

Elephant   Rhinoceros   Horse   Deer

Toes and hoofs Elephants have a broad foot with five toes encased in a fatty matrix. In the "true" ungulates, at least one toe has been lost and the remaining toes form a hoof. The hoofs of odd-toed ungulates have three toes (as in the rhinoceros) or a single toe (as in the horse), while even-toed ungulates have two or four toes, which may be fused to form a cloven hoof (as in deer).

Browsers and grazers Ungulates may be browsers, feeding on woody vegetation such as trees and shrubs; grazers, feeding on grasses; or mixed feeders. African elephants concentrate on savanna grasses during the wet season, but switch to the woody parts of trees and shrubs when the grasses die back in the dry season.

Advanced young Although pigs have several piglets in a litter, most female hoofed mammals bear a single well-developed young, which can stand, see, and hear soon after birth. An impala will isolate herself to calve, but will rejoin the herd with her calf within a day or two.

Fermenting stomachs Ungulates rely on internal microorganisms to break down plant cellulose. Many even-toed ungulates, such as deer, cattle, and sheep, are ruminants. Their complex stomachs include a rumen, a chamber in which food is fermented by microorganisms before being regurgitated and chewed a second time. A ruminant's digestive process can take up to four days and gains maximum nutrition from food. Odd-toed ungulates, such as horses, rhinoceroses, and tapirs, ferment their food in the hindgut (the cecum and large intestine). The two-day digestive process is less efficient than that of ruminants, requiring odd-toed ungulates to consume greater quantities of food.

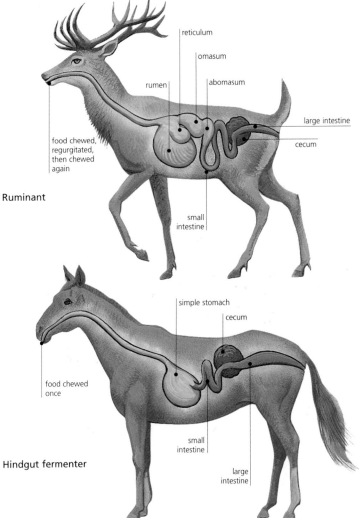

reticulum
omasum
rumen
abomasum
large intestine
cecum
food chewed, regurgitated, then chewed again
small intestine

Ruminant

simple stomach
cecum
food chewed once
small intestine
large intestine

Hindgut fermenter

ELEPHANTS

CLASS	Mammalia
ORDER	Proboscidea
FAMILY	Elephantidae
GENERA	2
SPECIES	2

Weighing up to 7 tons (6.3 tonnes), elephants are the world's largest land animals. Their massive bulk is supported by four pillar-like legs and broad feet. An enormous head features immense fan-shaped ears and a long, flexible trunk. Full of blood vessels, the ears help to disperse the animal's body heat and will be flapped on hot days. The trunk, a union of the nose and upper lip, has more than 150,000 muscle bands and can both pick up twigs and lift heavy logs. Elephants can live for up to 70 years, longer than any land mammal apart from humans. As well as their longevity and strength, elephants are very social, intelligent, and quick to learn, traits that encouraged their domestication.

REMOVING THE THREAT
While both elephant species are threatened by habitat loss, the African elephant's handsome tusks have also made it the prime target of the ivory trade. In an attempt to halt the trade, the Kenyan government burned piles of poached tusks (below) and the international sale of ivory was temporarily banned.

Asiatic elephant   African elephant

Flexible trunk Used for caressing, lifting, feeding, dusting, smelling, as a snorkel, as a weapon, and in sound production, the elephant's trunk is superbly dextrous. Asiatic elephants have one finger-like projection at the tip, while African elephants have two.

Close relations Female elephants usually bear a single young after a gestation period of 18–24 months—the longest of any mammal. The calf is weaned very gradually, and may continue to take milk from the mother for up to 10 years. Females remain with their mother's herd, while males leave at around 13 years.

Shared care
The female elephants in a herd share the care of the young and will form a circle around them to protect them from danger.

Rapid growth By the time it is 6 years old, a young elephant will weigh about 1 ton (1 tonne). Growth slows after about 15 years, but continues throughout its life.

MATRIARCHS AND MUSTH

The order Proboscidea appeared about 55 million years ago and once included immense mastodons and woolly mammoths. Its members inhabited all kinds of habitat, from polar regions to rain forest, and at some time lived on all continents apart from Australia and Antarctica. Today's elephants are restricted to the forests, savanna, grasslands, and deserts of Africa and Asia.

To fuel their immense bodies, elephants spend 18–20 hours a day foraging or traveling toward food. An adult may eat up to 330 pounds (150 kg) of vegetation and drink 40 gallons (160 l) of water per day.

The basic social unit of elephants is a family group of related females and offspring, led by a matriarch. Adult males visit these groups only to mate, spending the rest of their time alone or in bachelor bands. A number of family groups may form larger herds. To maintain their social ties, elephants rely on a range of communication methods, primarily using touch (greeting one another by intertwining trunks, for example), sound (some vocalizations are much lower than human ears can hear and travel up to 2½ miles, or 4 km), and postures (a raised trunk, for instance, can be used as a warning).

Mature male elephants have periods of musth, a time of high testosterone levels when the musth gland between the eye and ear secretes fluid. The animals become more aggressive and more likely to win fights, and wander greater distances in search of females.

Humped or level back

Asiatic elephant
Elephas maximus

Skin color often masked
by the dirt the animal
throws over itself or the
mud in which it wallows

Tusks usually
found only
on males

Runs from danger
with tail held up,
possibly as a
signal to fellow
herd members

Tusks used to remove
bark from trees, move
fallen branches, mark
trees, dig for water,
and for fighting

Four nails on
hindfoot

Larger ears than
Asiatic elephant

Heavier and taller
than Asiatic elephant

Concave back

Both males
and females
possess tusks

Trunk slightly
floppier than that
of Asiatic elephant

Young can
follow mother
a few days
after birth

African elephant
Loxodonta africana

Three nails on hindfoot

FACT FILE

Asiatic elephant Smaller than the
African elephant, this species is actually
more closely related to the extinct
woolly mammoth. The tusks are usually
absent in the female. Females live in
matriarchal herds of 8–40 mothers,
daughters, and sisters. Males may
spend much of their time alone or in
temporary groups of up to seven males,
but will join herds of females to mate.
Elephants are vocal communicators,
using low-frequency calls to maintain
contact over distance; high-frequency
calls to convey their mood; and loud
trumpeting calls to raise the alarm.

	Up to 21 ft (6.4 m)
	Up to 10 ft (3 m)
	Up to 6 tons (5.4 t)
	Variable
	Endangered

India to S.W. China, S.E. Asia

African elephant Occupying deserts,
forests, river valleys, and marshes as
well as savanna, this species survives
mainly in protected areas. Poaching has
led to a rapid decline in numbers, with
the Kenyan population dropping from
167,000 in 1970 to 22,000 in 1989. In
open habitat and especially during the
wet season, herds may come together
to form temporary aggregations of
hundreds of individuals. Elephants that
inhabit forests tend to be smaller in
size and live in small family groups.

	Up to 24½ ft (7.5 m)
	Up to 13 ft (4 m)
	Up to 7 tons (6.3 t)
	Variable
	Endangered

Sub-Saharan Africa

INSIDE THE HEAD

The elephant's massive skull contains
pockets of air to minimize its weight.
The tusks are elongated incisor teeth
emerging from deep sockets. The
molar teeth are replaced horizontally
in a conveyor-belt fashion, with new
teeth developing from behind and
slowly moving forward to take the
place of worn teeth.

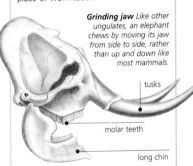

Grinding jaw Like other
*ungulates, an elephant
chews by moving its jaw
from side to side, rather
than up and down like
most mammals.*

tusks

molar teeth

long chin

DUGONG AND MANATEES

CLASS	Mammalia
ORDER	Sirenia
FAMILIES	2
GENERA	3
SPECIES	5

Credited with inspiring the myth of the mermaid, the aquatic mammals of the order Sirenia are languid, docile creatures that never venture onto land. They are the only mammals to feed primarily on grasses and plants in shallow waters, a unique niche that may explain the order's lack of diversity, since seagrasses are much less varied than terrestrial grasses. The four living sirenian species are all found in the warm waters of tropical and subtropical regions. The dugong is restricted to the sea, while the Amazonian manatee is found only in the fresh waters of the Amazon River. African and Caribbean manatees live in freshwater, estuarine, and marine habitats.

Shaped for the water The dugong has a streamlined body, with paddle-like flippers instead of front legs. Its tail is fluked like a dolphin's, whereas a manatee's tail is more like that of a beaver. While sirenians usually conserve energy by moving slowly through the water, they are capable of swimming swiftly to escape danger.

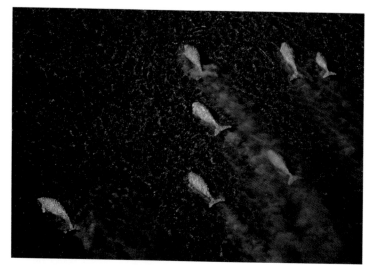

Slow mothers Sirenians have a very slow reproductive rate, which is a factor in their declining numbers. Females bear a single calf at a time, and wait at least 2 years and sometimes several before producing again. A calf nurses for up to 2 years, during which time it learns food sources and migration routes from its attentive mother.

Bristled feeder Like all sirenians, the Caribbean manatee swims slowly through the water, grazing on aquatic plants and seagrasses. It locates food with the aid of the sensitive, bristle-like hairs on its snout, and uses its muscular lips to grasp plants and pass them to the mouth.

GENTLE GRAZERS

Like other aquatic mammals, sirenians have a streamlined body, flippers, and flattened tail, and come to the surface to breathe through nostrils on top of the head. In keeping with a grazing lifestyle, the sirenian head is more like that of a pig and is used to root out grass rhizomes (underground stems) from the sediment. The angle of the dugong's snout restricts it to feeding from the bottom, but manatees can feed from all levels of the water.

The teeth of sirenians have become specialized in different ways for chewing great quantities of plant matter. The dugong uses rough horny pads in its mouth to crush its food, then grinds it with a few peg-like molars that continue to grow throughout its life. Manatees chew with their front molars, which are replaced by new teeth from behind when they wear out.

Sirenians have simple stomachs with extremely long intestines. Their plant food is broken down by microorganisms in the rear part of the digestive tract, making them hindgut fermenters like horses and other odd-toed ungulates. To counter the buoyancy provided by their gas-producing diet, sirenians have unusually dense, heavy bones.

Because their eyesight is generally poor, sirenians rely heavily on touch to find food. Their hearing is good underwater, with sounds transmitted through the skull and jaw bones. They emit squeaks to communicate, but how these sounds are produced is an enigma as they lack vocal cords.

Some sirenians live a solitary life, but more often, they are found in loosely structured groups of a dozen or so. Occasionally, such groups come together to form large herds of 100 or more. Social bonds are reinforced through playful nuzzling.

Because they have few natural predators, dugongs and manatees have developed no defenses other than their large size. This has made them easy targets for human hunters. Only about 130,000 individual sirenians remain, far fewer than in any other order of mammals.

Different skulls Manatees have a row of molar teeth, with new hind teeth constantly moving forward to replace the worn teeth. A dugong skull has an angled snout with only a few peg-like molars, which grow throughout life. In male dugongs, long incisors form tusks.

Manatee skull

Dugong skull

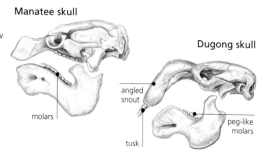

molars

angled snout

tusk

peg-like molars

African manatee
Trichechus senegalensis

Paddle-like tail

Less bulky than manatee

Fluked tail

Dugong
Dugong dugon

No nails on flippers

Thick, tough skin is often wrinkled

Nostrils closed by valves

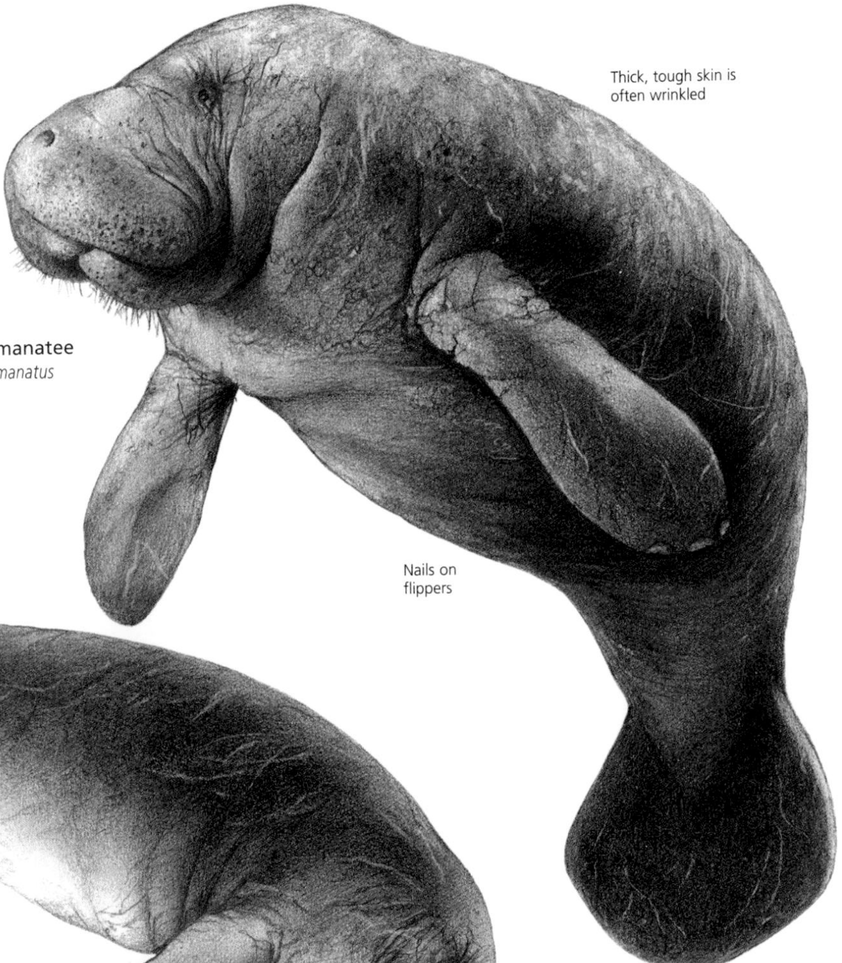

Caribbean manatee
Trichechus manatus

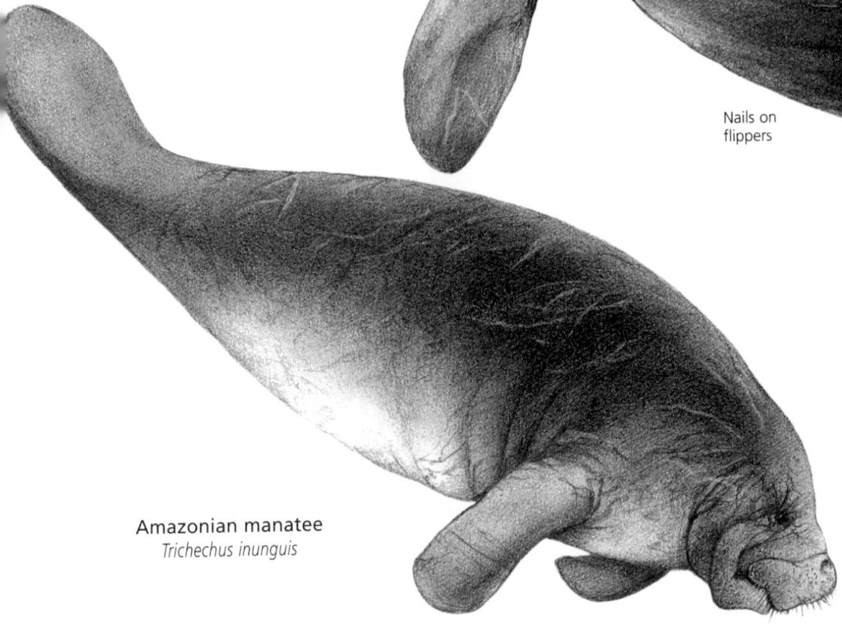

Nails on flippers

Amazonian manatee
Trichechus inunguis

Stiff bristles on large, mobile lips

FACT FILE

African manatee This poorly studied species is believed to be at least partly nocturnal. Found from coastal seas to rivers, it tends to feed from waters near or at the surface.

🐃 Up to 13 ft (4 m)
⚖ Up to 1,100 lb (500 kg)
🐾 Solitary, family group
⚡ Vulnerable

W. African coast, Niger River

Dugong Now much reduced, the original range of this marine species coincided with the distribution of seagrasses, its main food. The dugong grazes mostly on the ocean floor.

🐃 Up to 13 ft (4 m)
⚖ Up to 1,980 lb (900 kg)
🐾 Variable
⚡ Vulnerable

Red Sea to S.W. Pacific islands

Caribbean manatee This species moves freely between freshwater and marine habitats. When a female is ready to breed, a mating herd of up to 20 males will pursue her, competing for her attention for up to a month.

🐃 Up to 15 ft (4.5 m)
⚖ Up to 1,320 lb (600 kg)
🐾 Solitary
⚡ Vulnerable

Georgia & Florida to Brazil; Orinoco River

Amazonian manatee Few plants grow under the murky waters of the Amazon River, so this manatee feeds mainly on surface vegetation such as floating grasses and water hyacinths.

🐃 Up to 9 ft (2.8 m)
⚖ Up to 1,100 lb (500 kg)
🐾 Solitary
⚡ Vulnerable

Amazon basin

EXTINCT SEA COW

First sighted by Europeans in 1741, Steller's sea cow (*Hydrodamalis gigas*) became extinct in 1786, a victim of overhunting. It was the largest sirenian of all, weighing up to 11 tons (10 t).

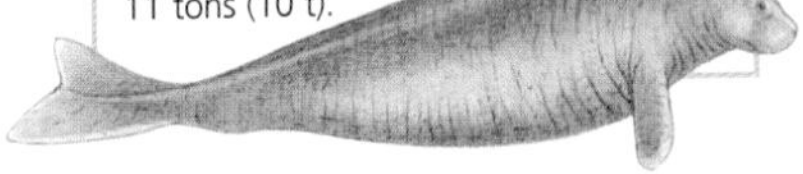

HORSES, ZEBRAS, AND ASSES

CLASS	Mammalia
ORDER	Perissodactyla
FAMILY	Equidae
GENUS	1
SPECIES	9

The horses, zebras, and asses of the family Equidae rely on their large size, swift running style, and herding behavior to evade predation. An equid bears its weight on the tip of a single toe on each foot, a stance that provides a springy gait. The slender legs lock when at rest rather than relying on muscle contraction, a mechanism that minimizes the energy used during the many hours that the animal feeds. The teeth are highly specialized for a diet of grasses and other plants, with incisors to clip the vegetation, and complex, ridged cheek teeth for grinding. Plant cellulose is fermented in the hindgut, allowing equids to live on the abundant low-quality food of arid lands.

Wild and feral Wild equids live in grasslands, savanna, and deserts in Africa and Asia. They generally congregate in herds over large territories. Hunted for meat or hides and persecuted as competitors for grazing land, all wild equids have declined and most are at risk of extinction. Feral herds of the domestic horse can be found on every continent except Antarctica.

EVOLVING EQUIDS

All equids are covered in thick hair, with a mane along their long neck. Most species have single-colored coats, but zebras can be instantly recognized by their striking black and white stripes. Eyes at the side of the head provide good all-round vision by night and day, while erect, mobile ears with acute hearing listen out for danger. Equids tend to run away from predators, but will kick and bite to defend themselves if necessary. These highly social animals may whinny, bray, nicker, or squeal to communicate. Visual cues involving tail, ear, and mouth positions also play a role in equid communication, as does scent.

The first horse-like animal, a dog-sized mammal that walked on the soft pads of its feet, appeared about 54 million years ago. North America was the center of equid evolution, which resulted in a single-toed horse-like creature by at least 5 million years ago. Equids migrated to Africa and Asia, where modern species of zebras and asses later emerged. By the end of the last ice age, horses had vanished from North America, and were only reintroduced there by Europeans.

Humans first domesticated asses in the Middle East before 3000 BC, but within 500 years the faster and stronger domestic horse had arrived from Central Asia. Domestic horses revolutionized agriculture, transport, hunting, and warfare. Today, virtually all wild horses are feral populations of the domestic horse.

Harem of horses Wild horses, like common and mountain zebras, live in permanent groups of females and their young led by a harem male. The females are usually unrelated, having been abducted from their family group. Adult asses and Grevy's zebras form temporary associations.

The last wild horse The domestic horse is descended from the tarpan (*Equus ferus*). The only tarpan to survive into the modern era is Przewalski's horse (*E. f. przewalskii*), a subspecies that once roamed Mongolia, but is now found only in zoos and a few reintroduced populations.

Grooming together Both domestic and wild horses rely on mutual grooming to cement social bonds. Here, two foals nibble each other's shoulders and withers. The nose-to-tail stance allows them to continue to watch for predators.

Kiang
Equus kiang

Upper coat red in summer, becomes browner and longer in winter

Largest wild ass

Onager
Equus onager

Ass (donkey, burro) domestic form
Equus africanus

Ass
wild form
Equus africanus

Coat can be grayish, brownish, or reddish, with white underparts

Some asses have bands on legs

Stallion curls back upper lip to help detect a female's reproductive status from her scent

Mongolian wild ass
Equus hemionus

Large head with short, erect mane and no forelock

Wild form is shorter and stockier than domestic form

Mongolian wild ass
Equus hemionus

✝
Horse
Equus ferus

FACT FILE

Zebra Like all zebra species, this equid has a coat pattern of black stripes on a white background. Suggestions that this provides camouflage or confuses predators have been discounted. The likeliest explanation is that the stripes perform a social role, possibly helping individuals to identify each other.

	Up to 8 ft (2.5 m)
	Up to 5 ft (1.5 m)
	Up to 850 lb (385 kg)
	Harem
	Common

E. & S. Africa

STRIPES AND MOODS

Like all equids, zebras use a range of visual signals to express their moods. Competing stallions will shake their head, arch their neck, and stamp their feet before resorting to physical contests in which they bite each other's neck and legs. Both mares and stallions will attempt to deter a predator with a kicking display.

Prepared to bite
Stallions will display a bite threat before making physical contact.

Ready to kick
A threatened equid will kick out its back legs in a defense display.

Getting a whiff
By curling back the upper lip, a stallion directs the scent of a mare's urine to the Jacobson's organ in the roof of his mouth, which can detect whether she is ready to mate.

Largest of the wild equids

Grevy's zebra
Equus grevyi

Finely divided pattern of black on white

Zebra southern form
Equus burchelli

Shorter ears than those on other zebra species

Zebra northern form
Equus burchelli

Wide stripes on body

Mountain zebra
Equus zebra

Each individual zebra's pattern of stripes is unique

Stripes wider on rump than on body

Foals can walk within an hour of birth

Mountain zebra foal

TAPIRS

CLASS Mammalia

ORDER Perissodactyla

FAMILY Tapiridae

GENERA 1

SPECIES 4

Appearing in the fossil record before either horses or rhinoceroses, tapirs as a group have changed little in the past 35 million years. These reclusive tropical-forest browsers are about the size of a donkey, with a squat, streamlined body for moving through the thick undergrowth, and a sensitive, prehensile trunk that is used for grasping food, detecting danger through scent, and as a snorkel. Tapirs emerge from the shelter of thickets at night to feed on the leaves, buds, twigs, and fruit of low-growing plants—by distributing seeds in their feces, they play an important ecological role in the forest. Being good swimmers and fond of water, they also consume aquatic vegetation. While their small eyes provide poor vision, the senses of hearing and smell are acute. Usually solitary and sparsely distributed, tapirs communicate through high-pitched whistles and scentmarking. Following a 13-month gestation period, a female tapir generally bears a single young. While the mother is foraging, a newborn will hide among the thickets, its patterned coat providing effective camouflage in the dappled light of the forest. After a week, it will accompany the mother, eventually going off on its own by the time it is 2 years of age.

Reduced range Found at various times throughout much of North America, Europe, and Asia, tapirs are now restricted to three species in Central and South America and a single species in Southeast Asia.

Water refuge Never far from water, tapirs spend much of their time submerged, often with just the trunk appearing above the surface as a snorkel. Water provides cover from predators and relief from the heat.

FACT FILE

Brazilian tapir When this tapir plunges into water to escape a jaguar, it risks being taken by a crocodile. Its chief predator, however, is the human hunter, who follows its clear foraging trails.

🐾	Up to 6½ ft (2 m)
🐾	Up to 3½ ft (1.1 m)
	Up to 550 lb (250 kg)
♟	Solitary
↯	Vulnerable

Tropical South America (E. of Andes)

Malayan tapir This species is the only living tapir in Asia. Courtship is marked by whistling pairs walking around in circles as they attempt to sniff each other's genitals.

🐾	Up to 8 ft (2.5 m)
🐾	Up to 4 ft (1.2 m)
	Up to 705 lb (320 kg)
♟	Solitary
↯	Vulnerable

Myanmar, Thailand, Malaya, Sumatra

⚡ CONSERVATION WATCH

Tapir numbers are in decline as their forest habitat is lost or fragmented, they are hunted for their meat, and livestock compete for their food. All four species are listed on the IUCN Red List, as follows:

2 Endangered

2 Vulnerable

Short, bristly mane

Infant tapirs are mottled and striped to help camouflage them when their mothers are absent

Brazilian tapir
Tapirus terrestris

Disruptive pattern of black and white camouflages animal in its shady rain-forest habitat

Malayan tapir
Tapirus indicus

RHINOCEROSES

CLASS	Mammalia
ORDER	Perissodactyla
FAMILY	Rhinocerotidae
GENERA	4
SPECIES	5

The snout of a rhinoceros features its most distinguishing characteristic—a great horn or two made up of fibrous keratin. Rhinoceroses use their remarkable horns to fight rivals, to defend their young against predators, to guide their young, and to push dung into piles as scented signposts. Although keratin is a common substance that is also found in human fingernails, rhino horns are attributed great potency in traditional Asian medicine. Such has been the demand that many thousands of rhinoceroses have been poached, and all five species are now at risk of extinction. Today, there are fewer than 15,000 wild rhinos in Africa, and no more than 3,000 in Asia.

High-speed charge The third heaviest of all mammal species (after the African and Asiatic elephants), the white rhinoceros is still capable of impressive bursts of speed when charging to scare off intruders.

Rhino standoff Both African rhino species use their horns to fight rivals, while Asian species use their sharp incisor or canine teeth. Before resorting to physical attack, rhinos will engage in a series of gestures, including pushing horns together, wiping horns on the ground, and spraying urine. Black rhinoceroses (pictured here) are especially aggressive, with half of all males and a third of females dying after fights.

MASSIVE HERBIVORES

The family Rhinocerotidae was once abundant, widespread, and diverse. The woolly rhinoceros, for example, roamed Europe until the end of the last ice age 10,000 years ago and appears in early cave art. A hornless rhinoceros, *Indricotherium*, was the largest animal ever to have lived on land. Today, only five species remain, two in Africa (the white rhinoceros and the black rhinoceros) and three in Asia (the Indian, the Javan, and the Sumatran).

Rhinoceroses have massive bodies supported by four stumpy legs. Three hoofed toes on each foot leave an "ace-of-clubs" print in their tracks. Their thick, wrinkled skin can be gray or brown, but its true color is often masked by dried mud, since rhinos like to wallow in muddy pools. Despite their names, there is no clear distinction in skin color between white and black rhinos: both are grayish, with the precise tinge determined by local soil color.

With a lifespan of up to 50 years, rhinos are slow breeders, a fact that has made them especially vulnerable to habitat loss and overhunting. After a gestation of about 16 months, a female bears a single young and nurses it for more than a year. Their association lasts for 2–4 years, until the next calf is born. Most adult rhinos are solitary, but breeding pairs may stay together for a few months, while females or immature males sometimes form temporary herds. White rhinos are known to form a circle around the young to protect them from predators.

Sharp hearing Erect ears swivel to pick up faint sounds.

Cropped horn To discourage poachers, the horn of this rhinoceros has been cropped.

Indian grazer A prehensile upper lip helps gather tall grasses, but can be folded away for eating short grasses.

Dim sight Small eyes at the side of the head provide poor vision.

Armor plating The Indian rhinoceros's distinctive dark gray skin falls into deep folds at the joints, resembling the plates of a coat of armor. This characteristic inspired Rudyard Kipling's famous story "How the Rhinoceros Got His Skin."

⚡ CONSERVATION WATCH

Coupled with extensive habitat loss, the trade in rhino horns (used in Asia for medicine and carvings) has had a devastating effect. Some conservationists advocate farming rhinos and cropping their horns to provide revenue for the local people and thus discourage poaching. All five rhinoceros species are listed on the IUCN Red List, as follows:

3	Critically endangered
1	Endangered
1	Near threatened

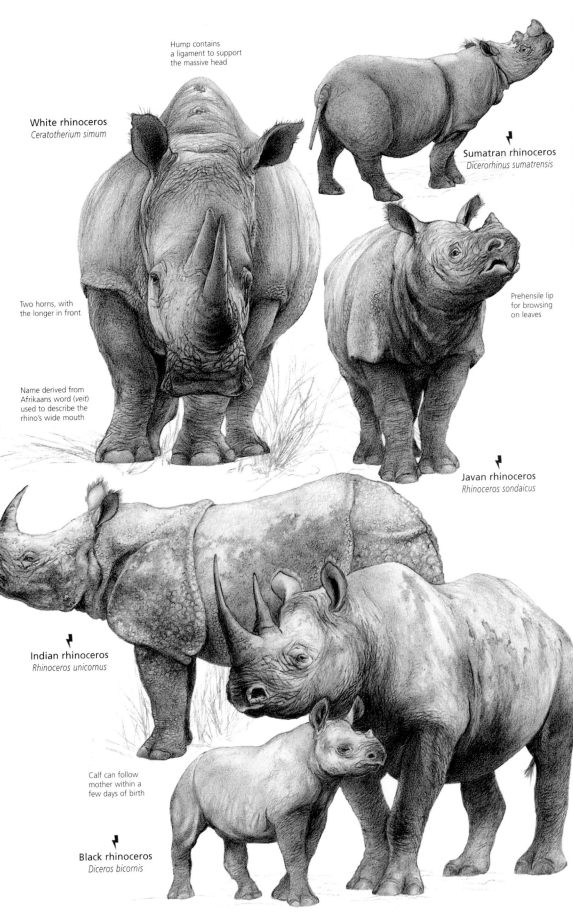

Hump contains a ligament to support the massive head

White rhinoceros
Ceratotherium simum

Sumatran rhinoceros
Dicerorhinus sumatrensis

Two horns, with the longer in front

Prehensile lip for browsing on leaves

Name derived from Afrikaans word (*veit*) used to describe the rhino's wide mouth

Javan rhinoceros
Rhinoceros sondaicus

Indian rhinoceros
Rhinoceros unicornus

Calf can follow mother within a few days of birth

Black rhinoceros
Diceros bicornis

HYRAXES

CLASS Mammalia
ORDER Hyracoidea
FAMILY Procaviidae
GENERA 3
SPECIES 7

About the size of a rabbit and looking rather like a large guinea pig, the hyraxes are often mistaken for rodents, but are in fact ungulates, with flattened hoof-like nails on their feet. Millions of years ago, hyraxes, some as large as tapirs, were the dominant grazing mammals of North Africa. They were displaced by larger ungulates such as antelopes and cattle. The surviving hyraxes are robust, agile creatures that scamper and leap along steep rocks and tree branches. The soles of their feet are uniquely equipped to provide traction, with soft pads kept moist by a glandular secretion, and muscles that retract the middle of the sole to form a suction pad. Hyraxes are gregarious animals, and some species live in colonies of up to 80 animals.

Africa to the Middle East Although they were once more diverse and widespread, hyraxes are now found only in Africa and the Middle East and there are just seven species in three genera. Rock hyraxes (*Procavia*) mostly live on rocky outcrops and cliffs in much of Africa and parts of the Middle East, but are also found in grassland and scrubby environments. Yellow-spotted hyraxes (*Heterohyrax*) occupy similar habitats, but are largely confined to East Africa. Tree hyraxes (*Dendrohyrax*) also make Africa their home, but tend to live in forests.

Warm huddle Most small mammals are nocturnal, but hyraxes are active by day. Unable to regulate their body temperature well, they conserve heat by huddling together and warm up by basking in the sun. Hyraxes live in family groups of several females and their offspring, headed by a territorial male. Females usually stay with their family for life, but males disperse at about 2 years of age. They may live on the edge of a family group, hoping to take over the territorial male's position.

Herbivorous hyraxes All hyrax species will feed both in trees and on the ground, and have been known to travel almost a mile (1.3 km) in search of food. Rock hyraxes mainly eat grasses, while yellow-spotted hyraxes and tree hyraxes tend to browse leafy plants. They rely on microorganisms in their multichambered stomach to digest the plant cellulose. Hyraxes are highly vocal creatures that sound like no other animal. The gregarious ground-dwellers chatter, whistle, and scream. At night, tree hyraxes begin a series of loud croaks that end in a scream. Hyrax vocabulary changes and expands throughout life. Young hyraxes utter long chatters that progressively intensify, but they make only a fraction of the sounds emitted by their adult relatives.

Tuft of hair covers scent gland on back

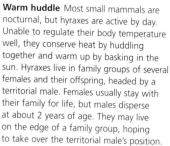

Southern tree hyrax
Dendrohyrax arboreus

Hoof-like nails

Yellow-spotted hyrax
Heterohyrax brucei

Rock hyrax
Procavia capensis

Large eyes provide sharp vision

Long upper incisors grow continuously

> ⚡ **CONSERVATION WATCH**
>
> **Hunted pelts** Tree hyraxes are losing habitat as forests disappear. At least one species, the eastern tree hyrax (*Dendrohyrax validus*), is extensively hunted for its pelt. Of the seven species of hyraxes, three are listed on the IUCN Red List as vulnerable.

Standing guard Hyraxes are prey for eagles, pythons, and leopards. While a group is busy foraging or resting, at least one member will keep watch for danger. A loud warning cry will send the animals scurrying for cover among the rocks.

MIXED LIVING

Hyraxes include one of the few examples of two mammal species cohabiting. The yellow–spotted hyrax and the rock hyrax have been found living together on the same kopje (rocky outcrop), sharing holes for shelter by night and then huddling together to warm up by day. They do not interbreed, but females tend to give birth at the same time, and members of both species show great interest in the young. The two species avoid competing with each other by eating different foods.

Different diets While the yellow-spotted hyrax is a browser, feeding mainly on leaves, the rock hyrax is a grazer and concentrates on grasses.

AARDVARK

CLASS	Mammalia
ORDER	Tubulidentata
FAMILY	Orycteropodidae
GENERA	1
SPECIES	1

A medium-sized pig-like animal with a stocky body, long snout, and large ears, the aardvark is the only living member of the order Tubulidentata, which evolved from an early hoofed mammal. Solitary and nocturnal, this highly specialized creature emerges from its burrow after dark to search for ants and termites, taking up to 50,000 insects a night. A keen sense of smell helps the aardvark to detect prey, while its powerful, clawed feet can excavate a termite mound in a few minutes. The hair-lined nostrils can be contracted and the ears folded back to keep out dirt. A long, sticky tongue snatches the insects, which are usually swallowed without chewing and ground up in the animal's muscular stomach. With its thick skin and sharp claws, an adult aardvark is vulnerable to few predators other than hyenas and humans.

Aardvark
Orycteropus afer

⚡ **CONSERVATION WATCH**

Specialized ardvarks Although the aardvark is not considered to be threatened, its highly specialized diet makes it vulnerable to habitat alteration. Grazing by both wild and domestic ungulates can benefit aardvarks, as termites flourish in the trampled ground. Crop farming, on the other hand, may cause a decline in aardvark numbers. Aardvarks are also hunted for their flesh.

Determined digger An aardvark uses the four shovel-shaped claws on each forefoot to dig out food and excavate burrows.

THE CATTLE FAMILY

The family Bovidae includes many millions of domesticated cattle, sheep, goats, and water buffalo. The 135 species of wild bovid are much more diverse, ranging from the dwarf antelope, just 10 inches (25 cm) tall and weighing as little as 5 pounds (2 kg), to massive animals such as bison, which can be more than 6 feet (2 m) high at the shoulder and weigh up to 1 ton (1 tonne). Cattle and their kin occur naturally throughout much of Eurasia and North America, and introduced species have formed feral populations in Australasia, but bovids reach their greatest numbers and variety in the grasslands, savanna, and forests of Africa.

CLASS	Mammalia
ORDER	Artiodactyla
FAMILY	Bovidae
GENERA	47
SPECIES	135

Wide spread Bovids are distributed from hot deserts and tropical forests to arctic and alpine regions. There are no native bovids in Australia or South America, but domestic species are found worldwide. There are more than a billion domestic cattle in the world, and all are descended from aurochs, wild cattle that were once widespread but became extinct in 1627.

A DIVERSE FAMILY

The cattle, buffalo, bison, antelopes, gazelles, sheep, goats, and other members of Bovidae are ruminants, with a four-chambered stomach for fermenting plant cellulose. This efficient digestive system has allowed bovids to make the most of low-nutrient foods such as grasses and colonize a wide array of habitats, from arid scrublands to arctic tundra. Grazing bovids tend to have a substantial, stocky build to house their large stomachs. Antelopes and other slender bovids are often more selective browsers.

All male and many female bovids have horns made up of a bony core covered in a layer of keratin, which is never shed. Always unbranched with pointed tips, bovid horns vary in size and can be straight, curved, or spiral. They may be used in contests between males or to fend off predators. The legs of bovids are adapted for running away from danger. Bovids bear their weight on the two middle toes of each foot, which form a cloven hoof. The main bones in the feet are fused to form the cannon bone, which helps to absorb the impact of running.

While some bovids are solitary or live in pairs, most are gregarious. Some species live in harems led by a male, while others have herds made up of females and young, with males being largely solitary or associating in bachelor herds. Group living not only reduces the risk of predation, but also allows members to share information about feeding sites.

Cattle and sheep were first domesticated several thousand years ago. Since then, most wild bovids have declined. Extensively hunted for food, hides, and sport, wild species have also lost much of their natural range to agriculture.

Ringed horns
When male gazelles wrestle with their horns, the ridges prevent the horns slipping and causing serious injury.

Safety in numbers Cape buffalo usually live in herds of 50–500 females and young but thousands, including males, may congregate during the wet season. Weak buffalo can survive within a herd, which will cooperate to drive off predators.

Two-tone
The Thomson's gazelle's two-toned coat is interrupted by black markings on its face and along its side.

Ever alert As prey species, bovids such as gazelles rely on their sharp senses to detect danger. Most have large, mobile ears; eyes on the side of the head for all-round vision; and a keen nose. The striking coloration of many species may help to camouflage them by breaking up their outline.

⚡ CONSERVATION WATCH

Of the 135 species of bovids, 83% are listed on the IUCN Red List, as follows:

4	Extinct
2	Extinct in the wild
7	Critically endangered
20	Endangered
25	Vulnerable
37	Conservation dependent
19	Near threatened

Abbott's duiker
Cephalophus spadix

Yellow-backed duiker
Cephalophus silvicultor

Duiker species vary in size but share distinctive body shape

Red-flanked duiker
Cephalophus rufilatus

Common duiker
Sylvicapra grimmia

Large gland beneath each eye produces secretion used in scentmarking

Banded duiker (zebra duiker)
Cephalophus zebra

Short, conical horns

White band breaks up animal's outline

Ader's duiker
Cephalophus adersi

Ogilby's duiker
Cephalophus ogilbyi

FACT FILE

Duikers Duikers are small, short-horned antelopes that browse at the forest edge, occasionally supplementing their diet with insects and small vertebrates. The name *duiker*—an Afrikaans word for "diver"—was inspired by this shy animal's habit of diving into the underbrush when frightened.

Banded duiker This muscular species is easily identified by the striking stripes of its coat. As in zebras, each banded duiker's pattern of stripes is unique. Banded duikers are diurnal creatures that usually live in pairs, with the pair bond reinforced by mutual grooming.

Up to 35½ in (90 cm)
Up to 19½ in (50 cm)
Up to 44 lb (20 kg)
Solitary, pair
Vulnerable

Liberia

Common duiker This nocturnal antelope lives at higher altitudes than any other African hoofed mammal. It relies on remarkable speed and stamina to escape its predators, which include big cats, dogs, baboons, pythons, crocodiles, and eagles.

Up to 45½ in (115 cm)
Up to 19½ in (50 cm)
Up to 46½ lb (21 kg)
Solitary
Common

Sub-Saharan Africa except rain-forest zone

Ader's duiker Active by day, this species tends to live in territorial pairs. It feeds mainly on flowers, fruit, and leaves from the forest floor, taking advantage of scraps discarded by monkeys and birds in the trees above. Like many other duikers, Ader's duiker has a soft, silky coat with a reddish crest on its head.

Up to 28½ in (72 cm)
Up to 12½ in (32 cm)
Up to 26½ lb (12 kg)
Solitary, pair
Endangered

Zanzibar, coastal S.W. Kenya

⚡ CONSERVATION WATCH

Dazzled prey Prized by the bushmeat trade and as trophies, duikers make easy targets at night when they are dazzled by the lights of hunters. The IUCN Red List classifies a total of 16 duiker species as threatened. The rarest of these species is the endangered Ader's duiker, which has fewer than 1,400 individuals remaining in the wild.

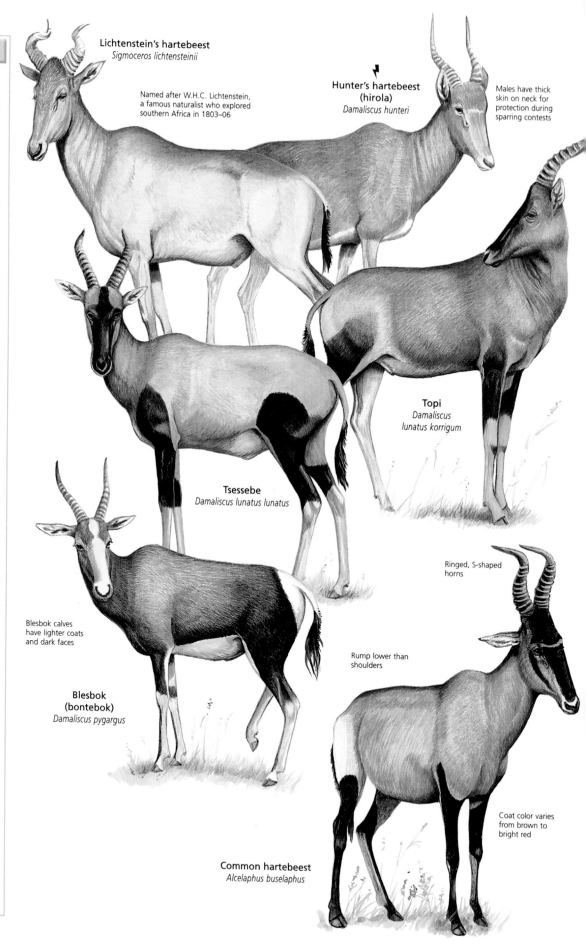

Lichtenstein's hartebeest
Sigmoceros lichtensteinii

Named after W.H.C. Lichtenstein, a famous naturalist who explored southern Africa in 1803–06

Hunter's hartebeest (hirola)
Damaliscus hunteri

Males have thick skin on neck for protection during sparring contests

Topi
Damaliscus lunatus korrigum

Tsessebe
Damaliscus lunatus lunatus

Ringed, S-shaped horns

Rump lower than shoulders

Blesbok calves have lighter coats and dark faces

Blesbok (bontebok)
Damaliscus pygargus

Coat color varies from brown to bright red

Common hartebeest
Alcelaphus buselaphus

FACT FILE

Lichtenstein's hartebeest Males of this savanna species are territorial, marking their claim in the ground with their horns. Rivals will fight for the right to mate, with the victor leading a harem of 3–10 females and their young. Weaker males live alone or form loose bachelor bands.

- Up to 7 ft (2.1 m)
- Up to 4¼ ft (1.3 m)
- Up to 375 lb (170 kg)
- Small group
- Conserv. dependent

S. Africa

Hunter's hartebeest One of the world's rarest mammals, this species declined from 14,000 to 300 animals in the wild between 1976 and 1995. A selective grazer, it eats only short, newly sprouting grasses and moves to another area if the grasses grow long or are disturbed by other grazers. Many scientists believe that Hunter's hartebeest represents the evolutionary link between true hartebeests and the genus *Damaliscus*, making its survival of the utmost importance to the study of antelope evolution.

- Up to 6½ ft (2 m)
- Up to 4¼ ft (1.3 m)
- Up to 355 lb (160 kg)
- Herd
- Critically endangered

Kenya–Somalia border region

Topi and tsessebe These large grazing antelopes breed once a year, usually calving at the end of the dry season. In a large herd, the calves are likely to be followers, protected within a group of adults. In smaller herds, the calves may be hiders, concealed in dense vegetation while the adults forage.

- Up to 8½ ft (2.6 m)
- Up to 4 ft (1.2 m)
- Up to 310 lb (140 kg)
- Herd
- Conserv. dependent

Savanna zone of sub-Saharan Africa

Common hartebeest With its sloping back, this large animal can appear awkward, but it is capable of reaching speeds of 50 miles per hour (80 km/h). It lives on open plains, preferring edge habitats near forest. Hartebeest may join other antelopes and zebras to form large aggregations.

- Up to 6¼ ft (1.9 m)
- Up to 4¼ ft (1.3 m)
- Up to 330 lb (150 kg)
- Herd
- Conserv. dependent

Sahel, Serengeti, Namibia to Botswana

Horns of a male
gemsbok can be up
to 5 feet (1.5 m) long

Gemsbok
Oryx gazella

Black-tufted tail

Dark markings on legs,
flanks, and face

S-shaped, ridged
horns only on male

Impala
Aepyceros melampus

Can leap up to 10 feet
(3 m) off the ground

Wildebeest
Connochaetes taurinus

Scent glands
beneath patches
of black hair on
rear feet

Vertical stripes
of longer hair
on back

**White-tailed gnu
(black wildebeest)**
Connochaetes gnou

RUTTING SEASON

During the mating season, which usually occurs at the end of the wet season, male impalas go into rut and will strongly defend their territory with scentmarking, defensive postures, and physical contests. Territoriality breaks down during the dry season, when impalas extend and overlap their ranges.

A noisy display
The male impala's roaring display begins with a few explosive snorts, followed by several resonant deep grunts.

PRONKING

Gazelles sometimes leap repeatedly, keeping their back arched and legs stiff and landing on all fours—a habit known as "pronking" or "stotting." This behavior may occur when an animal is excited or alarmed and possibly distracts predators or warns them that they have been detected.

Leaping springbok
Named after its habit of pronking, the springbok erects the white crest on its back as it leaps up to 13 feet (4 m) in the air.

Gerenuk
Litocranius walleri

Beira antelope
Dorcatragus megalotis

Can stand on hindlimbs to browse on leaves that are out of reach of most antelopes

Chiru
(Tibetan antelope)
Pantholops hodgsonii

Springbok
Antidorcas marsupialis

Blackbuck
Antilope cervicapra

Saiga
Saiga tatarica

Large, fleshy nose filters dust from air in summer and warms air in winter

Dibatag
Ammodorcas clarkei

Oribi
Ourebia ourebi

Scent gland
beneath eye

Steenbok
Raphicerus campestris

Klipspringer
Oreotragus oreotragus

Short, spike-like horns
on males

Guenther's dik-dik
Madoqua guentheri

Waterbuck
Kobus ellipsiprymnus

Peg-like hoofs for
traveling over rocks

Bohor reedbuck
Redunca redunca

Reedbucks usually
live near water

Mountain reedbuck
Redunca fulvorufula

FACT FILE

Steenbok Exclusively a browser, this swift antelope prefers nutritious new growth such as young leaves, flowers, fruit, and shoots. It is the only bovid known to scrape the ground before and after urinating and defecating.

🐂 Up to 33½ in (85 cm)
🐃 Up to 19½ in (50 cm)
⚖ Up to 24½ lb (11 kg)
🐾 Solitary, pair
🏃 Common

E. & S. Africa

Klipspringer This sure-footed cliff dweller has a thick, moss-like coat that protects it from bumps and scrapes. It lives in small family groups in which one member acts as sentinel, giving a shrill whistle to warn of danger.

🐂 Up to 35½ in (90 cm)
🐃 Up to 23½ in (60 cm)
⚖ Up to 28½ lb (13 kg)
🐾 Pair, family band
🏃 Conserv. dependent

Mountain & rocky areas in E. & S. Africa

Guenther's dik-dik The long, mobile snout of this shy antelope is thought to help control the animal's temperature, with blood diverted to the snout being cooled before it travels to the brain.

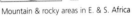

🐂 Up to 25½ in (65 cm)
🐃 Up to 15 in (38 cm)
⚖ Up to 12 lb (5.5 kg)
🐾 Pair
🏃 Common

N.E. Africa

Waterbuck Old, weak herd animals are usually easy targets for predators. As a waterbuck ages, however, the secretions released by its sweat glands gradually impart an unpleasant taste to the flesh, encouraging predators to look elsewhere for a meal.

🐂 Up to 8 ft (2.4 m)
🐃 Up to 4½ ft (1.4 m)
⚖ Up to 660 lb (300 kg)
🐾 Herd
🏃 Conserv. dependent

Savanna zone of sub-Saharan Africa

Mountain reedbuck This species can breed at any time, taking advantage of favorable conditions whenever they occur. Like many antelopes and deer, it has a white patch under the tail that is displayed as it runs away from danger.

🐂 Up to 4¼ ft (1.3 m)
🐃 Up to 28½ in (72 cm)
⚖ Up to 66 lb (30 kg)
🐾 Harem
🏃 Conserv. dependent

Mountains of C., E. & S. Africa

Cushioned fights
Sand gazelles sometimes have "air-cushion" fights, with the rivals charging at each other headfirst but stopping about 12 inches (30 cm) apart. If neither admits defeat after a series of such charges, physical fighting ensues.

Locking horns
Like other gazelles, sand gazelles fight by lowering the head, locking horns, and then twisting and pushing until one gives up and moves away.

MALE COMPETITION

During the breeding season, male gazelles will defend their territory and harem against challengers. Males mark their territory with secretions from the glands beneath each eye, as well as with urine and feces. Threat displays begin with the head raised so the horns lie along the back, and progress to the head tucked in with the horns vertical. Head lowered with horns pointed at the opponent is usually the last display before physical contact.

Time out
Gazelles may interrupt a fight for a spot of feigned grazing before resuming the contest.

Dangerous games
While bachelor gazelles will practice their fighting skills in harmless sparring bouts, territorial males sometimes inflict serious injury.

FACT FILE

Thomson's gazelle With 90 percent of its diet consisting of grass, this species is almost exclusively a grazer. Thousands of individuals congregate for annual migrations, moving to woodland in the dry season and to grassland in the wet. Mature males establish territories, which are crossed by small, loosely structured groups of foraging females and offspring.

- Up to 3½ ft (1.1 m)
- Up to 25½ in (65 cm)
- Up to 55 lb (25 kg)
- Herd
- Conserv. dependent

E. Africa

Grant's gazelle This large species has a unique courting ritual in which the male makes sputtering sounds as he follows the female with his head and tail raised. Highly adapted to its hot, dry environment, Grant's gazelle can stand on its hindlimbs to reach moisture-rich leaves.

- Up to 5 ft (1.5 m)
- Up to 37½ in (95 cm)
- Up to 175 lb (80 kg)
- Herd
- Conserv. dependent

E. Africa

Dama gazelle
Gazella dama

Thomson's gazelle
Gazella thomsonii

Male horns are thicker and longer than those of female

Grant's gazelle
Gazella granti

Spanish ibex
Capra pyrenaica

West Caucasian tur
Capra caucasica

Coat becomes
redder in summer

Long outer coat
protects warm,
dense underfur

Mountain goat
Oreamnos americanus

Flexible hoof
pads to grip
uneven ground

Chamois
Rupicapra rupicapra

Takin
Budorcas taxicolor

Serow
Capricornis sumatraensis

DEFENDERS AND GRAZERS

Sheep, goats, musk oxen, and their kin are collectively called goat-antelopes. They belong to the subfamily Caprinae, which first appeared in the tropics, but gradually moved into extreme locations such as deserts and mountains. Today's species range from resource defenders, such as serows, which live in productive habitats, to grazers, such as chamois, which tend to be gregarious, wide-ranging, and adapted to harsh conditions.

Alpine life *Living above the treeline in Europe and Western Asia, chamois announce danger with foot stamping and whistles and flee across uneven ground in great leaps until they reach an inaccessible spot.*

FACT FILE

Himalayan tahr During the summer rutting season, competing males walk with their mane erect and head down to display their horns, with the stronger blocking the other's path or chasing it away. The contest rarely escalates to head wrestling.

- 🐾 Up to 4½ ft (1.4 m)
- 📏 Up to 3¼ ft (1 m)
- ⚖ Up to 220 lb (100 kg)
- 🐾 Herd
- ⚡ Vulnerable

Himalayas

Nilgiri tahr Vast herds of this species once roamed the grass-covered hills of southern India, but hunting and habitat loss reduced its numbers to only about 100 animals. As a result of conservation efforts, the total population has now risen to about 1,000 animals. The Nilgiri tahr has a coarse coat, with a short, bristly mane.

- 🐾 Up to 4½ ft (1.4 m)
- 📏 Up to 3¼ ft (1 m)
- ⚖ Up to 220 lb (100 kg)
- 🐾 Herd
- ⚡ Endangered

Nilgiri Mts (S. India)

Markhor This animal has suffered from excessive hunting and now survives only in fragmented populations in isolated, rugged terrain above the treeline. Its spiraling horns are prized both by trophy hunters and in Chinese medicine. Competition with domestic goats for food is also taking its toll.

- 🐾 Up to 6 ft (1.8 m)
- 📏 Up to 3½ ft (1.1 m)
- ⚖ Up to 245 lb (110 kg)
- 🐾 Herd
- ⚡ Endangered

Turkmenistan to Pakistan

Saola Until the 1990s, no new large mammal species had been scientifically described for decades. Then in 1992, the saola was discovered in Vietnam, a country where a protracted war and limited international contact had impeded study of its fauna. Considered one of the world's rarest mammals, the saola is a nocturnal, forest-dwelling ox found only in remote mountainous regions. It travels in small groups of a few animals and appears to be a browser, feeding on fig leaves and other rain-forest vegetation.

- 🐾 Up to 6½ ft (2 m)
- 📏 Up to 35½ in (90 cm)
- ⚖ Up to 220 lb (100 kg)
- 🐾 Solitary, family band
- ⚡ Endangered

Laos, Vietnam

Male has thick mane of fur around neck and shoulders

Nilgiri tahr
Hemitragus hylocrius

Himalayan tahr
Hemitragus jemlahicus

Arabian tahr
Hemitragus jayakari

Inside edge of horns is sharp

Horns can be 5 feet (1.5 m) long

Markhor
Capra falconeri

Bezoar (wild goat)
Capra aegagrus

Saola (Vu Quang ox)
Pseudoryx nghetinhensis

Female horns have the same shape as male horns but are smaller

Aoudad (Barbary sheep)
Ammotragus lervia

Male's horn size determines rank

Bighorn sheep
Ovis canadensis

Ventral mane of long white hair

Siberian bighorn sheep
Ovis nivicola

Dall sheep
Ovis dalli

Male horns almost meet at top of head in a "boss;" female horns are smaller with no boss

Musk ox
Ovibos moschatus

Long guard hairs can almost reach the ground

FACT FILE

Aoudad Originating in the hills of the Sahara, the aoudad was introduced to Europe in the 1800s and to the southwest United States in the 1950s. Before mating takes place, the female aoudad licks the flanks of the male, and the pair may touch muzzles.

- Up to 5½ ft (1.7 m)
- Up to 3½ ft (1.1 m)
- Up to 320 lb (145 kg)
- Herd
- Vulnerable

N. Africa; introd. Europe & USA

Bighorn sheep To win the right to mate, male bighorn sheep will fight head to head, with contests sometimes lasting more than 24 hours. As well as their massive horns, which can weigh up to 30 pounds (14 kg), these sheep have a reinforced skull that is linked to the spine by a thick tendon.

- Up to 6 ft (1.8 m)
- Up to 4 ft (1.2 m)
- Up to 300 lb (135 kg)
- Herd
- Conserv. dependent

W. North America

ARCTIC SURVIVORS

Musk oxen must survive long winters with subzero temperatures and little light. Their long guard hairs shield a dense underfur that is shed in spring. To reach grasses and sedges under the snow, musk oxen clear circular feeding patches with their horns and feet. A herd will crowd around a young musk ox to protect it from a predator.

- Up to 7½ ft (2.3 m)
- Up to 5 ft (1.5 m)
- Up to 905 lb (410 kg)
- Herd
- Uncommon

Arctic Canada & Alaska; Greenland

FACT FILE

Anoa Usually solitary, this species lives in lowland forests and wetlands. It feeds on the plants of the understory.

	Up to 5½ ft (1.7 m)
	Up to 3¼ ft (1 m)
	Up to 660 lb (300 kg)
	Family band
	Endangered

Sulawesi

Bison North America was once home to about 60 million bison, but the species now survives in the wild only in two national parks. Bison live in groups of females, offspring, and a few older males. Other mature males may live alone or form bachelor bands.

	Up to 11½ ft (3.5 m)
	Up to 6½ ft (2 m)
	Up to 1 ton (1 t)
	Herd
	Conserv. dependent

Canada, N.W. USA

Gaur Herds of females and juveniles led by a single male emerge from the forest early in the day to graze on nearby grassy slopes. They return to the forest at night to sleep.

	Up to 11 ft (3.3 m)
	Up to 7¼ ft (2.2 m)
	Up to 1 ton (1 t)
	Herd
	Vulnerable

India to Indochina & Malaya

Yak While domesticated yaks are found throughout much of Asia, wild yaks are restricted to uninhabited alpine tundra and cold steppe regions.

	Up to 11 ft (3.3 m)
	Up to 6½ ft (2 m)
	Up to 1 ton (1 t)
	Herd
	Vulnerable

Tibet

⚡ CONSERVATION WATCH

Wild cattle and buffalo Hunting, habitat loss, and domestic livestock (which interbreed with wild species, pass on diseases, and compete for food) have had a devastating effect on wild cattle and buffalo species. Today, most depend on protected reserves for their survival and a few, including the kouprey and tamaraw, are near extinction. Conservation programs have had some success, notably with the wisent (European bison), which was declared extinct in the wild in 1919, but has since been reintroduced from zoo stocks.

Mountain anoa
Bubalus quarlesi

Anoa
Bubalus depressicornis

Horns can be held flat along back to avoid becoming entangled in forest undergrowth

Tamaraw
Bubalus mindorensis

Hump of raised muscle

Bison
Bison bison

Hair longer at front than at rear of body

Large lungs and high red blood cell count allow yaks to flourish at high altitudes

Yak
Bos grunniens

Gaur (seladang)
Bos frontalis

Wild males are three times heavier than wild females, and two to three times heavier than domestic males

Kouprey
Bos sauveli

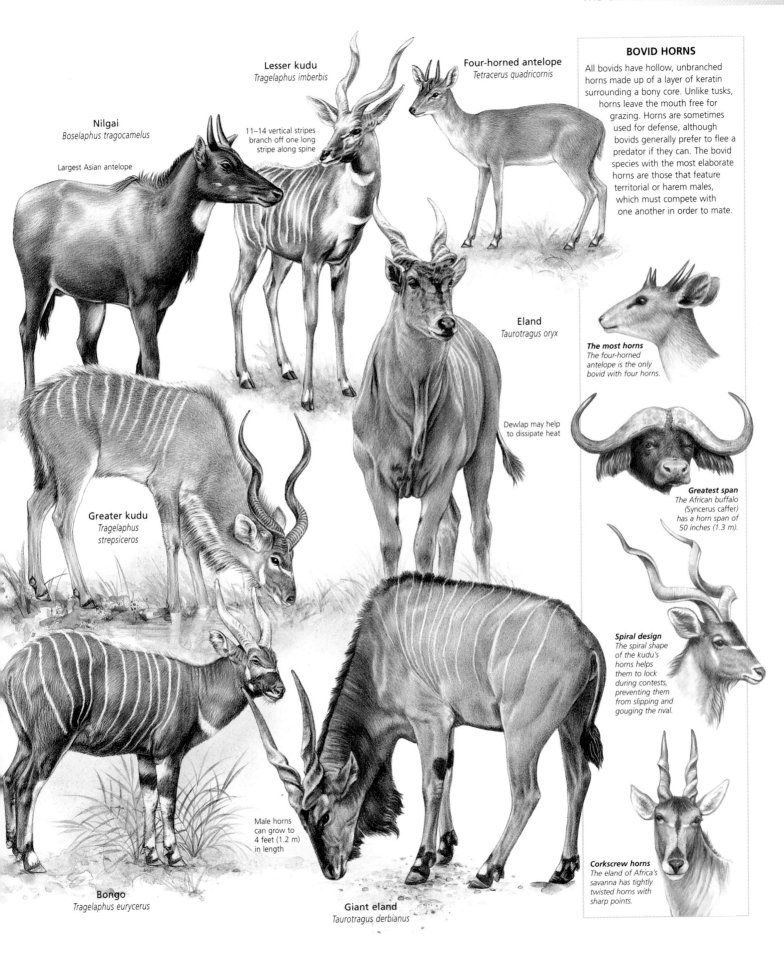

Nilgai
Boselaphus tragocamelus

Largest Asian antelope

Lesser kudu
Tragelaphus imberbis

11–14 vertical stripes
branch off one long
stripe along spine

Four-horned antelope
Tetracerus quadricornis

Eland
Taurotragus oryx

Dewlap may help
to dissipate heat

Greater kudu
*Tragelaphus
strepsiceros*

Male horns
can grow to
4 feet (1.2 m)
in length

Bongo
Tragelaphus eurycerus

Giant eland
Taurotragus derbianus

BOVID HORNS

All bovids have hollow, unbranched
horns made up of a layer of keratin
surrounding a bony core. Unlike tusks,
horns leave the mouth free for
grazing. Horns are sometimes
used for defense, although
bovids generally prefer to flee a
predator if they can. The bovid
species with the most elaborate
horns are those that feature
territorial or harem males,
which must compete with
one another in order to mate.

The most horns
*The four-horned
antelope is the only
bovid with four horns.*

Greatest span
*The African buffalo
(Syncerus caffer)
has a horn span of
50 inches (1.3 m).*

Spiral design
*The spiral shape
of the kudu's
horns helps
them to lock
during contests,
preventing them
from slipping and
gouging the rival.*

Corkscrew horns
*The eland of Africa's
savanna has tightly
twisted horns with
sharp points.*

INCREDIBLE JOURNEYS

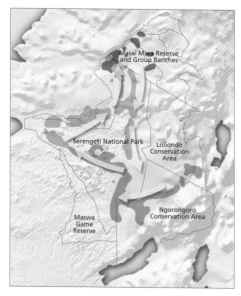

MOVING CLOCKWISE

Early in the year, thousands of wildebeest give birth and then spread out on the plains of the Serengeti, which offer rain-ripened, mineral-rich short grasses. By the end of May, as the wet season comes to an end, these plains are depleted and the animals travel west and north in small groups to a transitional zone, where they mate. During the rutting season of May to June, this region resounds with deep lowing as each dominant male defends a harem against the attentions of other males. By July, many thousands of animals have formed a single great herd that makes its way to the Masai Mara, where they feast on new grass growth and drink from permanent rivers through the dry season. They begin the return journey to their breeding grounds in late November.

● Wet season
● Transitional zone
● Dry season

Among the most astounding sights in nature are the seasonal migrations of large, hoofed grazers such as caribou in Canada and Alaska, gazelles in Mongolia, kob in southern Sudan, and wildebeest, zebras, and gazelles in East Africa. Migrating herds number in the thousands, and are prompted by changes in the weather to begin their mass movement. In colder climates, caribou and gazelles travel north to summer ranges, and south to winter ranges. In Africa, herd movement relates to the succession of wet and dry seasons. The greatest of all today's migrations is that of roughly 1.3 million wildebeest, accompanied by about 200,000 zebras and gazelles, from the Serengeti in Tanzania to the Masai Mara in Kenya—a clockwise route of more than 1,800 miles (2,900 km) per year. Some incredible journeys no longer take place, as ungulate numbers have dwindled and the land crossed by their migration routes has been developed. Hundreds of thousands of springbok once migrated over a vast range in southern Africa. In North America, as many as 4 million bison once traveled en masse through the Great Plains to reach the fresh grasses of their northern summer and southern winter ranges.

River crossing The most dangerous part of the great wildebeest migration is the crossing of the Mara River, which bisects the Masai Mara savanna. Swollen by the recent rains, the river can be a raging torrent. The migrating herd gathers at the banks until forced by the surge of animals to cross. Many wildebeest break their legs after leaping onto the rocky bottom. Others drown or are swept away by the powerful currents. Enormous crocodiles await, relying on the wildebeest prey for their main meal of the year. The animals that survive the crossing must then dodge the hungry lions on the other side.

Swimming caribou Each spring in Alaska and Canada, herds containing thousands of caribou migrate from their wintering grounds to their calving grounds farther north, traveling through deep snow and across icy rivers. After giving birth, the mothers feed on the tundra's nutritious new plant growth, which helps them produce rich milk. The calving grounds become cold and windswept as winter approaches, prompting the caribou to return south. In Europe, the same species has been domesticated and is known as the reindeer. The Sami herders follow the animals' natural migration routes.

Winter herd The world's largest remaining expanses of temperate grassland are found in eastern Mongolia. These steppes are home to the Mongolian gazelle (*Procapra gutturosa*). During summer, male and female gazelles form separate herds, and females give birth. Large mixed herds of several thousand animals gather for the winter migration, covering up to 180 miles (300 km) a day until they reach their southern mating grounds. Overall numbers of the species appear to be in decline. The impact of fires, epidemics, and other natural disasters has been exacerbated by such disruptions to the gazelles' migration routes as fencing along the border with China and a new railroad in Mongolia.

Annual cycle *The wildebeest mating season occurs in May to June. By the time of the perilous river crossing that occurs around July, more than 90 percent of all the adult female wildebeest may be pregnant.*

Fellow travelers *Zebras travel with or ahead of the great wildebeest herd, grazing on the tough longer grasses and exposing the sweet new growth for the wildebeest. Gazelles tend to follow the wildebeest herd.*

Survival of the fittest *Only the strongest members of the wildebeest herd make it from one side of the river to the other. They are rewarded by an abundance of lush green grass and spend their time building up reserves for the slow journey back to the Serengeti plains.*

Perilous journey About 400,000 wildebeest are born every year in the Serengeti during a 6-week period beginning in late January. Of these calves, two out of three will perish on their first migration to the Masai Mara, but enough will survive to replenish East Africa's great wildebeest herds.

Plenty for predators *The presence of the great herd in the Masai Mara attracts large numbers of predators. Crocodiles feast during the river crossing, while lions and hyenas target any stragglers.*

DEER

CLASS	Mammalia
ORDER	Artiodactyla
FAMILIES	4
GENERA	21
SPECIES	51

The largest family in the deer group, Cervidae contains deer and their allies, including moose, caribou, and elk. In many ways, deer resemble antelopes, with long bodies and necks, slender legs, short tails, large eyes on the side of the head, and high-set ears. They are distinguished, however, by the often spectacular antlers borne by the males of most species (and also by female caribou). Unlike horns, which are permanent and made of keratin, antlers are made of bone and are shed once a year. Growing antlers are covered in skin known as "velvet," which dies and is rubbed off once the antlers reach full size. Antlers can be small, simple spikes or enormous, branched structures.

Deer distribution Deer never moved into sub-Saharan Africa, but they occur naturally in northwest Africa, Eurasia, and the Americas, and some have been introduced elsewhere. The species in the family Cervidae fall into two groups according to their ancestry: Old World deer first evolved in Asia, while the New World group began in the Arctic.

Lying up Until they are strong enough to join the herd, newborn fawns such as this mule deer spend their days hiding among vegetation. The newborn of many deer species have a dappled coat to break up their outline and provide camouflage. The mother visits regularly to suckle the fawn.

Northern habitats Most deer are found in temperate or tropical forest, but some species have adapted to more extreme conditions. The moose (right) dwells in northern wetlands, where its summer diet includes the roots of aquatic plants. The caribou lives on the treeless arctic tundra.

LARGE AND SMALL

The family Cervidae ranges from the southern pudu, weighing as little as 17½ pounds (8 kg), to the moose, at 1,760 pounds (800 kg). A moose's antlers can have a span of 6½ feet (2 m), though even these are dwarfed by the antlers of the extinct Irish elk (*Megaloceros*), which spanned 12 feet (3.5 m). One cervid species, the Chinese water deer, has no antlers at all, but its elongated canine teeth form knife-like tusks. Southeast Asia's muntjacs have only spikes for antlers but also bear tusks.

As prey species, deer have evolved various escape strategies. Some leap and dodge into a hiding spot. Others rely on great speed and stamina to outrun the threat. The moose can easily trot over obstacles that slow down its shorter predators.

All deer are ruminants with a four-chambered stomach, but, unlike bovids, they are not adapted to a diet of coarse grasses and rely on more easily digested food such as shoots, young leaves, new grasses, lichens, and fruit. Even those species that do graze on grasses need large amounts of high-quality browse as well.

The deer family is comprised of Cervidae and three other families of ungulates that superficially resemble them. These include the chevrotains of Tragulidae and the musk deer of Moschidae, which both have long canine teeth rather than antlers. North America's pronghorn is in its own family, Antilocapridae.

Group grazers While some smaller deer species live alone or in small family groups, larger species such as fallow deer tend to form herds. Living in a group offers a better chance of avoiding predation, as predators are more likely to be detected and tend to target only the weakest members. In New Zealand and other places, deer have been introduced and are bred commercially.

Siberian musk deer
Moschus moschiferus

Muscular hindlegs allow animal to perform agile jumps

Like other members of the family Moschidae, this species lacks antlers but males have long canine teeth

Extensively hunted for musk produced by gland between navel and sex organs

Chinese water deer
Hydropotes inermis

Only species of Cervidae in which males lack antlers

Indian spotted chevrotain (Indian mouse deer)
Moschiola meminna

Water chevrotain
Hyemoschus aquaticus

Greater Malay chevrotain
Tragulus napu

Lesser Malay chevrotain
Tragulus javanicus

Spotted and striped coat helps camouflage animal amid forest foliage

FACT FILE

Sambar Native to Asia but introduced to Australia, New Zealand, and the United States, this nocturnal deer is most often found on forested hillsides. While several females with young may live together, males are usually solitary and aggressively defend their territories during the breeding season, when they mate with the females in their range.

🐂 Up to 8 ft (2.5 m)
📏 Up to 5¼ ft (1.6 m)
⚖ Up to 575 lb (260 kg)
🔀 Solitary, harem
🏃 Locally common
🏛 ❋ ⚘

India & Sri Lanka to S. China & S.E. Asia

IN A RUT

Known as wapiti or elk in North America and as red deer in Europe, *Cervus elaphus* is the noisiest of all deer species. After beginning courtship with a bugling call, males collect a harem, which they vigorously defend against other males throughout the rutting, or mating, season.

Vocal assault A harem male and his challenger will roar at each other for several minutes before engaging in a physical contest.

Antler lock After performing a ritual parallel walk, competing males will lock antlers and wrestle until one is pushed back and flees.

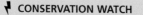

⚡ CONSERVATION WATCH

Of the 51 species in the four "deer" families, 76% are listed on the IUCN Red List, as follows:

  1 Extinct
  1 Critically endangered
  7 Endangered
 11 Vulnerable
  7 Near threatened
 12 Data deficient

Antlers can be up to 3 feet (1 m) long

When tail is raised, the white underside acts as a "follow me" signal

Sambar
Cervus unicolor

Barasingha
Cervus duvaucelii

Eld's deer
Cervus eldii

Rusa
Cervus timorensis

Fawn is spotted for camouflage

Roosevelt elk
Cervus elaphus roosevelti

Muscular hindlegs allow animal to perform agile jumps

Siberian musk deer
Moschus moschiferus

Like other members of the family Moschidae, this species lacks antlers but males have long canine teeth

Extensively hunted for musk produced by gland between navel and sex organs

Chinese water deer
Hydropotes inermis

Only species of Cervidae in which males lack antlers

**Indian spotted chevrotain
(Indian mouse deer)**
Moschiola meminna

Water chevrotain
Hyemoschus aquaticus

Greater Malay chevrotain
Tragulus napu

Lesser Malay chevrotain
Tragulus javanicus

Spotted and striped coat helps camouflage animal amid forest foliage

CHEVROTAINS

Also known as mouse deer, chevrotains are small, even-toed ruminants, but unlike true deer and cattle they lack horns or antlers. The males have long, continually growing canines. Shy and usually nocturnal, chevrotains tend to be solitary forest dwellers. They are classified in the family Tragulidae.

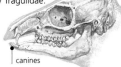

canines

Antlers can be up to
3 feet (1 m) long

When tail is raised, the
white underside acts as
a "follow me" signal

Sambar
Cervus unicolor

Barasingha
Cervus duvaucelii

Eld's deer
Cervus eldii

Rusa
Cervus timorensis

Fawn is spotted
for camouflage

Roosevelt elk
Cervus elaphus roosevelti

IN A RUT

Known as wapiti or elk in North America and as red deer in Europe, *Cervus elaphus* is the noisiest of all deer species. After beginning courtship with a bugling call, males collect a harem, which they vigorously defend against other males throughout the rutting, or mating, season.

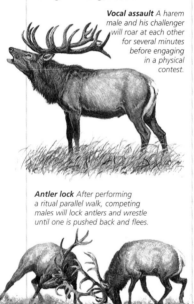

Vocal assault *A harem male and his challenger will roar at each other for several minutes before engaging in a physical contest.*

Antler lock *After performing a ritual parallel walk, competing males will lock antlers and wrestle until one is pushed back and flees.*

Critically endangered Chinese deer; disappeared from the wild around AD 200, but survived because a captive herd was maintained by the Chinese royal family and breeding pairs were reared in Europe; reintroduced into two Chinese national parks in the 1980s

Mesopotamian fallow deer
Dama mesopotamica

Père David's deer
Elaphurus davidianus

Palm-shaped antlers have numerous points

Chital
Axis axis

Fallow deer
Dama dama

Philippine hog deer
Axis calamianensis

Tufted deer
Elaphodus cephalophus

Tuft of hair hides male's antlers

Giant muntjak
Megamuntiacus vuquangensis

Tusk-like canine teeth on male

Indian muntjak
Muntiacus muntjak

FACT FILE

Chital This deer is mainly a grassland grazer, but will enter nearby forests to browse fallen fruit and leaves. Herds of females and young are followed by dominant males in the breeding season.

🐾 Up to 6 ft (1.8 m)
📏 Up to 3¼ ft (1 m)
⚖ Up to 245 lb (110 kg)
👣 Herd
🏃 Common

Sri Lanka, India, Nepal

Fallow deer Most populations have been translocated from their natural range. Adaptable foragers, fallow deer have been introduced to diverse habitats, from the tropics to mountains.

🐾 Up to 6 ft (1.8 m)
📏 Up to 3½ ft (1.1 m)
⚖ Up to 220 lb (100 kg)
👣 Herd
🏃 Locally common

Originally Mediterranean to S.W. Asia

Tufted deer The male tufted deer has simple, spiked antlers and long, tusk-like upper canines. The antlers are often hidden by the hair on its forehead.

🐾 Up to 5¼ ft (1.6 m)
📏 Up to 27½ in (70 cm)
⚖ Up to 110 lb (50 kg)
🏃 Solitary
🏃 Data deficient

E. Tibet & N. Myanmar to S.E. China

Giant muntjak First recorded in 1994, this secretive creature is about the size of a large dog, almost twice as big as the Indian muntjak.

🐾 Up to 3¼ ft (1 m)
📏 Up to 27½ in (70 cm)
⚖ Up to 110 lb (50 kg)
🏃 Solitary
🏃 Not known

Vietnam

Indian muntjak Also called the barking deer, this species may bark for more than an hour if it senses a predator. It is omnivorous, using kicks of its forelimbs and bites to subdue small prey.

🐾 Up to 3½ ft (1.1 m)
📏 Up to 25½ in (65 cm)
⚖ Up to 61½ lb (28 kg)
🏃 Solitary
🏃 Uncommon

Sri Lanka, India, Nepal to S. China, S.E. Asia

ANTLER CYCLE

Antlers are used in contests between male deer, but the reason they grow so large appears to be that they advertise the male's healthy genes to females. In deer species with the largest antlers, much of the courtship routine involves elaborate antler displays.

Spring
In temperate species, new antlers begin growing in late spring. They are covered in sensitive skin known as "velvet."

Summer
By late summer, the antlers are fully grown and have hardened. The velvet begins to dry and loosen.

Fall
The male rubs the velvet off on shrubs and small trees. The antlers are now ready for the contests and displays of the mating season.

Winter *Following the mating season, the two antlers are shed within days of each other.*

Male's massive antlers may have up to 20 points

Largest of all deer species; known as moose in North America and elk in Europe

Moose (elk)
Alces alces

Marsh deer
Blastocerus dichotomus

Female caribou is only female deer to possess true antlers

Male caribou has larger antlers than female

Clicking sound made during walking when tendons snap across bones in the feet

Caribou (reindeer)
Rangifer tarandus

Broad, flat feet can cross both snow and spongy tundra vegetation

Pampas deer
Ozotoceros bezoarticus

Mule deer (black-tailed deer)
Odocoileus hemionus

White-tailed deer
Odocoileus virginianus

Roe deer
Capreolus capreolus

Red brocket
Mazama americana

Both male and female bear horns, but the male's horns are longer and forked

Black markings only on male

Pronghorn
Antilocapra americana

Little red brocket
Mazama rufina

Peruvian guemal
Hippocamelus antisensis

Appears on Chile's coat of arms

Andean guemal
Hippocamelus bisulcus

Northern pudu
Pudu mephistophiles

Southern pudu
Pudu puda

Smallest of all deer in Cervidae family

FACT FILE

Red brocket An elusive creature that most often lives in dense, tropical forest, this small deer can disappear into undergrowth or swim to escape predators. It lives alone or in pairs and feeds on fruit, leaves, and fungi.

- Up to 5 ft (1.5 m)
- Up to 31½ in (80 cm)
- Up to 106 lb (48 kg)
- Solitary
- Data deficient

S. Mexico to N. Argentina

Southern pudu The smallest of all true deer species, the southern pudu will stand on its hindlegs to reach the leaves of trees or test the wind. Solitary except during the mating season, this territorial animal follows well-marked trails to its feeding and resting spots. Listed as vulnerable, this species suffers predation from domestic dogs, and competition for food from introduced species such as fallow deer.

- Up to 32½ in (83 cm)
- Up to 17 in (43 cm)
- Up to 28½ lb (13 kg)
- Solitary
- Vulnerable

S. Chile, S.W. Argentina

THE PRONGHORN

The sole species classified in the family Antilocapridae, the pronghorn is able to reach speeds of 40 miles per hour (65 km/h), making it one of the fastest mammals on land. It is distinguished by its unusual forked horns, which are made from keratin surrounding a bony core, like the horns of antelope. As with the antlers of deer, however, the keratin is shed annually.

- Up to 5 ft (1.5 m)
- Up to 3¼ ft (1 m)
- Up to 154 lb (70 kg)
- Herd
- Locally common

W. North America

Unique horns The head of a pronghorn buck features pronged horns, protruding eyes with long eyelashes, and a "mask" of black hair.

GIRAFFE AND OKAPI

CLASS	Mammalia
ORDER	Artiodactyla
FAMILY	Giraffidae
GENERA	2
SPECIES	2

With its head hovering up to 18 feet (5.5 m) above the ground, the giraffe is the tallest animal in the world. Along with its only close relative, the okapi, it is classified in the family Giraffidae. Both the giraffe and the okapi have a long neck, tail, and legs, with the forelimbs longer than the hindlimbs, creating a sloping back. Their small, constantly growing horns consist of bone covered by furred skin and are unique among mammals. The lips of giraffids are thin and mobile; the tongue is long, prehensile, and black; and the eyes and ears are large. Both species are found only in sub-Saharan Africa, where their strikingly patterned coats help them blend into their habitat—the giraffe's blotches mimic the dappled light of savanna woodland, while the okapi's rear stripes break up its outline amid the dense vegetation of the rain forest.

Dangerous drinking A giraffe gets most of the water it needs from its food, but will drink water when it is available. It must splay its forelegs to reach the water, making it vulnerable to predators at this time.

Necking rivals To establish rank, young male giraffes engage in ritualized necking contests. Much like arm wrestles, these involve two giraffes intertwining necks and pushing each other until one gives way. Older male giraffes head-butt each other in an attempt to push over the opponent. Competing okapi bulls also neck-wrestle before progressing to more aggressive contact.

STRIPED BEHIND

The okapi was first described only in 1901, when a British explorer went in search of a horse-like animal hunted by the local people. At first glance, the okapi looks more like a zebra than a giraffe. It shares distinctive features with the giraffe, however, including unusual fur-covered horns, specialized teeth and tongue, and a ruminating, four-chambered stomach. The stripes on the rump probably act as a "follow me" signal and allow a young okapi to keep track of its mother. With the stripes on the forelimbs, they also break up the animal's outline amid the dense vegetation of the forest.

DIFFERENT LIFESTYLES

Beyond their similarities, giraffes and okapi differ significantly. The most obvious contrast is in size and shape, with the okapi appearing rather horse-like, while the extreme elongation of the giraffe makes it instantly identifiable. The giraffe has only seven vertebrae in its neck, the same number as almost all other mammals, but each vertebra is lengthened. A specialized circulatory system powerfully pumps blood all the way up to its brain, but a series of valves adjusts the pressure when the animal leans down to drink.

The giraffe's extraordinary stature has allowed it to fully exploit the resources of its savanna woodland home. Because it is able to reach the leaves of tall acacia trees throughout the dry season, a giraffe can grow to a dramatic height and reproduce year-round. It is most vulnerable to lions and other predators when lying down or drinking. To avoid predation, a giraffe depends on its acute senses of vision, smell, and hearing. It may run away at speeds of more than 30 miles per hour (50 km/h) or deliver sharp kicks to the foe with its forefeet.

Living in dense, dark tropical forest, the okapi has poor vision but sharp hearing and a good sense of smell. It is extremely wary and will disappear into thick cover at the first hint of danger. Mostly solitary, this species marks its territory with urine or by rubbing its neck on trees.

The more open savanna habitat has encouraged giraffes to be social, and most live in small, loose herds of about a dozen animals. Young males may live in bachelor bands but tend to become solitary as they age. Males may fight each other for the right to mate, repeatedly swinging their long neck to deliver powerful head-butts to the rival's underbelly. A reinforced skull usually absorbs the impact of these blows, but occasionally an animal is knocked unconscious.

Horns on both males and females

Kenyan giraffe
Giraffa camelopardalis tippelskirschi

Southern giraffe
Giraffa camelopardalis giraffa

Short mane along neck

Long, tufted tail used to whisk away flies

Forelimbs longer than hindlimbs

Reticulated giraffe
Giraffa camelopardalis reticulata

Only males have horns

Okapi
Okapia johnstoni

SPECIALIZED BROWSERS

The giraffe and the okapi are almost exclusively browsers. They both have thin, muscular lips and a long, black tongue that is dextrous enough to pluck foliage or pull branches to the mouth. The tongue of the giraffe (right) is especially long and can measure up to 18 inches (46 cm). Both species strip leaves from branches using their unique lobed canine teeth, and then grind them with their molars. A four-chambered, ruminating stomach helps them gain maximum nutrition from their food, which is regurgitated and chewed a second time. Unlike other ruminants, giraffes can walk as they chew the cud, allowing them to spend more time feeding. A giraffe may spend between 12 and 20 hours a day feeding, consuming up to 75 pounds (34 kg) of vegetation in that time.

Giraffes can cope with a wide variety in the quality of their food. They prefer new growth, flowers, and fruit, but are able to switch to twigs and dried leaves. Their staple diet is acacia trees, which have chemical defenses that make the leaves toxic and unpalatable. In response, giraffes carefully select the least toxic foliage to eat and have thick, sticky saliva and a specialized liver function. Both the giraffe and the okapi appear to supplement their diet with minerals from other sources: the giraffe eats soil and chews bones discarded by scavengers, while the okapi licks the clay of riverbanks and eats charcoal from burned trees.

THE CAMEL FAMILY

CLASS	Mammalia
ORDER	Artiodactyla
FAMILY	Camelidae
GENERA	3
SPECIES	6

Famed for their humps and the ability to survive for long periods without drinking, the two species of camel are the single-humped dromedary, now found only in domesticated populations in northern Africa and the Middle East, and the Bactrian camel, domesticated in northern Asia, but also found in small numbers in the wild. Their relatives in the family Camelidae are the four camelids of South America—the wild guanaco and vicuña, and the domesticated llama and alpaca. Camelids first appeared some 45 million years ago in North America, but disappeared from there about 10,000 years ago at the end of the ice age. By then, they had dispersed to other parts of the world.

Old and new The two Old World camels occur in northern Africa and central Asia. The four South American species range from the foothills to the alpine meadows of the Andes Mountains. Domesticated camelids have been introduced in many places, including Australia, where feral dromedaries roam the central desert.

Precocious young In all camelid species, a single, well-developed young is born after a long gestation period, which lasts 11 months in South America's guanaco. A newborn guanaco can follow the mother within about 30 minutes of its birth.

ROBUST CAMELIDS

All camelids are adapted to arid or semiarid regions. A complex, three-chambered ruminating stomach extracts maximum nutrition from their main food of grasses. Their feet are unique among hoofed mammals in that only the front of the hoof touches the ground, and the animal's weight rests instead on a fleshy sole-pad. In camels, the feet are broad, helping it to travel over soft sand without sinking. The four South American species have a narrower foot for walking securely up rocky slopes. A thick double coat insulates against both heat and cold.

While Old World camelids are distinguished from the New World species by their much larger size and prominent humps, the overall anatomy is similar. All species have long, slender legs, a short tail, a long, curved neck, and a relatively small head with a split upper lip. When camelids walk, the front and back legs on the same side of the body move in unison, a distinctive gait known as pacing. Camelids are social animals and tend to live in harems of females and young led by a dominant male. Males without a harem may form bachelor bands.

By herding camelids, which provide meat, milk, wool, fuel, and transport, humans have been able to make a living in extreme locations, from the hot Sahara Desert to the cool high plains of the Andes. There are more than 20 million camelids in the world, but roughly 95 percent are domestic animals.

Water wise Dromedaries were introduced to northern Africa from Arabia several thousand years ago as domestic animals. Their hump stores fat, not water, but they can survive for months without drinking by feeding on desert plants. When water is available, however, they may gulp down the equivalent of a quarter of their weight.

DOMESTIC AND WILD

Until recently, both the domestic llama (right) and the domestic alpaca were assumed to be descendants of the wild guanaco. Molecular studies suggest, however, that the alpaca may be a cross between the llama and the wild vicuña. Both llamas and alpacas have been herded for many hundreds of years and there are now no wild individuals of these species. They vastly outnumber the wild guanaco and vicuña, but all South American camelids have been eclipsed by the introduced domestic sheep. Camelids were first domesticated in South America 4,000–5,000 years ago, and the llama was central to the success of the Inca Empire. Llamas and alpacas have now been introduced elsewhere, and are used for wool, to guard sheep, as pack animals for hikers, and as pets.

Bactrian camel
Camelus bactrianus

Fat stored in humps is used when food is scarce, causing humps to shrink

Long winter coat shed in summer

Narrow nostrils can close during dust storms

Long eyelashes keep desert dust out of eyes

Thick, tough lips can handle thorny vegetation

Dromedary
Camelus dromedarius

Guanaco
Lama guanicoe

Vicuña
Vicugna vicugna

FACT FILE

Bactrian camel Along with the dromedary, this species is unique among mammals because its blood cells are oval rather than round. This shape may help the cells travel through thick, dehydrated blood.

- Up to 11½ ft (3.5 m)
- Up to 7½ ft (2.3 m)
- Up to 1,540 lb (700 kg)
- Herd
- Critically endangered

Kazakhstan to Mongolia

Dromedary Although domesticated, many dromedary herds are unattended during the mating season, when they revert to living in harems.

- Up to 11½ ft (3.5 m)
- Up to 7½ ft (2.3 m)
- Up to 1,430 lb (650 kg)
- Herd
- Extinct in wild: domestic & feral only

N. Africa to India; introd. Australia

Guanaco Male guanacos, like all male South American camelids, have some sharp, hooked teeth that are used as weapons during fights with rival males.

- Up to 6½ ft (2 m)
- Up to 4 ft (1.2 m)
- Up to 265 lb (120 kg)
- Family band
- Locally common

S. Peru to E. Argentina & Tierra del Fuego

Vicuña Strictly a grazer, this small camelid has sharp, constantly growing incisors for snipping short grasses.

- Up to 6¼ ft (1.9 m)
- Up to 3½ ft (1.1 m)
- Up to 143 lb (65 kg)
- Family band
- Conserv. dependent

S. Peru to N.W. Argentina

CONSERVATION WATCH

Camelid threats Of the six species of camelids, two are listed on the IUCN Red List: the Bactrian camel, with about a thousand wild individuals left, is critically endangered, and the vicuña is conservation dependent. The dromedary has been extinct in the wild for many years but is prevalent as a domestic and feral species.

PIGS

CLASS	Mammalia
ORDER	Artiodactyla
FAMILY	Suidae
GENERA	5
SPECIES	14

Unlike most other ungulates, which are strictly herbivorous, the pigs, hogs, boars, and babirusa in the family Suidae are omnivores with a diet that includes insect larvae, earthworms, and small vertebrates, as well as a wide array of plants. The nostrils on a pig's prominent snout are enclosed in a disk of cartilage. Supported by a unique prenasal bone, this disk helps locate food by shoveling through leaf litter or dirt. The upper and lower canines in both males and females of most species form sharp tusks, which can be used as weapons. Occurring naturally in the forests of Africa and Eurasia, wild pigs have also been introduced in North America, Australia, and New Zealand.

Status symbols Thought to display status, the curved tusks of the male babirusa are elongated canine teeth. The upper tusks grow through the skin of the face.

Family ties Male wild pigs tend either to live alone or to belong to a bachelor band, while females and their offspring live in close-knit family groups that are known as sounders. These young warthogs are following their mother as she grazes.

> ⚡ **CONSERVATION WATCH**
>
> Feral pigs now threaten native fauna, including other pig species, in many places. Habitat loss has also contributed to the decline of some pigs. Of the 14 species in the family Suidae, 43% are listed on the IUCN Red List, as follows:
>
> 2 Critically endangered
> 1 Endangered
> 2 Vulnerable
> 1 Data deficient

Wild boar
Sus scrofa

Female has smaller tusks than male

Weighs 13–20 pounds (6–9 kg), making it the smallest species in Suidae

⚡ **Pygmy hog**
Sus salvanius

Piglets striped for camouflage; stripes fade with age

Warthog
Phacochoerus africanus

Padded knees allow for
kneeling during feeding

Bush pig
Potamochoerus larvatus

Upper tusks can grow to
14 inches (35 cm) long

Giant hog
Hylochoerus meinertzhageni

Large folds and
wrinkles in skin

Red river hog
Potamochoerus porcus

Mane and ear tassels can
be fluffed out to increase
the animal's apparent size

Lower canines
used in fighting

Babirusa
Babyrousa babyrussa

FACT FILE

Warthog This grassland grazer kneels on its forelimbs to feed, using its specialized incisor teeth to pluck new growth. When the grasses shrivel during the dry season, it digs out grass rhizomes (underground stems).

- Up to 5 ft (1.5 m)
- Up to 27½ in (70 cm)
- Up to 230 lb (105 kg)
- Mainly solitary
- Common

Sub-Saharan Africa

Bush pig The adaptable bush pig often follows monkeys to eat the fruit they drop, and will raid crops if they are available. Male rank is determined by head-to-head shoving contests.

- Up to 4¼ ft (1.3 m)
- Up to 35½ in (90 cm)
- Up to 255 lb (115 kg)
- Family band
- Locally common

E. & S.E. Africa; introd. Madagascar

Giant hog Around dusk, a mixed group of giant hogs will retreat through dense vegetation to a large sleeping nest. Females share the care of the piglets, suckling and protecting any in the group. Male contests are fierce, involving collisions at high speed that sometimes produce fractured skulls.

- Up to 7 ft (2.1 m)
- Up to 3¼ ft (1 m)
- Up to 520 lb (235 kg)
- Family band
- Uncommon

C. & W. Africa

Red river hog This species digs an underground burrow for sleeping. It rests by day and emerges at night to forage. Red river hogs live in harems of females and young led by a male.

- Up to 5 ft (1.5 m)
- Up to 3¼ ft (1 m)
- Up to 285 lb (130 kg)
- Family band
- Locally common

W. Africa

Babirusa With a diet of foliage, fruit, and fungi more specialized than that of other pigs, the babirusa also rarely uses its snout to root out food. Fossil studies suggest that it is the most primitive of all pig species.

- Up to 3½ ft (1.1 m)
- Up to 31½ in (80 cm)
- Up to 220 lb (100 kg)
- Family band
- Vulnerable

Sulawesi & nearby small islands

PECCARIES

CLASS	Mammalia
ORDER	Artiodactyla
FAMILY	Tayassuidae
GENERA	3
SPECIES	3

Although they resemble the pigs of Suidae in many ways, the three species of peccary in the family Tayassuidae can be distinguished by their long, slender legs, a more complex stomach, and a scent gland on the rump. They are omnivorous like pigs, but prefer fruit, seeds, roots, and vines, with the Chaco peccary depending largely on cacti. These gregarious animals live in herds ranging from 2–10 Chaco peccaries to 50–400 white-lipped peccaries. Social bonds are reinforced by herd members rubbing their cheeks on each other's scent glands. A few white-lipped peccaries will stay behind to fight a predator, allowing the rest of the herd to flee.

American pigs While pigs occur naturally only in Africa and Eurasia, peccaries are restricted to the Americas, where they range from southwest United States to northern Argentina. The collared peccary and white-lipped peccary are found in tropical forest, wooded savanna, and thorn scrub. The Chaco peccary is found mainly in semiarid thorn forest.

Twin peccaries A peccary litter most often contains two young, but can include up to four. Young collared peccaries depend on their mother for about 6 months.

⚡ **CONSERVATION WATCH**

Multiple threats Coupled with hunting for the bushmeat trade and diseases from introduced species, the rapid destruction of South America's tropical forests is having a serious impact on the three species of peccaries, which depend on large home ranges. The Chaco peccary, with only about 5,000 individuals remaining, is listed as endangered on the IUCN Red List.

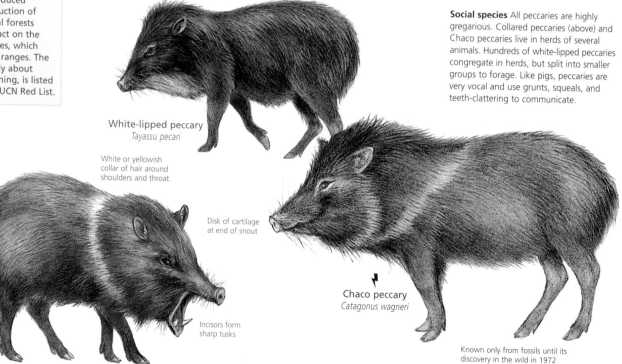

White-lipped peccary
Tayassu pecari

White or yellowish
collar of hair around
shoulders and throat

Social species All peccaries are highly gregarious. Collared peccaries (above) and Chaco peccaries live in herds of several animals. Hundreds of white-lipped peccaries congregate in herds, but split into smaller groups to forage. Like pigs, peccaries are very vocal and use grunts, squeals, and teeth-clattering to communicate.

Disk of cartilage
at end of snout

**Collared peccary
(javelina)**
Pecari tajacu

Incisors form
sharp tusks

⚡ **Chaco peccary**
Catagonus wagneri

Known only from fossils until its
discovery in the wild in 1972

HIPPOPOTAMUSES

CLASS Mammalia
ORDER Artiodactyla
FAMILY Hippopotamidae
GENERA 2
SPECIES 4

Now known to be more closely related to whales than to other ungulates, the two surviving species of hippopotamus lead a semiaquatic life, spending the day resting in water and emerging at night to forage on land. Their thick skin has only a thin outer layer, which rapidly dries out and cracks unless regularly moistened. Both species have large heads, a barrel-shaped body, and surprisingly short legs. There is, however, an enormous size disparity, with the grassland-grazing common hippopotamus being seven times heavier than the forest-foraging pygmy hippo. Because their weight is often borne by water, hippos conserve energy and need relatively little food.

Underwater ungulate Lacking sweat glands, the common hippopotamus relies on water to stay cool. It is a good swimmer and diver, and the density of its body allows it to walk along a river or lake bed and stay submerged for about 5 minutes at a time. It can float by filling its lungs with air. A hippo's feet are webbed; the nostrils and ears can close underwater; and the eyes, ears, and nostrils are positioned so that it can see, hear, and breathe with just the top of the head emerging above the surface. Young are born and suckled underwater. Herds of up to 40 hippos may spend the day together in water, devoting most of their time to sleeping or resting. At night, they leave the water to feed on land for about 6 hours.

Open wide Hinged far back in the skull, a hippo's jaw can open remarkably wide, achieving a gape of 150 degrees, more than 100 degrees wider than the human gape. A male's large lower canines are used as weapons in battles for mating rights.

Hippopotamus
Hippopotamus amphibius

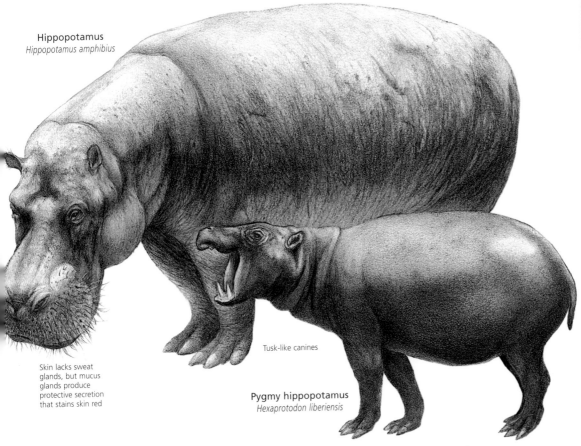

Skin lacks sweat glands, but mucus glands produce protective secretion that stains skin red

Tusk-like canines

Pygmy hippopotamus
Hexaprotodon liberiensis

FACT FILE

Hippopotamus Several females and their young spend the day together in water, but forage alone on land at night. Dominant males are territorial and mate with the females that enter their length of riverbank or lake shore. This species mainly eats grasses.

Up to 14 ft (4.2 m)
Up to 5 ft (1.5 m)
Up to 2¼ tons (2 t)
Herd
Locally common

Tropical & subtropical Africa

Pygmy hippopotamus Usually solitary, the pygmy hippo may spend the day hidden in a swamp or retreat to an otter's burrow in a riverbank. Its varied diet includes roots and fruit.

Up to 6½ ft (2 m)
Up to 35½ in (90 cm)
Up to 605 lb (275 kg)
Solitary, pair
Vulnerable

W. Africa

⚡ CONSERVATION WATCH

Hippo habitats While abundant in some areas, the common hippo is rare in western Africa. Its tendency to gather in large herds makes it easy prey for human hunters. Threatened by habitat loss and poaching, the pygmy hippo is listed as vulnerable on the IUCN Red List. Its dense habitat and solitary nature prevent an accurate tally of numbers.

CETACEANS

CLASS	Mammalia
ORDER	Cetacea
FAMILIES	10
GENERA	41
SPECIES	81

With their entirely aquatic lifestyle, the whales, dolphins, and porpoises of the order Cetacea are perhaps the most specialized of all mammals. They feed, rest, mate, give birth, and raise young in the water, yet they are warm-blooded and breathe air like other mammals. Gregarious and intelligent, cetaceans appear to be descended from the same land mammal that led to hippopotamuses, but their ancestors adapted to a watery life some 50 million years ago. Over time, they became as streamlined as fish, losing their hair and hindlimbs, modifying their arms into flippers, and developing a powerful fluked tail that makes some species the fastest creatures in the sea.

Ready to blow When a cetacean surfaces to breathe, it expels air and condensed moisture through nostrils modified to form a single or double blowhole on the top of the head. The blowhole closes underwater.

Staying warm and cool Being virtually hairless, a cetacean relies on a layer of blubber beneath the skin for insulation. A network of arteries and veins in the blubber, known as *retia mirabilia*, helps the animal to regulate its temperature.

melon | blowhole | heart | liver | dorsal fin

Aquatic mammal Although their body shape is highly modified for life in water, dolphins and other cetaceans are warm-blooded and breathe air through lungs. They have a four-chambered heart and a three-chambered stomach.

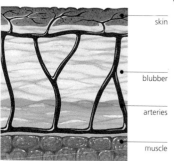

skin

blubber

arteries

muscle

pectoral fin

lung | stomach

A double fluke Like other cetaceans, the sperm whale propels itself with powerful up-and-down strokes of its fluked tail. The flippers are used for steering.

Bonded young After a long gestation period, a single calf is born tail-first underwater. Its mother and sometimes other members of the pod nudge it to the surface to take its first breath. Nourished by rich milk, the calf grows rapidly but stays with the mother for some years.

CETACEAN RECORDS

Cetaceans are found in all of the world's oceans and seas, and in some rivers and lakes. They are split into two living suborders: the toothed whales of Odontoceti, and the baleen whales of Mysticeti. Toothed whales, which include dolphins, porpoises, and sperm whales, have simple, conical teeth that can keep a firm grip on their slippery food of fishes and squids. Baleen whales include blue whales, humpbacks, gray whales, and right whales. They are filter feeders, straining great quantities of tiny plankton, other invertebrates, and small fish through bristled horny plates that hang from the roof of the mouth.

With water supporting their weight, some cetaceans have been able to reach enormous sizes. The blue whale is the largest animal that has ever lived, with a record weight of 209 tons (190 tonnes)—roughly equivalent to the weight of 35 elephants—and a record length of 110 feet (33.5 m).

Another cetacean, the sperm whale, boasts the deepest and longest dives of any mammal. Sperm whales are believed to descend to at least 10,000 feet (3,050 m), and their dives can last for more than 2 hours. When a cetacean dives, its heart rate slows by 50 percent and blood is directed away from the muscles to the vital organs, allowing the animal to survive on very little oxygen until it ascends to breathe.

Cetaceans have little or no sense of smell. Their relatively small eyes provide reasonable vision both above and beneath the water's surface. All species lack external ears, but their hearing is highly sensitive, allowing them to pick up distant calls from members of their species. To find prey and avoid obstacles, toothed whales use echolocation, emitting a series of clicks and whistles, and then analyzing the reflected sounds.

Sound is critical in cetacean communication. Blue whales and fin whales emit low-frequency pulses that carry across vast stretches of ocean and can reach 188 decibels—the loudest sound made by an animal. Male humpbacks produce

Clear breach Humpback whales may leap from the water a hundred or more times in a row. Such breaching appears to be used to communicate with other whales, but may also have other purposes.

the longest and most complex songs in the animal kingdom.

Because cetaceans spend most of their time underwater, accurate population statistics are difficult to compile. Nevertheless, it is certain that human activities have devastated cetacean numbers. Commercial whaling (now largely banned, but still pursued by Norway and Japan), driftnet fishing (which inadvertently traps cetaceans), and water pollution have all taken a heavy toll.

Social animals Almost all cetaceans are gregarious to some extent. Toothed whales tend to form larger groups than baleen whales and have more complex social structures. Hundreds or sometimes thousands of common dolphins travel together, swimming with great speed and leaping clear of the water. Members of a group usually feed at the same time and may hunt cooperatively, herding fish into clusters.

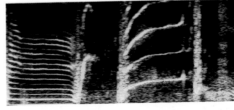

Orca song Each orca pod has a distinct dialect, a repetitive pattern of sounds used during travel and feeding that probably helps the animals to coordinate their activities. The calls become more varied during socializing.

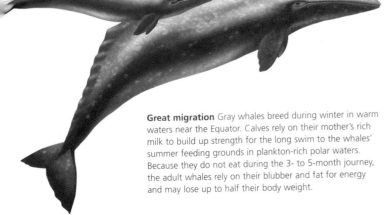

Great migration Gray whales breed during winter in warm waters near the Equator. Calves rely on their mother's rich milk to build up strength for the long swim to the whales' summer feeding grounds in plankton-rich polar waters. Because they do not eat during the 3- to 5-month journey, the adult whales rely on their blubber and fat for energy and may lose up to half their body weight.

Humpback song Male humpback whales sing complex songs that can be made up of nine themes and last half an hour. All males in an ocean basin sing the same song, but it may gradually change over time.

TOOTHED WHALES

CLASS	Mammalia
ORDER	Cetacea
FAMILIES	6
GENERA	35
SPECIES	68

About 90 percent of all cetaceans are toothed whales belonging to one of the six families in the suborder Odontoceti. In contrast to the enormous baleen whales, toothed whales tend to be medium-sized, although the largest of them, the sperm whale, is a massive creature. Their brains are relatively large, making them the most intelligent mammals other than primates. While some species are solitary, most are highly social and tend to be very vocal and playful. Members of a group may hunt cooperatively and help care for each other's young. Most toothed whales feed on fishes or squids, but one species, the orca, actively pursues warm-blooded prey such as seals and other whales.

Killer teeth Toothed whales have sharp, conical teeth. In fish-eating dolphins, these are small and numerous. Orcas, which hunt marine mammals, have fewer but larger teeth (above). Squid-eating beaked whales have just a single tooth per jaw.

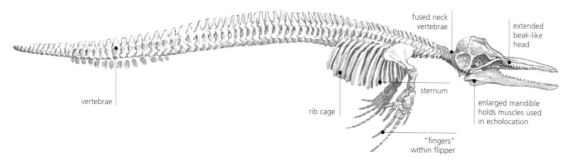

fused neck vertebrae

extended beak-like head

vertebrae

rib cage

sternum

"fingers" within flipper

enlarged mandible holds muscles used in echolocation

Long and narrow The skeleton of a toothed whale has been greatly modified from that of its land mammal ancestor. The hindlimbs have disappeared, while the forelimbs have become flippers, although the bones for five fingers remain. The head is usually long and narrow, forming a beak.

Serious playtime Dolphins, such as this dusky dolphin (*Lagenorhynchus obscurus*), may leap to impress mates, to herd fish, or simply for fun. Playful displays help to reinforce the social bonds among group members, providing the familiarity needed for successful cooperative hunting.

Gregarious species Sperm whales live in closeknit groups of about a dozen related females and their young. The adults look after each other's young and will protect an injured member from predators. Young males leave to form bachelor groups, but become less social as they age.

SOCIAL CETACEANS

The diverse members of Odontoceti include sperm whales; narwhals and belugas; beaked whales; dolphins, orcas, and pilot whales (grouped together in the family Delphinidae); porpoises; and river dolphins. Most have an elongated, beak-like head with sharp, conical teeth that can firmly seize prey but cannot chew it. Because there is only a single blowhole, the skull is asymmetrical. It supports a fluid-filled organ called the melon, which is thought to focus the clicks used in echolocation and communication. In sperm whales, the melon is greatly enlarged and filled with an oil known as spermaceti. This spermaceti organ may also help to focus sounds.

There is considerable variety in the social organization of toothed whales. Most groups are centered on the females, with males leaving the group at puberty. Orcas and pilot whales, however, never leave their birth group. River dolphins tend to form small groups or even live alone. Coastal dolphins form larger groups because their prey is concentrated in particular areas and they face more predators. In the deep ocean, small, closely related groups may form temporary herds containing thousands of dolphins.

Despite their reputation for being gentle and playful, dolphins do fight. Group living often involves rivalry for food or mates, and this may lead to physical clashes. Many toothed whales bear tooth rake marks as scars of such encounters.

⚡ CONSERVATION WATCH

Of the 68 species of toothed whales, 82% are listed on the IUCN Red List, as follows:

2	Critically endangered
2	Endangered
4	Vulnerable
10	Conservation dependent
38	Data deficient

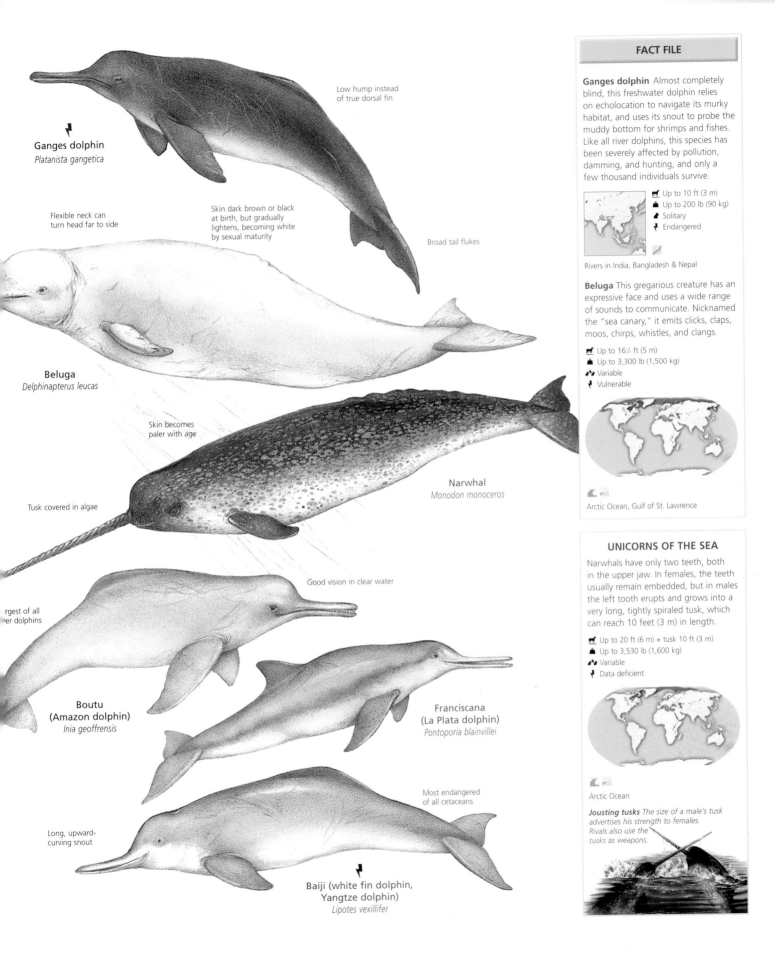

Ganges dolphin
Platanista gangetica

Low hump instead of true dorsal fin

Flexible neck can turn head far to side

Skin dark brown or black at birth, but gradually lightens, becoming white by sexual maturity

Broad tail flukes

Beluga
Delphinapterus leucas

Skin becomes paler with age

Tusk covered in algae

Narwhal
Monodon monoceros

rgest of all er dolphins

Good vision in clear water

Boutu (Amazon dolphin)
Inia geoffrensis

Franciscana (La Plata dolphin)
Pontoporia blainvillei

Most endangered of all cetaceans

Long, upward-curving snout

Baiji (white fin dolphin, Yangtze dolphin)
Lipotes vexillifer

FACT FILE

Ganges dolphin Almost completely blind, this freshwater dolphin relies on echolocation to navigate its murky habitat, and uses its snout to probe the muddy bottom for shrimps and fishes. Like all river dolphins, this species has been severely affected by pollution, damming, and hunting, and only a few thousand individuals survive.

Up to 10 ft (3 m)
Up to 200 lb (90 kg)
Solitary
Endangered

Rivers in India, Bangladesh & Nepal

Beluga This gregarious creature has an expressive face and uses a wide range of sounds to communicate. Nicknamed the "sea canary," it emits clicks, claps, moos, chirps, whistles, and clangs.

Up to 16½ ft (5 m)
Up to 3,300 lb (1,500 kg)
Variable
Vulnerable

Arctic Ocean, Gulf of St. Lawrence

UNICORNS OF THE SEA

Narwhals have only two teeth, both in the upper jaw. In females, the teeth usually remain embedded, but in males the left tooth erupts and grows into a very long, tightly spiraled tusk, which can reach 10 feet (3 m) in length.

Up to 20 ft (6 m) + tusk 10 ft (3 m)
Up to 3,530 lb (1,600 kg)
Variable
Data deficient

Arctic Ocean

Jousting tusks The size of a male's tusk advertises his strength to females. Rivals also use the tusks as weapons.

FACT FILE

Bottle-nosed dolphin This is the species made famous by the *Flipper* television series and is most often seen performing in marine parks. In the wild, it is found both inshore and offshore living in groups of about a dozen animals, which sometimes form herds of hundreds. It ranges widely for food, swimming at an average speed of 12 miles per hour (20 km/h).

- Up to 13 ft (4 m)
- Up to 605 lb (275 kg)
- Variable
- Data deficient

Temperate to tropical oceans & seas

Common dolphin The most abundant of all dolphins, this small species lives in herds of several hundred or even a few thousand individuals, which may be joined by white-sided or bottle-nosed dolphins during feeding.

- Up to 8 ft (2.4 m)
- Up to 187 lb (85 kg)
- Herd
- Common

Temperate to tropical oceans & seas

ENTANGLED VICTIMS

The vast nets used in commercial fisheries pose a great risk to dolphins, which follow their prey into the nets and become entangled. Unable to surface to breathe, the dolphins soon drown. Measures to make the nets more conspicuous have helped, but thousands of dolphins are still accidentally captured each year.

Tucuxi (river dolphin)
Sotalia fluviatilis

Found in both marine and freshwater environments

Striped dolphin
Stenella coeruleoalba

Short, stubby beak

Bottle-nosed dolphin
Tursiops truncatus

Largest of the beaked dolphins

Rough-toothed dolphin
Steno bredanensis

Criss-crossing scars from fights with squid prey or other dolphins

Risso's dolphin
Grampus griseus

Common dolphin
Delphinus delphis

Commerson's dolphin
Cephalorhynchus commersonii

Atlantic white-sided dolphin
Lagenorhynchus acutus

Two-tone coloration helps to camouflage animal in its marine environment

Pacific white-sided dolphin
Lagenorhynchus obliquidens

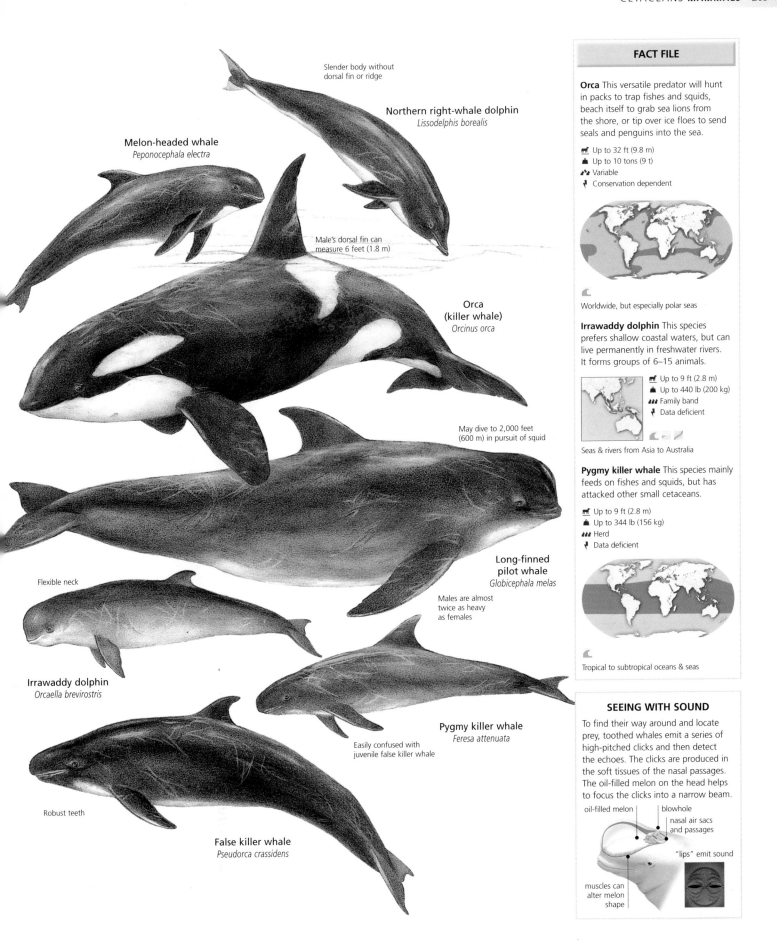

Slender body without dorsal fin or ridge

Northern right-whale dolphin
Lissodelphis borealis

Melon-headed whale
Peponocephala electra

Male's dorsal fin can measure 6 feet (1.8 m)

**Orca
(killer whale)**
Orcinus orca

May dive to 2,000 feet (600 m) in pursuit of squid

**Long-finned
pilot whale**
Globicephala melas

Males are almost twice as heavy as females

Flexible neck

Irrawaddy dolphin
Orcaella brevirostris

Pygmy killer whale
Feresa attenuata

Easily confused with juvenile false killer whale

Robust teeth

False killer whale
Pseudorca crassidens

FACT FILE

Orca This versatile predator will hunt in packs to trap fishes and squids, beach itself to grab sea lions from the shore, or tip over ice floes to send seals and penguins into the sea.

🐾 Up to 32 ft (9.8 m)
⚖ Up to 10 tons (9 t)
🐾 Variable
⚑ Conservation dependent

Worldwide, but especially polar seas

Irrawaddy dolphin This species prefers shallow coastal waters, but can live permanently in freshwater rivers. It forms groups of 6–15 animals.

🐾 Up to 9 ft (2.8 m)
⚖ Up to 440 lb (200 kg)
🐾 Family band
⚑ Data deficient

Seas & rivers from Asia to Australia

Pygmy killer whale This species mainly feeds on fishes and squids, but has attacked other small cetaceans.

🐾 Up to 9 ft (2.8 m)
⚖ Up to 344 lb (156 kg)
🐾 Herd
⚑ Data deficient

Tropical to subtropical oceans & seas

SEEING WITH SOUND

To find their way around and locate prey, toothed whales emit a series of high-pitched clicks and then detect the echoes. The clicks are produced in the soft tissues of the nasal passages. The oil-filled melon on the head helps to focus the clicks into a narrow beam.

oil-filled melon
blowhole
nasal air sacs and passages
"lips" emit sound
muscles can alter melon shape

FACT FILE

Spectacled porpoise Less acrobatic than many other small cetaceans, this porpoise moves slowly through the water. It usually lives alone or in pairs and feeds on fishes and squids.

- Up to 7 ft (2.1 m)
- Up to 255 lb (115 kg)
- Solitary, small group
- Data deficient

Off Argentina, Tasmania & sub-Antarctic islands

Burmeister's porpoise The species name of this animal (*spinipinnis*) means "spiny fin" and was inspired by the small bumps along the leading edge of its dorsal fin, a characteristic shared by most other porpoises.

- Up to 6 ft (1.8 m)
- Up to 154 lb (70 kg)
- Small groups
- Data deficient

Seas & coastal waters from Peru to Brazil

Gulf porpoise Found only in the northern part of the Gulf of California, this species has a more restricted range than any other cetacean. It appears to have evolved from South America's Burmeister's porpoise, but was isolated in the Northern Hemisphere when tropical waters became warmer.

- Up to 5 ft (1.5 m)
- Up to 121 lb (55 kg)
- Not known
- Critically endangered

Estuary of Colorado River; Gulf of California

Common porpoise A rounded shape and small flippers, tail, and fin minimize this cetacean's surface area. Along with a layer of blubber, this helps it survive in cooler waters despite its small size.

- Up to 6¼ ft (1.9 m)
- Up to 143 lb (65 kg)
- Variable
- Vulnerable

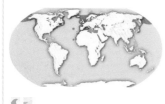

Temperate waters of Northern Hemisphere

Spectacled porpoise
Australophocaena dioptrica

Named for circles around the eyes

Dorsal fin set farther back than on any other small cetacean

Burmeister's porpoise
Phocoena spinipinnis

Gulf porpoise (vaquita)
Phocoena sinus

Common porpoise (harbor porpoise)
Phocoena phocoena

Less shy and slow than other porpoises

Tail sends up spray in the shape of a rooster's tail

Dall's porpoise
Phocoenoides dalli

Dorsal ridge rather than fin

Finless porpoise
Neophocaena phocaenoides

CETACEAN STRATEGIES

To reap a rich harvest from the ocean, cetaceans have developed remarkably varied physical characteristics and behavioral strategies. Baleen whales have enormous mouths that can engulf huge quantities of tiny animals, while toothed whales pursue individual prey, relying on echolocation to find it. Many species practise cooperative hunting, using a range of vocalizations to communicate the next move. The success of such efforts rests on a social structure that encourages close relationships.

Clicking pursuit Toothed whales track their food using echolocation. A dolphin may emit up to 600 clicks a second, analyzing the echoes to build up a picture of its surroundings, including the position of prey. Orcas use echolocation when pursuing fish, but rely more on vision to hunt other cetaceans or seals, which would be alerted by the clicks.

transmitted sound

returning echo

Humpback feeding Humpbacks (above) will synchronize their feeding, lunging at prey shoals together or herding scattered prey into clusters. In bubblenet feeding (below), a humpback spirals toward the surface while exhaling, producing a large "net" of bubbles that traps small prey. The whale lunges through the center of the net to capture its meal.

3. Lunge feeding Once the bubblenet has trapped the fishes, the humpback swims through the center, lunging to the surface with its mouth open to engulf the prey.

Bottom feeder A gray whale feeds in shallow waters on bottom-dwelling crustaceans, mollusks, and worms. It dives to the seafloor, turns on its side, and then sucks up a mouthful of sediment. Prey is filtered out by pushing the sediment through its baleen.

1. Slow exhale As a humpback whale spirals to the surface, it slowly lets out its breath, creating columns of bubbles. Small schooling fishes are trapped inside the net of bubbles.

2. Group effort Bubblenetting may be carried out by a single whale, or several whales may cooperate to create the net.

Together and alone Orcas feed mainly on the prey that are most abundant locally, which, in turn, determines their hunting methods. Where fishes such as salmon and herring are common, orcas tend to hunt cooperatively. In Argentina, a single whale will slide up onto the shore to grab a young sea lion (above).

FACT FILE

Sperm whale With a gullet large enough to swallow a human, the deep-diving sperm whale sometimes takes sharks and skates, but mainly feeds on giant squids, octopuses, and deepwater fishes. It is gregarious and lives in groups of 30–100 animals.

- Up to 61 ft (18.5 m)
- Up to 77 tons (70 t)
- Variable
- Vulnerable

Deep temperate to tropical oceans & seas

Baird's beaked whale Tightly knit social groups of 6–30 Baird's beaked whales live in deep offshore waters and are led by a dominant male. Rivalry for this position often leads to physical conflict, with most males bearing scars on their beak and back.

- Up to 43 ft (13 m)
- Up to 16½ tons (15 t)
- Variable
- Conservation dependent

North Pacific

TUSKED BEAKS

A diet of squids captured through suction has rendered beaked whales virtually toothless. In males, however, one or two pairs of teeth protrude from the mouth to form tusks, which appear to be used as weapons.

Wrap-arounds
In the male strap-toothed whale (Mesoplodon layardii), the tusks are especially long and wrap around the upper jaw. As a result, the mouth can only open an inch (2.5 cm) or so.

Pygmy sperm whale
Kogia breviceps

Sperm whale
Physeter catodon

Hump and ridges on back rather than dorsal fin

Males weigh three times as much as females

Baird's beaked whale
Berardius bairdii

Both males and females have two pairs of protruding teeth

Northern bottle-nosed whale
Hyperoodon ampullatus

Blainville's beaked whale
Mesoplodon densirostris

May spend 30 minutes plunging to depths of 3,300 feet (1,000 m)

Cuvier's beaked whale
Ziphius cavirostris

Female usually a little larger than male

Stubby beak

Sowerby's beaked whale
Mesoplodon bidens

BALEEN WHALES

ASS Mammalia

RDER Cetacea

MILIES 4

NERA 6

ECIES 13

The giants of the ocean, the baleen whales of the suborder Mysticeti feed on tiny prey, filtering aquatic invertebrates and small fishes through their sieve-like baleen plates. Their remarkable size is a great advantage in cooler waters, since, relative to body mass, they have a small surface area from which to lose heat. A thick layer of blubber provides insulation and can act as a food store for the epic annual migrations that many species undertake. Found in all the world's oceans, baleen whales include the gray whale, the right whales, the bowhead whale, and the rorquals—which comprise the blue whale, fin whale, sei whale, Bryde's whale, humpback whale, and minke whales.

Legal whaling For nutritional and cultural reasons, the Inuit (above) are permitted to hunt a small number of bowhead whales each year. The eastern bowhead population went into a severe decline as a result of the commercial whaling of the 19th century, but appears to be slowly stabilizing.

BALEEN AND BLUBBER

Baleen whales can be skimmers or gulpers. Right whales move slowly along the surface, skimming little animals from the water that crosses their long baleen plates. Rorquals lunge at shoals of prey with open mouths, gulp in water, and then force the water back out with their tongues, trapping krill and other creatures in their short plates. Gray whales are bottom feeders, filtering crustaceans and mollusks from the sediment with their heavy baleen.

Most of a baleen whale's prey species are minuscule, so it needs to consume vast quantities to stay alive. During summer, a large blue whale may eat 4½ tons (4 tonnes) of krill per day. It feeds very little during the rest of the year and lives off the fat and blubber laid down in summer.

Baleen and blubber, both crucial to the survival of these giants, also attracted commercial whalers. Since 1985, there has been a moratorium on all commercial whaling, but this is not observed by Norway or Japan.

Great gulp Like other rorquals, humpback whales have a pleated throat that expands into a great pouch as they gulp in water and plankton. The throat contracts again as they force the water back out and trap the prey in their bristled baleen plates.

Distinguishing marks Right whales are distinguished by the callosities on their head, patches of hardened skin that are often infested with parasites such as lice and barnacles. The callosities are slightly larger on male whales than on females, suggesting that they may be used as weapons by rivals.

Light bones Rather than supporting the animal's body weight, a cetacean's skeleton simply anchors the muscles. The bones are light, spongy, and filled with oil. The most dramatic feature of a baleen whale's skeleton is the massive head.

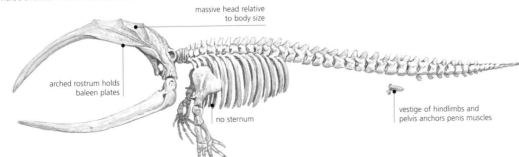

massive head relative to body size

arched rostrum holds baleen plates

no sternum

vestige of hindlimbs and pelvis anchors penis muscles

⚡ CONSERVATION WATCH

All 13 species of baleen whales are listed on the IUCN Red List, as follows:

5	Endangered
1	Vulnerable
4	Conservation dependent
1	Data deficient

FACT FILE

Blue whale Consuming about 4½ tons (4 tonnes) of krill every day during the main summer feeding season, the blue whale is the largest animal to have ever lived. A newborn measures at least 19½ feet (5.9 m) and guzzles 50 gallons (190 l) of milk per day, adding 8 pounds (3.6 kg) to its weight every hour. It may go on to live for 110 years. Blue whales were relentlessly hunted in the early to mid 20th century, and the total population now comprises only 6,000–14,000 individuals.

🐋 Up to 110 ft (33.5 m)
⚖ Up to 209 tons (190 t)
🐾 Variable
🏹 Endangered

🌊
All oceans except high Arctic

Fin whale Outsized only by the blue whale, the fin whale can attain speeds of 23 miles per hour (37 km/h), making it one of the fastest cetaceans. While groups of 300 or more may migrate together, the species is usually seen in pairs or small pods of several animals.

🐋 Up to 82 ft (25 m)
⚖ Up to 88 tons (80 t)
🐾 Variable
🏹 Endangered

🌊
All oceans except high Arctic

Northern right whale Full of oil and easy to catch, this species was named the "right" whale for hunting by early Basque whalers. There are now only a few hundred individuals left.

🐋 Up to 59 ft (18 m)
⚖ Up to 99 tons (90 t)
🐾 Variable
🏹 Endangered

🌊
North Pacific & W. North Atlantic

Blue whale
Balaenoptera musculus

50–90 throat pleats

Fin whale
Balaenoptera physalus

80 percent of fin whales were killed by 20th-century whalers

Differs from other right whales by having two throat grooves and a dorsal fin

Small dorsal fin positioned far along the back

Pygmy right whale
Caperea marginata

Bowhead whale
Balaena mysticetus

Longest baleen plates of any whale

24 inch (60 cm) layer of blubber keeps bowhead warm during total darkness of Arctic winter

"Necklace" of black spots

Only whale in temperate waters without dorsal fin

Callosities covered in whale lice

Northern right whale
Eubalaena glacialis

99 percent of blue whales were
killed by 20th-century whalers

Female larger than male

Fin whale's lower jaw has
asymmetrical color pattern:
white on the right side and
dark on the left

Sei whale's short
throat pleats and
fine baleen fringes
for skimming food
from water as whale
swims on its side

Minke whale
Balaenoptera acutorostrata

Mature gray whales
are covered in lice
and barnacles

Sei whale
Balaenoptera borealis

Gray whale
Eschrichtius robustus

Reduced dorsal fin forms
small hump on back

Humpback whale
Megaptera novaeangliae

97 percent of humpbacks
were killed by 20th-century
whalers

FACT FILE

Minke whale The smallest and most
abundant rorqual, this acrobatic whale
is found in all the world's oceans. It is
often seen in pairs, but up to 100 may
gather in a rich feeding area.

- Up to 36 ft (11 m)
- Up to 11 tons (10 t)
- Variable
- Near threatened

Most ocean areas except coldest regions

Gray whale Many gray whales make
a 12,400 mile (20,000 km) round-trip
from summer feeding grounds in Alaska
to winter breeding grounds in Mexico.

- Up to 49 ft (15 m)
- Up to 38½ tons (35 t)
- Variable
- Conservation dependent

Bering Sea to Gulf of Mexico & Sea of Okhotsk

Humpback whale This active animal is
famous for spectacular breaching, when
it leaps clear of the water. Other surface
behaviors include rolling, pec-slapping,
spyhopping (sticking its head vertically
out of the water), and tail splashing.

- Up to 49 ft (15 m)
- Up to 71½ tons (65 t)
- Variable
- Vulnerable

Breeds in tropics, feeds near Arctic & Antarctic

Pec-slapping
*A humpback may lie
on its side and loudly
slap its pectoral fins
against the water.*

RODENTS

CLASS	Mammalia
ORDER	Rodentia
FAMILIES	29
GENERA	442
SPECIES	2,010

With roughly 2,000 species, rodents account for more than 40 percent of all mammal species and have colonized almost every habitat on Earth. A key to their extraordinary success is the ability to reproduce quickly and abundantly, allowing species to survive harsh conditions and take full advantage of favorable ones. In addition, the small size of most rodents has helped them to exploit many microhabitats. Although rodents are among the earliest of placental mammals, with the oldest rodent fossils dating back some 57 million years, the largest family, Muridae (rats and mice), did not appear until 5 million years ago. It now contains almost two-thirds of all species in the order.

Successful spread Members of the order Rodentia are distributed throughout all the world's continents, except for Antarctica. Their association with humans has even helped them to reach isolated islands. They have adapted to a wide range of habitats, including arctic tundra, tropical forests, deserts, high mountains, and urban areas.

Persistent pests Being opportunistic feeders that can reproduce quickly, many rodents have thrived alongside humans. Pests such as black rats (left) not only eat vast quantities of crops and stored food, but also contaminate the remaining food with their droppings and spread disease.

Rapid reproduction Garden dormice usually mate in April or May after waking from their winter hibernation. Following a short gestation period of 22–28 days, the female gives birth to 2–9 young. The newborn are entirely dependent on the mother and will not open their eyes for another 21 days or so. They become independent by 6 weeks and then grow rapidly until their first hibernation. While some rodent species are ready to breed at 6 weeks, it takes about a year for garden dormice to become sexually mature. A captive garden dormouse lived for about 5½ years, a long lifespan for a rodent.

UNIFORM ANATOMY

Rodent size ranges from the tiny pygmy jerboa, less than 2 inches (5 cm) long and weighing just ⅕ ounce (5 g), to the substantial capybara, more than 50 inches (1.3 m) in length with a weight of 140 pounds (64 kg). Typically, however, rodents are small with squat bodies, short legs, and a tail.

The most distinguishing rodent feature is the arrangement of the teeth. All rodents have two pairs of extremely sharp incisors at the front of the mouth that can gnaw through seedpods, nut shells, and other tough matter to get at the nutritious food inside. The incisors grow continuously and are "self-sharpened" against each other. There are no canine teeth behind the incisors. Instead, a gap known as the diastema allows the lips to close during gnawing so inedible material is kept out of the mouth. At the back of the mouth, a series of molars grind up the plant matter that makes up most of a rodent's diet.

While a few rodents are mainly carnivorous, most are opportunistic feeders and eat leaves, fruit, nuts, and seeds, as well as caterpillars, spiders, and other small invertebrates. The indigestible cellulose in plant walls is broken down by bacteria in

Typical rodent The brown rat displays the typical rodent anatomy, with a compact body, short legs, clawed feet, long tail, and sensitive whiskers. A keen sense of smell and sharp hearing help rodents to find food and avoid predators.

Flexible feeder A resident of temperate forests, the red squirrel eats mainly seeds and nuts, but will also consume flowers, shoots, fungi, and small invertebrates. It builds up a cache of buried seeds and nuts that it can raid during the cold winter.

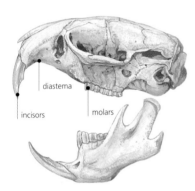

incisors diastema molars

Gnawing teeth A gap called the diastema separates a rodent's chewing molars from the sharp, continuously growing incisors. The diastema allows the lips to close behind the incisors and keep out inedible material when the animal is gnawing.

Ecosystem role As key prey species for medium-sized predators such as this barn owl, rodents play an important role in an ecosystem. Their distribution of the spores of fungi that nourish tree roots is also vital in North American and Australian forests.

A rodent's tail The diversity of rodent lifestyles is reflected in the various shapes and purposes of the tail (right). The northern flying squirrel (*Glaucomys sabrinus*) (above) uses its tail for steering and stability as it glides from tree to tree.

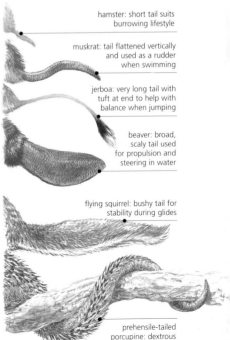

hamster: short tail suits burrowing lifestyle

muskrat: tail flattened vertically and used as a rudder when swimming

jerboa: very long tail with tuft at end to help with balance when jumping

beaver: broad, scaly tail used for propulsion and steering in water

flying squirrel: bushy tail for stability during glides

prehensile-tailed porcupine: dextrous tail acts as fifth limb

the large cecum (appendix) of the rodent digestive system. Some species then take this treated food from the anus and eat it again, gaining maximum nutrition from the meal before passing feces as dry pellets, a process known as refection.

Intelligent and resourceful, rodents have put their relatively uniform anatomy to diverse use. Many species are terrestrial, finding their food in forests, grasslands, deserts, or human settlements. Others spend most of their time in trees, scampering over branches and, in some cases, gliding from one tree to another. Some species make their life underground in extensive networks of burrows. And a few are excellent swimmers,

able to pursue a semiaquatic lifestyle. While a minority of rodents are solitary, most are highly social, a trait culminating in the townships that contain thousands of prairie dogs.

The order Rodentia was once split into three suborders according to the arrangement of jaw muscles: squirrel-like rodents, mouse-like rodents, and cavy-like rodents. These categories are still used informally for ease, but genetic evidence points to a division into just two suborders. The suborder Sciurognathi includes all the squirrel-like and mouse-like rodents, plus the gundis, a family of cavy-like rodents. The other suborder, Hystricognathi, includes all other cavy-like rodents.

⚡ CONSERVATION WATCH

Some rodent species have thrived alongside humans to the point of becoming serious pests. Many with limited ranges, however, have been threatened or even driven extinct by human activities. Of the 2,010 species of rodents, 33% are listed on the IUCN Red List, as follows:

32	Extinct
68	Critically endangered
95	Endangered
165	Vulnerable
5	Conservation dependent
255	Near threatened
49	Data deficient

SQUIRREL-LIKE RODENTS

CLASS	Mammalia
ORDER	Rodentia
FAMILIES	8
GENERA	71
SPECIES	383

The squirrels, beavers, and other animals collectively known as squirrel-like rodents share an arrangement of jaw muscles that gives them a strong forward bite. They have simple teeth and have retained one or two premolar teeth in each row, a characteristic not found in other rodents. Aside from the jaw muscles and premolar teeth, the families of squirrel-like rodents share few characteristics and probably diverged from each other early in rodent evolution. They include beavers (family Castoridae), mountain beaver (Aplodontidae), squirrels (Sciuridae), pocket gophers (Geomyidae), pocket mice (Heteromyidae), scaly-tailed squirrels (Anomaluridae), and springhare (Pedetidae).

Winter sleep From October until March or April, the woodchuck (*Marmota monax*) hibernates underground in burrows. During this time, its heartbeat slows, its body temperature falls, and it lives off body fat. It mates soon after emerging in spring.

Arboreal leaper When a tree squirrel such as this American red squirrel leaps from one tree to another, it stretches out its limbs, flattens its body, and slightly curves its tail to maximize its surface area. The long, bushy tail acts as a rudder.

Social species Prairie dogs live in large townships with a complex social structure. A township is made up of many coteries—groups of one male, several related females, and their offspring. Members of a coterie share burrows and food.

BURROWERS AND LEAPERS

The squirrels in the family Sciuridae make up about three-quarters of all squirrel-like rodents. Active by day, tree squirrels have long, lightweight bodies, sharp claws for clinging to bark, and excellent eyesight for judging distances. They move about by scampering along branches, climbing headfirst down trunks, or leaping from tree to tree. The nocturnal flying squirrels glide through the air, aided by a furred membrane along either side of the body. Arboreal squirrels mostly feed on fruit, nuts, seeds, shoots, and leaves, but may supplement this diet with insects. Ground-dwelling squirrels, which include ground squirrels, prairie dogs, marmots, and chipmunks, tend to favor grasses and herbs. Many of these terrestrial species are gregarious with complex social organization.

Scaly-tailed squirrels are only distantly related to true squirrels. Like tree squirrels, almost all species of scaly-tailed squirrels possess a membrane for gliding—an example of convergent evolution.

Beavers are superbly equipped for a life spent largely in water, with a streamlined body, flattened tail, and webbed feet. Their large incisor teeth allow them to cut down trees and build dams and lodges.

Pocket gophers, pocket mice, mountain beavers, and springhares are all burrowing rodents. The pocket gophers and pocket mice both carry food in cheek pouches on either side of the mouth.

Strong forward bite The chewing muscles are known as masseters. In squirrel-like rodents, the lateral masseter extends to the snout and pulls the jaw forward when biting. The deep masseter is short and direct and simply closes the jaw.

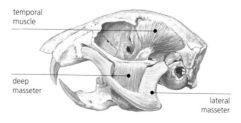

temporal muscle

deep masseter

lateral masseter

⚡ CONSERVATION WATCH

Of the 383 species of squirrel-like rodents, 21% are listed on the IUCN Red List, as follows:

- 8 Critically endangered
- 11 Endangered
- 3 Conservation dependent
- 58 Near threatened
- 2 Data deficient

Cheek pouches inside
mouth for carrying food

European souslik
Spermophilus citellus

Stands on hindlegs to
watch for predators

Spotted souslik
Spermophilus suslicus

Bobak marmot
Marmota bobak

Whistles to warn
other marmots
of danger

Alpine marmot
Marmota marmota

Hoary marmot
Marmota caligata

Sturdy, slightly curved
claws for digging

**Thirteen-lined
ground squirrel**
Spermophilus tridecemlineatus

13 stripes
alternating
between light
and dark fur

Poor eyesight but
keen hearing and
sense of touch

Black-tailed prairie dog
Cynomys ludovicianus

Mountain beaver
Aplodontia rufa

FACT FILE

Thirteen-lined ground squirrel
Active by day, this rodent feeds mainly
on grasses and seeds, transporting
some food in its cheek pouches to
underground caches.

- Up to 7 in (18 cm)
- Up to 5 in (13 cm)
- Up to 9½ oz (270 g)
- Solitary
- Locally common

Tallgrass prairies of C. North America

Black-tailed prairie dog Members
of this gregarious species play together,
groom one another, and communicate
with a range of calls. An alarm bark
warns of danger, while another type
of bark gives the "all clear" signal.

- Up to 13½ in (34 cm)
- Up to 3½ in (9 cm)
- Up to 3½ lb (1.5 kg)
- Family band, colonial
- Near threatened

Shortgrass prairies of W. North America

Mountain beaver Strictly terrestrial,
the mountain beaver lives in a network
of underground tunnels and chambers.
It is mostly solitary but spends limited
time with other members of its species.

- Up to 16½ in (42 cm)
- Up to 2 in (5 cm)
- Up to 2½ lb (1.2 kg)
- Solitary
- Near threatened

S.W. Canada to N. California

MARMOTS

Restricted to the Northern Hemisphere,
marmots are found mainly in mountain
habitats. They are true hibernators,
and spend the harsh winter at rest in
their burrows, living off body fat. All
marmots, except for the woodchuck,
live in family groups. Young female
marmots often stay with their parents
to help raise their siblings.

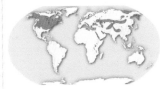

Face to face Like almost all other marmots,
Olympic marmots (Marmota olympus) are
highly social. Young remain dependent on
the mother for 2 years.

FACT FILE

Eastern gray squirrel In addition to maintaining a den in a hollow log, this squirrel builds a nest of twigs and leaves lined with grasses and shredded bark in the branches of a tree. The nest is used for resting and feeding, and may also serve as a nursery.

- Up to 11 in (28 cm)
- Up to 9½ in (24 cm)
- Up to 26½ oz (750 g)
- Solitary
- Common

S. Canada to Texas & Florida

Eurasian red squirrel Using its strong incisors, this squirrel can crack a tough nut in a few seconds. It spends much of the day collecting seeds and nuts, as well as fungi, birds' eggs, and tree sap.

- Up to 11 in (28 cm)
- Up to 9½ in (24 cm)
- Up to 10 oz (280 g)
- Solitary
- Near threatened

W. Europe to E. Russia, Korea & N. Japan

SQUIRREL LARDER

Like many tree squirrels in climates with harsh winters, the American red squirrel prepares for the cold months by storing food. It collects thousands of pine and spruce cones and caches them in a larder, known as a midden, which may be hidden under a log or inside a hollow stump. The territory surrounding the midden is vigorously defended.

⚑ CONSERVATION WATCH

Competing squirrels While the Eurasian red squirrel has remained common in much of central Europe, overhunting has led to a fall in numbers in eastern Europe. The species has now vanished from most parts of Great Britain, outcompeted for resources by the eastern gray squirrel, which was introduced from North America in 1902.

American red squirrel
Tamiasciurus hudsonicus

White band around eye

Variegated squirrel
Sciurus variegatoides

Eastern gray squirrel
Sciurus carolinensis

Long tufts on ears in winter

Variegated squirrel
Sciurus variegatoides

Relies on ponderosa pines for food and shelter

Tassel-eared squirrel (Abert's squirrel)
Sciurus aberti

Eurasian red squirrel
Sciurus vulgaris

Eurasian red squirrel coat can be red or black

Coat thickens in winter

Guayaquil squirrel
Sciurus stramineus

Nests high in forest canopy but feeds at lower levels

Prevost's squirrel
Callosciurus prevostii

Horse-tailed squirrel
Sundasciurus hippurus

Southern flying squirrel
Glaucomys volans

Gliding membrane, or patagium, is folded when squirrel is sitting

Siberian flying squirrel
Pteromys volans

Soil often tints coat

Red bush squirrel
Paraxerus palliatus

Handles food with dextrous forepaws

Striped ground squirrel
Xerus erythropus

Gambian sun squirrel
Heliosciurus gambianus

Sits up on hindlegs to eat or look for danger

Smallest squirrel in the world

African pygmy squirrel
Myoscuirus pumilio

Five black stripes on back

Eastern chipmunk
Tamias striatus

FACT FILE

Southern flying squirrel While this nocturnal glider eats mainly nuts and acorns, it also consumes many insects and young birds. It often lives in pairs, but larger groups may den together during the winter months.

🐾 Up to 5½ in (14 cm)
↔ Up to 4½ in (12 cm)
⚖ Up to 3 oz (85 g)
🐾🐾 Pair, small group
☘ Locally common

S. Canada to E. USA

Striped ground squirrel Like prairie dogs, this gregarious animal lives in colonies and is highly vocal, warning of danger with an alarm call.

🐾 Up to 16 in (40 cm)
↔ Up to 12 in (30 cm)
⚖ Up to 2 lb (1 kg)
🐾🐾 Colonial
☘ Locally common

W. Africa to Kenya

Eastern chipmunk This usually solitary species shelters in a burrow. When its cheek pouches are stuffed with food, they can be as large as its head.

🐾 Up to 6½ in (17 cm)
↔ Up to 4½ in (12 cm)
⚖ Up to 5½ oz (150 g)
🐾🐾 Solitary
☘ Locally common

S.E. North America

African pygmy squirrel About the size of a man's thumb, this species is the smallest of all squirrels. It lives in the hollow trunks of trees.

🐾 Up to 3 in (7.5 cm)
↔ Up to 2½ in (6 cm)
⚖ Up to ⅓ oz (17 g)
☘ Solitary
🏛 Vulnerable

Equatorial Africa

GLIDING

A flying squirrel can glide for more than 330 feet (100 m), using much less energy than climbing would require and allowing it to escape flightless predators. In most species, the gliding membrane that extends from wrist to ankle is tucked away when climbing.

Putting on the brakes
A flying squirrel brakes by raising the tail, and forms a parachute with its membrane by moving the limbs forward.

LUMBERJACK RODENTS

The great engineers of the animal world, beavers deliberately alter their environment by building dams, canals, and lodges. While often causing conflict with farmers and other humans, this construction work has an important ecological function, helping to reduce erosion and flooding and creating new habitats for aquatic species. Beavers live in family groups of a monogamous pair and several offspring. They communicate using various calls and postures and will slap their tails against the water to warn of danger. Although similar in appearance and behavior, the two species of beaver—the North American *Castor canadensis* and the Eurasian *C. fiber*—do not interbreed.

Chopping chisels Like all rodents, beavers have self-sharpening incisor teeth that never stop growing. The outer surface is protected by tough enamel, but the inner surface is softer and wears away as the beaver gnaws, creating a sharp, chiseled edge.

Winter quiet Beavers usually feed and build at night. Through snowy winters, however, they rarely emerge from their lodge, which provides a warm environment. For food, they rely on sticks and logs stored underwater, as well as the fat stored in the tail.

Lodges and dams A beaver colony may share a riverbank burrow system or they may build a lodge in the water. A lodge is a dome of sticks and mud with underwater entrances leading to a vegetation-lined living area above the water level. To create a calm pond for their lodge, beavers will construct dam walls that stop the flow of water. They will also dig out canals to link their dam to nearby sources of food and construction material. Several generations of beavers may maintain a dam, but eventually the pond silts up and the resident family must find a new location for their home.

Aquatic adaptations A beaver uses its flat, scaly tail and webbed rear feet to propel its streamlined body through the water. Clear eyelids shield the eyes underwater, while valved nostrils and ears keep out water. Thick, oil-coated fur insulates the animal in cold water.

Beaver babies Beaver litters contain an average of two to four kits, which are nursed for 6–8 weeks. The kits grow quickly, but will stay with their family group for up to 2 years so they can learn the craft of building dams and lodges.

Stopping the flow Beavers use mud, stones, sticks, and branches to construct a dam wall. The pond this creates acts as a moat around their lodge and deters most predators.

Rapid recovery Beavers prefer aspens, poplars, alders, and willows. These are all trees that grow rapidly and may even be reinvigorated after being felled by a beaver.

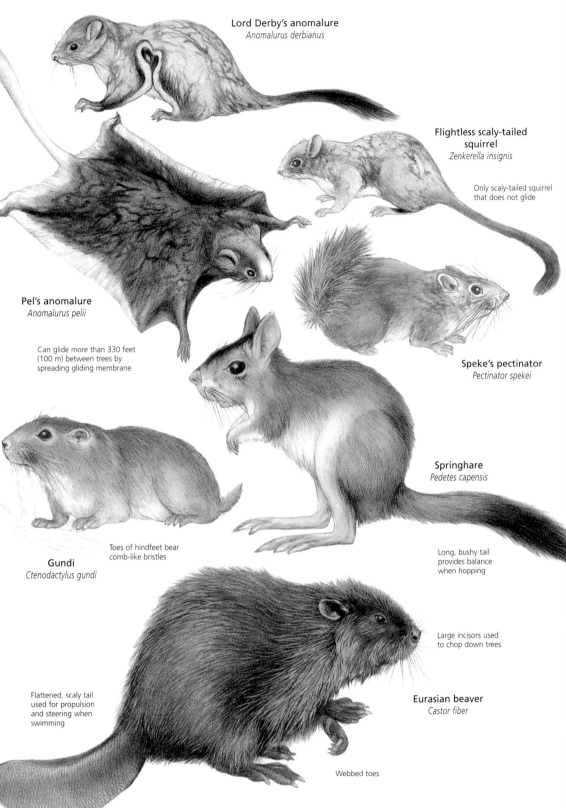

Lord Derby's anomalure
Anomalurus derbianus

Flightless scaly-tailed
squirrel
Zenkerella insignis

Only scaly-tailed squirrel
that does not glide

Pel's anomalure
Anomalurus pelii

Can glide more than 330 feet
(100 m) between trees by
spreading gliding membrane

Speke's pectinator
Pectinator spekei

Gundi
Ctenodactylus gundi

Toes of hindfeet bear
comb-like bristles

Springhare
Pedetes capensis

Long, bushy tail
provides balance
when hopping

Large incisors used
to chop down trees

Eurasian beaver
Castor fiber

Flattened, scaly tail
used for propulsion
and steering when
swimming

Webbed toes

FACT FILE

Gundi Once classified as a cavy-like rodent, but now grouped with squirrel-like and mouse-like rodents in the suborder Sciurognathi, the gundi eats a variety of desert plants and never drinks, extracting water from its food instead.

🐾 Up to 8 in (20 cm)
🐾 Up to 1 in (2.5 cm)
⬛ Up to 10 oz (290 g)
🐾 Family group, colonial
🌿 Locally common

N. Africa

Springhare Like a little kangaroo, the springhare has long hindlimbs that it uses for hopping. After sheltering from the heat of the day in its burrow, this rodent emerges at night to forage on grasses and crops.

🐾 Up to 17 in (43 cm)
🐾 Up to 18½ in (47 cm)
⬛ Up to 9 lb (4 kg)
🐾 Solitary
🌿 Vulnerable

E. & S. Africa

Eurasian beaver This semiaquatic rodent is strictly herbivorous, feeding mainly on water plants. It will also eat aspen trees, while it chops down alder and oak trees as building material.

🐾 Up to 31½ in (80 cm)
🐾 Up to 17½ in (45 cm)
⬛ Up to 55 lb (25 kg)
🐾 Pair, family band
🌿 Near threatened

W. Europe to E. Siberia

SCALY TAILS

The scaly-tailed squirrels of the Anomaluridae family are not directly related to the true squirrels of Sciuridae. All but one species moves about by gliding, an adaptation that developed independently in flying squirrels. The scales near the base of the tail help scaly-tailed squirrels cling to trees at the end of a glide and then climb back up the trunk.

🌿 CONSERVATION WATCH

Hollow sites Most of tropical Africa's scaly-tailed squirrels are considered near threatened. They depend on hollow trees for their nesting sites, but such trees are found only in old-growth forests, which are rapidly being cleared for agriculture.

FACT FILE

Heteromyidae This family includes pocket mice and kangaroo rats and mice. They are closely related to the pocket gophers of the Geomyidae family, with whom they share cheek pouches and a burrowing lifestyle.

Long-tailed pocket mouse This species is most often found in gravelly desert areas. During drought, females may avoid producing a litter.

🐁 Up to 4 in (10 cm)
🐀 Up to 4½ in (12 cm)
⚖ Up to 1 oz (25 g)
♦ Solitary
🗡 Common

Nevada & Utah to Baja California

Mexican spiny pocket mouse This species can breed at any time of year, allowing it to take advantage of favorable conditions when they arrive.

🐁 Up to 5 in (13 cm)
🐀 Up to 5 in (13 cm)
⚖ Up to 2 oz (60 g)
♦ Solitary
🗡 Uncommon
🏛 ♣♣

S. Texas to Mexico

Plains pocket gopher This solitary rodent digs an extensive burrow, with tunnels leading to a central chamber. During the mating season, a male may tunnel through to a female's burrow.

🐁 Up to 8 in (20 cm)
🐀 Up to 4½ in (12 cm)
⚖ Up to 9 oz (250 g)
♦ Solitary
🗡 Common

Tallgrass prairies from S. Canada to Texas

Big-eared kangaroo rat Like other kangaroo rats, this species has long hindlimbs and moves by hopping. The short forelimbs are used for feeding.

🐁 Up to 6 in (15 cm)
🐀 Up to 8 in (20 cm)
⚖ Up to 3 oz (90 g)
♦ Solitary
🗡 Not known

California

Desert kangaroo rat To conserve water in its arid environment, this rodent emerges from its burrow only at night, when humidity is highest, and concentrates its urine. It rarely drinks, obtaining most of its water from food.

🐁 Up to 6 in (15 cm)
🐀 Up to 8½ in (21 cm)
⚖ Up to 5½ oz (150 g)
♦ Solitary
🗡 Common

Nevada to N. Mexico

Long-tailed pocket mouse
Chaetodipus formosus

Tail longer than head and body

Mexican spiny pocket mouse
Liomys irroratus

Trinidad spiny pocket mouse
Heteromys anomalus

Harsh, bristly fur

Large, projecting teeth used for cutting roots or digging

Thick, ridged skull

Loose skin allows pocket gopher to maneuver in tight burrows

Enlarged claws used to dig burrows

Plains pocket gopher
Geomys bursarius

Botta's pocket gopher
Thomomys bottae

Usually moves in hops

Desert kangaroo rat
Dipodomys deserti

Big-eared kangaroo rat
Dipodomys elephantinus

Long tail provides stability when hopping

MOUSE-LIKE RODENTS

ᴀss Mammalia
ᴅᴇʀ Rodentia
ᴍɪʟɪᴇs 3
ɴᴇʀᴀ 306
ᴇᴄɪᴇs 1,409

More than a quarter of all mammal species are mouse-like rodents. Once grouped within their own suborder, these rodents share an arrangement of the chewing muscles that provides a versatile gnawing action. They also all have a maximum of three cheekteeth in each row. While their lifespans tend to be short, most are early and prolific breeders. The group is dominated by the Muridae family, which has more than 1,000 species, including Old World and New World rats and mice; voles and lemmings; hamsters; and gerbils. The other families of mouse-like rodents are the dormice of Myoxidae and the jumping mice, birchmice, and jerboas of Dipodidae.

On the scent As many as 50 house mice may live in a family group. They leave deposits of urine as scentmarks throughout their home territory, enabling them to recognize each other and detect intruders.

RAPID RADIATION

The first members of the Muridae family appeared only several million years ago, making it very young in evolutionary terms. Since that time, however, the family has diversified dramatically and now occupies almost every terrestrial habitat in the world, from polar regions to desert. The majority of its members are small, nocturnal, seed-eating ground-dwellers with a pointed face and long whiskers, but some spend much of their time in water or trees and others live underground.

There are more than 500 species of Old World rats and mice. These are highly varied but include the ubiquitous house mouse and brown rat, both of which are well known as urban pests. New World rats and mice range from climbing rats to fish-eating rats, but most live on the ground in forests or grasslands.

While rats and mice account for 80 percent of species in the family Muridae, voles and lemmings, hamsters, and gerbils form distinct subfamilies. Voles and lemmings, found throughout the Northern Hemisphere, have adapted to a diet of tough grasses. Many spend winter living in tunnels beneath the snow. Although popular as children's pets, the hamsters of Eurasia are decidedly solitary in the wild and will react very aggressively to intruders. Gerbils are found mainly in arid parts of Africa and Asia.

The Myoxidae and Dipodidae families are smaller and more specialized than Muridae. Dormice tend to live in trees and hibernate through cold winters. Jumping mice, birchmice, and jerboas all have long back feet and long tails that enable them to move by hopping. Jerboas have evolved to survive in some of the world's harshest deserts.

Fruit feeder Dormice (above) and most other mouse-like rodents are mainly herbivorous, existing on a diet of seeds, fruit, and buds supplemented by the occasional insect. Voles and lemmings have specialized to feed on grasses. A few species are more carnivorous. Water rats will sometimes add a turtle or bat to their diet of aquatic invertebrates, while brown rats may even attack poultry or rabbits.

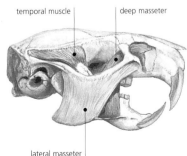

Big families Rats and mice are prolific breeders. Most mature quickly, have short gestation periods, and produce many large litters. In some species, a single pair and their offspring can produce thousands of animals in less than a year.

Versatile action The arrangement of their jaw muscles provides mouse-like rodents with a versatile gnawing action. The deep masseter extends onto the upper jaw and works together with the lateral masseter to pull the jaw forward for chewing.

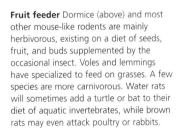

temporal muscle | deep masseter

lateral masseter

Golden hamster Now a popular pet and the best known of the hamsters, this species is endangered in the wild. It was introduced to the United States and England in the 1930s and has since proliferated in captivity.

Up to 7 in (18 cm)
Up to ¾ in (2 cm)
Up to 5½ oz (150 g)
Solitary
Endangered

Middle East, S.E. Europe, S.W. Asia

European hamster This solitary burrower hibernates through winter, waking once a week or so to feed on its massive hoard of seeds and roots. During warmer months, it stocks up this winter food supply, carrying plant matter in its cheek pouches.

Up to 12½ in (32 cm)
Up to 2½ in (6 cm)
Up to 13½ oz (385 g)
Solitary
Common

Belgium to Altai Mts of C. Asia

Eastern woodrat Nocturnal and solitary, this species falls prey to owls, weasels, and snakes. It protects its nest by building a shelter of sticks, bones, and leaves in a rock crevice or between tree roots.

Up to 10½ in (27 cm)
Up to 7 in (18 cm)
Up to 9 oz (260 g)
Solitary
Common

S.E. USA

Hispid cotton rat After a gestation of just 27 days, females of this species give birth to several fully furred young. The female is ready to mate again almost immediately, while the young are sexually mature within 40 days.

Up to 8 in (20 cm)
Up to 6½ in (16 cm)
Up to 8 oz (225 g)
Solitary
Common

S.E. USA to N. Venezuela & N. Peru

Deer mouse This small omnivore has adapted to diverse habitats, from boreal forest to desert. It breeds very quickly, with females producing up to four litters of 4–9 young each year.

Up to 4 in (10 cm)
Up to 4½ in (12 cm)
Up to 1 oz (30 g)
Solitary
Common

North America except tundra & far S.E. USA

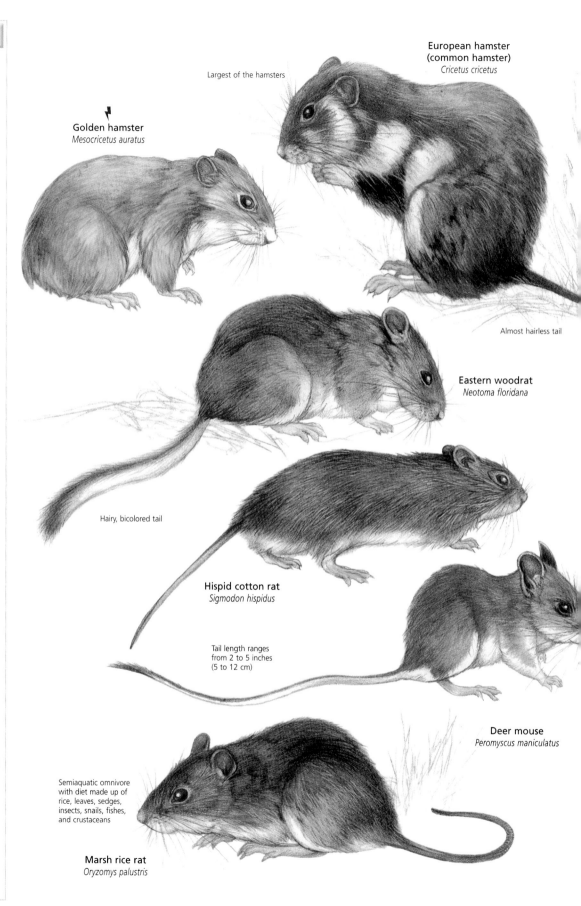

European hamster
(common hamster)
Cricetus cricetus

Largest of the hamsters

Golden hamster
Mesocricetus auratus

Almost hairless tail

Eastern woodrat
Neotoma floridana

Hairy, bicolored tail

Hispid cotton rat
Sigmodon hispidus

Tail length ranges from 2 to 5 inches (5 to 12 cm)

Deer mouse
Peromyscus maniculatus

Semiaquatic omnivore with diet made up of rice, leaves, sedges, insects, snails, fishes, and crustaceans

Marsh rice rat
Oryzomys palustris

Long-clawed mole-vole
Prometheomys schaposchnikowi

Long claws for digging burrows

Siberian collared lemming
(Arctic lemming)
Dicrostonyx torquatus

Lives farther north than any other rodent

White winter coat

Brown summer coat

European water vole
Arvicola terrestris

Bank vole
Clethrionomys glareolus

Great gerbil
Rhombomys opimus

Wood lemming
Myopus schisticolor

Norway lemming
Lemmus lemmus

Tail flattened vertically for use as rudder

Small webs between toes

Muskrat
Ondatra zibethicus

Largest of the voles

Libyan jird
Meriones lybicus

FACT FILE

Siberian collared lemming This tundra species spends summer in shallow burrows in high, rocky areas. In winter, it moves to lower meadows and shelters in tunnels under the snow.

- Up to 6 in (15 cm)
- Up to ½ in (1 cm)
- Up to 3 oz (90 g)
- Solitary
- Common

Arctic Eurasia

Great gerbil To survive cold winters, large groups of these gerbils huddle together in extensive burrows, not only providing each other with warmth but also protecting their stockpile of plants.

- Up to 8 in (20 cm)
- Up to 6½ in (16 cm)
- Not known
- Colonial
- Common

Caspian Sea to Mongolia, China & Pakistan

Muskrat This semiaquatic rodent swims with its large, webbed back feet and uses its naked tail as a rudder. Like the beaver, it lives in a group in a riverbank burrow or a lodge of twigs and mud.

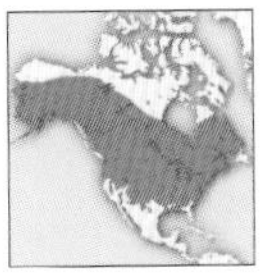

- Up to 13 in (33 cm)
- Up to 12 in (30 cm)
- Up to 4 lb (1.8 kg)
- Small to large group
- Common

USA & Canada except tundra; introd. Eurasia

FLUCTUATING LEMMINGS

Contrary to popular myth, Norway lemmings are not deliberately suicidal, but every 3 or 4 years, their number rises. Intolerant of one another at the best of times, the lemmings become highly aggressive. Such conflicts may trigger mass movements from the crowded alpine tundra to lower forests. When the lemmings meet obstacles such as rivers, panic can cause them to take flight, with some ending up in the sea.

Fighting techniques Norway lemmings may wrestle, box, or adopt dominating postures.

FACT FILE

Smooth-tailed giant rat This arboreal rat makes its nest in a tree hollow and is almost fully herbivorous, feeding mainly on shoots.

🐭	Up to 14½ in (37 cm)
🐀	Up to 16 in (41 cm)
⚖	Up to 3 lb (1.3 kg)
🐾	Solitary
♦	Common

Central highlands of New Guinea

Harvest mouse One of the smallest mice, this species lives among tall crops, reeds, bamboo, or long grass. For each litter, both parents spend days building a globe-shaped nest that is suspended between stalks above the ground.

🐭	Up to 1 in (2.5 cm)
🐀	Up to 1 in (2.5 cm)
⚖	Up to ¼ oz (7 g)
🐾	Solitary
♦	Near threatened

England & Spain to China, Korea & Japan

Striped grass mouse This savanna resident shelters in an underground burrow or an abandoned termite nest. It builds a roundish nest for its litters, which are born during the rainy season after a gestation of 28 days.

🐭	Up to 5½ in (14 cm)
🐀	Up to 6 in (15 cm)
⚖	Up to 2½ oz (68 g)
🐾	Solitary
♦	Common

Sub-Saharan Africa

House mouse Through its association with humans, this species has been able to spread throughout the world. It nests in buildings or nearby fields and will eat almost any human food as well as items such as glue and soap.

🐭	Up to 4 in (10 cm)
🐀	Up to 4 in (10 cm)
⚖	Up to 1 oz (30 g)
🐾	Variable
♦	Abundant, often regarded as pest

Worldwide except tundra & polar regions

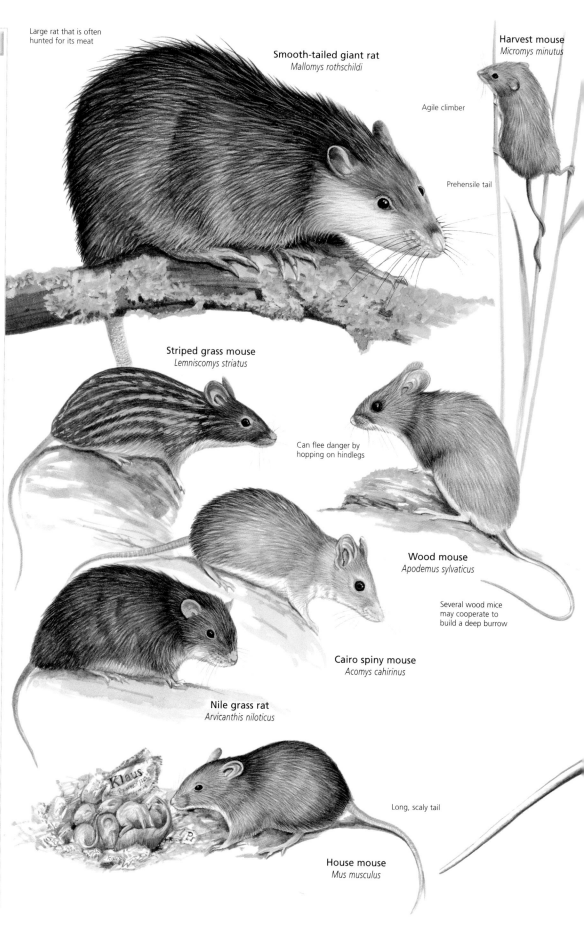

Large rat that is often hunted for its meat

Smooth-tailed giant rat
Mallomys rothschildi

Harvest mouse
Micromys minutus

Agile climber

Prehensile tail

Striped grass mouse
Lemniscomys striatus

Can flee danger by hopping on hindlegs

Wood mouse
Apodemus sylvaticus

Several wood mice may cooperate to build a deep burrow

Cairo spiny mouse
Acomys cahirinus

Nile grass rat
Arvicanthis niloticus

Long, scaly tail

House mouse
Mus musculus

Greater stick-nest rat
Leporillus conditor

Builds house of sticks as shelter from desert heat

Black rat is most often black but can be brown

Black rat
(roof rat, ship rat)
Rattus rattus

Brown rat
(Norway rat)
Rattus norvegicus

Short-tailed bandicoot rat
Nesokia indica

Feet adapted for climbing

Greater bandicoot rat
Bandicota indica

Natal multimammate mouse
Mastomys natalensis

Head and body can measure up to 17½ inches (45 cm)

Giant rat
Cricetomys gambianus

Cheek pouches for carrying food and nest material

FACT FILE

Greater stick-nest rat Once found throughout the shrublands of southern Australia, this small rat became extinct on the mainland when rabbits and sheep overgrazed their habitat. Various reintroduction campaigns are underway.

- Up to 10 in (26 cm)
- Up to 7 in (18 cm)
- Up to 16 oz (450 g)
- Family group
- Endangered

Franklin I. (Australia); reintrod. in several sites
● Former range

Short-tailed bandicoot rat This nocturnal burrowing species will raid crops and lawn grass for food. It stockpiles food in its burrow.

- Up to 8½ in (21 cm)
- Up to 5 in (13 cm)
- Up to 6 oz (175 g)
- Solitary
- Abundant

Egypt, Palestine & Syria to N. India

Greater bandicoot rat Found in both cultivated and urban areas, this species is considered a serious pest. It lives alone in a complex burrow system.

- Up to 14 in (36 cm)
- Up to 11 in (28 cm)
- Up to 3½ lb (1.5 kg)
- Solitary
- Abundant

S. Asia to China & S.E. Asia

RATS AS PESTS

Along with the house mouse, the black rat and the brown rat have had a close relationship with humans, living off crops and stored food. In the process, they have caused untold damage and spread serious disease, including the bubonic plague, which killed a third of Europe's population in the Middle Ages. Both rats live in large packs, with up to 60 black rats and 200 brown rats to a pack, and will vigorously defend their feeding area from intruders. Brown rats will eat almost anything and may even attack rabbits or human babies.

Disease carriers *By spreading disease, the brown rat has caused more deaths than all the wars in history.*

FACT FILE

Gray climbing mouse With its long, prehensile tail, this little mouse can easily climb the grasses and shrubs of its savanna home. It may dig a simple burrow to shelter from seasonal fires, but more often occupies a globe-shaped grass nest above the ground.

- 🐀 Up to 2¾ in (7 cm)
- 🐁 Up to 3 in (8 cm)
- ⚖ Up to ¼ oz (8 g)
- ♟ Solitary
- ♙ Common

Sub-Saharan Africa

Vlei rat Like voles and lemmings, this rat lives in moist grassland, and tunnels through the grass to its feeding areas. It will enter water to escape predators.

- 🐀 Up to 8½ in (22 cm)
- 🐁 Up to 4½ in (11 cm)
- ⚖ Up to 6½ oz (180 g)
- ♟ Solitary
- ♙ Locally common

S. Africa

Fawn hopping mouse If startled, this nocturnal burrower will bound away on its long hindlegs. Its urine is highly concentrated to help conserve water.

- 🐀 Up to 4¾ in (12 cm)
- 🐁 Up to 16½ in (16 cm)
- ⚖ Up to 1¾ oz (50 g)
- ♟ Family group
- ♙ Near threatened

Gibber desert of C. Australia

WATER RAT

The Australian water rat lives in burrows that run along the banks of rivers or lakes. Able to withstand pollution, it is often found in urban areas. It depends on fresh water for the bulk of its diet, which includes crustaceans, mollusks, and fish. Broad, webbed feet act as paddles in the water. Its coat is not waterproof, but a pad of fat helps keep the heart warm.

- 🐀 Up to 15½ in (39 cm)
- 🐁 Up to 12½ in (32 cm)
- ⚖ Up to 2½ lb (1.2 kg)
- ♟ Solitary
- ♙ Locally common

Papua New Guinea, Australia & Tasmania

Aquatic predator
The water rat often takes its catch to a preferred feeding platform.

Cloud rat
Phloeomys cumingi

Gray climbing mouse
Dendromus melanotis

Long tail is semiprehensile

Luzon bushy-tailed cloud rat
Crateromys schadenbergi

Long muzzle and small eyes resemble those of a shrew

Mount Data shrew rat
Rhynchomys soricoides

Vlei rat
Otomys irroratus

Australian water rat
Hydromys chrysogaster

Thick tail with white tip

Fawn hopping mouse
Notomys cervinus

Golden-backed tree rat
Mesembriomys macrurus

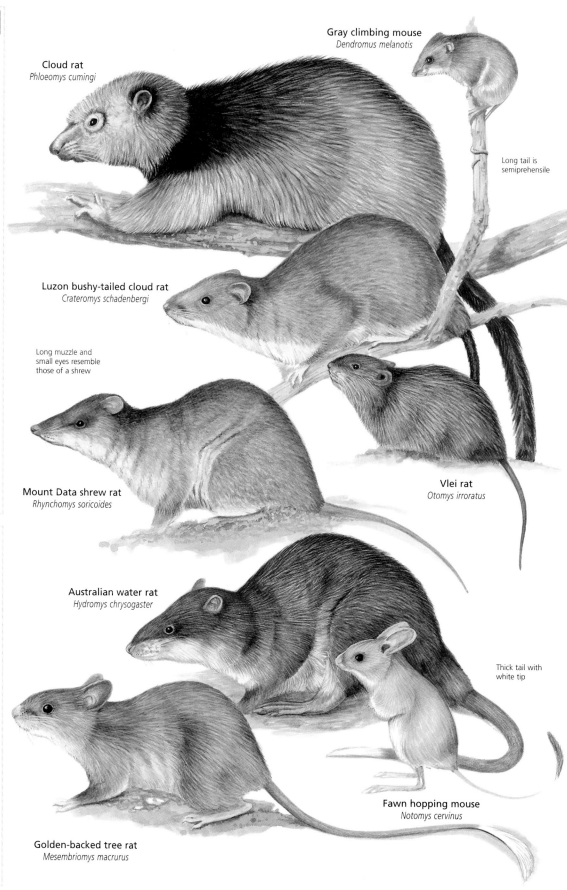

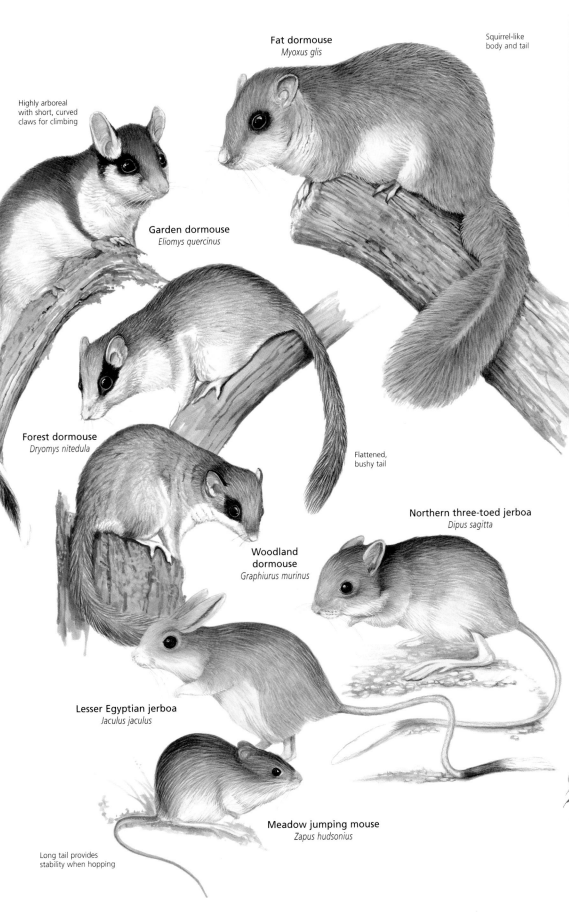

Fat dormouse
Myoxus glis

Squirrel-like body and tail

Highly arboreal with short, curved claws for climbing

Garden dormouse
Eliomys quercinus

Forest dormouse
Dryomys nitedula

Flattened, bushy tail

Woodland dormouse
Graphiurus murinus

Northern three-toed jerboa
Dipus sagitta

Lesser Egyptian jerboa
Jaculus jaculus

Meadow jumping mouse
Zapus hudsonius

Long tail provides stability when hopping

FACT FILE

Garden dormouse This noisy rodent lives in large colonies, making globe-shaped nests of leaves and grass in tree hollows, among shrubs, or in rock crevices. As well as eating acorns, nuts, and fruit, it hunts insects and small rodents and birds.

- Up to 6½ in (17 cm)
- Up to 5 in (13 cm)
- Up to 4 oz (120 g)
- Colonial
- Vulnerable

Europe

Lesser Egyptian jerboa With back legs four times longer than its front legs, this solitary desert animal will hop its way out of danger. It spends the day in a burrow, plugging the entrance with dirt during summer to create a cooler, humid microclimate.

- Up to 4 in (10 cm)
- Up to 5 in (13 cm)
- Up to 2 oz (55 g)
- Solitary
- Not known

Morocco & Senegal to S.W. Iran & Somalia

Meadow jumping mouse Although it usually moves in short hops, this mouse can leap up to 3 feet (1 m) in the air when startled. It breeds very soon after emerging from its winter hibernation.

- Up to 4 in (10 cm)
- Up to 5 in (13 cm)
- Up to 1 oz (30 g)
- Solitary
- Locally common

N. & E. North America

WINTER SLEEP

The dormice of Europe prepare for their long winter hibernation by laying down a layer of fat on their body and stockpiling food in their nest or burrow. Depending on the climate, they can spend up to 9 months of the year asleep. They start mating as soon as they emerge from hibernation.

CAVY-LIKE RODENTS

CLASS	Mammalia
ORDER	Rodentia
FAMILIES	18
GENERA	65
SPECIES	218

With their large heads, plump bodies, small legs, and short tails, the cavies of the Caviidae family—more commonly known as guinea pigs—are typical of cavy-like rodents. There are exceptions to this general body plan: some cavy-like rodents, such as the spiny rats of the Echimyidae family, look more like common mice and rats. All cavy-like rodents share a distinctive arrangement of the jaw muscles that gives them a strong forward bite. Unlike most other rodents, they also tend to have small litters of well-developed young. There are cavy-like rodents in both the Old World and the New World, but their relationship continues to be debated.

Spiky young New World porcupines, like other cavy-like rodents, give birth to well-developed young and have low infant mortality rates. Newborns have open eyes, can walk almost straight away, and are able to climb trees within a few days.

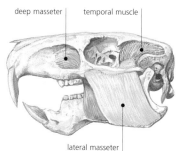

Powerful bite Like squirrel-like rodents, cavy-like rodents have a strong forward bite, but it is produced by a different arrangement of the jaw muscles. The lateral masseter closes the jaw, while the deep masseter extends past the eye and pulls the jaw forward for biting.

deep masseter | temporal muscle
lateral masseter

Rat-like form The spiny rats of the family Echimyidae look more like the common mice and rats of Muridae than most other cavy-like rodents. If grabbed by a predator, a spiny rat's tail will break off, allowing it to quickly escape.

Living in pairs The mara or Patagonian cavy (*Dolichotus patagonum*) is unusual among mammals in maintaining a lifelong monogamous pair bond. One member of the pair watches for danger while the other feeds. Different pairs of maras rarely interact, but they do share a communal nursery den, where young are visited daily by their parents for nursing.

CAVIES AND THEIR KIN

Although controversy persists as to whether the South American cavy-like rodents came from North America or rafted over from Africa, most of today's cavy-like rodents are found in Central and South America. The cavies of Caviomorpha include not only guinea pigs and similar species, but also the deer-like mara, a long-legged grazer. The semiaquatic capybara is more than 3 feet (1 m) long, making it the largest of all rodents. Chinchillas and viscachas live mainly at high elevations, kept warm by a thick, soft coat. Agoutis have long, slender limbs that allow them to run swiftly from danger. While most of South America's cavy-like rodents are terrestrial, tuco-tucos dig complex burrows. Other South American cavy-like rodents include degus, hutias, pacas, coypu, and pacarana.

New World porcupines, found in both North and South America, are arboreal and climb trees with agility, aided in some species by a prehensile tail. They share many features with the Old World porcupines of Africa, Asia, and Europe, but the latter are mostly ground-dwelling.

As well as porcupines, the Old World's cavy-like rodents include the African mole-rats, cane rats, and dassie rat. The gundis of North Africa (family Ctenodactylidae), have cavy-like jaw muscles, but are now classified in the suborder Sciurognathi. All other cavy-like rodents are now grouped together in the suborder Hystricognathi.

⚡ **CONSERVATION WATCH**

Of the 218 species of cavy-like rodents, 33% are listed on the IUCN Red List, as follows:

12	Extinct
8	Critically endangered
3	Endangered
15	Vulnerable
24	Near threatened
9	Data deficient

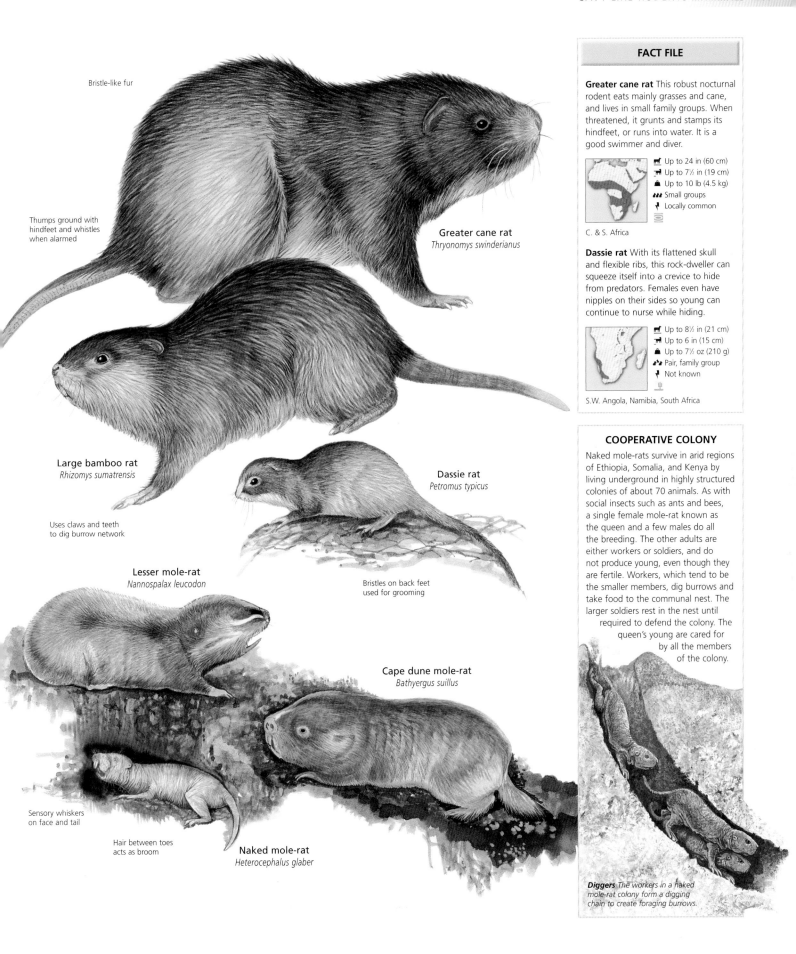

Bristle-like fur

Thumps ground with hindfeet and whistles when alarmed

Greater cane rat
Thryonomys swinderianus

Large bamboo rat
Rhizomys sumatrensis

Uses claws and teeth to dig burrow network

Dassie rat
Petromus typicus

Lesser mole-rat
Nannospalax leucodon

Bristles on back feet used for grooming

Cape dune mole-rat
Bathyergus suillus

Sensory whiskers on face and tail

Hair between toes acts as broom

Naked mole-rat
Heterocephalus glaber

FACT FILE

Greater cane rat This robust nocturnal rodent eats mainly grasses and cane, and lives in small family groups. When threatened, it grunts and stamps its hindfeet, or runs into water. It is a good swimmer and diver.

🐂 Up to 24 in (60 cm)
🐀 Up to 7½ in (19 cm)
⚖ Up to 10 lb (4.5 kg)
♦♦♦ Small groups
🕯 Locally common
👁

C. & S. Africa

Dassie rat With its flattened skull and flexible ribs, this rock-dweller can squeeze itself into a crevice to hide from predators. Females even have nipples on their sides so young can continue to nurse while hiding.

🐂 Up to 8½ in (21 cm)
🐀 Up to 6 in (15 cm)
⚖ Up to 7½ oz (210 g)
♦♦ Pair, family group
🕯 Not known
🕯

S.W. Angola, Namibia, South Africa

COOPERATIVE COLONY

Naked mole-rats survive in arid regions of Ethiopia, Somalia, and Kenya by living underground in highly structured colonies of about 70 animals. As with social insects such as ants and bees, a single female mole-rat known as the queen and a few males do all the breeding. The other adults are either workers or soldiers, and do not produce young, even though they are fertile. Workers, which tend to be the smaller members, dig burrows and take food to the communal nest. The larger soldiers rest in the nest until required to defend the colony. The queen's young are cared for by all the members of the colony.

Diggers The workers in a naked mole-rat colony form a digging chain to create foraging burrows.

FACT FILE

Old World porcupines These porcupines are generally terrestrial and their quills are embedded in clusters, while New World species are arboreal with singly embedded quills. The Old World species can be divided into brush-tailed porcupines (with long, slender tails that rustle when shaken), and crested porcupines (with long black and white spines and a short tail that rattles).

Crested porcupine This porcupine has been known to kill lions, hyenas, and humans. When threatened, it raises its quills to make itself seem larger, but if this display fails to deter, it will then back itself into the predator so that the quills become embedded.

🐀 Up to 27½ in (70 cm)
🐀 Up to 4½ in (12 cm)
🏋 Up to 33 lb (15 kg)
🐾 Family group
🏃 Near threatened

Italy, Balkans, Africa

African brush-tailed porcupine This species lives in a family group made up of a breeding pair and several offspring. They spend the day together hiding in caves, crevices, or logs, emerging at night to forage alone for roots, leaves, fruit, and bulbs.

🐀 Up to 22½ in (57 cm)
🐀 Up to 9 in (23 cm)
🏋 Up to 9 lb (4 kg)
🐾 Family group
🏃 Common

Equatorial Africa

Long-tailed porcupine Unlike other Old World porcupines, this forest-dweller cannot bristle or rattle its quills, but the long tail can break off from the body if snatched by a predator. It climbs trees to reach fruit and other food.

🐀 Up to 19 in (48 cm)
🐀 Up to 9 in (23 cm)
🏋 Up to 5 lb (2.3 kg)
🏃 Not known
🏃 Not known

Malaya, Sumatra, Borneo

⚡ CONSERVATION WATCH

Spiky protection With their spines forming effective armor, porcupines have few natural predators. They are, however, killed by humans for food, sport, and pest control. Most Old World species are common, but the Malayan porcupine is listed as vulnerable, and the crested porcupine and the thick-spined porcupine (*Hystrix crassispinis*) are considered near threatened.

Dark quills along neck can be raised into a crest

Crested porcupine
Hystrix cristata

Indian porcupine
Hystrix indica

Sumatran porcupine
Hystrix sumatrae

Tail quills rattle when shaken

Malayan porcupine
Hystrix brachyura

Quills cannot bristle or rattle

Scaly tail with bristled tip

African brush-tailed porcupine
Atherurus africanus

Long-tailed porcupine
Trichys fasciculata

Partially webbed feet

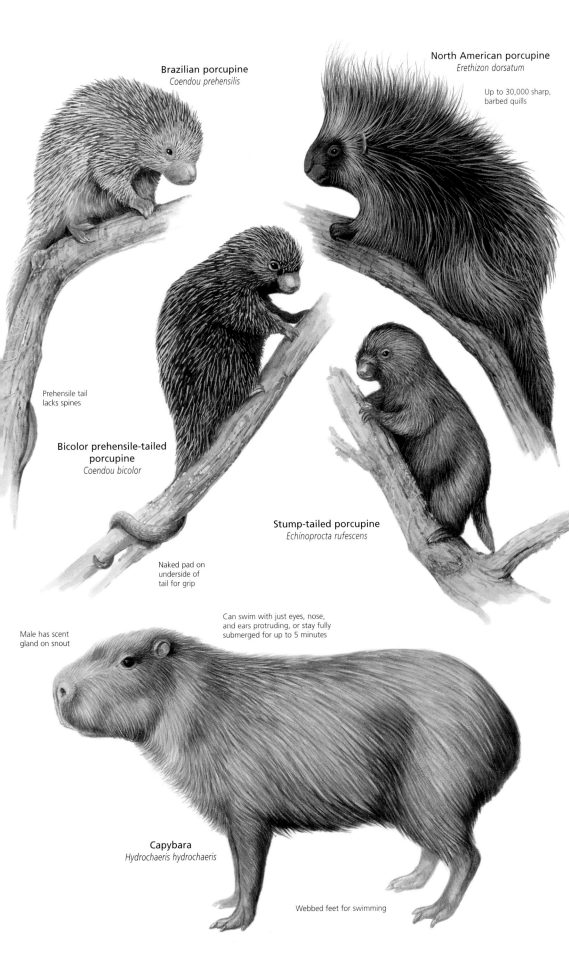

Brazilian porcupine
Coendou prehensilis

North American porcupine
Erethizon dorsatum

Up to 30,000 sharp, barbed quills

Prehensile tail lacks spines

Bicolor prehensile-tailed porcupine
Coendou bicolor

Naked pad on underside of tail for grip

Stump-tailed porcupine
Echinoprocta rufescens

Male has scent gland on snout

Can swim with just eyes, nose, and ears protruding, or stay fully submerged for up to 5 minutes

Capybara
Hydrochaeris hydrochaeris

Webbed feet for swimming

FACT FILE

New World porcupines Aided by large, gripping feet with strong claws and naked soles, these porcupines can climb trees. They have poor eyesight, but good senses of smell and hearing.

North American porcupine This species dens in a cave, rock crevice, or fallen log, emerging at night to browse on the bark of trees and shrubs.

🐾 Up to 3½ ft (1.1 m)
🐀 Up to 10 in (25 cm)
🏋 Up to 39½ lb (18 kg)
🐾 Solitary
🕴 Locally common

N. & W. North America

Bicolor prehensile-tailed porcupine A long, gripping tail helps this species clamber through the middle and upper layers of the forest. It climbs down to the ground only occasionally.

🐾 Up to 19½ in (49 cm)
🐀 Up to 21 in (54 cm)
🏋 Up to 10½ lb (4.7 kg)
🐾 Pair
🕴 Locally common

E. foothills of Andes from Colombia to Bolivia

Stump-tailed porcupine This little-studied species has a short, hairy tail. Its spines become thicker and shorter toward the rear.

🐾 Up to 14½ in (37 cm)
🐀 Up to 6 in (15 cm)
🏋 Not known
🐾 Not known
🕴 Not known

Andes in Colombia

THE BIGGEST RODENT

Capybaras are barrel-shaped grazers that feed mainly on grasses growing in or near water. They enter the water to seek refuge from the midday heat, to escape predators, and to mate. These social animals usually live in family groups of one male, several females, and their young, but in dry times may form larger temporary herds of up to a hundred individuals.

🐾 Up to 4½ ft (1.3 m)
🐀 Up to ¾ in (2 cm)
🏋 Up to 143 lb (65 kg)
🐾 Family group
🕴 Locally common

Panama to N.E. Argentina

AGOUTIS AND ACOUCHIS

The young of agoutis and acouchis are born after a relatively long gestation period of approximately 100 days. They emerge fully furred and with their eyes open. Within hours, they are able to run and will nibble on green plants. Even so, they continue to drink their mother's milk for some weeks. Their potential lifespan is long, up to 17 years, but most do not make it through their first year, falling prey to predators such as coatis or starving during the dry season.

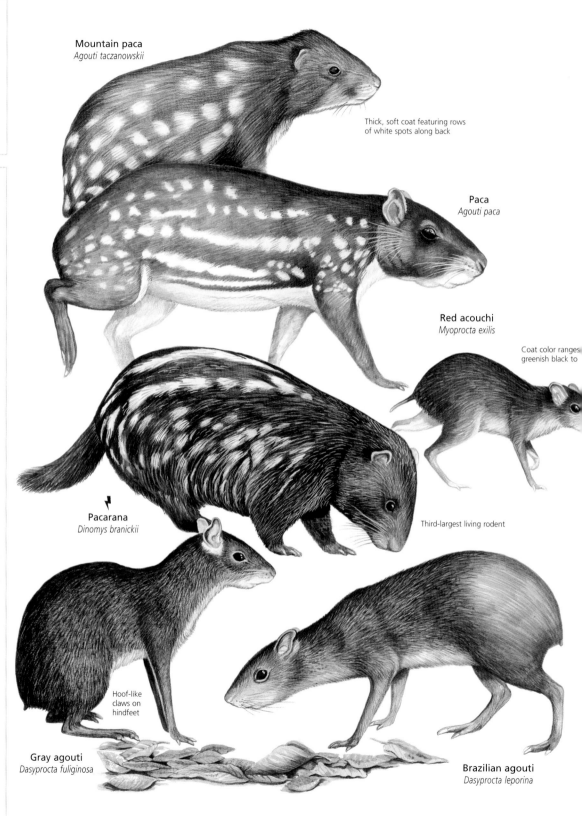

Mountain paca
Agouti taczanowskii

Thick, soft coat featuring rows of white spots along back

Paca
Agouti paca

Red acouchi
Myoprocta exilis

Coat color ranges greenish black to

Pacarana
Dinomys branickii

Third-largest living rodent

Gray agouti
Dasyprocta fuliginosa

Hoof-like claws on hindfeet

Brazilian agouti
Dasyprocta leporina

Demarest's hutia
Capromys pilorides

Hispaniolan hutia
Plagiodontia aedium

Scaly tail

Hairy tail

Prehensile-tailed hutia
Mysateles prehensilis

Brown's hutia
Geocapromys brownii

Coypu (nutria)
Myocastor coypus

Fringes of bristles
on hindfeet used
for grooming

White-bellied tuco-tuco
Ctenomys colburni

FACT FILE

Demarest's hutia While this species has strong claws and easily climbs trees, it spends more time on the ground than most other hutias. It shares with other hutias a stomach divided into three compartments, the most complex stomach of any rodent.

- Up to 24 in (60 cm)
- Up to 12 in (30 cm)
- Up to 18½ lb (8.5 kg)
- Pair
- Common, declining

Cuba & adjacent offshore islands

Brown's hutia This terrestrial animal usually lives alone, but can be found in family groups of up to 10 animals. It hides in rock crevices by day, but waddles out at night to search the forest for leaves, roots, bark, and fruit.

- Up to 17½ in (45 cm)
- Up to 2½ in (6 cm)
- Up to 4½ lb (2 kg)
- Not known
- Vulnerable

Jamaica

Coypu Equally at home in salt and fresh water, this semiaquatic species swims with its webbed hindfeet and can stay submerged for 5 minutes. It eats aquatic plants and shellfish.

- Up to 25 in (64 cm)
- Up to 16½ in (42 cm)
- Up to 22 lb (10 kg)
- Pair, family group
- Common

Bolivia & Brazil to Patagonia; introd. elsewhere

White-bellied tuco-tuco This robust burrower digs tunnels with the strong claws on its forefeet, cutting through any roots with its prominent incisors. Considered a pest, it has been hunted in great numbers.

- Up to 6½ in (17 cm)
- Up to 3 in (8 cm)
- Not known
- Not known
- Not known

Known from only two localities in Argentina

⚡ CONSERVATION WATCH

Vanishing hutias Living only in the West Indies, the hutias of the Capromyidae family are hunted intensively for food. They also fall prey to birds, snakes, and introduced domestic animals. In recent times, six species of hutia have become extinct and a further six are critically endangered, four are vulnerable, and two are near threatened.

FACT FILE

Coruro Living in complex burrows, this social species communicates through a variety of calls, including a long-distance musical trilling that can last for 2 minutes.

📏	Up to 6½ in (17 cm)
🐁	Up to 1½ in (4 cm)
⚖	Up to 4 oz (120 g)
👥	Colonial
🌱	Common, declining

C. Chile

Chinchilla-rat This little rodent has soft fur like that of the chinchilla but a body and head like that of a rat. It is nocturnal, resting by day in a burrow or rock crevice.

📏	Up to 7½ in (19 cm)
🐁	Up to 2¾ in (7 cm)
⚖	Not known
👥	Colonial
🌱	Not known

S.W. Peru, N. Chile & N.W. Argentina

Chinchilla Most commonly found in barren, rocky mountain areas, this nocturnal species lives in large colonies of up to 100 animals. Popular as pets, chinchillas are now rare in the wild.

📏	Up to 9 in (23 cm)
🐁	Up to 6 in (15 cm)
⚖	Up to 1 lb (500 g)
👥	Colonial
🌱	Vulnerable

Andes of N. Chile

DEGUS AND DOGS

Degus are the ecological equivalent of North America's prairie dogs (left). Both are diurnal rodents that live in large colonies in extensive burrow systems and communicate through a range of vocalizations. Degus and prairie dogs are only distantly related. Their similarities are the result of convergent evolution, in which both groups adapted in the same way to their semiarid habitats.

🌱 CONSERVATION WATCH

Hunted chinchillids Prized for their soft fur, the chinchillas and viscachas of the Chinchillidae family have been hunted in large numbers. In 1900 alone, 500,000 chinchilla pelts were exported from Chile. The short-tailed chinchilla (*C. brevicaudata*) is now critically endangered, while the chinchilla is considered vulnerable.

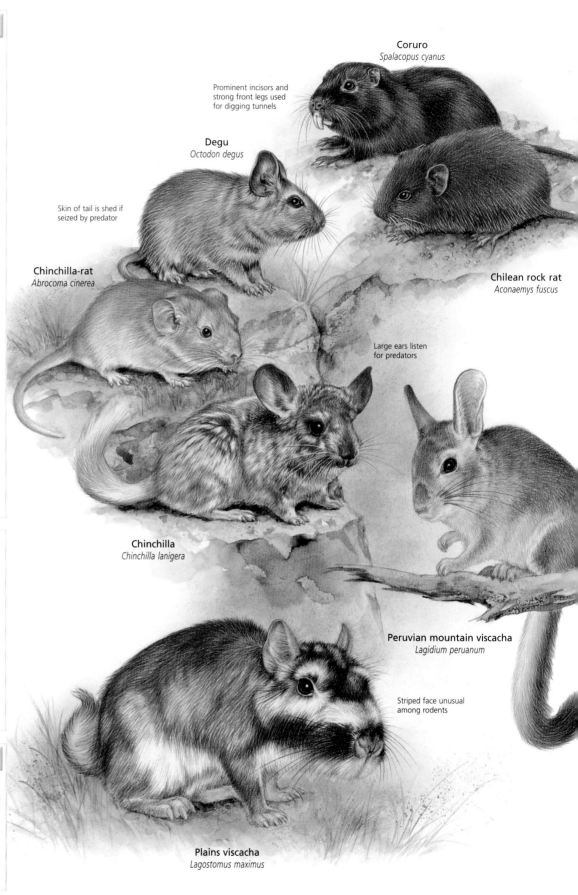

Coruro
Spalacopus cyanus

Prominent incisors and strong front legs used for digging tunnels

Degu
Octodon degus

Skin of tail is shed if seized by predator

Chinchilla-rat
Abrocoma cinerea

Chilean rock rat
Aconaemys fuscus

Large ears listen for predators

Chinchilla
Chinchilla lanigera

Peruvian mountain viscacha
Lagidium peruanum

Striped face unusual among rodents

Plains viscacha
Lagostomus maximus

LAGOMORPHS

The rabbits, hares, and pikas of the order Lagomorpha were once considered a suborder of Rodentia, and they do bear some resemblance to large rodents. They are gnawing herbivores, with large, constantly growing incisors, no canine teeth, and a gap between the incisors and molars. Like rodents, they can close their lips in this gap and gnaw on material without taking it into the mouth. Unlike rodents, they have a second, smaller pair of upper incisors, known as peg teeth, behind the first pair. All lagomorphs are terrestrial. They are found almost worldwide in a diverse range of habitats, from snowy arctic tundra to steamy tropical forest and scorching desert.

ʟASS Mammalia
ʀDER Lagomorpha
ᴀMILIES 2
ᴇNERA 15
ᴘECIES 82

In human company While their close association with humans has allowed lagomorphs to spread almost worldwide, they are absent from southern South America and many islands. Where they have been introduced, as in Australia and New Zealand, lagomorphs have often had a devastating effect, outcompeting both native animals and livestock for food.

Harvesting pika During summer and fall, most pikas prepare for winter, spending up to a third of their time gathering grasses, leaves, and flowers in their mouth and adding them to a haystack sheltered under an overhanging rock.

Fleet of foot With their long, powerful hindlimbs, hares are designed to outrun their predators and have been known to exceed speeds of 45 miles per hour (70 km/h). Even at full speed, these agile animals can abruptly change direction.

RABBITS, HARES, AND PIKAS

Lagomorphs are divided into two families: the rabbits and hares of Leporidae, and the pikas of Ochotonidae. As key prey species for many birds and carnivores, all lagomorphs have eyes on the side of the head that provide a wide field of vision, and relatively large ears that contribute to their sharp hearing. In pikas, the ears are short and round, while rabbits and hares have very long ears.

Many lagomorphs are social species, and all use their scent glands in communication. In pikas, these are supplemented by a variety of vocalizations, which inspired their nickname of "calling hares."

To escape predators, both rabbits and hares have long hindlegs that allow them to run at speed—rabbits tend to flee to cover, while hares try to outrun the danger over open ground. Pikas have shorter legs, but usually live in rocky terrain where they can quickly slip into a crevice.

In spite of their anti-predator tactics, all lagomorph species are an important food source for other animals and suffer a high mortality rate. To cope with this, they tend to be prolific breeders. Gestation periods are short, usually lasting just 30–40 days, and litters are often large. Many species reach sexual maturity quickly—female European rabbits can breed at only 3 months. The female's eggs are released in response to copulation, and she is able to become pregnant almost immediately after giving birth. In some species, the female can conceive a second litter while still pregnant with the first. These reproductive strategies have allowed some species, such as the European rabbit, to become so successful that they are considered pests.

Boxing hares Throughout the mating season, male Arctic hares compete for access to females, while each female fends off any males that she is not interested in. Arctic hares most often live in small family groups, but larger bands of more than a hundred individuals can occur, especially on cold, northern islands.

FACT FILE

Northern pika Active through long, cold winters, this species stays on the surface until the snow cover is about 12 inches (30 cm) deep, and then retreats into tunnels under the snow.

🐾 Up to 8 in (20 cm)
🐾 None
⚖ Up to 7 oz (200 g)
♦ Solitary
✦ Not known

Mongolia, Siberia

Royle's pika Although this pika normally nests in natural rock piles, it has been known to live in the rock walls of huts built by humans. Because it forages throughout the year, it does not create a haystack as other pikas do.

🐾 Up to 8 in (20 cm)
🐾 None
⚖ Up to 7 oz (200 g)
♦ Solitary
✦ Not known

Himalayas of Pakistan, India, Nepal & Tibet

Daurian pika This steppe-dweller is a highly social burrower, living in large colonies made up of family groups. Family members communicate through a repertoire of calls, groom each other, rub noses, and play together.

🐾 Up to 8 in (20 cm)
🐾 None
⚖ Up to 7 oz (200 g)
🐁 Colonial
✦ Common

Steppes of Mongolia & S. Siberia

American pika Each American pika defends an area of rock, with males and females occupying adjacent but separate territories.

🐾 Up to 8½ in (22 cm)
🐾 None
⚖ Up to 6 oz (175 g)
♦ Solitary
✦ Locally common

W. North America

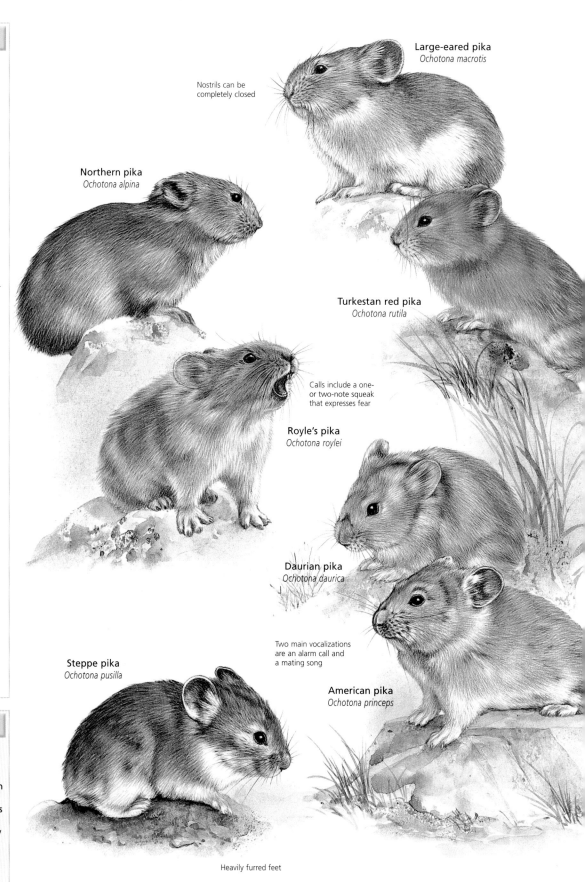

Large-eared pika
Ochotona macrotis

Nostrils can be completely closed

Northern pika
Ochotona alpina

Turkestan red pika
Ochotona rutila

Calls include a one- or two-note squeak that expresses fear

Royle's pika
Ochotona roylei

Daurian pika
Ochotona daurica

Two main vocalizations are an alarm call and a mating song

American pika
Ochotona princeps

Steppe pika
Ochotona pusilla

Heavily furred feet

Arctic hare
Lepus timidus

Black ear tips

Brown summer coat blends in with tundra vegetation

Grows white coat and eats barks and buds in winter

Snowshoe hare
Lepus americanus

White winter coat provides camouflage against snow cover

Grows brown coat and eats green plants and berries in summer

Antelope jackrabbit
Lepus alleni

Asiatic brown hare
Lepus tolai

Black-tailed jackrabbit
Lepus californicus

Coat lightens in summer

Can flee predators at speeds of 35 miles per hour (56 km/h)

White-tailed jackrabbit
Lepus townsendii

Brown hare
Lepus europaeus

FACT FILE

Arctic hare To survive the harsh Arctic winter, this usually solitary hare may gather in flocks of several hundred individuals, and smaller groups may cooperate to build a protective wall of snow. The Arctic hare also changes color with the season, growing a white winter coat to blend in with the snow, then shedding this for a brown coat to match the summer tundra vegetation.

- Up to 24 in (60 cm)
- Up to 3 in (8 cm)
- Up to 13 lb (6 kg)
- Solitary
- Common

Iceland, Ireland, Scotland, N. Eurasia

Antelope jackrabbit Like an antelope, this large hare is able to make great leaps. A desert species, it is nocturnal and can survive without drinking water, obtaining all its water from plants.

- Up to 24 in (60 cm)
- Up to 3 in (8 cm)
- Up to 13 lb (6 kg)
- Solitary
- Common, declining

S. Arizona to N. Mexico

Brown hare Female brown hares can produce four litters a year. For the first month, the young are left in a form, a shallow depression in long grass, and visited once a day for feeding.

- Up to 27 in (68 cm)
- Up to 4 in (10 cm)
- Up to 15½ lb (7 kg)
- Solitary
- Common, declining

Europe to Middle East; introd. widely

COOL EARS

A resident of North America's deserts, the black-tailed jackrabbit relies on its long ears to keep cool. Hundreds of tiny blood vessels radiate heat to the surface of the ears, allowing the blood to cool before it returns to the heart. A jackrabbit spends the hottest part of the day resting in the shade of a shrub or long grass.

FACT FILE

Eastern cottontail This solitary rabbit spends the daytime resting in a hollow beneath a log or brush. Although blind and naked at birth, young cottontails can leave the nest at 2 weeks of age, disperse at 7 weeks, and are sexually mature at only 3 months.

- Up to 19½ in (50 cm)
- Up to 2½ in (6 cm)
- Up to 3½ lb (1.5 kg)
- Solitary
- Common

E. & S. North America

Pygmy rabbit The smallest rabbit, this species lives amid dense sagebrush, which forms much of its diet, and digs its own burrow. It emits a distinctive whistle to warn neighbors of danger.

- Up to 11 in (28 cm)
- Up to ¼ in (2 cm)
- Up to 16 oz (460 g)
- Solitary
- Near threatened

W. USA

European rabbit Popular as a game animal, this species was deliberately introduced in many places, often with devastating effects on native fauna.

- Up to 18 in (46 cm)
- Up to 3 in (8 cm)
- Up to 5 lb (2.2 kg)
- Family group
- Abundant

British Isles; Spain to Balkans; introd. widely

Hispid hare Hunting and domestic dogs have taken their toll, but this animal is most threatened by deliberate burning of its grassland habitat.

- Up to 19½ in (50 cm)
- Up to 1½ in (4 cm)
- Up to 5½ lb (2.5 kg)
- Solitary, pair
- Endangered

Foothills of Himalayas in Nepal & N. India

RABBIT WARREN

The European rabbit is one of the few rabbit or hare species to dig its own burrow, and is the only one to live in stable, territorial breeding groups. Large numbers of young are raised in the shelter of an underground warren.

Forest rabbit
Sylvilagus brasiliensis

Eastern cottontail
Sylvilagus floridanus

Pygmy rabbit
Brachylagus idahoensis

White tail displayed when running

Brush rabbit
Sylvilagus bachmani

European rabbit
Oryctolagus cuniculus

No visible tail

Volcano rabbit
Romerolagus diazi

Coarse, bristly outer fur, with softer underfur

Thumps hindleg on ground to warn others of danger

Hispid hare
Caprolagus hispidus

Sumatran rabbit
Nesolagus netscheri

The rarest lagomorph, with just one record from 1972 and a photograph taken by remote camera in 1998

Central African hare
Poelagus marjorita

ELEPHANT SHREWS

Little studied until recent years and categorized at different times with insectivores, ungulates, tree shrews, and lagomorphs, elephant shrews are so distinctive that they are now placed in their own order, the Macroscelidea. Their long, mobile snout suggested an elephant's trunk to naturalists, inspiring their common name. They are ground-dwellers, relying on keen hearing and vision to detect danger, and on the speed provided by their long, slender legs to escape predators. Some smaller species move in leaps when alarmed, resembling miniature antelopes. Elephant shrews are insect-eaters, but have larger, more developed brains than the insectivores.

ᴀss	Mammalia
ʀᴅᴇʀ	Macroscelidea
ᴀᴍɪʟʏ	Macroscelididae
ᴇɴᴇʀᴀ	4
ᴇᴄɪᴇs	15

African habitats Elephant shrews are found throughout much of Africa, but are missing from West Africa and the Sahara Desert. They occupy diverse habitats, including rocky outcrops, desert, savanna, grasslands, thornbrush plains, and tropical forest. Although terrestrial and often active by day, these secretive, swift mammals are not common and are rarely seen.

INSECT-EATERS
An elephant shrew can spend up to 80 percent of its waking hours foraging. Although elephant shrews have a large cecum similar to that of herbivores and may include some fruit, seeds, and other plant matter in their diet, they are largely insectivorous, feeding mainly on invertebrates such as spiders, beetles, termites, ants, centipedes, and earthworms. Their long, sensitive, mobile snout can root around in leaf litter, detecting the small prey by smell. Some species use their claws and teeth to raid the tunnel systems of ants and termites. Like other elephant shrews, the bushveld elephant shrew (*Elephantulus intufi*) has a long tongue (shown above) that is able to speedily flick insects into its mouth.

Checkered elephant shrew
Rhynchocyon cirnei

Highly sensitive, flexible snout

Together and alone Like other elephant shrews, rufous elephant shrews live in monogamous pairs, usually mating for life. The pairs rarely meet, but share and defend exactly the same territory. They rely on scentmarking to communicate, and maintain a network of trails that allows fast escape from predators. Trespassing male elephant shrews are driven off by the male of the pair, while the female deters female intruders.

Rufous elephant shrew (spectacled elephant shrew)
Elephantulus rufescens

Back legs longer than front legs

Large eyes and ears

Four-toed elephant shrew
Petrodromus tetradactylus

Rat-like, bristled tail

BIRDS

BIRDS

Descended from reptiles that developed the ability to fly, birds are among the most mobile of all animals alive today. Most numerous in marshes, woodlands, and rain forests, they have also adapted to live in large cities, inhospitable deserts, and even the North and South poles. Some never stray from home; others cross entire oceans or continents, sometimes in a single flight. In size, they range from the tiny bee hummingbird to the imposing ostrich. Among the more than 9,700 species are birds of almost every color and a dazzling array of patterns. While some hover on the verge of extinction, others, such as domestic chickens, outnumber humans.

PHYLUM	Chordata
CLASS	Aves
ORDERS	29
FAMILIES	194
GENERA	2,161
SPECIES	9,721

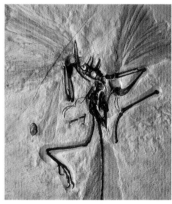

Early bird The first birds evolved from dinosaurs more than 150 million years ago. The oldest-known bird is the Archaeopteryx (above), the first fossil of which was discovered in Germany in 1861. It was a swift, two-legged predator whose feather-covered body was probably capable of powered flight.

ON THE WING

Birds' flying and singing abilities have awed humans through the millennia. These distinctive animals have been the inspiration not only for the invention of mechanical flying machines, but also for myths, musical works, and other art forms.

Although some species of birds have lost the ability to fly, they all have feathers. In general, no other types of animals can travel so far or so fast under their own power.

Birds can perform an astonishing range of aerial maneuvers, from swooping and soaring to hovering or even, in some cases, flying backwards. They may fly to search for food and more favorable climates, or else to ward off predators or communicate with each other.

Some birds seek out territories for temporary breeding, roosting, or feeding purposes, while others spend their entire lives in one area. Many species of birds migrate over great distances to escape harsh winters before returning to their starting places when the weather begins to improve again. They are guided by instinct and experience.

Not all types of birds sing, but more than half of the species are considered songbirds. Some of these have very large repertoires, which are used primarily by males to court females and deter other males from their territories.

Some birds are solitary; others live in large flocks and may even help raise other parents' young.

More than 30 species of birds became extinct during the 20th century, mostly due to human activities. Today, one out of every nine species of birds is threatened, and many more species are in decline. Past species extinctions were mostly due to overhunting, as in the case of the passenger pigeon.

Today, habitat loss and destruction are by far the most serious factors in the decline of birdlife worldwide. Pesticides and other pollutants also take their toll, as do introduced species such as cats and rats, especially on populations of island birds that never evolved defenses against such animals.

Although many types of birds appear on the threatened list, parrots feature very prominently, because their bright feathers attract poachers and because they tend to live in tropical rain forests, which are disappearing at a fast rate.

Spreading out
The wings are fully spread and thrust forward. The feathers overlap and the feet are tucked against the body.

Pushing forward
The wings drop and the feathers dig into the air. This pulls the robin forward.

Tucking in
The wings are tucked well into the body throughout the upstroke to reduce air resistance.

Ending
The feathers twist apart as the downstroke ends.

Beginning
At the start of the upstroke the robin's feathers are separated to reduce air resistance and save energy.

On the wing The wingbeat of a European robin (above) is typical of flying birds: a smooth alternation between the wings moving upward (upstroke or recovery stroke) and downward (downstroke or powerstroke). Penguins (right) are in the minority of flightless birds.

Power in the air Like other raptors or birds of prey, the white-tailed sea eagle (*Haliaeetus albicilla*) (far right) uses its formidable talons and beak to catch and immobilize fishes or other animals.

A slow start All birds hatch from eggs (left). One or both parents typically sit on the eggs to incubate them, keeping them warm enough to allow the embryos to develop. When it is ready, a chick frees itself by pecking a hole through the shell.

Courtship Most birds are monogamous. They attract mates and then cement their bonds in a variety of ways, including singing, displaying special breeding plumage, performing aerial acrobatics, or dancing, like these cranes (below).

FEATHER TYPES

There are three kinds of feathers. Closest to the body are the fluffy down feathers that protect a bird from the cold. Over these are the contour feathers. These are short and round, and give the bird its streamlined shape. The longer feathers on a bird's wing and tail enable the bird to take off, fly, maneuver, and land.

Flight feather The vane of a flight feather is made up of fine strands that interlock to create a smooth surface.

hooklets

barb

shaft

shaft | vane

barbules

Purpose-built Feathers provide warmth, protection, color, and shape, as well as the dynamics that make flight possible.

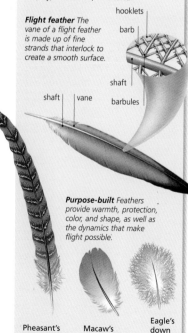

Pheasant's tail feather | Macaw's body feather | Eagle's down feather

Bills of fare A bird's bill can provide important clues to its lifestyle. Above, from left to right: a puffin's serrated bill helps it hold fishes; a macaw's can easily shell nuts and seeds; and an egret's is well suited to catching pond life in the rushes.

ENGINEERING MARVELS

Birds come in all shapes and sizes. Yet, although external features vary markedly among species, all birds share a remarkably similar anatomy. All birds are descended from flying ancestors and have front limbs in the form of wings. A flying bird's body is designed to enable it to move efficiently through the air. For example, most of a bird's weight is in the center of its body, which is light and strong, with fewer bones than the bodies of reptiles or mammals. Flight is a very demanding activity that requires a high metabolism, so the respiratory system and heart of birds are very efficient.

A bird's digestive tract is fairly typical of vertebrates, although some birds have a crop for the temporary storage of food (for subsequent grinding in the gizzard and for absorption through the intestine walls, or for regurgitation to feed the young).

Some avian features—such as the scales on their legs and feet and the laying of eggs—are reminiscent of their reptilian ancestors.

However, because birds are warm-blooded, they must spend a large part of their waking time searching for food. Their diets range from seeds and fruits to insects, small mammals, and even other birds.

Bills serve similar functions for birds as hands do for humans; their primary function is food gathering. There is enormous variation in the structure and shape of birds' bills and feet, reflecting the ecological niches species have adapted to fill.

The forms of a bird's wings and tail determine the amount of lift, thrust, and maneuverability it will have in flight. An improvement in one of these aspects can generally be achieved only

with a sacrifice of performance in the other two. Small birds that need to fly off quickly to evade predators tend to have short wings; large birds such as albatrosses that regularly fly very long distances have long, thin wings for gliding.

Eyesight is birds' most well-developed sense, with hearing ranking second, despite the lack of any external ear structure.

Birds' feathers come in many colors (which may vary with the seasons) and are

Looming large The ostrich (*Struthio camelus*) (right) is the world's largest bird. Unlike most birds, it has shaggy feathers and cannot fly. The group to which the ostrich belongs is also unusual among birds for having a flat sternum.

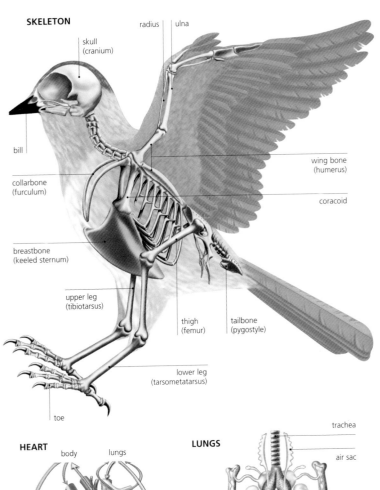

SKELETON

radius | ulna

skull (cranium)

bill

collarbone (furculum)

wing bone (humerus)

coracoid

breastbone (keeled sternum)

upper leg (tibiotarsus)

thigh (femur)

tailbone (pygostyle)

lower leg (tarsometatarsus)

toe

Controlled chaos This flock of snow geese (*Chen caerulescens*) (above) may look chaotic, but the birds can sense their neighbors and maneuver quickly to avoid bumping into each other, very much like human commuters on busy city streets.

Keeping warm Birds such as the northern cardinal (*Cardinalis cardinalis*) (left) that don't live in or migrate to warmer climates struggle to find enough food and maintain a constant body temperature during the winter months. Fluffing up their feathers traps air and helps insulate their bodies.

Adapted to flight A bird's skeleton (right) is typical of vertebrates, with modifications to support powered flight. Parts of the backbone are fused into a sturdy frame. The collarbone is fused into a furculum (commonly known as a wishbone). As the bird flies, this bone acts like a spring. The breastbone, or sternum, is broad and curved to provide a secure anchor for the strong flight muscles. Lungs continuously transfer oxygen to the blood, which the powerful heart pumps to the muscles.

Disappearing act One of the hundreds of threatened birds is the harpy eagle (*Harpia harpyja*) (right), the world's most powerful bird of prey, which is in dangerous decline in the tropical forests of Central and South America. The majority of threatened bird species live in developing countries.

HEART

body | lungs

right side | left side

LUNGS

trachea

air sac

lung

wing bone (humerus)

air sac

frequently used for social displays as well as for flying. Males tend to be more colorful than females.

To keep feathers in fine shape, birds preen regularly, running their beaks along the length of their feathers to smooth them out and to remove foreign particles. But even the most well-tended bird's feathers eventually become old and tattered, so at least once a year birds shed their feathers and grow new ones in a process known as molting. Birds also clean themselves by bathing in water or dust.

Most birds build nests to house their young. They lay from one to a dozen or more eggs, either once or several times a year.

Some species are born well-developed enough to fend for themselves right away; others are born blind and helpless. Until they are strong enough to fly and be independent, they remain in the nest, relying on one or both of their parents for food and for protection from predators. This nestling period can range from about a week to more than 5 months.

⚡ CONSERVATION WATCH

The 2,139 species of birds on the IUCN Red List are categorized as follows:

129	Extinct
3	Extinct in the wild
182	Critically endangered
331	Endangered
681	Vulnerable
3	Conservation dependent
731	Near threatened
79	Data deficient

RATITES AND TINAMOUS

CLASS	Aves
ORDERS	5
FAMILIES	6
GENERA	15
SPECIES	59

Ratites are flightless birds with flat sternums, or breastbones, that lack the prominent, keel-like sternums of flying birds. Like other flightless birds, they are believed to have lost the ability to fly because they either lacked predators or could evade them by less energy-intensive means. This category includes two groups of large extinct birds, the moa of New Zealand and the elephant bird of Madagascar. Most ratites are believed to have evolved along similar lines rather than from one distinct ancestor. Tinamous are a related order of flying birds with keeled sternums that share some unusual anatomical characteristics—such as a distinctive palate structure—with ratites.

Southern exposure Ratites and tinamous are found only on southern Gondwanan continents. The various species are adapted to grasslands, jungles, woodlands, or high mountain ranges. The emu is the most versatile, roaming in diverse habitats.

Courting A reversal of the usual sex roles is apparent among tinamous (above), with the females taking the more aggressive lead role in courtship.

Fathers lead Like most other ratites and tinamous, the male cassowary (above, right) is responsible for incubating and raising the young. Male emus lives off their fat reserves for the entire 8-week incubation period, not taking a break to eat, drink, or even defecate during that time.

Great tinamou
Tinamus major

GIANTS AND RUNNERS

Adult ratites and tinamous—among them the largest of all birds, the ostrich, which may exceed 9 feet (2.8 m) in height—resemble overgrown chicks, with underdeveloped, stubby wings and soft flight feathers.

Powerful runners, ratites have as few as two toes per foot, rather than the standard four toes of other birds. Ostriches can run faster than a racehorse and sometimes defend themselves by kicking. Rheas, sometimes called South American ostriches, occasionally raise one of their wings while running, but in this case the front limb serves the function of a sail rather than a wing.

The kiwis of New Zealand have vestigial wings. These three species of nocturnal birds—the country's national symbol—live in burrows and find their invertebrate prey by scent, using their long, sensitive bills. They are one of the few birds with a good sense of smell.

Tinamous are an ancient order of diverse, abundant smaller birds with plump bodies, short, rounded wings, and loose plumage. Some of the 47 species roost in trees, others on the ground. Although tinamous can fly, they usually escape predators by stealing through cover or by freezing.

Ornate tinamou
Nothoprocta ornata

Dwarf tinamou
Taoniscus nanus

Elegant crested tinamou
Eudromia elegans

Variegated tinamou
Crypturellus variegatus

PADDING ALONG

Rheas are large, heavy birds that can reach 5 feet (1.5 m) in height. Their powerful legs bear a lot of weight when they run, but extra flesh on their feet helps absorb some of the impact. The bird's three forward-facing toes are also well adapted to running.

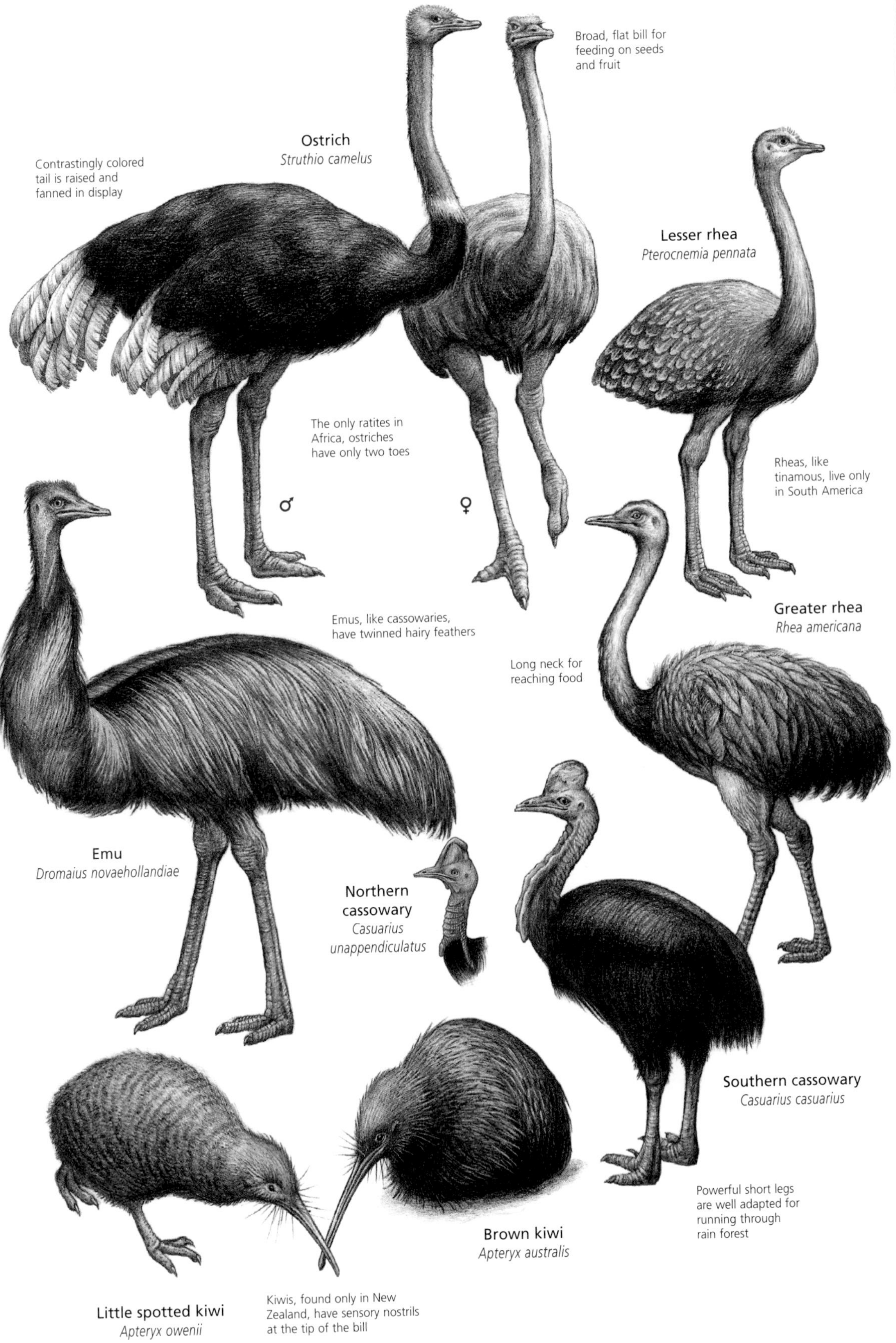

Broad, flat bill for feeding on seeds and fruit
Ostrich
Struthio camelus
Contrastingly colored tail is raised and fanned in display
Lesser rhea
Pterocnemia pennata
The only ratites in Africa, ostriches have only two toes
♂
♀
Rheas, like tinamous, live only in South America
Emus, like cassowaries, have twinned hairy feathers
Greater rhea
Rhea americana
Long neck for reaching food
Emu
Dromaius novaehollandiae
Northern cassowary
Casuarius unappendiculatus
Southern cassowary
Casuarius casuarius
Powerful short legs are well adapted for running through rain forest
Brown kiwi
Apteryx australis
Little spotted kiwi
Apteryx owenii
Kiwis, found only in New Zealand, have sensory nostrils at the tip of the bill

FACT FILE

Ostrich This large bird once ranged to the Middle East and Asia, but is now limited mainly to national parks in eastern and southern Africa. Males keep a temporary harem when breeding and scrape nests on the ground.

- Up to 9½ ft (2.9 m)
- 5–11 (major hen)
- Sexes differ
- Nomadic
- Locally common

C., E. & S. Africa

Emu This species is the sole surviving member of its family. Adaptable birds, emus live in a variety of habitats. They form temporary pairs to mate, and walk vast distances in search of fruit, seeds, insects, and young shoots.

- Up to 6½ ft (2 m)
- 7–11
- Sexes alike
- Nomadic
- Locally common

Australia, Tasmania (extinct)

Greater rhea The larger of two species of rhea, the greater rhea has a wide, flat beak that is well adapted for grazing on the plains where it roams. It is a very agile runner, able to twist and turn abruptly to elude its pursuers. Males fight for territory.

- Up to 5¼ ft (1.6 m)
- 13–30
- Sexes alike
- Sedentary
- Near threatened

E., S.E. & C.W. South America

Southern cassowary Females of this secretive, solitary rain-forest species are larger and more brightly colored than the males, with taller casques atop their heads. Males rear the chicks.

- Up to 6½ ft (2 m)
- 3–5
- Sexes alike
- Sedentary
- Vulnerable in Australia

Ceram, S. New Guinea, N.E. Australia

CONSERVATION WATCH

Ratites IUCN lists two of the three species of cassowary as vulnerable, and all three species of kiwi as well. The emu is common enough on mainland Australia; but two pygmy species that were alive on King and Kangaroo Islands, off southeastern Australia, at the time of European settlement are now extinct.

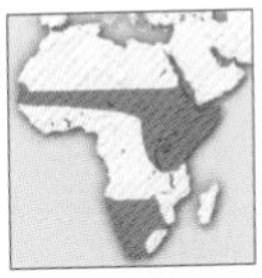

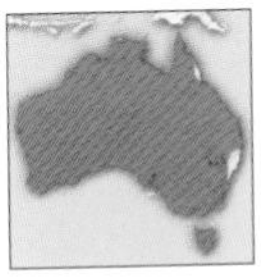

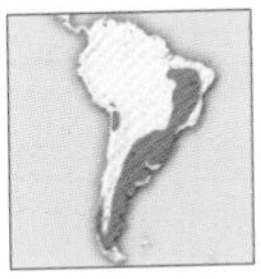

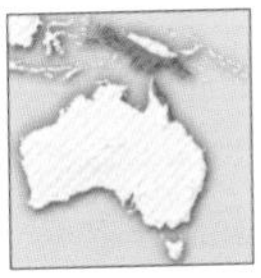

GAMEBIRDS

CLASS	Aves
ORDERS	1
FAMILIES	5
GENERA	80
SPECIES	290

These familiar birds include chickens (domesticated forms of central Asian jungle fowl) and turkeys. Humans also hunt and eat pheasants, partridges, grouse, and quail. Peafowl are sometimes kept in captivity because of their beautiful plumage and ornate displays. These birds vary widely in size, but they all have stocky frames, relatively small heads, and short, broad wings. They typically fly low and fast. They are a favorite prey animal of many wild predators. To elude them, gamebirds rely on the camouflage effect of their dull-colored plumage, or else they fly or scurry away quickly. They have large clutches of up to 20 eggs, but species' populations tend to fluctuate greatly.

Far and wide Gamebirds live in a variety of climatic zones and, depending on the species, prefer either forests, scrub, open habitats, or grasslands. Some types of gamebirds are much more widespread than others: for example, the quails and partridges are found on several continents across the world, whereas turkeys occur only in North America.

FACT FILE

Australian brush-turkey This bird incubates its eggs by placing them in a mound of heat-generating decaying matter, adding or removing material to keep the mound's temperature constant.

⬆	Up to 27 in (70 cm)
●	15–27
⫻	Sexes alike
⊘	Sedentary
⬥	Locally common

E. Australia

Great curassow This slender, forest-dwelling bird spends much of its time on the ground, foraging for fallen fruit and seeds that it picks up with its sturdy bill. It roosts and seeks refuge in trees, and also builds nests of plant matter among the branches.

⬆	Up to 36 in (92 cm)
●	2
⫻	Sexes differ
⊘	Sedentary
⬥	Locally common

E. Mexico to far N.W. South America; Cozumel I.

White-crested guan This handsome, secretive bird is confined to a fairly small area south of the Amazon River. It has a call that reaches far through the forest, thanks to a windpipe that is adapted for amplifying sound.

⬆	Up to 32 in (83 cm)
●	3–4
⫻	Sexes alike
⊘	Sedentary
⬥	Near threatened

N.E. Amazonian Brazil

Check me out Many gamebirds stage elaborate courtship displays. Sage grouse males (*Centrocercus urophasianus*) (above) gather at strutting grounds known as leks. After each male performs, the fortunate dominant male mates with dozens of females.

Australian brush-turkey
Alectura lathami

Brush-turkeys live in rain-forest scrubs. They have huge feet to scrape up litter for their nest mounds

Most guams, including the white-crested, are localized species in tropical South America

Great curassows live in heavy rain forests at low altitudes, and feed on the forest floor singly, in pairs, or in small groups

Great curassow
Crax rubra

White-crested guan
Penelope pileata

On show A male common pheasant (above) displays by rapidly beating its wings while uttering its characteristic "crowing" call. Some species dwell primarily in trees, but most gamebirds spend much time on the ground and can run well on their strong legs. All male pheasants are brilliantly colored.

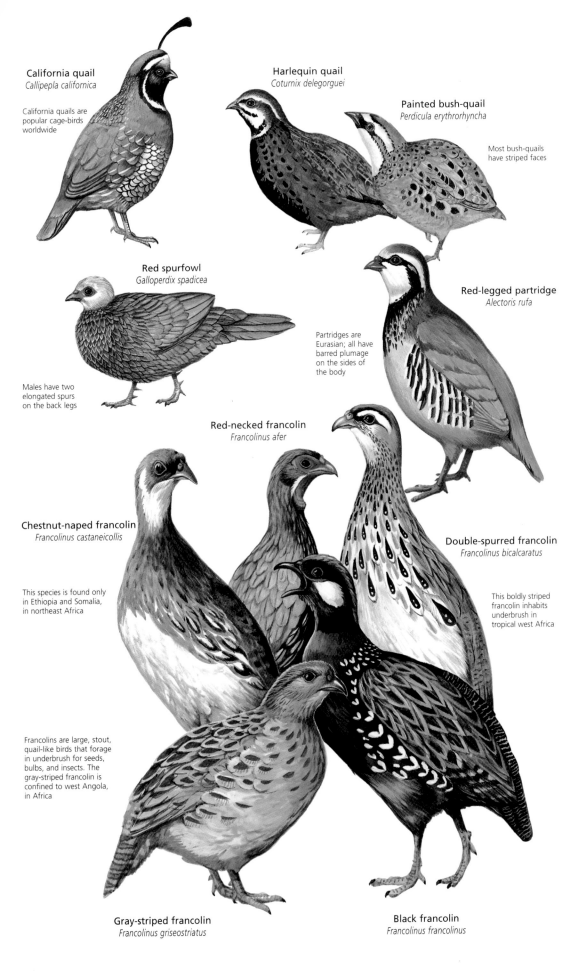

California quail
Callipepla californica

California quails are popular cage-birds worldwide

Harlequin quail
Coturnix delegorguei

Painted bush-quail
Perdicula erythrorhyncha

Most bush-quails have striped faces

Red spurfowl
Galloperdix spadicea

Red-legged partridge
Alectoris rufa

Partridges are Eurasian; all have barred plumage on the sides of the body

Males have two elongated spurs on the back legs

Red-necked francolin
Francolinus afer

Chestnut-naped francolin
Francolinus castaneicollis

This species is found only in Ethiopia and Somalia, in northeast Africa

Double-spurred francolin
Francolinus bicalcaratus

This boldly striped francolin inhabits underbrush in tropical west Africa

Francolins are large, stout, quail-like birds that forage in underbrush for seeds, bulbs, and insects. The gray-striped francolin is confined to west Angola, in Africa

Gray-striped francolin
Francolinus griseostriatus

Black francolin
Francolinus francolinus

FACT FILE

Black grouse This bird lives in a variety of habitats, from wooded areas to marshy ground. The black grouse nests in a hollow that it digs in the ground, where the female incubates her eggs. It has a largely herbivorous diet.

↥ Up to 23 in (60 cm)
● 6–11
✎ Sexes differ
⊘ Sedentary
↟ Locally common

Britain & N. Eurasia to E. Siberia & N. Korea

Rock ptarmigan In the summer, the male and female's upper parts are dusky gray and brown, respectively, but in winter both sexes are almost entirely pure white. Feathers on the bird's legs and toes enable it to walk on deep snow without sinking.

↥ Up to 15 in (38 cm)
● 5–8
✎ Sexes differ
↺ Partial migrant
↟ Locally common

N. North America, Europe & N. Eurasia

Red junglefowl This colorful bird was domesticated as the chicken at least 5,000 years ago in the Indus valley. The meat and eggs of chickens are now eaten by people the world over.

↥ Up to 30 in (75 cm)
● 4–9
✎ Sexes differ
⊘ Sedentary
↟ Locally common

N. India & S.E. Asia to W. Lesser Sundas

SPURRED ON

In the junglefowl (and some of its close relatives within this order), the male bird has spurs on the back of its legs, just above its toes. These are used as weapons when fighting other males. The sport of cockfighting, using domesticated birds, has been practiced for millennia. It is illegal in most countries but still takes place in others. The spurs themselves don't kill, but when sharp blades are attached to the legs as well, fighting can be lethal.

Blue and white eared-pheasants belong to a small group of round-tailed pheasants with projecting white ear tufts. They live only in China

Blue eared-pheasant
Crossoptilon auritum

White eared-pheasant
Crossoptilon crossoptilon

The only member of its group with whitish plumage; considered vulnerable

Female red junglefowl carry out all the brooding of eggs and chicks

Red junglefowl
Gallus gallus

Males crow to advertise their territory

Koklass pheasant
Pucrasia macrolopha

Eurasian black grouse
Lyrurus tetrix

Rock ptarmigan
Lagopus muta

Summer breeding plumage

Winter plumage for camouflage

Ocellated turkey
Meleagris ocellata

Vulturine guineafowl
Acryllium vulturinum

All guineafowl have spotted plumage and bare, colored skin on the head

Helmeted guineafowl
Numida meleagris

Mikado pheasants live only on Taiwan; they are considered vulnerable

This is a black male. Note the red facial skin

Mikado pheasant
Syrmaticus mikado

This species, endemic to northeast China, has perhaps the longest tail of all pheasants

Reeves' pheasant
Syrmaticus reevesii

Pheasants have powerful feet for scratching in litter on the ground to find their food of seeds and invertebrates

In display, the male raises his huge tail of ocellated feathers in a giant fan

Indian peafowl
Pavo cristatus

Feeds on dropped fruits on the forest floor

♂

Great argus
Argusianus argus

♀

FACT FILE

Reeve's pheasant This beautiful bird lives only in the hill forests of central eastern Asia. For many centuries, the Chinese used its impressively long tail feathers and other plumage in decorative, ceremonial, or religious motifs.

- ⚊ Up to 7 ft (2.1 m)
- ● 6–9
- Sexes differ
- ⊘ Sedentary
- ⚑ Vulnerable

N. & C. China & Mongolia

Helmeted guineafowl This ancestor of the domesticated guineafowl nests on the ground amid vegetation. After they are reared by the female, nestlings sometimes remain together in groups.

- ⚊ Up to 25 in (63 cm)
- ● 6–12
- Sexes alike
- ⊘ Sedentary
- ⚑ Locally common

Sub-Saharan Africa

Great argus During this pheasant's courtship display, the male clears a hilltop in a forest. Dancing around the female, he opens his wings to reveal "eye" patterns that appear three-dimensional. As with most pheasants, the male's breeding role ends after mating.

- ⚊ Up to 6½ feet (2 m)
- ● 2
- Sexes differ
- ⊘ Sedentary
- ⚑ Near threatened

Malay Peninsula, Sumatra, Borneo

FLIGHT MUSCLES

Gamebirds tend to have powerful flight muscles, which they use to launch themselves off the ground quickly in order to escape from their many predators. All flying birds develop a keel on the breast bone, to separate the muscle attachments for each wing. This area of bone is enlarged in game birds, in concert with enlarged breast muscles.

Lift-off
A Reeve's pheasant takes advantage of its powerful wings to launch its stocky body off the ground.

WATERFOWL

CLASS Aves	
ORDERS 1	
FAMILIES 3	
GENERA 52	
SPECIES 162	

Among the first birds to be domesticated, ducks and geese were raised for food more than 4,500 years ago. Swans have also been kept in captivity, because of their beauty and grace. Several species are flightless, but the rest are powerful fliers; many northern species migrate in family groups over great distances. In the air, they flap their wings continuously and can attain maximum speeds greater than 68 miles per hour (122 km/h). Some have been observed flying at an altitude of 28,000 feet (8,485 m), near the summit of Mount Everest. Some doze on water; others come ashore to rest. Their calls range from quacks to barks, hisses, whistles and even trumpeting sounds.

Citizens of the world Waterfowl predominate in the Northern Hemisphere (with the largest number of species being in North America), but they can be found worldwide, except in Antarctica. They occur in almost every type of wetland, from city ponds to Arctic sea inlets. Some species spend a great deal of time at sea.

AT HOME ON THE WATER
All species are remarkably similar in form, with short legs, webbed feet, relatively long necks, and flattened, broad bills. Most are excellent swimmers, although several species have adapted to life on land and have less webbing on their feet.

To protect them from the cold water, these birds rely on their dense, waterproof feathers and a thick coat of insulating down. Their plumage is often brightly colored and patterned.

Domestic waterfowl breeds are descended from mallards, muscovies, graylags and swan geese. Mallards, native to the Northern Hemisphere, have been widely introduced in other parts of the world, which has led to unwanted hybridization and genetic dilution of local species.

Many waterfowl consume grass, seeds, grain, and other vegetation, but some species eat fishes, insects, mollusks, and crustaceans.

The South American screamers bear little superficial resemblance to other waterfowl, but they all share certain anatomical similarities.

Primed for take-off The largest and most majestic of waterfowl, swans (above) reach a height of almost 5 feet (1.5 m). They must run on water and vigorously flap their wings to build up enough speed to launch their heavy bodies into the air.

Follow the leader Young shelducks swim behind their mother (above, top). Ducklings can feed themselves, but they imprint on a parent, following the adult in order to learn various behaviors. The image of their mother also later influences what they look to when seeking a mate.

Whooper swan
Cygnus cygnus

Mute swan
Cygnus olor

Black-necked swan
Cygnus melanocoryphus

Coscoroba swan
Coscoroba coscoroba

Southern screamer
Chauna torquata

Vulturine bill is adapted for grasping and pulling

The three species of screamers all live in tropical and subtropical South America. They feed on succulent vegetation on stream sides

Magpie geese are found only in the swamps of northern Australia and New Guinea. They breed in huge colonies on floating vegetation

Magpie goose
Anseranas semipalmata

Although its feet are only part-webbed, this bird is a good swimmer

Snow goose
Anser caerulescens

Snow geese are migratory, breeding in the north American tundra and wintering on the east and west coasts of the United States

White-faced whistling-duck
Dendrocygna viduata

This species has an unusual distribution, with widespread populations in both Africa and South America

Red-breasted goose
Branta ruficollis

Canada goose
Branta canadensis

The only goose with a black head and neck ornamented with a white bib

Bean goose
Anser fabalis

Bean geese breed in the high arctic zone of Eurasia and fly south to winter in China and the Mediterranean

FACT FILE

Southern screamer This bird lives in marshlands and along watercourses. It typically lives in groups and is a fine swimmer, despite the lack of webbing on its feet. Some indigenous people raise domesticated southern screamers.

↟ Up to 37 in (95 cm)
● 3–5
⫽ Sexes alike
⊘ Sedentary
↡ Locally common

C. & E. South America

Snow goose This bird has two color forms, one all white with black wing feathers, the other white-headed with a blue-gray body and wings. It breeds in the Arctic tundra and can be heard cackling as it flies.

↟ Up to 32 in (80 cm)
● 4–5 usually
⫽ Sexes alike
↻ Migrant
↡ Common

Arctic & S. North America

Canada goose There are about a dozen races of Canada goose, which vary in size, color, and geographic distribution. Originally native to North America, where it is a common sight in city parks, the Canada goose has been introduced widely around the world.

↟ Up to 3¾ ft (1.15 m)
● 4–7 usually
⫽ Sexes alike
↻ Migrant
↡ Common

North America, N. Europe, N.E. Asia

V-FORMATION

Canada geese are among the species of geese that migrate flying in wedge formation. The leader breaks the air and causes a stream-lining effect, stirring up updrafts that help the birds behind it conserve energy by reducing their air resistance. The birds take turns flying in the lead, and do so as they fly by both day and night to reach summer or winter quarters.

FACT FILE

Common shelduck This pugnacious bird looks heavy in flight and beats its wings slowly. It often breeds in old rabbit burrows or in gaps under old buildings. The male shelduck whistles and the female growls.

- ⚓ Up to 25 in (65 cm)
- ● 8–10
- ⫻ Sexes differ
- ↻ Migrant
- ⚑ Common

W. Europe, C. Asia, N.W. Africa

Comb duck This bird is also known as the knob-billed duck, due to the prominent fatty comb atop the drake's bill. This comb enlarges considerably prior to the breeding season.

- ⚓ Up to 30 in (76 cm) (female)
- ● 6–20
- ⫻ Sexes differ
- ⊘ Mainly sedentary
- ⚑ Locally common

Sub-Saharan Africa, S. Asia, N. & E. South America

Wood duck This handsome duck nests in tree cavities. Nestboxes have helped to reestablish this threatened species. After nesting, the male morphs into drab, female-like plumage, but retains a distinctive red bill. The wood duck makes a nasal squealing sound.

- ⚓ Up to 20 in (51 cm)
- ● 9–15
- ⫻ Sexes differ
- ↰ Partial migrant
- ⚑ Locally common

C. & S. North America & W. Cuba

SPATULATE BEAK

The northern shoveler gets its name from its distinctive, heavy, shovel-like bill, which is longer than its head. It feeds in shallow water, extending its neck and dabbling its bill just below the water surface. Comb-like serrations along the sides of the bill help it strain food items from the water.

Feeding tool
This bird's long bill helps it feed on a variety of plants, mollusks, and crustaceans in the water.

Torrent duck
Merganetta armata

Lives only in the Andes of South America, and feeds by diving in swift-flowing mountain streams

Northern shoveler
Anas clypeata

Muscovy duck
Cairina moschata

Freckled duck
Stictonetta naevosa

A rare duck of inland Australia that feeds on microplankton in shallow, freshly flooded swamps

Orinoco goose
Neochen jubata

Lives along the forested sides of rivers in tropical South America

Common shelduck
Tadorna tadorna

Male has an orange knob above the base of the bill; the female's knob is white

Comb duck
Sarkidiornis melanotos

Males are twice as big as females

Magellanic steamer duck
Tachyeres pteneres

Mallard
Anas platyrhynchos

Wood duck
Aix sponsa

Red-crested pochard
Netta rufina

**Oldsquaw
(long-tailed duck)**
Clangula hyemalis

Common pochard
Aythya ferina

Pochards feed mostly by
diving and dabbling. They
paddle to the floors of lakes
and swamps for mainly
vegetable matter and seeds

Despite their distinctively
long tails, oldsquaws dive
almost exclusively for
their food of crustaceans
and mollusks in the Arctic

Common mergansers dive
for their food of fishes
and aquatic invertebrates

Common merganser
Mergus merganser

Musk duck
Biziura lobata

This southern
Australian species
has a large lobe of
skin under the bill

Eiders are marine, living in
often rough seas around
rocky coasts, and diving for
mollusks and crustaceans

Common eider
Somateria mollissima

♀

♂

White-headed duck
Oxyura leucocephala

Males have an eclipse
plumage, but are
never as dull as the
brown females

Male in full
breeding plumage

FACT FILE

Red-crested pochard This diving duck
feeds mainly at night. To fly, it has
to run over the surface of the water
before it can gain enough momentum
to lift itself into the air. The male may
bring food or twigs to the female as
part of its courtship feeding ritual.

- Up to 23 in (58 cm)
- 6–14
- Sexes differ
- Partial migrant
- Locally common

C. & S. Europe to S. Asia, N. Africa

Common eider The down of the eider
is thick and heavy, and has the best
thermal insulating properties of any
natural material. From a young age,
the eider spends much of its time at
sea, often in rough water. This bird
often nests in tern colonies.

- Up to 27 in (69 cm)
- 3–6
- Sexes differ
- Partial migrant
- Common

N. & arctic North America, N. Europe & N. Asia

White-headed duck This rare duck
breeds in shallow, brackish wetlands
with abundant vegetation and reeds.
It has declined due to hunting, habitat
loss, and hybridization. The male's
bright blue, swollen bill becomes
smaller and grayer in winter.

- Up to 18 in (46 cm)
- 5–8
- Sexes differ
- Partial migrant
- Endangered

S. Europe, Middle East, C. Asia to N. India

WEBBED PROPELLERS

Most waterfowl are excellent
swimmers, due to their webbed front
toes, which act as paddles to propel
them through the water. They also
help the birds walk on mud. But
because the legs are set so far back
on the body for propulsion in
water, waterfowl can do little
more than
waddle
awkwardly
on land.

PENGUINS

CLASS	Aves
ORDERS	1
FAMILIES	1
GENERA	6
SPECIES	17

The most aquatic of all birds, penguins have remained unchanged in form for at least 45 million years. Although they evolved from flying birds, none of the 17 species of penguins can fly. Very highly specialized, social seabirds, they take advantage of their streamlined bodies (which minimize drag) and small, flipper-like wings to travel underwater at speeds of up to 15 miles per hour (24 km/h). They spend up to three-quarters of their life in the sea, staying underwater for as long as 20 minutes or more at a time, and coming ashore only to breed and molt. Penguins eat fish, krill, and other invertebrates.

South seas dwellers Penguins are widely distributed throughout the cooler waters of the world's southern oceans. The greatest diversity of species occurs in and around New Zealand and the Falkland Islands. The northernmost penguin lives almost exactly on the equator, in the Galápagos Islands.

Flying through the water Penguins are ungainly on land, but underwater they are efficient swimmers (left). They often enter the water by tobogganing down icy shores on their stomachs and plunging in.

GOING TO EXTREMES

Penguins endure a remarkably wide range of temperatures, from –80°F (–63°C) in the southern polar regions to 100°F (37°C) in the tropics. Their dense, fur-like feathers protect them by trapping warm air in an insulating layer. Penguins also benefit from a layer of blubbery fat that provides warmth as well as neutral buoyancy.

Chicks are born with insulating down, but cannot enter the water until they have grown their first adult layer of waterproof feathers. As adults, they then lose this layer when they molt and must remain ashore for 3 to 6 weeks to regrow it; during this period, they may lose a third or more of their body weight.

Penguins are superb swimmers and divers, propelling themselves with their wings, which have evolved into stiff, flat, paddle-like flippers. They twist their flippers in such a way as to provide thrust on both the upstroke and downstroke. Their bodies also have numerous special physiological features that allow them to regulate their body temperatures and oxygen levels in cold waters.

Warm spot Emperor penguin chicks (left) depend on their parents for warmth and protection. Some penguins nest on the ground or in burrows, but emperor penguins are among those that incubate eggs on their feet. After 6 months, the chicks molt and mature enough to go to sea.

A crowded life Penguins raise their young in crowded colonies. A lone adult king penguin stands out among a sea of chicks (above), which are covered in monochromatic gray-brown down before growing more colorful adult feathers. The parents recognize their own chicks by voice.

Emperor penguin
Aptenodytes forsteri

Emperor penguin chicks are grayish white with faces patterned in black

King penguin
Aptenodytes patagonicus

Jackass penguin
Spheniscus demersus

King penguins can dive to depths of over 650 feet (200 m) chasing their food of fish and some cephalopods

Adelie penguin
Pygoscelis adeliae

Snares penguin
Eudyptes robustus

Snares penguins live off the southern end of New Zealand, and capture their food (crustaceans and cephalopods) by pursuit-diving

Little penguin
Eudyptula minor

Royal penguin
Eudyptes schlegeli

This species breeds only around Macquarie Island, in the Southern Ocean

Confined to southwest New Zealand and considered vulnerable

Yellow-eyed penguin
Megadyptes antipodes

FACT FILE

Emperor penguin The largest penguin, the emperor is unique in that it breeds in the middle of winter. These penguins usually do so on annual fast ice, making them the only birds to not normally set foot on solid ground.

- Up to 4 ft (1.2 m)
- 1
- Sexes alike
- Sedentary
- Locally common

Seas & coasts of Antarctica

King penguin The young of this second-largest (and most colorful) species spend the winter in large groups known as creches. They are fed sporadically and many perish. Breeding colonies can exceed 100,000 pairs.

- Up to 3¼ ft (1 m)
- 1
- Sexes alike
- Locally nomadic
- Locally common

Sub-Antarctic seas and islands

Little penguin It would take up to 30 of the world's smallest penguin species to equal the weight of one emperor penguin. They have indigo blue and slate gray feather colors. Also known as fairy penguins, they are popular with tourists at several sites.

- Up to 18 in (45 cm)
- 2
- Sexes alike
- Sedentary
- Locally common

Coasts of S. Australia & New Zealand

GENTOO PENGUIN

These penguins, whose main colony is on the Falkland Islands, nest on rocky shores and inland grasslands. They build skimpy nests using pebbles and molted feathers (in Antarctica) or vegetation (on sub-Antarctic islands). They can be very aggressive and will often fight over these materials as they walk around gathering them.

DIVERS AND GREBES

CLASS	Aves
ORDERS	2
FAMILIES	2
GENERA	7
SPECIES	27

Although both are aquatic birds that use their webbed feet for propulsion underwater, divers (known as loons in North America) and grebes are only distantly related to each other. Their similarities are believed to be the result of convergent evolutionary paths, over the course of which they both developed features suited to diving for fishes and aquatic invertebrates. Physically, loons resemble stout ducks and cormorants that ride low in the water and have pointed bills. They are thought to be descended from wing-propelled swimming ancestors and therefore may be related to penguins and petrels. Some grebes are slim and elegant; smaller ones resemble ducklings.

Watery world Loons and grebes spend most of their lives on water, coming ashore only to nest (grebes even nest on the water, on floating platforms of vegetation). Loons live on freshwater lakes; grebes inhabit wetlands in general.

Dancing duo The courtship rituals of the Western grebes (above) are very highly ritualized. They initially perform a "weed dance," holding strands of vegetation in their bills. Then they rise and run across the water surface together. Rival males also dance to defend their territories.

Feathers aren't just for flying Great crested grebes (above) and several other grebe species eat feathers in large quantities. These form soft balls in their stomachs, perhaps protecting the birds from ingested fish bones. Because they swim much more than they fly, grebes can molt all their wing feathers at the same time.

SKILLED SWIMMERS

Loons are shy birds that cannot walk properly. They nest on the edges of lakes, to make it easier to slip into the water, where they can swim rapidly and dive to depths of more than 200 feet (61 m) chasing fishes. They spend their winters at sea.

Grebes can also be elusive. They appear to lack tails; their bodies end in a powder-puff effect of loose feathers. Females often mount males during the breeding season.

Loon and grebe chicks can swim and dive as soon as they are born, but because they are sensitive to cold water, they prefer to hitch rides on their parents' backs or shelter under their parents' wings.

These birds do not fly much on a daily basis and usually avoid danger by diving underwater.

LOON CALL

A loon (above) gives an aggressive territorial call. The loud, yodelling calls—audible for a mile (1.6 km) or more—of returning migratory loons are a familiar sign of summer in boreal North American and Eurasia. These shy birds are most easily seen during migrations.

Common loon
Gavia immer

Pointed bill and streamlined head are adapted to cleaving water in the chase for food

Yellow-billed loon
Gavia adamsii

The rarest and least known of the loons

The common loon breeds in northern North America and migrates to spend winter in middle North America and Europe

Arctic loon
Gavia arctica

Distinguishing gray heads and black throats

Red-throated loon
Gavia stellata

All loons have streamlined bodies and webbed feet set far back on the body—superb adaptations for swimming

White-tufted grebes feed mainly on the surface of the water on fishes and aquatic invertebrates

White-tufted grebe
Rollandia rolland

New Zealand grebe
Poliocephalus rufopectus

One of two large species of grebes in North America

Western grebe
Aechmophorus occidentalis

Great grebe
Podiceps major

Named for its mostly black head

Hooded grebe
Podiceps gallardoi

Great crested grebe
Podiceps cristatus

Face plumes, called tippets, are flared in sexual display. Like other grebes, this species builds a rough nest of vegetable matter floating in water

Eared grebe
Podiceps nigricollis

Horned grebe
Podiceps auritus

If forced to leave the nest, this grebe first covers its eggs with wisps of vegetation to hide them

Little grebe
Tachybaptus ruficollis

Lobed toes on feet placed far back on the body are much better adapted for swimming than walking

FACT FILE

Western grebe This elegant bird is famous for its spectacular courtship displays. It nests in colonies on lakes and migrates to the coast. A nocturnal feeder, it belongs to the only group of grebes that spear fish with the bill.

⚲ Up to 30 in (76 cm)
● 3–4
∥ Sexes alike
↻ Migrant
↯ Locally common

W. & C. North America

Great crested grebe This elegant waterbird, the best-known grebe, was almost hunted to extinction in some countries for its ornate head feathers.

⚲ Up to 25 in (64 cm)
● 3–5
∥ Sexes alike
↻ Partial migrant
↯ Locally common

Eurasia, S. Africa, S. Australia, New Zealand

Eared grebe This grebe inhabits mostly marshes, ponds, and lakes. Its feathers were once popular adornments for clothing, and its eggs were collected for food. Like all grebes, pairs perform ritual courtship and bonding dances.

⚲ Up to 13 in (33 cm)
● 3–5
∥ Sexes alike
↻ Migrant
↯ Common

North America, Europe, Asia, Africa

Little grebe This small, dumpy bird appears to have a fluffy rear end. If disturbed, it quickly submerges, reappearing some distance away. It makes loud, distinctive whinnying trills.

⚲ Up to 11 in (28 cm)
● 3–7
∥ Sexes alike
↻ Partial migrant
↯ Common

Sub-Saharan Africa, W. & S. Eurasia to N. Melanesia

ALBATROSSES AND PETRELS

CLASS Aves	
ORDERS 1	
FAMILIES 4	
GENERA 26	
SPECIES 112	

These seabirds, collectively known as tubenoses, are highly adapted to life at sea. They can glide for hours without beating their wings, and it is not uncommon for them to fly hundreds of miles (km) in search of squid, fishes, and the large zooplankton that comprise their diets. They seldom come within sight of land, except to breed. The biggest tubenose is the wandering albatross, which has the largest wingspan—up to 11 feet (3.3 m)—of any bird. Giant petrels are the size of some albatrosses, but the smallest storm petrels have an average wingspan of only about 12 inches (30 cm). Diving petrels have small, rigid wings as much suited for underwater propulsion as for flying.

Ocean-going birds Albatrosses are most frequently associated with the windswept expanses of southern oceans.

Nesting instinct Two Laysan albatross (*Diomedea immutabilis*) parents tend their week-old chick (above). Adults regurgitate a fattening mixture of half-digested food and stored oil to feed their young, which take up to 9 months to fledge.

Greeting display Wandering albatrosses (*D. exulans*) (left) greet each other with outstretched wings. Up to eight males may take part in courtship ceremonies.

EXPRESSING THEMSELVES

Tubenoses are found throughout the world's oceans. They have large, external, tubular nostrils and a relatively well-developed sense of smell that they may use to locate food, breeding sites, and each other. All species have a musty body odor that persists for decades in museum exhibits.

Most tubenoses store large quantities of oil in their stomachs, which they regurgitate for their young or eject to deter predators.

The male and female perform spectacular courtship displays on the ground, facing each other with their wings spread and their tails fanned, throwing their heads back as they gurgle and bray. They renew their bonds with their life partners with complex head movements and rituals that can last for days.

Tubenoses are long-lived birds, but do not begin breeding until they are at least 10 years old. Some only breed every other year.

Northern fulmar
Fulmarus glacialis

Huge long wings enable albatrosses to wander the oceans by effortlessly gliding on wind currents

Royal albatross
Diomedea epomophora

Cape petrel
Daption capense

Yellow-nosed albatross
Thalassarche chlororhynchos

⚑ CONSERVATION WATCH

Albatrosses and other large tube-noses are seriously threatened by long-line fishing. The birds follow fishing ships, dive for bait on lines thrown overboard, get hooked, and drown. Another albatross, the short-tailed albatross (*Phoebastria albatrus*), was hunted almost to extinction for the feather trade in the early 20th century. The 78 species of Procellariiformes on the IUCN Red List are categorized as follows:

3	Extinct
13	Critically endangered
14	Endangered
31	Vulnerable
13	Near threatened
4	Data deficient

Gray petrel
Procellaria cinerea

Fiji petrel
Pseudobulweria macgillivrayi

A rare species known only from around Fiji

Jouanin's petrel
Bulweria fallax

Bulwer's petrel
Bulweria bulwerii

Manx shearwater
Puffinus puffinus

Band-rumped storm-petrel
Oceanodroma castro

Longer, more pointed wings, forked tails, and short feet distinguish northern storm-petrels from those in Southern Hemisphere waters

Diving-petrels have small, chunky, bullet-shaped bodies adapted to plunge-diving for food

Common diving-petrel
Pelecanoides urinatrix

This transequatorial migrant is the most widespread storm-petrel in the world

Wedge-tailed shearwater
Puffinus pacificus

Wilson's storm-petrel
Oceanites oceanicus

FACT FILE

Wilson's storm-petrel Like other storm-petrels, this bird is colonial and nests in rock crevices or underground, mostly on isolated islands. It makes long migrations, traveling between the Antarctic and the subarctic.

⚖ Up to 7½ in (19 cm)
🥚 1
⫽ Sexes alike
↻ Migrant
🐾 Common

Antarctica, all oceans to north of equator

Common diving petrel This bird dives for its food, traveling through water and air with equal ease. Its tubular nostrils open upward rather than forward to prevent water entering.

⚖ Up to 10 in (25 cm)
🥚 1
⫽ Sexes alike
⊘ Sedentary to locally nomadic
🐾 Common

Southern Ocean of S.E. Australia & New Zealand, S.W. Africa & S.W. South America

---

THE LONG WAY HOME

Several shearwaters are examples of birds that migrate in a loop pattern. They skirt the coastline of one continent in the spring and a different one in the fall, in a large figure-eight pattern. Loop migrations may be undertaken due to food availability, ocean currents, prevailing winds, and temperature.

Long flight
A shearwater's long wings help it migrate over very large distances across open oceans.

MIGRATIONS

Nearly half of the world's birds divide their time between two main localities. Most migrate because of seasonal fluctuations in the availability of food. Many species travel alone, but others prefer to migrate in flocks comprising one or more species. They travel either by day or night, and either overland or across oceans. Most break the journey into short hops of a couple of hundred miles (km). But terrestrial birds that cross oceans and cannot land on water must complete the entire journey in one epic flight. One such bird is the American golden plover, which flies between Alaska and Hawaii, a journey it undertakes twice a year. To prepare for such energy-intensive journeys, birds load up on food almost equal to their regular weight before they leave.

Travelers Garden warblers (*Sylvia borin*) (above) spend their summers in Europe and their winters in Africa.

Epic journey The arctic tern (*Sterna paradisaea*) (below) migrates over a greater distance than any other bird species. Twice a year, these birds fly between the two polar regions.

Look up Loud flocks of migrating snow geese (*Chen caerulescens*) (above) are a breathtaking sight in parts of North America.

Breeding range
Arctic terns spend their summers in the world's far northern latitudes. The birds nest there in colonies on the ground.

On the wing
An Arctic tern's migration route depends on which part of the Arctic its particular population is starting from.

A record
The Arctic tern's biannual journey eclipses that of any other bird species, covering 12,400 miles (20,000 km) each way.

Meals en route
Arctic terns can eat fish along their migration routes. They hover, then plunge into the water to catch their quarry.

SUN AND MIGRATION
Environmental cues such as day lengths and temperatures tell birds when it is time to migrate. They also navigate by other cues, including landmarks, the position of the sun and stars, and the Earth's magnetic field. Birds have a keen time sense as well that helps them judge journey lengths.

A bird innately knows it should fly at a certain angle to the sun.

In a cage where mirrors deflect the sun's angle, the bird reorients.

No matter how the sun's angle is altered, the bird flies relative to it.

FLAMINGOS

These distinctively beautiful birds are easily recognized by their bright pink or red and white plumage, their long legs and neck (proportionately longer than on any other bird), and their oddly depressed bills. There are five species of flamingo; the largest is the greater flamingo, which reaches a height of nearly 5 feet (1.5 m). Vast flamingo flocks that congregate on the lakes of the Great Rift Valley are one of the famous natural spectacles of Africa. The reddish color of flamingo plumage comes from carotenoid proteins in the birds' diet of plant and animal microplankton. Liver enzymes break down these proteins into usable pigments that are deposited in both skin and feathers.

CLASS	Aves
ORDERS	1
FAMILIES	1
GENERA	3
SPECIES	5

Old salts Flamingos were once found on every continent, but have disappeared from Australasia. These mainly tropical birds live in shallow lakes and coastal regions, preferring salty or brackish water. They inhabit some isolated islands and can also be seen at high altitudes in the Andes.

All together now Greater flamingos (right) are gregarious birds that build conical nests out of mud and sand that they scrape together in shallow waters in Central Africa and the Caribbean.

Unusual angle
The long neck and downturned bill are adapted to feeding by immersing the head deep under water for long periods.

SPECIALIZED FEEDERS
The evolutionary lineage of flamingos is still something of a mystery, though they may represent the link between herons and their allies, and waterfowl.

The flamingo's strongly hooked bill is well designed for filter feeding on small shellfish, insects, single-celled animals, and algae. To feed, the bird bends forward, turns its head upside down (looking backward between its legs) and drags its opened bill through the water. After closing its bill, it uses its lower jaw and tongue (which has tooth-like projections) to pump water and mud out through the slits lining the upper jaw. The bird then swallows the food that remains.

The shallow waters where flamingos live sometimes drain away, forcing them to travel great distances to more abundant feeding grounds. The flocks travel at night, honking as they fly.

Flamingos nest on lakes and in coastal regions, laying one or two eggs per breeding season. The young can run and swim well at an early age, leaving the nest to follow their parents around by the time they are about 4 days old, and flying by the age of 70–80 days.

Adults sometimes swim rather than walk while they feed in deep waters, often at night.

Andean flamingo
Phoenicoparrus andinus

Greater flamingo
Phoeniconaias ruber

Lesser flamingo
Phoenicopterus minor

Chilean flamingo
Phoenicopterus chilensis

Long, slender, bare-legged feet adapted for wading in water 3¼ feet (1 m) or more deep

Feet have a short but distinct hind toe missing from other species

⚡ CONSERVATION WATCH

Salty habitats As flamingos live in saline waters, their wetland habitats have not been as affected as those of other water birds. Even so, two species found only in the South American Andes, the Andean flamingo and Puna flamingo (*Phoenicoparrus jamesi*), are considered threatened.

HERONS AND ALLIES

CLASS	Aves
ORDERS	3
FAMILIES	5
GENERA	41
SPECIES	118

This group of long-legged wading birds includes herons, storks, ibises, and spoonbills. They can step through water without getting their plumage wet as they search for fishes, insects, and amphibians to eat. Some types of herons are known as egrets; they were thus named for their special, filamentous breeding plumage, which was highly sought after by 19th-century hat-makers. Many species within this group are gregarious, and large groups comprising several species can sometimes be seen feeding, roosting, or nesting together. The white stork, a migratory bird that often nests in pairs on chimneys, has long been associated with human births in folk tales.

Freshwater dwellers Herons and their allies live worldwide, except near the poles. They are typically found in or near various types of freshwater habitats, including swamps, marshes, rivers, streams, lakes, and ponds. However, several species have adapted to drier environments.

A safe perch Although they spend most of their waking time down on the ground, scarlet ibises (*Eudocimus ruber*) (above) roost in large groups in trees, out of harm's way. Some other species in this group nest on the ground.

A catch A young great blue heron (above, right) holds a fish it has just caught. Most herons have long bills with which to stab at fishes. The bills of other birds in this group vary, depending on their diets and feeding styles.

WADING IN

All birds in this group have short tails and long beaks and necks, as well as long legs. Nonetheless, there is considerable variation in size, color, plumage pattern, and feeding behavior among species.

Some species have specialized to fill very specific niches. The cattle egret, for example, follows grazing animals such as buffalo, eating the insects disturbed. Some herons camouflage themselves in marshes; if approached, they try to blend in with the reeds by pointing their heads skyward, compressing their bodies, and swaying with the vegetation. Unlike their close relatives, herons fly with their heads folded back on their shoulders, making them easy to recognize.

Like some other types of birds, herons have specialized patches of feathers called powder-down. These are never molted, but grow continually. As the tips fray, they turn into a fine powder, which the bird picks up in its bill and uses to remove slime and oil from its feathers when grooming. Some storks and ibises have bare necks, perhaps to prevent fouling plumage while eating carrion. Most species in this group migrate long distances, maybe because their large bodies would require a great deal of energy to warm in winter.

Eurasian bittern
Botaurus stellaris

Little bittern
Ixobrychus minutus

Boat-billed heron
Cochlearius cochlearius

White-crested bittern
Tigriornis leucolopha

American bittern
Botaurus lentiginosus

Black-headed heron
Ardea melanocephala

Long, articulated neck for stabbing strikes at prey

Gray heron
Ardea cinerea

Whistling heron
Syrigma sibilatrix

Long, bare legs and feet for wading

Dagger-shaped bill for grasping prey from strikes

Great blue heron
Ardea herodias

Goliath heron
Ardea goliath

Purple heron
Ardea purpurea

Cattle egret
Bubulcus ibis

Cattle egret populations erupted around the world in the 20th century. This species associates in small flocks with domestic and wild stock, often perching on the backs of cattle

Capped heron
Pilherodius pileatus

Chinese egret
Egretta eulophotes

FACT FILE

Gray heron This colonial bird generally nests high up in trees, often with other species. Both parents look after the young, which remain in the nest for almost 2 months. The gray heron's diet sometimes includes small birds.

- Up to 3¼ feet (1 m)
- 3–5
- Sexes alike
- Partial migrant
- Common

Sub-Saharan Africa, C. & S. Eurasia to Indonesia

Great blue heron This bird has both white and gray phases in different regions. The most familiar large wading bird in North America, it is often seen stalking the edges of lakes or marshes.

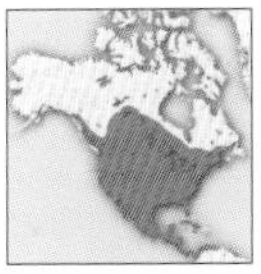

- Up to 4½ ft (1.4 m)
- 3–7
- Sexes alike
- Partial migrant
- Common

Mid-North to Central America, Galápagos Is.

Purple heron Less sociable than the larger gray heron, the purple heron nests alone or in colonies among reeds. A common bird, it sometimes eats small birds and small mammals, in addition to its staple diet of amphibians, fishes, and invertebrates.

- Up to 35 in (90 cm)
- 2–5
- Sexes alike
- Partial migrant
- Locally common

S. & C. Europe to Middle East, Sub-Saharan Africa, Madagascar, S. & E. Asia to Sunda Is.

SHADOW HUNTING

The black heron (*Egretta ardesiaca*) spreads its wings to create a cowl over the water. This reduces reflections on the surface, improving the bird's vision, and may also attract fishes into the shade. Some species of heron stand still, then grab any prey that comes near. Other species actively pursue their quarry.

FACT FILE

Hamerkop Named for its hammer-shaped head, the hamerkop generally lives alone or in pairs. It flies in a slow, undulating way and has a raucous call. It feeds on aquatic animals, primarily at twilight or at night.

⤓ Up to 22 in (56 cm)
● 3–6
⫽ Sexes alike
⊘ Sedentary
🏋 Locally common

Sub-Saharan Africa, Madagascar

Shoebill This rather odd-looking bird is thus named because it appears to be wearing a clog on its face. Its large bill, which it usually rests on its breast, is well adapted for catching the slippery lungfish found in its wetland habitats.

⤓ Up to 4 ft (1.2 m)
● 1–3
⫽ Sexes alike
⊘ Sedentary
🏋 Near threatened

C. Africa

Sacred ibis This distinctive bird was very prominent in ancient Egyptian mythology, symbolizing Thoth, the god of writing and wisdom. The Egyptians mummified these birds. Sacred ibis eventually became extinct in Egypt, although they still thrive elsewhere.

⤓ Up to 35 in (90 cm)
● 2–3
⫽ Sexes alike
♁ Partial migrant
🏋 Common

Sub-Saharan Africa & W. Madagascar

Madagascan crested ibis This large terrestrial bird feeds on moist ground in forests and scrubs. When disturbed, it prefers to run away rather than fly, dodging through trees along the way.

⤓ Up to 20 in (50 cm)
⫽ Sexes alike
● 2–3
⊘ Sedentary
🏋 Near threatened

E. & W. Madagascar

⚡ CONSERVATION WATCH

Waning wetlands The draining and pollution of wetlands worldwide have caused marked contractions in the populations of many species of herons and allies that rely on these habitats for food. Several species of Asian storks, known as the Asian adjutants, have also collapsed from loss of food and poisoning.

Madagascan crested ibis
Lophotibis cristata

Color pattern of rusty brown body with white wings is unique to this species

Crested ibis
Nipponia nippon

Eurasian spoonbill
Platalea leucorodia

Shoebill
Balaeniceps rex

Hamerkop
Scopus umbretta

Roseate spoonbill
Platalea ajaja

Sacred ibis
Threskiornis aethiopicus

Black stork
Ciconia nigra

Long, bare legs and feet are adapted for wading in shallow waters

Wood stork
Mycteria americana

These South American birds use their open bills to detect fish in muddy water

PELICANS AND ALLIES

The distinctive pelicans (which have been in existence since mid-tertiary times, up to 30 million years ago) are related to four other families of water birds: tropicbirds; gannets and boobies; cormorants and anhingas; and frigatebirds. They all have webbed feet that allow them to move easily through water, and the webbing extends uniquely across all four toes. Many have large, naked throat sacs that are used to hold fishes or as a sexual attractant in courtship displays. Unusually extensive air-sac systems in their chests and lower necks make those areas well-cushioned (hence protective in diving) and buoyant. They eat primarily fishes, as well as squid and other invertebrates.

CLASS	Aves
ORDERS	1
FAMILIES	6
GENERA	8
SPECIES	63

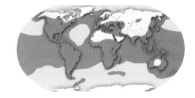

A varied range Pelicans and their close relatives are found in all types of water environments, from open oceans and sea coasts to lakes, swamps, and rivers. Most species live in tropical or temperate areas.

EXPERT FISHERS

These birds are much better suited to water than land. Tropicbirds cannot even walk because their legs are situated too far back on their bodies, and so they have to push themselves forward on their bellies. One species of cormorant in the Galápagos Islands cannot fly. But pelicans are surprisingly graceful in the air, despite being among the largest of flying birds in terms of body weight. Frigatebirds, on the other hand, are very lightweight and can remain in the air for days.

Pelicans and their allies are skilled at catching fish. Some species dive from great heights. For centuries, Chinese fishermen have attached roped collars to cormorants, letting them out to catch (but not swallow) fishes, before pulling them back to the boats and taking their prey.

Cormorants and darters lack waterproofing in their wings. This allows them to dive deeper and in general to move through water more efficiently. But their plumage eventually becomes waterlogged, forcing them to spend considerable time ashore waiting for it to dry.

All of the group breed in large colonies with other species, but may vigorously defend their individual patches. All stages of the breeding cycle may be synchronized within a colony. Many of these birds reuse the same nest sites year after year.

Pelican party Most pelicans (above) feed as they sit on the surface of the water, dipping down to capture fishes in their pouches. Groups sometimes herd fishes into shallow water where they can be more easily caught. Pelicans can often be seen scavenging near fishing boats and piers.

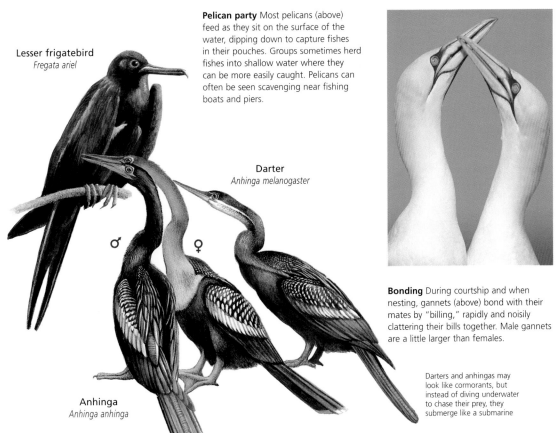

Lesser frigatebird
Fregata ariel

Darter
Anhinga melanogaster

♂ ♀

Anhinga
Anhinga anhinga

Bonding During courtship and when nesting, gannets (above) bond with their mates by "billing," rapidly and noisily clattering their bills together. Male gannets are a little larger than females.

Darters and anhingas may look like cormorants, but instead of diving underwater to chase their prey, they submerge like a submarine

⚑ CONSERVATION WATCH

Under threat Although many species of pelicans and their allies live at sea and nest on remote islands, they have still been vulnerable to massed killing at their breeding colonies and to disturbance from guano extraction. Two endangered species breed only on Christmas Island.

FACT FILE

Great white pelican Because of its size (it is among the heaviest flying birds in the world), the great white pelican relies as much as possible on thermals. For reasons of balance, it cannot fly with a full pouch.

- Up to 5¾ feet (1.75 m)
- 1–3
- Sexes alike
- Partial migrant
- Locally common

S.E. Europe, Africa, S. & S.C. Asia

Northern gannet This bird is distinctive in flight, with its pointed tail and beak, and long, narrow wings. It nests in colonies on rocky coasts, and is renowned for its spectacular high dives in search of fish.

- Up to 36 in (92 cm)
- 1
- Sexes alike
- Partial migrant
- Locally common

N. Atlantic, Mediterranean Sea

Blue-footed booby The name booby is thought to be derived from the Spanish word for "clown." This second-rarest of boobies dives for fish, attaining such a high speed that it passes its prey and must then snatch it from below as it resurfaces.

- Up to 33 in (84 cm)
- 1–3
- Sexes alike
- Partial migrant
- Locally common

N.W. Mexico to N. Peru, Galápagos Is.

PRE-DIGESTED FOOD

Before pelican chicks are old enough to hunt for themselves, they stick their bills down their parents' throats, prompting them to regurgitate a half-digested fish meal. Unlike adult pelicans, which are essentially voiceless, chicks use their voices to loudly beg for food.

Lunch box
The bill pouch of a pelican does more than snare fishes; it also serves as a receptacle for regurgitated food for its chick.

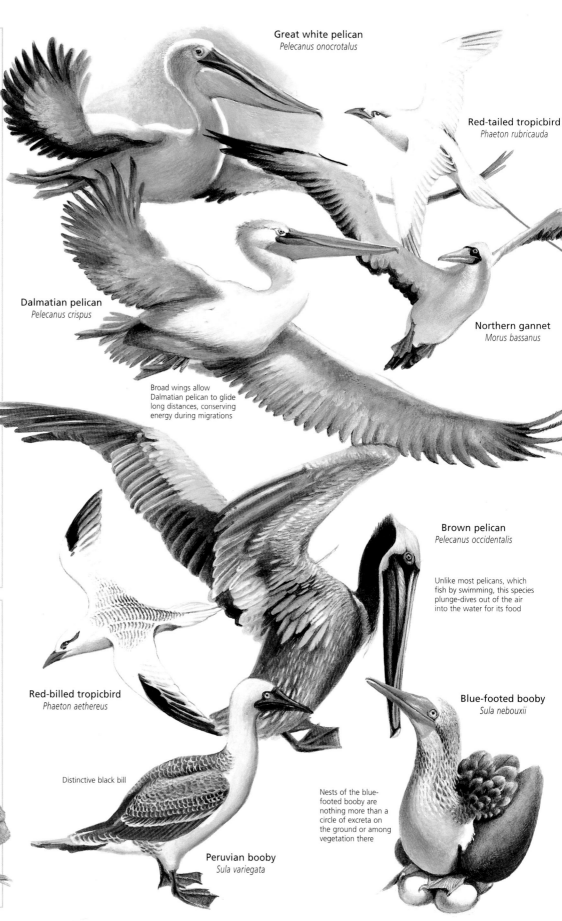

Great white pelican
Pelecanus onocrotalus

Red-tailed tropicbird
Phaeton rubricauda

Dalmatian pelican
Pelecanus crispus

Northern gannet
Morus bassanus

Broad wings allow Dalmatian pelican to glide long distances, conserving energy during migrations

Brown pelican
Pelecanus occidentalis

Unlike most pelicans, which fish by swimming, this species plunge-dives out of the air into the water for its food

Red-billed tropicbird
Phaeton aethereus

Blue-footed booby
Sula nebouxii

Distinctive black bill

Peruvian booby
Sula variegata

Nests of the blue-footed booby are nothing more than a circle of excreta on the ground or among vegetation there

Double-crested cormorant
Phalacrocorax auritus

Adults have short crests
on either side of the
head and bright orange
skin under the bill

Great cormorant
Phalacrocorax carbo

The most widespread
species of cormorant

White patches
develop on the flanks
in breeding plumage

Kerguelen shag
Phalacrocorax verrucosus

Breeding adult
European shags
develop a short
crest on the head

European shag
Phalacrocorax aristotelis

Pelagic cormorant
Phalacrocorax pelagicus

Imperial shag
Phalacrocorax atriceps

Tail is flattened out
when bird swims and
can be used as a rudder

Rough-faced shag
Phalacrocorax carunculatus

GANNET DIVE

Gannets regularly plunge from a height of about 100 feet (30 m). Their three-dimensional vision (a result of having both eyes positioned toward the front of the head) helps them pinpoint prey.

Missiles away
Gannets plunge-dive head-first into a school of fishes. After catching their prey, they surface to eat it or fly off with it.

BIRDS OF PREY

CLASS Aves	
ORDERS 1	
FAMILIES 3	
GENERA 83	
SPECIES 304	

These skilled hunters are collectively known as raptors, from the Latin word meaning "one who seizes and carries away." They comprise one of the avian world's largest orders, with members ranging from the world's fastest birds to its ugliest scavengers, and varying in standing height from 6 inches (15 cm) to more than 4 feet (1.2 m). The group includes eagles, kites, falcons, buzzards, vultures, and hawks. The raptors' hunting prowess has awed humans throughout history, making them common features on military insignia, national crests, and business logos. They all have large eyes and powerful beaks and claws, but there is considerable variety in behavior among the groups.

Widespread predators Raptors are found in most habitats, from Arctic tundra to tropical rain forests, deserts, marshes, fields, and cities. Because they need space for their hunting, their presence is determined by physical environment rather than by type of vegetation.

WINGING IT

Some raptors have long, broad wings suited for soaring while looking for carrion or live prey on the ground far below. Others have more pointed wings that allow them to fly quickly and change direction rapidly.

***Andean condor** – 9½ feet (2.9 m)*

***Bearded vulture** – 8 feet (2.5 m)*

***Secretarybird** – 6¾ feet (2.1 m)*

***White-bellied sea eagle** – 6½ feet (2 m)*

***Rough-legged buzzard** – 5½ feet (1.5 m)*

***Peregrine falcon** – 2¼ feet (0.7 m)*

***Lesser kestrel** – 2¼ feet (0.7 m)*

***Little sparrowhawk** – 1¼ feet (0.4 m)*

Got it Some raptors such as this osprey (above, top) specialize in swooping upon prey, which they grab and carry off with their talons. Many species can fly far and fast in search of prey. Scavengers do not need to be as agile and instead may soar for long periods on thermals looking for a kill.

A meaty treat Like this bald eagle (above), most raptors breed in trees, although some nest among vegetation on the ground or in small hollows scraped on cliff edges. There is usually a clear division of labor, with mothers being the ones who offer torn-up prey to nestlings.

Sizing up Female raptors are typically larger than their male counterparts, as in the case of this pair of African fish eagles (*Haliaetus vocifer*) (above). This dimorphism is less pronounced among species that eat slower-moving or immobile prey.

SUPREME HUNTERS

Raptors have sharp, hooked bills adapted for tearing flesh; powerful feet and talons for grabbing prey; and large eyes for spotting their quarry in daylight. Their diets vary from species to species and include insects, birds, mammals, fishes, and reptiles. Anatomical features vary accordingly. Long toes help falcons grab airborne prey. Their sturdy legs allow them to hit birds hard enough to incapacitate them. Large legs and talons help forest eagles capture monkeys, sloths, and other large, tree-dwelling mammals.

Raptors are renowned for their keen eyesight and aerial prowess. A rabbit that a human can barely see at a distance of 1,640 feet (500 m), a wedge-tailed eagle can see from a mile (1.6 km) away. Eagles, hawks, and ospreys often swoop suddenly when they spot a likely target.

Many scavenging raptors have bare skin on their heads and necks, perhaps to prevent messing up their feathers when they stick their heads into carcasses, or else to assist with regulating body temperature.

Andean condor
Vultur gryphus

World's largest
bird of prey

California condor
Gymnogyps californianus

Black vulture
Coragyps atratus

Like most vultures,
these birds are
scavengers and
have weak talons

Turkey vulture
Cathartes aura

King vulture
Sarcoramphus papa

Osprey
Pandion haliaetus

The ultimate fish-
catching raptor,
the osprey ranges
around the coasts
of the world

Black baza
*Aviceda
leuphotes*

African cuckoo-hawk
Aviceda cuculoides

Cuckoo-hawks are inoffensive
raptors that hunt low over tree
tops in search of large insect
food, or dive on small reptiles
from set perches

Hook-billed kite
Chondrohierax uncinatus

FACT FILE

California condor This scavenger has
the longest nestling period—5 months
—of any bird. During this time, the
young are totally dependent on the
parents. The California condor is now
being bred in captivity and released
into the wild in efforts to save it.

- ⚖ Up to 4⅓ ft (1.3 m)
- ↔ Up to 9 ft (2.7 m)
- ● 1
- ⊘ Sedentary
- ⚑ Critically endangered

S.W. USA

Osprey This bird lives near lakes, rivers,
and seashores. It builds huge, crude
nests of dry twigs, branches, and other
plant matter that it accumulates over
years. It feeds almost exclusively on fishes.

- ⚖ Up to 23 in (58 cm)
- ↔ Up to 5½ ft (1.7 m)
- ● 2–4
- ⚓ Partial migrant
- ⚑ Common

N. & S. America, Eurasia, Africa, Australia

King vulture This distinctive bird is still
quite common. It typically nests high
up in trees and feeds on carrion. It uses
its strong beak to tear apart and eat
the flesh of carcasses ahead of its rivals.

- ⚖ Up to 32 in (81 cm)
- ↔ Up to 6½ ft (2 m)
- ● 1
- ⊘ Sedentary
- ⚑ Locally common

Central & N. & C.E. South America

POWERFUL TOOLS

Ospreys drop feet first into the water
to catch fishes. Their reversible
outer toe, long curved talons,
and rough, spiny toes
(with thorny
growths called
"spicules") help
them grab and carry
prey. Like all raptors,
they have strong,
curved beaks to
tear flesh.

FACT FILE

Snail kite This raptor eats only water snails, which it collects in freshwater lowland marshes. Snail kites frequent open places, where they sometimes gather in large flocks. Nests are built among grasses and aquatic bushes.

- Up to 17 in (43 cm)
- Up to 3½ ft (1.1 m)
- 2–3
- Partial migrant
- Locally common

S.E. USA, C. America, N.E. South America

Bald eagle This bird became the national symbol of the United States due to its fierce look. It hunts mainly fishes, though it may also feed on ducks or carrion. It has also been known to steal food from other birds.

- Up to 38 in (96 cm)
- Up to 6½ feet (2 m)
- 1–3
- Migrant
- Locally common

North America, S. to N. Mexico

Short-toed eagle This eagle prefers to nest in evergreen trees. Its habitat includes terrain with scrubby vegetation and woods with large clearings. It eats mainly reptiles, especially snakes and lizards. It even eats venomous snakes.

- Up to 26½ in (67 cm)
- Up to 73 in (1.85 m)
- 1
- Partial migrant
- Locally common

N.W. Africa, W. to C. Eurasia, W. China & India

PAIR BONDS

Like most raptors, bald eagles are monogamous. Pairs renew their bonds with spectacular acrobatic displays. The male usually dives toward the female, who flies below. She then rolls and raises her legs. She might then grasp his feet as the two of them tumble together.

Flying united
A pair of bald eagles clasp talons in mid air during an aerial display fight.

Mississippi kite
Ictinia mississippiensis

Mississippi kites hunt on the wing, stooping to feed mainly on insects flushed by grazing animals or fires

Distinguished by its deeply forked tail and elegant gray form

Europ

Scissor-tailed kite
Chelictinia riocourii

Snail kite
Rostrhamus sociabilis

Congregate to roost in flocks of up to 1,000

Distinctive red feet and facial skin

Black kite
Milvus migrans

Bald eagles are sea eagles, and have huge talons and feet bare of feathers that could otherwise become waterlogged

Bald eagle
Haliaeetus leucocephalus

Short-toed eagle
Circaetus gallicus

Specializes in preying on small reptiles, by plunging from the sky and swallowing them whole

White-tailed eagle
Haliaeetus albicilla

Lappet-faced vulture
Torgos tracheliotus

Lammergeier
Gypaetus barbatus

These huge vultures feed mainly on the large bones of animal carcasses, which they break open on rocks

Cinereous vulture
Aegypius monachus

Eurasian griffon
Gyps fulvus

Egyptian vulture
Neophron percnopterus

Hooded vulture
Necrosyrtes monachus

African white-backed vulture
Gyps africanus

Palm-nut vulture
Gypohierax angolensis

FACT FILE

Eurasian griffon The largest type of vulture in Europe, this bird prefers to nest, roost, and soar in mountainous terrain, but then moves to open plains to feed. It eats carrion, mainly from large mammals such as sheep.

- Up to 3½ ft (1.1 m)
- Up to 9 ft (2.8 m)
- 1
- Sedentary
- Locally common

N. & S. Africa, S. Europe to Middle East & Caucasia

Egyptian vulture This scavenger builds crude nests using small branches and rubbish, which it typically places in holes and nooks in rocks. Its diet includes rotten fruit, rubbish, carrion, and dung.

- Up to 27 in (69 cm)
- Up to 5½ ft (1.7 m)
- 2
- Partial migrant
- Locally common

S. Europe, N. & E. Africa, S.W. Asia to India

Hooded vulture This vulture cannot compete with larger scavengers at carcass sites, so it circles around the periphery, picking at scraps. It is the only member of this species group that can live in areas of high rainfall.

- Up to 27 in (69 cm)
- Up to 6 ft (1.8 m)
- 1
- Sedentary
- Common

Sub-Saharan Africa excl. Congo Basin

SCAVENGING

Like other griffon vultures, Ruppell's griffons are gregarious birds. Several hundred may descend to feed on a carcass that one of them has spotted.

⚡ CONSERVATION WATCH

Vulture populations collapse Across southern Asia, vulture populations collapsed at the close of the 20th century. The reason was found to be poisons concentrated in carrion and other sources, which caused the vultures' kidneys to fail.

FACT FILE

Gabar goshawk This bird has two very different color forms, one gray and one mostly black. It hunts birds, mammals, lizards, and insects by swiftly flying out from trees. It prefers lower rainfall areas.

- ⏚ Up to 14 in (36 cm)
- ⏛ Up to 24 in (60 cm)
- ● 2–4
- ⏱ Partial migrant
- ⚑ Common

Sub-Saharan Africa excl. Congo, S. Yemen

Congo serpent-eagle Very little is known about the distinctively long-tailed Congo serpent-eagle. Its large eyes help it see in the dimly lit forest understory. It eats chameleons, lizards, and other animals aside from snakes.

- ⏚ Up to 20 in (51 cm)
- ⏛ Up to 3½ ft (1.1 m)
- ● Unknown
- ⊘ Sedentary
- ⚑ Locally common

Liberia to Congo Basin

Northern harrier Formerly called a marsh hawk, the northern harrier is widespread in grasslands and open fields, especially in marshy areas. Unlike other hawks, it rarely perches on anything higher than a fence post.

- ⏚ Up to 20 in (51 cm)
- ⏛ Up to 4 ft (1.2 m)
- ● 3–6
- ↻ Migrant
- ⚑ Common

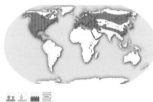

North & Central America, N. & C. Eurasia

FLEDGLINGS

Most raptors breed in trees, but northern harriers nest on the ground among tall vegetation. The female feeds the young with food brought by the male until they are ready to fly and hunt for themselves, about a month after hatching.

Lizard buzzard
Kaupifalco monogrammicus

Congo serpent-eagle
Dryotriorchis spectabilis

Gabar goshawk
Micronisus gabar

Crested serpent-eagle
Spilornis cheela

Hunts from exposed perches, diving to pounce on tree snakes and other reptiles

Dark chanting-goshawk
Melierax metabates

Black harrier
Circus maurus

Harriers hunt low over open areas and grassland, gliding on upraised wings while looking for prey

Northern harrier
Circus cyaneus

Western marsh-harrie
Circus aeruginosus

Harriers have long, slender feet and talons adapted for reaching into vegetation and grasping animal prey from shallow dives or drops

Martial eagle
Polemaetus bellicosus

Lives in open habitats in sub-Saharan Africa, and hunts on the wing

Rough-legged hawk
Buteo lagopus

Hunts from vantage perches; favored prey is lemmings, voles, and other small mammals

Eurasian buzzard
Buteo buteo

White hawk
Leucopternis albicollis

Common black-hawk
Buteogallus anthracinus

Lives in open country in central South America

Crowned eagle
Harpyhaliaetus coronatus

Mangrove black-hawk
Buteogallus subtilis

Harris's hawk
Parabuteo unicinctus

Ferruginous hawk
Buteo regalis

Sparrowhawks and goshawks hunt by stealth, watching from perches under cover, then flashing out to ambush their quarry

White phase; there are also dusky phases. Even in the normal phase, this species varies from gray-and-white to rufous-and-brown.

Variable goshawk
Accipiter novaehollandiae

Eurasian sparrowhawk
Accipiter nisus

FACT FILE

Rough-legged hawk This bird is one of the few raptors that breed in the Arctic. After raising its young above the treeline on open Arctic tundra, it migrates to its winter quarters: marshes and farmlands of North America, Europe, and Asia.

⤒ Up to 24 in (60 cm)
⤢ Up to 5 ft (1.5 m)
● 3–5
↻ Migrant
⚑ Common

North America, N. & C. Eurasia

Eurasian sparrowhawk This raptor lives in forested areas interspersed with open spaces. It hunts its prey by flying low along the edge of the woods, catching mostly birds, as well as small mammals and insects.

⤒ Up to 15 in (38 cm)
⤢ Up to 29 in (74 cm)
● 3–6
↻ Partial migrant
⚑ Uncommon

Europe, far N. Africa, N. to S. Asia

FIGHTING BIRDS

Raptors sometimes squabble over prey. Eurasian buzzards attack each other with their powerful feet and sharp talons. Beaks, no matter how dangerous looking, are never used.

Foot fighting
Two competing hawks try to get the upper hand the only way they know.

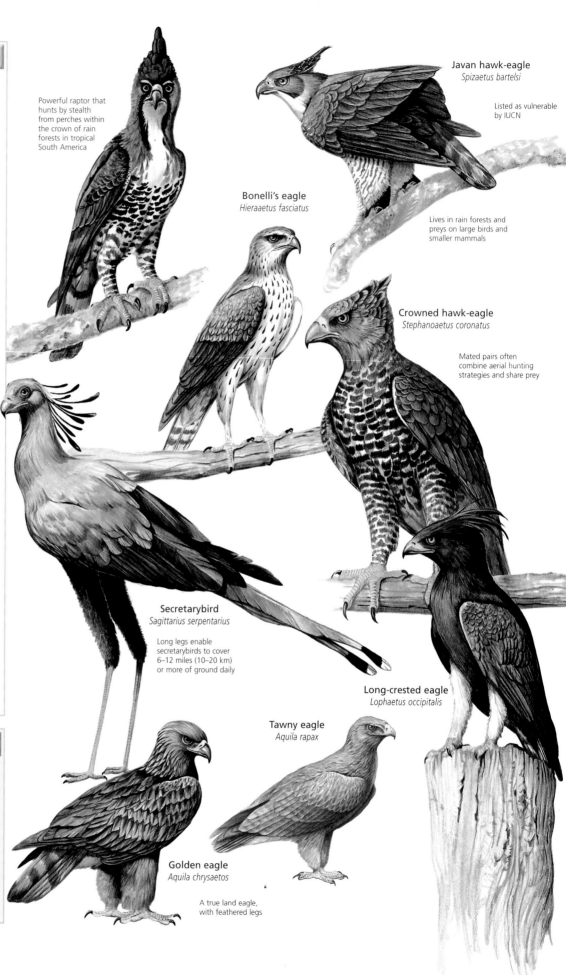

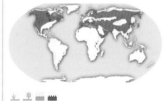
Powerful raptor that hunts by stealth from perches within the crown of rain forests in tropical South America

Javan hawk-eagle
Spizaetus bartelsi

Listed as vulnerable by IUCN

Lives in rain forests and preys on large birds and smaller mammals

Bonelli's eagle
Hieraaetus fasciatus

Crowned hawk-eagle
Stephanoaetus coronatus

Mated pairs often combine aerial hunting strategies and share prey

Secretarybird
Sagittarius serpentarius

Long legs enable secretarybirds to cover 6–12 miles (10–20 km) or more of ground daily

Long-crested eagle
Lophaetus occipitalis

Tawny eagle
Aquila rapax

Golden eagle
Aquila chrysaetos

A true land eagle, with feathered legs

HUNTING METHODS

R aptors hunt and kill their prey in a variety of ways, although their powerful feet and sharp talons are their main weapons. Some pursue airborne prey; others capture terrestrial reptiles and mammals. Hawks kill with their strong grips, squeezing their victims to death. Some vultures drop tortoises until they break, then swoop down to eat the flesh inside. Sea eagles and ospreys snatch fishes out of the water. The unusual secretarybirds subdue their prey by kicking it. And the African harrier-hawk has extra- ordinarily flexible legs which it can bend at extreme angles to grope inside tree hollows for nestling birds and other small animals.

Stages in prey capture Falcons strike and take their prey in mid air. Here we see the sequence of attack by a peregrine falcon: circling to scout prey (top left); the attacking dive (left); the hit, a Eurasian oystercatcher (below).

CRACKING UP

Egyptian vultures sometimes use tools such as twigs to open ostrich eggs. They may also drop eggs from the air, or hurl stones at eggs that are too large to drop. Lammergeiers also drop large bones onto favored rocks ("ossuaries") to break them up for eating.

Outfoxing prey Eagles (above) use their powerful legs and sharp talons to catch and kill mammals such as foxes. Fishes can be slippery prey, but fish eagles have particularly large talons with spiny soles to cope with this problem.

ALL IN THE FAMILY

The family Falconidae includes true falcons and caracaras. Unlike true falcons, caracaras walk about on the ground, and they eat insects, fruit, and seeds, or scavenge for flesh.

Contrasting ways
Falcons typically hunt prey in the air; but their closest relatives, the South American caracaras, scavenge on the ground.

Barred forest-falcon
Micrastur ruficollis

Laughing falcon
Herpetotheres cachinnans

Spot-winged falconet
Spiziapteryx circumcincta

Collared falconet
Microhierax caerulescens

Caracaras are scavengers, walking about on the ground in search of worms, maggots, and other food

Carunculated caracara
Phalcoboenus carunculatus

Yellow-headed caracara
Milvago chimachima

Yellowish-white head is unique among caracaras

Common kestrel
Falco tinnunculus

♂   ♀

Red-footed falcon
Falco vespertinus

CRANES AND ALLIES

This ancient bird order, whose members are sometimes known as the misfits of the avian world, comprises a variety of predominantly ground-living birds that prefer walking and swimming to flying. Some species have, in fact, altogether lost the ability to fly. Descended from a ground-dwelling shorebird, cranes and their relatives have filled a variety of ecological niches around the world. They typically nest on the ground or on platform nests in shallow water. Most have loud calls, and in some cases the male and female perform duets. In parts of Asia, cranes are symbols of good luck and long life; one captive crane is known to have lived to the age of 83.

LASS	Aves
RDERS	1
AMILIES	11
ENERA	61
PECIES	212

Widespread At least one species of this group of birds can be found on every continent except Antarctica and on many islands. They live in wetlands, deserts, grasslands, and forests. Trumpeters and limpkins occur only in the New World; the majority of bustards inhabit Africa. Some unusual gruids have very narrow ranges.

Pick me Two crowned-cranes (*Balearica regulorum*) (left) perform an elaborate courtship dance, involving bowing and head shaking. Other related species have even more bizarre mating rituals, featuring such displays as inflated throat sacs.

Surprise discovery The once-abundant takahe (above) was rediscovered in an inaccessible valley in New Zealand in the 1940s, after not having been sighted for 50 years. This secretive, flightless bird is one of many unusual species in this group.

White-naped crane
Grus vipio

Siberian crane
Grus leucogeranus

Individuals dance together, bugling as they do, probably to attract and strengthen pair bonds

All crowned-cranes have a distinct tuft of straw-like plumes on the head

Black crowned-crane
Balearica pavonina

Migratory, like most Northern Hemisphere cranes; it breeds across central Asia and winter in north sub-Saharan Africa and India

Demoiselle crane
Anthropoides virgo

Red-crowned crane
Grus japonensis

FACT FILE

White-naped crane This crane, which stands out with its colorful head, builds flat nests made of dry grass, which it places on slightly elevated ground among marshes. It can often be found in cultivated rice fields.

- Up to 5 ft (1.5 m)
- 2–3
- Sexes alike
- Migrant
- Vulnerable

N.E. Asia

Siberian crane Unlike most cranes, the Siberian crane has a flute-like voice and often wades in shallow waters. It does not breed until it is 5 to 7 years old. It is very wary and difficult to approach. Nests are built near water.

- Up to 4½ ft (1.4 m)
- 2
- Sexes alike
- Migrant
- Critically endangered

N.E. Siberia to Iran, N.W. India & China

Demoiselle crane This bird is often seen foraging and nesting close to human settlements. The demoiselle crane eats vegetable matter, insects, and other invertebrates. The young remain with both parents until they are thoroughly independent.

- Up to 35 in (90 cm)
- 1–2
- Sexes alike
- Migrant
- Rare to locally common

C. Eurasia to N.E. Africa & N. India

FACT FILE

Barred buttonquail Although this bird generally lives at lower elevations, it can be found in mountains up to 7,500 feet (2,273 m) in the Himalayas. It often forages for food in sugarcane, tea, and coffee plantations.

- Up to 7 in (17 cm)
- 3–5
- Sexes differ
- Sedentary
- Common

S., S.E. & E. Asia to Philippines & Sulawesi

Hoatzin This prehistoric-looking bird has an extraordinarily large crop. Just a few days after hatching, chicks can climb trees using their feet, bills, and specially adapted wings; the wings have "claws" that later disappear.

- Up to 27 in (70 cm)
- 2–4
- Sexes alike
- Sedentary
- Locally common

N. South America

Limpkin The sole species within its family, the limpkin, which is closely related to cranes, uses its long, curved bill to extract snails from their shells. It builds large, bulky nests of rushes and sticks. Before it was protected, hunting had reduced its numbers.

- 27 in (70 cm)
- 5–7
- Sexes alike
- Sedentary
- Locally common

C. America to N.E. South America, West Indies

CRANE'S TRACHEA

Cranes can make a wide range of calls, from purrs to screams. Their very long tracheas are coiled and fused with their sternums. In that region, the bony rings of the trachea are like thin plates that vibrate, amplifying the sounds produced in the voicebox and allowing the voice to carry for more than 1 mile (1.6 km) in some circumstances. Crane species with better developed tracheas make higher-pitched calls.

folded trachea

sternum

Small buttonquail
Turnix sylvatica

Despite their quail-like appearance, buttonquail have only three toes and females are more brightly plumaged than males

Barred buttonquail
Turnix suscitator
♀

Hoatzin
Opisthocomus hoazin

After attracting a mate, a female buttonquail leaves him to incubate and rear the young, and goes off to find another

Barred buttonquail
Turnix suscitator
♂

The relationships of the strange, fowl-sized hoatzin are unclear; some research even associates them with cuckoos

White-breasted mesite
Mesitornis variegatus

Subdesert mesite
Monias benschi

Gray-winged trumpeter
Psophia crepitans

Limpkin
Aramus guarauna

Inaccessible rail
Atlantisia rogersi

Swamphens flick their tails
to flash the white underside
as they walk about feeding
in wetlands and on verges;
this is a social signal to
others in their group

Coots are the most
aquatic of all rails

Horned coot
Fulica cornuta

Purple swamphen
Porphyrio porphyrio

White-breasted waterhen
Amaurornis phoenicurus

Little crake
Porzana parva

Corn crake
Crex crex

Takahes live on high moors in
the mountains of southeast New
Zealand, digging out and eating
the fleshy bases of tussocks of
grass and sedges

Gray-necked wood-rail
Aramides cajanea

Takahes have such
vestigial wings that
they cannot fly

Takahe
Porphyrio hochstetteri

Wood-rails have very
slender bodies and long
legs, enabling them to
run very quickly through
dense undergrowth

RAILS AND PREDATORS

A high proportion of island-based rail
species cannot fly and are therefore
vulnerable to over-predation by
introduced animals such as rats.
Reintroduction programs have helped
some species recover.

Back from the brink
*The Lord Howe
rail had fallen
to 10 breeding
pairs, but has
been bred in
captivity and
released
successfully.*

Kagu
Rhynochetos jubatus

Red-legged seriema
Cariama cristata

Sunbittern
Eurypyga helias

Long legs adapted for walking and feeding on the ground

Great bustard
Otis tarda

Colorful neck plumes are flared in display

Colorful tail is raised in display. Male bustards are much larger and more colorfully plumaged than females

Denham's bustard
Neotis denhami

One of 20 of the 26 species of bustards that occur in Africa; 18 are only found here

Houbara bustard
Chlamydotis undulata

Bustards have long legs adapted for walking through open habitats to feed; when the birds rise to fly, the feet are tucked up under the body

Lesser florican
Sypheotides indica

African finfoot
Podica senegalensis

WADERS AND SHOREBIRDS

CLASS Aves	
ORDERS 1	
FAMILIES 16	
GENERA 86	
SPECIES 351	

The world's shallow waters and shorelines teem with marine and terrestrial creatures that this order of generally sociable birds has evolved to hunt. The groups within the order exhibit significant diversity, allowing them to exploit different resources within aquatic habitats. Typical waders such as sandpipers and plovers patrol shallow waters and shorelines. Gulls scavenge along shorelines, too, but they are also adapted to swim out to feed on the surface of deeper water. Terns venture even further from shore and dive for food. Auks swim underwater after their prey, much like penguins. The eyes of many waders and shorebirds are set on the sides of the head to scan for predators.

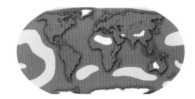

Watery world Some species in this group live next to oceans, along estuaries or seashores; others can live far inland in arid climates, in mid deserts.

A WORLD OF DIFFERENCES

Birds within this order exhibit many anatomical differences, as befits their respective niches. Those that scavenge along the mud flats and in the shallows, probing for food, tend to have long, thin legs and long necks and bills. Those that patrol the surf are usually shorter and can scurry out of the way of incoming water. Birds that swim out and scoop up food on the water's surface tend to be stouter and to have webbed feet.

Species that search for marine prey further out over the water are more capable fliers: they have shorter legs and smaller feet, but long, narrow wings. Terns are very agile, and have long, forked tails for quick maneuvering in the air.

Auks have webbed feet set well back on their compact bodies, and use their wings as flippers underwater.

The diets of these birds range from insects and worms to fishes and crustaceans. Some are scavengers.

Walking on water Jacanas (above) are distinctive birds that can walk across the surface of still waters by stepping on large lily leaves with their specially adapted feet.

Best foot forward *Jacanas have splayed, long-nailed toes that distribute their body weight as they walk on floating vegetation.*

Holes in the ground Like many waders and shorebirds, puffins (right) often nest in colonies on the shore. Few birds in this order build elaborate nests; like puffins, many lay eggs in holes that they scrape in the ground. They bring food to their chicks until they can fend for themselves.

Pirate The pomarine jaeger (*Stercorarius pomarinus*) (above) is the largest of the gull-like jaegers or skuas. They are big, sturdy, broad-winged birds that scavenge around seabird colonies. They also steal food by forcing other species to disgorge their stomach contents as they fly.

FACT FILE

African jacana Also known as a lily-trotter, this bird uses a high-stepping gait to walk on floating vegetation without stumbling on its long toes. It probes under the plants for insects, snails, and other organisms.

⬧ Up to 12 in (30 cm)
● 4
⫻ Sexes alike
⊘ Sedentary
⫞ Common

Sub-Saharan Africa

Eurasian oystercatcher This elegant bird feeds on mollusks and other invertebrates that it catches along seashores. It gathers in large flocks. A strong flier, it can also swim and dive.

⬧ Up to 18 in (46 cm)
● 2–5
⫻ Sexes alike
↻ Migrant
⫞ Common

Europe; W., S.W. & E. Asia; N.W., N. & E. Africa

Beach stone curlew This bird's extremely large yellow eyes help it find crabs and other shellfish on the reefs, muddy seashores, and sandy beaches it patrols at night. Its call is harsh and eerie. It scrapes a nest in the sand.

⬧ Up to 22 in (56 cm)
● 1
⫻ Sexes alike
⊘ Sedentary
⫞ Vulnerable

Malay Peninsula to Philippines, New Guinea & N. Australia

BILLS OF FARE

Bills vary with diet and feeding habits. Puffins have big bills with serrated edges that help them catch and carry fish underwater. Wading birds such as greenshanks have long, slender bills for probing and picking up insects and other small prey. Gulls have stout bills with tips suited for tearing carrion.

tufted puffin

common greenshank

great black-headed gull

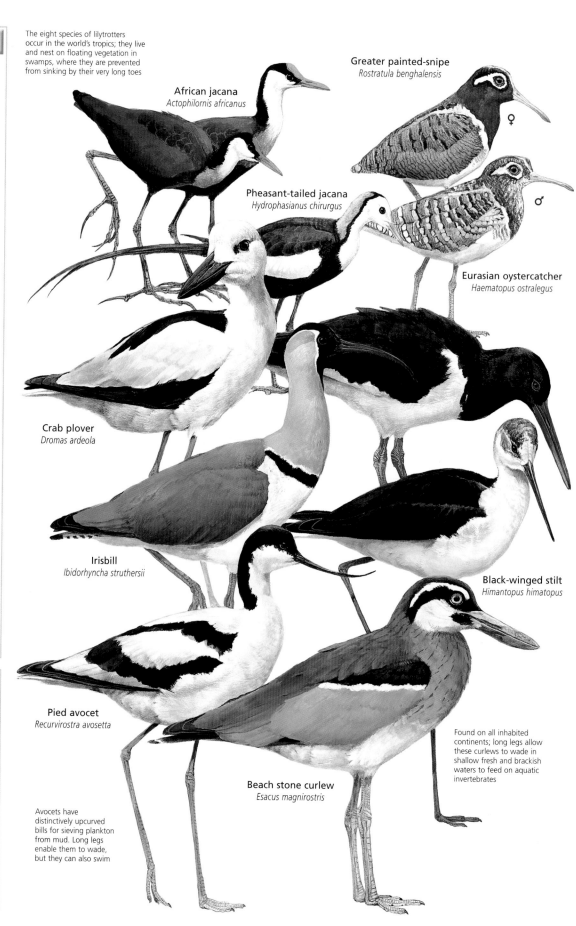

The eight species of lilytrotters occur in the world's tropics; they live and nest on floating vegetation in swamps, where they are prevented from sinking by their very long toes

African jacana
Actophilornis africanus

Greater painted-snipe
Rostratula benghalensis

♀

♂

Pheasant-tailed jacana
Hydrophasianus chirurgus

Eurasian oystercatcher
Haematopus ostralegus

Crab plover
Dromas ardeola

Black-winged stilt
Himantopus himatopus

Irisbill
Ibidorhyncha struthersii

Pied avocet
Recurvirostra avosetta

Beach stone curlew
Esacus magnirostris

Found on all inhabited continents; long legs allow these curlews to wade in shallow fresh and brackish waters to feed on aquatic invertebrates

Avocets have distinctively upcurved bills for sieving plankton from mud. Long legs enable them to wade, but they can also swim

Southern lapwing
Vanellus chilensis

Only gray lapwing with a crest

Black-tailed godwit
Limosa limosa

Rusty breeding plumage is lost when bird migrates to winter quarters

Breeds across northern Eurasia and winters from the Mediterranean and Africa to Australia

Spotted redshank
Tringa erythropus

Eurasian curlew
Numenius arquata

Large, fowl-sized waders with very long downcurved bills

Common redshank
Tringa totanus

Collared pratincole
Glareola pratincola

Collared pratincoles feed in flocks on the wing, mostly at dawn and dusk

Common snipe
Gallinago gallinago

Live in marshlands, where they dig under cover for food

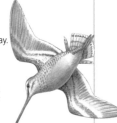

FACT FILE

Green sandpiper This bird's name derives from the fact that its underparts have a greenish sheen in summer. It makes piping sounds in flight. It feeds by probing deeply into mud along inland waterways.

- Up to 9 in (24 cm)
- 3–4
- Sexes alike
- Migrant
- Locally common

W.C. to E.C Eurasia, C. Africa, S. Asia

Red-necked phalarope The smallest of the three phalarope species, this bird flies swiftly and erratically in flocks. It winters at sea, and swims gracefully, carrying its head high and nodding.

- Up to 8 in (20 cm)
- 3–4
- Sexes alike
- Migrant
- Locally common

Arctic, North & Central America, Eurasia

Ruff In spring, the ruff can be seen in flocks of hundreds. The female, or reeve, builds a nest out of fine grasses, either among thick grass or in clumps of sedge or rushes.

- Up to 13 in (32 cm)
- 3–4
- Sexes differ
- Migrant
- Locally common

Arctic to W. & S. Eurasia, Africa

RUFFING IT

During the spring breeding season, the male ruff has colorful ear tufts and a raised collar of feathers. Males assemble on selected hillocks each morning to display to the females.

Common greenshank
Tringa nebularia

Green sandpiper
Tringa ochropus

Black marks on this bird's back and its speckled breast indicate breeding plumage

This species breeds in the swampy woodlands of Eurasia and migrates to Africa and Southeast Asia for winter

Little stint
Calidris minuta

Curlew sandpiper
Calidris ferruginea

Other phalaropes spend winter at sea, but this species migrates from its breeding grounds in the North American midwest to the lakes and mudflats of Andean South America

Wilson's phalarope
Phalaropus tricolor

Red-necked phalarope
Phalaropus lobatus

Ruff in full breeding regalia. Ruffs congregate in leks at the beginning of breeding to attract and mate with females

Reeves look like ruffs in non-breeding plumage; this is the plumage they wear at their winter quarters from Africa to Southeast Asia

♂

♀

Ruff
Philomachus pugnax

Gray-breasted seedsnipe
Thinocorus orbignyianus

Seedsnipes browse on the tips of low marshy vegetation in the South American highlands

Black-faced sheathbill
Chionis minor

A fatty layer keeps the bird warm and gives it a chubby look

A rudimentary spur on the wing is used as a weapon

White-eyed gull
Larus leucophthalmus

Jaegers are parasitic; they live in cold seas and chase other seabirds for scraps

Breeding birds have black heads; out of breeding, the head molts to white

Bonaparte's gull
Larus philadelphia

Long-tailed jaeger
Stercorarius longicaudus

Plains-wanderer
Pedionomus torquatus

Once thought to be a buttonquail, this Australian bird is now know to be a wader related to the South American seedsnipes

Great black-backed gulls are omnivorous, and usually breed in small colonies

Great black-backed gull
Larus marinus

Herring gull
Larus argentatus

In Europe and North America, herring gulls can be distinguished from other large gulls by their combination of a pale gray back with black-tipped wings, pink legs, and pale eyes

FACT FILE

Long-tailed jaeger Like other skuas, the long-tailed jaeger has acrobatic courtship flights. Its nest is an unlined shallow hollow on the ground. This bird eats lemmings at nesting grounds.

⚓ Up to 21 in (53 cm)
● 2
⫻ Sexes alike
↻ Migrant
⚡ Common

Arctic, Antarctic, N. & S. Pacific & Atlantic oceans

Herring gull This large, familiar gull can be seen at harbors and on beaches, as well as at garbage dumps and in ploughed fields. It flies inland to bathe in fresh water, and roosts communally.

⚓ Up to 26 in (66 cm)
● 2–3
⫻ Sexes alike
⚡ Partial migrant
⚡ Common

North & Central America, W., N. & E. Eurasia

JAEGER TAIL FEATHERS

The three species of jaeger are easily recognized by their elongated central tail feathers that taper or are rounded at the tip. The smallest of these birds is the long-tailed jaeger. It grows its central tail feathers—which can project up to 10 inches (25 cm) beyond the other tail feathers—when it is an adult. The function of these feathers is thought to be connected with display. They seem to assist in display flight. Pomarine jaegers even bite them off after breeding because they increase drag in normal flight.

Untold tails
Why long-tailed jaegers have projecting central tail feathers is still not really known.

FACT FILE

Fairy tern This tern nests only on isolated sandy beaches or spits above high-tide marks and is vulnerable to disturbance. It is colonial in Australia but more solitary in New Zealand.

⚓ Up to 10½ in (27 cm)
🥚 1–2
🪶 Sexes alike
🧭 Partial migrant
🚩 Vulnerable

New Caledonia, Coral Sea, W. & S. Australia, Tasmania, N. North I. (New Zealand)

Little tern This is one of the smallest of the terns, a group of birds whose species look very similar to each other. It lives on sea beaches, bays, and large rivers.

⚓ Up to 11 in (28 cm)
🥚 2–3
🪶 Sexes alike
🧭 Migrant
🚩 Locally common

Europe, Africa, Asia to Australasia, Indian & W. Pacific oceans

Common tern This small bird inhabits lakes, oceans, bays, and beaches in the Northern Hemisphere and migrates to the Southern Hemisphere. It nests in colonies on sandy beaches and small islands.

⚓ Up to 15 in (38 cm)
🥚 2–4
🪶 Sexes alike
🧭 Migrant
🚩 Locally common

Worldwide

DIVING TERNS

Terns resemble gulls but are specially adapted for catching fishes by plunge diving. They have compact, streamlined bodies; heavy heads with strong, thin, tapering bills; long, narrow wings; and forked tails for fast braking and maneuvering. They fly low over oceanic or inland waters, hover briefly, then plunge for food.

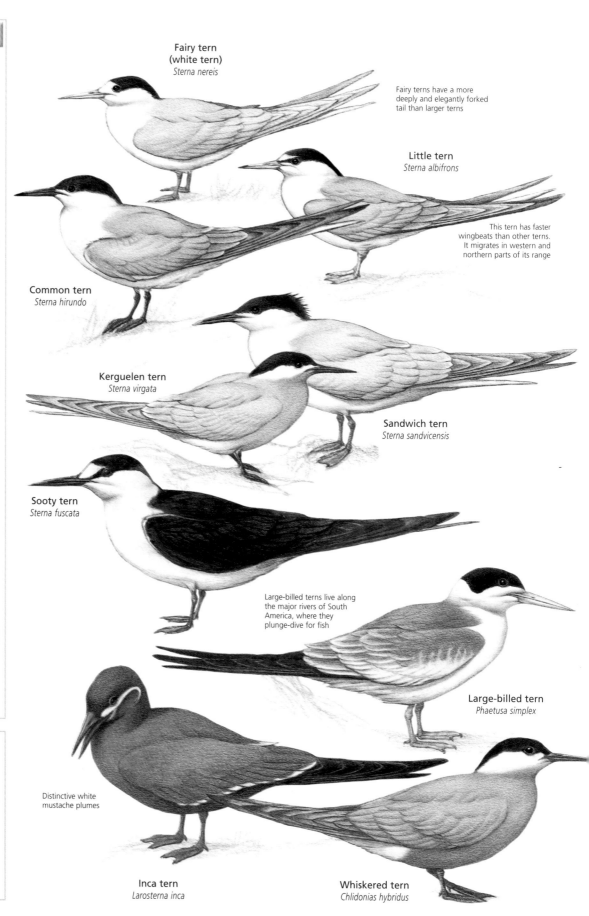

Fairy tern (white tern)
Sterna nereis

Fairy terns have a more deeply and elegantly forked tail than larger terns

Little tern
Sterna albifrons

This tern has faster wingbeats than other terns. It migrates in western and northern parts of its range

Common tern
Sterna hirundo

Kerguelen tern
Sterna virgata

Sandwich tern
Sterna sandvicensis

Sooty tern
Sterna fuscata

Large-billed terns live along the major rivers of South America, where they plunge-dive for fish

Large-billed tern
Phaetusa simplex

Distinctive white mustache plumes

Inca tern
Larosterna inca

Whiskered tern
Chlidonias hybridus

Common murre
Uria aalge

Murres live in arctic and subarctic seas, sit on the water in large flocks, and dive for a minute or more to chase and catch fish; their feet are webbed

Ancient murrelet
Synthliboramphus antiquus

Parakeet auklet
Cyclorrhynchus psittacula

Nests colonially like other auks around the north Pacific, but it nests in burrows

Rhinoceros auklet
Cerorhinca monocerata

Crested auklet
Aethia cristatella

Tufted puffin
Fratercula cirrhata

Atlantic puffin
Fratercula arctica

Black skimmer
Rhynchops niger

In skimmers, the lower bill is longer than the upper. They feed by flying just above rather still water with the lower bill immersed, ready to snap up any fishes that come in contact

PIGEONS AND SANDGROUSE

CLASS Aves	
ORDERS 1	
FAMILIES 3	
GENERA 46	
SPECIES 327	

Pigeons and sandgrouse are quite dissimilar and may not even be related. Pigeons and doves are commonplace, tree-dwelling birds that eat fruits and seeds. They have a close association with humans, who have used pigeons for carrying messages. They vary in color from the drab bluish-gray of the familiar street pigeon to the riot of hues that characterize the fruit doves of the Indo-Pacific region. Pigeons feed their young with a milky substance produced in their crops. They also have specialized bills that enable them to suck up water when they drink. By contrast, sandgrouse are dull-colored, fast-flying desert dwellers that are well adapted to arid climates.

Abundant Pigeons and doves are found worldwide, except in polar regions. Sandgrouse only inhabit Africa and Eurasia.

FACT FILE

Rock dove Also known as a feral pigeon, this bird is familiar to urban dwellers across the world. Originally from Eurasia and North Africa, where it nests on cliffs, it has readily adapted to living on building ledges in cities.

- ⊥ Up to 13 in (33 cm)
- ● 2
- ✕ Sexes alike
- ⊘ Sedentary
- ↟ Common

S. Europe, Middle East, S.W. & C.E. Asia, N. Africa

Pallas's sandgrouse This rare vagrant bird breeds on open steppes in Central Asia. Each year large numbers fly for long distances outside their main range, to grasslands and beaches, to avoid winter snows that cover feeding areas.

- ⊥ Up to 16 in (40 cm)
- ● 2–3
- ✕ Sexes differ
- ↺ Partial migrant
- ↟ Locally common

S. Urals & Transcaspia to Mongolia

Emerald dove This bird is commonly found in rain forests or nearby dense vegetation, where it forages on the forest floor, alone or in pairs. It can also be seen flying swiftly over open ground between foraging areas.

- ⊥ Up to 11 in (27 cm)
- ● 2
- ✕ Sexes differ
- ⊘ Sedentary
- ↟ Common

India & S.E. Asia to E. Australia & Melanesia

Emerald dove
Chalcophaps indica

Emerald doves feed on fallen fruits on the floor of rain forests

Namaqua dove
Oena capensis

Pallas' sandgrouse
Syrrhaptes paradoxus

Rock dove
Columba livia

Crowning achievement Three species of spectacular crowned pigeons (above) inhabit the rain forests of New Guinea. The world's largest pigeons, they are distinguished by their filmy crests. Pigeon courtship rituals include bowing and special display flights.

Side by side Like other species in this group, green-pigeons (above, left) roost communally. Highly social birds, pigeons, doves, and sandgrouse often flock in very large groups numbering in the thousands.

Madagascar green-pigeon
Treron australis

The Indo-African green-pigeons are arboreal fruit-eaters that have a gizzard modified for grinding seeds as well as fruit

Black-backed fruit-dove
Ptilinopus cinctus

Fruit-doves have thin-walled gizzards to allow the ingestion of fruit and the passing of seeds

Seychelles blue pigeon
Alectroenas pulcherrima

Bruce's green pigeon
Treron waalia

The turkey-sized crowned pigeons of New Guinea have a uniquely fan-like crest

Victoria crowned pigeon
Goura victoria

Mourning dove
Zenaida macroura

Gray-headed dove
Leptotila plumbeiceps

Zebra dove
Geopelia striata

PARROTS

CLASS	Aves
ORDERS	1
FAMILIES	1
GENERA	85
SPECIES	364

Parrots and cockatoos form an ancient and highly distinct order of birds. They are easily recognized by their short, blunt bills, which have downcurved upper mandibles, and also by their feet, which have two toes pointing forward and two pointing backward. Although some species are dull, most have brilliantly colored plumage, predominantly in shades of green accented by splashes of red, yellow, and blue. Their visual appeal is one reason for their popularity as pets over the centuries. Another is their antics: they can perform acrobatics, hanging on to perches with their feet or bills. They can also mimic human voices.

Southerly distibution Parrots live primarily in the Southern Hemisphere. They are most common in tropical rain forests, especially lowland tropical rain forests, but some species prefer open, arid regions. The highest concentrations of species occur in Australasia and South America. The most southerly parrot inhabits Tierra del Fuego.

A SOCIABLE GROUP

Most parrots eat seeds and nuts (which they crack open with their heavy bills) as well as fruit. They forage among the treetops or on the ground. Lorikeets, on the other hand, are strictly arboreal; they eat soft fruit, and harvest pollen and nectar from blossoms.

Although parrots' basic features differ little among species, there is considerable variation in size and shape. Wings can be narrow and pointed, or broad and rounded. Similarly, tails may be long and pointed or short and squarish. Some have ornate feathers. Cockatoos, a separate family from "true parrots," have prominent, erectile head crests. The sexes are usually alike.

Parrots are very social birds. They squawk loudly and frequently, and are heard more readily than observed in the wild. They can be difficult to glimpse as they fly through the forest canopy, camouflaged by their green plumage. That has not stopped many of them from falling prey to poachers, who sell them to the worldwide pet market. That and habitat destruction have made parrots one of the most common groups of birds on lists of threatened species.

Group dynamics This colorful gathering of orange-checked (*Pionopsitta barrabandi*) and blue-headed (*Pionus menstruus*) parrots (above) is typical of these very gregarious birds. Parrots often forage and roost in small groups or large flocks comprising many pairs or family groups.

Fancy feathers Red-and-green macaws (*Ara chloroptera*) (below), though difficult to see among the forest canopy, can be recognized by their colorful plumage and long tails. Their tapered wings allow them to fly faster than would otherwise be expected for birds of their size.

Open wide Parrots' bills have a greater range of motion and are more powerful than the bills of other birds. A well-developed hinge on the upper mandible of a parrot's jaw (below) provides leverage that enables the bird to use its bill to climb branches, and a strongly muscled, cutting-edged lower mandible to crack open large, hard-shelled nuts.

Inner view The cutaway shows the adaptations to jaw and bill.

upper bill when jaw is open

upper hinge

lower hinge

lower bill when jaw is open

base of bill for cracking food

hook for grabbing food

Sounding off Parrots can use their feet like hands, to handle objects. This palm cockatoo (*Progosciger aterrimus*) (above) creates mechanical sounds by beating a stick against a hollow tree.

Aloft Two brilliantly colored red-and-green macaws display a rainbow of hues when they fly. They are among the largest parrots.

Buff-faced pygmy parrot
Micropsitta pusio

Red-rumped parrot
Psephotus haematonotus

Eclectus parrot
Eclectus roratus

♀

♂

Males are green, with red on the sides of the breast; females are richer red and blue

Inhabits woodlands in inland southeastern Australia, and feeds on seeds on the ground

Pesquet's parrot
Psittrichas fulgidus

Eastern rosella
Platycercus eximius

Breast barring is present in some races (New Guinea) but not others (Australia)

Rainbow lorikeet
Trichoglossus haematodus

The delicate tone to the gray on a galah's back comes from powder down on the rump; if washed out, the back turns dull dark gray

Galah
Eolophus roseicapilla

Kea
Nestor notabilis

FACT FILE

Alexandrine parakeet This bird normally nests in holes it gnaws in trees, or in naturally occurring hollows, or even in chimneys. It lives in small groups that join together to form larger groups for the night. Only the male has a prominent black collar around its neck. The species can be found in forested and cultivated lands.

⚊ Up to 24 in (62 cm)
● 3
⫽ Sexes differ
⊘ Sedentary
▮ Locally common

S. and S.E. Asia

Blue-crowned hanging parrot This small parrot is quite common in the forested lowlands of Southeast Asia. It eats buds and flowers, as well as fruits, nuts, and seeds. It gets its name from its habit of sleeping upside down and is part of a group known as the bat parrots. It wanders in small groups and is rarely seen by itself.

⚊ Up to 5 in (12 cm)
● 3–4
⫽ Sexes differ
⊘ Sedentary
▮ Common

Malay Peninsula, Borneo, Sumatra & nearby islands

BLUE VARIATIONS

Alexandrine parakeets are among those parrot species that also come in a mutant blue color variety due to the suppression of yellow pigmentation. They are popular pets because of their excellent mimicking skills and affectionate behavior. They are named after the legendary Alexander the Great.

Spread out
An Australian eastern rosella in flight shows the wide range of colors and patterns in its plumage.

CONSERVATION WATCH

Fischer's lovebird This beautiful, once-common species of parrot, which lives in the wooded steppes and forests of East Africa, is caught in large numbers for the pet trade. Even so, numbers are still up to a million birds in the wild, at least half of them protected in reserves and parks. "Lovebirds" are so named for the close bond between pairs of the species.

Blue-crowned hanging parrot
Loriculus galgulus

Alexandrine parakeet
Psittacula eupatria

Hanging parrots are among the few groups of parrots with short, blunt tails; they may rest and sleep hanging upside-down in foliage

Swift parrot
Lathamus discolor

Vernal hanging-parrot
Loriculus vernalis

A common feature of the Afro-Asian parakeets is a ringed neck

Plum-headed parakeet
Psittacula cyanocephala

Senegal parrot
Poicephalus senegalus

Yellow-collared lovebird
Agapornis personatus

Fischer's lovebird
Agapornis fischeri

Ground parrot
Pezoporus wallicus

Thick-billed parrot
Rhynchopsitta pachyrhyncha

Blue-and-yellow macaw
Ara ararauna

Hyacinth macaw
Anodorhynchus hyacinthinus

Macaws use their
massive bills to
open hard-shelled
nuts and fruits

**Scarlet
macaw**
Ara macao

**White-eared
parakeet**
Pyrrhura leucotis

Military macaw
Ara militaris

Nanday parakeets feed
on seeds and fruit,
mainly on the ground

Nanday parakeet
Nandayus nenday

Burrowing parakeet
Cyanoliseus patagonus

BURROWING PARROT

Most parrots nest in tree hollows or
in cavities in structures such as termite
mounds. The burrowing parrot of
northern Argentina, on the other hand,
excavates deep burrows in cliffs and
banks along rivers or
near the ocean. These
offer protection
from predators.

FACT FILE

Kakapo The world's heaviest parrot is nocturnal and cannot fly because it lacks a sternal keel. It chews leaves or stems to extract their juices. Males court females by booming with their inflated gular air-sacs.

⚖	Up to 25 in (64 cm)
⬤	1–3
⫽	Sexes alike
⊘	Sedentary
⚡	Extinct in natural range

S. W. South Island, New Zealand; introd. to Little Barrier, Maud, Codfish & Pearl Is.

Blue-fronted parrot This arboreal, climbing parrot lives in forests and builds its nests in hollows in trees. It lays a clutch of two eggs, which are incubated mainly by the female over the course of about 25 days.

⚖	Up to 15 in (37 cm)
⬤	2–4
⫽	Sexes alike
⊘	Sedentary
⚡	Common

N.E. & C. South America

Monk parakeet This highly adaptable South American species is found in city parks, on farms and in gardens as well as in savannas, forests, and palm groves. Unlike other species of parrots, it builds its own rather complex nests of twigs..

⚖	Up to 12 in (29 cm)
⬤	1–11
⫽	Sexes alike
⊘	Sedentary
⚡	Common

C. & S.E. South America

STRIKING RUFF

The hawk-headed parrot (*Deroptyus accipitrinus*) of Amazonia has a striking ruff of long, colorful erectile feathers on the nape. It raises them when it is excited or angry. It is a noisy, sociable, and conspicuous parrot that usually inhabits jungle interiors and edges, often along watercourses.

Head on
A raised collar

In profile
Ruffled feathers

Red-fan parrot
Deroptyus accipitrinus

White-crowned parrot
Pionus senilis

Monk parakeet
Myiopsitta monachus

Canary-winged parakeet
Brotogeris versicolurus

Orange-winged parrot
Amazona amazonica

Cuban parrot
Amazona leucocephala

Blue-fronted parrot
Amazona aestiva

Kakapo
Strigops habroptilus

Flightless; has cryptic plumage to help it hide in its scrubby habitat, but this has not stopped introduced stoats, rats, and feral cats from exterminating it in its natural range

Like other parrots, the kakapo's feet have two toes forward and two toes back

CUCKOOS AND TURACOS

These two old families of birds are related by virtue of some molecular similarities, but otherwise differ markedly in development, anatomy, and other characteristics. Cuckoos are notorious as parasitical birds that trick other species into raising their chicks. However, less than half of the 140 or so species of cuckoos actually engage in such behavior. All cuckoos have feet with two toes pointing forward and two pointing backward, but otherwise they differ considerably among species. Turacos are more homogeneous: with one exception, they are rather slender-necked birds with long tails, short, rounded wings, and erectile, laterally compressed crests.

CLASS Aves
ORDERS 1
FAMILIES 3
GENERA 42
SPECIES 162

Here, there and everywhere Turacos inhabit savannas and forests. They live only in Africa, south of the Sahara. Cuckoos thrive in a variety of habitats—from open moorland to tropical rainforest—and are virtually cosmopolitan, although they predominate in the tropics and subtropics.

Doing away with the competition
A common cuckoo chick (left) hatches before the eggs of its host parent, in this case a reed warbler. Within three or four days, the young cuckoo is strong enough to evict the other eggs or hatchlings so that it can monopolize the foster parent's attention and food. Different females within one species of cuckoo may each be adapted to parasitize one species of host.

ROBBING THEM BLIND

Some cuckoo species have evolved remarkably devious strategies to get away with their brood parasitism, such as laying eggs that closely resemble those of their host.

Most "true cuckoos" are drab, although the bronze-cuckoos of the Old World tropics are brightly colored. Some cuckoo species are solitary; others breed communally and maintain group territories. Some, such as the roadrunners and ground cuckoos, rarely fly and prefer to run.

Turacos are gregarious, noisy birds whose harsh, barking alarm calls can be heard from afar. They gather in small groups and glide or flap from tree to tree. They are more agile when running among the branches than when flying.

The savanna-based turaco species have dull plumage, but those that favor forests tend to be brightly colored, with special blue-green pigments that are water-soluble. Turacos are primarily herbivorous, but they also eat insects.

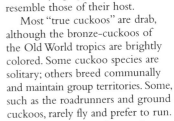

Great blue turaco
Corythaeola cristata

As big as a small turkey, this is the largest of all turacos; it lives in tropical Africa, and has a distinctive orange belly

Hartlaub's turaco
Tauraco hartlaubi

Has the green plumage found in most turaco; lives in equatorial east Africa

All turacos have long, slender tails

Jacobin cuckoo
Clamator jacobinus

Cuckoos of the Afro-Asia *Clamator* group are the only ones with crests

Great spotted cuckoo
Clamator glandarius

Violet turaco
Musophaga violacea

Violet-blue plumage is characteristic of violet and Ross's turaco species; both live in central west African riverine forests

Look at me Unlike most other birds, whose bright plumage colors are caused by refraction of light, turacos such as the Knysna turaco (*Tauraco corythaix*) (below) of South Africa have true pigmentation.

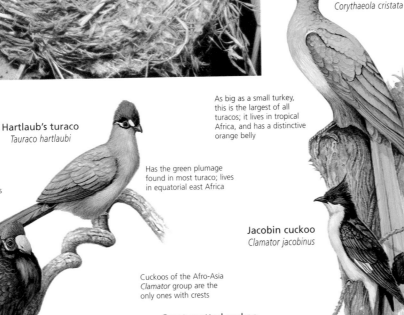

FACT FILE

Common cuckoo This well-known bird inhabits woodland clearings and cultivated fields, where it feeds on insects. It has a familiar call from which its name arose. Its females are polyandrous and parasitic.

- Up to 13 in (33 cm)
- Up to 12, even 20, singly in host nests
- Sexes differ
- Migrant
- Common

Eurasia except S.W., N.W. & S. Africa

Greater coucal This non-parasitic bird has terrestrial habits. It is often mistaken for a game bird due to its size and voluminous tail. It has a distinctive dull, booming call, as well as a variety of croaks and chuckles.

- Up to 20 in (52 cm)
- 2–4
- Sexes alike
- Sedentary
- Common

S. & S.E. Asia, Greater Sundas, some islands off Philippines

Greater roadrunner This ground-loving bird is the basis for the famous cartoon character. It inhabits arid regions and deserts with shrubs and cacti. It uses its long tail as a rudder to change direction when it runs quickly. Its diet includes snakes and other reptiles.

- Up to 22 in (56 cm)
- 2–6
- Sexes alike
- Sedentary
- Common

S.W. USA to C. Mexico

FURRY FEAST

Old World cuckoos eat insects and their larvae. They are especially fond of hairy caterpillars, which are shunned by most other birds. Their diet forces European cuckoos to migrate to Africa for the winter.

Common koel
Eudynamys scolopaceces

Dideric cuckoo
Chrysococcyx caprius

One of the bronze cuckoos, this species has the characteristic brilliantly glossed green back of the group

Greater coucal
Centropus sinensis

Common cuckoo
Cuculus canorus

A female common cuckoo about to place her egg in the nest of another bird—in this case, a reed warbler

Yellow-billed cuckoo
Coccyzus americanus

Feeds on insects on the floor of rain forests, but roosts and nests in trees

Bornean ground-cuckoo
Carpococcyx radiceus

Rufous-vented ground-cuckoo
Neomorphus geoffroyi

High-ridged bill for parting vegetation as bird forages for foliage

Smooth-billed ani
Crotophaga ani

Greater roadrunner
Geococcyx californianus

Omnivorous, and an efficient predator on small reptiles; also eats insects, mice, and the eggs of other small birds

OWLS

Solitary creatures of the night, owls are instantly recognizable by their forward-facing eyes, face masks, and stout silhouettes. There are two families of owls: barn owls, with heart-shaped faces and elongate bills, and "true owls," with rounded heads and hawk-like bills. They usually roost in out-of-the-way spots. Even when out in the open, they tend to perch stoically, camouflaged by their mottled, earth-toned plumage, so they are more often heard than seen. Their mysterious nocturnal habits, coupled with their far-reaching hoots and other sometimes eerie-sounding calls, have given rise to many folk superstitions. In lifestyle, however, they are no more than nocturnal birds of prey.

LASS	Aves
RDERS	1
AMILIES	2
ENERA	29
PECIES	195

Cosmopolitan birds Both families of owls are widespread; some species such as the barn owl are among the most widely distributed of all birds. Most species inhabit woodlands and forest edges, but some prefer treeless habitats instead.

Hungry mouths to feed A great gray owl (above) feeds a rodent to its chicks. Some owls vary their breeding seasons and clutch sizes according to availability of food. Hatching may be asynchronous, so that if not enough food is available for all offspring, the more-developed chicks can survive by eating their younger siblings.

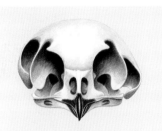

STEREO HEARING

Many owls have ears that are asymmetrical in both size and shape. This structure enables them to more precisely locate their prey, by noting the subtle difference between the sound signals reaching each ear.

Stealthy flight The eastern screech owl (*Otus asio*) (right) has soft, loose plumage. Its flight feathers have frayed rather than smooth edges, which slows the flow of air over the wings and silences flight.

Silent flier
The eastern screech owl's wings allow it to make less noise when it swoops down to strike unsuspecting prey.

NIGHT HUNTERS

Owls, which typically become active at twilight, are well adapted to a nocturnal predatory lifestyle. Their forward-facing eyes give them binocular vision that helps them judge distances well. They can also rotate their heads to see behind them. They can see well in poor light, thanks to their tubular eyes and abundance of light-sensitive rods in their eyes. They also rely on their acute hearing: some species even have facial masks to help them capture sound waves more effectively.

Owls also have sharp, hooked bills and strong legs with talon-tipped feet. They remain on the alert for prey as they perch on branches or ledges, then swoop down to catch the unwary mammal or insect on the ground. They may also snatch insects or mammals from trees, or catch insects in the air.

When they capture sizable prey with their feet, owls sometimes kill it by reaching down to bite it. Small prey is often lifted to the beak with one foot, in the manner of parrots, and swallowed whole. Larger prey is held with the feet and torn apart with the bill before being eaten.

Owls flatten their plumage to become less conspicuous when they are disturbed. Several species have erectile feathers that resemble "ears," which they raise in emotion.

In most species, females are larger than the males, sometimes weighing twice as much as their mates. The young often leave the nest well before they can fly; their parents continue to feed them until they can fend for themselves.

All owls call, especially as the breeding season approaches, and species are often named for the typical sounds they make. A few owl species almost sing.

FACT FILE

Great gray owl This large bird sometimes uses the abandoned nests of other birds. It may not lay any eggs in years when food is particularly scarce, whereas in good years, it may lay up to nine eggs.

- Up to 27 in (70 cm)
- 2–9
- Sexes alike
- Sedentary
- Uncommon

N. North America, N. Eurasia

Snowy owl This bird migrates during the winter, when it may be found on lake shores, in marshland, or by the sea. It perches on rocks or in trees, and kills its prey—invertebrates, small mammals, or birds—by pouncing in flight.

- Up to 27 in (70 cm)
- 3–11
- Sexes differ
- Partial migrant
- Uncommon

N. Eurasia, N. Canada, Arctic

PYGMY OWL DISGUISE

Owls protect themselves from predators primarily by camouflaging themselves against their surroundings and by roosting out of sight. Their nocturnal habits also put them on a different schedule than most bird predators. As further protection, many pygmy owls have spots on the back of their heads that resemble eyes. These are thought to deceive predators into thinking that the owl is already aware of their presence.

Eyes in the back of its head?
Left: real eyes in the face of a pygmy owl; right: false eyes, from feather marks, in the back of the head of the same owl.

Ural owl
Strix uralensis

Ural owls range across the boreal forests of Eurasia; their pale gray plumage is distinctive

Great gray owl
Strix nebulosa

Facial disk assists hearing by conducting sound to the ears, like a satellite dish

Black-banded owl
Ciccaba huhula

Eurasian pygmy-owl
Glaucidium passerinum

Barred owl
Strix varia

Northern hawk owl
Surnia ulula

This predator of forest tundra and taiga preys on mammals and birds, specializing in voles; it nests mostly in hollows and has clutches of 6–10 eggs

Elf owl
Micrathene whitneyi

Snowy owl
Nyctea scandiaca

Great horned owl
Bubo virginianus

Verreaux's eagle-owl
Bubo lacteus

The size of a buzzard, and one of the largest owls; confined to sub-Saharan Africa, it preys on medium-sized mammals and rather large birds

Tropical screech-owl
Otus choliba

Barn owl
Tyto alba

Comb-like central toenail is used for preeening

Spectacled owl
Pulsatrix perspicillata

Has the shortest "ears"—twin crests above the eyes—of its group; lives in open country and hunts mostly small mammals

Short-eared owl
Asio flammeus

Northern long-eared owl
Asio otus

Boreal owl
Aegolius funereus

The palest member of a group of small, mostly New World owls with incomplete facial disks

Burrowing owl
Athene cunicularia

Northern saw-whet owl
Aegolius acadicus

FACT FILE

Barn owl One of the most widespread birds in the world, the barn owl can often be seen hunting rodents in the grassy shoulders and medians of highways. Barn owls have exceptional hearing, and can catch prey in total darkness, guided by sound alone. They also occasionally hunt in daylight.

- Up to 17 in (44 cm)
- 4–7
- Sexes alike
- Sedentary
- Common

S. North & South America, sub-Saharan Africa, W. Eurasia to Australia

Northern long-eared owl This bird sometimes nests in squirrels' nests. The young are fed by the female, although it is the male that does most of the hunting and delivers their diet of small mammals, birds, and invertebrates. Females are generally darker and larger than males. Eye color varies with the region, from yellow to golden.

- Up to 16 in (40 cm)
- 5–7
- Sexes alike
- Partial migrant
- Locally common

C. & S. North America, W. to E. temperate Eurasia

LONG-EARED OWL

To discourage would-be predators from approaching her nest, the female long-eared owl spreads out her wings, flares her flight feathers, and lowers her head, making herself appear larger.

NIGHTJARS AND ALLIES

CLASS	Aves
ORDERS	1
FAMILIES	5
GENERA	22
SPECIES	118

Similar to owls in nocturnal lifestyle, this order of birds—which comprises oilbirds, frogmouths, potoos, owlet-nightjars, and nightjars—is thought to be distantly related to them. Like owls, nightjars and their allies are active at twilight and at night. They have soft plumage in patterns and shades that make them difficult to distinguish from the trees or ground. They tend to have rather large heads and large eyes, though the latter are less forward-facing in this group than in owls. Tubular in shape, their eyes are adapted to see well in low light, and they have a keen sense of hearing.

Diverse habitats The oilbird is found only in tropical South American caves. Potoos inhabit open woodlands in Central and South America. Frogmouths and owlet-nighjars live in or near arboreal areas in Australasia. Nightjars live in a variety of warm-climate habitats across the world.

Trailing behind Nightjars (above) have long tails and long, pointed wings that help them fly fast. In ancient times, nightjars were accused of stealing milk from goats, hence their nickname of goatsuckers. In reality, it is the insects that gather around livestock that they pursue.

Spot the bird The tawny frogmouth (above) is well camouflaged against its background. These nocturnal birds try to remain inconspicuous during the day. Here one sits motionless on its nest, its shape and color imitating the branch of a tree.

MASTERS OF DISGUISE

These rather unusual-looking night-birds are experts at blending in and often strike curious poses that make them resemble broken-off tree limbs.

They have broad, flattened bills for catching insects. The bristles that surround the bills help funnel food (insects, fruit, or other animals) into their large mouths.

The oilbird is the only member of its family. It has a fan-like tail and long, broad wings. It is a cave dweller that can navigate in total darkness by relying on echolo-cation, in a similar way to bats.

Frogmouths are odd-looking birds whose bodies taper from their very large, flat, shaggy heads. Their stout bills, hardened like bones, snap shut to trap their prey.

Potoos resemble frogmouths but have thinner bills and fewer bristles. They eat insects while flying.

Nightjars comprise about half the species in this order. They fly swiftly but have weak legs and feet, which they rarely use. They open and display their brightly colored mouths when they feel threatened.

Owlet-nightjars look like a cross between nightjars and owls. Their broad, flat bills are almost hidden by bristles. They have longer legs and run more than the other birds in this group, and hunt by perch-pouncing.

The only regular fruit-eater among nightjar-like birds; feeds mainly on palms, and regurgitates hard seeds

Oilbird
Steatornis caripensis

Perch-and-pounce hunters; they perch motionless on a vantage perch, then, sighting prey, dive down onto it

Tawny frogmouth
Podargus strigoides

Spotted nightjar
Eurostopodus argus

Common potoo
Nyctibius griseus

Camouflages itself at roost by imitating a dead branch; plumage pattern is cryptic

---

⚡ **CONSERVATION WATCH**

In reasonable shape Only three species of nightjars and their allies are listed as critically endangered by IUCN, and five as vulnerable, but more may be under threat because so little information is available on these cryptic birds. All the critically endangered species are nightjars found in Jamaica, Puerto Rico, and Central and South America.

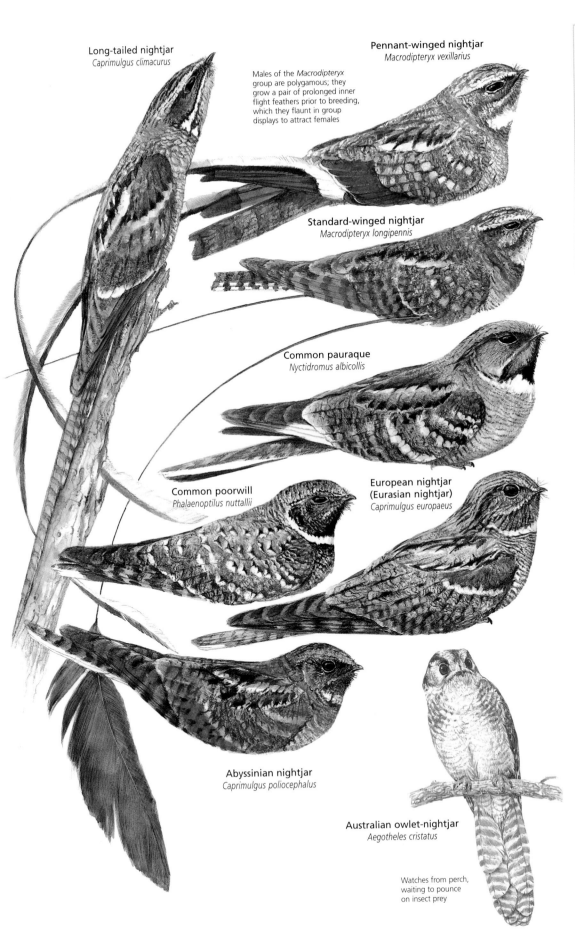

Long-tailed nightjar
Caprimulgus climacurus

Pennant-winged nightjar
Macrodipteryx vexillarius

Males of the *Macrodipteryx* group are polygamous; they grow a pair of prolonged inner flight feathers prior to breeding, which they flaunt in group displays to attract females

Standard-winged nightjar
Macrodipteryx longipennis

Common pauraque
Nyctidromus albicollis

Common poorwill
Phalaenoptilus nuttallii

European nightjar
(Eurasian nightjar)
Caprimulgus europaeus

Abyssinian nightjar
Caprimulgus poliocephalus

Australian owlet-nightjar
Aegotheles cristatus

Watches from perch, waiting to pounce on insect prey

FACT FILE

Oilbird This bird lives in caves, mostly in mountains but also sometimes along rocky coasts. Its cup-like nests are made of regurgitated fruit. It hovers while plucking fruits and seeds, which it locates by sight and by scent.

- Up to 20 in (50 cm)
- 1–4
- Sexes alike
- Sedentary
- Locally common

N. & N.W. South America

Tawny frogmouth This sociable bird can be seen perching in pairs or in family groups. It searches for food—invertebrates, small mammals, and rodents—on the ground, flying into the trees at any hint of danger.

- Up to 21 in (53 cm)
- 1–3
- Sexes alike
- Sedentary
- Common

Mainland Australia & Tasmania

European nightjar This bird's eggs hatch asynchronously in a nest made in a shallow, unlined scrape in the ground. Eggs are color-camouflaged. The adult uses a special display to distract and drive away any predators that approach the nest.

- Up to 11 in (28 cm)
- 1–2
- Sexes differ
- Migrant
- Common

W. & C. Eurasia, W. & S.E. Africa

OPEN-MOUTHED FEEDING

Common nightjars fly quickly through the darkened air, their capacious mouths fully open in order to catch moths, beetles, crickets, and other insects that they pursue with their keen eyesight. Bristles on either side of their mouths help trap their prey.

Open wide
A nightjar catches an insect while flying with its mouth open.

HUMMINGBIRDS AND SWIFTS

CLASS	Aves
ORDERS	1
FAMILIES	3
GENERA	124
SPECIES	429

Any common ancestry between these dissimilar birds would have to be very old indeed. Nevertheless, hummingbirds and swifts do have significant anatomical similarities, such as the relative length of their wing bones. This is related to two distinguishing features for both types of birds: their very rapid wing beats and flight behavior. Hummingbirds are also known for their tiny size, bright iridescent colors, and hovering flight. The average weight of these birds is less than one-third of an ounce (8.3 g); the bee hummingbird is the smallest known species of bird, at a mere one-tenth of an ounce (2.5 g). Swifts are relatively bigger birds that are the fastest fliers of the avian world.

Prevalent throughout Swifts are widely distributed, but are most numerous in the tropics. Hummingbirds are confined to the New World; most live in its tropical belt.

FREQUENT FLIERS

Swifts spend most of their time in the air, taking advantage of their narrow, swept-back wings to maneuver after insects, especially swarming species such as mayflies and termites. They also eat bees and wasps. Several species make long migrations over land and stretches of ocean to reach their Southern Hemisphere wintering grounds.

Predominantly dark colored, swifts have short legs with strong claws. Some cave swiftlets nest and roost in total darkness, deep in caves. They are among the few bird species capable of navigating by echolocation, using clicking calls.

Hummingbirds have long, very narrow bills that they poke into flowers for nectar, hovering in front of the blossoms while they collect the sweet substance with their brush-tipped tongues. They also supplement their diet with insects that they eat for protein.

Swifts and hummingbirds can become torpid to conserve energy, lowering their metabolism and body temperatures while resting.

Compact caretaker A green hermit hummingbird (*Phaethornis guy*) (above) feeds her chicks. The female usually builds the nest and incubates the chicks. Her nest is a small cup of plant material held together with spider web and sometimes anchored to the underside of a leaf.

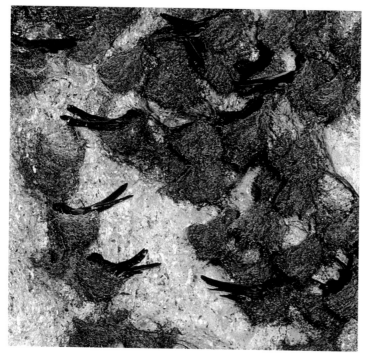

Spit it out The salivary glands of many swifts (left) enlarge during the breeding season. They use secretions from these glands to glue together sticks to form their nests, which they then attach to vertical walls of hollow-tree nesting sites. The nests of some species of cave-dwelling swifts are made entirely of saliva and are highly sought after in some Asian cuisines.

Eating on the fly Hummingbirds can hover while they feed (right) thanks to the anatomy of their wings. They can turn the wings completely over on the back-stroke as well as on the forestroke, thereby counteracting the tendency to move in any one direction. Their broad tails help them make precise maneuvers.

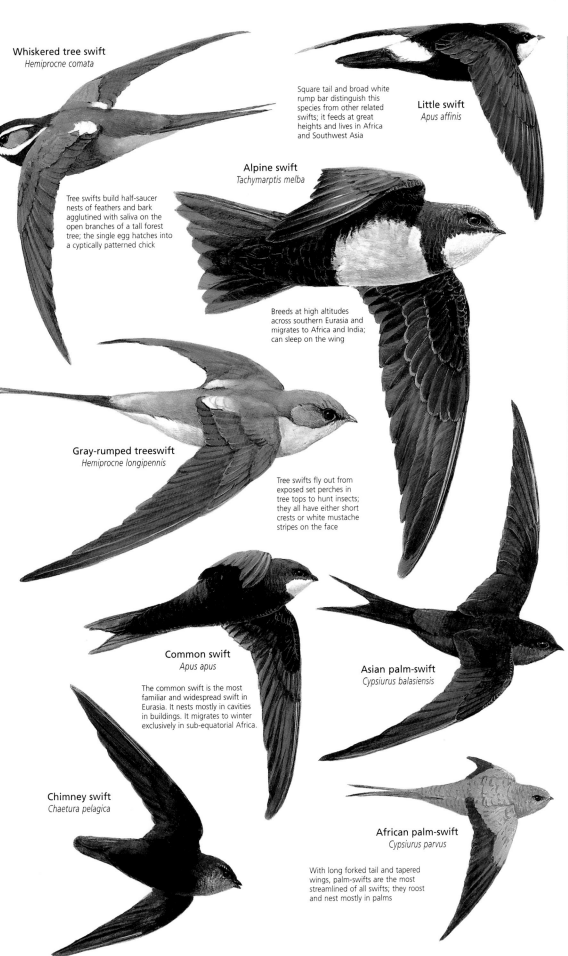

Whiskered tree swift
Hemiprocne comata

Tree swifts build half-saucer nests of feathers and bark agglutinated with saliva on the open branches of a tall forest tree; the single egg hatches into a cyptically patterned chick

Square tail and broad white rump bar distinguish this species from other related swifts; it feeds at great heights and lives in Africa and Southwest Asia

Little swift
Apus affinis

Alpine swift
Tachymarptis melba

Breeds at high altitudes across southern Eurasia and migrates to Africa and India; can sleep on the wing

Gray-rumped treeswift
Hemiprocne longipennis

Tree swifts fly out from exposed set perches in tree tops to hunt insects; they all have either short crests or white mustache stripes on the face

Common swift
Apus apus

The common swift is the most familiar and widespread swift in Eurasia. It nests mostly in cavities in buildings. It migrates to winter exclusively in sub-equatorial Africa.

Asian palm-swift
Cypsiurus balasiensis

Chimney swift
Chaetura pelagica

African palm-swift
Cypsiurus parvus

With long forked tail and tapered wings, palm-swifts are the most streamlined of all swifts; they roost and nest mostly in palms

FACT FILE

Common swift This bird spends most of its time in flight, in urban and rural areas. It is highly sociable and can often be seen flying in chirruping groups. Nestlings are reared by both parents and can fly within 2 months.

⤓ Up to 7 in (17 cm)
● 1–4
∥ Sexes alike
↻ Migrant
⚑ Common

W. & C. Eurasia, S. Africa

Chimney swift This bird can often be heard making high-pitched twittering calls as it flies high overhead. It can often be seen in the company of swallows, a species it resembles but to which it is not closely related.

⤓ Up to 5 in (13 cm)
● 2–7
∥ Sexes alike
↻ Migrant
⚑ Common

E. North America, N.W. South America

African palm-swift This bird's nest is a small, flat pad of seed down that it glues to a leaf. The female then uses saliva to glue one or two eggs to the pad. The parents incubate the eggs in a vertical position. Chicks hook themselves to the pad until they can fly.

⤓ Up to 6 in (16 cm)
● 2
∥ Sexes alike
⊘ Sedentary
⚑ Locally common

Sub-Saharan Africa, Madagascar

AERIAL EXISTENCE

Swifts—easily recognized in the sky thanks to their compact bodies and distinctive wing shapes—spend most of their time darting around, perching only at their overnight roosts. They can even copulate in mid-air. The female signals her interest by holding her wings in a V-shape and gliding downward. The male pursues her, then gently lands on her back. They then mate as they glide downward.

Mating
A male and female swift mate as they glide together.

FACT FILE

Crimson topaz This visually stunning bird is the second-largest humming-bird. It lives in the Amazon region, where it is often found in the forest canopy, making it difficult to observe. The male's long tail feathers cross halfway down its tail.

- Up to 9 in (22 cm)
- 2
- Sexes differ
- Sedentary
- Locally common

N. South America

Giant hummingbird The largest bird in this order, the giant hummingbird is the size of a large swift. Its plumage is mostly brown, and quite dull compared to that of most hummingbirds. It has a long, slender bill and a forked tongue.

- Up to 9 in (23 cm)
- 1–2
- Sexes differ
- Sedentary
- Locally common

Andean W. South America

Sword-billed hummingbird This bird's bill is longer than its body, which makes it the longest bill relative to body length of any bird. The sword-billed hummingbird has co-evolved with the climbing passion flower.

- Up to 9 in (23 cm)
- Unknown
- Sexes differ
- Sedentary
- Locally common

Andean N.W. South America

Black-throated mango This small bird prefers to feed on nectar from red flowers, though it will eat from others also. During courtship flight displays, its wings beat twice as fast as normal.

- Up to 5 in (12 cm)
- 2
- Sexes differ
- Partial migrant
- Common

N.W., N.E. and E. C. South America

CONSERVATION WATCH

Habitat crises IUCN lists 7 species of hummingbirds as critically endangered. All occur in Central America and north South America, where habitat destruction has proved disastrous. Another 7 hummingbird species are endangered, and 11 vulnerable. Swifts, however, have suffered less: only 7 species are considered vulnerable at worst.

Crimson topaz
Topaza pella

Great sapphirewing
Pterophanes cyanopterus

Giant hummingbird
Patagona gigas

Collared inca
Coeligena torquata

Lives in the cloud forests of the northern South American Andes, and can be identified by its combination of blue head, white breast, and white sides to the tail

Sword-billed hummingbird
Ensifera ensifera

Green-backed firecrown
Sephanoides sephanoides

Buff-winged starfrontlet
Coeligena lutetiae

Hummingbirds have been given the most imaginative names in the world of birds; a buff wing patch identifies this north Andean species

White-tipped sicklebill
Eutoxeres aquila

Polygamous; males group to display, and females build pendent, often cone-shaped nests attached on the undersides of leaves or twigs

Black-throated mango
Anthracothorax nigricollis

Uniquely upturned tip on the bill for accessing nectar; curled, bifurcate tongue snakes out into the nectar in the base of a flower and takes it up by capillary action

Fiery-tailed awlbill
Anthracothorax recurvirostris

Males hold a flower-centered territory year-round, but females only do so when not breeding; the sexes are alike in plumage

Purple-throated carib
Eulampis jugularis

Festive coquette
Lophornis chalybeus

Fork-tailed woodnymph
Thalurania furcata

One of a group of fork-tailed hummingbirds, this species is widespread in rain forests across northern South America

Ruby topaz
Chrysolampis mosquitus

Frilled coquette
Lophornis magnificus

Identifiable by its glistening golden gorget. Iridescent plumage in hummingbirds comes from the physical structure of platelets in the feather barbules

Green-throated carib
Eulampis holosericeus

Racket-tailed coquette
Discosura longicauda

Racket-tipped tail feathers are present in several hummingbirds; tail feathers play an important role in flight maneuvers

FACT FILE

Purple-throated carib This bird lives in the Lesser Antilles, in high altitude forests and various other habitats. The female's bill is slightly longer and more curved than the male's. Its cup-shaped nest is camouflaged with lichens.

- Up to 5 in (12 cm)
- 2
- Sexes alike
- Sedentary
- Locally common

Lesser Antilles

Festive coquette This hummingbird inhabits humid forests and scrub-covered areas. It can be found in the lowlands, at altitudes up to 3,300 feet (1,000 m), east of the Andes from Colombia down to Argentina.

- Up to 3¼ in (8.5 cm)
- 2
- Sexes differ
- Sedentary
- Indeterminate

N.W. South America

Green-throated carib This primarily green hummingbird has a thin, curved black bill. It lives mainly in dry low-altitude areas, though it can also be found in higher regions, as well as in mangroves. It sometimes builds its nests quite high up in the branches.

- Up to 5 in (13 cm)
- 2
- Sexes alike
- Locally nomadic
- Locally common

E. Puerto Rico, Lesser Antilles, Grenada

HIGH METABOLISM

The hummingbird's energy-intensive flying style requires it to have a highly efficient system for obtaining oxygen, as well as the ability to quickly process large amounts of food. It has a huge heart, too, to pump blood to distribute energy to the wing muscles; the heart is proportionally twice the size of that in songbirds. Some species migrate long distances to remain near sources of sweet nectar.

Thirsting for energy
A hummingbird dips its long bill into a flower to suck nectar.

HUMMINGBIRD FLIGHT

HUMMINGBIRD FLIGHT

Flying forward Hummingbirds flap their wings up and down to move forward.

Hovering Hummingbirds move their wings in a rapid figure-eight motion to hover in place.

Flying upward By altering the wing angle of the movements, many directional changes are possible. Here, a more up-and-down angle allows a vertical rise.

Flying backward Hummingbirds flap their wings above and behind their heads to move backward.

Free movement The wings of this scintillant hummingbird are shown turned backward for a split second as it hovers to suck nectar. A unique joint system at the shoulder allows such free movement.

Most birds can only fly forward, but hummingbirds can also fly backward, sideways, and straight up or down. They can even switch direction without turning their bodies. They can accomplish this thanks to their specialized physiology and anatomy, with wings that can be turned through 180 degrees. To maintain hovering, they can beat their wings up to 90 times per second. They can also stop and accelerate very quickly. Their flight feathers take up almost the entire wing, and their breastbones are very strong.

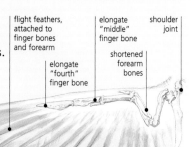

flight feathers, attached to finger bones and forearm

elongate "middle" finger bone

shoulder joint

shortened forearm bones

elongate "fourth" finger bone

Positioning agent The scintillant hummingbird (*Selasphorus scintilla*) (below) can position itself at just the right angle to reach the nectar in the flower. Then its wings hover at blurring speed to keep it in position, the body quite still.

Hummingbird wing structure
This arrangement transfers enormous power to the flight feathers.

MOUSEBIRDS

CLASS Aves
ORDERS 1
FAMILIES 1
GENERA 2
SPECIES 6

The mousebird gets its name from its odd habit of creeping and crawling among bushes and clinging upside down, with its long tail held high. Herbivores, their diets include wild or cultivated fruit and even seedlings. They are therefore considered pests by gardeners and farmers. They build nests in bushes and sometimes cannibalize their young if they stray from the nest. They dislike rain and cold, huddling together and sometimes becoming torpid. Their feathers are replaced randomly and have long aftershafts.

Only in Africa Mousebirds live in a wide range of African habitats, from almost-dry bushland to the edge of forests, south of the Sahara.

Spread out Blue-naped mousebirds (*Urocalius macrourus*) (left) demonstrate one of the unusual characteristics of this bird family: a perching style in which the feet are kept almost level with the shoulders.

This way or that Mousebirds have unique feet with two reversible outer toes that can point either forward or backward. Because of this, they can imitate the toe arrangements of all other birds. It helps when scrambling and clinging to vegetation.

All points of the compass
The center toes of mousebirds can point in any direction.

White-headed mousebird
Colius leucocephalus

Endemic to the horn of Africa; little is known of its habits but, like other mousebirds, it clumps together to roost in family groups

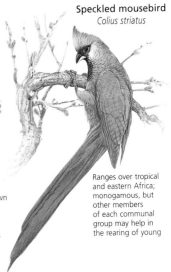

Speckled mousebird
Colius striatus

Ranges over tropical and eastern Africa; monogamous, but other members of each communal group may help in the rearing of young

TROGONS

CLASS Aves
ORDERS 1
FAMILIES 1
GENERA 6
SPECIES 39

The best known of these brilliantly colored birds is the resplendent quetzal, the national bird of Guatemala, which was considered divine by the Aztecs. Female trogons are generally duller than males, and the Asian species are not as colorful. Trogons are secretive and territorial, perching while they scan for food. They eat insects and small lizards; some species also eat fruit. Males sometimes stage displays that include chases through the trees.

Pan-tropical beauties Trogons are forest dwellers. They can be found in the tropical regions of several continents. Arboreal birds, they live in rain forests and monsoon scrubs.

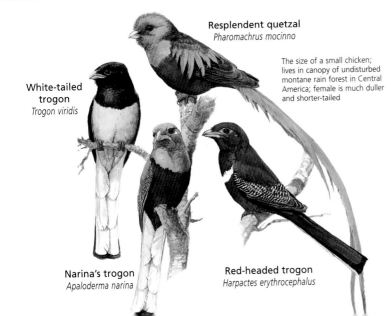

Resplendent quetzal
Pharomachrus mocinno

The size of a small chicken; lives in canopy of undisturbed montane rain forest in Central America; female is much duller and shorter-tailed

White-tailed trogon
Trogon viridis

Narina's trogon
Apaloderma narina

Red-headed trogon
Harpactes erythrocephalus

Sticking out A male resplendent quetzal (above) peers from its nest in a tree hollow. Trogons nest in such hollows or excavate nest holes in dead trees. Although male trogons may group to display to several females, they are monogamous and share nesting duties.

KINGFISHERS AND ALLIES

CLASS Aves

ORDERS 1

FAMILIES 11

GENERA 51

SPECIES 209

Kingfishers are linked to todies, motmots, bee-eaters, rollers, hornbills, the hoopoe, and wood-hoopoes due to certain anatomical and behavioral similarities. All of these birds have small feet with three fused forward toes; they also have affinities in their ear bones and egg proteins. Many species have brilliantly colored plumage, and all nest in holes that they dig in soil or in rotten trees with their bills. Kingfishers are recognized by their short legs and their large, robust, long, and straight bills, which typically have sharply pointed or slightly hooked tips. The former are best suited for striking at and grasping fishes and other prey, and the latter for holding and crushing prey.

Waters and woods Kingfishers and their relatives can be found in a variety of aquatic and wooded environments around the world. The regions with the most species are Africa and Southeast Asia. Ground rollers are found only in Madagascar. "Non-fishing" kingfishers favor tropical rain forest and woodland.

Mother's helpers Carmine bee-eaters (*Merops nubicus*) (above) congregate at a nesting colony in Botswana. Many species of kingfishers and their allies are cooperative breeders, helping to raise other adults' chicks. This dramatically increases fledging success.

DIFFERENT HABITS

Most kingfisher species are generalized predators that eat a variety of terrestrial and aquatic invertebrates and vertebrates. Their typical foraging technique involves sitting quietly on a perch from which they can quietly survey their surroundings. Spotting prey, they then swoop or dive down and seize it in their bills. After bringing it back to the perch, they immobilize it by striking it repeatedly against a branch before eating it.

Some kingfishers, however, have markedly different diets and exhibit a variety of foraging techniques, such as shoveling into moist soil in search of earthworms.

Todies—very small, stocky birds with flattened bills—capture insect prey from the undersides of leaves; they also catch insects in flight.

Bee-eaters eat stinging insects, squeezing out and wiping off the sting before swallowing their prey.

Caught in the act A common kingfisher (*Alcedo atthis*) (above) catches a fish. Despite their name, most species of kingfishers eat a variety of foods, including insects.

Giant kingfisher
Megaceryle maxima

Straight, stout bill is typical of kingfishers; this species, widespread in sub-Saharan Africa, is the largest of all kingfishers and as big as a chicken

Belted kingfisher
Megaceryle alcyon

Pied kingfisher
Ceryle rudis

Widespread in sub-Saharan Africa and southern Asia; lives along rivers, lakes, and waterways and dives for fish from perches along the banks

Short, wedge-shaped tail, characteristic of many kingfishers

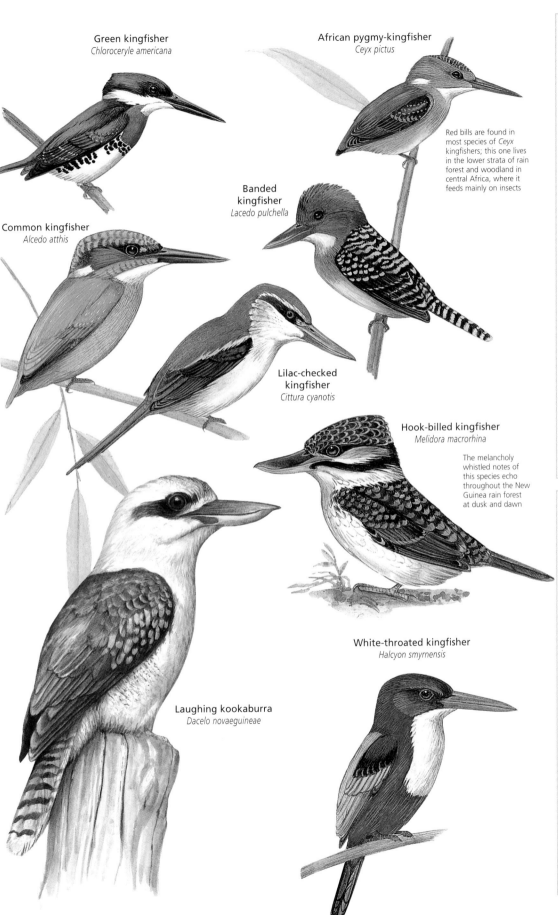

Green kingfisher
Chloroceryle americana

African pygmy-kingfisher
Ceyx pictus

Red bills are found in most species of *Ceyx* kingfishers; this one lives in the lower strata of rain forest and woodland in central Africa, where it feeds mainly on insects

Common kingfisher
Alcedo atthis

Banded kingfisher
Lacedo pulchella

Lilac-checked kingfisher
Cittura cyanotis

Hook-billed kingfisher
Melidora macrorhina

The melancholy whistled notes of this species echo throughout the New Guinea rain forest at dusk and dawn

White-throated kingfisher
Halcyon smyrnensis

Laughing kookaburra
Dacelo novaeguineae

FACT FILE

Common kingfisher This bird's upper parts can appear either brilliant blue or emerald green, depending on the refraction of the light. It flies low over the water, and nests inside tunnels on river banks.

⚖ Up to 6 in (16 cm)
● 4–10
✱ Sexes alike
ᖶ Partial migrant
➹ Common

Europe, N. Africa, W. Eurasia to N. Melanesia

Laughing kookaburra This bird's unmistakable loud cry sounds like laughter. It nests in hollow trunks or holes in trees, or sometimes in emptied-out termite nests. The young remain in the nest for over a month.

⚖ Up to 17 in (43 cm)
● 1–4
✱ Sexes alike
⊘ Sedentary
➹ Common

E. Australia; introd. S.W. Australia, Tasmania

DIVING ACTION

Young kingfishers are taught to fish by their parents, who drop dead fish into the water for the young to retrieve. After the young have mastered the technique and can catch their own live food, they are chased away from the territory by the parents.

Kingfishers typically look for prey either while sitting on a tree, wire, or other high perch overlooking the water, or while hovering overhead. Some kingfishers have polarizing filters in their eyes that eliminate water reflections so they can better see their prey beneath the surface. Their eyes are protected by a membrane. They snap their bills shut when they feel the fish in their mouths.

A trained eye
The kingfisher flies above the water looking for fish.

Plunging in
After spotting its prey, it drops straight down to grab it with its bill.

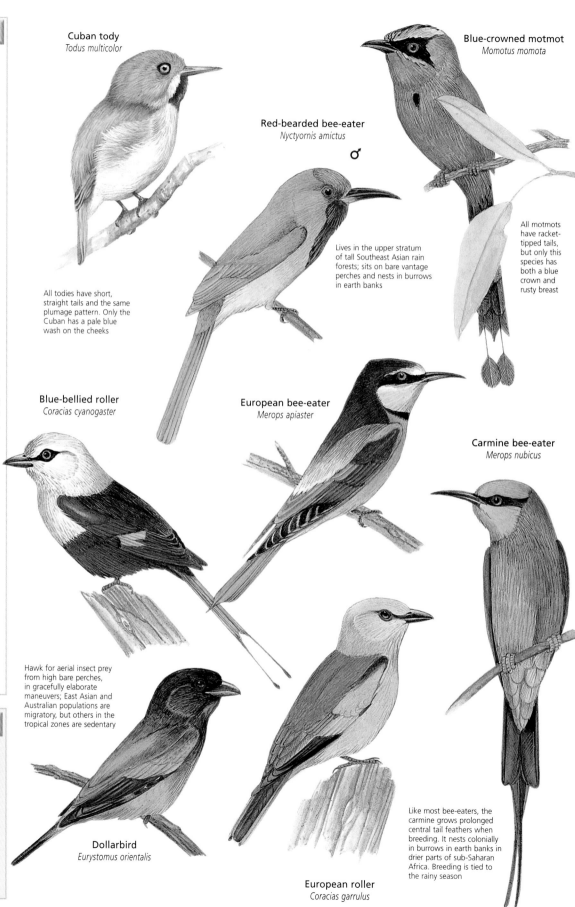

Cuban tody
Todus multicolor

Blue-crowned motmot
Momotus momota

Red-bearded bee-eater
Nyctyornis amictus
♂

Lives in the upper stratum of tall Southeast Asian rain forests; sits on bare vantage perches and nests in burrows in earth banks

All motmots have racket-tipped tails, but only this species has both a blue crown and rusty breast

All todies have short, straight tails and the same plumage pattern. Only the Cuban has a pale blue wash on the cheeks

Blue-bellied roller
Coracias cyanogaster

European bee-eater
Merops apiaster

Carmine bee-eater
Merops nubicus

Hawk for aerial insect prey from high bare perches, in gracefully elaborate maneuvers; East Asian and Australian populations are migratory, but others in the tropical zones are sedentary

Dollarbird
Eurystomus orientalis

European roller
Coracias garrulus

Like most bee-eaters, the carmine grows prolonged central tail feathers when breeding. It nests colonially in burrows in earth banks in drier parts of sub-Saharan Africa. Breeding is tied to the rainy season

Long-tailed ground-roller
Uratelornis chimaera

Rufous-headed ground-roller
Atelornis crossleyi

Common hoopoe
Upupa epops

Cuckoo-rollers give loud whistling and cackling cries as they glide over Madagascan forests; the calls are among the most characteristic for the island

Cuckoo-roller
Leptosomus discolor

♂

Green wood-hoopoe
Phoeniculus purpureus

Red-billed hornbill
Tockus erythrorhynchus

♂

♂

Northern ground-hornbill
Bucorvus abyssinicus

Feeds on the ground, walking about in pairs or small groups, picking up a range of vertebrate and invertebrate prey

Great hornbill
Buceros bicornis

Monogamous, territorial birds that nest in big hollows in tall rain-forest trees; the male seals the female in to incubate by blocking up the entrance hole with mud, then feeds his mate and chicks through a small hole in the entrance for the period of incubation and rearing, up to about 100 days

NESTS

In most birds, as soon as a breeding pair forms, one or both adults will begin to build a nest. The structure is used to cradle the eggs, house the developing chicks, and help keep them safe from predators. Selecting the nest site and building the nest are tasks that may be performed by either sex, alone or as a pair. Each adult's involvement, the choice of site, technique and building materials, and the energy that is invested in nest-building are characteristics that vary among species. The widely recognized cup-shaped nest is typically built using twigs and other plant matter that is threaded or twisted together to prevent the structure from falling apart. Other nest types include stick platforms, excavated tree hollows, scrapes or burrows in the ground, mud "ovens," and rotting mounds of vegetation. Some birds reuse the nests of others. A few species build nests for purposes other than to house chicks, or else build no nests at all. Some use nests at non-breeding times of the year, for sleeping or for shelter; such dormitory nests rarely house their young. Bowerbird males build elaborate nests, but only to attract females, who, after mating, then go off alone to build the actual nest that will house eggs and young. A few birds build no nests: some penguins, for example, simply incubate their eggs and hold their developing young on top of their feet.

A ready-made home The eastern screech owl (*Otus asio*) (above) makes its nest in available tree hollows. Owls also raise their young in holes in cliffs or in old buildings, and they sometimes reuse the stick nests of crows or hawks. Other birds such as kingfishers also nest in cavities, which they dig in riverbanks. Some birds have been known to use old rabbit holes.

A prickly abode The cactus wren (*Campylorhynchus brunneicapillus*) (right) uses sticks to make a somewhat bulky, domed nest, which it prefers to place among the spines of a cactus. Wrens also build nests for sleeping.

A nest with a view European white storks (*Ciconia ciconia*) (below) often nest on man-made structures, in this case the ruins of a church in Algeria. These birds, which mate for life, pile branches into massive nests that they use for many years.

Easy access A red-throated loon (*Gavia stellata*) (above) joins a chick in a nest built on the edge of a lake in Alaska. Loon nests are usually nothing more than shallow scrapes in boggy ground, although some pairs use reeds and water weeds. They are often placed on peninsulas or small islands.

One among many A sociable weaver (*Philetairus socius*) (above) peeks out of its hole in a communal nest. Weavers are renowned for building intricately woven grass nests. They push and pull material into loops and knots in a technique that is similar to basket weaving.

A multi-family dwelling The very large, communal nests (right) of sociable weavers can weigh more than 1 ton (1 tonne) and house hundreds of birds. A small chamber houses each pair. Nests are used for years.

On the rocks A large colony of king cormorants (*Phalacrocorax albiventer*) (left) sit on their nests in the Falkland Islands. This species of cormorant builds its nest on cliff-top slopes using mud and vegetation. Other species use seaweed, guano, or old bones. Those cormorants that nest in trees use twigs.

Home in a cup A vireo (below) sits on her nest in an oak tree. The cup shape is ideal for preventing eggs from rolling out. Fragments of spider web may serve to bind, and lichen on the outside may serve as camouflage.

Spiders' web for binding

Careful construction *Vireos (right) build cup-shaped nests using fine grasses, spider's silk and strips of bark. They suspend them from forks in branches in the mid- to upper levels of trees.*

Strips of bark for camouflage

WOODPECKERS AND ALLIES

CLASS	Aves
ORDERS	1
FAMILIES	5
GENERA	68
SPECIES	398

These six families of birds—woodpeckers, honeyguides, jacamars, puffbirds, barbets, and toucans—look different but share certain anatomical features, such as zygodactyl feet, with two toes in front and two at the back. They also generally lack down feathers and lay white eggs. Most species are colorful; some are even gaudy. They are primarily tropical birds that nest in cavities in trees, in termite mounds, or in the ground. Woodpeckers and barbets typically excavate their own cavities; after they are abandoned, the holes are often used by other species. Woodpeckers sometimes live within sight of large cities and are frequent visitors to bird-feeding stations.

Wide ranges Woodpeckers live in forests; toucans, jacamars, and puffbirds in the New World tropics. Honeyguides are mainly African. Barbets are more widely dispersed.

Hanging on An acorn woodpecker (*Melanerpes formicivorus*) (far left) clings to a pine tree studded with acorns. These birds store their acorns in holes in special storage trees, which they fiercely defend. They then rely on these food stores during the lean winter months. They live in oak and mixed oak woodland in western America and may be monogamous or polygamous, depending on the area.

Sticking out A toco toucan (*Rhamphastotos toco*) (left) eats a goiaba fruit. These odd-looking birds have frilled tongues and often serrated bills, which they use to adroitly pick fruits from trees. They also sometimes prey on untended eggs or nestlings. They are the only toucans that live outside rain forest, frequenting riverside growth and savannas.

SPECIALIZED FEATURES

Woodpeckers are often heard before they are seen. They use their strong, tapering, often chisel-tipped bills to peck at the bark of trees, probing for insects. Bony, muscle-cushioned skulls help protect them from the impact. They use their strong, long-clawed toes to grip the bark and their straightened tail feathers to prop themselves against the trees. Most of them hop rather than walk.

Despite their large, gaudy-colored bills, which can be up to a third or more of their body length, toucans are often surprisingly inconspicuous as they perch on branches amid foliage. They feed mainly on fruit, as do the brightly colored and patterned barbets, which have relatively stout, sometimes notched bills.

Jacamars and puffbirds eat insects, flying forth from perches to catch them. They nest in tunnels that they excavate; some species cover the entrance with a pile of sticks. The typically inconspicuous and inactive puffbirds are named for their unusually loose, fluffy plumage.

Honeyguides also eat insects, as well as wax from beehives. Their name comes from the fact that one species leads humans to beehives.

Some species within these families are solitary; others, such as the nunbirds and puffbirds, are social and even breed cooperatively. Some African barbets nest in pair groups of 100 or more birds, all residing in one dead tree. The greater honeyguide is a brood parasite, laying one egg in the nest of each host bird.

Barbets tend to be highly vocal; some are called "brain-fever birds" because of their unceasing, monotonous calls. Uniquely among birds, woodpeckers have developed instrumental drumming on wood. The tone and pattern are species-specific.

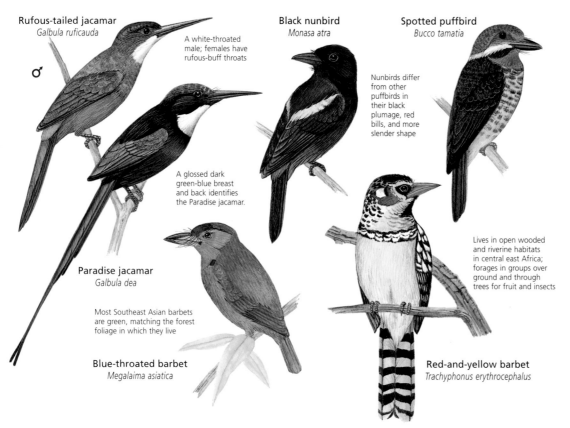

Rufous-tailed jacamar
Galbula ruficauda

A white-throated male; females have rufous-buff throats

♂

A glossed dark green-blue breast and back identifies the Paradise jacamar.

Paradise jacamar
Galbula dea

Most Southeast Asian barbets are green, matching the forest foliage in which they live

Blue-throated barbet
Megalaima asiatica

Black nunbird
Monasa atra

Nunbirds differ from other puffbirds in their black plumage, red bills, and more slender shape

Spotted puffbird
Bucco tamatia

Lives in open wooded and riverine habitats in central east Africa; forages in groups over ground and through trees for fruit and insects

Red-and-yellow barbet
Trachyphonus erythrocephalus

Greater honeyguide
Indicator indicator

Yellow-rumped honeyguide
Indicator xanthonotus

This Himalayan honeyguide is one of only two species of the group to occur outside Africa

Honeyguides are rather small gray-green birds with distinctive white flashes at the sides of the tail; wax from insects or their nests is their staple diet

Lesser honeyguide
Indicator minor

Black-necked aracari
Pteroglossus aracari

Curl-crested aracari
Pteroglossus beauharnaesii

Emerald toucanet
Aulacorhynchus prasinus

Toco toucan
Ramphastos toco

Distinctively striped bill; this species lives in montane rain forest in the northern Andes

Gray-breasted mountain-toucan
Andigena hypoglauca

Zygodactyl toes with two forward and two back, an arrangement better fitted for clasping perches than for running

Channel-billed toucan
Ramphastos vitellinus

Moderately alternated tail that is often cocked temporarily at perches or during vocalizations

Identified by its broad crimson breast band and dark bill. It lives in the lowland rain forests of tropical South America. It eats palm nuts, fruits, and insects and drinks from bromeliads or by holding the bill open in rain

---

FACT FILE

Greater honeyguide This bird is known for leading humans to the hives of wild honeybees. The humans harvest the comb and leave the remains for the bird, which is particularly fond of the wax.

- Up to 8 in (20 cm)
- Sets of up to 5
- Sexes alike
- Sedentary
- Common

Sub-Saharan Africa excl. Congo basin & S.W.

Lesser honeyguide Like other honeyguides, the lesser honeyguide is a brood parasite. Newly hatched young use a hook on their bills to kill the host's chicks and push them out of the nest, or they break the eggs before hatching.

- Up to 6½ in (16 cm)
- Sets of 2–4
- Sexes alike
- Sedentary
- Locally common

Sub-Saharan Africa excl. Congo basin & S.W.

Emerald toucanet The male and female of this small green toucan are alike in color but differ in size, the male being slightly larger. The emerald toucanet sometimes appropriates the previously used nesting holes of other species, or excavates its own.

- Up to 14½ in (37 cm)
- 1–5
- Sexes alike
- Sedentary
- Locally common

Mexico, S. Central & N.W. South America

Toco toucan The toco toucan lives in family groups or flocks. It nests in holes in trees and often pecks noisily at tree branches. Two birds will sometimes clash their bills together.

- Up to 24 in (60 cm)
- 2–4
- Sexes alike
- Sedentary
- Locally common

N.E. South America

⚡ CONSERVATION WATCH

Woodpecker wake The IUCN lists 14 species of woodpeckers and their allies as threatened: 7 as vulnerable, 4 endangered, and 3 critical. The critically endangered species are all woodpeckers, one of which, the ivory-billed, may be extinct. It had disappeared from southeast mainland United States by the 1970s, but persisted in Cuba until the 1990s.

FACT FILE

Acorn woodpecker This unmistakable bird is common in oak woods. It is a social bird that can be heard making chattering and rhythmic laughing calls. The birds eat insects (pried from tree trunks or caught in flight) and acorns.

- Up to 9 in (23 cm)
- 4–6
- Sexes alike
- Sedentary
- Common

W. North to N.W. South America

Yellow-bellied sapsucker The male and female of this species have different colored throat patches. At nesting grounds, the bird makes distinctive drumming sounds, with alternating rapid and slow thumps.

- Up to 8 in (21 cm)
- 4–7
- Sexes differ
- Migrant
- Common

C.N. to S.E. North America, Central America

Ground woodpecker This species feeds almost entirely on the ground. It is specialized to eat ants by probing crannies and digging its bill into ants' nests, head down. It also nests on the ground, in 3¼ ft (1 m) long burrows, dug mostly by the male.

- Up to 12 in (30 cm)
- 2–5
- Sexes alike
- Sedentary
- Locally common

Cape Province to Transvaal & Natal (Africa)

SUCKING SAP

The yellow-bellied sapsucker is one of several North American woodpeckers that peck rows of holes in the trunks of certain trees. They feed on the sap that accumulates in the holes, as well as on the insects that are attracted to the sap. These sap wells, dug in late summer and fall, are also exploited by other birds and mammals.

Hunting sap
A sapsucker searches the bark crevices on a birch tree.

Soft, cryptic plumage; one of only three migratory woodpeckers

Northern wryneck
Jynx torquilla

Feeds mostly on ants, hopping around mainly on the ground; curved bill is used to open anthills

Acorn woodpecker
Melanerpes formicivorus

Acorn woodpeckers are well known for their habit of storing nuts in holes in trees for winter

A white rump helps to identify this species

Gray woodpecker
Dendropicos goertae

Yellow-bellied sapsucker
Sphyrapicus varius

Feeds by pecking on the middle branches of trees on ants, termites, and other insects

Red-headed woodpecker
Melanerpes erythrocephalus

Feeds on both insects and seeds in forests, and stores nuts and seeds in cracks and crevices for winter

Golden-tailed woodpecker
Campethera abingoni

Brown-capped woodpeckers live in woodlands and forage on trees for ants and other insects

Brown-capped woodpecker
Dendrocopos moluccensis

Ground woodpecker
Geocolaptes olivaceus

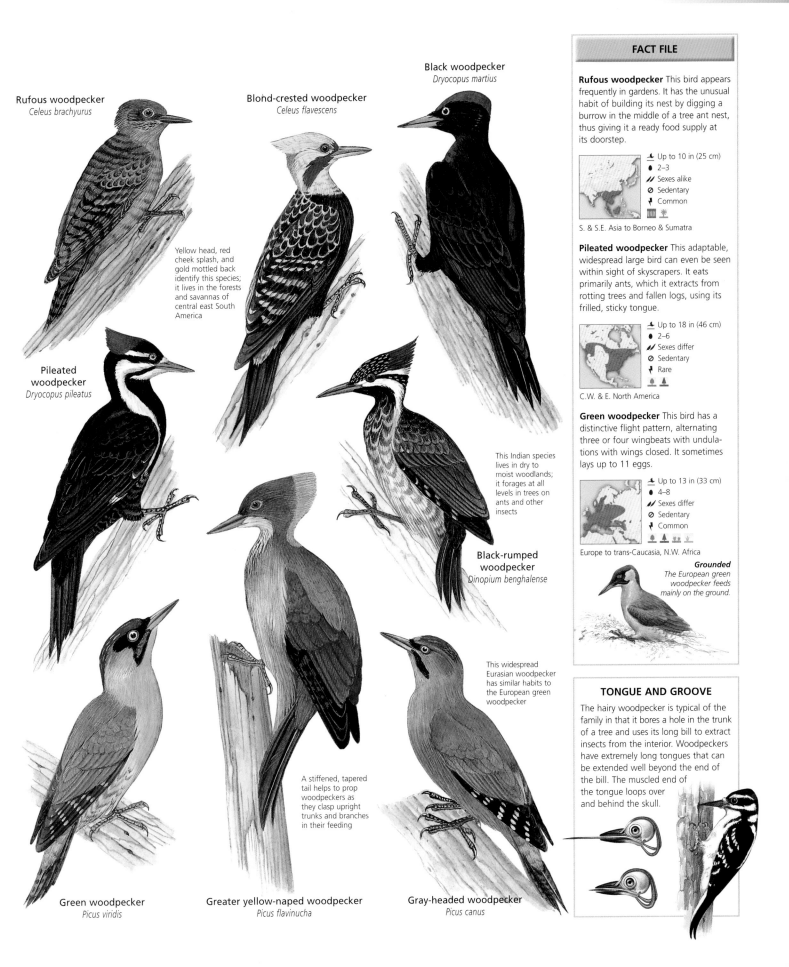

Rufous woodpecker
Celeus brachyurus

Blond-crested woodpecker
Celeus flavescens

Black woodpecker
Dryocopus martius

Yellow head, red cheek splash, and gold mottled back identify this species; it lives in the forests and savannas of central east South America

Pileated woodpecker
Dryocopus pileatus

This Indian species lives in dry to moist woodlands; it forages at all levels in trees on ants and other insects

Black-rumped woodpecker
Dinopium benghalense

This widespread Eurasian woodpecker has similar habits to the European green woodpecker

A stiffened, tapered tail helps to prop woodpeckers as they clasp upright trunks and branches in their feeding

Green woodpecker
Picus viridis

Greater yellow-naped woodpecker
Picus flavinucha

Gray-headed woodpecker
Picus canus

FACT FILE

Rufous woodpecker This bird appears frequently in gardens. It has the unusual habit of building its nest by digging a burrow in the middle of a tree ant nest, thus giving it a ready food supply at its doorstep.

- Up to 10 in (25 cm)
- 2–3
- Sexes alike
- Sedentary
- Common

S. & S.E. Asia to Borneo & Sumatra

Pileated woodpecker This adaptable, widespread large bird can even be seen within sight of skyscrapers. It eats primarily ants, which it extracts from rotting trees and fallen logs, using its frilled, sticky tongue.

- Up to 18 in (46 cm)
- 2–6
- Sexes differ
- Sedentary
- Rare

C.W. & E. North America

Green woodpecker This bird has a distinctive flight pattern, alternating three or four wingbeats with undulations with wings closed. It sometimes lays up to 11 eggs.

- Up to 13 in (33 cm)
- 4–8
- Sexes differ
- Sedentary
- Common

Europe to trans-Caucasia, N.W. Africa

Grounded
The European green woodpecker feeds mainly on the ground.

TONGUE AND GROOVE

The hairy woodpecker is typical of the family in that it bores a hole in the trunk of a tree and uses its long bill to extract insects from the interior. Woodpeckers have extremely long tongues that can be extended well beyond the end of the bill. The muscled end of the tongue loops over and behind the skull.

PASSERINES

CLASS	Aves
ORDER	Passeriformes
FAMILIES	96
GENERA	1,218
SPECIES	5,754

Passeriformes is the largest order of birds by far, encompassing more than 5,700 species. At the same time, this order includes a disproportionately small percentage of bird families, meaning that each passerine family has a relatively high number of species. These two facts together attest to the great evolutionary success of the passerines. Relative newcomers—they are believed to have evolved about 75 million years ago on the southern super-continent Gondwana—they have proven themselves remarkably adaptable, having spread to every continent except Antarctica. Sometimes known as perching birds or songbirds, they are distinguished by their distinctive palates, voice box, and feet.

SKILLED VOCALIZERS

The passerines are typically small or medium-sized birds, which nevertheless vary greatly in shape, size, and plumage colors (the last ranges from very drab to brilliant).

Their feet, unlike those of other bird orders, have four toes, each of which joins the leg at the same level; three of the toes are directed forward, whereas the innermost one is directed backward. This foot structure and the flexibility of the toes are ideally suited to grasping perches, whether they be trees, shrubs, or even blades of grass.

Most passerines have wings that taper to a point at the outer tip. This shape makes them well suited for rapid take-offs and good in-air maneuverability, which is useful for catching prey and also for avoiding predators. They are less capable of sustaining high flight speeds.

Most passerines are songbirds, or oscines. Unlike mammals, which produce sound in their larynx, birds make sound in a voice box known

Small but powerful singer The yellow warbler (*Dendroica petechia*) (above) sings from a branch. Its bright song has been phrased as "sweet sweet sweet, I'm so sweet." The warblers are members of the suborder Oscines, whose members have the most elaborately muscled voice boxes and are therefore the best singers.

as a syrinx. This unique structure at the base of the windpipe, or trachea, consists of two chambers and relies on about half a dozen pairs of muscles to control the tension of its three elastic, vibrating membranes. The bird's trachea, interclavicular air sac and mouth modify the sound that is produced. The two chambers allow the bird effectively to sing a duet by itself.

One large group of passerines includes birds with poorly muscled voice boxes and simple, stereotyped songs, among them the pittas and broadbills of the Old World tropics, and the ovenbirds, antbirds, tyrant flycatchers, and cotingas of Central and South America.

🐦 CONSERVATION WATCH

The 1,039 species of Passeriformes on the IUCN Red List are categorized as follows:

40	Extinct
73	Critically endangered
168	Endangered
326	Vulnerable
1	Conservation dependent
389	Near threatened
42	Data deficient

On the hunt A blue tit (*Parus caeruleus*) (above) hovers beside an oak tree as it scans the branches for insects. Hardy and enterprising, many tits vary their diet with the seasons, eating seeds during winter when other prey is harder to find. In spring, they turn to insect protein for breeding.

A good match The iiwi (*Vestiaria coccinea*) (right) is a Hawaiian honeycreeper whose long, curved, delicate bill matches the sickle shape of the flower blossoms on which it feeds. Three species in this small subfamily, which has played an important role in Hawaiian culture, are now extinct.

Mealtime A noisy pitta (*Pitta versicolor*) (left), an Australian native, sits on a perch after a successful attempt to catch some worms. Pittas are terrestrial forest inhabitants and are among the most colorful of birds. Some other insect-eating passerines hunt their prey in flight.

Hungry mouths to feed A marsh wren (*Cistothorus palustris*) (right) feeds its chicks at a nest on Long Island, in New York, United States. All passerine young are altricial, meaning they are born naked, helpless, and blind. The nestlings gape to indicate that they want to be fed. Parents have such a strong feeding instinct that they occasionally even feed other species.

In the past, many families in this order were grouped together due to morphological similarities. These were in many cases later attributed to convergent evolution rather than very similar genetics; for example, the wrens of Australia are not close relatives of the wrens that live in the Northern Hemisphere.

Most passerines eat plants or insects or both. During each breeding season, adults and young alike may consume more protein.

Most of the birds in this order are monogamous, and both parents help rear the young, although the female tends to do more brooding.

Wading in An American dipper (*Cinclus mexicanus*) (left) stands in a waterhole. The five species of dippers, which do not have webbed feet but use their wings to hunt for insects underwater, are the only truly aquatic birds in the passerine order.

Small passengers A group of yellow-billed oxpeckers (*Buphagus africanus*) (below) sits on the back of a buffalo in Africa. Also known as tickbirds, they can be seen riding on different species of large mammals, pecking for ticks on their hides. Some related starlings eat parasites on livestock, but also swoop down to eat insects disturbed by the grazers.

Displaying A male three-wattled bellbird (*Procnias tricarunculatus*) (below) gets close to a female in a cloud forest in Costa Rica. Bellbirds are among the world's loudest birds. During the breeding season, the male stands on a display perch high in a tree canopy and makes a sound akin to a sledgehammer hitting an anvil, which can be heard up to ½ mile (1 km) away.

Black-bellied gnateater
Conopophaga melanogaster,
family Conopophagidae

Barred antshrike
Thamnophilus doliatus,
family Thamnophilidae

Red-billed scythebill
Campylorhamphus trochilirostris,
family Dendrocolaptidae

Slaty spinetail
Synallaxis brachyura,
family Furnariidae

Slaty breasts and rufous wings characterize the species of *Synallaxis*

Black-throated huet-huet
Pteroptochos tarnii,
family Rhinocryptidae

Gleans for insects on tree branches

Ruddy treerunner
Margarornis rubiginosus,
family Furnariidae

Rufous-winged antwren
Herpsilochmus rufimarginatus,
family Thamnophilidae

Rufous hornero
Furnarius rufus,
family Furnariidae

Horneros build ovenlike nests

Streak-chested antpitta
Hylopezus perspicillatus,
family Formicariidae

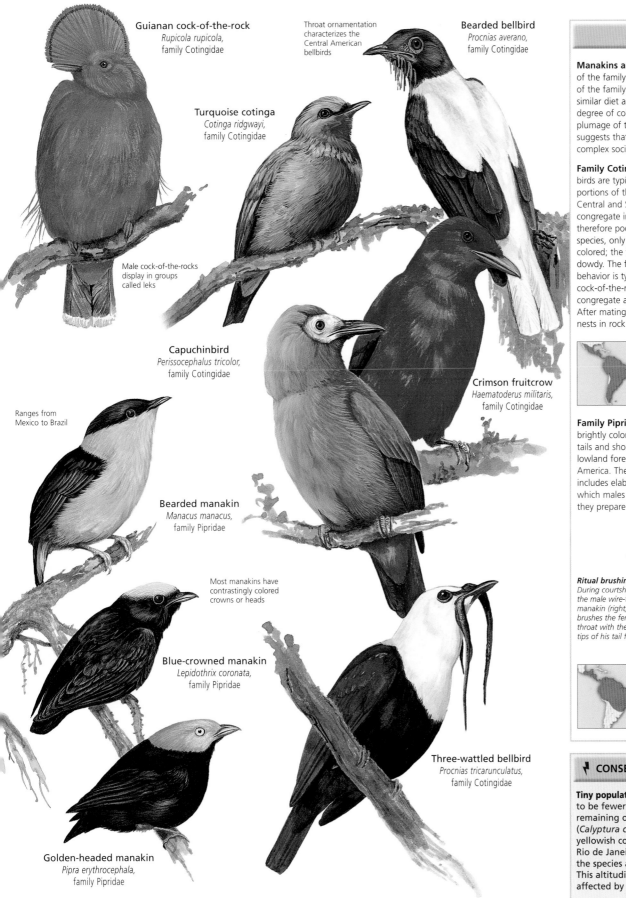

Guianan cock-of-the-rock
Rupicola rupicola,
family Cotingidae

Throat ornamentation
characterizes the
Central American
bellbirds

Bearded bellbird
Procnias averano,
family Cotingidae

Turquoise cotinga
Cotinga ridgwayi,
family Cotingidae

Male cock-of-the-rocks
display in groups
called leks

Capuchinbird
Perissocephalus tricolor,
family Cotingidae

Crimson fruitcrow
Haematoderus militaris,
family Cotingidae

Ranges from
Mexico to Brazil

Bearded manakin
Manacus manacus,
family Pipridae

Most manakins have
contrastingly colored
crowns or heads

Blue-crowned manakin
Lepidothrix coronata,
family Pipridae

Three-wattled bellbird
Procnias tricarunculatus,
family Cotingidae

Golden-headed manakin
Pipra erythrocephala,
family Pipridae

FACT FILE

Tyrant flycatchers These birds of the Americas are unrelated to the Old World flycatchers, although they exploit similar ecological niches. Some catch insects by sallying forth from perches; others are ground feeders.

Family Tyrannidae This largest of the passerine families includes the tyrant flycatchers as well as phoebes, flatbills, elaenias, kingbirds, and wood-peewees. The sexes usually look alike, with green, brown, yellow, and white plumage.

Genera 98
Species 400

South America, West Indies, North America

DIET SPECIALIZATION

Different species within the Tyrannidae family exploit very particular niches within their environments, allowing dozens of species to coexist in some areas. In some cases this is due to different food preferences, such as for fruit as opposed to insects. But in the majority of cases it can be attributed to slightly different combinations of prey size, habitat, vegetation type, foraging position, and capture technique.

Ground feeders *The white-browed ground-tyrant (below, right) is similar to the northern wheatear in terms of its long legs, upright stance, and similar plumage pattern.*

Foliage gleaners *The tufted tit-tyrant (below, left) bears a strong resemblance to its Old World counterpart, the crested tit, with which it also shares a similar feeding strategy.*

⚡ CONSERVATION WATCH

Fragmented range The known range of a newly described species that belongs to the Tyrannidae family, the mishima tyrranulet (*Zimmerius villarejoi*), falls primarily within a protected area in Peru. But the very small, severely fragmented nature of its range has caused the species to be listed as vulnerable.

Scissor-tailed flycatcher
Tyrannus forficatus,
family Tyrannidae

Great kiskadee
Pitangus sulphuratus,
family Tyrannidae

Scarlet crown plumage spread in display

Royal flycatcher
Onychorhynchus coronatus,
family Tyrannidae

Flattened, slightly hooked bill for catching insects in flight

Eastern kingbird
Tyrannus tyrannus,
family Tyrannidae

Olive-sided flycatcher
Contopus cooperi,
family Tyrannidae

Sulfur-bellied flycatcher
Myiodynastes luteiventris,
family Tyrannidae

Erectile crest is raised in excitement

Tufted flycatcher
Mitrephanes phaeocercus,
family Tyrannidae

Sallies out to catch insects on the wing

Black phoebe
Sayornis nigricans,
family Tyrannidae

Long tail for maneuverability in flight

COOPERATIVE BREEDING

About 3 percent of all bird species are classified as cooperative breeders, meaning that additional birds—either juveniles or adults—help the parents raise their young. Recent studies have shown that this phenomenon is more prevalent among passerines than was previously thought. There are two forms of cooperative breeding. One involves non-breeding birds helping parents protect and rear their nestlings. The other, called communal breeding, involves some shared parentage (whether shared maternity, paternity, or both) of the offspring that a group of adults are raising together.

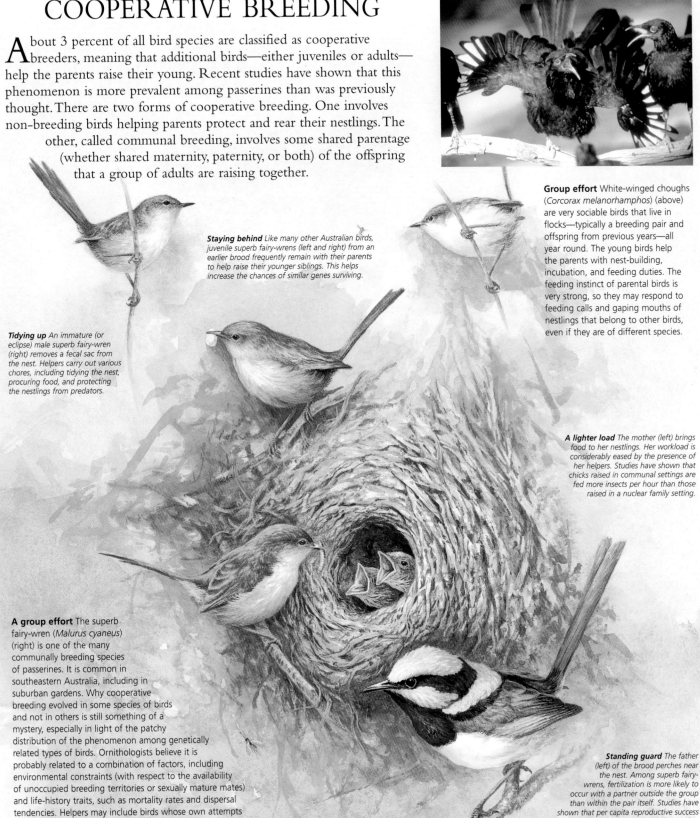

Group effort White-winged choughs (*Corcorax melanorhamphos*) (above) are very sociable birds that live in flocks—typically a breeding pair and offspring from previous years—all year round. The young birds help the parents with nest-building, incubation, and feeding duties. The feeding instinct of parental birds is very strong, so they may respond to feeding calls and gaping mouths of nestlings that belong to other birds, even if they are of different species.

Staying behind *Like many other Australian birds, juvenile superb fairy-wrens (left and right) from an earlier brood frequently remain with their parents to help raise their younger siblings. This helps increase the chances of similar genes surviving.*

Tidying up *An immature (or eclipse) male superb fairy-wren (right) removes a fecal sac from the nest. Helpers carry out various chores, including tidying the nest, procuring food, and protecting the nestlings from predators.*

A lighter load *The mother (left) brings food to her nestlings. Her workload is considerably eased by the presence of her helpers. Studies have shown that chicks raised in communal settings are fed more insects per hour than those raised in a nuclear family setting.*

A group effort The superb fairy-wren (*Malurus cyaneus*) (right) is one of the many communally breeding species of passerines. It is common in southeastern Australia, including in suburban gardens. Why cooperative breeding evolved in some species of birds and not in others is still something of a mystery, especially in light of the patchy distribution of the phenomenon among genetically related types of birds. Ornithologists believe it is probably related to a combination of factors, including environmental constraints (with respect to the availability of unoccupied breeding territories or sexually mature mates) and life-history traits, such as mortality rates and dispersal tendencies. Helpers may include birds whose own attempts at nesting have failed that season.

Standing guard *The father (left) of the brood perches near the nest. Among superb fairy-wrens, fertilization is more likely to occur with a partner outside the group than within the pair itself. Studies have shown that per capita reproductive success does not increase with group size.*

FACT FILE

Tropical and Southern Hemisphere families The greatest concentration of bird species occurs in those parts of the world with the greatest degree of biodiversity in general, which gives birds more ecological niches to exploit. These areas tend to be concentrated in the tropics. The generally milder climes of the Southern Hemisphere also support many bird species. Many birds in these regions are brightly colored and often green, like surrounding foliage.

Family Eurylaimidae Known as broadbills, the members of this family are stout birds with broad heads; broad, flattened bills; and short legs. They have striking plumage and, in some cases, brilliant red, green, or yellow eyes. Most species eat insects, as well as small lizards, frogs, and fruit.

Genera 9
Species 14

Tropical Africa, S.E. Asia, Gr. Sundas, S. Philippines

Family Pittidae Pittas are thrush-sized, insect-eating birds with short tails. They are brilliantly colored forest-dwellers that prefer to hop or run rather than fly. Their large nests, built on the ground or in low vegetation, resemble piles of plant debris. All give loud, melodious double whistles.

Genera 1
Species 30

C. Africa, S.E. Asia to N. & E. Australia, Melanesia

Stealthy bird
The banded pitta, though colorful, is surprisingly inconspicuous on the forest floor. It feeds quietly and seldom calls.

⚡ CONSERVATION WATCH

Dwindling mangroves The IUCN lists 13 members of the Pittidae family, including the mangrove pitta (*Pitta megarhyncha*). It is considered near threatened because its mangrove habitat in southern Asia is being cleared for construction materials, fuel, and charcoal production.

Green broadbill
Calyptomena viridis,
family Eurylaimidae

Cream head-spots shine in the forest, possibly for social signalling

Sharpbill
Oxyruncus cristatus,
family Cotingidae

Long-tailed broadbill
Psarisomus dalhousiae,
family Eurylaimidae

Red-bellied pitta
Pitta erythrogaster,
family Pittidae

Red-bellied pittas range from the Philippines to Australia

Rufous-tailed plantcutter
Phytotoma rara,
family Cotingidae

Rifleman
Acanthisitta chloris,
family Acanthisittidae

Velvet asity
Philepitta castanea,
family Philepittidae

Noisy scrub-bird
Atrichornis clamosus,
family Atrichornithidae

A lyrebird singing in display posture

Superb lyrebird
Menura novaehollandiae,
family Menuridae

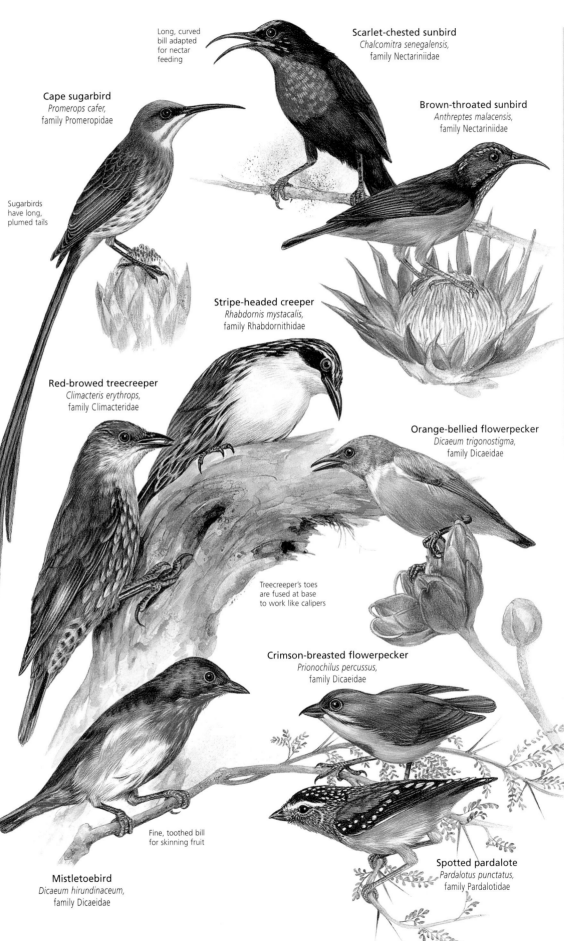

Long, curved bill adapted for nectar feeding

Scarlet-chested sunbird
Chalcomitra senegalensis,
family Nectariniidae

Cape sugarbird
Promerops cafer,
family Promeropidae

Brown-throated sunbird
Anthreptes malacensis,
family Nectariniidae

Sugarbirds have long, plumed tails

Stripe-headed creeper
Rhabdornis mystacalis,
family Rhabdornithidae

Red-browed treecreeper
Climacteris erythrops,
family Climacteridae

Orange-bellied flowerpecker
Dicaeum trigonostigma,
family Dicaeidae

Treecreeper's toes are fused at base to work like calipers

Crimson-breasted flowerpecker
Prionochilus percussus,
family Dicaeidae

Fine, toothed bill for skinning fruit

Mistletoebird
Dicaeum hirundinaceum,
family Dicaeidae

Spotted pardalote
Pardalotus punctatus,
family Pardalotidae

FACT FILE

Sunbirds, tree-creepers, and flower-peckers This group of small, tree-dwelling birds is found throughout the Old World tropics and south temperate zones. Families native to Australasia are the most distinctive.

Family Nectariniidae Sunbirds are strongly associated with flowers. They are lively, colorful birds with long, slender, decurved, and finely serrated bills. Their tubular, deeply cleft tongues are adapted for sucking nectar.

Genera 16
Species 127

Sub-Saharan Africa, S. Asia to Melanesia & N.E. Australia

Family Dicaeidae The flowerpeckers are dumpy birds with pointed wings, stubby tails, and short, conical bills. Their cleft tongues are adapted for feeding on nectar as well as fruit.

Genera 2
Species 44

India, S.E. Asia & its archipelagos to Melanesia & Australia

Family Climacteridae This small family of bark-climbing birds known as tree-creepers is found only in Australia and New Guinea. They have slender bills, short, square tails, and strong legs.

Genera 2
Species 7

Mainland Australia excl. deserts, New Guinea

Family Pardalotidae These common, multicolored birds closely resemble flowerpeckers in size and proportions. They feed almost entirely on larvae, and on an exudate produced by the larvae, of leaf-eating insects in foliage.

Genera 1
Species 4

Australia, Tasmania

Cavity dweller *Pardalotes, such as the spotted pardalote, typically nest in tree cavities or in tunnels in the ground; they have also been known to nest in buildings.*

This species calls like a reed-warbler

Brown honeyeater
Lichmera indistincta,
family Meliphagidae

Red-headed honeyeater
Myzomela erythrocephala,
family Meliphagidae

Pied honeyeater
Certhionyx variegatus,
family Meliphagidae

A nomadic honeyeater found in Australian deserts

Most Myzomela honeyeaters have glistening red plumage on the belly

Lewin's honeyeater
Meliphaga lewinii,
family Meliphagidae

Black-chinned honeyeater
Melithreptus gularis,
family Meliphagidae

Silver-eye
Zosterops lateralis,
family Zosteropidae

White eye-rings are found in most white-eyes

Oriental white-eye
Zosterops palpebrosus,
family Zosteropidae

Blue-faced honeyeater
Entomyzon cyanotis,
family Meliphagidae

Asian paradise-flycatcher
Terpsiphone paradisi,
family Monarchidae

Chestnut-breasted whiteface
Aphelocephala pectoralis,
family Acanthizidae

Crimson chat
Epthianura tricolor,
family Meliphagidae

Penduline tit
Remiz pendulinus,
family Remizidae

Black-throated tit
Aegithalos concinnus,
family Aegithalidae

Penduline tits are found
in Eurasia, Africa, and
southern North America

Yellow-bellied fantail
Rhipidura hypoxantha,
family Rhipiduridae

Crested tit
Parus cristatus,
family Paridae

Gray shrike-thrush
Colluricincla harmonica,
family Colluricinclidae

This species gleans over
ground and trees in the
Australian bush

FACT FILE

Tits and flycatchers These small birds feed primarily on insects. A number of them are distinguished by disproportionately long fanned tails.

Family Monarchidae Known as monarch flycatchers, the members of this family of small songbirds have a predilection for flicking their tails and for sallying for food on the wing. They prefer wooded habitats and often forage in mixed-species parties.

Genera 15
Species 87

Australia & Melanesia to Oceania,
S.E. & S. Asia, & Africa

Family Rhipiduridae Fantails get their common name from their habit of fanning or waving their long tails. They also pirouette through the air and appear hyperactive. Their broad bills are surrounded by bristles that aid in catching insect prey.

Genera 1
Species 43

Australasia & Melanesia,
Oceania, S.E. Asia & India

Family Paridae Tits, chickadees, and titmice tend to have black caps, some with crests, and white cheeks. Most are sedentary, forest-dwelling birds that use their short, straight bills to feed on insects and seeds in foliage.

Standing out
The colorful great tit
inhabits woods, parks,
and gardens. It nests in
trees and holes, including
artificial nesting boxes.

Genera 3
Species 54

North America, Africa & Eurasia
to Japan & L. Sundas

FACT FILE

Family Sittidae Nuthatches are found throughout the Northern Hemisphere and spend their whole lives in trees. These bark-climbing birds are superficially similar to the Australasian sittellas in appearance, but the two families are not closely related, representing an example of evolutionary convergence, a response to similar environmental factors.

Eurasian nuthatch
Both nuthatches and sittellas can clamber up and down tree trunks as they forage for insects.

Genera 2
Species 25

North America, Eurasia to Japan, Philippines & Gr. Sundas

Family Maluridae Like the northern wrens to which they are not closely related, the fairy-wrens, grass-wrens, and emu-wrens that make up this family are small birds that cock their tails. They are insectivores; most forage on the ground and in shrubbery.

Genera 5
Species 28

Australia, Tasmania, New Guinea, Aru Is.

Standing out
The brilliantly colored male fairy-wrens stand out, none more so than the splendid fairy-wren, a species that lives in Australia's western and interior scrublands.

⚡ CONSERVATION WATCH

Sparse sightings Several nuthatch species are considered threatened; a number of others are rarely seen. The endangered Algerian nuthatch (*Sitta ledanti*) has only been found in a handful of localities in Algeria; all are close to each other and one is threatened by logging.

Wallcreeper
Tichodroma muraria, family Sittidae

Wallcreepers nest in rock crevices in cliff faces

Chinspot batis
Batis molitor, family Platysteiridae

Males have black breast band

Varied sitella
Neositta chrysoptera, family Neosittidae

Variegated fairywren
Malurus lamberti, family Maluridae

Iridescent blue head characterizes many malurid wrens

Eurasian treecreeper
Certhia familiaris, family Certhiidae

White-breasted nuthatch
Sitta carolinensis, family Sittidae

White-browed scrubwren
Sericornis frontalis, family Acanthizidae

Yellow-rumped thornbill
Acanthiza chrysorrhoa, family Acanthizidae

Fairy gerygone
Gerygone palpebrosa, family Acanthizidae

Gerygones, scrub-wrens, and thornbills live in Australia and New Guinea

White-browed woodswallow
Artamus superciliosus,
family Artamidae

Brubru
Nilaus afer,
family Malaconotidae

Red-backed shrike
Lanius collurio,
family Laniidae

Greater
racket-tailed drongo
Dicrurus paradiseus,
family Dicruridae

Eurasian golden oriole
Oriolus oriolus,
family Oriolidae

Saddleback
Philesturnus carunculatus,
family Callaeatidae

Magpie-lark
Grallina cyanoleuca,
family Monarchidae

Tail streamers

This species carols
in duet with others
of its flock

Strong feet adapted
for ground feeding

Australian magpie
Gymnorhina tibicen,
family Cracticidae

FACT FILE

Shrikes and allies These families of primarily insect-eating birds can be found throughout the world, although most species are in the Old World.

Family Artamidae Wood-swallows have strong, pointed wings, blue-gray bills with black tips, and short legs. They forage in all types of habitats, hawking on the wing. Sociable birds, they perch side by side and preen each other.

Genera 1
Species 10

Australia & Melanesia
to S.E. Asia & India

Family Oriolidae Not to be confused with New World orioles, the orioles in this family are found in Old World forests and woodlands. Sturdy birds with often yellow-plumaged males, they tend to spend most of their lives high in the treetops, so are often heard but rarely seen.

An unusual bird The figbird differs from other orioles in several ways. It is gregarious, noisy, and has bare skin around the eye. Its diet consists almost entirely of fruit, primarily figs.

Genera 2
Species 29

Australia & New Guinea to temperate Eurasia,
sub-Saharan Africa

Family Laniidae Shrike numbers have been decreasing because pesticides are wiping out the large insects they favor. True shrikes with hooked bills often impale surplus food on thorns. Vangas are related birds that have radiated into a variety of habitats in Madagascar, similar to the finches of the Galápagos.

Genera 4
Species 30

North America, Afric, Eurasia to montane
New Guinea

Perching The long-tailed shrike is a solitary, aggressive bird of the open country. It perches conspicuously on poles or trees as it looks for food.

Singing A marsh wren (*Cistothorus palustris*) (left) sings while perching on a common cattail in Alberta, Canada. This tiny bird is a perennial singer, except during the molting season. The size of song repertoires varies by population.

Well-researched singers Zebra finches (*Taeniopygia guttata*), widespread in the dry regions of Australia, are among the best-studied songbirds. To make their songs heard far afield, birds may sing from the highest perches available. This is also one reason that many species prefer to sing in the morning, when the clear air helps carry the sound further. Birds can produce sounds with higher frequencies than the human voice.

Volume matters Male zebra finches (far left) sing far more loudly when other birds are present, in order to prompt females (left) to respond to their advances and to ward off other males.

SONG

Bird songs have inspired musicians, poets, and other artists through the ages. The syrinx, the resonating chamber in which bird sounds are produced, is a complex structure that is best developed in the group of passerines known as oscines. Different species can be distinguished by their songs, although some birds, such as the village indigobird, are very skilled at mimicking other birds. Songs are generally partly or entirely learned, and different dialects often exist among neighboring populations. In most species, it is the males that sing, to attract females or to warn other males to stay away from an established territory. Studies have shown that males can distinguish the songs of their neighbors from the songs of other males. Sometimes birds sing together in duets.

Repertoires matter Most male songbirds can sing two or more different songs. Female zebra finches have been shown to consistently prefer males with more complex songs.

Stimulating sounds Songs help coordinate the reproductive cycles of mates. They can prompt females to ovulate, build nests, and lay eggs.

Begging to be fed Four hungry baby northern mockingbirds (*Mimus polyglottos*) (right) beg for food. Parents respond not only to gaping mouths but also to feeding calls. Calls are shorter and less complex than songs and can signal threats, hunger, sexual interest and various other elements of social behavior.

Common green magpie
Cissa chinensis,
family Corvidae

Red-billed blue-magpie
Urocissa erythrorhyncha,
family Corvidae

This species is split in
its distribution between
southwestern Europe
and eastern Asia

Eurasian nutcracker
Nucifraga caryocatactes,
family Corvidae

Azure-winged magpie
Cyanopica cyanus,
family Corvidae

Characteristic
long tail

Clark's nutcracker
Nucifraga columbiana,
family Corvidae

Red-billed chough
Pyrrhocorax pyrrhocorax,
family Corvidae

Alpine chough
Pyrrhocorax graculus,
family Corvidae

Distinctively
massive bill

Rufous treepie
Dendrocitta vagabunda,
family Corvidae

White-necked raven
Corvus albicollis,
family Corvidae

FACT FILE

Crows The ancestors of this very
successful branch of the passerines
most likely originated in Australia.
Extensive evolution after they got to
the Northern Hemisphere resulted in
the many different forms of crows,
choughs, nutcrackers, magpies, jays,
and treepies of today.

Family Corvidae These medium to
large birds have bristle-covered nostrils
and fairly long legs. Coloring varies from
the raven's deep black to the brilliant
reds and greens of the Asian magpies.
They may eat berries, insects or seeds,
or scavenge meat from carcasses.

Nest robber *The North
American scrub jay is notorious
for taking the eggs or young of
many songbirds.*

Genera 24
Species 117

Worldwide except S. South America
& polar regions

FOOD STORAGE HABITS

Many corvids hide their surplus food
in caches, to which they return long
afterward. Their ability to recall the
exact location of their caches has
intrigued scientists, who believe that
corvids rely on landmarks and possess
a notably well-evolved spatial memory.
Here, a Clark's nutcracker buries nuts for
later in a shallow hole in the ground.

⚑ CONSERVATION WATCH

Dwarf jay The dwarf jay (*Cyanolyca
nana*) is a small, slender, agile, slate-
blue bird that is endemic to forests
in southeast Mexico. Although its
status has been downgraded from
endangered to vulnerable due to
new sightings in areas where it was
feared extinct, its habitat is being
destroyed and fragmented.

Birds of paradise and bowerbirds
These two groups of birds are found only in New Guinea, Australia, and the Moluccas. The males court females by building elaborate bowers for display (in the case of bowerbirds) or by displaying their bizarre plumage in elaborate courtship rituals (in the case of birds of paradise).

Family Paradisaeidae The birds of paradise are stout- or long-billed and strong-footed birds similar to crows in size and appearance. But they are distinguished by their gaudy plumage, which the males show off singly or in groups to attract females. They live in mountain forests.

Genera 16
Species 40

New Guinea & satellite islands, E. Australia, N. Moluccas

Family Ptilonorhynchidae Most bowerbirds live in forested areas, although others prefer the more open woodlands of the Australian bush. Most species can mimic bird calls or other sounds, and several species are called catbirds because of their cat-like calls. They eat mostly fruit and other plant matter, although their nestlings are sometimes fed insects or even other birds' nestlings.

Genera 8
Species 18

New Guinea & satellite islands, Australia

Bower on display Male bowerbirds build elaborate bowers, which they decorate with various small objects.

Blue beauty The uniquely plumed blue bird of paradise (*Paradisaea rudolphi*), which lives in montane forests in New Guinea, is one of several vulnerable members of the Paradisaeidae family. Threats to its survival include loss of habitat and hunting for its feathers. Parts of its range are still inaccessible.

Arfak astrapia
Astrapia nigra,
family Paradisaeidae

Velvet sheen gorget

Western parotia
Parotia sefilata,
family Paradisaeidae

Wire-like head plumes are twirled in display

Magnificent riflebird
Ptiloris magnificus,
family Paradisaeidae

Wallace's standardwing
Semioptera wallacii,
family Paradisaeidae

White plumes are raised in display

Twelve-wired bird-of-paradise
Seleucidis melanoleucus,
family Paradisaeidae

Great bowerbird
Chlamydera nuchalis,
family Ptilonorhynchidae

Satin bowerbird
Ptilonorhynchus violaceus,
family Ptilonorhynchidae

Always decorates his bower with blue or violet-colored objects

Golden bowerbird
Prionodura newtoniana,
family Ptilonorhynchidae

White-eyed river martin
Pseudochelidon sirintarae,
family Hirundinidae

White-winged swallow
Tachycineta albiventer,
family Hirundinidae

White-banded swallow
Atticora fasciata,
family Hirundinidae

Black lark
Melanocorypha yeltoniensis,
family Alaudidae

This large lark lives
in the deserts of
northern Africa and
southwestern Asia

Horned lark
Eremophila alpestris,
family Alaudidae

Greater hoopoe-lark
Alaemon alaudipes,
family Alaudidae

Greater short-toed lark
Calandrella brachydactyla,
family Alaudidae

Black-crowned sparrow-lark
Eremopterix nigriceps,
family Alaudidae

FACT FILE

Larks and others This group—larks, martins, swallows, wagtails, and cuckooshrikes—was once thought to comprise related families. Molecular research has since revealed that they are quite disparate.

Family Hirundinidae Typical swallows are often recognized by their outer tail-quills, which extend into long, narrow filaments. However, some species have squared tails. Many are strong migrants and are usually silent except near the nest. Like swallows, martins feed on insects taken during flight.

Genera 20
Species 84

Cosmopolitan except polar regions

Family Alaudidae Of the lark family, the skylark has the most impressive vocalizations; because of these, it has been introduced to many parts of the world. Other larks also sing well while flying or perching. Larks have long legs and hind toes and walk rather than hop. They eat insects, other invertebrates, seeds, and grain.

Genera 19
Species 92

Cosmopolitan except C. & S. South America, S.W. Australia

Courtship
*The courtship of the male horned lark (*Eremophila alpestris*) involves some complicated acrobatics.*

⚡ CONSERVATION WATCH

Under threat The Bahama swallow (*Tachycineta cyaneoviridis*) is a vulnerable native of the Caribbean. A member of the Hirundinidae family, it often builds nests near human settlements. It is forced to compete with introduced predators in its breeding range.

FACT FILE

Family Campephagidae Birds in this family are known as cuckooshrikes because their bustle of feathers on the rump and undulating flight feathers are reminiscent of the cuckoos. The coloration of cuckooshrikes varies markedly among the different genera. These mostly secretive birds forage for insects and fruit in trees, and build tiny, saucer nests high up in the branches.

Genera 7
Species 81

Australia & Melanesia to S. & E. Asia, Madagascar, Africa

Family Motacillidae These small, slender, often long-tailed birds are known as wagtails and pipits. They mostly nest on the ground and feed on insects. The typical wagtails are often found near running water, on riverbanks, and in moist grassland areas. Pipits are very widespread, and prefer open country.

Genera 5
Species 64

Cosmopolitan (only one species in North America)

Gray wagtail
The gray wagtail (*Motacilla cinerea*) is seldom found far from water.

⚡ CONSERVATION WATCH

Small range The Sharpe's longclaw (*Macronyx sharpei*) is a sedentary, terrestrial, pipit-like member of the Motacillidae family. Endemic to Kenya, it is listed as endangered because its habitat is very small and highly fragmented. The remaining habitat is expected to decrease by at least half over the next decade.

Scarlet minivet
Pericrocotus flammeus, family Campephagidae

Bar-bellied cuckoo-shrike
Coracina striata, family Campephagidae

Black-faced cuckoo-shrike
Coracina novaehollandiae, family Campephagidae

Varied triller
Lalage leucomela, family Campephagidae

Ranges nomadically through the woodlands of Australia

Ashy minivet
Pericrocotus divaricatus, family Campephagidae

Wagtails are so called because of their habit of waving their tails up and down

Rosy pipit
Anthus roseatus, family Motacillidae

Madagascan wagtail
Motacilla flaviventris, family Motacillidae

Red-throated pipit
Anthus cervinus, family Motacillidae

Palmchat
Dulus dominicus,
family Dulidae

This West Indian
endemic prefers
groves of royal palms,
where it eats fruit

Orange-bellied leafbird
Chloropsis hardwickei,
family Chloropseidae

Long-tailed silky-flycatcher
Ptilogonys caudatus,
family Bombycillidae

Bohemian waxwing
Bombycilla garrulus,
family Bombycillidae

White-capped dipper
Cinclus leucocephalus,
family Cinclidae

Gray hypocolius
Hypocolius ampelinus,
family Bombycillidae

Marsh wren
Cistothorus palustris,
family Troglodytidae

Makes cheerful calls

Gray catbird
Dumatella carolinensis,
family Mimidae

Red-whiskered bulbul
Pycnonotus jocusus,
family Pycnonotidae

DISPLAYS AND LEK BEHAVIOR

Because birds respond strongly to visual signs, much of their communication revolves around displays. These may serve a number of functions, including courtship, greeting, threat, submission, or distraction of predators. Courtship displays themselves vary widely, and may involve ritualized flying, constructing elaborate structures, demonstrating skills, or performing such singular actions as the male manakin's brushing of the female's throat with his tail feathers. In some cases, only the male displays, and the female is merely a spectator; in other cases, the male and female perform together. Many courtship displays involve showing off plumage, especially that found on the most visible parts of the body, such as the head, neck, breast, upper wings, or tail. In some cases, many males gather together in staging areas called leks, where they compete to attract females. Most birds are monogamous, forming pairs for one or more breeding seasons.

Trying to impress A male great bowerbird (*Chlamydera nuchalis*) (above) displays in front of a female who inspects his handiwork in the bower he has constructed. The species of bowerbirds in which the males are relatively drab build more elaborate bowers than those in which the males are more colorful. Bowers may take different shapes and may be decorated with various small objects. After mating, female bowerbirds go away to build separate, cup-like nests.

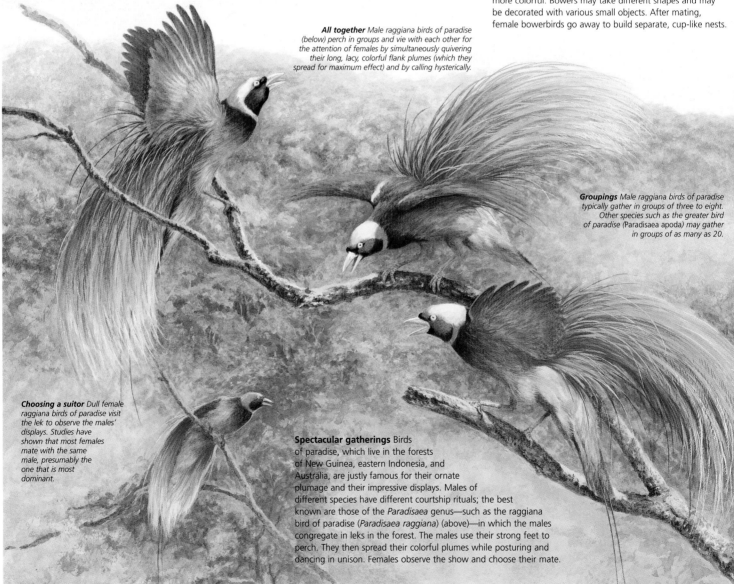

All together *Male raggiana birds of paradise (below) perch in groups and vie with each other for the attention of females by simultaneously quivering their long, lacy, colorful flank plumes (which they spread for maximum effect) and by calling hysterically.*

Groupings *Male raggiana birds of paradise typically gather in groups of three to eight. Other species such as the greater bird of paradise (Paradisaea apoda) may gather in groups of as many as 20.*

Choosing a suitor *Dull female raggiana birds of paradise visit the lek to observe the males' displays. Studies have shown that most females mate with the same male, presumably the one that is most dominant.*

Spectacular gatherings Birds of paradise, which live in the forests of New Guinea, eastern Indonesia, and Australia, are justly famous for their ornate plumage and their impressive displays. Males of different species have different courtship rituals; the best known are those of the *Paradisaea* genus—such as the raggiana bird of paradise (*Paradisaea raggiana*) (above)—in which the males congregate in leks in the forest. The males use their strong feet to perch. They then spread their colorful plumes while posturing and dancing in unison. Females observe the show and choose their mate.

Elegance in the forest A male superb lyrebird (*Menura novaehollandiae*) (left), which lives in the forests of southeastern Australia, is reknowned for having one of the most stunning of avian displays. To attract a mate, it makes use of its unique tail, the two outer feathers of which are shaped like a Greek lyre, hence its name. The bird stands on a mound and spreads its tail into a spectacular fan. It then throws its tail forward over its head and vibrates the plumes as it dances and sings.

Going solo The male western parotia (*Parotia sefilata*) (right) of New Guinea may perform in dispersed leks or by itself. It stands in a small, cleared performing ground in the forest and dances to attract females. Other types of passerines, such as the superb bird of paradise and the riflebirds, also perform by themselves, spreading their wings while perched.

Saving the best display for last When a visiting female raggiana bird of paradise (below, right) shows enough interest in a courting male to land on his perch, he begins his most impressive show. He leans forward until his body is nearly inverted, then extends his colorful wings and clasps them together, giving the female a better view of his brilliant plumage. This is a very strenuous exercise for the male. After mating, the female will fly off and the male will resume his displays in the quest for another mate. The males are not involved in the raising of offspring.

Inspiring humans The display dances of male raggiana birds of paradise have inspired the native people of eastern New Guinea to adorn themselves with the birds' plumes and perform ceremonial dances of their own. The raggiana bird of paradise is the national bird of Papua New Guinea. Trade in them is controlled, and non-nationals cannot possess a bird of paradise or its plumes.

Different strokes Male birds of paradise have different routines, depending on their species. King birds of paradise (Cicinnurus regius) invert themselves even more than the raggiana, by hanging under branches.

Showing interest The female raggiana bird of paradise observes her suitor's private display, gauging his desirability as a mate. She also casually pecks at his beak. By primarily choosing to mate with the same male—the one who outshines his rivals—the females optimize the quality of male genes they pass on to their offspring.

Babblers, thrushes, and others This grouping has been extensively restructured by DNA analyses. It includes some of the world's best-known birds, such as the European and American robins, the common nightingale, and the dunnock.

Family Muscicapidae The Old World flycatchers (such as the pied flycatcher, left, with the female at the nest and the male below) are small, insect-eating birds that live in wooded areas. They often sit on low perches in dense cover, then fly off to catch insects in flight.

Genera 48
Species 275

W. Alaska, N.E. Canada, Africa, Eurasia to W. Melanesia

Family Petroicidae Although known as robins, these common Australasian birds are not related to European nor American robins, only resembling them in shape, size, and nature. They live in forests and woodlands.

Red-capped robin
The red-capped robin (Petroica goodenovii) catches insects by pouncing down from perches.

Genera 13
Species 45

Australia, New Guinea, New Zealand, S.W. Pacific

Family Prunellidae These birds, known as accentors, look like slender-billed sparrows. They prefer high-altitude habitats, and forage for seeds, insects, or berries. They insulate their ground-level nests with feathers.

Genera 1
Species 13

N.W. Africa, extra-tropical Eurasia

Black-throated accentor
Prunella atrogularis,
family Prunellidae

A shrubbery bird of western Europe

White-browed shortwing
Brachypteryx montana,
family Turdidae

Rufous scrub-robin
Cercotrichas galactotes,
family Muscicapidae

European
Erithacus rub...
family Muscic...

White-starred robin
Pogonocichla stellata,
family Muscicapidae

Siberian rubythroat
Luscinia calliope,
family Muscicapidae

Northern scrub-robin
Drymodes superciliaris,
family Petroicidae

This scrub-robin is superficially thrush-like

Bluethroat
Luscinia svecica,
family Muscicapidae

Malay rail-babbler
Eupetes macrocerus,
family Eupetidae

Arabian babbler
Turdoides squamiceps,
family Timaliidae

Large wren-babbler
Napothera macrodactyla,
family Timaliidae

Forages on the forest
floor in Southeast
Asia's rain forests

Eastern whipbird
Psophodes olivaceus,
family Eupetidae

Black-throated thrush
Turdus atrogularis,
family Turdidae

American robin
Turdus migratorius,
family Turdidae

Fluffy-backed tit-babbler
Macronous ptilosus,
family Timaliidae

Chestnut-crowned babbler
Pomatostomus ruficeps,
family Pomatostomidae

Builds dormitory
nests to sleep in

FACT FILE

Family Turdidae Thrushes live in most parts of the world. Familiar members of this family are the Eurasian blackbird and the American robin. They vary in color and habitat preferences. The sexes are usually alike. They eat fruits and animal prey, mostly on the ground.

Fine singer The white-rumped shama (Copsychus malabaricus) has a fine voice that carries far and rivals that of the nightingale in its richness.

Genera 24
Species 165

Cosmopolitan except W. & C. Australia,
New Zealand

Family Timaliidae Babblers are mainly Old World birds that live in warm areas and eat insects. They prefer wooded habitats, where they live in under-growth and foliage. They are small, highly social birds. Most species are a nondescript brown.

Genera 50
Species 273

W. coastal USA, Africa, tropical to temperate
Eurasia, E. Asian is.

Family Pomatostomidae These birds are known as Australian babblers. Isolated in Australia and New Guinea, they have evolved convergently with the true babblers, imitating them in appearance and behavior. Most species are gray, white, and ruddy.

Genera 2
Species 5

Australia, lowland
New Guinea

*Anyone home?
An Australian gray-crowned babbler visits one of its nests, a bulky dome of twigs.*

FACT FILE

Old World warblers and others This group of birds is characterized by the impressive singing abilities of many of its species. Warblers and their allies are for the most part insectivores, but some species also eat fruit and other items. The strange African bald crows are anomalous here, and the gnat-catchers are exclusively American.

Family Sylviidae Most warblers are residents of the Old World. They live in woodland, scrub, and forest. Many breed at high latitudes or high altitudes but migrate over sometimes very long distances to warmer winter climates. Their navigational abilities appear to have a strong genetic component.

Genera 48
Species 265

Africa, Eurasia to Australasia & Oceania, W. Alaska

Tiny songster
The golden-crowned kinglet (Regulus satrapa) inhabits coniferous forests. The tiny, plump warblers in its genus are characterized by a patch of vivid red or yellow on the crown.

Family Picathartidae The bald crows, or rockfowl, are unusual birds that look like long-legged thrushes. Their bare heads are colored either orange-yellow or blue and pink. They inhabit dense forests in West Africa, where they build bowl-like mud nests. They eat insects, amphibians, and fruit.

Genera 1
Species 2

Equatorial W. Africa (Guinea to Gabon)

Family Polioptilidae The insect-eating gnatcatchers are very small birds, even when their long tails are taken into account. Most species are found only in Central and South America. Notable is the long-billed gnatwren, whose bill is equal to more than one-third of the bird's body length.

Genera 3
Species 14

Temperate North America to subtropical South America

Icterine warbler
Hippolais icterina,
family Sylviidae

African yellow warbler
Chloropeta natalensis,
family Sylviidae

Gives clicking calls in display flights

Zitting cisticola
Cisticola juncidis,
family Cisticolidae

Sedge warbler
Acrocephalus schoenobaenus,
family Sylviidae

White-necked picathartes
Picathartes gymnocephalus,
family Picathartidae

Goldcrest
Regulus regulus,
family Regulidae

Tropical gnatcatcher
Polioptila plumbea,
family Polioptilidae

Gray-headed parrotbill
Paradoxornis gularis,
family Timaliidae

Some colonies number in the millions

Red-billed quelea
Quelea quelea,
family Ploceidae

Southern red bishop
Euplectes orix,
family Ploceidae

Long-tailed widowbird
Euplectes progne,
family Ploceidae

Spot-winged starling
Saroglossa spiloptera,
family Sturnidae

Amethyst starling
Cinnyricinclus leucogaster,
family Sturnidae

Red-billed buffalo weaver
Bubalornis niger,
family Ploceidae

Metallic starling
Aplonis metallica,
family Sturnidae

Builds globular hanging nests in trees in large colonies

Gray-headed social weaver
Pseudonigrita arnaudi,
family Passeridae

This weaver is a member of the Old World sparrow family

FACT FILE

Starlings and weavers Starlings are found throughout the Old World. They have also been deliberately introduced to (and prospered in) North America and in countries outside their ancestral ranges. Weavers are finch-like birds restricted mostly to Africa.

Family Sturnidae The starling family includes the common starling, one of the most familiar of garden birds. This stumpy, relatively short-tailed bird has a confident strut and a pugnacious air. Although some starlings inhabit open country and eat insects, most species live in rain forests and eat fruit.

Safari bird *The superb starling* (Lamprotornis superbus) *is often seen by safari-goers in East Africa. It commonly visits campsites. Its plumage makes it one of the more beautiful of the starlings.*

Genera 25
Species 115

Africa, temperate & tropical Eurasia to N.E. Australia & Oceania

Family Ploceidae The weaver birds are notable for their nests. The males of some species build their own nests by looping and knotting blades of grass in a process akin to basket-weaving. Other species build large communal nests of thorns or sticks. Red-billed queleas live in huge flocks and feed on grain.

Genera 11
Species 108

Sub-Saharan Africa, S. & S.E. Asia, Gr. Sundas (except Borneo)

⚡ CONSERVATION WATCH

Weaver woes The critically endangered Mauritius fody (*Foudia rubra*) is a medium-sized forest weaver of the Ploceidae family. It lives only in southwest Mauritius, where the clearance of upland forests and the introduction of predators such as rats have led to a dramatic decline in its numbers.

FACT FILE

Waxbills, whydahs, and finches
This group of finch-like birds includes some of the most popular cage birds. They are, for the most part, seed eaters with cone-shaped bills. They can be found in a variety of habitats around the world, from parks to grasslands to arid areas. Many have fine songs.

Family Fringillidae Almost all finches within this family have red or yellow in their plumage. They are migratory birds that inhabit temperate climates. They have only nine large primary feathers in each wing; the tenth is vestigial. They rarely eat insects, even during nesting.

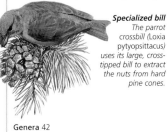

Specialized bill
The parrot crossbill (Loxia pytyopsittacus) uses its large, cross-tipped bill to extract the nuts from hard pine cones.

Genera 42
Species 168

The Americas, Africa, Eurasia to S.E. Asia

Family Viduidae Whydahs get their name from the Portuguese word for a widow's veil, thanks to the males' very long tails. Each species of whydah is a brood parasite of a particular waxbill species, and their young even have the same mouth markings and begging calls as the foster species' own young.

Genera 2
Species 20

Sub-saharan Africa (except rainforested Congo & Namibian desert)

VIREO NEST

The vireos—small gray-green birds of the Americas that sing compulsively and eat insects—build their nests at various elevations in trees or in low shrubs. Their nests are almost always woven cups, suspended by their rims from forked branches.

European goldfinch
Carduelis carduelis,
family Fringillidae

Common redpoll
Carduelis flammea,
family Fringillidae

Introduced to Australia
and New Zealand

Zebra finch
Taeniopygia guttata,
family Estrildidae

Widespread in dry
regions of Australia

Eastern paradise whydah
Vidua paradisaea,
family Viduidae

Pin-tailed whydah
Vidua macroura,
family Viduidae

Straw-tailed whydah
Vidua fischeri,
family Viduidae

Gouldian finch
Erythrura gouldiae,
family Estrildidae

Blue-faced parrotfinch
Erythrura trichroa,
family Estrildidae

American goldfinch
Carduelis tristis,
family Fringillidae

Black-capped vireo
Vireo atricapilla,
family Vireonidae

Scarlet tanager
Piranga olivacea,
family Thraupidae

A common,
gregarious forest bird

Red-legged honeycreeper
Cyanerpes cyaneus,
family Thraupidae

Green honeycreeper
Chlorophanes spiza,
family Thraupidae

Giant conebill
Oreomanes fraseri,
family Thraupidae

Golden-hooded tanager
Tangara larvata,
family Thraupidae

Swallow-tanager
Tersina viridis,
family Thraupidae

Bananaquit
Coereba flaveola,
family Coerebidae

Paradise tanager
Tangara chilensis,
family Thraupidae

Their brilliant color makes
tanagers attractive to the
live bird trade

FACT FILE

Tangers and bananaquits These small to medium birds are found only in the New World, mainly in tropical and mountainous regions of South America. Most eat fruit, though some species specialize in catching insects or drawing nectar from flowers. Most of them prefer either forests or semi-open areas, but some species have no distinct habitat preference.

Family Thraupidae The Andes are the center of radiation for the tanagers. A few species of tanager are drab and secretive, but the family also includes some of the most colorful of birds. The honeycreepers have long, thin, delicate bills for extracting nectar. The insect specialists have heavy, notched bills. Most tanagers build cup-shaped nests among moss or dead leaves. Females carry out most of the nesting tasks.

Genera 50
Species 202

North America to
South America

Superb plumage
The seven-colored tanager (Tangara fastuosa) lives in a small sector of Brazil. Its numbers have declined greatly due to trapping for the pet trade and habitat loss.

Family Coerebidae
The bananaquit is the only member of this family. It lives in the tropics of South America, north to Mexico and the Caribbean, and it sometimes ventures to the southern tip of the United States. This very small bird has a slender, curved bill adapted for taking nectar, which it sometimes extracts by piercing the flower from the side, without pollinating it. It is often seen in gardens.

Genera 1
Species 1

S. Mexico to
W.C. South America

⚡ CONSERVATION WATCH

Azure-rumped tanager *Tangara cabanisi* is one of more than half a dozen endangered members of the Thraupidae family. It is a highly social bird that lives in Southwest Mexico and nearby Guatemala, where the conversion of its forest habitat to coffee plantations is causing a decline in its numbers.

SEXUAL PLUMAGE

Most New World warblers are brightly colored and patterned. Among the North American species, the breeding males are often more brightly colored than females, but both sexes usually molt into more drab plumage before their autumn migrations.

Variety The range of sexual plumage dimorphism in this group is exemplified by the similarity of the sexes in the orange-crowned warbler (Vermivora celata); the slight overlap in the Magnolia warbler (Dendroica magnolia); and the huge differences in the American redstart (Setophaga ruticilla).

Golden-winged warbler
Vermivora chrysoptera,
family Parulidae

Breeds in eastern United States and migrates as far as northern South America for winter

Northern parula
Parula americana,
family Parulidae

MacGillivray's warbler
Oporornis tolmiei,
family Parulidae

Most New World warblers build simple cup-shaped nests. Only the female builds and incubates

Unlike its relatives, may nest in areas without wooded vegetation

Canada warbler
Wilsonia canadensis,
family Parulidae

Common yellowthroat
Geothlypis trichas,
family Parulidae

Yellow-rumped warbler
Dendroica coronata,
family Parulidae

Elegant patterns of red and yellow on head and breast characterize wood warblers in breeding plumage

Collared redstart
Myioborus torquatus,
family Parulidae

Nests in cavities in trees

Prothonotary warbler
Protonotaria citrea,
family Parulidae

Red-winged blackbird
Agelaius phoeniceius,
family Icteridae

One of the most common
North American birds

Eastern meadowlark
Sturnella magna,
family Icteridae

Males jump up in
display to show
breast marks

Shiny cowbird
Molothrus bonariensis,
family Icteridae

Large size difference
between males and
females in this family

Breast feathers
are fluffed in display

Giant cowbird
Molothrus oryzivorus,
family Icteridae

White-crowned sparrow
Zonotrichia leucophrys,
family Emberizidae

Common cactus-finch
Geospiza scandens,
family Emberizidae

Baltimore oriole
Icterus galbula,
family Icteridae

Black-thighed grosbeak
Pheucticus tibialis,
family Cardinalidae

McKay's bunting
Plectrophenax hyperboreus,
family Emberizidae

Thick, cone-shaped
bill for picking and
shelling seeds

FACT FILE

Buntings, cardinals, and troupials
These birds are either confined to the
New World or else originated in the
Americas and spread to the Old World
in successive waves of colonization.
They are quite diverse, but most eat
seeds and use plant matter to build
cup-shaped nests.

Family Icteridae This family, often
misleadingly referred to as "American
blackbirds," is the most ecologically
and morphologically diverse group of
New World songbirds. The family is
predominantly tropical. Very few of its
species are true forest birds, preferring
swamps, marshes, savannas, arid scrub,
or open woodlands. Males in this
family are much larger than females.

Genera 26
Species 98

North America to South
America, West Indies

Family Emberizidae Included in this
family are New World sparrows and
finches and Old World buntings. These
small birds all have short, conical bills
adapted to eating seeds, but they tend
to feed their young with insects. They
are primarily terrestrial and most have
brown-streaked plumage. The snow
bunting has the distinction of breeding
further north (in northern Greenland)
than any other land bird. These birds
tend to have small song repertoires.

Genera 73
Species 308

Cosmopolitan except Madagascar,
Indonesia & Australasia

A standout
Cardinals, such as the painted
bunting (Passerina ciris), tend to
be much more brightly colored
than other buntings. The male
in particular has notably multi-
hued plumage.

🏴 CONSERVATION WATCH

Black news The saffron-cowled
blackbird (*Xanthopsar flavus*) is one
of several vulnerable members of the
Icteridae family. It is trapped for the
pet trade, due to its striking looks.
Also, pesticide use and other factors
are threatening its grassland habitat.

EVOLUTION OF BILLS

The size and shape of birds' bills vary, adapted to their diets and food-gathering methods. It was Charles Darwin who first offered convincing proof that such physical features are not immutable, and that species change over time in response to their environments. He theorized that Galápagos Islands finches, which differed most noticeably in terms of bill size and shape, were all descended from a common ancestor and had evolved to exploit different ecological niches that were not being filled by other animals. Genetic mutations that helped certain birds thrive were likely to be passed on to successive generations. This adaptive radiation is most noticeable on remote islands, where outside influences and competitor species are almost absent.

A productive relationship The brilliantly colored iiwi (above), one of the Hawaiian honeycreepers, has a bill that allows it to easily reach nectar in the tropical blooms on which it feeds. It can often be seen hanging upside down while feeding.

Isolated habitats Left to right: Hawaii; Galápagos Islands; Madagascar.

Rapid spread Specialized species that also eat insects (H1 and H3) or seeds (H2) developed rapidly—within a few hundred thousand years—to fill Hawaii's untapped ecological niches.

Special skills One Darwin finch species (G1) has learned to extract insects from under bark or cracks in wood using a cactus spine or twig. The largest of the ground finches (G2) uses its sturdy bill to eat large seeds.

Different appetites Nectar specialists such as H4 evolved long, curved bills that allowed them to probe and extract nectar, whereas seed-eaters such as H5 were better served by strong, stout bills that could crack hard shells.

Limitations Despite having bills that allowed them to eat well on the Hawaiian islands, many honeycreepers such as H6 went extinct. Evolutionary change could not keep pace with introduced predators and other threats.

A rainbow of differences The Hawaiian honeycreepers belong to the subfamily Drepanidinae, which includes more than 30 species (many of them now extinct) that look and behave very differently from each other, despite the fact that they all trace their lineage back to a single species. The honeycreepers' ancestors were most likely Eurasian rosefinches (*Carpodacus carpodacus*) that flew 2,500 miles (4,000 km) across open ocean, possibly with the aid of storm winds, to settle on this volcanic archipelago millions of years ago. Many honeycreepers evolved bills that fit into certain types of flowers. The pictured Hawaiian honeycreepers (above) are: H1: anianiau (*Hemignathus parvus*); H2: Laysan finch (*Telespyza cantans*); H3: crested honeycreeper, or akohekohe (*Palmeria dolei*); H4: iiwi (*Vestiaria coccinea*); H5: Maui parrotbill (*Pseudonestor xanthophrys*); H6: black mamo (*Drepanis funerea*, extinct).

Diversity on a large island The vangas are a diverse family (called Vangidae) of passerines found only on the island of Madagascar. The 22 species all prey on invertebrates or small reptiles, but they have evolved to fill many different niches on a long-isolated island whose very different habitats span a latitudinal range equal to that of California. The vangas pictured are: M1: Sickle-billed vanga (*Falculea palliata*); M2: Pollen's vanga (*Xenopirostris polleni*); M3: Helmet vanga (*Euryceros prevostii*); M4: Red-shouldered vanga (*Calicalicus rufocarpalis*); M5: Blue vanga (*Cyanolanius madagascarinus*).

M2

M3

M1

M4

Niche filler A Chabert's vanga (*Leptopterus chabert*) (left) perches on a branch in Madagascar. This bird is an insectivore that catches its prey by flying forth from a perch, or else by behaving like a woodswallow.

A variety of roles
The vangas have specialized so much that there are almost as many genera as species in this group. The long bill of M1 is ideal for feeding in spiny forests. M2 probes in dead wood; M4 gleans insects from bushes and in flight.

M5

Similarities to other birds
Individual vanga species have filled niches similar to those of birds elsewhere and have thus grown to resemble them. The M3, for example, recalls a small hornbill.

Blood-sucker One finch (G5) is also known as the "vampire finch" due to its habit of pecking seabirds in order to drink their blood.

G4

Versatile bird One of the ground finches (G6) evolved a longer bill, which allows it to feed on cactus flowers and fruits as well as seeds. Diet specialization is especially important in lean times.

G6

G5

G3

Spawning a scientific revolution Modern genetic analyses have proven Darwin right, showing that the 14 species of Galápagos-based birds now known as Darwin's finches are indeed descended from one species, a seed-eating finch that arrived there from the South American mainland. The Galápagos birds pictured are: G1: woodpecker finch (*Camarhynchus pallidus*); G2: large ground finch (*Geospiza magnirostris*); G3: warbler finch (*Certhidea olivacea*); G4: masked booby (*Sula dactylatra*); G5: sharp-beaked ground finch (*Geospiza difficilis*); G6: common cactus-finch (*Geospiza scandens*).

Few predators A Galápagos hawk (*Buteo galápagoensis*) (right) soars above Fernandina Island. It is one of the few land birds, apart from the finches, that has managed to colonize these remote islands. It is one of three raptor species found there.

REPTILES

REPTILES

PHYLUM	Chordata
CLASS	Reptilia
ORDERS	4
FAMILIES	60
GENERA	1,012
SPECIES	8,163

Reptiles are often thought of as relics of an age gone by, leftovers from the age of the dinosaurs. But in fact reptiles are continuously evolving. Modern species are the product of over 300 million years of evolution since the basal amniotes split into two lineages: one leading to mammals (Synapsida) and the other to birds and reptiles (Diapsida). The diapsid reptiles diverged into two main groups, the Lepidosaura (scaly reptiles—lizards, snakes, amphisbaenas, and tuataras) and the Archosauria (ruling reptiles—crocodiles, pterosaurs, dinosaurs, and ancestral birds) in the Triassic. Turtles appeared in the fossil record about 210 million years ago. Even though current skeletal shapes look similar to species living 200 million years ago, today's reptiles are highly specialized.

Parental care Some species of crocodiles guard their nests and carry their young to the water after they hatch. Some stay with their pod for over a year.

Basking turtles Butterflies are attracted to the salts excreted from the eyes of yellow-spotted Amazon river turtles (*Podocnemis unifilis*). The turtles bask to raise their body temperature and to absorb ultraviolet radiation to help in the synthesis of vitamin D. The process also dries the shell to rid it of algal colonies.

THE CLEDOIC EGG

The closed system of the egg—where water, nutrients, and waste are stored until hatching—allowed reptiles to be independent of water and invade land. All eggs need to respire; they have tiny pores that allow oxygen to enter. Some eggs are very permeable, and if they are not buried in a moist place will dehydrate. Other species have developed very thick-shelled eggs to avoid dehydration. Eggs are sometimes retained in many snakes and lizards, and development progresses in the female. Variations on this theme led to viviparity evolving in lizards and snakes, whereby large, yolked, shell-less eggs were maintained in the oviduct until hatching. Where there was nutrient exchange between the female and her embryos this evolved into the formation of a placenta.

Internal fertilization was another mechanism that freed reptiles from an aquatic life. Lizards and snakes have two functional hemipenes. Tuataras abut their cloacas, and turtles and crocodilians have a penis for internal sperm transmission.

Development of impermeable skin allowed reptiles the chance to live permanently out of water without dehydrating. The epidermis radiated in diverse directions—the skin beads on gila monsters, the crests on iguanas, the keeled scales in snakes, the rattle on rattlesnakes, the scutes on turtle shells, and the armored plates on crocodiles. Reptile skin varies in its permeability to water and gas exchange.

On land, temperatures fluctuate more rapidly than in the water, so reptiles had to develop behavioral mechanisms to control their core body temperatures so that enzyme processes would function. Since, unlike mammals, most reptiles do not burn calories to maintain body temperatures, they use behavioral means, sunning themselves in the early morning hours, remaining in the shade or underground in the heat of the day, or becoming nocturnal in hot environments. Many behavioral, structural, and physiological mechanisms evolved to allow reptiles to be more efficient in using energy than mammals.

Reptile egg The embryo absorbs oxygen through the blood vessels lying adjacent to the pores of the shell. Waste products are stored in the allantois. The amnion maintains the water balance in the egg, and acts as a shock absorber. The embryo grows from the energy supplied in the yolk sac, the oxygen that filters through the shell, and the water absorbed through the leathery shell.

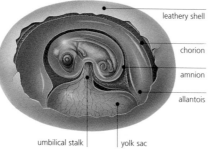

leathery shell

chorion

amnion

allantois

umbilical stalk | yolk sac

Thorny or spiny? The thorny devil is the porcupine of the lizard world. Because every part of its body is covered in spines, it is safe from many predators, including snakes and most lizards, though some monitor lizards (family Varanidae) have managed to swallow them and live.

Matamata characterized by its long snorkel nose, which allows it to lie in shallow water and reach the surface to respire without frightening its prey

Matamata
Chelus fimbriatus

Twist-necked turtle
Platemys platycephala

Common snake-necked turtle
Chelodina longivollis

Twist-necked turtle is the only species of turtle known to have triploid chromosomes in some populations

Helmeted turtle
Pelomedusa subrufa

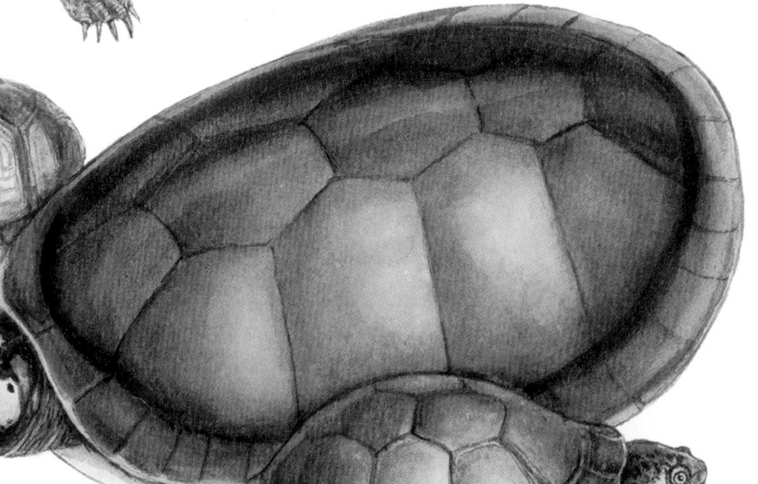

Giant Amazon river turtle
Podocnemis expansa

Chelidae Fitzroy turtle
Rheodytes leukops

Serrated turtle
Pelusios sinuatus

Victoria short-necked turtle
Emydura victoriae

Hilaire's toadhead turtle
Phrynops hilarii

FACT FILE

Matamata The matamata, relying on its camouflage, lies waiting for small fishes to pass. With lightning speed, it opens its mouth and thrusts out its neck simultaneously, creating a sucking force that pulls fishes into its mouth.

- Up to 18 in (46 cm)
- Aquatic
- Unknown
- 12–28
- Common

Amazonia (N. South America)

Giant Amazon river turtle Selection for fast incubation was influenced by the unpredictability of rising river levels. The eggs hatch in 45 days, and nest temperatures often reach 104°F (40°C), the highest known for any turtle.

- Up to 42 in (107 cm)
- Aquatic
- TSD
- 60–150
- Conserv. dependent

Amazon & Orinoco basins (N. South America)

BEGINNING OF LIFE

Sea turtles lay their eggs on the same beaches that their ancestors have used for millennia. Once the nest has been covered with sand, the eggs are left to chance to go undetected by predators. Hatchlings must race to the sea, running the risk of predation by hungry mammals and birds, to the dubious safety of waters filled with even more predators. Survival rates for some species may be as low as 1 in 50,000.

Hatchling flatback turtles
Hatchling turtles are on their own. Their only chance of survival is to scamper to the sea in a large group.

FACT FILE

Green turtle Green turtles migrate huge distances across the open seas between their feeding grounds and nesting beaches. They graze in shallow waters on submerged vegetation. Adults are primarily herbivorous, while juveniles are more carnivorous.

- Up to 5 ft (1.5 m)
- Aquatic
- ♀♂ TSD
- 50–240
- Endangered

W. Atlantic from N. USA to Argentina

Olive Ridley turtle This species is characteristic of turtles that nest in arribadas. Most of the population comes to nest within a 2–3 day period on the same beach. Arribadas in Orissa, India, have numbered more than 100,000 turtles nesting each year.

- Up to 31 in (79 cm)
- Aquatic
- ♀♂ TSD
- 30–168
- Endangered

Pacific and Indian Oceans, Atlantic in West Indies, N. South America & West Africa

SHELL SHAPES

Tortoises are high-domed to protect them from predators and store water. Pond turtles are more streamlined so that they can swim with less resistance. Semi-terrestrial turtles have higher domes for predator protection. Sea turtles are designed for least resistance to glide through the water.

Land tortoise

Semi-terrestrial

Sea turtle

Pond turtle

Front feet modified as flippers; swim with a figure-eight stroke

Loggerhead
Caretta caretta

Green turtle
Chelonia mydas

Olive Ridley turtle
Lepidochelys olivacea

Kemp's Ridley turtle
Lepidochelys kempi

Leatherback turtle
Dermochelys coriacea

Carapace scutes used for making combs and hair ornaments

Flatback turtle
Natador depressus

Hawksbill turtle
Eretmochelys imbricata

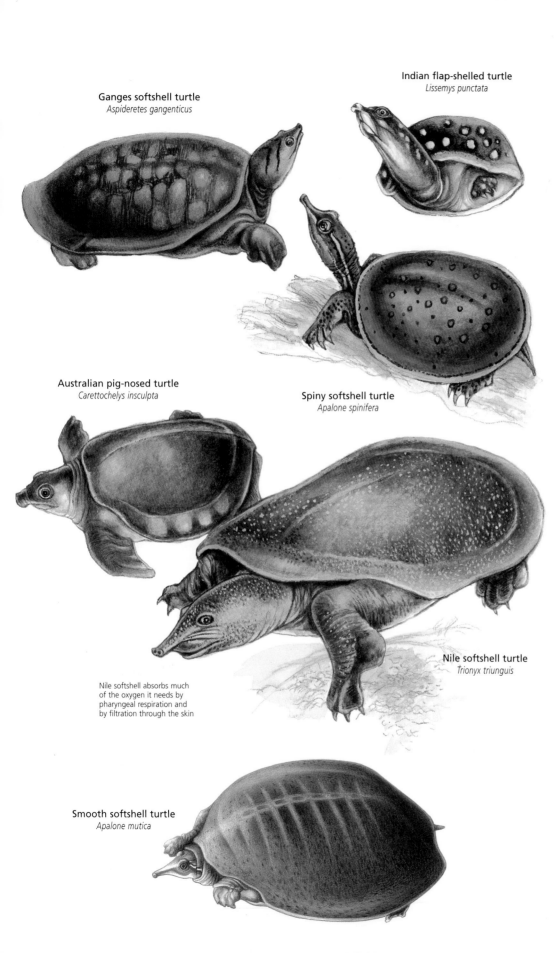

Ganges softshell turtle
Aspideretes gangenticus

Indian flap-shelled turtle
Lissemys punctata

Australian pig-nosed turtle
Carettochelys insculpta

Spiny softshell turtle
Apalone spinifera

Nile softshell turtle
Trionyx triunguis

Nile softshell absorbs much
of the oxygen it needs by
pharyngeal respiration and
by filtration through the skin

Smooth softshell turtle
Apalone mutica

FACT FILE

Ganges softshell turtle An important scavenger, this species helps to lower pollution levels in the Ganges River by consuming partially cremated human corpses that are thrown into the river in traditional funeral rites.

🐢 Up to 28 in (71 cm)
🌊 Aquatic
♀♂ Unknown
● 25–35
🏹 Vulnerable

N. India, N.W. Pakistan, Bangladesh & Nepal

HEAT AND SEX

Sex in some species of turtles, as in most vertebrates, is genetic (referred to as GSD, or genetic sex determination). In most turtles studied, some lizards, all crocodiles, and the tuatara, sex is controlled by the incubation temperature (referred to as TSD, or temperature sex determination). The temperature during the middle third of the incubation period determines what the sex of the hatchling turtle will be. Females are produced at extreme low and high temperatures and males at intermediate temperatures. Thus a female can control the sex of her offspring by where she lays her eggs: nests out in the full sunlight produce females, while nests in the shade produce males; the first nests in summer produce females, while the last nests in the fall produce males.

Wood turtle hatching
Wood turtles are one of the few North American turtles in the family Emydidae to have GSD; most others have TSD.

🏹 CONSERVATION WATCH

The 198 species of testudines on the IUCN Red List are categorized as follows:

7	Extinct
1	Extinct in the wild
25	Critically endangered
46	Endangered
57	Vulnerable
1	Conservation dependent
41	Near threatened
13	Data deficient
7	Least concern

FACT FILE

Diamond-back terrapin This is the only turtle to specialize in living in brackish water marshes. It was once driven to near extinction because of its popularity for the gourmet table.

- Up to 9½ in (24 cm)
- Aquatic
- ♀♂ TSD
- 4–18
- Near threatened

E. seaboard & Gulf Coast (USA)

Ringed sawback This species thrives in fast-moving water, where it forages for aquatic insect larvae. It is endangered due to pollution of the Pearl River. The pet trade and target practice have also reduced its numbers.

- Up to 8½ in (21 cm)
- Aquatic
- ♀♂ TSD
- 4–8
- Endangered

Pearl River, Mississippi (USA)

European pond turtle Only males hatch at incubation temperatures of 75–82°F (24–28°C), while at 86°F (30°C), 96 percent of the hatchlings are females. At higher temperatures, only females are produced.

- Up to 8 in (20 cm)
- Aquatic
- ♀♂ TSD
- 3–16
- Common

S. Europe & W. Asia

Painted turtle In spring, this cold-adapted species can be seen swimming under the ice and mating in iceflows. Hatchlings overwinter in the nest, super-cooling yet not freezing at –20°F (–4°C).

- Up to 10 in (25 cm)
- Aquatic
- ♀♂ TSD
- 4–20
- Common

E. & C. USA

River cooter A complete herbivore, this species is often seen basking by the dozens on floating logs in order to raise its body temperature to speed digestion. Basking is also important for ridding the shell and limbs of fungal and algal colonies.

- Up to 17 in (43 cm)
- Aquatic
- ♀♂ TSD
- 6–28
- Common

S.E. USA

Diamond-back terrapin
Malaclemys terrapin

Ringed sawback
Graptemys oculifera

Wood turtle
Clemmys insculpta

Eastern box turtle
Terrapene carolina

Red-eared slider
Trachemys scripta

Cagle's map turtle
Graptemys caglei

Red-bellied turtle
Pseudemys rubiventris

European pond turtle
Emys orbicularis

Muhlenberg's turtle
Clemmys muhlenbergi

Scorpion mud turtle
Kinosternon scorpioides

Painted turtle
Chrysemys picta

River cooter
Pseudemys concinna

Furrowed wood turtle
Rhinoclemmys areolata

Indian roofed turtle
Kachuga tecta

Caspian turtle
Mauremys caspica

Malayan box turtle
Cuora amboinensis

Malayan snail-eating turtle
Malayemys subtrijuga

One of the few freshwater turtles to nest on ocean beaches. The hatchlings can live for at least 2 weeks in sea water

Spotted pond turtle
Geoclemys hamiltoni

Black-breasted leaf turtle
Geoemyda spengleri

River terrapin
Batagur basca

Painted terrapin
Callagur borneoensis

Feeds almost entirely on terrestrial plants

FACT FILE

River terrapin During the breeding season, from September to November, the male's head coloration changes. The nostrils become blue, the iris white, the head deep black, and the neck and front limbs deep red and black. Sand mining is destroying the preferred nesting beaches of this species.

- ⬡ Up to 24 in (60 cm)
- ⬡ Aquatic
- ♀♂ Unknown
- ● 12–34
- ⚑ Critically endangered

E. India & Bangladesh to Burma, Thailand, Cambodia & Indonesia

Painted terrapin Painted terrapins are exploited for their eggs, which fetch five times the price of chicken eggs. Habitat destruction is also exacerbating the decline of the species. Only one or two rivers harbor more than a hundred nesting females.

- ⬡ Up to 30 in (76 cm)
- ⬡ Aquatic
- ♀♂ Unknown
- ● 10–15
- ⚑ Critically endangered

S. Thailand, Malaysia, Sumatra & Borneo

LEG ADAPTATIONS

From the shape of the leg, foot, and claws, it is possible to deduce what habitat the turtle was adapted for, but not necessarily where it is living today.

Sea turtle The forelimbs of sea turtles are aero-dynamically designed to fly through water. They are tapered from tip to base like the wings of an albatross. There is no webbing, as the toes are fused together within the flippers.

Land tortoise This animal is designed for walking with a load on land, not for swimming. The legs have armored plates, the unwebbed toes are elephantine, and the feet are flat to support the tortoise's weight on land.

Pond turtle The legs are only slightly paddle-shaped to navigate through aquatic vegetation and to walk on land. The toes are webbed, with long, claw-like toenails for traction when crawling up logs to bask.

FACT FILE

Leopard tortoise During courtship, the male trails the female, butting her into submission. After mounting, he extends his neck and releases a grunt-like bellow. Females lay 5–7 clutches of 5–30 eggs from May to October.

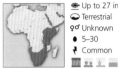

👁 Up to 27 in (68 cm)
🌓 Terrestrial
♀♂ Unknown
● 5–30
🏹 Common

S. Sudan & Ethiopia to Natal & South Africa

African tent tortoise Nesting takes place from September to December, when up to three ellipsoidal eggs are laid in a single yearly clutch. The eggs hatch between April and May. The hatchlings are 1 inch (2.5 cm) long.

👁 Up to 5½ in (16 cm)
🌓 Terrestrial
♀♂ Unknown
● 1–3
🏹 Common

S.W. Africa to Cape (South Africa)

Texas tortoise This species feeds on the pads, flowers, and fruit of cactus. In the Chihuahuan Desert, they are active early in the morning, and rest in the shade or in burrows during the heat of the day.

👁 Up to 8½ in (22 cm)
🌓 Terrestrial
♀♂ TSD
● 1–4
🏹 Vulnerable

S. Texas (USA) to N. Mexico

COURTSHIP

Courtship occurs in turtles that have sympatric congeners to ensure intra-specific mating. Males come face to face with the female, and present head bobs or titillations with the foreclaws to the face of the female. The number of bobs or vibrations and the length of the bout are species-specific.

Titillating a female Red-eared slider males are smaller than females. A courting male vibrates his foreclaws against the sides of the female's head. If the number of beats per minute is correct, it identifies him as a suitable male to breed with.

Leopard tortoise
Geochelone pardalis

Dark blotches on carapace

Bell's hinge-back tortoise
Kinixis belliana

Marginated tortoise
Testudo marginata

Radiated tortoise
Geochelone radiata

Dark lines on carapace

Speckled cape tortoise
Homopus signatus

Elongated tortoise
Indotestudo elongata

Bright red scales on legs

South American red-footed tortoise
Geochelone carbonaria

African tent tortoise
Psammobates tentorius

African pancake tortoise
Malacochersus tornieri

Texas tortoise
Gopherus polyphemus

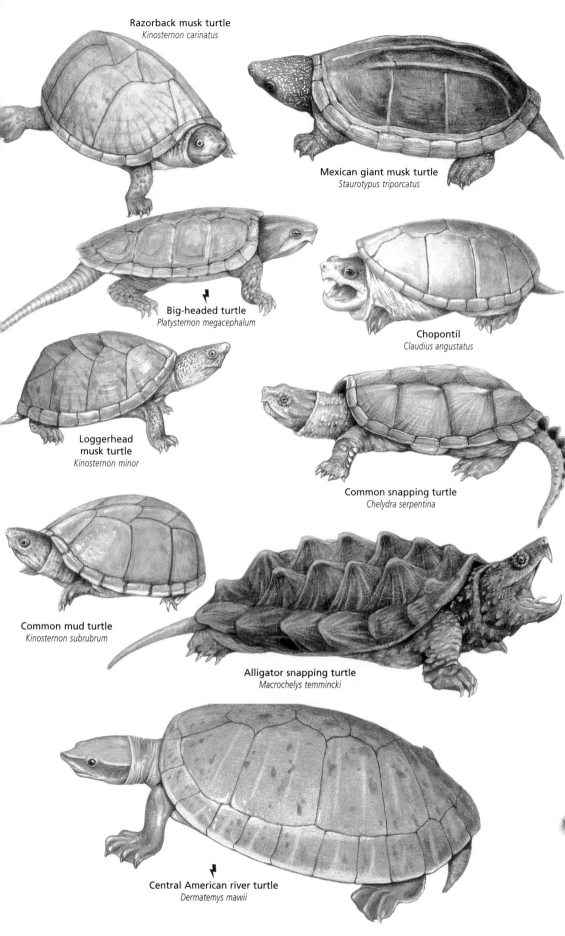

Razorback musk turtle
Kinosternon carinatus

Mexican giant musk turtle
Staurotypus triporcatus

Big-headed turtle
Platysternon megacephalum

Chopontil
Claudius angustatus

Loggerhead
musk turtle
Kinosternon minor

Common snapping turtle
Chelydra serpentina

Common mud turtle
Kinosternon subrubrum

Alligator snapping turtle
Macrochelys temmincki

Central American river turtle
Dermatemys mawii

CAMOUFLAGE

The alligator snapping turtle is a master of disguise. With flaps of loose skin hanging on its neck and filamentous algae growing on its shell, it blends in well with the logs and plants among which it sits waiting to ambush its prey. Only its pink lure looks out of place.

Alligator snapping turtle
This turtle sits with its mouth open at the bottom of a river or lake, fishing with its own bright pink lure.

DEATH ROW

During the last 5 years, the decimation of turtle populations has accelerated rapidly. The Turtle Survival Alliance (TSA) has been formed to monitor and attempt to reverse this trend. TSA shortlisted 25 of the most critically endangered species to raise awareness of the problem. Most of these species occur in hotspots for high biodiversity—areas that are important for the critical habitat of a number of other taxa as well. The turtles are in trouble because of overexploitation for food or traditional remedies, the pet trade, restricted range, and habitat degradation. Only five of these are suffering due to habitat degradation—15 of the 25 are on the brink of extinction because of uncontrolled exploitation by humans.

Bogged down The endangered bog turtle (*Clemmys muhlenbergii*) is endemic to spring-fed meadows and acidic wetlands of the Piedmont and Appalachian mountains, United States. Habitat destruction and fragmentation and the pet trade have decimated this species. Creating reserves may slow this turtle's extinction rate.

THE WORLD'S MOST ENDANGERED TURTLES

1. MESOAMERICA HOTSPOT
Central American river turtle
Dermatemys mawii

2. CHOCÓ-DARIÉN-WESTERN ECUADOR HOTSPOT
Dahl's toad-headed tortoise
Batrachemys dahli

3. MEDITERRANEAN BASIN HOTSPOT
Egyptian tortoise
Testudo kleinmanni

4. MADAGASCAR AND INDIAN OCEAN ISLANDS HOTSPOT
Madagascar big-headed turtle
Erymnochelys madagascariensis

Ploughshare tortoise
Geochelone yniphora

Flat-tailed tortoise
Pyxis planicauda

5. SUCCULENT KAROO HOTSPOT
Southern speckled padloper tortoise
Homopus signatus cafer

6. CAPE FLORISTIC REGION HOTSPOT
Geometric tortoise
Psammobates geometricus

7. INDO-BURMA HOTSPOT
Striped narrow-headed softshell turtle
Chitra chitra

Chinese three-striped box turtle
Cuora trifasciata

Arakan forest turtle
Heosemys depressa

Burmese star tortoise
Geochelone platynota

Burmese roofed turtle
Kachuga trivittata

Leaf turtle
Mauremys annamensis

Yangtze giant softshell turtle
Rafetus swinhoei

8. SUNDALAND HOTSPOT
River terrapin
Batagur baska

Painted terrapin
Callagur borneoensis

9. PHILIPPINES HOTSPOT
Philippine forest turtle
Heosemys leytensis

10. WALLACEA HOTSPOT
Roti snake-necked turtle
Chelodina mccordi

Sulawesi forest turtle
Leucocephalon yuwonoi

11. SOUTHWESTERN AUSTRALIA HOTSPOT
Western swamp turtle
Pseudemydura umbrina

Reaping death There are now fewer than 400 ploughshare tortoises (below) in the wild; they are now restricted to the bamboo scrub near Baly Bay in northeastern Madagascar. Local consumption, the pet trade, and habitat degradation have led to their downfall. The tortoise's name comes from the long gular scute that extends under its head and is used in combat by the males.

Rapid decline The Malayan river turtle (above) is extinct over much of Indochina due to harvesting of eggs and adults. A small population has been protected in Cambodia since 2001 and 30 hatchlings were released in 2002. However, the Mary River turtle (*Elusor macrurus*) of Australia, which was also on the verge of extinction, has been saved by conservation efforts.

Lonely giants Once there were 15 distinct island forms of Galápagos tortoises; some are extinct and now only one Abingdon Island tortoise (*Geochelone nigra*) remains. Human predation and feral mammals have had a devastating impact.

CROCODILIANS

ᴸASS Reptilia

ᴿᴅER Crocodilia

ᴬMILIES 3

ᴱNERA 8

ᴾECIES 23

Alligators, caimans, crocodiles, and gharials are all crocodilians, belonging to the lineage of archosaurs, which includes the dinosaurs and birds. Crocodilians are much more closely related to birds than they are to other reptiles, and have existed for 220 million years, since the Triassic period. Part of their success is a result of being the top aquatic predator in their domain. As the body form of crocodilians has stayed the same, they are often called living fossils, but these beasts have been evolving for millions of years and are very different from their ancestors in the age of the dinosaurs. Unlike most reptiles, crocodilians are very vocal, especially during courtship.

Distribution Crocodilians are found in tropical, subtropical, and temperate zones worldwide: Gavialidae in South Asia, and Alligatoridae in eastern North America, Central and South America, and eastern China. Crocodylidae inhabit estuaries and freshwater streams of Africa, India, Indonesia, Australia, northern South America, Central America, and the West Indies.

Nile crocodile Despite their ferocious look, these animals feed primarily on fishes. Those over 3 feet (1 m) mainly eat fishes that are predatory on human food fishes. Maintaining populations of this species in the wild may help the local fishing industry.

Nile crocodile baby When about to hatch, the young call from the egg. Their mother responds by scraping away the material covering the eggs. When they hatch, she gently gulps them into a pouch that has developed in the floor of her mouth, then transports them to their wetland nursery.

CANNIBALISM AND CARE

Crocodilians have an elongated, cylindrical body with short, muscular limbs and a laterally compressed tail. The massive skull is set on a short neck and features strongly toothed jaws. Crocodilians are aquatic, yet bask and nest on shorelines.

Crocodilians are oviparous, with internal fertilization. Clutches usually contain 12–48 eggs; all species studied have temperature-controlled sex determination, in which females are produced at high and low incubation temperatures and males only at a narrow range of intermediate temperatures.

The eggs are laid in nests made of mounds of vegetation or in nests dug in the sand. Nest site selection can control the sex of the young. Both males and females of some species guard the nest. Females of some species are known to respond to the grunts of the nearly hatched embryos by opening the nests for them. The female will also carry the hatchlings in her mouth to the water, where she has bulldozed a nursery pond for them.

Females will protect their young for the first 2 months. However, males are also known to protect and maintain their territories and food supplies by killing and eating any young male crocodilian in their territory, so the female is often protecting her young from their own father. Female American alligators often stay with their young near the nesting site for 1–2 years.

⚑ CONSERVATION WATCH

The 14 species of crocodilians on the IUCN Red List are categorized as follows:

- 4 Critically endangered
- 3 Endangered
- 3 Vulnerable
- 2 Conservation dependent
- 1 Data deficient
- 1 Least concern

Walk
Crocodilians walking on land hold the limbs vertically beneath the body.

Crawl
Crocodiles crawl low to the ground when they are trying to conceal themselves from potential prey or predators.

Gallop
Australian freshwater crocodiles often gallop over rough terrain.

Crocodile gaits Although crocodiles are primarily aquatic, they use their legs for walking on land and the tail for swimming. On land, the use of ball-and-socket vertebrae allows crocodilians to have varied gaits.

FACT FILE

American alligator Populations of this species were low in the 1950s and protected as endangered in 1967. Populations rebounded in 20 years to over 800,000 animals. Hunting of some populations is now allowed, in order to control their numbers.

⤢	Up to 19 ft (5.8 m)
⊖	Aquatic
○	Oviparous
●	10–40
⚑	Common

S.E. USA

Black caiman This species has recently been reclassified from endangered in Brazil. Conservation efforts over more than 10 years were highly successful, with populations recuperating rapidly. A program to sustainably harvest this species is being enacted.

⤢	Up to 20 ft (6 m)
⊖	Aquatic
○	Oviparous
●	35–50
⚑	Conserv. dependent

Amazon Basin (N. South America)

Chinese alligator This species spends the majority of its life in a complex of underground burrows. The alligators build these systems with pools of water above and below ground as well as air holes for breathing.

⤢	Up to 6½ ft (2 m)
⊖	Aquatic
○	Oviparous
●	10–40
⚑	Critically endangered

Yangtze Valley (China)

ALLIGATOR ACTIVITY

Alligators frequently forage for fishes beneath waterbird rookeries. On occasion, for a change in diet, they will launch themselves at birds as well, using just the propulsion of their tails. They can reach amazing speeds and heights, and it often appears as if they are walking on their tails.

American alligator
Alligator mississippiensis

Black caiman
Melanosuchus niger

Cuvier's dwarf caiman
Paleosuchus palpebrosus

African dwarf crocodile
Osteolaemus tetraspis

Chinese alligator
Alligator sinensis

False gharial
Tomistoma schlegeli

Mugger
Crocodylus palustri

Saltwater crocodile
Crocodylus porosus

Orinoco crocodile
Crocodylus intermedius

Spectacled caiman
Caiman crocodilus

Black tail bands
are characteristic
of the caiman

Siamese crocodile
Crocodylus siamensis

Five clawed digits
are present on
the forelimb

American crocodile
Crocodylus acutus

Gharial
Gavialis gangeticus

Only the male has
a distinct boss on
the tip of the snout

Nile crocodile
Crocodylus niloticus

ANATOMY

Crocodilians have various adaptations to an aquatic life. The position of the eyes, ear openings, and nostrils at the highest part of the body allows them to lie concealed just below the surface of the water when stalking prey. A secondary palate lets them breathe with the mouth closed. A flap of skin in the throat prevents water from entering the throat when the jaws are being opened to capture prey.

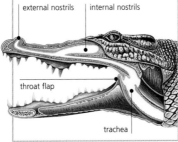

external nostrils   internal nostrils

throat flap

trachea

⚑ CONSERVATION WATCH

Fashion victims Skins from 15 species of crocodilians have been sold as luxury leather products. This has had a negative effect on wild populations of crocodilians. Crocodile farms, meanwhile, are vanishing due to high costs, lower demand, and lower prices. Sustainable use of wild populations seems to be more feasible in the long term.

TUATARA

The tuatara is often called a "living fossil" and is now only to be found on the islands of New Zealand. It is the only survivor of a large group of reptiles that roamed with the dinosaurs over 225 million years ago, the rest of which became extinct 60 million years ago. Its tooth arrangement is unique: a single row in the lower jaw fits between two rows of teeth in the upper jaw. Lizards have visible ear openings but tuatara do not.

| CLASS Reptilia |
| ORDER Rhynchocephalia |
| FAMILY Sphenodontidae |
| GENUS Sphenodon |
| SPECIES 2 |

Distribution About 400 *Sphenodon guntheri* live on North Brother Island, New Zealand. More than 60,000 *S. punctatus* live on some 30 islands off the northeast coast of New Zealand's North Island.

Tuatara
Sphenodon punctatus

Tuatara The name Tuatara is from Maori, meaning "peaks on the back," referring to the dorsal crest. Tuatara keep growing until they are 35 years old, and may live for more than 100 years. Sex is determined by temperature at the time of incubation.

AMPHISBAENIANS

These legless squamates have reduced pectoral and pelvic girdles. They have an annulated pattern of scutes, and short tails. Amphisbaenians are built for burrowing, with heavily ossified skulls. Their brain is surrounded by the frontal bones. The right lung of amphisbaenians is reduced in size, while in other limbless lizards and snakes the left lung is smaller. Three of the four families of worm lizards have no limbs whatsoever, while the remaining family has enlarged forelimbs, which help it with locomotion and digging.

| CLASS Reptilia |
| ORDER Squamata |
| SUBORDER |
| Amphisbaenia |
| FAMILIES 4 |
| GENERA 21 |
| SPECIES 140 |

Distribution Amphisbaenians are found in tropical and subtropical regions of southern North America, South America to Patagonia, West Indies, Africa, the Iberian Penisula, Arabia, and western Asia.

Shovel-snouted worm lizard The illustration below shows how the worm lizard pushes its head against the ceiling of the tunnel to widen it. Worm lizards have large, interlocking upper and lower teeth, which allow them to grasp prey and drag it into their tunnel.

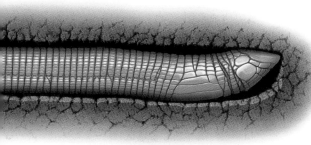

Two-headed legless lizard Amphisbaenians are often erroneously called two-headed snakes, as the tail is designed to mimic the head to confuse potential predators. A wound to the tail would be easier to survive than one to the head.

HOW THEY BURROW

The rings of scales behind the head are close together at the start of the stroke. During penetration the rings separate, pushing the head forward. To widen the tunnel, the head is lifted against the ceiling of the tunnel.

round heads

shovel heads

keel heads

chisel heads

LIZARDS

Today, lizards occupy almost all landmasses except for Antarctica and some Arctic regions. At the end of the Cretaceous period, some 65 million years ago, lizards survived when the dinosaurs and other large reptiles died out. With more than 4,000 species, they are the largest group of living reptiles. Although the largest lizard, the Komodo dragon, reaches an impressive length of 10 feet (3 m), few lizards exceed 12 inches (30 cm) and this is one reason for their continued success. Lizards tend to be restricted to specific habitat niches, as mountains and water are significant barriers to their movement.

ASS Reptilia	
DER Squamata	
BORDER Sauria	
MILIES 27	
NERA 442	
ECIES 4,560	

Distribution Lizards are found from New Zealand to Norway, and from southern Canada to Tierra del Fuego. They are also endemic to many islands in the world's oceans. The only continent they have not colonized is Antarctica.

The lizard assumes a threatening stance on all fours

The back is arched and the eyes stare firmly forward

The back feet leave the ground in an aggressive jump

Territorial defense A defined territory provides an area in which an adult male lizard can hunt for food and find a female for mating. If a rival invades its territory, a collard lizard does a series of "push-ups" to make itself look bigger and more threatening to the unwelcome interloper.

Lizard features The cone-like spines on the crest of the forest dragon make it appear to be larger and more formidable than it actually is, and its multicolored scales provide perfect camouflage in the mottled forest light.

Prehensile-tailed skink The Solomon Island skink (*Corucia zebrata*) is arboreal and herbivorous. It uses its prehensile tail to hold onto branches as it feeds on fruits and leaves in the treetops.

DEFENSE AND ESCAPE

Lizards are preyed upon by spiders, scorpions, other lizards, snakes, birds, and mammals. The Gila monster and the Mexican beaded lizard are the only two venomous species, but even these will resort to scare tactics at the start of a confrontation. Many others have developed an impressive array of tactics to defend themselves or to escape an attacker.

Most lizards are extremely well camouflaged and may keep absolutely still until a predator passes by. Chameleons, in particular, are well known for their ability to change color to blend in with their environment. Other lizards surprise or distract a predator to give themselves a chance to escape. The Australian frilled lizard, for example, opens its mouth, hisses loudly, and flourishes its neck frill before scampering away.

Some species have sharp spines that can injure a predator's mouth, or slippery scales that make them hard to grip. The armadillo girdle-tailed lizard curls itself into a ball and protects itself with a prickly fence of spikes, while the basilisk escapes by skimming on water before diving in to safety.

Diurnal dragon Boyd's forest dragon forages by day near streams or forest edges in tropical rain forests.

Dragon lizard Boyd's forest dragon (*Hypsilurus boydii*), a chisel-tooth lizard from the rain forests of northeastern Australia, is one of the largest dragon lizards in the world. It can puff out its dewlap—a flap of skin on its throat—to communicate with other forest dragons.

Australian shingleback

tree-living chameleon

Lizard tails Some lizards have tails that mimic leaves; others mimic heads. Some are used to hold onto branches, and many are expendable. If a predator grabs a skink's tail, the tail stays in the predator's mouth and the lizard escapes.

skink

leaf-tailed gecko

Underwater grazers Marine iguanas on the Galápagos Islands are herbivores from the time they hatch. They dive into the ocean to forage on algae. Before and after diving, they bask to increase their body temperatures to help them more efficiently digest their food, and also to increase swimming stamina in the usually cold seas.

Lizards use their tails for defense. Monitors and iguanas beat their attackers with their tails. Skinks and other small lizards may give up a little piece of tail in return for their lives. These lizards often have brightly colored tails that they will wave from side to side, inducing the predator to attack the tail and not the head. The tail continues to wriggle after the lizard has escaped. Shed tails grow back after time; the lizard loses some stored energy but remains alive to reproduce. In territorial bouts, some geckos attack and eat the tails of their opponents.

Horned lizards, when attacked by foxes and coyotes, will squirt a stream of bad-tasting blood out of their eyes to distract and discourage potential predators. Some lizards also attempt to scare off predators by sticking out their tongues. Australian skinks hiss and thrust their brightly colored tongues out to startle predators. Even though the Gila monster is venomous and brightly colored to warn predators of the risk of attacking, if molested it will display its bright purple tongue and hiss at its aggressor.

Rain-forest ranch The common green iguana (above) is completely herbivorous after hatching. The low-cholesterol meat is eaten as a delicacy in Meso-America.

Predatory dragon The Indonesian Komodo dragon (left) tests the air for the scent of warm-blooded prey. These lizards grow to be more than 10 feet (3 m) in total length and feed on large mammals such as deer, pigs, goats, and even water buffalo.

Fiji banded iguana
Brachylophus fasciatus, family Iguanidae

West Indian iguana
Iguana delicatissima, family Iguanidae

Marine iguana
Amblyrhynchus cristatus,
family Iguanidae

Galápagos
land iguana
Conolophus subcristatus,
family Iguanidae

Rhinoceros iguana
Cyclura cornuta,
family Iguanidae

Black iguana
Ctenosaura similis,
family Iguanidae

The warmer male black
iguana (top) is lighter
in color; the other male
is colder and darker

Green iguana
Iguana iguana, family Iguanidae

FACT FILE

Family Iguanidae This family contains terrestrial, rock-dwelling, marine, and arboreal species. They range from the 5½ inch (14 cm) snout–vent length *Dipsosaurus* to the more than 27½ inch (70 cm) *Cyclura*. Young iguanas of some species eat insects, but later become predominantly herbivorous, feeding on leaves, fruit, and even algae. All species are oviparous.

Genera 8
Species 36

USA to Paraguay,
Galápagos Is., Fiji,
West Indies

Old salts
*Marine iguanas spend
so much time at sea that
they have a special gland to
exude salt from their bodies.*

Long time coming
*The Fiji crested
iguana (*Brachylophus
vitiensis*) has a 30-week
incubation—three times
that of other iguanas.*

LIZARD TONGUES

The long forked tongues of monitors are used to capture scent information from the air and to transfer it to the Jacobson's organ for analysis. The long, sticky tongues of chameleons are used to capture prey, while the blue-tongued skink (below) flashes its tongue to deter attacking birds.

⚡ CONSERVATION WATCH

The 15 species of iguanas on the IUCN Red List are categorized as follows:

5   Critically endangered
2   Endangered
7   Vulnerable
1   Near threatened

Family Crotaphytidae Collard and leopard lizards are medium-length diurnal lizards active in deserts and other rocky, arid areas. They feed on invertebrates, lizards, and other small vertebrates. They employ squealing vocalizations when stressed, and can use a form of bipedalism when moving among rocks. They are oviparous, laying three to eight eggs per clutch.

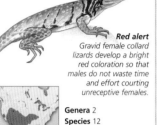

Red alert
Gravid female collard lizards develop a bright red coloration so that males do not waste time and effort courting unreceptive females.

Genera 2
Species 12

S.W. North America

Family Phrynosomatidae The spiny lizards, *Sceloporus*, are the most diverse group in this family, with 70 species. They have the generalized body form of a sit-and-wait predator. They predominate in desert regions, where there are both terrestrial and arboreal species. Horned lizards are terrestrial with a flattened body. Zebra-tailed lizards are a rapid bipedal desert lizard.

Defensive spines
The regal horned lizard has the largest crown of spines of any species. These are effective in avoiding predation by many species of snakes.

Genera 9
Species 110

Southern Canada & USA to Panama

Lizard losses One species of Crotaphytidae, the blunt-nosed leopard lizard (*Gambelia sila*), is listed as endangered on the IUCN Red List due to habitat encroachment. One species of Phrynosomatidae, the Coachella Valley fringe-toed lizard (*Uma inornata*) of North America, is on the IUCN Red List as endangered. It lives in moving dune areas which have now been stabilized by vegetation and construction. Populations of collard and leopard lizards, however, are abundant in the southwest deserts of the United States.

Merrem's Madagascar swift
Oplurus cyclurus, family Opluridae

Collard lizard
Crotaphytus collaris, family Crotaphytidae

Side-blotched lizard
Uta stansburiana, family Phrynosomatidae

Chuckwala
Sauromalus obesus, family Phrynosomatidae

Leopard lizard
Gambelia sila, family Crotaphytidae

Regal horned lizard
Phrynosoma solare, family Phrynosomatidae

Short-horned lizard
Phrynosoma douglass, family Phrynosomatidae

Desert spiny lizard
Sceloporus magister, family Phrynosomatidae

Colorado desert fringe-toed lizard
Uma notata, family Phrynosomatidae

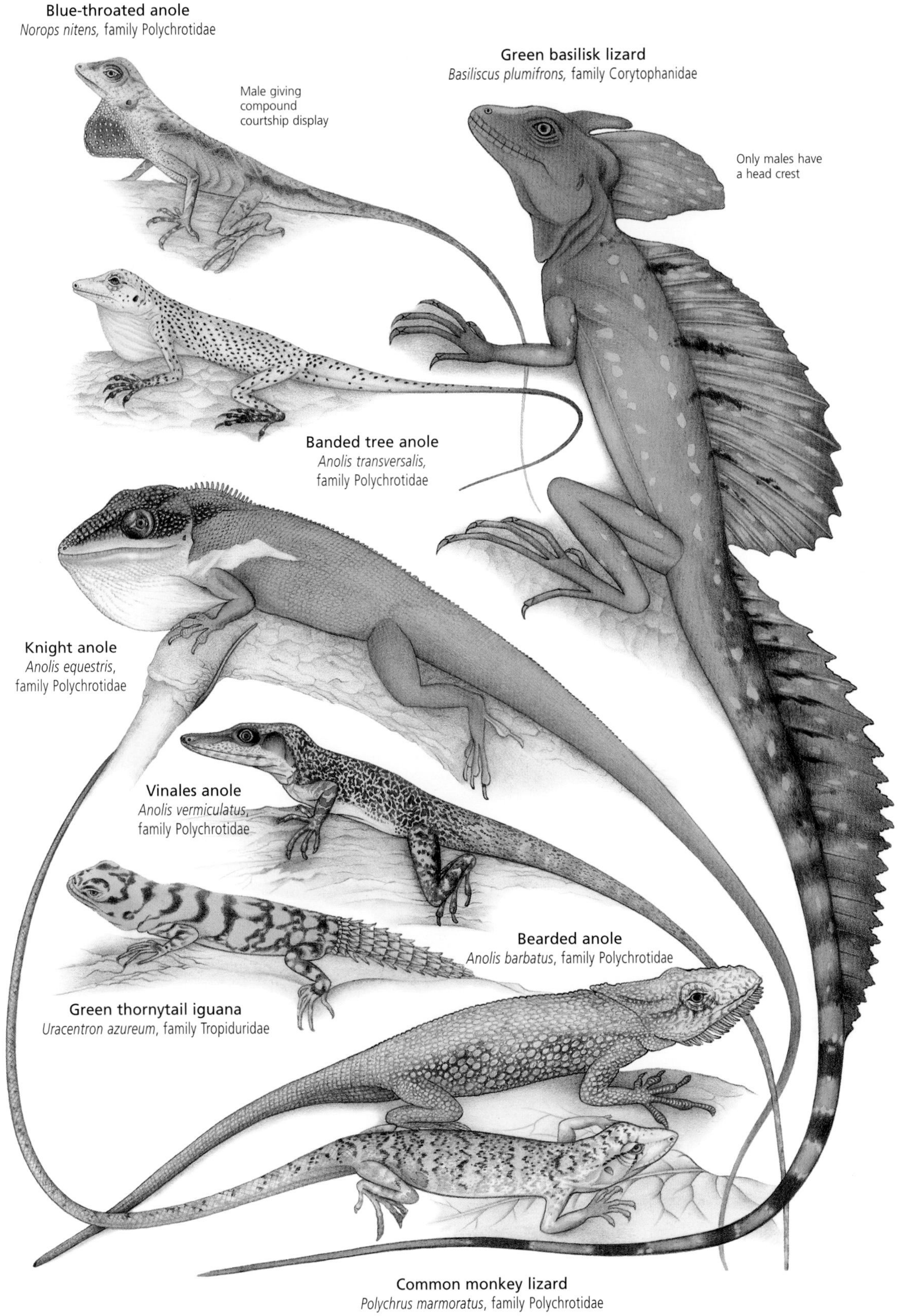

Blue-throated anole
Norops nitens, family Polychrotidae

Male giving
compound
courtship display

Green basilisk lizard
Basiliscus plumifrons, family Corytophanidae

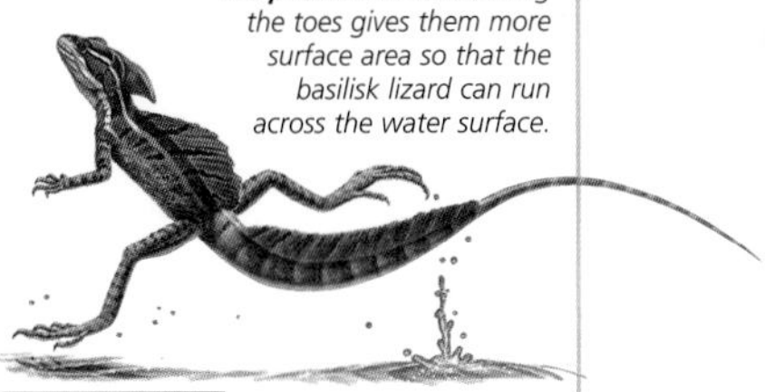

Only males have
a head crest

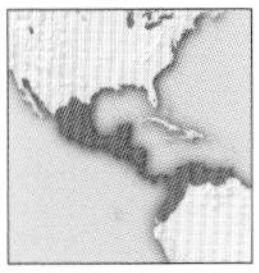

Banded tree anole
Anolis transversalis,
family Polychrotidae

Knight anole
Anolis equestris,
family Polychrotidae

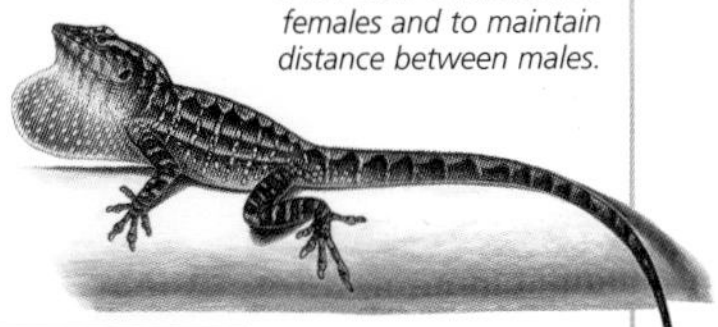

Vinales anole
Anolis vermiculatus,
family Polychrotidae

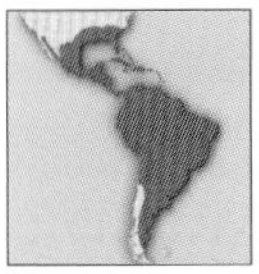

Green thornytail iguana
Uracentron azureum, family Tropiduridae

Bearded anole
Anolis barbatus, family Polychrotidae

Common monkey lizard
Polychrus marmoratus, family Polychrotidae

FACT FILE

Family Agamidae Chisel-tooth lizards are from Africa, Asia, and Australia. There are 52 genera, ranging from 2 to 14 inches (5 to 35 cm) in length from snout to vent. Forest species are green; desert species are brown, gray, or black. Sexual dimorphism in coloration is common; some can change coloration rapidly. The head is usually large, and distinct from the neck; the body form may be compressed, depressed, or cylindrical. The thick tongue is notched.

Genera 50
Species 420

Australia, Indonesia, Asia, Africa

Frilled lizard This arboreal lizard is active diurnally in dry woodlands. When escaping, it runs bipedally. If threatened, it erects its huge frilled collar to appear much larger than it is.

➤ Up to 11¼ in (28 cm)
⊕ Arboreal
○ Oviparous
● 8–23
➤ Common

N. Australia &
S. New Guinea

Carnivores
Frilled lizards feed on cicadas in the trees and will descend to the ground to feed on ants and crickets.

Thorny devil A hygroscopic system of grooves on this lizard's skin leads to the corners of its mouth, allowing it to drink the dew that falls on its back. It feeds only on ants, similar to the North American horned lizards.

➤ Up to 4½ in (11 cm)
⊖ Terrestrial
○ Oviparous
● 3–10
➤ Common

W. Australia

⚡ CONSERVATION WATCH

The five species of Agamidae on the 2003 IUCN Red List are categorized as follows:

  2 Endangered
  1 Vulnerable
  2 Data deficient

1   Frilled lizard
      Chlamydosaurus kingii, family Agamidae

2   Banded agama
      Laudakia stellio, family Agamidae

3   Thorny devil
      Moloch horridus, family Agamidae

4   Eastern bearded dragon
      Pogona barbata, family Agamidae

5   Ocellated mastigure
      Uromastyx ocellata, family Agamidae

6   Sinai agama
      Pseudotrapelus sinaitus,
      family Agamidae

7   Iranian toad-headed agama
      Phrynocephalus persicus,
      family Agamidae

8   Persian agama
      Trapelus persicus, family Agamidae

9   Moroccan dabb lizard
      Uromastyx acanthinura,
      family Agamidae

10   Common agama
      Agama agama, family Agamidae

FACT FILE

Sailfin lizard This semiaquatic lizard forages and basks along the edges of streams. When danger approaches, it runs bipedally across the surface of the water with the aid of a fringe on the toes of its hindlimbs. Populations are declining due to over-collection for the pet trade and its popularity as food.

⟀ Up to 40 in (100 cm)
◉ Variable
○ Oviparous
● 6–12
♟ Common

S.E. Asia, New Guinea

Chinese water dragon These lizards are semiaquatic, living along the shores of rivers, climbing into the lower branches of trees, and sharing burrows in colonies of one dominant male and several females. Males have larger crests than females. Incubation time for water dragon eggs is 60 days.

⟀ Up to 12 in (30 cm)
◉ Variable
○ Oviparous
● 7–12
♟ Common

Thailand, E. Indochina, Vietnam, S. China

Common green forest lizard This arboreal lizard inhabits most forest areas to an altitude of 4,900 feet (1,500 m), beyond which they are replaced by other species. Males, the largest lizards in Sri Lanka, have very vivid coloration.

⟀ Up to 12 in (30 cm)
⬆ Arboreal
○ Oviparous
● 10–20
♟ Common

India & Sri Lanka

TREE DRAGON

The semiarboreal tree dragon (*Diporiphora superba*) is one of the most slender agamids. It measures 3 inches (8 cm) in snout–vent length and its tail is up to four times its body length.

Treetop maneuvers
The tree dragon uses its long tail for balance while climbing and scampering through the branches.

1	**Armored pricklenape** *Acanthosaura armata,* family Agamidae	
2	**Five-lined flying dragon** *Draco quinquefasciatus,* family Agamidae	
3	**Sailfin lizard** *Hydrosaurus amboinensis,* family Agamidae	
4	**Giant forest dragon** *Gonocephalus grandis,* family Agamidae	
5	**Chinese water dragon** *Physignathus cocincinus,* family Agamidae	
6	**Sumatra nose-horned lizard** *Harpesaurus beccarii,* family Agamidae	
7	**Indo-Chinese forest lizard** *Calotes mystaceus,* family Agamidae	
8	**Common green forest lizard** *Calotes calotes,* family Agamidae	
9	**Tropical forest dragon** *Gonocephalus liogaster,* family Agamidae	

LIZARD REPRODUCTION

Reproductive patterns in lizards vary widely. Some lizard species mature rapidly, are short lived, and are continuously producing an egg. Other species take several years to mature and then lay large clutches of large eggs for many years. Between these two extremes there are all possible variations of numbers of clutches per year and viviparity. Some species are egg layers at low altitudes but populations at high altitudes or latitudes are viviparous. The female acts as an incubator, moving with the developing embryos into optimum incubation temperatures. Equally amazing are some species that are stuck in evolutionary inertia and produce no more than two eggs per clutch regardless of their size or the amount of stored energy.

Clutch size Geckos (top), regardless of size or nutrient condition, produce two eggs per clutch. Some species can vary clutch size and egg size (center). Large species (bottom) have large clutches of large eggs and the number and size of the eggs is proportionate to the size and physical condition of the female.

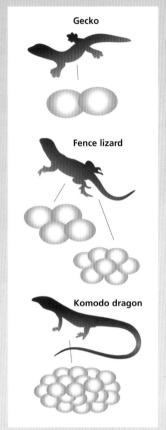

Gecko

Fence lizard

Komodo dragon

Continuous ovulation Anoles of all sizes only produce one egg per clutch; they have continuous ovulation, first from one ovary then the other.

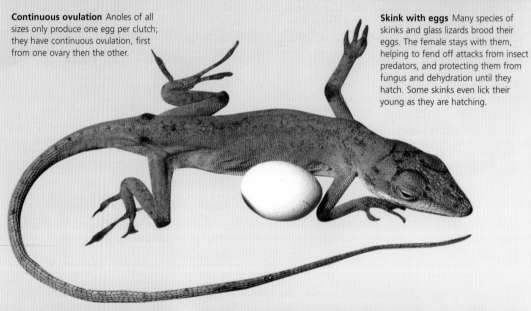

Skink with eggs Many species of skinks and glass lizards brood their eggs. The female stays with them, helping to fend off attacks from insect predators, and protecting them from fungus and dehydration until they hatch. Some skinks even lick their young as they are hatching.

Sceloporus with eggs Fence lizards (Sceloporus) are able to vary the number and size of their eggs according to the time of year and their physical condition. They put their energy into large eggs for the first clutches and many smaller eggs for later clutches that will have less time to develop before the onset of winter.

Iguana with eggs Iguanas normally dig tunnels in sandy beaches and deposit between 25 and 40 eggs, depending on the size of the female. The eggs hatch in 3 months and the hatchlings seek out other iguanas in order to eat their feces, thereby inoculating their guts with cellulose-digesting microbes.

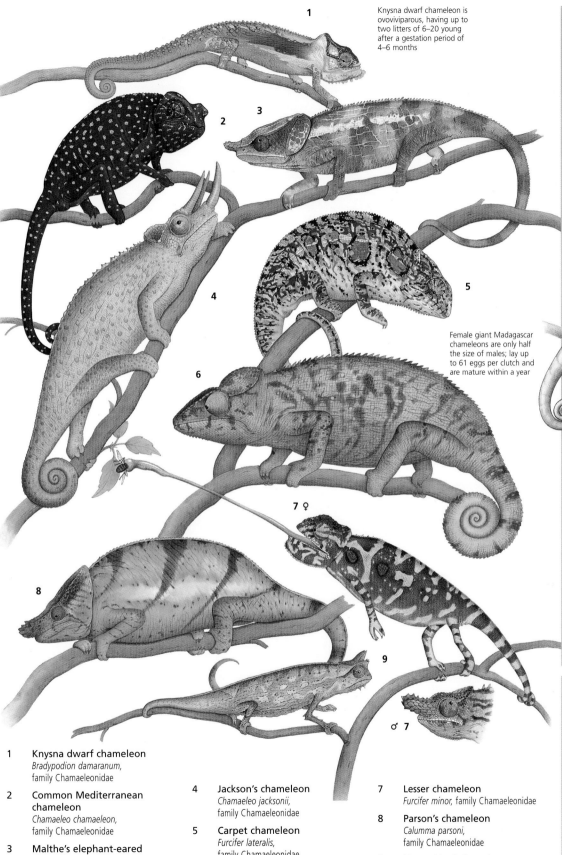

Knysna dwarf chameleon is ovoviviparous, having up to two litters of 6–20 young after a gestation period of 4–6 months

Female giant Madagascar chameleons are only half the size of males; lay up to 61 eggs per clutch and are mature within a year

1

2

3

4

5

6

7 ♀

8

9

♂ 7

FACT FILE

Family Chamaeleonidae Chameleons are famous for their ability to change colors. Their tail is prehensile and two or three of their toes are fused to form grasping pads. Their eyes can move independently and they have three-dimensional vision when they look ahead. This allows them to focus on the prey in front of them. Their tongue is elongated, allowing them to catch prey at a distance of one body length. They grow up to 25¼ inches (63 cm) long and are largely terrestrial.

Genera 6
Species 135

Africa, Madagascar, S. Europe, Middle East, India & Sri Lanka

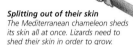

Splitting out of their skin
The Mediterranean chameleon sheds its skin all at once. Lizards need to shed their skin in order to grow.

Giant Madagascar chameleon This species inhabits warm, humid coastal lowlands as well as drier forests. It has a high casque without occipital lobes, horns, or rostral processes. Gular and dorsal crests are composed of specialized, prominent conical scales. Males are larger and have broader tail bases than females.

Up to 24 in (60 cm)
Arboreal
Oviparous
1–61
Common

Madagascar

CONSERVATION WATCH

Habitat loss is the prime reason for the decline in chameleon numbers. But they are also threatened by the pet trade: from 1993 to 1998, 476,000 chameleons were exported from Africa, Yemen, the Seychelles, and Madagascar. The eight species of chameleons on the IUCN Red List are categorized as follows:

1 Critically endangered
1 Endangered
4 Vulnerable
2 Near threatened

1 **Knysna dwarf chameleon**
Bradypodion damaranum,
family Chamaeleonidae

2 **Common Mediterranean chameleon**
Chamaeleo chamaeleon,
family Chamaeleonidae

3 **Malthe's elephant-eared chameleon**
Calumma malthe,
family Chamaeleonidae

4 **Jackson's chameleon**
Chamaeleo jacksonii,
family Chamaeleonidae

5 **Carpet chameleon**
Furcifer lateralis,
family Chamaeleonidae

6 **Giant Madagascar chameleon**
Furcifer oustaleti, family Chamaeleonidae

7 **Lesser chameleon**
Furcifer minor, family Chamaeleonidae

8 **Parson's chameleon**
Calumma parsoni,
family Chamaeleonidae

9 **Horned leaf chameleon**
Brookesia superciliaris,
family Chamaeleonidae

FACT FILE

Family Gekkonidae The geckos are the most speciose family of lizards, split into four subfamilies. Covered with small granular scales, they vary in size from ¾ inch to 13 inches (1.5 to 33 cm). The tails are usually fragile and break easily.

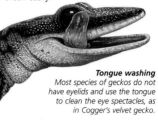

Tongue washing *Most species of geckos do not have eyelids and use the tongue to clean the eye spectacles, as in Cogger's velvet gecko.*

Genera 109
Species 970

Pan-tropical, S. North America, South America, Africa, S. Europe, S. Asia, Indo-Australia

Common leopard gecko Leopard geckos have moveable eyelids and lack toe pads. The sex is determined by incubation temperature.

🦎 Up to 10 in (25 cm)
🥚 Terrestrial
◯ Oviparous
● 2
🌿 Common

Afghanistan, Pakistan, W. India, Iraq, Iran

Banded gecko This nocturnal gecko has large, moveable eyelids, no toe pads and pre-anal pores in a continuous row. Tails are easily lost.

🦎 Up to 4 in (10 cm)
🥚 Terrestrial
◯ Oviparous
● 2
🌿 Common

S.W. USA to Mexico & Panama

ADAPTED FOR SAND

Fine free-flowing sand requires specialized feet to efficiently traverse the terrain. The African web-footed gecko (*Pallmatogecko rangeri*) uses its feet like snow shoes on the sand to keep it on the surface.

Sand shoes *Some lizards have fringes along the digits to allow them better traction on sand; others have paddle-like webbed feet.*

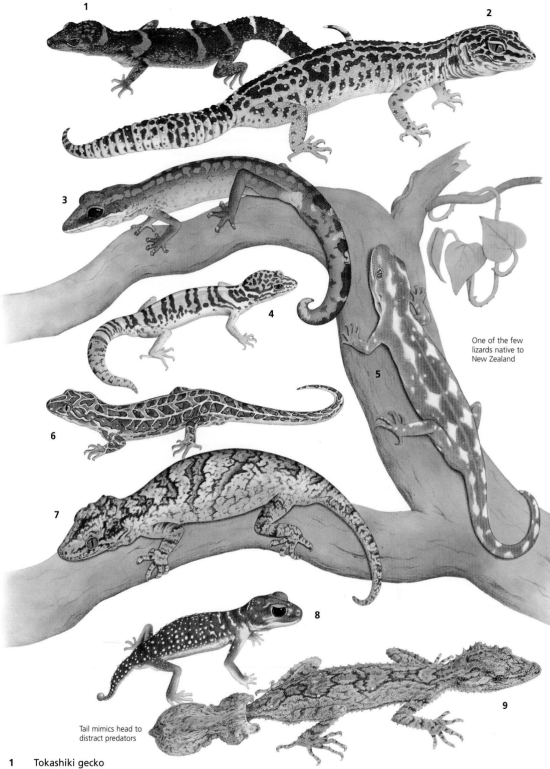

One of the few lizards native to New Zealand

Tail mimics head to distract predators

1 Tokashiki gecko
Goniurosaurus kuroiwae,
family Gekkonidae

2 Common leopard gecko
Eublepharis macularius,
family Gekkonidae

3 Prehensile tailed gecko
Aeluroscalabotes felinus,
family Gekkonidae

4 Banded gecko
Coleonyx variegatus, family Gekkonidae

5 Green tree gecko
Naultinus elegans, family Gekkonidae

6 Thomas's sticky-toed gecko
Hoplodactylus rakiurae,
family Gekkonidae

7 New Caledonia bumpy gecko
Rhacodactylus auriculatus,
family Gekkonidae

8 Stellate knob-tail
Nephrurus stellatus, family Gekkonidae

9 Northern leaf-tailed gecko
Saltuarius cornutus, family Gekkonidae

Tail is flattened

Two-thirds of
length is tail

Enormous eyes,
bulbous head,
short tail

1 **Lined gecko**
Gekko vittatus, family Gekkonidae

2 **Henkel's flat-tailed gecko**
Uroplatus henkeli, family Gekkonidae

3 **Tokay gecko**
Gekko gecko, family Gekkonidae

4 **Northern spiny-tailed gecko**
Diplodactylus ciliaris, family Gekkonidae

5 **Marbled velvet gecko**
Oedura marmorata, family Gekkonidae

6 **Marble-faced worm lizard**
Delma australus, family Pygopodidae

7 **Burton's snake-lizard**
Lialis burtonis, family Pygopodidae

8 **Gray's bow-fingered gecko**
Cyrtodactylus pulchellus,
family Gekkonidae

9 **Common wonder gecko**
Teratoscincus scincus,
family Gekkonidae

FACT FILE

Lined gecko This gekko has a distinct white dorsal stripe. The tail is cross-banded. They are nocturnal and have large distinct laminae on the toes for climbing. Males have a V-shaped row of enlarged pre-anal pores and hemi-penal bulges at the base of the tail.

✱ Up to 10 in (25 cm)
⬭ Terrestrial
◯ Oviparous
● 2
🍗 Common

Indo-Australian archipelago

Gray's bow-fingered gecko This gecko has a flattened, compact body. The toes are long and thin, bending upward and at the ultimate phalange turning downwards, terminating with small lamellae. Granular dorsal scales become spine-like on the tail.

✱ Up to 8 in (20 cm)
⬭ Terrestrial
◯ Oviparous
● 2
🍗 Common

Central Asia, S.E. Asia

Family Pygopodidae This family is closely related to the geckos. Forelimbs are lacking, and the hindlimbs are represented by a scaly flap just anterior to the cloaca. Their tails are fragile and easily broken. The eyes are snake-like, as they lack lids and are covered with an immovable spectacle. Some species have external ear openings. They breed in summer and most lay two eggs. The majority are insectivorous.

Rudimentary feet
The hindlimb of the Burton's snake lizard has been reduced to a flap-like scale, useful for traction in flowing sand.

Genera 8
Species 36

Australia except Tasmania, Aru Islands, New Guinea, New Briton

---

🔖 CONSERVATION WATCH

The seven species of Pygopodidae listed on the IUCN Red List are categorized as follows:

6 Vulnerable

1 Near threatened

FACT FILE

Mourning gecko This arboreal gecko is covered with small scales. The tail is long and depressed with a lateral fringe of small, spinose scales.

- ⟷ Up to 2 in (5 cm)
- 🌲 Arboreal
- ○ Oviparous
- ● 2
- 🗡 Common

N.E. Australia, Malaysia to Oceania

Common wall gecko This gecko is strong and heavily built, with rows of keeled, tubercular scales. Males are territorial. The eggs take 10 weeks to hatch and the young 2 years to mature.

- ⟷ Up to 6 in (15 cm)
- ◠ Arboreal
- ○ Oviparous
- ● 2
- 🗡 Common

Mediterranean & S. European coast

Kuhl's flying gecko This terrestrial species lays clutches of two eggs about 30 days apart. The eggs attach to bark or rocks and hatch in about 60 days.

- ⟷ Up to 6 in (15 cm)
- 🌲 Arboreal
- ○ Oviparous
- ● 2
- 🗡 Common

S.E. Asia

FLYING GECKO

At high speeds, the cutaneous flaps allow flying geckos to slow down and achieve shallower glide angles. The webbed feet function in gliding and their aerial locomotion is more similar to "flying" tree frogs than to other reptilian gliders.

Parachuting
The skin flaps along the body and tail flare out in flight and allow the Kuhl's flying gecko to break its fall.

Broad-tailed day gecko is native to Madagascar but is well established on Hawaii

1 Madagascar day gecko
Phelsuma madagascariensis,
family Gekkonidae

2 Broad-tailed day gecko
Phelsuma laticauda, family Gekkonidae

3 Mourning gecko
Lepidodactylus lugubris,
family Gekkonidae

4 Common wall gecko
Tarentola mauritanica,
family Gekkonidae

5 Israeli fan-fingered gecko
Ptyodactylus puiseuxi, family Gekkonidae

6 Kuhl's flying gecko
Ptychozoon kuhli, family Gekkonidae

7 Anderson's short-fingered gecko
Stenodactylus petrii, family Gekkonidae

8 Ruppell's leaf-toed gecko
Hemidactylus flaviviridis,
family Gekkonidae

9 Cradock thick-toed gecko
Pachydactylus geitje, family Gekkonidae

10 Wiegmann's striped gecko
Gonatodes vittatus, family Gekkonidae

Long-tail whip lizard is more than two-thirds tail; the tail will break off, if grasped by a predator, so that the lizard can escape

1 **Rough-scaled plated lizard**
Gerrhosaurus major,
family Gerrhosauridae

2 **Madagascar girdled lizard**
Zonosaurus madagascariensis,
family Gerrhosauridae

3 **Long-tail whip lizard**
Tetradactylus tetradactylus,
family Gerrhosauridae

4 **Lesser flat lizard**
Platysaurus guttatus, family Cordylidae

5 **Karoo girdled lizard**
Cordylus polyzonus, family Cordylidae

6 **Black crag lizard**
Pseudocordylus melanotus,
family Cordylidae

7 **Mole skink**
Eumeces egregius, family Scincidae

8 **Schneider's skink**
Novoeumeces schneideri, family Scincidae

9 **Sand fish**
Scincus scincus, family Scincidae

FACT FILE

Family Scincidae Skinks are usually covered with overlapping smooth scales. The adults vary from 1 to 14 inches (2.5 to 35 cm) in length. The body shape varies from robust species to species with no external limbs. Limbless species are usually burrowers and the rest are terrestrial or arboreal. Tails are long to moderately long. Tail autonomy is present in most species. Skinks are active foragers, using both visual and olfactory cues to find prey. Most species lay eggs, but viviparity evolved in the family at least 25 times.

Genera 116
Species 993

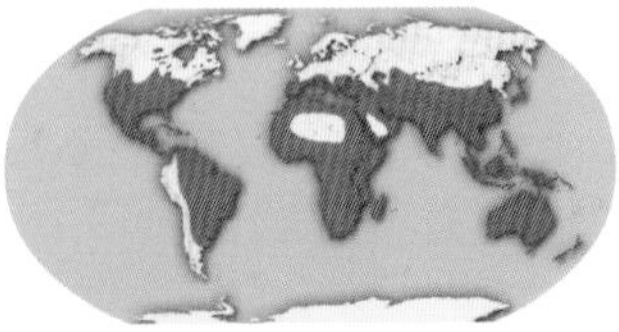

Cosmopolitan in most terrestrial habitats

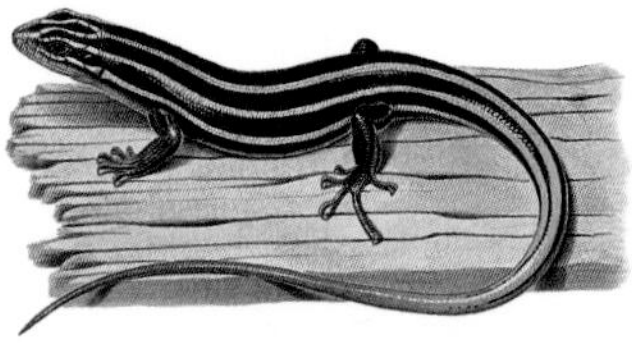

Telltale tails *Territorial male five-lined skinks (Eumeces fasciatus) recognize females and juveniles by their bright blue tails and allow them access to their territories.*

Hosmer's spiny-tailed skink The base of this skink's tail lacks enlarged and expanded scales, and long, rugose ear lobules almost conceal the ear. This is a diurnal rock-dwelling lizard found in rock outcrops and stony hillsides, where it lives in crevices, under boulders, and on tumbled rock slides.

- Up to 7 in (18 cm)
- Terrestrial
- Oviparous
- 2
- Common

N.E. Australia

Centralian blue-tongued lizard This lizard has a bright blue tongue, relatively short five-toed limbs, and a short tail. It thrives in stony areas in desert and semidesert areas where it feeds on insects and other invertebrates. The tail is about half the body length.

- Up to 12 in (30 cm)
- Terrestrial
- Viviparous
- 10 live young
- Common

N. & N.W. Australia

1
Solomon Island skink can grasp a branch and hang by its prehensile tail
2
3
4
5
6
7

1 **Solomon Island skink**
Corucia zebrata, family Scincidae

2 **Emerald skink**
Dasia smaragdina, family Scincidae

3 **Hosmer's spiny-tailed skink**
Egernia hosmeri, family Scincidae

4 **Centralian blue-tongued lizard**
Tiliqua multifasciata, family Scincidae

5 **Shingleback lizard**
Tiliqua rugosa, family Scincidae

6 **Three-lined burrowing skink**
Androngo trivittatus, family Scincidae

7 **Otago skink**
Oligosoma otagense, family Scincidae

1 **Bridled mabuya**
   Mabuya vittata, family Scincidae

2 **Christmas Island grass skink**
   Lygosoma bowringii, family Scincidae

3 **Boulenger's legless skink**
   Typhlosaurus vermis, family Scincidae

4 **Red-sided ctenotus**
   Ctenotus pulchellus, family Scincidae

5 **Northwestern sandslider**
   Lerista bipes, family Scincidae

6 **Cape York mulch skink**
   Glaphyromorphus crassicaudum,
   family Scincidae

7 **Lined fire-tailed skink**
   Morethia ruficauda, family Scincidae

8 **Juniper skink**
   Ablepharus kitaibelii, family Scincidae

9 **Desert rainbow skink**
   Carlia triacantha, family Scincidae

10 **Six-lined burrowing skink**
   Scelotes sexlineatus, family Scincidae

LIZARD COURTSHIP

Courtship displays have evolved in lizards to ensure that gametes are not wasted on mating with the wrong species or inferior individuals. Males often maintain territories in resource-rich areas and will fight other males to maintain them. Females are allowed to enter males' territories and are encouraged to mate by energetic courtship displays. When there are many similar species or species in the same genus in the same habitat the courtship behavioral patterns are more complex. Courtship displays are most complex and have been most intensively studied in the family Polychrotidae. Courtship in monitor lizards (Varanidae) involves tongue flicks by males along different parts of the female's body during the initiation of courtship. Communication in skinks is primarily chemical.

Anole courtship displays Small, colorful arboreal lizards are most adapted to using visual communication for courtship. The males have species-specific, bright dewlap colors and head bobs, push-ups, or other motions that attract females to them for copulation. Variation exists between individuals, so that females can select a male by the way he courts her. The male's vigor in courtship is the component most often used by females in their selection of a mate.

Bobbing rapidly *The carpenter anole (Norops carpenteri) (above) has a simple display: the dewlap, which is uniformly orange-colored, is extended and held open, while the head is bobbed at regular frequency. The male has a series of over a dozen rapid head bobs in his courtship display, while holding his dewlap extended the entire time.*

Variable bobbing *The silky anole (N. sericeus) (above) has a compound display: the dewlap has a central color that contrasts with the basal color. When the dewlap is extended only partway, the spot does not show. The head bob pattern is independent of the dewlap extension rhythm. The male begins bobbing rapidly for the first five bobs, then bobs in slow motion for the remainder of the sequence; the dewlap is extended and retracted twice.*

Slow bobbing *The display of the lichen anole (N. pentaprion) (above) is complex. The dewlap has a red basal color with blue lines: the number of blue lines exposed varies according to the extension of the dewlap. The slow bobbing sequence is synchronized with the extension of the dewlap—bob and extend, retract, bob and extend.*

Larger than life *When the frill is extended fully, it can reach 14 inches (35 cm) across, longer than its body length.*

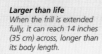

Rain-maker *Australian Aborigines will not hunt the frilled lizard because they believe that it calls the rain to come; without rain, the country falls into drought.*

Pseudo-copulation In the whiptails (Cnemidophorus) of the southwestern deserts of the United States, females reproduce without males or their sperm (parthenogenetic reproduction). The genetic component of the female is identical to that of her offspring. Pseudo-copulation occurs in some species of these clones of all female lizards. One female behaves like a male and attempts to mount with another female. The courtship and pseudo-copulation stimulate ovulation.

Larger than life *By expanding its folded neck skin, the frilled dragon can appear much larger when threatening other males or attempting to attract females.*

Expanding frill Dominant male frilled lizards show off different display sequences during courtship of females or territoriality displays toward other males. Over 75 different sequences have been recorded, including head bobbing, push-ups, beard erection, color changes, body inflations, head licking, and jaw gaping. Apart from the changes in shape and size of the lizard, changes in color and flashes of color as they open the mouth are a major part of their repertoire.

1 European green lizard
Ablepharus kitaibelii, family Lacertidae

2 Giant Canary Island lizard
Gallotia stehlini, family Lacertidae

3 Zagrosian lizard
Timon princeps, family Lacertidae

4 Tiger lizard
Nucras tessellata, family Lacertidae

5 Jeweled lizard
Timon lepidus, family Lacertidae

6 Sand lizard
Lacerta agilis, family Lacertidae

7 Gallot's lizard
Gallotia galloti, family Lacertidae

8 Balkan emerald lizard
Lacerta trilineata, family Lacertidae

FACT FILE

Family Xantusiidae Even though
Xantusidae are called night lizards,
many species are diurnal. They are
small lizards, up to 4 inches (10 cm)
long, with small granular dorsal scales
and large ventral scales. All species
are viviparous. Most species feed
on invertebrates, although one cave-
dwelling species feeds on figs. Night
lizards are terrestrial and arboreal. In
Mexico, Lepidophyma, a species of
night lizard, are called scorpions and
are mistakenly thought to be venomous.
Like many lizards, their tails will break
off if grasped; a new tail will regenerate
but will be of a uniform color.

Liberated from males *The yellow-
spotted night lizard (Lepidophyma
flavimaculata), is nocturnal and
viviparous. Some populations are
parthenogenetic.*

Genera 3
Species 20

W. USA & E. Mexico,
Central America &
N. South America

Common wall lizard This is a colonial
species. It is a good climber and basks
on rocks. Wall lizards feed on invertebrates: beetles, flies, butterflies, and
spiders. They are prey items for raptors
and snakes, particularly young horned
vipers or large whip snakes.

⚘ Up to 4 in (10 cm)
🌐 Terrestrial, arboreal
○ Oviparous
● 2–8
🍃 Common

Southern Europe & Balkans

🍃 CONSERVATION WATCH

Restricted range One species of
Xantusiidae, the North American
island night lizard (*Xantusia
riversiana*), is listed on the IUCN Red
List as vulnerable. The distribution
of this lizard is restricted to San
Clemente, Santa Barbara, and the
San Nicholas Islands off the coast of
southern California, United States.
They inhabit grasslands, cactus
clumps, cliffs, and rocky beaches.
This lizard is vulnerable because
of habitat destruction and development of its island habitats.

1	Italian wall lizard *Podarcis sicula*, family Lacertidae	**4**	Desert night lizard *Xantusia vigilis*, family Xantusiidae	**7**	Radd's rock lizard *Lacerta raddei*, family Xantusiidae
2	Common wall lizard *Podarcis muralis*, family Lacertidae	**5**	Granite night lizard *Xantusia henshawi*, family Xantusiidae	**8**	Uzzell's rock lizard *Lacerta uzzeli*, family Xantusiidae
3	Menorca wall lizard *Podarcis perspicillata*, family Lacertidae	**6**	Viviporus lizard *Lacerta vivipara*, family Xantusiidae	**9**	Milo's wall lizard *Podarcis milensis*, family Xantusiidae

FACT FILE

Egyptian fringe-fingered lizard
The toes on these lizards have a fringe enabling better traction over windblown sand dunes. They also have shovel noses and countersunk lower jaws to allow them ease in plowing through sand. Ants make up most of their diet.

✦ Up to 8 in (20 cm)
⬭ Terrestrial
○ Oviparous
● 3–5
⚑ Common

Algeria, Egypt, Israel, Jordan & Libya

Blue-throated keeled-lizard These diurnal lizards are the smallest lacertids in Europe, and forage for insects in vineyards and buildings. They emerge from hibernation and breed in April; breeding males have a bright blue throat and orange-red ventral surface.

✦ Up to 8 in (20 cm)
⊛ Terrestrial, arboreal
○ Oviparous
● 2–3
⚑ Common

N.E. Italy to Gulf of Corinth & Ionian islands

Sawtail lizard Its flattened shape and striking coloration make this species unmistakable. This lizard uses its flat body and broad tail to glide through the air from tree to tree. They have long, stout trunks and tails and relatively strong hindlimbs.

Long jumper
The sawtail lizard jumps up to 33 feet (10 m) from tree to tree, its flattened body and tail assisting the feat.

✦ Up to 5 in (12.5 cm)
⊕ Arboreal
○ Oviparous
● 2
⚑ Common

C., E. & W. Africa & Mozambique

Snake-eyed lizard This lizard gets its name from the large transparent disks that cover its eyes, so large that the eyelids no longer cover the eye. When escaping from a predator, they scamper from bush to bush and then pop up to look for the aggressor. They can also run short distances bipedally.

✦ Up to 2 in (5 cm)
⬭ Terrestrial
○ Oviparous
● 4–5
⚑ Common

N. Africa, S.E. Europe to Middle East & India

1 **Egyptian fringe-fingered lizard**
Acanthodactylus pardalis,
family Lacertidae

2 **Rapid racerunner**
Eremias velox, family Lacertidae

3 **Algerian psammodromus**
Psammodromus algirus,
family Lacertidae

4 **Blue-throated keeled-lizard**
Algyroides nigropunctatus,
family Lacertidae

5 **Small-spotted lizard**
Mesalina guttulata, family Lacertidae

6 **Asian grass lizard**
Takydromus sexlineatus,
family Lacertidae

7 **Sawtail lizard**
Holaspis guentheri, family Lacertidae

8 **Mourning racerunner**
Heliobolus lugubris, family Lacertidae

9 **Slender sand lizard**
Meroles anchietae, family Lacertidae

10 **Snake-eyed lizard**
Ophisops elegans, family Lacertidae

FACT FILE

Family Gymnophthalmidae The microtiids are small, oviparous lizards, up to 2½ inches (6 cm) long, with a wide variation of scale patterns. Some species have reduced limbs. Tails are autonomous. Most of the lizards in this diverse family live within the forest-floor litter. A few species are semiaquatic. All species consume invertebrates.

Genera 36
Species 160

S. Central America & N.W. South America

Family Teiidae Whiptail lizards and tegus, which are oviparous, vary in adult size from 2 to 16 inches (5 to 40 cm) in snout–vent length. Dorsal and lateral scales are small and granular, while the ventral scales are larger, rectangular, and in rows. The tails are long and autonomic. Smaller whiptails prefer deserts, grasslands, and open areas in the forest where they prey on invertebrates. The larger tegus prefer open forests and are omnivorous.

Sexless clones
The checkered whiptail (Cnemidophorus tesselatus) reproduces asexually; only females are present in the population.

Genera 9
Species 118

N. USA to South America

Golden tegu This heavy-bodied lizard is one of the main predators of turtle eggs in the Amazon Basin. To protect their own eggs from predation, they lay them in arborial termite nests.

✳ Up to 12 in (30 cm)
◗ Terrestrial
○ Oviparous
● 4–32
⚑ Common

N. South America

Heliothermic species that feeds on ants and spiders in the leaf litter

1   **Ocellated tegu**
Cercosaura ocellata,
family Gymnophthalmidae

2   **Thomas's bachia**
Bachia panoplia,
family Gymnophthalmidae

3   **Two-ridged neusticurus**
Neusticurus bicarinatus,
family Gymnophthalmidae

4   **Rainbow lizard**
Cnemidophorus lemniscatus,
family Teiidae

5   **Green calango**
Kentropyx calcarata, family Teiidae

6   **Four-toed tegu**
Teius teyou, family Teiidae

7   **Slow worm**
Anguis fragilis, family Anguidae

8   **Common worm lizard**
Ophiodes intermedius, family Anguidae

9   **Southern alligator lizard**
Elgaria multicarinata, family Anguidae

FACT FILE

Family Xenosauridae The xenosaurian or knob-scaled lizards are covered with granular scales and large tubercules. These insectivorous lizards inhabit mountainous tropical rain forests in southern Mexico and Guatemala, where they live in rock crevices or tree holes. The Chinese crocodile lizard, their widely disjunct relative, forages for fishes and tadpoles in mountain streams.

Genera 2
Species 5

S. China & S. Mexico & Guatemala

Family Anguidae Anguids have heavily armored scales with underlying osteoderms. Most have limbs, but a large number of them are limbless with long tails that are two-thirds of their total length. They are commonly called glass snakes, because when attacked by a predator the tail breaks into several pieces.

Prehensile tails
Cope's arboreal alligator lizard (Abronia aurita) is adapted for living high in the canopy of tropical rain forests; they hold on with their prehensile tails.

Genera 13
Species 101

S. Asia, S.W. Asia, Europe & Americas

Lives in the low branches of trees above the water; the tail is high and flattened dorso-laterally for swimming

1 Giant ameiva
Ameiva ameiva, family Teiidae

2 Banded galliwasp
Diploglossus fasciatus, family Anguidae

3 Chinese crocodile lizard
Shinisaurus crocodilurus,
family Xenosauridae

4 Golden tegu
Tupinambis teguixin, family Teiidae

5 European glass lizard
Pseudopus apodus, family Anguidae

6 Slender glass lizard
Ophisaurus attenuatus
family Xenosauridae

7 Crocodile tegu
Crocodilurus lacertinus, family Teiidae

8 Paraguay Caiman lizard
Dracaena paraguayensis

⚡ CONSERVATION WATCH

The 10 species of Anguidae listed on the IUCN Red List are categorized as follows:

1 Extinct
3 Critically endangered
1 Endangered
1 Vulnerable
1 Near threatened
3 Data deficient

FACT FILE

Family Helodermatidae The Gila monster and the Mexican beaded lizard are the only known venomous lizards. They have broad, flattened heads, and stocky bodies, strong limbs, and thick tails that are used for fat storage. Both species are slow-moving, oviparous, and diurnal. They feed on nestlings and lizard or birds' eggs.

Genera 2
Species 2

S.W. USA to Guatemala

Family Lanthanotidae These earless monitors have ptergoid teeth and lack a parietal eye. They are related to the Varanidae and Helodermatidae. The species is nocturnal and semi-aquatic, living along and in forest streams. Monitors forage in the water and on land for small vertebrates and invertebrates. They are oviparous with clutches of up to six eggs.

Genera 1
Species 1

Borneo

SHEDDING SKIN

The outer layer of a lizard's skin (the epidermis) is composed of keratin; scales are thickenings of this layer. As the lizard grows, the outer layer of keratin is shed in large flakes.

Splitting out
All lizards need to shed their skin periodically in order to grow.

⚡ CONSERVATION WATCH

Added protection Both species of Helodermatidae are listed on the IUCN Red List as vulnerable. The Gila monster is given complete protection in Arizona, United States, to keep it from wanton slaughter and being engulfed by the pet trade. Mexico lists both species, so the lizards cannot be legally sold on the international pet market and cannot be killed, stuffed, and then sold as curios for the tourist trade.

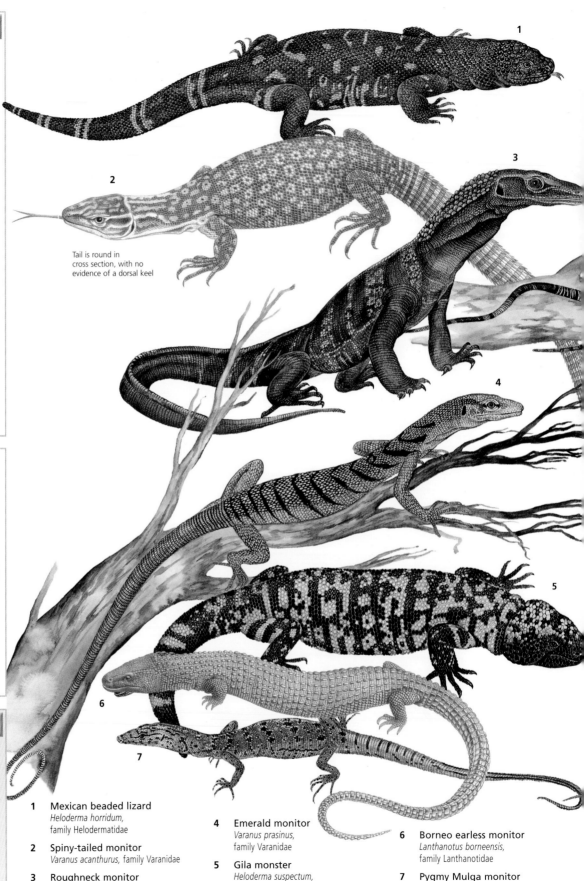

Tail is round in cross section, with no evidence of a dorsal keel

1. **Mexican beaded lizard**
   Heloderma horridum,
   family Helodermatidae

2. **Spiny-tailed monitor**
   Varanus acanthurus, family Varanidae

3. **Roughneck monitor**
   Varanus rudicollis, family Varanidae

4. **Emerald monitor**
   Varanus prasinus,
   family Varanidae

5. **Gila monster**
   Heloderma suspectum,
   family Helodermatidae

6. **Borneo earless monitor**
   Lanthanotus borneensis,
   family Lanthanotidae

7. **Pygmy Mulga monitor**
   Varanus gilleni, family Varanidae

JACOBSON'S ORGAN

A large number of lizards sense the air around them with their tongues, picking up chemical cues regarding food, conspecifics, likely mates, and predators. The Jacobson's organ has a pair of cavities lined with sensory cells on the roof of the mouth. The particles collected on the lizard's tongue are transported to ducts leading to these cavities by the tongue, explaining why the tongue is forked. The vomeronasal nerve then transmits the information collected to the brain for action.

Tongue testing *Monitors and many other lizards pick up airborne scents on their forked tongues.*

Komodo dragons have an extremely virulent symbiotic bacteria living in their mouths; within a few hours of being bitten, wounded prey will lose energy due to a fever caused by the bacteria, then drop to the ground. The dragon trails the animal and eats it when it is down

1 Crocodile monitor
Varanus salvadorii, family Varanidae

2 Perentie
Varanus giganteus, family Varanidae

3 Gray's monitor
Varanus olivaceus, family Varanidae

4 Desert monitor
Varanus griseus, family Varanidae

5 Sand monitor
Varanus gouldii, family Varanidae

6 Nile monitor
Varanus niloticus, family Varanidae

7 Komodo dragon
Varanus komodoensis, family Varanidae

SNAKES

CLASS	Reptilia
ORDERS	Squamata
FAMILIES	17
GENERA	438
SPECIES	2,955

At nearly 3,000 species, the variety in snakes is enormous. They range from the tiny burrowing blind snakes of a few inches (10 cm) long to huge constrictors over 33 feet (10 m) in length. Snakes have evolved special methods of locomotion, different ways of gathering environmental cues, and diverse venom delivery systems. Lizards appeared in the fossil record before snakes, and the presence of vestigial pelvic girdles and spurs in some primitive snakes indicates that snakes have evolved from them; the left lung is reduced or absent in snakes and legless lizards, but the right lung is reduced in amphisbaenians. Most of the organs of snakes are reduced in girth and elongated.

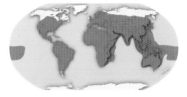

Tropic lovers Snakes are found on all continents except Antarctica, and ranges of some even pass above the Arctic circle. Some islands are also devoid of snakes, such as New Zealand, Ireland, and Iceland, and sea snakes are absent from the Atlantic. Snake biodiversity is highest in the tropics.

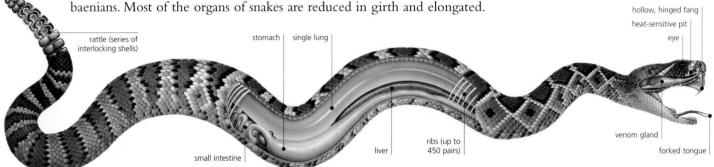

rattle (series of interlocking shells) · stomach · single lung · hollow, hinged fang · heat-sensitive pit · eye · venom gland · forked tongue · small intestine · liver · ribs (up to 450 pairs)

Snake anatomy As in the example of the Western diamondback rattlesnake (above), most snake organs have been reduced in diameter and have increased in length. Lungs in most species have been reduced to one, but males have two functional reproductive organs.

Defensive display Most species of cobras, such as the Cape cobra (*Naja nivea*) (right), have a hood that they can spread to look larger than life in order to frighten enemies or to transfix prey. The dorsal surface of the hood of some species is marked with large eye-like spots to suggest the eyes of a much larger animal.

Hidden fangs When pit vipers, including timber rattlesnakes (below), threaten with an open mouth they keep their fangs relaxed, so they do not obstruct the functioning of their pit organs. These scan for infrared signals from predators or prey.

AN EAR TO THE GROUND

Snakes are very close to their environment. In fact, their bellies are in contact with the substrate most of the time—only when they are gliding between trees, swimming, or climbing trees are they free from this close contact with the ground. It is through this earthy contact that they pick up sound vibrations and follow the scent trails of prey and receptive females.

The shape of the snake determines where it will live and what it will do. Shovel-nosed snakes and worm snakes remain within the ground with modified snouts for burrowing and feeding on their subterranean prey. The slender, thin-necked, arboreal snakes are adapted for gliding through the branches and stretching out to reach the next branch only a nose away. They have small clutches of elongate eggs and eat thin, elongate prey, such as lizards or frogs. Short, thick-bodied snakes, like some vipers, are sit-and-wait predators and often gorge on a meal that weighs more than they do. Their awkward shape does not affect them, as they do not move much anyway, and many are venomous. Long-bodied, agile, fast-moving snakes are the athletes in the group, chasing down lizards and other snakes, and using their speed for escape. It is amazing how fast mambas and whip snakes can move

SNAKE SKIN TYPES

Snake skin is made up of scales, which are not separate items, but a thickened part of the skin connected to other scales by thinner areas of elastic skin. Fine, granular scales are common in boas, the smooth, glossy over-lapping scales are common in racers, and strongly keeled scales are characteristic of rattlesnakes.

Granular scales

Smooth scales

Keeled scales

Poised to strike
This young green tree python is coiled, half in the air, ready to strike.

venom duct

rear grooved fangs

Colubridae

Fang varieties Rear grooved fangs are associated with Duvernoy's gland secretions in Colubridae. Elapidae (cobras, coral snakes, sea snakes, and taipans) have fixed short front fangs. The vipers all have hinged front fangs.

venom duct

fixed front hollow fangs

Elapidae

venom duct

swinging front hollow fangs

Viperidae

Infrared sensors
Green tree pythons have heat-sensing receptors embedded in the labial scales, which enable them to receive a visual image of light and heat.

Green constrictors The green tree python is an arboreal predator, feeding on birds, mammals, and lizards. Grasping a bird or bat on the wing with its long teeth, it simultaneously throws a coil around the prey and constricts it. The hatchlings of the green tree python are 11–14 inches (28–35 cm) long and are colored differently from the adults. The bright juvenile color pattern transforms to a vivid green in 6–8 months.

Keen vision The king cobra has large, round pupils adapted for acute diurnal vision. They are designed for detecting the slightest motion of their preferred prey—other snakes—within the vegetation.

through the forest canopy: they do not slither, but flow like a stream of water. Sea snakes have modified their shape to have a paddle-like tail, useful for a pelagic life. Many species never touch land, but feed, breed, and give birth at sea.

All male snakes have two hemipenes in the base of the tail; they are usually used alternately, each receiving sperm from only one of the testicles. This system perhaps evolved from the selective advantage of having a fast recovery time when participating in mass spring orgies following emergence from hibernation. Sperm can be stored for months or more than a year before it is utilized to fertilize eggs.

Females have evolved a range of modified behavior patterns to enhance their genes. Most snakes lay eggs, but many of the more recent snakes have evolved viviparity and some parental care. Through muscle

contractions, pythons can raise their body temperatures when brooding their eggs. Female cobras guard their nest of eggs in a loose construction of leaves built over the nest.

Some of the most fantastic adaptations of modern snakes are the mechanisms they have developed for finding, capturing, killing, and swallowing prey. Highly developed chemical receptors on the roof of the mouth help snakes to identify mates, enemies, and prey. Infrared receptors possessed by pit vipers, boas, and pythons allow them to visualize warm-blooded prey in terms of heat. The venom produced by Viperidae, Elapidae, and Colubridae allow them to stop prey in its tracks and help in swallowing and digesting the meal. Obviously some snakes have evolved more than others, and many are now losing ground—humans are changing the environment faster than snakes can evolve.

Hatchlings emerge Most pythons guard and incubate their eggs. However, once their first week outside the shell is over, the young shed their skins, leave the protective coil of the mother, and disperse, like the young green tree python below.

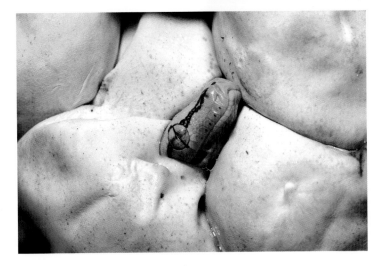

BOA AND PYTHON SKULLS

The kinetic jaws of snakes are designed to swallow objects up to five times their diameter. Teeth are curved so that by first moving one side and then the other, the food is pushed down the throat.

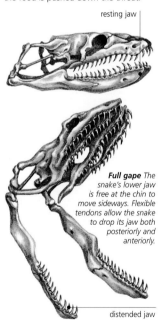

resting jaw

Full gape The snake's lower jaw is free at the chin to move sideways. Flexible tendons allow the snake to drop its jaw both posteriorly and anteriorly.

distended jaw

Male Mexican rosy boas have well-developed spurs; those of the female are smaller

Tatar sand boas are sold for high prices in central Asia for medicinal purposes

1   **Mexican rosy boa**
    Charina trivirgata, family Boidae

2   **Wood snake**
    Tropidophis melanurus,
    family Tropidophiidae

3   **New Guinea ground boa**
    Candoia aspera, family Boidae

4   **Calabar ground python**
    Calabaria reinhardtii, family Boidae

5   **Rubber boa**
    Charina bottae, family Boidae

6   **East African sand boa**
    Gongylophis colubrinu, family Boidae

7   **Schaefer's dwarf boa**
    Xenophidion schaeferi,
    family Tropidophiidae

8   **Isthmian dwarf boa**
    Ungaliophis continentalis,
    family Tropidophiidae

9   **Tatar sand boa**
    Eryx tataricus, family Boidae

10 **Feick's dwarf boa**
    Tropidophis feicki, family Tropidophiidae

Because of its gentle disposition and beautiful pattern, the Indian python has been prized in the pet trade for decades; now most individuals sold are captive-born

Heat-sensing pits are located in the labial scales

Tongue is used to collect airborne scents

1 Common boa constrictor
Boa constrictor, family Boidae

2 Indian python
Python molarus, family Boidae

3 Madagascar ground boa
Acrantophis madagascariensis, family Boidae

4 African rock python
Python sebae, family Boidae

5 Reticulate python
Python reticulatus, family Boidae

6 Anaconda
Eunectes murinus, family Boidae

7 Emerald tree boa
Corallus caninus, family Boidae

FACT FILE

Family Boidae This family includes the boas, pythons, and sand boas—the largest living snakes—but some species are less than 20 inches (50 cm) long. The largest may reach 33 feet (10 m) and feed on large mammals, capybaras, deer, and caimans. Boas and sand boas are viviparous; pythons are oviparous.

Egg brooding
The children's python (Antaresia childreni) coils around its eggs to conceal and protect them. It increases body temperature by muscle contractions.

Genera 20
Species 74

Worldwide except Europe and Antarctica

Common boa constrictor Boas are heavy-bodied arboreal snakes that kill their prey by constriction. They have vestigial legs as spurs on either side of the cloaca, which suggests that they are of ancient origin.

☀ Up to 14 ft (4.2 m)
⬤ Terrestrial, arboreal
⬤ Viviparous
● 30–50
⬤ Common

S. Mexico to Argentina

Reticulated python One of the two largest snakes in the world, this python is more massive than the anaconda. Its prey includes large reptiles, such as lizards and crocodilians, and medium to large mammals, even humans.

☀ Up to 33 ft (10 m)
⬤ Terrestrial
○ Oviparous
● 80–100
⬤ Common

S.E. Asia

⚡ CONSERVATION WATCH

The 10 species of Boidae on the IUCN Red List are categorized as follows:

1 Extinct
2 Endangered
4 Vulnerable
3 Near threatened

FACT FILE

Green tree python This nocturnal constrictor feeds on mammals and birds. It is an example of parallel evolution: the emerald tree boa (*Corallus caninus*) of South America evolved in the same habitat and is similar in color and form.

- ✂ Up to 6½ ft (2 m)
- 🦅 Arboreal
- ○ Oviparous
- ● 12–18
- ♟ Uncommon
- ▥

N.E. Australia, New Guinea

Polymorphic colors
The young differ in color from the bright-green adults; young are bright yellow to brick red when they hatch.

Cook's tree boa This boa feeds on small lizards when young, progressing to rodents and birds as it grows. Prey is ambushed: in one motion, the boa bites and wraps several coils of its body around its victim while suspended from a branch by its prehensile tail.

- ✂ Up to 6¼ ft (1.9 m)
- 🦅 Arboreal
- 🐾 Viviparous
- ● 30–80 live young
- ♟ Common
- ▥

Panama, N. South America, West Indies

Black-headed python This python is nocturnal in warmer months and diurnal in cooler months. Its diet consists of monitor lizards, frogs, birds, mammals, and snakes—even death adders. Dingos are its main predator.

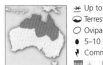

- ✂ Up to 8½ ft (2.6 m)
- ◠ Terrestrial
- ○ Oviparous
- ● 5–10
- ♟ Common
- ▥ ⋎ 🌵

N. Australia

Blood python Blood pythons are stocky, robust, and semiaquatic, spending much time underwater in swamps or streams. They feed primarily on mammals and small birds. Blood python populations are diminishing due to the skin trade: the colorful skin is prized for the luxury leather market.

- ✂ Up to 10 ft (3 m)
- 🌐 Terrestrial, aquatic
- ○ Oviparous
- ● 18–30
- ♟ Common
- ▥ ▱ ◉

S.E. Asia

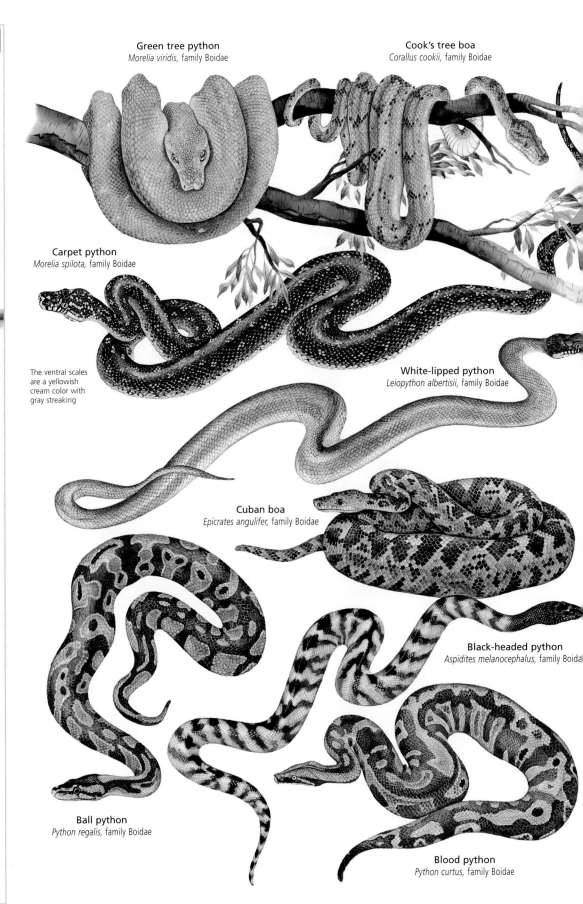

Green tree python
Morelia viridis, family Boidae

Cook's tree boa
Corallus cookii, family Boidae

Carpet python
Morelia spilota, family Boidae

The ventral scales are a yellowish cream color with gray streaking

White-lipped python
Leiopython albertisii, family Boidae

Cuban boa
Epicrates angulifer, family Boidae

Black-headed python
Aspidites melanocephalus, family Boidae

Ball python
Python regalis, family Boidae

Blood python
Python curtus, family Boidae

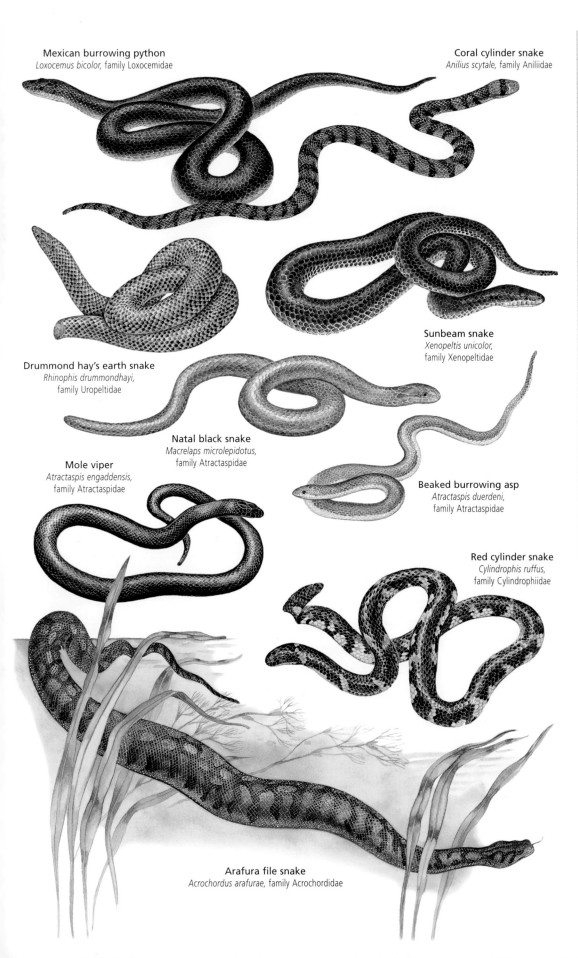

Mexican burrowing python
Loxocemus bicolor, family Loxocemidae

Coral cylinder snake
Anilius scytale, family Aniliidae

Drummond hay's earth snake
Rhinophis drummondhayi,
family Uropeltidae

Sunbeam snake
Xenopeltis unicolor,
family Xenopeltidae

Natal black snake
Macrelaps microlepidotus,
family Atractaspidae

Mole viper
Atractaspis engaddensis,
family Atractaspidae

Beaked burrowing asp
Atractaspis duerdeni,
family Atractaspidae

Red cylinder snake
Cylindrophis ruffus,
family Cylindrophiidae

Arafura file snake
Acrochordus arafurae, family Acrochordidae

FACT FILE

Mexican burrowing python This nocturnal python forages for small mammals and reptiles, and sea turtle and lizard eggs. Its pointed snout is useful for burrowing into reptile nests. The large plates on the head suggest those of more recent colubrid snakes, rather than the closely related boas.

Up to 4¼ ft (1.4 m)
Terrestrial, burrowing
Oviparous
2–4
Rare

S. Mexico to Costa Rica

Family Atractaspidae These oviparous snakes are called stiletto vipers because they have a large semi-erect fang on each maxillary bone (upper jaw). They prey on burrowing rodents: the tight spaces mean it is impossible for them to gape their mouths to strike, so they wiggle alongside their prey, shift their jaw to the opposite side from the prey, and expose their long fangs. They then proceed to envenomate their prey, stabbing it with a backslash motion.

Genera 1
Species 18

Africa & Middle East

Family Aniliidae These brightly colored coral-snake mimics live in leaf litter and feed on earthworms, caecilians, eels, amphisbaenians, and snakes. They are on occasion found in the water, but are probably burrowers. When handled, they roll up into a ball.

Genera 1
Species 1

Amazonia (South America)

Family Acrochordidae These snakes have small, strongly keeled scales. File snakes are one of the most aquatic snakes, and are almost incapable of moving on land. They are nocturnal, and feed on fishes and crustaceans.

Genera 1
Species 3

India, S.E. Asia, Australia

Paddle tail
*The laterally compressed tail of the little file snake (*Acrochordus granulatus*) is an adaptation for swimming.*

SNAKE DEFENSE STRATEGIES

Looping in alarm The bandy-bandy (*Vermiculla annulata*) is a nocturnal, venomous elapid snake. It throws its body into loops off the ground when it is alarmed. It is possible that the snake is attempting to appear larger than it is by doing this, or that it is striking in a new direction to catch the predator off guard.

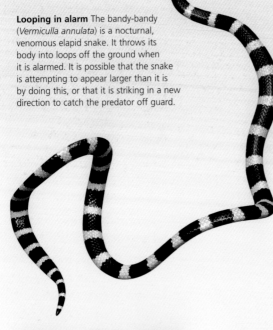

Snakes have many different strategies to avoid becoming a meal. Camouflage—utilizing color, pattern, and shape—is the most common tactic. Many arboreal snakes have long, slender heads and necks like twigs so that they blend in with their favored habitat. Escape works if you are very fast, like the red racer (*Masticophis flagellum*), which is built for speed. Many defensive maneuvers were designed to be effective against birds or mammals. The Indian cobra (*Naja naja*) raises half its body off the ground, spreads its hood, and displays large eye spots so that it appears to be a much larger animal. The Arizona coral snake (*Micruoides euryxanthus*) raises its tail and pops its cloaca to distract predators from attacking its head. The prairie ringneck snake (*Diadophis punctatus*) is black dorsally; when threatened, it hides its head in its coils and raises a coiled tail which is bright orange ventrally to startle its attacker. Many snakes will hiss, thrash, bite, and defecate to dissuade unwanted attention.

Rattling good tail When disturbed, a rattlesnake will vibrate its tail and make a loud rattling sound. The noise and the movement distract the potential prey or predators, who end up watching the tail and not the striking head. Many species of non-venomous snakes vibrate their tails as well.

Master of disguise The eyelash viper (*Bothriechis schlegelii*), an arboreal pit viper, is perfectly camouflaged. It chooses an appropriate perch on which to wait for passing mice or frogs. Its color and shape blend in with the vegetation so that predatory birds are not alerted to its presence. The "eyelashes" may help to protect its eyes from abrasion as it slips through the rain-forest canopy.

Surprise under the sand Waiting for a meal to pass, the Peringuey's sidewinding adder (*Bitis peringueyi*) (below) becomes invisible as the shifting sand of the Namibian desert blows over it. The adder's pattern and coloration blend in with the golden sand; only its eyes and the black tip of its tail remain visible. Hiding is a good defense against canids and raptors.

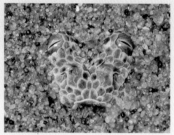

Mimicking danger Many non-venomous snakes have evolved color patterns similar to those of venomous species in their region, to reap the benefit this warning coloration has on predators. This adaptation is called Batesian mimicry. The long-tailed false coral snake (*Pliocercus elapoides*) of the Yucatán peninsula, Mexico, is one of these imitators (far right). It closely resembles the Mayan coral snake (*Micrurus hippocrepis*) (right).

Feigning death When molested, the Eastern hognose snake (above) hisses and strikes with a closed mouth. If this fails to deter a predator, the snake writhes and contorts its body; its mouth is open, and it discharges fetid cloacal fluids. After a few minutes it will roll onto its back and lie limply with its mouth open as though dead. The impact is lost if you turn the snake over, as it will immediately roll onto its back again.

Aging rattles The rattle grows one button each time the snake sheds its skin. This can be four to six times a year.

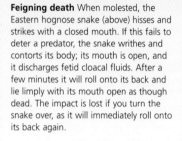

Snake in the grass When threatened, a bull snake (*Pituophis melanoleucus*) raises its head off the ground and expels air from its trachea; vibrating a cartilage at the tracheal opening creates a loud hissing sound. The tail is vibrated simultaneously, which produces a rattling sound in dry leaves or grasses. These effects combine to scare off most predators.

Threat display The venomous Florida cottonmouth (*Agkistrodon piscivorous*) opens its jaws to reveal a startlingly white mouth. This alerts predators to its presence and to the imminent danger. Sympatric nonvenomous water snakes have a similar color pattern and mimic this behavior for their defense.

FACT FILE

Family Colubridae About 63 percent of all species of snakes are included in this family, which utilizes all available reproductive styles and habitats. There are six loosely defined groups within the colubrids: Natricinae are small aquatic or terrestrial snakes; the largest group, Colubrinae, exploit all snake habitats except marine; the primarily oviparous Xenodontinae contains small, New World tropical and terrestrial species; Dipsadinae have diversified in Central and South America and occur in all available habitats; Homalopsinae contains 11 genera of diverse aquatic snakes occurring from northern Australia to Asia; and Aparallactinae are small snakes that range from sub-Saharan Africa to the Middle East.

Genera 320
Species 1,800

Worldwide, except Antarctica & polar regions

Snake trade
The brightly colored corn snake (Elaphe guttata) has been sought after in the pet trade for over 50 years because of its attractive colors and mild disposition. Captive breeding colonies are now providing most pet snakes sold today.

COLUBRID VENOM

People often ignore the fact that most species of colubrids produce venom in the Duvernoy's gland. The venom is released and flows along grooved erect fangs in the back of the mouth. Since the venom is not forced through hollow fangs under pressure, as in front-fanged venom delivery (common to vipers and elapids), only about half enters the wound.

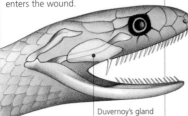

Duvernoy's gland

Non-lethal venom *When dispensed via a wound from the rear teeth, the venom from colubrids can be deadly. Herpetologists have died after under-estimating the venom's potency.*

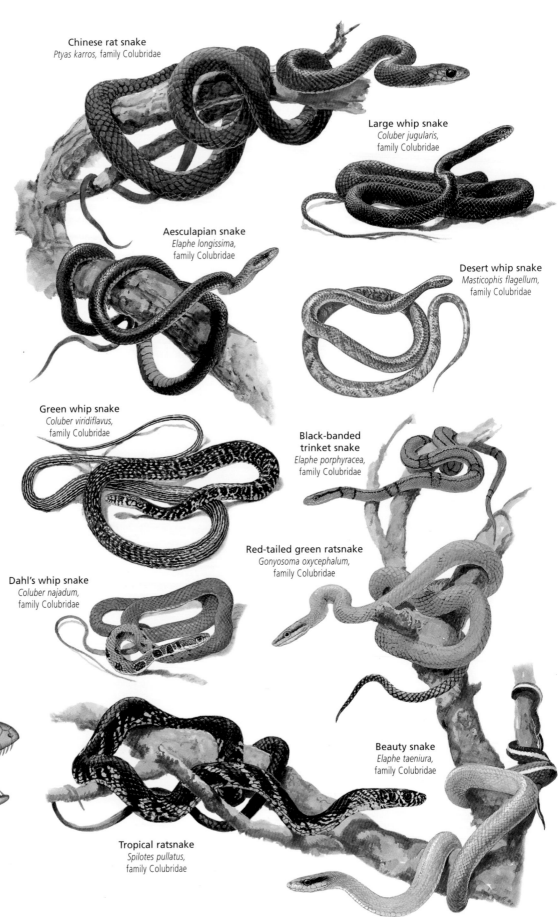

Chinese rat snake
Ptyas karros, family Colubridae

Large whip snake
Coluber jugularis, family Colubridae

Aesculapian snake
Elaphe longissima, family Colubridae

Desert whip snake
Masticophis flagellum, family Colubridae

Green whip snake
Coluber viridiflavus, family Colubridae

Black-banded trinket snake
Elaphe porphyracea, family Colubridae

Red-tailed green ratsnake
Gonyosoma oxycephalum, family Colubridae

Dahl's whip snake
Coluber najadum, family Colubridae

Beauty snake
Elaphe taeniura, family Colubridae

Tropical ratsnake
Spilotes pullatus, family Colubridae

Indigo snake
Drymarchon corais,
family Colubridae

Spotted bush snake
Philothamnus semivariegatus,
family Colubridae

Chinese slug snake
Pareas chinensis,
family Colubridae

Milk snake
Lampropeltis triangulum,
family Colubridae

Gray-banded king snake
Lampropeltis alterna,
family Colubridae

Aesculapian false coral snake
Erythrolamprus aesculapii,
family Colubridae

Milk snake
(juvenile color pattern)
Lampropeltis triangulum,
family Colubridae

Rhombic egg-eating snake
Dasypeltis scabra,
family Colubridae

Montpelier snake is rear-fanged
and has venom potentially
dangerous to humans

Schokari sand racer
Psammophis schokar,
family Colubridae

Montpelier snake
Malpolon monspessulanus,
family Colubridae

SPECIALIZED EGG-EATERS

African egg-eating snakes have evolved vertebrae with blunt spines that pierce the egg as it is forced down by muscle contraction. The contents are squeezed out and the eggshell regurgitated.

Egg room By not swallowing the eggshell, the snake has more space in its stomach for high-protein egg yolk.

Western ground snake This snake is secretive, inhabiting areas with loose alluvial sand in dry washes, creosote bushes, desert flats, and rocky hillsides. It prefers to hunt for prey at night and consumes centipedes, spiders, crickets, grasshoppers, and insect larvae.

- ↔ Up to 20 in (50 cm)
- ◗ Terrestrial
- ○ Oviparous
- ● Unknown
- ⚊ Uncommon

S.W. USA & N.W. Mexico

Common bronze-back snake Bronze-backs are primarily arboreal in rain forest, coconut plantations, and urban areas. Frogs and lizards, including flying lizards (Draco), are consumed.

- ↔ Up to 3¼ ft (1 m)
- ◗ Terrestrial, arboreal
- ○ Oviparous
- ● Unknown
- ⚊ Common

India, Burma, W. Malaysia, Indonesia, S. China

Ring-neck snake When attacked, these small woodland snakes hide their head and raise a coiled tail exposing the bright orange ventral surface. They feed on slugs, beetles, frogs, salamanders, and small snakes.

- ↔ Up to 28 in (71 cm)
- ◗ Terrestrial
- ○ Oviparous
- ● 1–7
- ⚊ Common

S.E. Canada, USA, N. Mexico

GARTER SNAKE

In Canada, garter snakes (*Thamnophis sirtalis*) hibernate en masse in deep crevices below the frost line. In spring, thousands emerge simultaneously to bask and breed. Males trail receptive females by the scent trails they lay down; up to a hundred males can be in a ball courting the same female. After copulating, the male leaves a gelatinous sperm plug in the female, giving his sperm time to fertilize the embryos before other males' sperm can enter.

Different fathers
DNA studies have shown that litters often have multiple paternities.

Striped kukri snake
Oligodon octolineatus,
family Colubridae

Striped kukri snake is so-named because the shape of its rear fangs resembles the kukri knife used by Gurkha soldiers

Western ground snake
Sonora semiannulata, family Colubridae

Common bronze-back snake
Dendrelaphis pictus,
family Colubridae

Red-sided garter snake
Thamnophis sirtalis,
family Colubridae

Laotian wolf snake
Lycodon laoensis, family Colubridae

Ring-neck snake
Diadophis punctatus,
family Colubridae

Hong Kong dwarf snake
Calamaria septemtrionalis,
family Colubridae

Asia Minor dwarf racer
Eirenis modestus,
family Colubridae

Crowned leaf-nosed snake
Lytorhynchus diadema, family Colubridae

New Guinea bockadam is tolerant of salt water

Rainbow water snake is important to the snake-skin leather industry

1   **Blue-necked keelback**
     Macropisthodon rhodomelas,
     family Colubridae

2   **Red-necked keelback**
     Rhabdophis subminiatus,
     family Colubridae

3   **European grass snake**
     Natrix natrix, family Colubridae

4   **Northern water snake**
     Nerodia sipedon, family Colubridae

5   **Queen snake**
     Regina septemvittata, family Colubridae

6   **Masked water snake**
     Homalopsis buccata, family Colubridae

7   **New Guinea bockadam**
     Cerberus rynchops, family Colubridae

8   **Tentacle snake**
     Erpeton tentaculatum,
     family Colubridae

9   **Rainbow water snake**
     Enhydris enhydris, family Colubridae

FACT FILE

European grass snake This is one of the few reptiles to live inside the arctic circle and above 7,000 feet (2,121 m). Frogs are one of their main dietary items. Declining frog populations mean grass snake numbers are falling.

Dead or alive?
When grass snakes are attacked, they inflate with air, hiss, bite, release a foul-smelling liquid from the cloaca, and writhe around with their mouth open before lying limp as if dead.

- ⟷ Up to 6½ ft (2 m)
- Terrestrial, aquatic
- ○ Oviparous
- ● 15–35
- ⚑ Common

Europe, W. Asia, N.W. Africa

Queen snake The queen snake is endangered in parts of its range due to habitat destruction and pollution. This water snake inhabits cool, clear streams, where it forages for newly molted crayfish under flat rocks.

- ⟷ Up to 36½ in (93 cm)
- Aquatic
- Ovoviviparous
- ● 5–23 live young
- ⚑ Status

S.E. Canada, E. USA

Masked water snake This nocturnal aquatic snake feeds on fishes and frogs. It has well-developed Duvernoy's glands and grooved, posterior maxillary teeth. Duvernoy's secretions are used to incapacitate prey animals, which makes them easier to eat.

- ⟷ Up to 4 ft (1.2 m)
- Aquatic
- Viviparous
- ● Unknown
- ⚑ Common

S.E. Asia

Rainbow water snake This tropical water snake feeds primarily on fresh-water fish and is not tolerant of salt water. It is the main aquatic snake in most wetland habitats in Southeast Asia, but is at risk from the intrusion of salt water, associated with El Niño, into its habitats.

- ⟷ Up to 32 in (81 cm)
- Aquatic
- Viviparous
- ● Unknown
- ⚑ Common

S.E. Asia

FACT FILE

Mangrove snake This rear-fanged snake has mildly potent venom, and is capable of killing humans. They are semiarboreal, feeding on small mammals, tree shrews, birds, and other snakes in mangrove swamps.

- ⟷ Up to 8¼ ft (2.5 m)
- Aquatic, arboreal
- ○ Oviparous
- ● 7–14
- ⚑ Common

S.E. Asia

Boomslang This venomous snake is the most dangerous colubrid. It is a rear-fanged snake, but its venom is more potent than that of cobras or vipers and causes internal bleeding. Boomslangs are active snakes that feed mainly on chameleons and other tree lizards, nestling birds, and eggs.

- ⟷ Up to 6 ft (1.8 m)
- Arboreal
- ○ Oviparous
- ● 10–25
- ⚑ Common

Sub-Saharan Africa

Cape twig snake This is a venomous rear-fanged colubrid, capable of inflicting fatal bites on humans. The venom causes internal bleeding. The twig snake has special cartilage in the throat that allows it to puff its neck up greatly in a threat display.

- ⟷ Up to 5½ ft (1.6 m)
- Arboreal
- ○ Oviparous
- ● 4–13
- ⚑ Common

Sub-Saharan Africa

BROWN VINE SNAKE

These diurnal snakes forage in the lower branches of neotropical forests, where they easily strike and capture passing ground-dwelling lizards. They also feed on arboreal lizards, frogs, small birds, and small mammals.

Look, no hands
The brown vine snake is a rear-fanged snake and kills its prey so that it is easier to manipulate and has less chance of dropping to the ground.

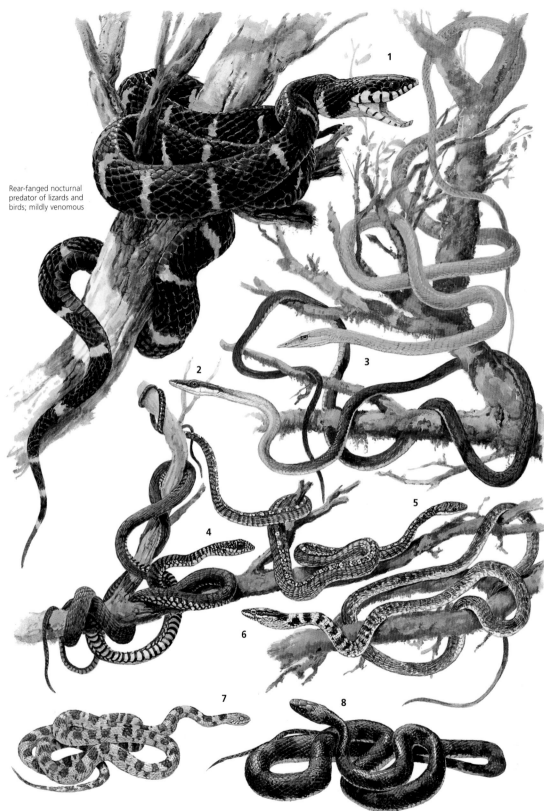

Rear-fanged nocturnal predator of lizards and birds; mildly venomous

1 **Mangrove snake**
Boiga dendrophila, family Colubridae

2 **Brown vine snake**
Oxybelis aeneus, family Colubridae

3 **Long-nosed tree snake**
Ahaetulla nasuta, family Colubridae

4 **Boomslang**
Dispholidus typus, family Colubridae

5 **Paradise flying snake**
Chrysopelea paradisi, family Colubridae

6 **Cape twig snake**
Thelotornis capensis, family Colubridae

7 **Mediterranean cat snake**
Telescopus fallax, family Colubridae

8 **Forest flame snake**
Oxyrhopus petola, family Colubridae

SNAKES AS PREDATORS

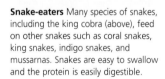

Heat vision Carpet pythons (Morelia spilota), with their labial heat-sensing pits, are able to detect and identify the size of a warm-blooded object before seeing it optically.

All snakes are carnivorous. The variety of prey sizes and taxa ranges from termites to crocodiles. Small snakes that feed on invertebrates search for their prey by sight or smell, often attacking them in their nests. Snakes use their tongues to collect scent data on where their prey has been and often trail them or wait for them to return; many of the large vipers and constricting snakes wait along rodent or game trails for prey to come to them. The large-eyed diurnal racers and whipsnakes are visual predators and chase down lizards and other snakes. The venom that vipers, elapids, and some colubrids inject to kill their prey also helps in the digestion of the animal, since venom is an enzyme. Many large snakes eat a meal only a few times a year, but sometimes this meal has a greater mass than they do.

Strike force The eyelash viper (Bothriechis schlegelii) (above) is an arboreal sit-and-wait predator, ready to rapidly identify prey through visual, infrared, and olfactory cues. It strikes on a moment's notice, as a bird, bat, lizard, frog, or in this case, rodent, passes by.

Snake-eaters Many species of snakes, including the king cobra (above), feed on other snakes such as coral snakes, king snakes, indigo snakes, and mussarnas. Snakes are easy to swallow and the protein is easily digestible.

Egg-eaters The egg-eater snake (Dasypeltis scabra) has a gland behind each eye that may supply extra fluids to the mouth to lubricate egg shells for swallowing. It can swallow an object about three times its diameter.

Gone fishing The mild venom of the Duvernoy's gland is sufficient to paralyze a fish. The grass snake (Natrix natrix) (above) can then more easily manipulate it for swallowing. All snakes swallow their prey whole.

Suffocation Many snakes, not just pythons, boas, and anacondas, constrict their prey to kill it. The amethyst python (Morelia amethistina) (left) captures prey in its teeth and simultaneously wraps several coils around it. The coils get tighter as the animal exhales, and it dies from suffocation. Normally no bones are broken in this process.

Catesby's snail eater These nocturnal snakes stalk snails and slugs. They extract the snail meat from the shell with their toothed mandible.

⬩ Up to 28 in (71 cm)
⬩ Terrestrial, arboreal
○ Oviparous
⬩ 1–5
⬩ Common

Amazonia (N. South America)

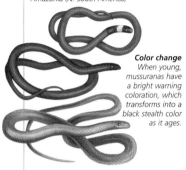

Color change *When young, mussuranas have a bright warning coloration, which transforms into a black stealth color as it ages.*

Amazon false ferdelance When alarmed, this snake will flatten its head and neck and distend its neck to make a hood. It is a rear-fanged snake, so the bite is potentially dangerous.

⬩ Up to 4 ft (1.2 m)
⬩ Terrestrial
○ Oviparous
⬩ 9–26
⬩ Common

Amazonia (N. South America)

False water cobra These snakes are not highly dangerous, but their bites have caused serious side-effects in humans, so they should be treated with care. Adults both bite and constrict mammals and birds, but swallow frogs and fishes directly.

⬩ Up to 9 ft (2.8 m)
⬩ Aquatic
○ Oviparous
⬩ 20–36
⬩ Common

N. South America

Catesby's snail eater
Dipsas catesbyi, family Colubridae

Mountain sipo
Chironius monticola,
family Colubridae

Amazon false
ferdelance
Xenodon severus,
family Colubridae

Blunthead tree snake
Imantodes cenchoa, family Colubridae

Mussurana
Clelia clelia, family Colubridae

Ringed hognose snake
Lystrophis semicinctus,
family Colubridae

Eastern hognose snake
Heterodon platyrhinos,
family Colubridae

Rainbow snake
Farancia erytrogramma, family Colubridae

False water cobra
Hydrodynastes gigas, family Colubridae

Egyptian cobra
Naja haje, family Elapidae

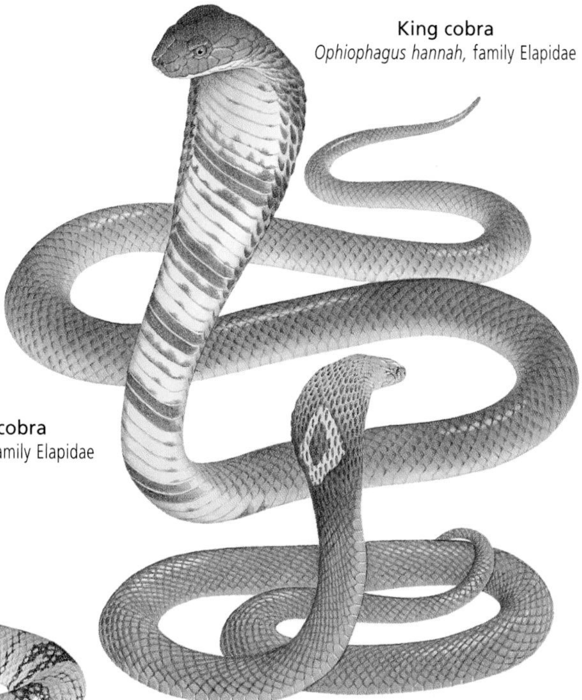

King cobra
Ophiophagus hannah, family Elapidae

Ringed water cobra
Boulengerina annulata, family Elapidae

Monocled cobra
Naja kaouthia, family Elapidae

Venom can cause blindness if not treated

Spitting cobra
Naja nigricollis, family Elapidae

Western green mamba
Debdroaspis viridis,
family Elapidae

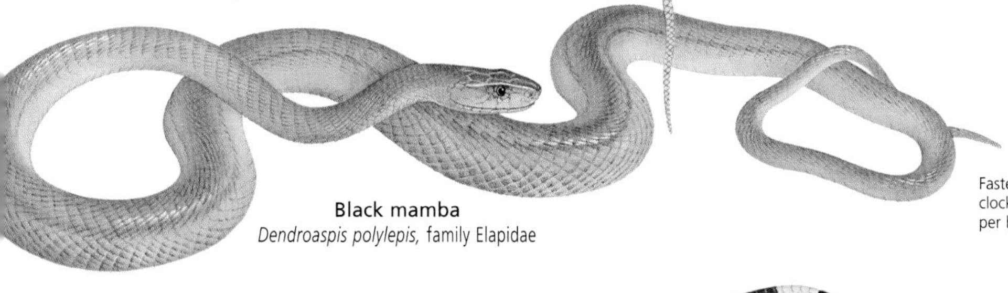

Fastest snake known: clocked at up to 12½ miles per hour (20 km/h)

Black mamba
Dendroaspis polylepis, family Elapidae

Blue Malaysian coral snake
Maticora bivirgata, family Elapidae

Banded krait
Bungarus fasciatus,
family Elapidae

FACT FILE

Family Elapidae This family comprises the cobras, kraits, seasnakes, coral snakes, death adders, and allies. Elapids are venomous, with anterior erect fangs on each maxilla. Only the mambas and tree cobras are arboreal. Many are burrowing or leaf-litter snakes and have bright coloration. Most terrestrial species are oviparous or viviparous. All true sea snakes are viviparous, giving birth in the water. The sea kraits lay eggs on land.

Genera 62
Species 300

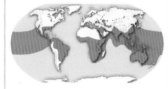

Worldwide except Antarctica & Atlantic

Spitting cobra Although this species usually does not bite, it can squirt its venom accurately up to 9 feet (2.6 m), aiming for its attacker's eyes.

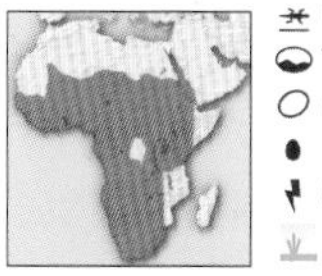

✴ Up to 7¼ ft (2.2 m)
◯ Terrestrial
◯ Oviparous
● 10–22
🗡 Common

Sub-Saharan Africa

Black mamba This large snake carries enough neuro-toxic venom to kill up to 10 adult humans. It is one of the fastest and most fearless snakes known. It can strike out to 40 percent of its length (most snakes only manage 25 percent).

✴ Up to 14 ft (4.2 m)
◐ Terrestrial, arboreal
◯ Oviparous
● 12–14
🗡 Common

África

SPITTING COBRA

Spitting cobras have an opening on the anterior surface of the fang above the tip. For better accuracy, the inside of the fangs have spiral grooves that put a spin on the projected venom.

Safe spitting
Spitting cobras avoid contact with potential predators by spitting venom in their eyes; this avoids injuries to their fangs and mouths.

Eastern coral snake Even though it has neuro-toxic venom, this snake is not very dangerous to humans because of its small size. Several non-venomous snakes mimic the color pattern.

✸ Up to 4 ft (1.2 m)
◯ Terrestrial
◯ Oviparous
● 3–13
♟ Common

S.E. USA & E. Mexico

Redtail coral snake Eggs are laid in January and hatch in March. Hatchlings are 6¾ inches (17 cm) long. Cope's false coral snake (*Pliocercus euryzona*) mimics the color pattern in Costa Rica.

✸ Up to 3½ ft (1.1 m)
◯ Terrestrial
◯ Oviparous
● 15–18
♟ Uncommon

S. Central America & N.W. South America

Common death adder These nocturnal snakes spend the day buried in sand or leaf litter, often at the base of trees or shrubs. Death adders have large fangs and toxic venom.

✸ Up to 3¼ ft (1 m)
◯ Terrestrial
◉ Ovoviparous
● 10–20
♟ Common

Australia & Papua New Guinea

SNAKE EYES

The habits of a snake can be seen in their eyes. Small degenerate eyes suggest burrowers. Vertical elliptical pupils are nocturnal species. Large round eyes are active diurnal predators that chase down their prey.

Sea snake status Crocker's sea snake (*Laticauda crockeri*) is listed on the IUCN Red List as vulnerable in Oceania. It is the only species of Hydrophiinae listed and the only species of sea snake to be found primarily in fresh water. It is vulnerable because of its limited distribution in the Solomon Islands.

Eastern coral snake
Micrurus fulvius, family Elapida

Arizona coral snake
Micruroides euryxanthus, family Elapidae

Redtail coral snake
Micrurus mipartitus, family Elapidae

Many-banded coral snake
Micrurus multifasciatus, family Elapidae

Loveridge's garter snake
Elapsoidea loveridgei,
family Elapidae

Loveridge's garter snake is nocturnal and secretive and feeds on lizards; although venomous, there are few recorded bites on humans

Cape coral snake
Aspidelaps lubricus, family Elapidae

Common death adder
Acanthophis antarcticus,
family Elapidae

Curl snake
Suta suta,
family Elapidae

Black-striped burrowing snake
Simoselaps calonotus,
family Elapidae

Taipan
Oxyuranus scutellatus, family Elapidae

Black tiger snake
Notechis ater, family Elapidae

Mulga
Pseudechis australis, family Elapidae

Blue-lipped sea krait
Laticauda laticauda, family Elapidae

Western brown snake
Pseudonaja nuchalis, family Elapidae

Olive-brown sea snake
Aipysurus laevis, family Elapidae

Yellow-belly sea snake
Pelamis platurus, family Elapidae

Turtle-headed sea snake
Emydocephalus annulatus,
family Elapidae

Ornate reef snake
Hydrophis ornatus, family Elapidae

Scales at center of
ornate reef snake's body
are hexagonal in shape

FACT FILE

Taipan The taipan is the world's deadliest snake, due to the toxicity of its venom. It inhabits wet tropical forests as well as dry forests and open savanna, and feeds on small mammals.

  Up to 6½ ft (2 m)
  Terrestrial
  Oviparous
  3–20
  Common

S. New Guinea, Indonesia & Australia

Blue-lipped sea krait Unlike other sea snakes, sea kraits are oviparous and lay their eggs on land. They do not have reduced ventral scales or a well-developed paddle-like tail. They often come to shore to digest large meals, another point of difference with other sea snakes, and possess muscles that are well developed for moving on land.

  Up to 3¼ ft (1 m)
  Aquatic
  Oviparous
  1–10
  Common

Indian & Pacific oceans

SEA SNAKES

Sea snakes were once grouped with Hydrophiidae, but are now part of the family Elapidae. They have many similar characteristics: all are venomous, and all except the sea krait (*Laticauda*) are viviparous and give birth at sea. All of these snakes have laterally compressed bodies, paddle-like tails, reduced ventral scales, and are incapable of moving well on land. There are 15 genera and 70 species distributed in Papua New Guinea, Australia, and tropical Pacific and Indian oceans.

No table *Sea snakes evolved a very potent neuro-toxic venom so that they could stun fishes rapidly, and then easily manipulate them into their mouths without dropping them to the ocean floor.*

⚑ CONSERVATION WATCH

The nine species of Elapidae on the IUCN Red List are categorized as follows:

  7 Vulnerable
  2 Near threatened

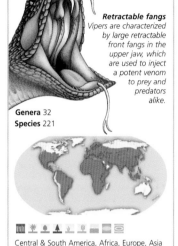

Retractable fangs
Vipers are characterized by large retractable front fangs in the upper jaw, which are used to inject a potent venom to prey and predators alike.

Genera 32
Species 221

Central & South America, Africa, Europe, Asia

SIDEWINDER

The sidewinder (*Crotalus cerastes*) moves in loose sand by sidewinding. The head and neck are held off the sand and thrown sideways while the body remains in place. As soon as the head and neck hit the sand, the body and tail are hurled after it. As the tail touches the ground, the head and neck become airborne again, resulting in a continuous looping motion.

Sidewinding
Vipers in African deserts and side-winder rattlesnakes in North American deserts evolved this skill separately for moving across windblown sand.

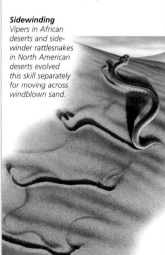

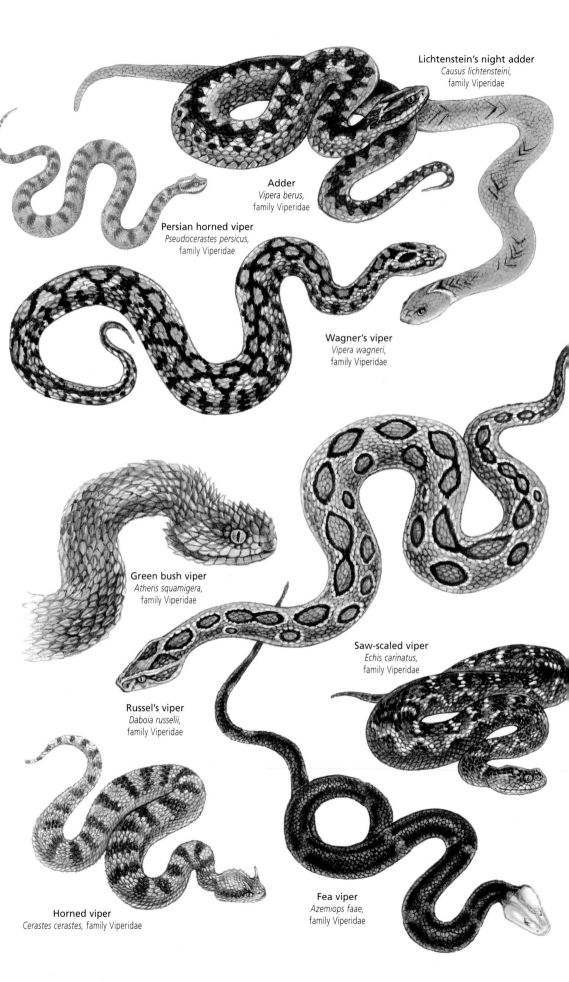

Lichtenstein's night adder
Causus lichtensteini,
family Viperidae

Adder
Vipera berus,
family Viperidae

Persian horned viper
Pseudocerastes persicus,
family Viperidae

Wagner's viper
Vipera wagneri,
family Viperidae

Green bush viper
Atheris squamigera,
family Viperidae

Saw-scaled viper
Echis carinatus,
family Viperidae

Russel's viper
Daboia russelii,
family Viperidae

Horned viper
Cerastes cerastes, family Viperidae

Fea viper
Azemiops faae,
family Viperidae

Levantine viper
Macrovipera lebetina, family Viperidae

Rhinoceros viper
Bitis nasicornis,
family Viperidae

Mauritanic viper
Macrovipera mauritanica,
family Viperidae

Tropical rattlesnake
Crotalus durissus,
family Viperidae

Gaboon adder
Bitis gabonica, family Viperidae

Pygmy rattlesnake
Sistrurus miliarius,
family Viperidae

**Banded rock
rattlesnake**
Crotalus lepidus,
family Viperidae

Copperhead
Agkistrodon contortix,
family Viperidae

Cantil
Agkistrodon bilineatus,
family Viperidae

**Western
diamondback
rattlesnake**
Crotalus atrox
family Viperidae

FACT FILE

Rhinoceros viper This heavy-bodied viper has two to three horns on the tip of its snout and is venomous but not aggressive. It is a sit-and-wait predator and mostly preys on mammals.

✦	Up to 4 ft (1.2 m)
◯	Terrestrial
⚲	Viviparous
●	6–35 live young
⅄	Common

C. & W. Africa

Pygmy rattlesnake The genus *Sistrurus* is distinguished by having nine large scales on the dorsal surface of the head. Young pygmy rattlesnakes have a bright yellow tail tip, which they wiggle as a lure to attract small prey.

✦	Up to 31½ in (80 cm)
◯	Terrestrial
⚲	Viviparous
●	6–10 live young
⅄	Common

S.E. USA

Western diamondback rattlesnake One of the most abundant rattlesnakes in the United States, it is responsible for more snake bites there than any other snake. Diet is primarily rodents, but the young prey on lizards.

✦	Up to 7½ ft (2.3 m)
◯	Terrestrial
⚲	Viviparous
●	4–25 live young
⅄	Common

S.W. USA & N.W. Mexico

RATTLE

Rattlesnakes have a unique rattle at the end of their tails. It is made up of interlocking segments of keratin; each time the snake sheds its skin, a new segment is added. When the tail is vibrating, the hollow segments bang against each other, making the anti-predator rattling sound.

⚡ CONSERVATION WATCH

The 20 species of Viperidae on the IUCN Red List are categorized as follows:

7	Critically endangered
4	Endangered
7	Vulnerable
1	Data deficient
1	Least concern

FACT FILE

Tiger rattlesnake This small nocturnal rattlesnake can often be found resting in pack rat (*Neotoma*) mounds in the daytime. At night it feeds on rodents and lizards. Although small, it has long fangs and potent venom.

- Up to 36 in (92 cm)
- Terrestrial
- Viviparous
- 2–5 live young
- Common

Sonoran Desert (S.W. USA & N.W. Mexico)

Massasauga Massasauga means great river mouth in the Chippeawa language. The preferred habitat for this species is river-bottom forest in the floodplains of large rivers. They hibernate in crayfish burrows after foraging for rodents and frogs during the summer. When vibrated, their small rattle sounds like the buzz of a cricket.

- Up to 30 in (76 cm)
- Terrestrial
- Viviparous
- 8–20 live young
- Rare

S.E. Canada, N.E. USA to N.W. USA, N. Mexico

Black-tailed rattlesnake In the spring, males follow the chemical trails of females to court them. Courtship and mating often last several days. Females give birth in August and stay with their young until they have shed for the first time. They find prey by following animal scent trails to burrows or nests and waiting for their chosen prey to return.

- Up to 4¼ ft (1.3 m)
- Terrestrial
- Viviparous
- 3–13 live young
- Common

S.W. USA & N.W. Mexico

Timber rattlesnake These rattlesnakes have communal dens in rock outcroppings, usually on the south side of hills or mountains. Breeding usually takes place in the spring at the dens before the snakes disperse. The young follow the scent trails of the adults to reach the dens in the fall. Although this species is primarily terrestrial, young have been found in the lower branches of trees and recent studies have found adults in the treetops hunting squirrels.

- Up to 5 ft (1.5 m)
- Terrestrial
- Viviparous
- 6–15 live young
- Common

N.E. USA

Tiger rattlesnake
Crotalus tigris, family Viperidae

Massasauga
Sistrurus catenatus, family Viperidae

Named for its solid black unbanded tail

Black-tailed rattlesnake
Crotalus molossus, family Viperidae

Twin-spotted rattlesnake
Crotalus pricei, family Viperidae

Ridge-nosed rattlesnake
Crotalus willardi, family Viperidae

Mojave rattlesnake
Crotalus scutulatus, family Viperidae

Speckled rattlesnake
Crotalus mitchelli, family Viperidae

Prairie rattlesnake
Crotalus viridis, family Viperidae

Timber rattlesnake
Crotalus horridus, family Viperidae

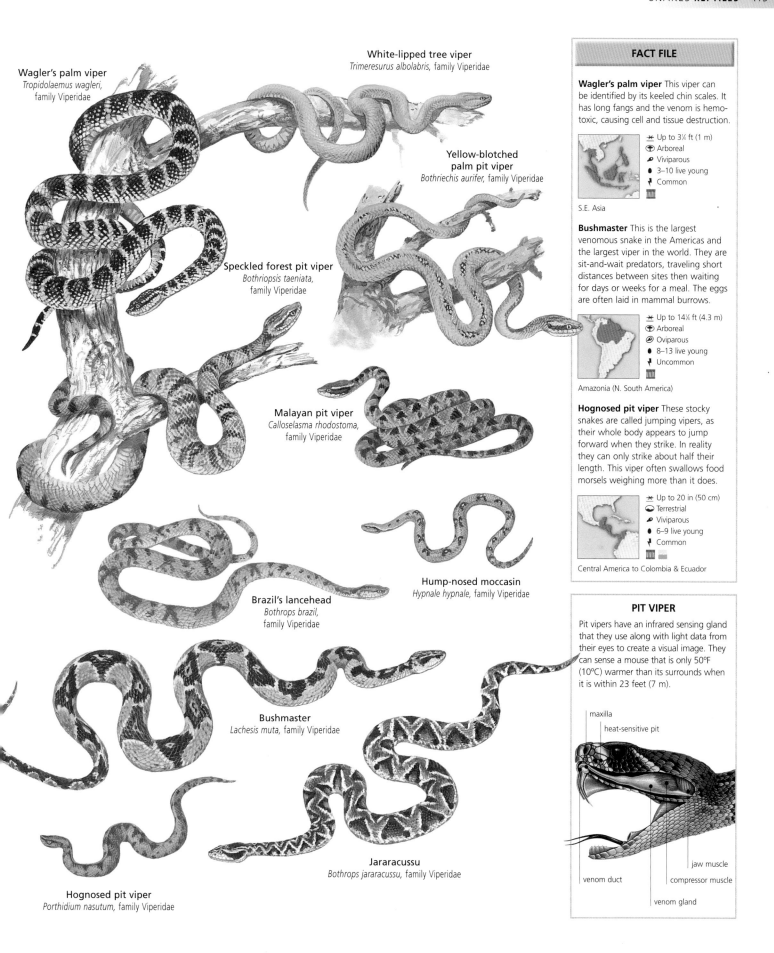

Wagler's palm viper
Tropidolaemus wagleri,
family Viperidae

White-lipped tree viper
Trimeresurus albolabris, family Viperidae

Yellow-blotched
palm pit viper
Bothriechis aurifer, family Viperidae

Speckled forest pit viper
Bothriopsis taeniata,
family Viperidae

Malayan pit viper
Calloselasma rhodostoma,
family Viperidae

Hump-nosed moccasin
Hypnale hypnale, family Viperidae

Brazil's lancehead
Bothrops brazil,
family Viperidae

Bushmaster
Lachesis muta, family Viperidae

Jararacussu
Bothrops jararacussu, family Viperidae

Hognosed pit viper
Porthidium nasutum, family Viperidae

FACT FILE

Wagler's palm viper This viper can be identified by its keeled chin scales. It has long fangs and the venom is hemotoxic, causing cell and tissue destruction.

- ⵌ Up to 3¼ ft (1 m)
- Arboreal
- Viviparous
- 3–10 live young
- Common

S.E. Asia

Bushmaster This is the largest venomous snake in the Americas and the largest viper in the world. They are sit-and-wait predators, traveling short distances between sites then waiting for days or weeks for a meal. The eggs are often laid in mammal burrows.

- ⵌ Up to 14¼ ft (4.3 m)
- Arboreal
- Oviparous
- 8–13 live young
- Uncommon

Amazonia (N. South America)

Hognosed pit viper These stocky snakes are called jumping vipers, as their whole body appears to jump forward when they strike. In reality they can only strike about half their length. This viper often swallows food morsels weighing more than it does.

- ⵌ Up to 20 in (50 cm)
- Terrestrial
- Viviparous
- 6–9 live young
- Common

Central America to Colombia & Ecuador

PIT VIPER

Pit vipers have an infrared sensing gland that they use along with light data from their eyes to create a visual image. They can sense a mouse that is only 50°F (10°C) warmer than its surrounds when it is within 23 feet (7 m).

maxilla
heat-sensitive pit
jaw muscle
compressor muscle
venom gland
venom duct

AMPHIBIANS

AMPHIBIANS

PHYLUM	Chordata
CLASS	Amphibia
ORDERS	3
FAMILIES	44
GENERA	434
SPECIES	5,400

Amphibians appeared 360 million years ago. Descended directly from early lobe-finned fishes, they were the first vertebrates to become established on land, where they gave rise to the reptiles. Amphibians are cold-blooded, with moist skin and no scales or claws. There are three orders. Caecilians resemble earthworms, being limbless with bullet-shaped heads and tails. Salamanders (Caudata, meaning "with tail") have cylindrical bodies, long tails, distinct heads and necks, and well-developed limbs. Frogs and toads (Anura, meaning "lacking tail") differ from all other vertebrates in having stout, tailless bodies with a continuous head and body and well-developed limbs.

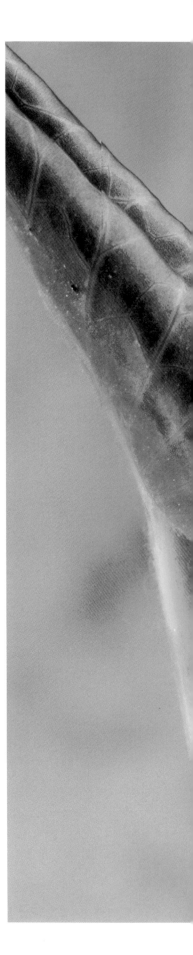

Caecileans The ringed caecilian (*Siphonops annulatus*) is a typical fossorial form built for a burrowing life, with reduced eyes, a segmented body, and a fortified, bullet-shaped head and tail. Although there are aquatic species as well, all are tropical, secretive, and rarely seen.

Amphibian skin The skin is moistened by mucus gland secretions. It maintains water balance, respires, and protects the body. Many species possess antibiotic or protective substances, including poisons, in their skin secretions. These secretions are antipredatory mechanisms, being distasteful and often deadly.

Leaping for their lives The red-eyed tree frog spends most of its life in the canopy of neotropical rain forests. It feeds, calls, breeds, and even lays its eggs in the trees. When these frogs are threatened by a predator high in the tree-tops, they have only one choice—go airborne and hope they reach another tree branch.

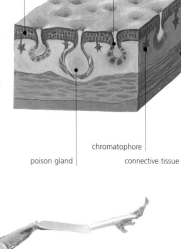

Flash colors
The red-eyed tree frog appears all green until the shocking bright blue inner surfaces of the thighs and groin are displayed while leaping.

A DOUBLE LIFE

The word amphibian is from the Greek *amphibios*, meaning "a being with a double life." This refers to the reproductive cycle of most amphibians, in which they have an aquatic larval form and a terrestrial adult life. Most species breed and lay eggs in the water. In these cases, the young have a free-swimming larval stage with external gills, and metamorphose later into a miniature replica of the terrestrial adult. There are many exceptions to this. Some species have direct development on land and no larval stage. Others are entirely aquatic and viviparous. Still others are neotenic.

Some of the distinctive features that amphibians have developed to allow them to invade terrestrial habitats are: the tongue, which allows them to moisten and move

stratified epithelium | mucus gland

poison gland | chromatophore | connective tissue

food; eyelids, which along with adjacent glands wet and protect the cornea; an outer layer of dead skin cells that is regularly sloughed; ears; a sound-producing structure called the larynx; and the Jacobson's chemosensory organ in the nasal cavity, which gives them a sense of smell and taste.

An amphibian's skin is active in controlling water balance. This can be illustrated by looking more closely at frogs. The permeability of the skin is dynamic, changing with the activity cycle of the frog; when the frog is foraging away from water, the skin is highly permeable to absorb water. When the animal is aquatic, the permeability of the skin is reduced. Some terrestrial frogs have a special patch of skin in the pelvic region that is highly vascularized, allowing them to absorb water from any moist area. When frogs enter hibernation in the water, they must change their ionic balance and skin permeability to lower the absorption of water by osmosis. Some desert toads are able to absorb water from dry soils by retaining urea in their urine, thus creating an osmotic gradient across their skin.

Adults of all amphibians are carnivores, although a few species of frogs have been reported to eat some fruits. The larvae of salamanders are carnivorous, while those of most anurans are herbivores. The tadpoles of some species of frogs are cannibalistic, as are the adults of many species.

Male display This male great crested newt (*Triturus cristatus*) (right) is in breeding condition. The males' dorsal and tail crests increase in size during the breeding season. After courtship in the water, the eggs are fertilized internally and deposited singly on vegetation. The larvae are carnivorous.

SALAMANDERS AND NEWTS

CLASS Amphibia	
ORDER Caudata	
FAMILIES 10	
GENERA 60	
SPECIES 472	

Salamanders, newts, mudpuppies, waterdogs, and sirens belong to the order Caudata. All have tails and most have four legs. Most species are very secretive, living terrestrially in leaf litter or rotting logs, underground, in arboreal bromeliads, or underwater. Some live terrestrially but return to the water to reproduce. Although seldom seen, they are not uncommon; astoundingly, in some deciduous forests in North America, the biomass of salamander populations is greater than that of the birds and mammals. All salamanders and their larvae are carnivorous, and they in turn become prey for many larger carnivores, such as snakes, birds, and mammals.

Distribution Salamanders are widely distributed in North and Central America, northern South America, Europe, northern Africa, and Asia.

Warning colors The red salamander (*Pseudotriton ruber*) is semiaquatic, living in or under moss or flat rocks in or near springs or spring creeks. It prefers cool, clear water that flows through woods or meadows. Many species of salamanders are brightly colored to warn their principal predators—birds—that they are distasteful.

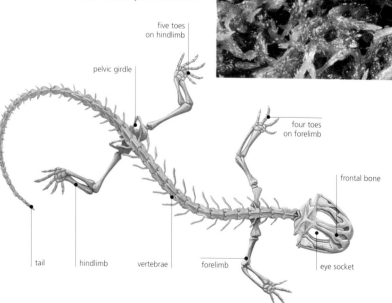

five toes on hindlimb

pelvic girdle

four toes on forelimb

frontal bone

tail   hindlimb   vertebrae   forelimb   eye socket

Four-toed salamander *Hemidactylium scutam* lays up to 24 eggs under logs or in moss near clear-water creeks. The female guards them until they hatch; then the larvae enter the water to finish development.

Salamander skeleton A salamander's vertebral column is rigid to support the head, pelvic and pectoral girdles, and viscera, but is also flexible, permitting lateral and dorsoventral movements. Salamanders have 10–60 presacral vertebrae and a variable number of postsacral vertebrae: 2–4 precaudals and a number of caudal vertebrae.

SILENT CARNIVORES

Salamanders evolved to live on land, with four legs and a tail. Some species have secondarily returned to live permanently in the water and have reduced limbs; these are known as sirens.

Salamanders more closely resemble early fossil amphibians than do frogs and toads. Unlike frogs, they do not emphasize vocal communications during breeding. The skin of salamanders is generally smooth, moist, and flexible, with oxygen exchange taking place through the skin. Many species have lungs as well, but some families lack lungs and respire entirely cutaneously. As the salamander grows, the skin is regularly shed. Because it is made up of the salamander's own cells and is easy to digest, the animal usually devours its own skin as it is sloughed.

Salamanders and their larvae are completely carnivorous. Adults consume insects, spiders, snails, slugs, worms, and other invertebrates. The larvae devour mosquito larvae and other insects. As they grow, they eat larger prey, including tadpoles and the larvae of the dragonfly that was once their feared predator. Some species of salamanders are able to extend their tongues rapidly for half the length of their body to capture prey.

Salamanders have moist, slippery skin, while newts, both in their aquatic and terrestrial phases, have rough, dry skin. Some species of newts live on land as adults, coming to the water only to breed, while the adults of some species are aquatic. Red-spotted newts are equally at home in the water or on land. The aquatic larvae metamorphose into a terrestrial red eft, and after 1–2 years on land return to the water to breed.

Hellbender
Cryptobranchus alleganiensis

Greater siren
Siren lacertina

Chinese giant salamander
Andrias davidianus

Lacks eyelids

Lesser siren
Siren intermedia

Two-toed amphiuma
Amphiuma means

Dwarf siren
Pseudobranchus striatus

Feathery protrusions
near head are
external gills

FACT FILE

Hellbender Hellbenders live under rocks and thrive in streams, where they forage for crayfish, mollusks, and small fishes. They are harmless but extremely slimy; in order to pick them up, they must be grasped firmly around the head.

- Up to 29½ in (74 cm)
- Aquatic
- Fall
- Common

E.C. USA

Greater siren Although most salamanders are silent, greater sirens when first caught will sometimes yelp. They also make clicking sounds, which perhaps may be used in intraspecific communication.

- Up to 38½ in (98 cm)
- Aquatic
- Unknown
- Common

S.E. USA

Chinese giant salamander Breeding takes place in an underwater cavity where the female lays up to 500 eggs in a cylindrical string. The male fertilizes the eggs externally and guards them until they hatch, in 50–60 days.

- Up to 6 ft (1.8 m)
- Aquatic
- Fall
- Data deficient

E.C. China

SALAMANDER REPRODUCTION

During courtship, males of species that fertilize their eggs internally deposit sperm packets, or spermatophores, which are picked up by the cloacal lips of the females. Species with direct development lay eggs in moist areas, and the females guard the eggs until a miniature version of the adult emerges. Eggs laid in ponds or streams and attached to vegetation hatch within 1–2 weeks into free-swimming carnivorous larvae. Some species do not develop past their final larval stage.

Courtship consequences
After courtship, the male deposits his spermatophore below the female. She picks it up with her cloacal lips and the eggs are fertilized internally.

FACT FILE

Mudpuppy These neotenic salamanders are characterized by their large, red external gills. They feed nocturnally on crustaceans, insect larvae, fishes, worms, mollusks, and other amphibians. Females guard and aerate their eggs.

▶ Up to 19 in (48 cm)
◖ Aquatic
⇌ Fall
ͳ Common

N.C. USA

Olm Olms live in flooded underground caves and are blind, with only vestigial eyes occurring beneath the pigmentless skin. They are neotenic. They lay up to 70 eggs, or may retain eggs and produce fewer living young.

▶ Up to 12 in (30 cm)
◖ Aquatic
⇌ Spring
ͳ Vulnerable

Adriatic seaboard & N.E. Italy

Chinese salamander This terrestrial salamander with short limbs migrates to streams or ponds for breeding. Males are attracted to females by chemical and visual cues.

▶ Up to 4 in (10 cm)
◖ Terrestrial
⇌ Spring
ͳ Common

Hubei Province (S. China)

⚡ CONSERVATION WATCH

Going, going... Three species of salamander are listed on the IUCN Red List as critically endangered. The Lake Lerma salamander (*Ambystoma lermaense*) has not been seen for years and is now presumed to be extinct. It was endemic to Lake Lerma, Mexico, which was affected by dam construction and pollution.

The Sardinian brook salamander (*Europroctus platycephalus*) is the most threatened salamander in Europe, due to pesticide (DDT) treatment of water bodies in the battle against malaria; the introduction of trout, which feed on and compete with the salamanders for food; and the reduction of water levels due to agriculture.

The desert slender salamander (*Batrachoseps major aridus*) only lives in one canyon in Riverside County, California, United States. As the water table lowers, this species' habitat is slowly vanishing.

Mudpuppy
Necturus maculosus

Extremely slimy skin

Olm
Proteus anguinus

Vestigial eyes

Siberian salamander
Ranodon sibiricus

Fischer's clawed salamander
Onychodactylus fischeri

Dybowski's salamander
Salamandrella keyserlingii

Sichuan salamander
Liua shihi

Chinese salamander
Hynobius chinensis

Paghman mountain salamander
Batrachuperus mustersi

Luschan's salamander
Mertensiella luschani

European fire salamander
Salamandra salamandra

Olympic torrent salamander
Rhyacotriton olympicus

Golden striped salamander
Chioglossa lusitanica

Pyrenees mountain salamander
Euproctus asper

Californian giant salamander
Dicamptodon ensatus

Red-spotted newt
Notophthalmus viridescens

Common newt
Triturus vulgaris

Male in breeding
condition

Luristan newt
Neurergus kaiseri

Banded newt
Triturus vittatus

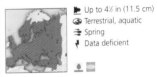
NEOTENY

Neoteny is the developmental state where the aquatic larval form with gills is maintained throughout adulthood. This usually occurs in species which live in oxygen-deficient water. Some species are permanently neotenic. The Mexican axolotl is neotenic if iodine is absent in the environment.

Fire salamander
"Salamander" is from the Greek word for "fire lizard," because they were seen crawling out of logs that had been thrown onto campfires.

FACT FILE

Ringed salamander This species has narrow crossbands, a slate-colored belly, and a light gray stripe along the sides of its thin body, unlike the marbled and tiger salamanders.

- ➤ Up to 9 in (23 cm)
- ⊕ Aquatic, burrowing
- ⇌ Fall
- ⚑ Common

S.C. USA

Tiger salamander These salamanders are fossorial for most of the year, only emerging to breed once the early spring rains begin. After breeding in ponds or marshes, the adults retreat underground.

- ➤ Up to 13 in (33 cm)
- ⊕ Aquatic, burrowing
- ⇌ Early spring
- ⚑ Common

S. Canada to Mexico

Marbled salamander This species congregates to lay up to 230 eggs in dry depressions from September to December. Females remain wrapped around the eggs, protecting them from predation and desiccation, until fall rains fill the pools.

- ➤ Up to 5 in (13 cm)
- ⊕ Terrestrial, aquatic
- ⇌ Fall
- ⚑ Common

S.E. USA

Japanese firebelly newt Males of this species differ from females by having swollen cloacas. The tails of breeding males develop a bluish iridescent sheen and a small filament at the tip.

- ➤ Up to 5¼ in (13.2 cm)
- ⊕ Aquatic
- ⇌ Spring
- ⚑ Common

Japan

Vietnam warty newt This, the largest species of the genus, is characterized by a distinct belly pattern of orange-yellow or reddish blotches, each with black or dark brown mottling.

- ➤ Up to 8 in (20 cm)
- ⊕ Aquatic
- ⇌ Spring
- ⚑ Vulnerable

N. Vietnam

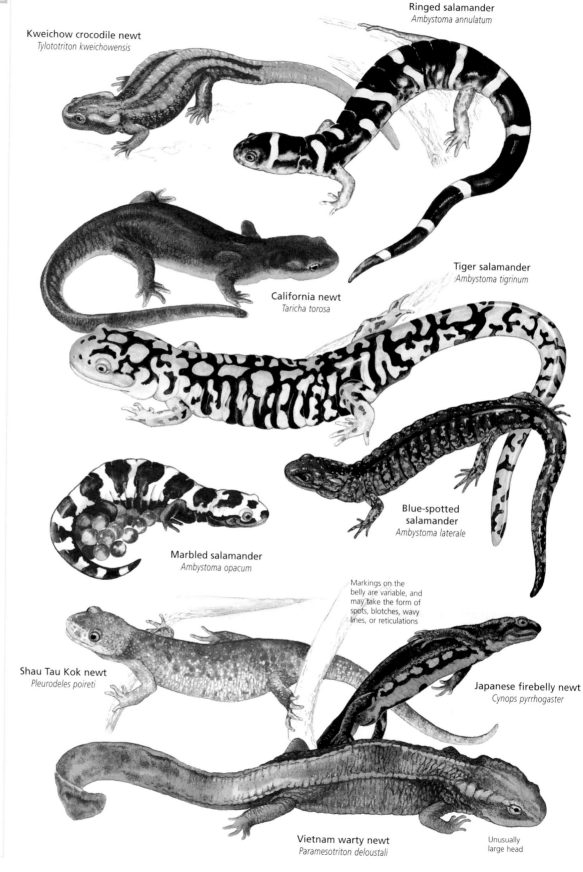

Ringed salamander
Ambystoma annulatum

Kweichow crocodile newt
Tylototriton kweichowensis

California newt
Taricha torosa

Tiger salamander
Ambystoma tigrinum

Marbled salamander
Ambystoma opacum

Blue-spotted salamander
Ambystoma laterale

Markings on the belly are variable, and may take the form of spots, blotches, wavy lines, or reticulations

Shau Tau Kok newt
Pleurodeles poireti

Japanese firebelly newt
Cynops pyrrhogaster

Vietnam warty newt
Paramesotriton deloustali

Unusually large head

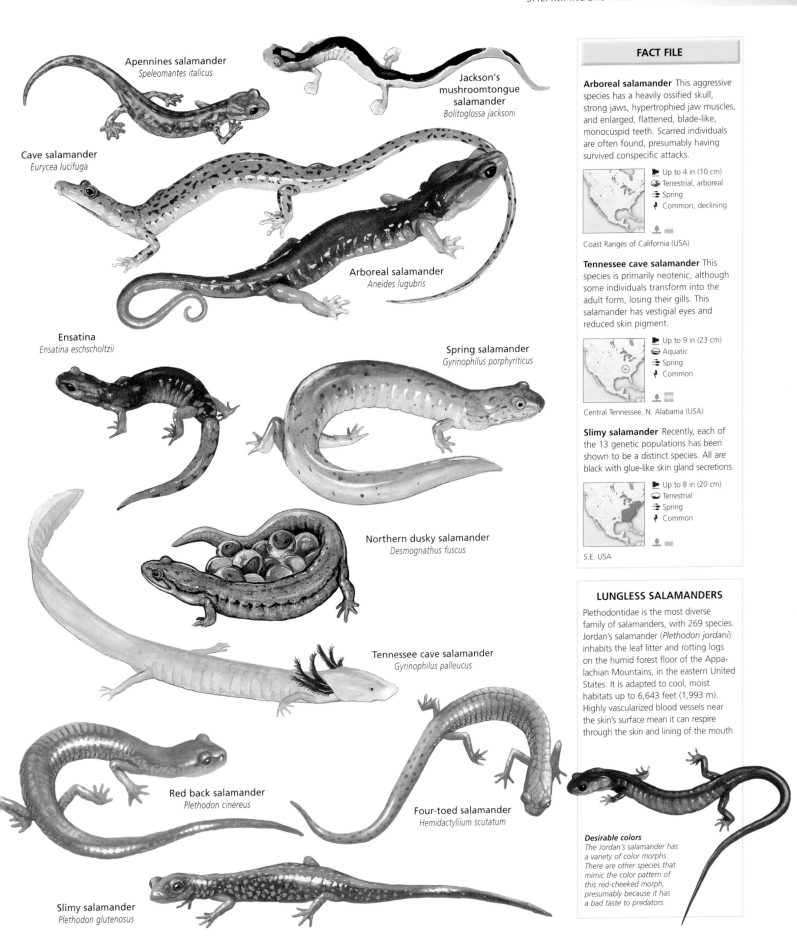

Apennines salamander
Speleomantes italicus

Jackson's mushroomtongue salamander
Bolitoglossa jacksoni

Cave salamander
Eurycea lucifuga

Arboreal salamander
Aneides lugubris

Ensatina
Ensatina eschscholtzii

Spring salamander
Gyrinophilus porphyriticus

Northern dusky salamander
Desmognathus fuscus

Tennessee cave salamander
Gyrinophilus palleucus

Red back salamander
Plethodon cinereus

Four-toed salamander
Hemidactyliium scutatum

Slimy salamander
Plethodon glutenosus

FACT FILE

Arboreal salamander This aggressive species has a heavily ossified skull, strong jaws, hypertrophied jaw muscles, and enlarged, flattened, blade-like, monocuspid teeth. Scarred individuals are often found, presumably having survived conspecific attacks.

- Up to 4 in (10 cm)
- Terrestrial, arboreal
- Spring
- Common, declining

Coast Ranges of California (USA)

Tennessee cave salamander This species is primarily neotenic, although some individuals transform into the adult form, losing their gills. This salamander has vestigial eyes and reduced skin pigment.

- Up to 9 in (23 cm)
- Aquatic
- Spring
- Common

Central Tennessee, N. Alabama (USA)

Slimy salamander Recently, each of the 13 genetic populations has been shown to be a distinct species. All are black with glue-like skin gland secretions.

- Up to 8 in (20 cm)
- Terrestrial
- Spring
- Common

S.E. USA

LUNGLESS SALAMANDERS

Plethodontidae is the most diverse family of salamanders, with 269 species. Jordan's salamander (*Plethodon jordani*) inhabits the leaf litter and rotting logs on the humid forest floor of the Appalachian Mountains, in the eastern United States. It is adapted to cool, moist habitats up to 6,643 feet (1,993 m). Highly vascularized blood vessels near the skin's surface mean it can respire through the skin and lining of the mouth.

Desirable colors
The Jordan's salamander has a variety of color morphs. There are other species that mimic the color pattern of this red-cheeked morph, presumably because it has a bad taste to predators.

AMPHIBIAN LIFE-CYCLES

M ost amphibians have a dual life-cycle involving a court–ship ritual, the deposition of sperm and eggs, external fertilization, an aquatic larval stage, and transformation into the adult life form. Modifications made to this cycle involve internal fertilization, direct development, viviparity, neoteny, or parental care. There is a trade-off between multitudes of small eggs with no parental care and a few large eggs with parental care.

Swollen egg sacs The egg sac of the Jefferson salamander (*Ambystoma jeffersonianum*) is much larger than the adult after the eggs have absorbed water. This species does not brood or guard its eggs.

Salamander life-cycle Pheromones stimulate courtship behavior in sala-manders. Males then deposit sperm as spermatophores on the substrate below the female. She retrieves the spermatophores with her cloacal lips, fertilizing the eggs internally as she deposits them in water or in a moist substrate. Aquatic larvae hatch from the eggs in the water or direct develop-ment may occur in terrestrial eggs.

terrestrial adult

larva with fully developed limbs and gills

egg

larva with gill buds

larva with developing gills and forelimbs, and hindlimb buds

1. Mating *The male clasps the female around the middle. The female carries the male until she is ready to lay her eggs. The male then sheds his sperm over the eggs as they are laid.*

5. Change complete *After 6 weeks, the metamorphosis from tadpole to adult frog is complete, and the young frog begins to hunt insects.*

2. Egg development *The eggs float in large masses while the embryos develop inside. The eggs are covered in a jelly-like substance that swells when it comes into contact with water.*

Safe development Marsupial frogs (*Gastrotheca*) have one or two pouches, similiar to those of marsupials, which open dorsally to the cloaca. Eggs are placed in the pouch as the male fertilizes them. Thus the eggs develop in a moist and safe environment. The young emerge as tadpoles or froglets, depending on the species.

Frog life-cycle Frogs and toads have a complex life-cycle. Males call to attact females, and shed sperm over the eggs as the female deposits them in a selected site. The eggs absorb water and in a few days develop into free-swimming tadpoles, which are normally herbivorous. After a few weeks, months, or years (depending on the species) the tadpoles meta-morphose into identical small replicas of the adults. It then takes months or years of a carnivorous diet before sexual maturity is reached.

4. Tadpole development *Lungs begin to form after 3 weeks, and legs after 4 weeks, the hindlimbs developing first. The tadpole's tail shrinks and its gills are reabsorbed.*

3. Tadpoles emerge *After 2 weeks, the eggs hatch into tadpoles. When the mouth develops, the tadpoles feed on algae. They breathe through external gills.*

CAECILIANS

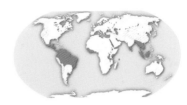

Caecilians have radiated to a variety of aquatic and terrestrial habitats, with specialized characteristics for a fossorial life not present in any other vertebrates—chemo-sensing tentacles, an extra set of jaw-closing muscles, and skin that is ossified to the skull. Adults range in size from just under 3 inches (7 cm) to 5¼ feet (1.6 m). Fertilization is internal. Some species are oviparous; others have direct development; and about half of the known species are viviparous. They feed on earthworms, beetle larvae, termites, and crickets.

CLASS Amphibia
ORDER Gymnophiona
FAMILIES 6
GENERA 33
SPECIES 149

Distribution Caecilians have a limited pantropical distribution in India, southern China, Malaysia, the Philippines, Africa, Central America, and South America. No species are known from Europe, Australia, or Antarctica.

Caecilian skull Caecilians are built for burrowing, with massive, bullet-shaped skulls. They are streamlined, but reinforced for burrowing through soil or protruding into the substrate in search of invertebrate prey. They move through the soil by undulating the body: the muscles move in a wave from the head to the tail. The body curves resist the soil or water, and forward motion is achieved.

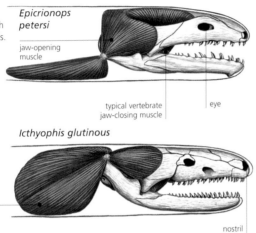

Epicrionops petersi

jaw-opening muscle

typical vertebrate jaw-closing muscle

eye

Icthyophis glutinous

strong jaw-closing muscles

nostril

Gymnophis multiplicate All species in the family Caeciliidae are chacterized by distinct annuli, which have caused some people to confuse them with earthworms. Caecilians have a retractable tentacle on each side of the head, which helps them to search for prey by transmitting chemical clues to the nasal cavity.

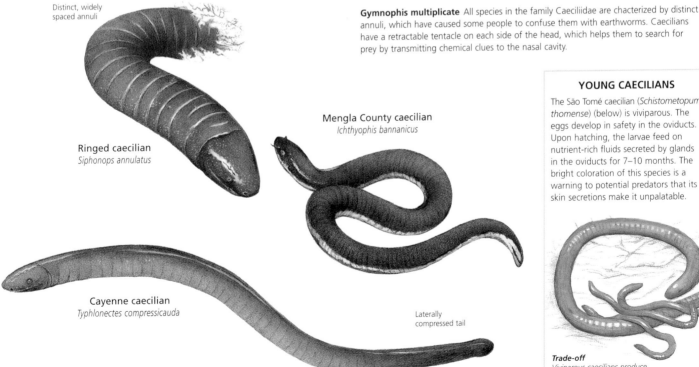

Distinct, widely spaced annuli

Ringed caecilian
Siphonops annulatus

Mengla County caecilian
Ichthyophis bannanicus

Cayenne caecilian
Typhlonectes compressicauda

Laterally compressed tail

YOUNG CAECILIANS

The São Tomé caecilian (*Schistometopum thomense*) (below) is viviparous. The eggs develop in safety in the oviducts. Upon hatching, the larvae feed on nutrient-rich fluids secreted by glands in the oviducts for 7–10 months. The bright coloration of this species is a warning to potential predators that its skin secretions make it unpalatable.

Trade-off
Viviparous caecilians produce fewer young than oviparous species but provide better care.

FROGS AND TOADS

CLASS	Amphibia
ORDER	Anura
FAMILIES	28
GENERA	338
SPECIES	4,937

Anuran is the collective term for this group of tailless amphibians. This is the largest order of amphibians, with nearly 5,000 known species. Anurans come in all shapes and sizes, from the tiny Brazilian Izecksohn's toad (*Psyllophryne didactyla*), which is less than half an inch (1 cm) long, to the African goliath frog (*Conraua goliath*), which is 12 inches (30 cm) long and weighs 7⅓ pounds (3.3 kg). Frogs and toads are easily identified by their long hindlegs, short trunk, moist skin, and lack of tail. Anurans are the only amphibians that are extremely vocal when breeding. The largest genus of frogs—*Eleutherodactylus* (rain frogs)—is also the largest genus of all vertebrates.

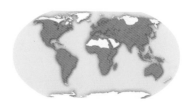

Distribution Frogs and toads occur on all continents except Antarctica. More than 80 percent of species are tropical; at one site alone, 67 species have been recorded. Two species extend north of the Arctic Circle.

Flying frogs The Java flying frog (*Rhacophirus reinwardtii*) has broad webs of skin between its toes, allowing it to parachute between trees or guiding it in a controlled fall to another branch. This adaptation is important for life in the rain-forest canopy.

SINGING FOR THEIR LIVES

A notable feature of frogs and toads is the voices of the males calling for mates. These calls herald spring in the temperate zone and the onset of the rainy season in the tropics. Each species of frog has a specific call that distinguishes it from all other species. Females can assess the size of the male, and thus his fitness, by the tone of his call. Only males call, and females come to the calling male to mate. These calls also have the disadvantage of attracting some predators, notably bats.

In some species, calling sites are at a premium and males defend them against other calling males. To avoid confrontations, satellite males wait silently near the calling male in hopes of intercepting a gravid female before she reaches him. This lessens their chance of being eaten by a bat, and reduces the energy lost by calling or by fighting with the dominant male.

The above are characteristics of frogs with long breeding seasons. Many other species are explosive breeders, coming to the breeding ponds with the first heavy rains, breeding by the thousands for 2–3 days, and then retreating until the following year. These species

Tongue lashing Frogs and toads have long, sticky tongues, like the Japanese toad (*Bufo japonicus*) (above), which they use to capture prey. Once the insect is inside the mouth, the toad blinks its eyes and the eyes push the food out of the mouth and down the throat. Males also use their tongues in combat, lashing them into the eyes of other males in territorial battles or in bouts over access to a female.

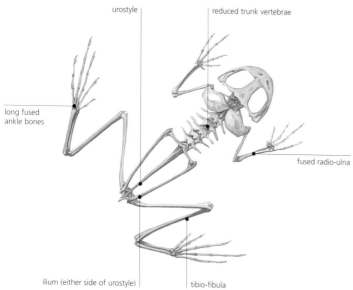

urostyle

reduced trunk vertebrae

long fused ankle bones

fused radio-ulna

ilium (either side of urostyle)

tibio-fibula

Frog skeleton Anurans have a shortened vertebral column of nine or fewer vertebrae. The epipodial elements of both fore and hindlimbs are fused; also, the ankle is elongated and made up of two fused elements. The postsacral vertebrae are uniquely fused into a rod-shaped urostyle. The long hindlimbs give frogs the propulsive force to propel themselves forward. A short, compact body is easier to move forward. The reinforced pectoral girdle and forelimbs are built to absorb the shock of landing.

Calling for a mate During the breeding season, male frogs and toads call to attract the females to mate. Each species has a distinct attraction call. This natterjack toad (*Bufo calamita*) (above) has his gular sac inflated and is making a trilling sound.

Energy transfers The northern leopard frog (*Rana pipiens*) (right) leaps to escape a predator. Frogs and tadpoles transform the energy in their food (algae, detritus, and insects) into larger packets of energy that are then available to bigger predators.

usually have loud, carrying calls and lay thousands of small eggs. To avoid predation of their eggs by aquatic predators, many species of tree frogs (*Hylidae*) lay their eggs in vegetation overhanging their breeding ponds. When the eggs hatch, the tadpoles fall into the water, where they complete their development. Other species of casque-headed tree frogs (*Osteocephalus*) lay a few large fertilized eggs in tree holes. The female returns regularly to deposit unfertilized eggs in the nest to feed her tadpoles. Many species of poison frogs (*Dendrobatidae*) guard their eggs in terrestrial nests and, upon hatching, carry their tadpoles to the water, or in some cases up trees and into bromeliads. Many species of *Eleutherodactylus* lay 15–25 large yolked eggs in terrestrial nests, the eggs having direct development. Midwife toads, Surinam toads, and gastric-brooding frog females carry the eggs with them until they hatch into small frogs. Other *Leptodactylid* and *Rhacophorid* frogs build foam nests to deposit their eggs and rear their tadpoles.

Frogs are indicators of environmental pollution. Since the 1970s, there has been growing worldwide evidence that frog populations are declining on a massive scale and that many species are going extinct, even though their populations appear to be within pristine areas. The decline may be the result of a number of factors, including pollution, pesticides (such as DDT), herbicides (such as Atrizine), road salts, UV radiation, global warming, lower rainfall, or lower humidity. At present, the fungus chytridiomycosis seems to be the greatest threat. The fungus, which probably originated in Africa, invades the skin and lowers the frogs' immune system so they are more susceptible to diseases.

⚡ CONSERVATION WATCH

The 342 species of anurans on the IUCN Red List are categorized as follows:

7	Extinct
27	Critically endangered
27	Endangered
64	Vulnerable
1	Conservation dependent
15	Near threatened
67	Data deficient
134	Least concern

FACT FILE

Brown New Zealand frog Males of this species do not have a prominent breeding call. Call notes are made by resonance frequencies in the head and body, not by the vibration frequency of the vocal cords.

- Up to 2 in (5 cm)
- Terrestrial
- Spring
- Vulnerable

Stephens & Maud Islands (New Zealand)

Oriental firebelly toad This toad is one of the most common in its range, constituting up to a third of the total anurans at some localities. The density at breeding sites can reach three toads per square foot (nine toads per sq. m).

- Up to 2½ in (6 cm)
- Terrestrial
- Summer
- Common

Russia, Korea, China

Tailed frog This frog has no breeding call. Its "tail"—used for internal fertilization in fast-moving streams— is unique among anurans. Eggs are deposited in strings under rocks in streams. Tadpoles take 1–4 years to metamorphose to juveniles, and then 7–8 years to first breeding.

- Up to 2 in (5 cm)
- Terrestrial, aquatic
- Summer
- Common

N.W. USA

ANURAN FERTILIZATION

Fertilization of the eggs usually occurs while the frogs are in amplexus. As the female deposits her eggs, males release sperm above them in the water. The coqui and some other terrestrial species achieve internal fertilization by cloacal apposition; only the tailed frog has an intermittent organ.

Frogs in amplexus
The female carries the male to an appropriate spot to deposit the eggs.

Brown midwife toad
Alytes cisternasii

Yellowbelly toad
Bombina variegata

Brown New Zealand frog
Leiopelma hamiltoni

Olive midwife toad
Alytes obstetricans

Hochstetter's
New Zealand frog
Leiopelma hochstetteri

Oriental firebelly toad
Bombina orientalis

Painted frog
Discoglossus pictus

Belly and underside of
limbs are flashed as a
warning to predators

Corsica painted frog
Discoglossus montalentii

Tailed frog
Ascaphus truei

Malayan horned frog
Megophrys nasuta

Shaping frog
Atympanophrys shapingensis

Common parsley frog
Pelodytes punctatus

Surinam toad
Pipa pipa

Eggs embedded
in back

Mexican spadefoot toad
Spea multiplicata

Mid-dorsal
red-orange stripe

Mueller's clawed frog
Xenopus muelleri

Mexican burrowing toad
Rhinophrynus dorsalis

Cape clawed frog
Xenopus gilli

Syrian spadefoot toad
Pelobates syriacus

FACT FILE

Turtle frog This frog burrows head-first with its strong forelimbs, often into termite nests to eat the inhabitants. Males call from the soil surface or with only their heads uncovered. Up to 40 large eggs are laid underground.

- Up to 2½ in (6 cm)
- Terrestrial, burrowing
- Rainy season
- Uncommon

S.W. Australia

Crucifix toad This fossorial toad spends most of its life underground and only emerges to breed after heavy rains in temporary pools. Males call while floating on the surface. Small eggs are deposited in the water.

- Up to 2¼ in (5.5 cm)
- Terrestrial, burrowing
- Rainy season
- Common

S.E. Australia

Giant banjo frog Adults spend day-light hours and drier months buried beneath the surface. Males call while floating on vegetation or concealed in burrows at the water's edge. Oviposition take place in flooded burrows.

- Up to 3½ in (9 cm)
- Aquatic
- Spring, summer
- Uncommon

S.E. Australia

GASTRIC BROODERS

Fertilized eggs or larvae are swallowed by the female gastric-brooding frog (*Rheobatrachus silus*) and develop in her stomach. The production of hydro-chloric acid in the female's stomach ceases during this time. Tadpoles develop from the nutrients in the egg yolk, so the labial teeth are absent.

Up they come The young crawl out into the world after developing in their mother's stomach for 6–7 weeks.

Red-crowned toadlet
Pseudophryne australis

Skeleton ghost frog
Heleophryne rosei

Spatulate toes used for climbing rocks

Thomasset's frog
Nesomantis thomasseti

Turtle frog
Myobatrachus gouldii

Crucifix toad
Notaden bennettii

Natal ghost frog
Heleophryne natalensis

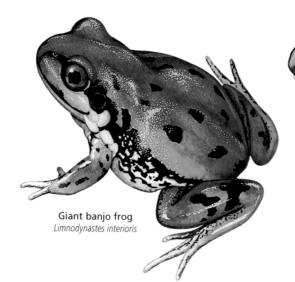

Giant banjo frog
Limnodynastes interioris

Differs from the giant burrowing frog (*Helioporus australicacus*) as lacks black spines on the back and throat

Western marsh frog
Heleioporus barycragus

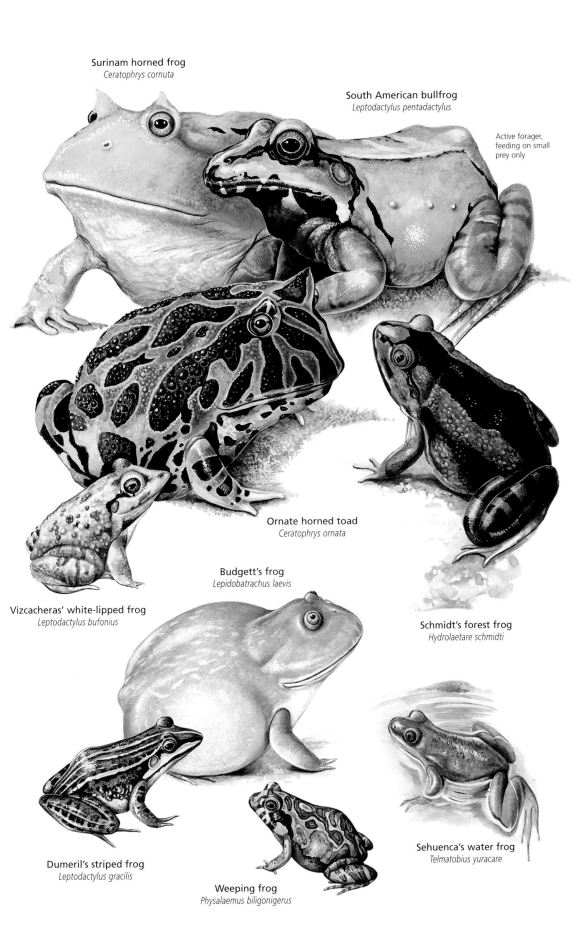

Surinam horned frog
Ceratophrys cornuta

South American bullfrog
Leptodactylus pentadactylus

Active forager, feeding on small prey only

Ornate horned toad
Ceratophrys ornata

Schmidt's forest frog
Hydrolaetare schmidti

Vizcacheras' white-lipped frog
Leptodactylus bufonius

Budgett's frog
Lepidobatrachus laevis

Dumeril's striped frog
Leptodactylus gracilis

Weeping frog
Physalaemus biligonigerus

Sehuenca's water frog
Telmatobius yuracare

FACT FILE

Surinam horned frog When prey ventures nearby, this frog lurches forward, opening the mouth that dominates the entire top of its head, and swallowing the prey whole.

- ▶ Up to 4¾ in (12 cm)
- ◯ Terrestrial
- ⇌ Fall rainy season
- ⚡ Common

Amazonia (South America)

South American bullfrog The call is a single, loud, whistle-like hoot. Up to 3,000 eggs are laid in a depression in the forest floor. The cannibalistic tadpoles are dispersed to nearby streams or swamps by heavy rains.

- ▶ Up to 7 in (18 cm)
- ◯ Terrestrial
- ⇌ Fall
- ⚡ Common

Honduras to Bolivia

Budgett's frog This is a large, stout, aquatic frog with a flattened body. The head is large and robust, one-third of the total body length, and broad to accommodate its very wide jaws.

- ▶ Up to 4 in (10 cm)
- ◯ Aquatic
- ⇌ Summer
- ⚡ Common

Gran Chaco (Argentina, Bolivia, Paraguay)

DROUGHT RESISTANCE

During the dry winter months, ornate horned toads are inactive underground, encased in a hard shell composed of layers of unshed skin. This "cocoon" protects the animal from excessive water loss and allows it to survive until the rains arrive. These signal the beginning of the wet summer months in South America, which last from October to February. The heavy rains flood the Gran Chaco and create transient pools for foraging and breeding.

Digging in Ornate horned toads will remain buried until rains signal the beginning of the wet summer months.

FACT FILE

Coqui This species gets its name from its high-pitched chirp, which sounds like "co qui." Eggs are laid on land with direct development in the egg. Coquis have been introduced into Florida and Hawaii, United States.

- Up to 2¼ in (5.5 cm)
- Terrestrial, arboreal
- Year round
- Common

Puerto Rico (Caribbean)

Barking frog The breeding call of this species sounds like a dog barking when heard at a distance; close up, it is a guttural "whurr." Females make a screech when grasped. When captured, their bodies puff up immensely.

- Up to 3¾ in (9.5 cm)
- Terrestrial
- Spring
- Common

S.W. USA to C.W. Mexico

Peru robber frog This species has direct development. The eggs are deposited in moist areas in the leaf litter. The males, which are considerably smaller than the females, call day and night from the forest floor.

- Up to 1¼ in (3 cm)
- Terrestrial
- Fall rainy season
- Common

W. Amazonia (Brazil, Bolivia, Peru, Ecuador)

Lowland tropical frog Females deposit up to 20 unpigmented eggs in a foam nest. The tadpoles developing in the nest subsist entirely on the egg yolk. Males call from mudflats.

- Up to 1¼ in (3 cm)
- Terrestrial
- Spring
- Common

Amazonia (N. South America)

Gold-striped frog The flash coloration —bright red spots in the groin and on the rear of the thighs—is displayed when the frog moves. Its coloration mimics that of the spotted-thighed poison frog (*Epipedobates femoralis*). Eggs are deposited in foam nests.

- Up to 2 in (5 cm)
- Terrestrial
- Winter
- Common

N. South America

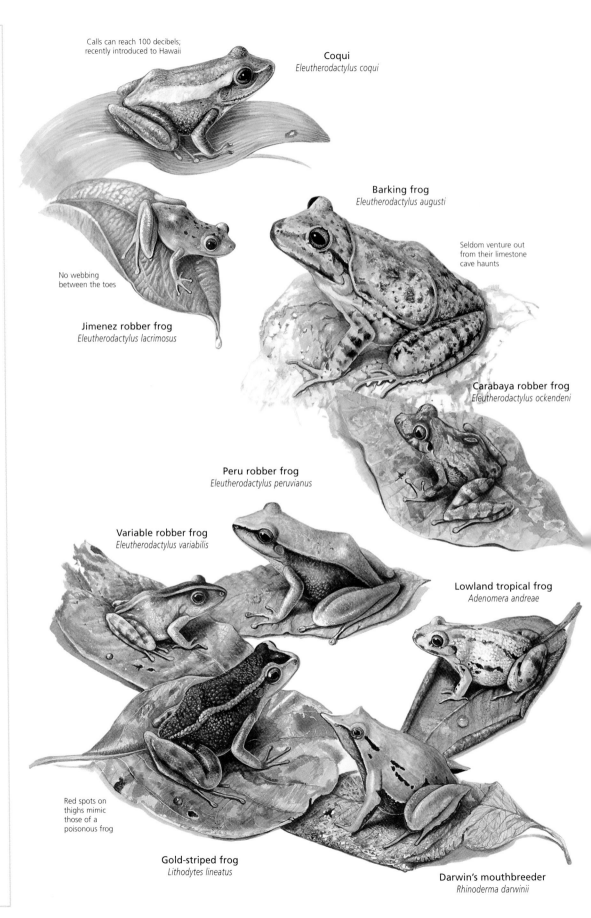

Calls can reach 100 decibels; recently introduced to Hawaii

Coqui
Eleutherodactylus coqui

Barking frog
Eleutherodactylus augusti

Seldom venture out from their limestone cave haunts

No webbing between the toes

Jimenez robber frog
Eleutherodactylus lacrimosus

Carabaya robber frog
Eleutherodactylus ockendeni

Peru robber frog
Eleutherodactylus peruvianus

Variable robber frog
Eleutherodactylus variabilis

Lowland tropical frog
Adenomera andreae

Red spots on thighs mimic those of a poisonous frog

Gold-striped frog
Lithodytes lineatus

Darwin's mouthbreeder
Rhinoderma darwinii

ADVANCE OF THE CANE TOAD

Successful stories of the cane toad (*Bufo marinos*) (also known as the marine toad) being introduced into the sugarcane fields of Puerto Rico to control insect pests convinced the sugarcane growers in Queensland, Australia, to import and breed 100 adult toads. In 1935, they released 62,000 subadults in Queensland to control the grayback cane beetle (*Dermolepida albohirtum*). Since the neotropical cane toad has no natural predators in Australia, populations have reached plague proportions. The toads adapted well to feeding on all animals smaller than themselves, even competing with dogs for the rations in their bowls. Studies in Brazil are yet to come up with a solution.

paratoid gland

Toad nightmare The cane fields did not have proper daytime cover so the introduced toads dispersed into the countryside. They are now so abundant that gardens become a moving mass of toads at night and the roads are slippery with splattered bodies.

Cane toad poison When cane toads are threatened, they inflate their bodies and begin sweating a milky, latex-like liquid from the paratoid glands. In extreme cases, the secretions can be projected up to 3 feet (92 cm) toward the aggressive attacker, but are not toxic unless ingested. The poison has been used as a hallucinogenic drug.

Breeding The cane toad breeds year-round in Australia in temporary pools, stock tanks, lakes, and streams. The prolific toads are capable of producing up to 13,000 eggs at a time. Their tadpoles transform rapidly and have minimal habitat requirements: algae and water. Although cane toads take several years to mature, they can live for over 20 years. Calling cane toads interfere with the breeding choruses of native frogs.

Indiscriminate eaters The cane toad grows up to 9 inches (23 cm) long and can weigh over 2¼ pounds (1 kg). It eats nearly anything smaller than itself, including many beneficial frogs. The cane toad above is swallowing a pygmy possum.

Fatal meal The cane toad's paratoid glands contain a secretion that is fatal to some animals. The cane toad pictured right has angled its paratoid glands toward the approaching snake. Populations of Australian snakes that normally consume native anurans are diminishing because they die from swallowing cane toads. This is also a problem for other indigenous animals. Birds, dingos, monitor lizards, and other anurans see the cane toad as possible prey. Fishes die from ingesting the toxic tadpoles. And the problem is not restricted to areas into which it has been introduced: in South America, a group of Peruvian Indians is believed to have perished after eating a soup made from cane toad eggs.

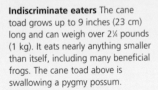

FACT FILE

Colorado river toad The skin glands of these toads produce a toxin that can cause intoxication and hallucination in humans, and is a controlled substance in the United States, where people have been arrested for milking toads.

🐾 Up to 8 in (20 cm)
⬭ Terrestrial
⇆ Summer
🕯 Common

S.W. USA & N.W. Mexico

Common Asian toad Once common, this species is declining due to habitat alteration and loss, caused by drought, deforestation, drainage of habitat, pesticides, fertilizers, and pollutants.

🐾 Up to 6 in (15 cm)
⬭ Terrestrial
⇆ Summer
🕯 Declining

S.E. Asia

AMPHIBIAN EGG

Amphibian embryos have gelatinous membranes around them, but lack the protective amnion present in reptiles. Since eggs also lack shells, they must be protected from desiccation. Eggs contain nutrients for the tadpoles to develop into free-swimming larvae. Eggs of some species have enough yolk to nourish the embryos until they hatch as miniature adults.

Tadpole developing in egg
Anurans must either deposit their eggs in water or in moist places to avoid desiccation.

White chest and abdomen

Raised ridges of warts

1 Cururu toad *Bufo paracnemis*	**4** Cameroon toad *Bufo superciliaris*	**7** Colorado river toad *Bufo alvarius*
2 Colombian giant toad *Bufo blombergi*	**5** Square-marked toad *Bufo regularis*	**8** Berber toad *Bufo mauritanicus*
3 Cane toad *Bufo marinus*	**6** Tschudi's Caribbean toad *Bufo pelticephalus*	**9** Common Asian toad *Bufo melanostictus*

Everett's Asian tree toad
Pedostibes everetti

Horned toad
Bufo ceratophrys

Sonoran green toad
Bufo retiformis

Red toad
Schismaderma carens

South American common toad
Bufo margaritifer

Oak toad
Bufo quercinus

Red-spotted toad
Bufo punctatus

Two color morphs
of harlequin frogs

Green toad
Bufo viridis

Harlequin frog
Atelopus varius

FACT FILE

Horned toad This diurnal toad lives on the forest floor where it forages for termites and breeds in forest streams.

- Up to 3¼ in (8.2 cm)
- Terrestrial
- Unknown
- Common

N.W. Amazonia (South America)

Green toad Green toads form dense populations in areas altered by people, and these are often much higher than in adjacent natural habitats. They prefer burrows and live in the burrows of rodents. Abundance can reach more than 100 individuals per 1,075 square feet (100 sq. m).

- Up to 4¼ in (12 cm)
- Terrestrial
- Spring
- Common

Europe to Asia

Harlequin frog No courtship has been reported for this species. Amplexus is prolonged; females carry males on their backs from a few days to over a month until about 20 eggs are laid in water.

- Up to 2 in (5 cm)
- Terrestrial, aquatic
- Fall
- Declining

S. Costa Rica, Panama, N. Colombia

TOADS VERSUS FROGS

In Europe and North America, toads (family Bufonidae) have short legs for hopping, dry warty skin, and are terrestrial. Frogs (family Ranidae) have long, slender legs for leaping great distances, moist skin, and are aquatic. However, the use of the terms "frog" and "toad" depends on the region of the world you are in. In Africa, the smooth and moist-skinned aquatic Cape clawed frog (*Xenopus gilli*) is called a clawed toad.

To kiss a toad
Even though South Africans call Bufo pardalis *(above) the leopard frog, it belongs to the toad family Bufonidae.*

FACT FILE

Australian lace-lid frog This frog has huge eyes with vertically elliptical pupils and a lower eyelid with a golden reticulation. It is found on rocks and vegetation in fast-flowing rocky streams in rain forest up to 3,900 feet (1,200 m).

- Up to 2½ in (6 cm)
- Arboreal
- Spring, summer
- Endangered

N.E. Australia

Southern bell frog Males call while floating on the water. The call is a low growl lasting about a second. The eggs are deposited in a floating jelly that later sinks. They feed on other frogs and even their own species.

- Up to 4 in (10 cm)
- Terrestrial, aquatic
- Summer
- Locally common

S.E. Australia & Tasmania

Short-footed frog This is a robust, burrowing frog with striking and highly variable markings. The back is dark brown with silver-brown blotches. The frog usually also has a silver-brown stripe down the back and a white belly.

- Up to 2 in (5 cm)
- Terrestrial, burrowing
- Unknown
- Common

N.E. Australia

Paradox frog This frog has large webbed feet and extremely slippery skin. The tadpoles are three times as large as the adults, up to 10 inches (25 cm) in length.

- Up to 3 in (7.5 cm)
- Aquatic
- Rainy season
- Common

Amazonia (South America) to N. Argentina

Water-holding frog This species spends the long dry periods, which can be up to several years long, in burrows deep underground. It surrounds itself with a cocoon-like chamber, formed with sloughed dead layers of its skin.

- Up to 2½ in (6 cm)
- Terrestrial, burrowing
- Variable
- Common

Central Australia

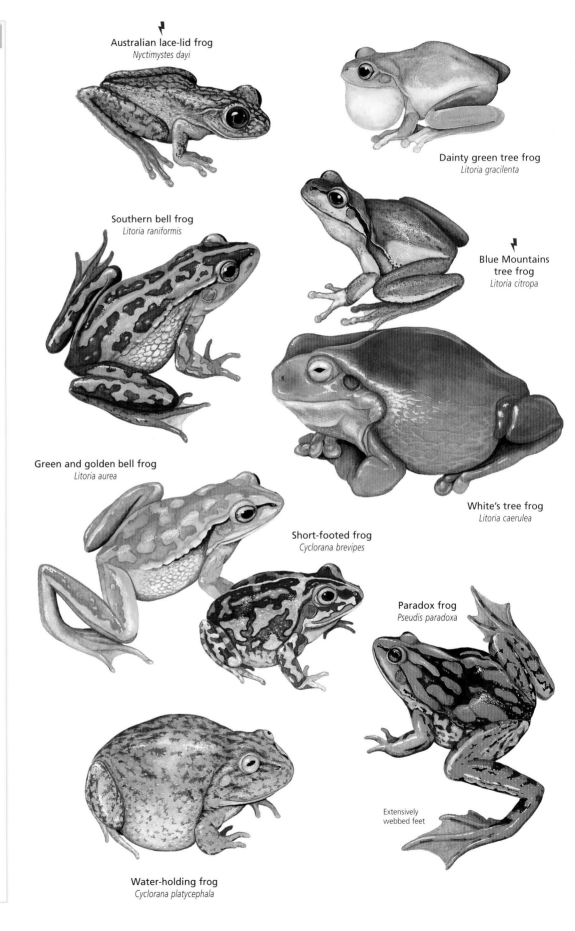

Australian lace-lid frog
Nyctimystes dayi

Dainty green tree frog
Litoria gracilenta

Southern bell frog
Litoria raniformis

Blue Mountains tree frog
Litoria citropa

Green and golden bell frog
Litoria aurea

Short-footed frog
Cyclorana brevipes

White's tree frog
Litoria caerulea

Paradox frog
Pseudis paradoxa

Extensively webbed feet

Water-holding frog
Cyclorana platycephala

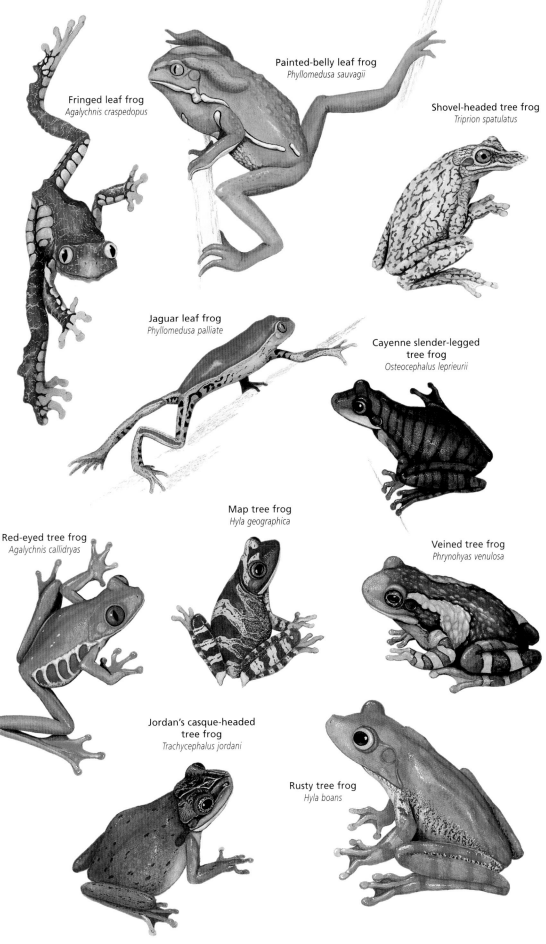

Fringed leaf frog
Agalychnis craspedopus

Painted-belly leaf frog
Phyllomedusa sauvagii

Shovel-headed tree frog
Triprion spatulatus

Jaguar leaf frog
Phyllomedusa palliate

Cayenne slender-legged
tree frog
Osteocephalus leprieurii

Red-eyed tree frog
Agalychnis callidryas

Map tree frog
Hyla geographica

Veined tree frog
Phrynohyas venulosa

Jordan's casque-headed
tree frog
Trachycephalus jordani

Rusty tree frog
Hyla boans

FACT FILE

Painted-belly leaf frog This tree frog coats itself with a waxy secretion to keep in moisture. It lays eggs on a leaf and then rolls it up with this waxy skin secretion.

📏	Up to 3¼ in (8.5 cm)
⊕	Arboreal
☷	Rainy season
🏹	Common

Gran Chaco (N. Paraguay, N. Argentina, E. Bolivia, S. Brazil)

Shovel-headed tree frog The skin of this frog's head is fused to the cranial bone. When in retreats, such as holes in trees or rock crevices, the frog plugs the hole with its head to prevent desiccation and predation.

📏	Up to 3 in (7.5 cm)
⊕	Terrestrial, arboreal
☷	Summer
🏹	Common

W. Mexico

Red-eyed tree frog This species lays its green eggs on vegetation or rocks overhanging temporary ponds. The tadpoles hatch and fall into the water to complete development. The males become sexually mature after a year.

📏	Up to 3 in (7.5 cm)
⊕	Arboreal
☷	Summer
🏹	Common

S. Mexico to Colombia

FROG FEET

Frogs have four digits on the forelimbs and five on the hindlimbs. The shapes vary depending on their habits: fully webbed for swimming, or toe pads for climbing, or modified claws with extra tubercules for digging.

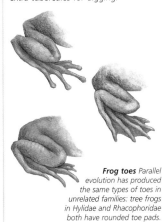

Frog toes *Parallel evolution has produced the same types of toes in unrelated families: tree frogs in Hylidae and Rhacophoridae both have rounded toe pads.*

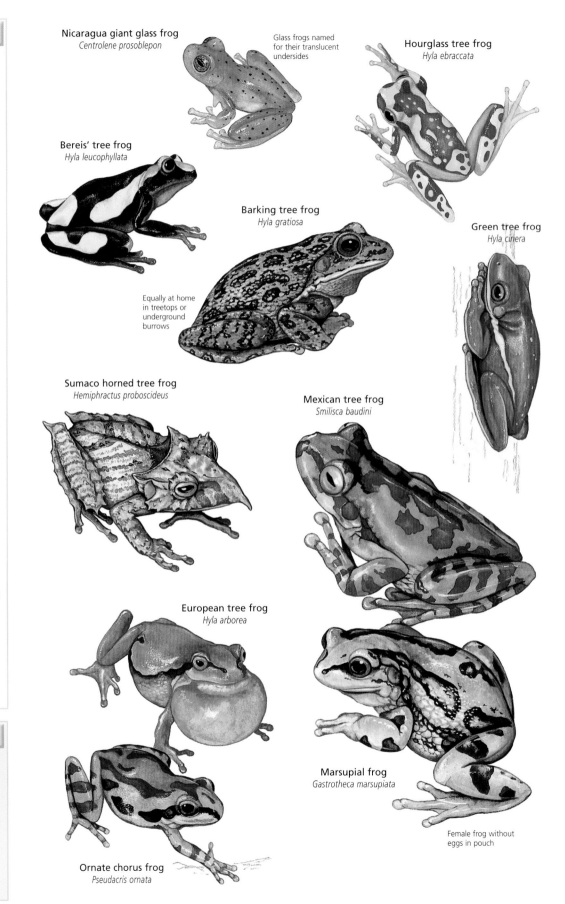

Nicaragua giant glass frog
Centrolene prosoblepon

Glass frogs named for their translucent undersides

Hourglass tree frog
Hyla ebraccata

Bereis' tree frog
Hyla leucophyllata

Barking tree frog
Hyla gratiosa

Green tree frog
Hyla cinera

Equally at home in treetops or underground burrows

Sumaco horned tree frog
Hemiphractus proboscideus

Mexican tree frog
Smilisca baudini

European tree frog
Hyla arborea

Marsupial frog
Gastrotheca marsupiata

Female frog without eggs in pouch

Ornate chorus frog
Pseudacris ornata

TREE FROGS

The family Hylidae—tree frogs—comprises 855 species in 42 genera, distributed primarily in North America, South America, and Australia, and to a lesser extent in Europe and Asia. Tree frogs range in size from the half an inch (12 mm) adult javelin frog (*Litoria microbelos*) to the 5½ inch (14 cm) Hispaniola tree frog (*Hyla vasta*). Most hylids are arboreal, but some are terrestrial, aquatic, or burrowing. Most possess adhesive toe pads, which allow them to cling to any surface. Almost all are very flattened and streamlined, with long legs that are useful for leaping from branch to branch. Tree frogs have horizontally elliptical pupils, with the exception of the Phyllomedusids, which have vertical pupils.

Striking skin The Bereis' tree frog has a variable color pattern resembling the markings of a giraffe; other tree frogs are less striking. Large toe pads and loose belly skin allow this frog to stick to leaves and other smooth surfaces when climbing and searching for prey.

Burrowing habits The burrowing tree frog (*Pternohyla fodiens*) is found in many habitats subject to seasonal inundation. They are well adapted for burrowing and spend the dry season in a dormant state beneath the surface. They will only breed after heavy rain.

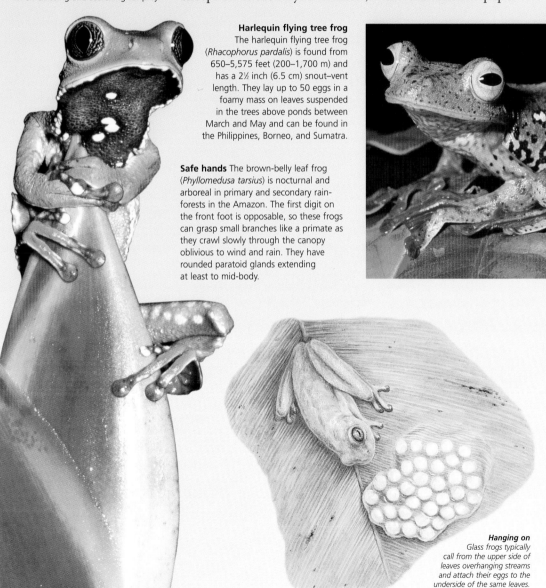

Harlequin flying tree frog The harlequin flying tree frog (*Rhacophorus pardalis*) is found from 650–5,575 feet (200–1,700 m) and has a 2½ inch (6.5 cm) snout–vent length. They lay up to 50 eggs in a foamy mass on leaves suspended in the trees above ponds between March and May and can be found in the Philippines, Borneo, and Sumatra.

Safe hands The brown-belly leaf frog (*Phyllomedusa tarsius*) is nocturnal and arboreal in primary and secondary rain-forests in the Amazon. The first digit on the front foot is opposable, so these frogs can grasp small branches like a primate as they crawl slowly through the canopy oblivious to wind and rain. They have rounded paratoid glands extending at least to mid-body.

Hanging on
Glass frogs typically call from the upper side of leaves overhanging streams and attach their eggs to the underside of the same leaves.

Reproduction in the trees Males have species-specific vocalizations that they use to attract gravid females. Amplexus stimulates the females to deposit eggs, usually in water, where they hatch into free-swimming tad-poles. Many species deposit their eggs in overhanging vegetation; when the eggs hatch, the tadpoles fall into the water to finish their development. Some species deposit their eggs in bromeliads and the tadpoles develop there. Parental care in the casque-headed tree frog (*Osteocephalus oophagus*) involves males guarding their nests after oviposition and females depositing yolked unfertilized eggs for their tadpoles developed from eggs previously deposited in bromeliads or tree holes.

FACT FILE

Nosy Be giant tree frog This species' call is a single whistling note that lasts for only 55–65 milliseconds. The call is repeated over a hundred times a minute. Their hands lack webbing and the feet are barely webbed.

⚑ Up to 1¼ in (3 cm)
🌐 Terrestrial, arboreal
⇄ Unknown
⚐ Common

N.W. Madagascar

Dotted humming frog This frog exists mutualistically with tarantulas in their burrows, feeding on ants that prey on tarantula eggs. At night, the frogs sit between the legs of the spiders at the mouths of their burrows.

⚑ Up to 1 in (2.5 cm)
🌐 Terrestrial
⇄ Summer
⚐ Common

E. Ecuador, S.E. Peru, N.W. Bolivia

Tomato frog The bright red coloration of this frog acts as a clear warning to potential predators that it is a toxic meal. A white, sticky fluid secreted from the skin deters snakes and can produce an allergic reaction in humans.

⚑ Up to 4 in (10 cm)
🌐 Terrestrial, aquatic
⇄ Summer
⚐ Vulnerable

N.E. Madagascar

RAIN FROGS OF AFRICA

Rain frogs (genus *Breviceps*) have short heads and stout bodies. They appear above ground only during torrential rain. Males have loud, bellowing calls that can be heard from great distances, and sticky skin secretions that attach them to the females during amplexus. They lay their eggs underground, and the young frogs develop without any need for water.

Puffed up like a balloon
Desert rain frog males (Breviceps maacrops) inflate their bodies enormously when calling for mates.

Ornate rice frog
Microhyla ornata

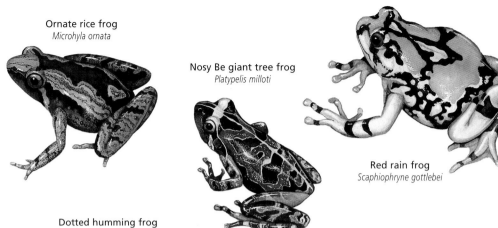

Nosy Be giant tree frog
Platypelis milloti

Red rain frog
Scaphiophryne gottlebei

Dotted humming frog
Chiasmocleis ventrimaculata

Red-banded frog
Phrynomantis bifasciata

Shown in defensive posture, with the head lowered and the rear legs extended

Muller's termite frog
Dermatonotus muelleri

Great Plains
narrow-mouthed toad
Gastrophryne olivacea

Malaysian
narrow-mouthed toad
Kaloula pulchra

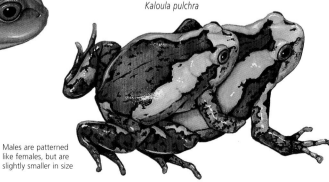

Tomato frog
Dyscophus antongilii

Males are patterned like females, but are slightly smaller in size

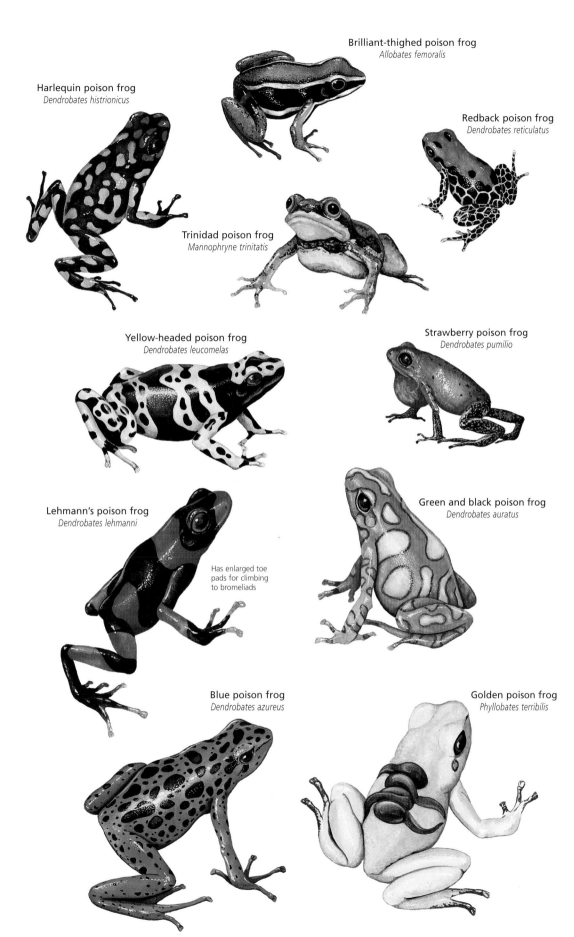

Brilliant-thighed poison frog
Allobates femoralis

Harlequin poison frog
Dendrobates histrionicus

Redback poison frog
Dendrobates reticulatus

Trinidad poison frog
Mannophryne trinitatis

Yellow-headed poison frog
Dendrobates leucomelas

Strawberry poison frog
Dendrobates pumilio

Lehmann's poison frog
Dendrobates lehmanni

Has enlarged toe
pads for climbing
to bromeliads

Green and black poison frog
Dendrobates auratus

Blue poison frog
Dendrobates azureus

Golden poison frog
Phyllobates terribilis

FACT FILE

Harlequin poison frog The male calls
from branches up to 3 feet (92 cm) high.
The male often sits on the back of the
female on the way to the oviposition
site. Their complex courtship ritual,
which lasts for 2–3 hours, involves a
sequence of sitting, bowing, crouching,
touching, and circling behavior patterns.

Up to 1½ in (4 cm)
Terrestrial, arboreal
Rainy season
Common

Colombia, Ecuador

Redback poison frog Females lay
either two or three eggs, ⅛ inch (2 mm)
in diameter. Although they are primarily
found on the forest floor, males hop
up tree trunks carrying one or two
tadpoles to deposit in bromeliads.

Up to ¾ in (2 cm)
Terrestrial, arboreal
Rainy season
Common

N.E. Peru, W. Brazil

Blue poison frog Females lay eggs
in water for the males to fertilize. The
males usually guard the eggs until the
tadpoles develop after about 12 days.
At 12 weeks, the tadpoles metamorphose into grown frogs.

Up to 2 in (5 cm)
Terrestrial
Rainy season
Uncommon

Suriname

POISON-ARROW FROGS

Indigenous tribes coat their dart tips
with the secretions of Dendrobatids,
or poison-arrow frogs. This toxin blocks
neural transmission at the acetylcholine
receptor of the neuro-muscular junc-
tion, which causes paralysis or even
death. To produce these skin toxins,
the frogs need to consume ants that
produce formic acid.

Madagascar reed frog
Heterixalus madagascariensis

Striped spiny reed frog
Afrixalus dorsalis

Distinctive yellow stripe on body, bordered on both sides with black

Yellow-striped reed frog
Hyperolius semidiscus

Natal forest tree frog
Leptopelis natalensis

Seychelles reed frog
Tachycnemis seychellensis

Marbled reed frog
Hyperolius marmoratus

Weal's running frog
Semnodactylus wealii

Tulear golden frog
Mantella expectata

Runs instead of hopping, jumping, or crawling

Red-legged kassina
Kassina maculata

Madagascar golden frog
Mantella madagascariensis

Red-eared frog
Rana erythraea

African bullfrog
Pyxicephalus adspersus

European common frog
Rana temporaria

Pickerel frog
Rana palustris

Malaysian frog
Rana luctuosa

Marsh frog
Rana ridibunda

Pig frog
Rana grylio

Dorsum has two rows of square chocolate-colored blotches, while the leopard frog, for which it has been mistaken, usually has round blotches

Amazon river frog
Rana palmipes

Bullfrog
Rana catesbeiana

MASS SPAWNING

The European common frog mates en masse in 3 days in spring, each laying about 400 black eggs in gelatinous clusters. The eggs absorb UV radiation, which increases the water temperature and thus the eggs' developmental rate.

Group sex
Mass breeding reduces the chance of predation by swamping predators with more than they could possibly eat.

ANURAN DEFORMITIES

Like the canaries in the coal mines, frogs are whistling a warning about chemical pollution in the environment when they begin appearing with grossly deformed bodies. Part of the chemical pollution is hormonal from the progesterone used in birth control pills not being filtered out at sewage-treatment plants.

Multi-legged frog
Notice how chemical pollution in the breeding ponds caused this frog to develop with extra legs.

Dahaoping sucker frog
Amolops viridimaculatus

Borneo splash frog
Staurois tuberilinguis

Banded stream frog
Strongylopus bonaspei

Cape dainty frog
Cacosternum capense

Mascarene Ridge frog
Ptychadena mascareniensis

Bulbous glands on the back and sides

Striped sand frog
Tomopterna cryptotis

Singapore wart frog
Limnonectes malesianus

African ornate frog
Hildebrandtia ornata

Spotted snout-burrower
Hemisus guttatus

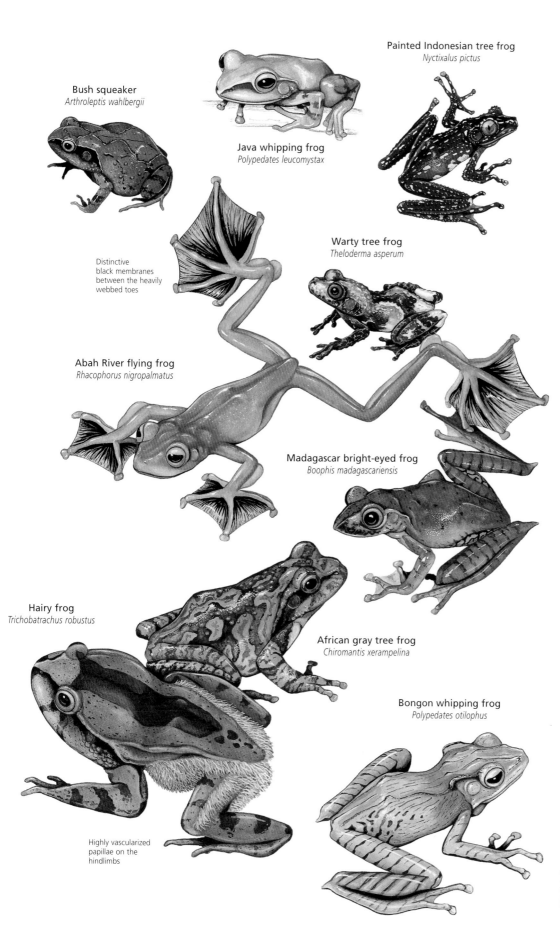

Bush squeaker
Arthroleptis wahlbergii

Painted Indonesian tree frog
Nyctixalus pictus

Java whipping frog
Polypedates leucomystax

Distinctive
black membranes
between the heavily
webbed toes

Warty tree frog
Theloderma asperum

Abah River flying frog
Rhacophorus nigropalmatus

Madagascar bright-eyed frog
Boophis madagascariensis

Hairy frog
Trichobatrachus robustus

African gray tree frog
Chiromantis xerampelina

Bongon whipping frog
Polypedates otilophus

Highly vascularized
papillae on the
hindlimbs

CONTROLLED FALLING

Frogs possessing expanded webs of the feet are able to fall in a controlled fashion to avoid predation or move about. They can adjust the impact and control the direction of their fall.

Expanded webbing *The Java flying frog* (Rhacophorus reinwardtii) *uses its expanded webbed feet to guide its leaps from tree to tree.*

FISHES

FISHES

PHYLUM	Chordata
SUBPHYLUM	Vertebrata
CLASSES	5
ORDERS	62
FAMILIES	504
SPECIES	25,777

With more than 25,000 described so far, and perhaps several thousand more still to be identified, fishes comprise over half of all living vertebrate species. They first evolved about 500 million years ago in fresh water and now exploit almost every aquatic habitat from polar seas to tropical ponds. Some even live briefly on land. This diverse group includes tiny gobies ⅖ inch (1 cm) long and massive 40 foot (12 m) whale sharks, drab wobbegongs and vibrantly colored butterfly fish, fiercely predatory great white sharks and placid algae-grazing parrotfish. Five classes survive today: hagfishes, lampreys, cartilaginous fishes, lobe-finned fishes, and ray-finned fishes.

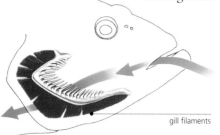

gill filaments

Gill-breathing Water enters the mouth and passes over feather-like gill filaments that are richly endowed with capillaries. Each filament has many folds, creating an enormous area for gas exchange. As blood and water pass closely in opposite directions, oxygen diffuses into the capillaries as carbon dioxide is expelled.

AQUATIC ADAPTATIONS

Many adaptations in the fishes have been driven by water's physical and chemical properties. The basic streamlined torpedo fish body shape, for example, has evolved to move through this medium, which is 800 times denser than air.

Forward motion in most species is created by oscillations of the tail fin and body, the other fins providing stability and maneuverability. The demands of such movements mean swimming muscles make up about half the body weight of most fishes. Maneuverability is enhanced in water by neutral buoyancy (weightlessness), achieved in most fishes by a swim bladder, an internal gas-filled sac.

The aquatic medium has also propelled the evolution of specialist sensory equipment, such as the lateral line organs. These usually run the length of the body and detect tiny changes in the surrounding water pressure, assisting in both prey location and avoiding obstacles.

Some fishes reproduce by means of internal fertilization and produce live young. Water, however, provides for the ready mixing of sex cells and dispersal of offspring and so most fishes lay eggs that are fertilized and hatched outside the female.

Fishes are the only vertebrates with true fins and most use them to swim. In some, however, these appendages allow for "walking," while in others they accommodate brief bouts of "flight."

One of the most important early evolutionary events in fishes, and indeed all vertebrates, was the development of jaws. This is thought to have first occurred in the fishes about 450 million years ago. Jaws probably evolved from gill arches. It is thought one of the anterior arches became fused to the skull, its upper section developing into the top jaw, the lower section developing into the bottom jaw.

The earliest fishes were filter feeders, but the evolution of jaws expanded options considerably and underpinned the enormous diversification of the group that followed.

Fishes-in-waiting Many fishes emerge from externally laid eggs as larvae to exploit different habitats to their parents before metamorphosing into adult form. In others, such as the striped blenny (*Chasmodes bosquianus*) (above), young pass their larval stage in the egg and emerge with the appearance of tiny adults.

Daredevils Some of the largest fishes are among the devil rays (right), a group which includes species that can attain widths of 23 feet (7 m) and weigh up to 2,200 pounds (1,000 kg). Despite their size, some can leap clear of the water.

Leaping devil
The acrobatic lesser devil ray (Mobula hypostoma) is a small Atlantic species of up to 4 feet (1.2 m) wide.

Extreme specialist The honeycomb cowfish (*Acanthostracion polygonius*) (right) belongs to an advanced group that has lost the typical torpedo-shaped streamlining. The plate-like armor prevents body movement but the fins are able to facilitate finely controlled "hovering."

HERMAPHRODITISM

Most fishes are dioecious; they are either male or female from an early age and remain that way throughout life. Hermaphroditism (where individuals possess both male and female sex organs during their lives) is, however, widespread in the group, much more so than among other vertebrates. In some species, individuals are sequential hermaphrodites, changing gender once they reach a certain size or when there is a shortfall of one sex or the other. Less common are simultaneous hermaphrodites, which can perform as either sex at the same time. Hermaphroditism tends to be more prevalent among fishes of lower latitudes. The fish life on tropical coral reefs, in particular, is noted for the phenomenon.

Sex-changing eel Ribbon morays (*Rhinomuraena quaesita*) are protandrous hermaphrodites: males become females. This male will turn bright yellow as it matures into a female.

Dominance changeover Among spotbreast angelfish (*Genicanthus melanospilos*), which are protogynous hermaphrodites, a single dominant male is associated with a small harem of females. In response to the death or removal of the male, the largest female grows larger and then changes color before becoming the dominant male.

Juvenile: not sexually active

Initial phase: usually female

Terminal phase: always a mature male

Replaceable males Hermaphroditism is common in parrotfishes. A dominant male often maintains a harem of females but, if he disappears, the largest and most aggressive female undergoes a sex change accompanied by a dramatic color change, all in a matter of weeks.

1. Harem life *Female spotbreast angelfish are yellow on their upper surface and pale blue below, with strong black lines defining their tails. Harems usually contain between three and five females.*

4. Quick change *The sexual transformation is completed within about 14 days and the new male begins spawning with the members of his harem.*

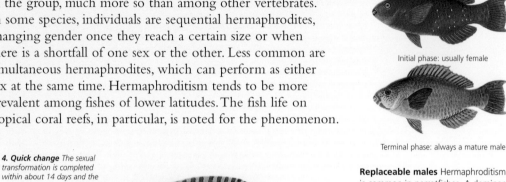

3. Girl power *As her sexual transformation continues, her body reabsorbs its eggs and begins sperm production. She begins behaving more aggressively and displays courtship behavior toward the harem's remaining females.*

2. Changing colors *When the dominant male disappears, the harem's largest female begins to grow and develop the distinctive coloration of a male: a pale blue body with black stripes.*

JAWLESS FISHES

SUPERCLASS Agnatha	
CLASSES 2	
ORDERS 2	
FAMILIES 2	
SPECIES 105	

These were the first fishes and most became extinct by about 360 million years ago. The two surviving groups—hagfishes and lampreys—are probably only distantly related to each other. They comprise 105 species. All lack scales and jaws and have cartilaginous skeletons. True fins are either absent or poorly developed. Hagfishes are eel-like in appearance and produce copious quantities of mucus from slime glands along the length of the body, possibly for defensive reasons. These scavengers of dead and dying fishes and invertebrates have photoreceptors but do not have true eyes or a larval phase. Lampreys have functional eyes and a long larval phase. Larval lampreys are filter feeders but most adults are external parasites on other fishes.

Widespread curiosities Hagfishes and lampreys are found in temperate waters of the Northern and Southern hemispheres, and in cool, deep waters in parts of the tropics. Although lampreys are found in both fresh and salt water, hagfishes are exclusively marine. Larval lampreys burrow into soft stream and river substrates where they feed by filtering algae, detritus, and microorganisms from the water.

Serious suckers Lampreys parasitize other fishes by attaching to their bodies with a toothed oral disk (right). They mainly live on their host's body fluids but sometimes also consume pieces of flesh and even internal body organs. They often cause the death of their hosts. The arrangement of the teeth lining the oral disk and mouth is an important aid to identifying lampreys, which look very similar.

> ⚡ **CONSERVATION WATCH**
>
> **Lamprey losses** The IUCN Red List records that the Lombardy brook lamprey (*Lethenteron zanandreai*) is endangered and the Greek brook lamprey (*Eudontomyzon hellenicus*) and non-parasitic lamprey (*Mordacia praecox*) are vulnerable. Habitat destruction is a major threat.

REPRODUCTION

Hagfish reproduction remains largely a mystery. These fishes begin life as hermaphrodites, later becoming either female or male. They are thought to spawn repeatedly during their lives, each time producing a small number of large eggs—about 1 inch (2.5 cm) long—in toughened cases.

Lampreys only spawn once, producing large numbers of much smaller eggs before dying. They spend their first few years as filter-feeding larvae called ammocoetes. Metamorphosis into young adults usually occurs when they are 3–6½ inches (7.5–16.5 cm) long and takes 3–6 months. Some migrate downstream to the sea.

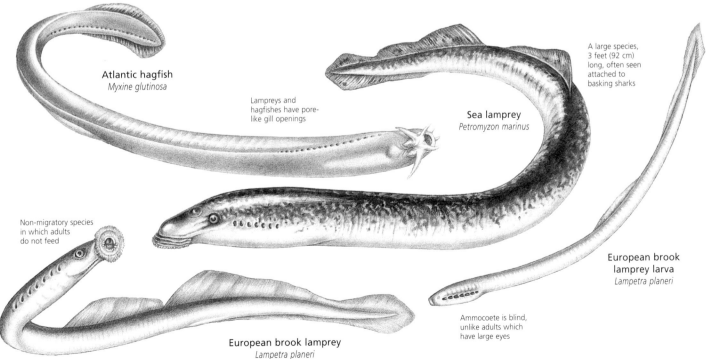

Atlantic hagfish
Myxine glutinosa

Lampreys and hagfishes have pore-like gill openings

Sea lamprey
Petromyzon marinus

A large species, 3 feet (92 cm) long, often seen attached to basking sharks

Non-migratory species in which adults do not feed

European brook lamprey larva
Lampetra planeri

Ammocoete is blind, unlike adults which have large eyes

European brook lamprey
Lampetra planeri

SHARKS, RAYS, AND ALLIES

CLASS Chondrichthyes	
SUBCLASSES 2	
ORDERS 12	
FAMILIES 47	
SPECIES 999	

Having skeletons of cartilage, not bone, sharks and rays are known collectively as cartilaginous fishes. Shark evolutionary history extends back some 400 million years, while rays probably first appeared 200 million years ago. All feed on other animals and most are marine. Teeth are embedded in connective tissue and replaced throughout life; they usually have five, but sometimes six or seven, external gill slits on each side of the mouth; and all have internal fertilization. Together with skates (a type of ray) and chimaeras (a mainly deepwater group, including spookfishes, ghost sharks, and elephant-fishes), sharks and rays today comprise about 1,000 species.

Shark napping Caribbean reef sharks (above) occur on reefs around islands in the Caribbean Sea. They are known for the way they rest motionlessly in caves and on the ocean bottom as if sleeping.

Gentle giant The largest living fish, the whale shark (left) is about 25 years old and 30 feet (9 m) long before it reaches sexual maturity. It roams the Indian, Pacific, and Atlantic oceans, mouth agape, filtering mainly plankton from surface waters.

ANCIENT FEATURES

Unlike jawless fishes (which also have skeletons of cartilage), sharks, rays, and their close relatives have well-developed jaws, paired nostrils, and paired pectoral and pelvic fins. In addition to their cartilaginous skeletons, they differ from the bony fishes in having dermal denticles and teeth that are replaced throughout their lives or which are fused into continuously growing bony plates.

The class Chondrichthyes, to which the cartilaginous fishes belong, is usually divided into two subclasses. Elasmobranchii is a large group that includes the wide-ranging sharks, skates, and rays. The smaller Holocephali group contains the relatively "primitive" chimaeras. These bizarre-looking and mostly bottom-dwelling fishes differ from the elasmobranchs in having only a single gill cleft and four gills on each side of the head, mainly naked skin, teeth fused into plates, and the upper jaw fused to the skull.

Shark anatomy Basic shark attributes (left) have changed little since the group's early evolution. Body shape is designed for hydrodynamic efficiency and to suit a predatory lifestyle. Fins are thick, stiff, and usually lack spines. All sharks have paired pelvic and pectoral fins and between two and four unpaired anal, caudal, and dorsal fins. Gill openings are visible externally.

upper lobe

precaudal pit

caudal fin

free rear tip

first dorsal fin

spiracle

second dorsal fin

eye

snout

subterminal notch

mouth

nostril

gill openings

rear margin

anal fin

pelvic fin (paired)

pectoral fin (paired)

lower lobe

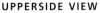

Excellent eyesight Large eyes on short stalks provide this reef-dwelling ray with excellent vision to the front and sides.

UPPERSIDE VIEW

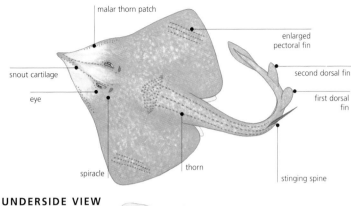

malar thorn patch

snout cartilage

eye

enlarged pectoral fin

second dorsal fin

first dorsal fin

spiracle

thorn

stinging spine

UNDERSIDE VIEW

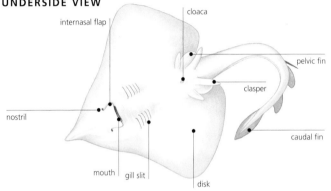

internasal flap

cloaca

nostril

pelvic fin

clasper

caudal fin

mouth | gill slit

disk

REPRODUCTION

All of the cartilaginous fishes have internal fertilization, and elaborate courtship rituals are believed to be widespread in the group. Males do not possess a penis but have claspers instead. These stiffened, fin-like rods, which extend parallel to the body from behind the pelvic fins, are inserted into the female's cloaca to guide sperm. Pup numbers range from 2 to 300 per pregnancy, depending on the species.

Most fishes adopt a strength-in-numbers approach to reproduction, dispersing sometimes millions of tiny eggs at a time so that at least some reach adulthood. In contrast, sharks and rays produce very few offspring but invest far more energy in their early development.

Some sharks and rays lay large yolky eggs that nourish embryos for months before hatching. In most sharks and rays, however, mothers retain young internally to give birth after a lengthy gestation. Either way, baby sharks and rays are born as miniature versions of their parents.

Parental devotion almost always ends with birth. In fact, the pups of many shark species probably separate quickly from their mothers to avoid ending up as prey.

Powerful swimmers Spotted eagle rays (*Aetobatus narinari*) are highly social fish that frequently frolic in groups near the ocean's surface. They move through the water with great strength and grace and have been reported to escape predators by leaping clear of the water.

Feeding migrations The blue-spotted ribbontail ray (*Taeniura lymma*) (above) moves in small groups on the rising tide to graze on invertebrates living on reefs. It returns with the falling tide to deeper water to hide under ledges and in caves.

Ray anatomy Rays have flattened, disk-like bodies with eyes on the dorsal surface and a ventrally located mouth (left). They also have five or six pairs of gill slits located on the underside of the body, just behind the mouth. Rays take in water for respiration through large openings on the upper surface of the head called spiracles, as well as or instead of through the mouth. The long slender tail is often equipped with sharp spines to deter predators and is not used for swimming. Instead, rays propel themselves using their greatly enlarged wing-like pectoral fins. The caudal and dorsal fins are often reduced or lacking.

SHARKS

CLASS Chondrichthyes	
SUBCLASS Elasmobranchii	
ORDERS 8	
FAMILIES 31	
SPECIES 415	

Although less than 2 percent of living fish species are sharks, they are of crucial ecological importance in marine ecosystems because they are apex predators. As a result they occur at naturally low levels of abundance compared with most bony fishes. This, combined with low reproductive rates, makes sharks particularly vulnerable to overexploitation. Most only reach sexual maturity after 6 years, and some not until 18 years or more. They produce few young and embryos undergo long periods of development before birth. Chimaeras are related to the sharks but belong to the subclass Holocephali, which contains a single order with 3 families and 37 species.

Widespread distribution Sharks are found throughout the world's oceans, although very few species can tolerate polar waters. Most prefer shallow marine habitats but a small number of species, such as dogfish sharks, dwell at great depths.

"Mermaid purses" Shark eggs are large and yolky. The purse-like outer layer is made of a tough keratinoid (horny) protein and can be up to 6¾ inches (17 cm) long. Each contains a single embryo that can take up to 15 months to mature.

Gatherings Sharks don't school in the same highly coordinated way that bony fishes often do. But many species form aggregations. In the Sea of Cortez, large day-time groups of mainly juvenile female hammerheads form regularly in areas of high productivity (left). Reasons for this remain unclear.

PREDATORY LIVES

Because most sharks are highly active predators, they tend to be strong, agile swimmers, and some species make long migrations covering many hundreds of miles (km) in search of food. As they have slower metabolisms than most fishes, sharks do not need to feed as frequently and, despite their reputation as irrepressible killers, only hunt when necessary. Prey typically includes small fishes and invertebrates but the larger sharks also hunt sea turtles and marine mammals.

They have no internal swim bladder but other adaptations help to increase buoyancy and enhance their swimming capabilities. The cartilaginous skeleton, which is lighter than true bone, contributes. And large livers with high levels of oil also help keep many afloat. They do need, however, to keep swimming to avoid dropping to the ocean floor.

Sharks reduce water loss by retaining urea in their tissues.

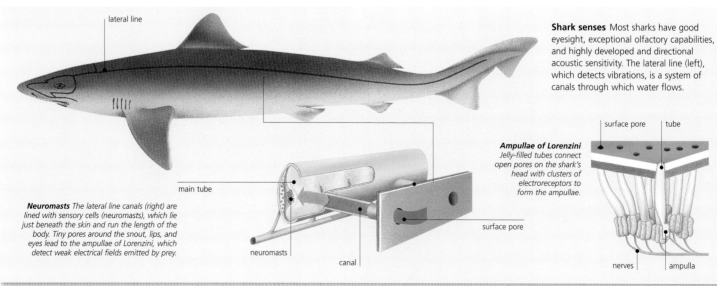

Shark senses Most sharks have good eyesight, exceptional olfactory capabilities, and highly developed and directional acoustic sensitivity. The lateral line (left), which detects vibrations, is a system of canals through which water flows.

lateral line

Ampullae of Lorenzini Jelly-filled tubes connect open pores on the shark's head with clusters of electroreceptors to form the ampullae.

surface pore

tube

main tube

surface pore

Neuromasts The lateral line canals (right) are lined with sensory cells (neuromasts), which lie just beneath the skin and run the length of the body. Tiny pores around the snout, lips, and eyes lead to the ampullae of Lorenzini, which detect weak electrical fields emitted by prey.

neuromasts

canal

nerves

ampulla

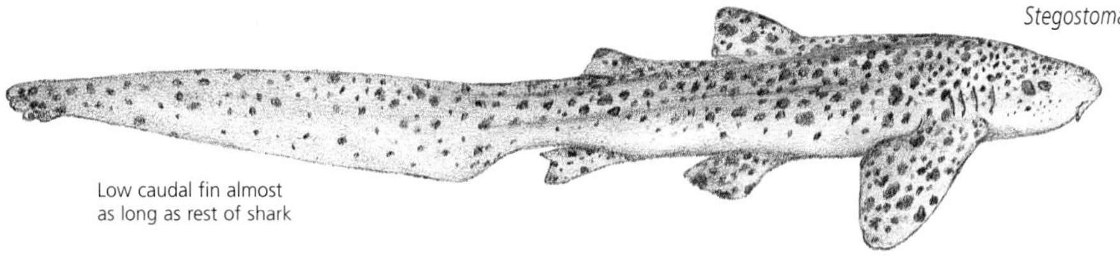
Low caudal fin almost
as long as rest of shark

Zebra shark
Stegostoma fasciatum

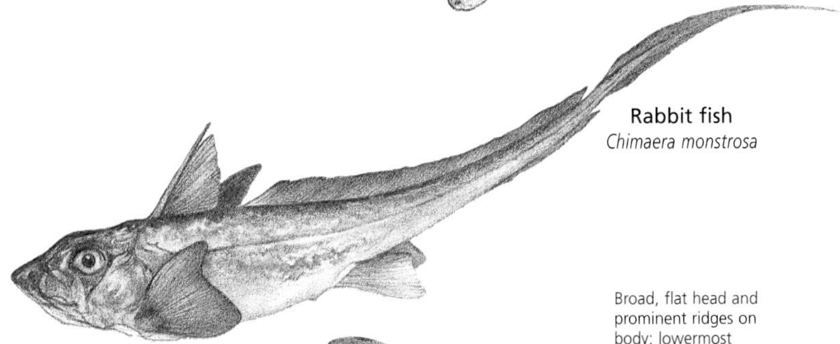
Males have
a clasper on
their forehead,
in addition to
pelvic claspers

Rabbit fish
Chimaera monstrosa

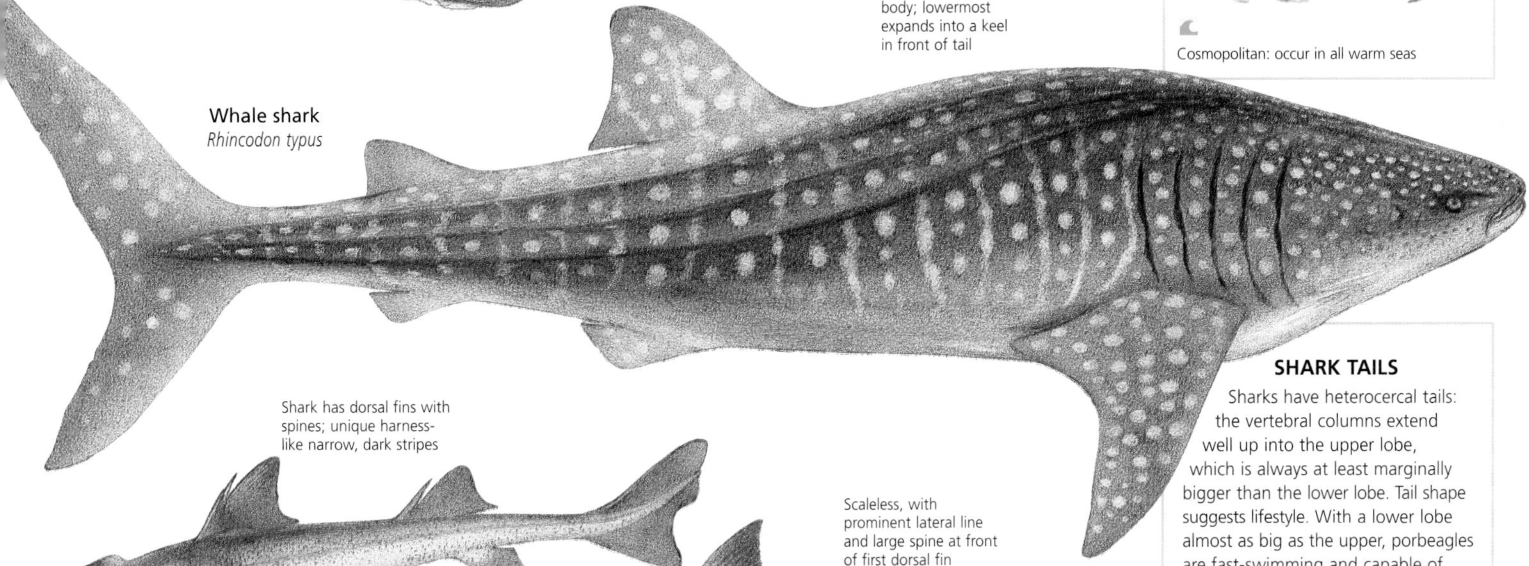
Broad, flat head and
prominent ridges on
body; lowermost
expands into a keel
in front of tail

Whale shark
Rhincodon typus

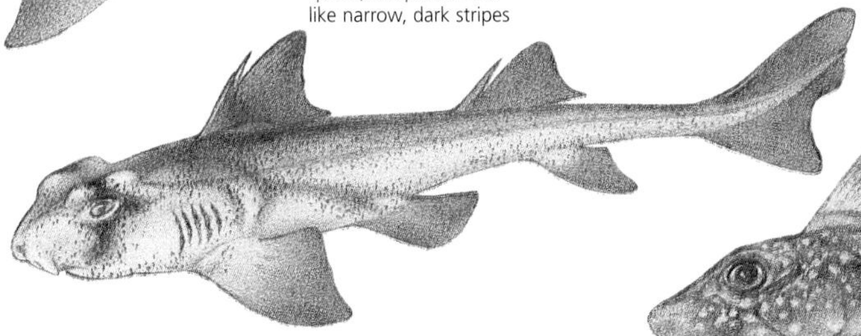
Shark has dorsal fins with
spines; unique harness-
like narrow, dark stripes

Port Jackson shark
Heterodontus portusjacksoni

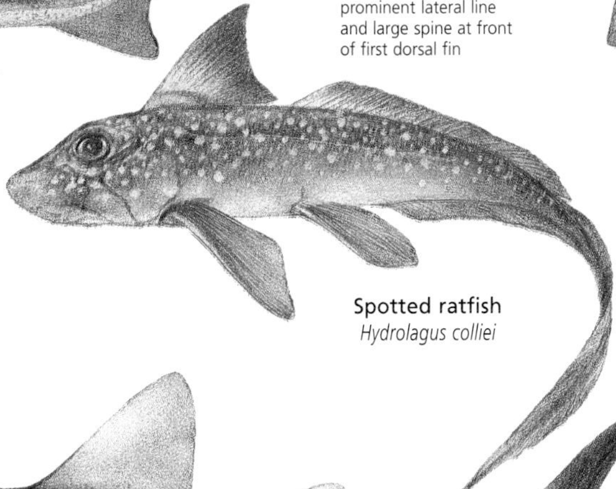
Scaleless, with
prominent lateral line
and large spine at front
of first dorsal fin

Spotted ratfish
Hydrolagus colliei

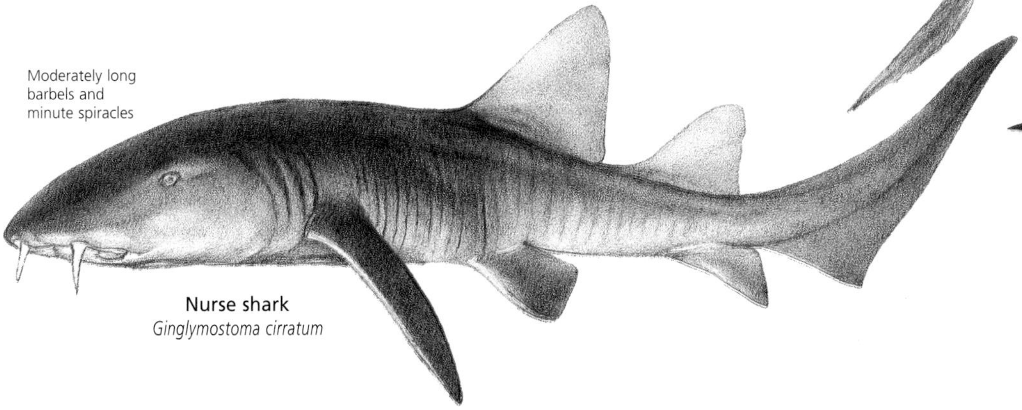
Moderately long
barbels and
minute spiracles

Nurse shark
Ginglymostoma cirratum

FACT FILE

Whale shark Massive, filter-feeding whale sharks are believed to be highly migratory. They sometimes form loose aggregations of a hundred or more individuals but it is thought that most lead solitary lives.

- Up to 59 ft (18 m)
- Up to 44,100 lb (20,000 kg)
- Ovoviviparous
- Male & female
- Vulnerable

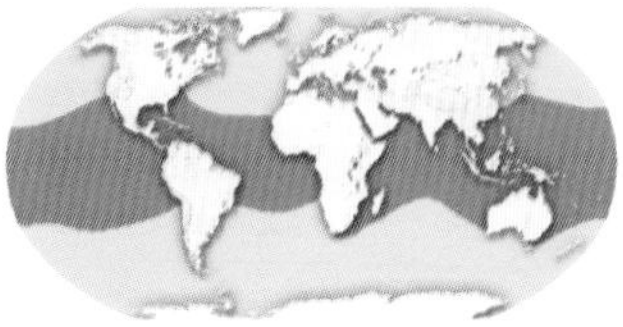

Cosmopolitan: occur in all warm seas

SHARK TAILS

Sharks have heterocercal tails: the vertebral columns extend well up into the upper lobe, which is always at least marginally bigger than the lower lobe. Tail shape suggests lifestyle. With a lower lobe almost as big as the upper, porbeagles are fast-swimming and capable of rapid acceleration. The marked tail asymmetry in the tiger shark indicates a slower swimmer.

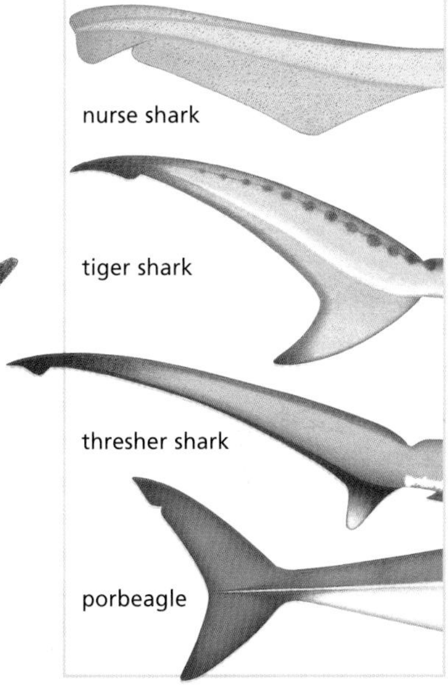
nurse shark
tiger shark
thresher shark
porbeagle

FACT FILE

Smooth hammerhead This mostly temperate species feeds on bony fishes, small sharks and rays, crustaceans, and squid. Groups migrate to lower latitudes.

- Up to 16½ ft (5 m)
- Up to 880 lb (400 kg)
- Viviparous
- Male & female
- Near threatened

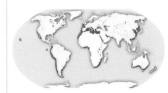

Widespread in temperate and tropical seas

Blue shark Courtship in the wide-ranging blue shark is so aggressive that, to cope with the nips and bites of mates, females have evolved skin more than twice as thick as that of the males.

- Up 13 ft (4 m)
- Up to 452 lb (205 kg)
- Viviparous
- Male & female
- Near threatened

Widespread in temperate and tropical seas

AGONISTIC DISPLAY

When threatened, grey reef sharks indicate their readiness to attack by raising the snout, dropping the pectoral fins, and holding the tail sideways as the back is arched and flexed. They swim like this in a figure-of-eight pattern with increasing intensity until they make either a rapid attack or retreat.

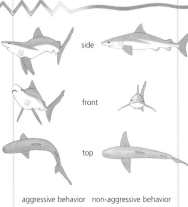

side

front

top

aggressive behavior   non-aggressive behavior

Lower caudal lobe strong; posterior margin of anal fin deeply notched

Hammer-shaped head enhances maneuverability and prey capture and might also improve sensory capabilities

Smooth hammerhead
Sphyrna zygaena

Tiger shark
Galeocerdo cuvier

Long snout and large eyes

Blue shark
Prionace glauca

Bull shark
Carcharhinus leucas

One of the few species that can enter fresh water, bull sharks have been found as far as 2,600 miles (4,200 km) up the Amazon River

Smooth hound
Mustelus mustelus

Compressed precaudal tail and long anal fin

Blackmouth catshark
Galeus melastomus

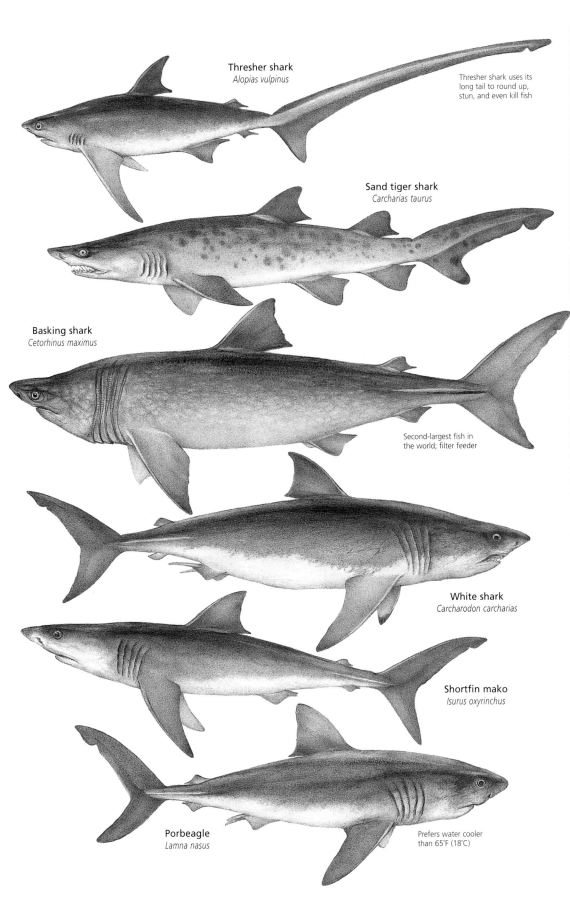

Thresher shark
Alopias vulpinus

Thresher shark uses its long tail to round up, stun, and even kill fish

Sand tiger shark
Carcharias taurus

Basking shark
Cetorhinus maximus

Second-largest fish in the world; filter feeder

White shark
Carcharodon carcharias

Shortfin mako
Isurus oxyrinchus

Porbeagle
Lamna nasus

Prefers water cooler than 65°F (18°C)

FACT FILE

Sand tiger shark The embryos of sand tiger sharks cannibalize each other until only one in each uterus survives. As a result, these sharks can only ever produce two pups per pregnancy.

- Up to 10½ ft (3.2 m)
- Up to 348 lb (158 kg)
- Viviparous
- Male & female
- Vulnerable

Widespread in warm seas except E. Pacific

Basking shark These huge sharks filter tiny animals from surface waters and can process a body of water equivalent in size to an Olympic pool every hour.

- Up to 32 ft (9.8 m)
- Up to 8,820 lb (4,000 kg)
- Ovoviviparous
- Male & female
- Vulnerable

Cosmopolitan in temperate and tropical seas

PLACENTAL VIVIPARITY

Developing young in some sharks—including blue and hammerhead sharks—are nourished, like mammals, by maternal nutrients. Embryos hatch in the oviducts and live initially off egg sac yolk after which empty sacs develop into placentae that interweave with the uterine walls. Yolk stalks become "umbilical cords" extending from between the embryos' pectoral fins. Sharks with placental viviparity can produce up to a hundred offspring with every pregnancy, depending on the species.

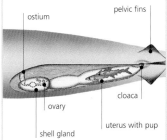

ostium

pelvic fins

cloaca

ovary

uterus with pup

shell gland

THE BIG BITE

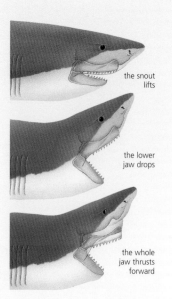

the snout lifts

the lower jaw drops

the whole jaw thrusts forward

A daytime predator of fish, squid, stingrays, turtles, and marine mammals, the white shark is believed to have exceptional eyesight and to see in color. This huge carnivore has evolved to consume large meals infrequently and is thought to undertake long migrations to favored hunting grounds. The white shark is one of just 27 shark species that is known to have attacked either people or boats.

Nocturnal hunter The gray reef shark (*Carcharhinus amblyrhynchos*) has serrated triangular teeth and is a mostly night-time predator of fishes, mollusks, and crustaceans.

Menacing mouth The mouth of the white shark can readily sever large pieces of flesh or chunks of blubber from large animals with its powerful jaws and massive bite. The multiple rows of razor-sharp, dagger-shaped teeth are serrated like steak knives and are constantly replaced.

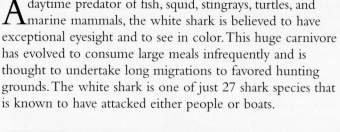

Protrusible jaws As in other sharks, white shark jaws are connected to each other at their outer corners but hang loosely beneath the skull. The snout lifts as the bite begins, the lower jaw drops and, as the mouth opens, the whole jaw arrangement thrusts forward, providing extra reach.

Garbage guts The tiger shark (*Galeocerdo cuvier*) (left) will attack and eat floating rubbish, such as tin cans.

SHARK TEETH

Teeth reveal much about shark prey preferences: white sharks have large, triangular-shaped teeth with serrated edges for tearing flesh; the flattened hind teeth of the hornshark crush hard-shelled invertebrates; blue shark teeth are finely serrated for catching fishes and squid; and the needle-like mako teeth are adept at grasping large, slippery prey.

horn shark

blue shark

great white shark

shortfin mako

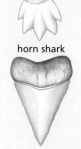

Breaching behavior White sharks around seal colonies off South Africa have been observed making sudden powerful and spectacular leaps, sometimes clear of the water, to snatch unsuspecting prey.

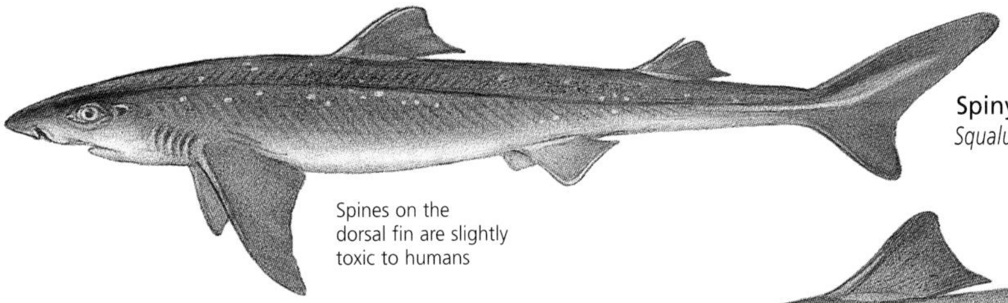

Spiny dogfish
Squalus acanthius

Spines on the
dorsal fin are slightly
toxic to humans

Six gill slits give this
shark its name

Sixgill shark
Hexanchus griseus

Sharpnose sevengill shark
Heptranchias perlo

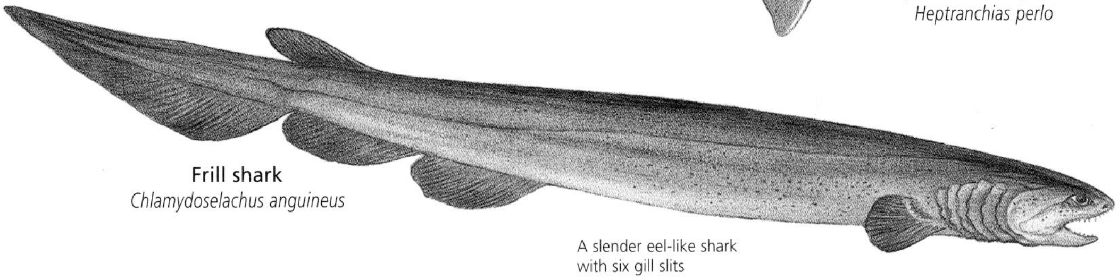

Frill shark
Chlamydoselachus anguineus

A slender eel-like shark
with six gill slits

Bramble shark
Echinorhinus brucus

Common name comes
from the unusual and
prominent thorn-like
scales covering the body

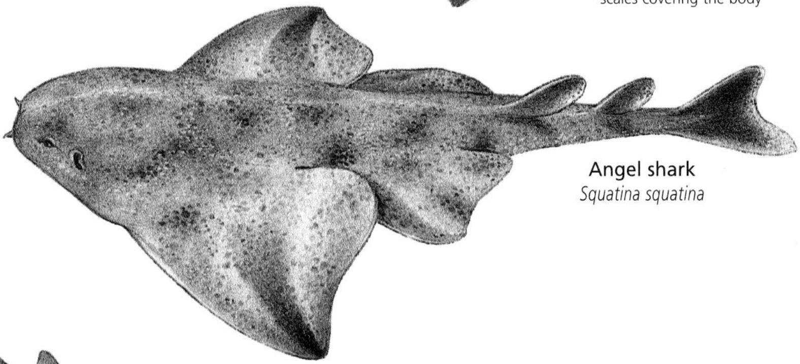

Angel shark
Squatina squatina

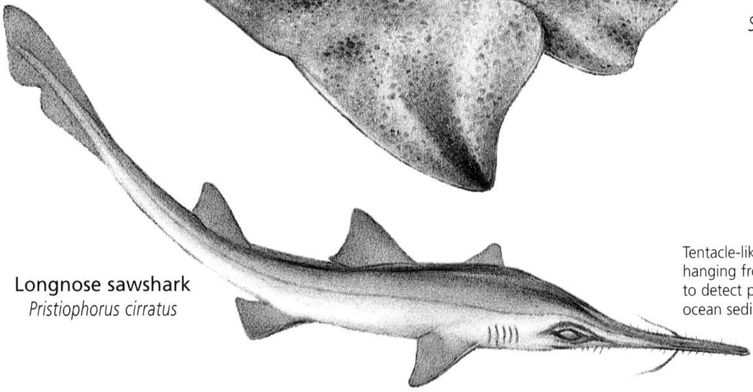

Longnose sawshark
Pristiophorus cirratus

Tentacle-like sensory barbels
hanging from snout are used
to detect prey buried in
ocean sediments

FACT FILE

Spiny dogfish The spiny dogfish has
one of the longest gestation periods of
any cartilaginous fish—up to 24 months.
With a life expectancy of about 70 years,
it is one of the longest-lived sharks.
Although a mostly deepwater species,
it can form shallow-water coastal groups.

- Up to 5¼ ft (1.6 m)
- Up to 20 lb (9 kg)
- Ovoviviparous
- Male & female
- Near threatened

Cosmopolitan in temperate seas

Angel shark By day, this ambush
predator lies buried in sediment with
only its eyes uncovered. It bursts
rapidly up to capture unsuspecting prey
such as bony fishes, squid, skates, and
crustaceans as they swim overhead.

- Up to 8 ft (2.4 m)
- Up to 175 lb (80 kg)
- Ovoviviparous
- Male & female
- Vulnerable

E. North Atlantic & Mediterranean Sea

FEEDING BEHAVIOR OF COOKIE-CUTTER SHARKS

Cookie-cutters attach to prey with their
suctorial lips, then spin around to carve
deep round plugs of flesh using large
triangular-shaped lower teeth. They
remain attached with small hook-like
upper teeth. It is thought that light-
emitting organs on their undersides
make them appear to be much smaller
fishes when viewed from beneath. The
illusion is revealed when would-be
predators launch attacks and become
prey instead. Cookie-
cutter scars have
been reported on
large fishes, marine
mammals, and
even the rubber
sonar domes
of submarines.

large teeth and
very powerful jaws

RAYS AND ALLIES

Mostly benthic Skates and rays are found in most benthic marine communities. A small number of ray families live in the open ocean and a few stingray and sawfish species survive in brackish estuaries and freshwater rivers and lakes, ranging from the tropics to polar waters. The guitarfishes, too—found in the temperate and tropical waters of the Atlantic, Pacific, and Indian oceans—are mostly marine.

CLASS	Chondrichthyes
SUBCLASS	Elasmobranchii
ORDERS	3
FAMILIES	13
SPECIES	547

Body shape is the principal difference between these fishes and other cartilaginous fishes. Known as batoids, skates and rays are dorsoventrally flattened, an adaptation for a bottom-dwelling lifestyle. Their greatly enlarged pectoral fins extend from near the snout to the base of the tail. Together with the body, and often the head, they form the "disk," which can be triangular, round, or diamond-shaped. Most batoids take in water for gill ventilation through spiracles, openings located on the tops of their heads that are often mistaken for the eyes. Teeth are often plate-like and used for crushing prey, which ranges from bottom-dwelling invertebrates to pelagic fishes.

BATOID DIVERSITY

Almost half the living batoid species are skates, all of which are bottom-dwelling egg-layers with large flat disks and small tails. The dorsal surface usually bears "thorns" for defence against predators and these are also used by males to grip females during mating. Although some species can attain lengths over 8 feet (2.4 m), most are less than 3¼ feet (1 m). Most skates prefer shallow water but some are found to depths of 9,000 feet (2,750 m).

The rays are often categorized into four major groups: the electric rays, sawfishes, stingrays and their allies, and guitarfishes.

Sawfishes have distinctive flat elongated snouts with conspicuous "teeth" along their edges. These can account for as much as a third of the body length in adults. They are thrashed about in schools of fishes to stun and kill prey. Some of the seven known living species attain lengths of over 23 feet (7 m).

Electric rays stun prey using electric organs located behind the eyes. There are more than 40 species worldwide, all of which tend to be rather lethargic bottom dwellers.

Many of the 150-plus species of stingrays and their allies have slender tails armed with serrated spines and are powerful swimmers. They tend to be bottom dwellers but three families have adopted an open ocean lifestyle. The largest, the manta, attains widths in excess of 21 feet (6.4 m).

Guitarfishes comprise about 50 species and have well-developed tails and either a poorly developed or absent disk. They are all ovo-viviparous and tend to have more shark-like features than the other battoids, including well-developed tails bearing two dorsal fins.

claspers (modified pelvic fins); the male copulatory organ

pectoral fin or wing

fleshy fins supported by horny fin rays

Typical skate Most skates have slender tails bearing two small dorsal fins. The nose is often pointed and spiny thorns on the dorsal surface are common. The disk of most species is diamond-shaped.

Clever fishes Rays are inquisitive, often sociable animals with complex behaviors. Although they are usually seen alone, many species will aggregate into loosely formed groups, particularly for the purposes of mating or migration.

Out of sight Taking water through spiracles located just behind the eyes enables batoids, such as the thornback ray (above), to breathe when their mouths are buried. Most batoids have good vision.

DANGEROUS TAIL

The tail of the thorntail stingray (*Dasyatis thetidis*) can be twice as long as its body and makes a formidable weapon against would-be predators. Thick at its base, but whip-like toward the end, it is spiked with small sharp thorns and armed with one or two sharp, serrated, and venomous barbs. The venom is dangerous to humans.

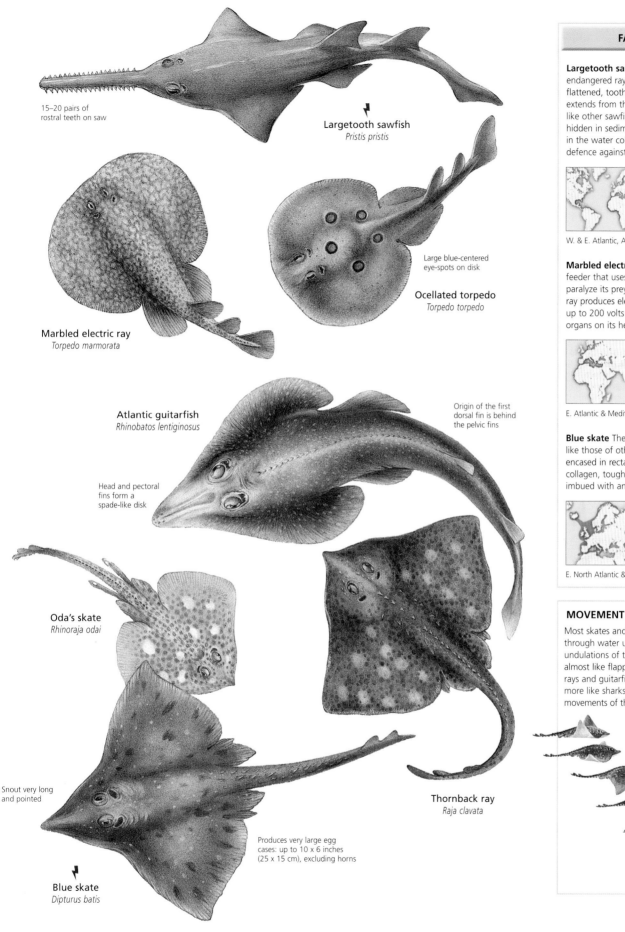

15–20 pairs of
rostral teeth on saw

Largetooth sawfish
Pristis pristis

Large blue-centered
eye-spots on disk

Ocellated torpedo
Torpedo torpedo

Marbled electric ray
Torpedo marmorata

Atlantic guitarfish
Rhinobatos lentiginosus

Origin of the first
dorsal fin is behind
the pelvic fins

Head and pectoral
fins form a
spade-like disk

Oda's skate
Rhinoraja odai

Thornback ray
Raja clavata

Snout very long
and pointed

Produces very large egg
cases: up to 10 x 6 inches
(25 x 15 cm), excluding horns

Blue skate
Dipturus batis

FACT FILE

Largetooth sawfish This critically
endangered ray uses the long,
flattened, tooth-studded "saw" that
extends from the front of its head,
like other sawfishes, to slash at prey
hidden in sediments or swimming
in the water column as well as for
defence against predators.

- Up to 15 ft (4.5 m)
- Up to 1,000 lb (454 kg)
- Ovoviviparous
- Male & female
- Critically endangered

W. & E. Atlantic, Amazon River

Marbled electric ray A nocturnal
feeder that uses shock tactics to
paralyze its prey, the marbled electric
ray produces electrical discharges of
up to 200 volts from two large electric
organs on its head.

- Up to 23½ in (60 cm)
- Up to 6.5 lb (3 kg)
- Viviparous
- Male & female
- Uncommon

E. Atlantic & Mediterranean Sea

Blue skate The eggs of the blue skate,
like those of other skate species, are
encased in rectangular capsules of
collagen, toughened with keratin and
imbued with antibacterial sulphur.

- Up to 9¼ ft (2.85 m)
- Up to 250 lb (113 kg)
- Oviparous
- Male & female
- Endangered

E. North Atlantic & W. Mediterranean Sea

MOVEMENT THROUGH WATER

Most skates and rays propel themselves
through water using wave-like vertical
undulations of their pectoral fins—
almost like flapping birds. The electric
rays and guitarfishes, however, swim
more like sharks, using horizontal
movements of their tails and caudal fins.

Graceful swimmers
As the pectorals flap
vertically, wave-like ripples
pass across them
horizontally.

BONY FISHES

SUPERCLASS	Gnathostomata
CLASSES	2
ORDERS	48
FAMILIES	455
SPECIES	24,673

In terms of both species and total numbers, this is by far the most successful living vertebrate group. Bony fishes first appeared about 395 million years ago, the fossil record indicating that the earliest forms inhabited fresh water. There are two distinct evolutionary lineages. The lobe-finned fishes (the sarcopterygians) are now represented by only a small number of living species. They are critically important in evolutionary terms because their ancestors gave rise to the earliest tetrapods—four-limbed land vertebrates—which, in turn, ultimately led to all other vertebrates. The overwhelming majority of living bony fishes, however, are the ray-finned fishes—the actinopterygians.

Successful radiation The bony fishes are found in virtually every available marine, freshwater, and brackish habitat throughout the world, and occasionally even desiccated environments. Species diversity increases nearer to the tropics and decreases toward the poles. Diversity tends to be highest close to coastlines and lowest in the open ocean.

EFFECTIVE EVOLUTION

Bony fishes are distinguished, as the name suggests, by a lightweight internal skeleton that is strengthened entirely or partially by true bone.

The fins of bony fishes are supported by more complex skeletal and muscle arrangements than those of cartilaginous fishes, giving bony fishes much finer control of swimming movements. As a result, many can move backward and even hover mid-water.

Maneuverability is also enhanced by the capacity to precisely and immediately adjust their buoyancy. This is achieved with a gas-filled sac known as a swim bladder. In some species, this is connected to the gullet and emptied and filled via the mouth, requiring fishes to gulp air at the water's surface. Mostly, however, there is no external connection and the swim bladder's content is controlled by the transfer of gases between adjacent blood vessels.

As they have a flap covering the gills (an operculum) and additional bony supports in the gill chamber known as branchiostegal rays, bony fishes can pump water over their gills and do not need to move forward to breathe.

About 90 percent of all bony fishes expel their reproductive cells from the body, making use of the watery environments in which they live for fertilization and the distribution of young.

Although cartilaginous fishes sometimes form aggregations, they don't school in the same highly coordinated way as many bony fishes, behavior made possible by a well-developed lateral line system and exceptional hearing and vision. Schools confuse predators by making lone targets difficult to select.

Safety in numbers Blue-and-gold fusiliers (*Caesio teres*) (above) have elongated bullet-shaped bodies and iridescent coloring. Like all members of the family Caesionidae, these fish form huge, fast-moving, mid-water schools that feed on zooplankton by day but shelter by night on the outer slopes of reefs. Their aggregations sometimes incorporate other fusilier species.

Symbiotic harmony This cleaner wrasse (*Labroides* sp.) (right) gains a feed as the much larger potato grouper (*Epinephelus tukula*) is picked free of parasites in a relationship of mutual benefit. The attentions of this wrasse species are not only tolerated, but actively sought out by a range of large fishes, including those that would normally eat small fishes.

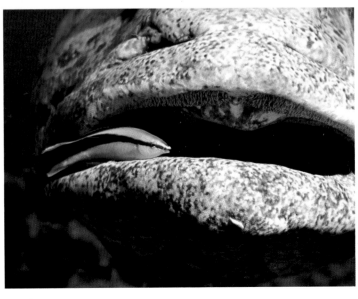

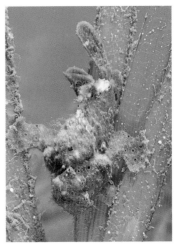

Fish lure The heads of anglerfishes (below) are equipped with a bony rod complete with a fleshy "bait" at the tip to attract prey. This is often waved around to enhance the ruse but the strategy doesn't always work as planned. Anglerfishes have been found with missing lures, suggesting their prey sometimes escapes with the bait.

SUCCESSFUL MOVEMENTS

Migration has been another significant behavioral adaptation among bony fishes that has helped underpin the group's success. Mass, predictable movements are undertaken by many species to exploit changing food resources, avoid predators, or for the purposes of mating and spawning.

Such movements can be vertical, from deep to shallower waters and back, and measured in just meters. Horizontal migrations, however, can cover many hundreds, and even thousands, of miles (kilometers). When this involves migration from fresh water to ocean and back, and most of a species' life is spent at sea, it is termed anadromy. This is typical of the salmon family. The reverse—ocean to fresh water and back—is termed catadromy and is typical of freshwater eels.

Spawning migrations are common among the bony fishes because they allow adults and young to exploit different niches or even habitats.

Bony fish anatomy The gills of nearly all adult bony fishes are covered by an operculum. Most species have thin, flexible scales covered by a thin mucus-secreting skin layer. Teeth are fixed to the upper jaw. The upper and lower lobes of the tail are usually symmetrical and the vertebral column stops before the fin. Most bony fishes have at least one dorsal fin, one anal fin, and paired pectoral fins. Except for a number of viviparous species, males do not have external organs to aid reproduction.

Deadly stalker Venom glands at the base of the fin spines of red lionfish (*Pterois volitans*) (above) deliver a poison that is potentially fatal to humans. This highly aggressive and mostly solitary fish tends to hide by day and hunt by night, stalking prey such as small fishes and crustaceans, which it corners by stretching out and expanding its fan-like pectoral fins.

Arduous journeys Like most tuna species, the northern bluefin (*Thunnus thynnus*) is highly migratory. It forms large schools of like-sized fish to make seasonal movements thought to be influenced by spawning, water temperature, and prey aggregations. One study found that northern bluefins migrated across the Atlantic at an average of 40 miles (65 km) per day, covering 4,800 miles (7,700 km) in just 119 days.

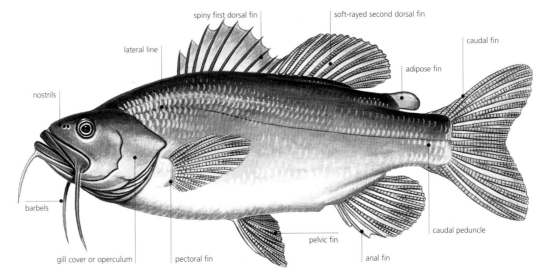

spiny first dorsal fin
soft-rayed second dorsal fin
lateral line
caudal fin
adipose fin
nostrils
barbels
gill cover or operculum
pectoral fin
pelvic fin
anal fin
caudal peduncle

LUNGFISHES AND ALLIES

CLASS	Sarcopterygii
SUBCLASSES	2
ORDERS	3
FAMILIES	4
SPECIES	11

Long-extinct ancestors of this relict group, known as the Sarcopterygii or "lobe-finned fishes," gave rise to the earliest land animals. Today, just nine lungfish and two coelacanth species survive. All have fleshy fins with skeletal structures and musculature more reminiscent of tetrapod limbs than they are of the fan-like fins of most other living fishes. In fact, 19th-century scientists mistakenly classified lungfishes as both reptiles and amphibians before realizing that they were fishes. Larval lungfishes breathe through gills but, in all but one species, adults rely on lungs to survive, allowing them to exist in waters with low oxygen levels. They occur in tropical freshwater rivers, lakes, and flood-plains, while the gill-breathing coelacanths are exclusively marine.

Restricted distribution Lungfishes survive only in Africa, South America, and Australia. One of the two coelacanth species (*Latimeria chalumnae*) is known from a small number of specimens caught off coastlines in the Indian Ocean around the Comoros Islands. The other (*L. menadoensis*) was discovered off Sulawesi in Indonesian waters in 1999. These are the only two known sites that support living coelacanth populations.

Surviving desiccation The African lungfish (*Protopterus annectens*) (above) estivates during the dry season in a vertical burrow (right), protected by a cocoon formed from dried mucus. It breathes air through a tube that connects with the surface. It can exist in this dormant state for years.

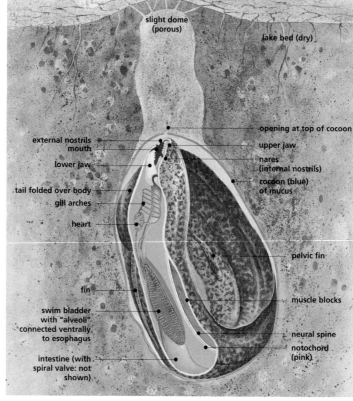

slight dome (porous)

lake bed (dry)

external nostrils
mouth

opening at top of cocoon

upper jaw

lower jaw

nares (internal nostrils)

cocoon (blue) of mucus

tail folded over body

gill arches

heart

pelvic fin

fin

muscle blocks

swim bladder with "alveoli" connected ventrally to esophagus

neural spine

notochord (pink)

intestine (with spiral valve: not shown)

AIR-BREATHERS

African and American lungfishes have two lungs and thread-like paired pectoral and pelvic fins. Spawning occurs at the onset of the rainy season to aid survival of the gill-breathing larvae. Adults endure protracted dry spells by estivating in burrows, extracting oxygen from air via their lungs.

The Australian lungfish has a single lung and paddle-shaped fins. Although it can breathe air, it retains functional gills and cannot survive complete desiccation.

Living fossils Coelacanths were thought to have been extinct since the time of the dinosaurs, 65 million years ago, until one was caught in 1938 off the coast of Madagascar by some fishermen.

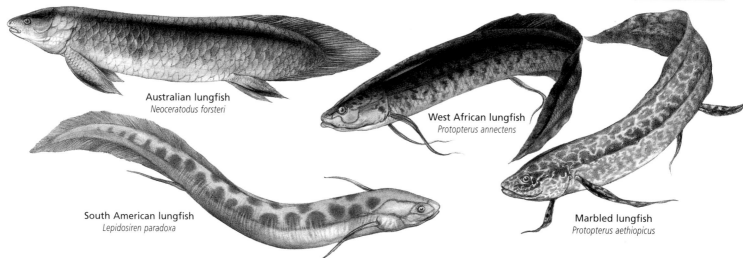

Australian lungfish
Neoceratodus forsteri

West African lungfish
Protopterus annectens

South American lungfish
Lepidosiren paradoxa

Marbled lungfish
Protopterus aethiopicus

BICHIRS AND ALLIES

CLASS	Actinopterygii
SUBCLASS	Chondrostei
ORDERS	2
FAMILIES	3
SPECIES	47

During the Permian period (285–245 million years ago) the direct ancestors of fishes such as the bichirs, reedfish, sturgeons, and paddlefishes—known collectively as chondostreans—were widespread and numerous both in species and total numbers. Today, these most primitive of the living ray-finned fishes have restricted distributions and are largely considered relics with archaic features. Most have rhombic-shaped scales hardened with a coating of an enamel-like substance called ganoine, not seen in modern species. Spiracles and intestinal spiral valves, characteristics more typical of cartilaginous fishes, are other common features. Most also have a single dorsal fin and a swim bladder, used for breathing air, connected to the gut.

Relict distribution Although fossil records reveal this group once occurred worldwide, its distribution is now restricted to Europe, Asia, Africa, and North America. One paddlefish species occurs in the eastern United States, the other in China. Fossils of the freshwater bichir and reedfish family (Polypteridae) have been uncovered in North Africa but it is now confined to tropical Africa and the Nile River system.

Fish delicacy Caviar (unfertilized sturgeon eggs) has become a symbol of opulence and indulgence. When the ripe females are stripped of their roe and released alive, it is a renewable resource. Too often, however, these fishes are killed for their roe.

Sensitive snouts The elongated snouts of paddlefishes, which account for as much as half their adult length, have a sensory function and are richly endowed with electroreceptors. These fishes are largely freshwater and feed by filtering zooplankton from the water. Although once thought mistakenly to be sharks, their closest living relatives are sturgeons.

STURGEON FEATURES

Sturgeons have been better studied than the other chondostreans, because of their high economic value. Found in both fresh and coastal marine waters, they are very large, long-lived fishes that take many years to reach sexual maturity. Females only spawn every few years.

Like the paddlefishes, sturgeons have several shark-like features, such as partly cartilaginous skeletons and heterocercal tails, where the vertebral column extends well into the tail's upper lobe.

They are unique among the living actinopterygians in having five rows of bony plates running along the body. They also have long flattened snouts and several tentacle-like barbels surrounding a ventrally located protractile mouth.

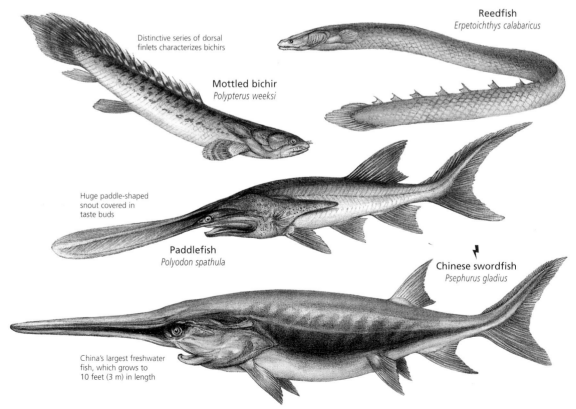

Distinctive series of dorsal finlets characterizes bichirs

Reedfish
Erpetoichthys calabaricus

Mottled bichir
Polypterus weeksi

Huge paddle-shaped snout covered in taste buds

Paddlefish
Polyodon spathula

Chinese swordfish
Psephurus gladius

China's largest freshwater fish, which grows to 10 feet (3 m) in length

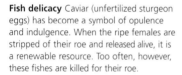

⚡ CONSERVATION WATCH

Sturgeon losses Overharvesting of caviar, habitat degradation, and overfishing have decimated sturgeon populations since the late 20th century. Most species are included on the IUCN Red List, including five that are classified as critically endangered. As part of efforts to halt the decline, sturgeon products were placed under the Convention on International Trade in Endangered Species of Wild Fauna and Flora in 1998.

White sturgeon The largest of the freshwater fishes in North America, the white sturgeon grows to 20 feet (6 m) in length and attains weights in excess of 1,500 pounds (680 kg). Sturgeons are typically very long-lived and this species is thought to have a life expectancy of at least a century.

- Up to 20 ft (6 m)
- Up to 1,800 lb (820 kg)
- Oviparous
- Male & female
- Near threatened

N.W. North America

Stellate sturgeon This fish has all the characteristic sturgeon features: five rows of bony plates on the dorsal surface of its body, a flattened snout, a protractile, ventrally located toothless mouth, and sensitive, fleshy, tentacle-like projections known as barbels in front of the mouth below the snout. Sturgeons drag these over bottom sediments in search of prey such as small fishes and invertebrates.

- Up to 7¼ ft (2.2 m)
- Up to 175 lb (80 kg)
- Oviparous
- Male & female
- Endangered

Basins of Black, Azov, Caspian seas; Adriatic Sea

Beluga Often described as the "most expensive fish in the world," the beluga is the most sought-after of the caviar sturgeons because of both the quality and quantity of its roe. A 13 foot (4 m) female can yield 400 pounds (180 kg).

- Up to 13 ft (4 m)
- Up to 1,760 lb (800 kg)
- Oviparous
- Male & female
- Endangered

Basins of Black & Caspian seas; Adriatic Sea

STERLET

The sterlet (*Acipenser ruthenus*), a caviar sturgeon, is designated as vulnerable on the IUCN Red List. Endemic to freshwater tributaries feeding the Black and Caspian seas, it survives cold winter months like other sturgeons in a type of "suspended animation," resting at depth without eating. Adults rise in spring to migrate upriver and spawn.

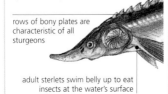

rows of bony plates are characteristic of all sturgeons

adult sterlets swim belly up to eat insects at the water's surface

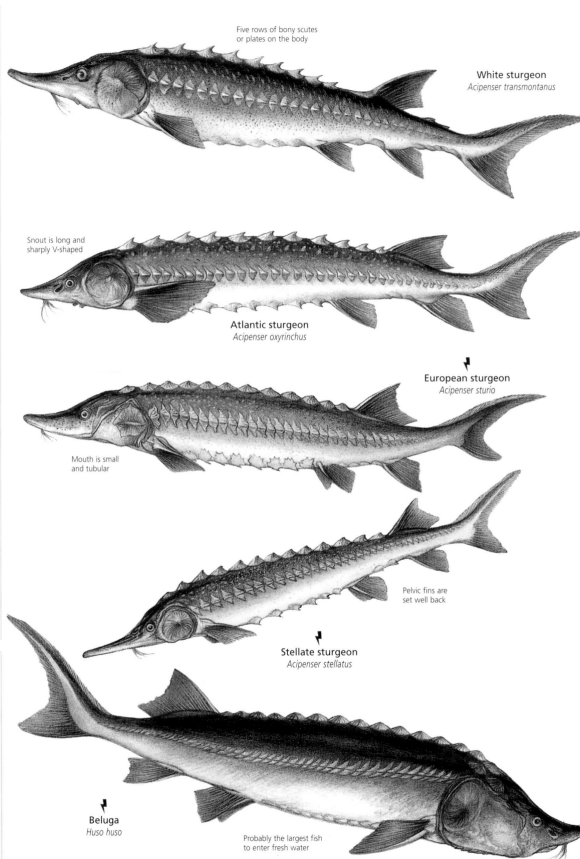

Five rows of bony scutes or plates on the body

White sturgeon
Acipenser transmontanus

Snout is long and sharply V-shaped

Atlantic sturgeon
Acipenser oxyrinchus

European sturgeon
Acipenser sturio

Mouth is small and tubular

Pelvic fins are set well back

Stellate sturgeon
Acipenser stellatus

Beluga
Huso huso

Probably the largest fish to enter fresh water

PRIMITIVE NEOPTERYGII

CLASS Actinopterygii	
SUBCLASS Neopterygii	
ORDERS 2	
FAMILIES 2	
SPECIES 8	

The Neopterygii are the second of the two living ray-finned fish groups. They arose from primitive fishes—probably early chondostreans—some 250 million years ago and, compared with their predecessors, had more mobile mouth parts, more symmetrical compact tails, and a simpler fin structure. Such advances enhanced feeding and swimming capabilities and the group ultimately gave rise to the extraordinarily successful teleosts, which comprise the majority of living fish species. Features of the earliest Neopterygii can still be seen today in a few surviving primitive representatives of the group: the bowfin and the seven species of gars, all of which are agile and voracious predators.

Northern Hemisphere inhabitants The bowfin is mostly restricted to still bodies of fresh water in temperate eastern North America. Five gar species are found in eastern North America. The remaining two species occur in Central America.

ANCIENT FISHES

Gars and the bowfin have elongated body shapes, abbreviated shark-like heterocercal tails, and numerous sharp teeth. They inhabit mostly still-water swamps and backwaters where they can survive low levels of dissolved oxygen by using lung-like gas bladders to breathe air. Gars have primitive, interlocking ganoid scales, while the bowfin has cycloid scales, the type common in more modern fishes.

Rapacious carnivore The longnose gar (above) is, like all other living gars, an aggressive ambush predator of other fishes and crustaceans. It is capable of rapid acceleration and armed with long rows of sharp teeth in an elongated "beak."

Bayou dweller Like other semionotiforms, the alligator gar (*Atractosteus spatula*) (above) is an aggressive carnivore with lightning reflexes that prefers slow-moving, poorly oxygenated, swampy backwaters. Growing to almost 10 feet (3 m) in length, it is the largest of the seven gar species.

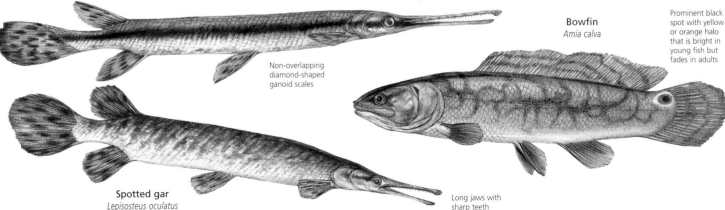
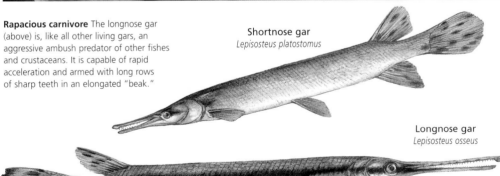

Shortnose gar
Lepisosteus platostomus

Longnose gar
Lepisosteus osseus

Non-overlapping diamond-shaped ganoid scales

Bowfin
Amia calva

Prominent black spot with yellow or orange halo that is bright in young fish but fades in adults

Spotted gar
Lepisosteus oculatus

Long jaws with sharp teeth

BONYTONGUES AND ALLIES

CLASS Actinopterygii	
SUBDIVISION Osteoglossomorpha	
ORDER Osteoglossiformes	
FAMILIES 6	
SPECIES 221	

These fishes, the osteoglossomorphs, are considered the most primitive of the modern teleosts. The term bonytongue describes a trait shared by all members of the group: well-developed, tooth-like tongue bones that bite against teeth on the roof of the mouth. Beyond this unifying trait—and the fact that all are exclusively freshwater, even though the fossil record indicates that some past forms may have endured brackish water—these fishes exhibit an extraordinary diversity of form and behaviors. They are found on all continents except Europe and Antarctica, with most species occurring in Africa and just one family containing two species in North America.

Tropical endemics True bonytongues occur in South America, Africa, Australia, Malaysia, Borneo, Sumatra, Thailand, and New Guinea. Featherbacks are restricted to Asia and Africa. All elephantfish species are found only in Africa.

Odd-shaped head The arawana (right) has a distinctive appearance. It has two forward-pointing sensory barbels on the chin and a large trap-like mouth in which the lower jaw edge opens almost vertically.

Skillful swimmer With rippling movements of its very long anal fin, the African knifefish (*Xenomystus nigri*) (above) can swim backward with as much ease as it moves forward.

FRESHWATER CURIOSITIES

There are three main groups in the osteoglossomorph subdivision. True bonytongues range from a small butterfly fish that seemingly glides in air using its well-developed pectoral fins, to the arapaima, one of the largest freshwater fishes.

Featherbacks have almost no tail fin but their anal fin runs two-thirds of their body length.

The elephantfishes, the third main group, includes species with elongated snouts that look like elephant trunks. Anal fin shape differs between the sexes in these bizarre-looking fishes. It is thought these fins are brought together in spawning pairs to form a cup into which eggs and sperm are ejected and mixed.

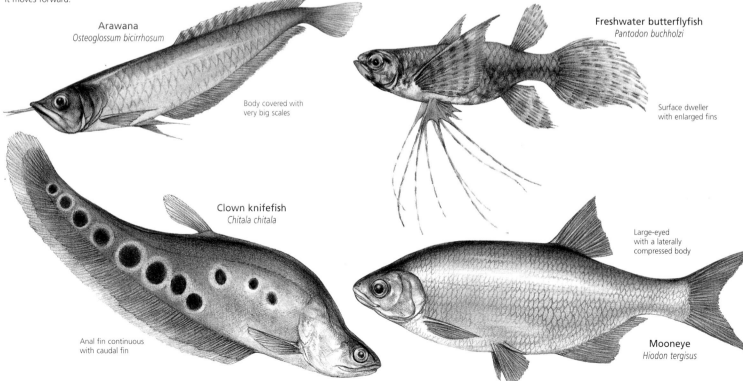

Arawana
Osteoglossum bicirrhosum

Body covered with very big scales

Freshwater butterflyfish
Pantodon buchholzi

Surface dweller with enlarged fins

Clown knifefish
Chitala chitala

Anal fin continuous with caudal fin

Large-eyed with a laterally compressed body

Mooneye
Hiodon tergisus

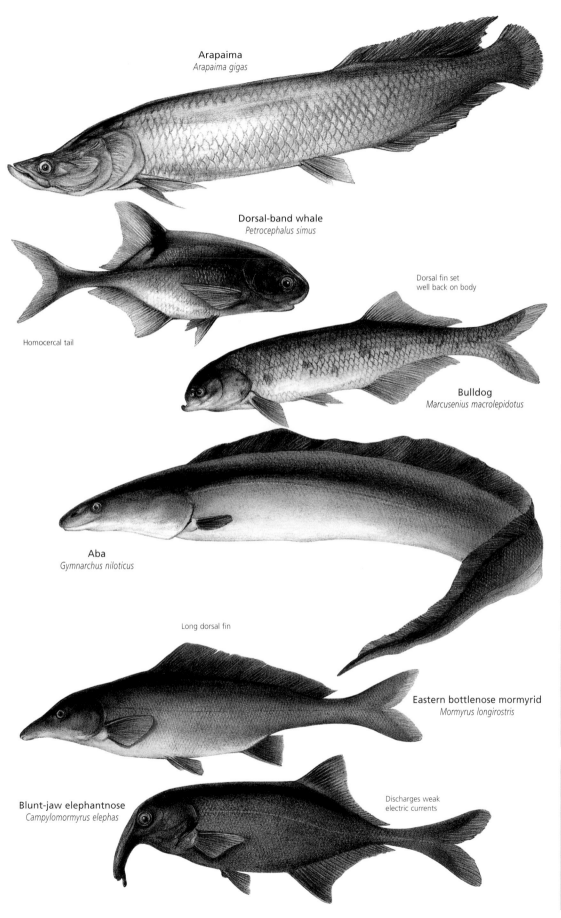

Arapaima
Arapaima gigas

Dorsal-band whale
Petrocephalus simus

Homocercal tail

Dorsal fin set
well back on body

Bulldog
Marcusenius macrolepidotus

Aba
Gymnarchus niloticus

Long dorsal fin

Eastern bottlenose mormyrid
Mormyrus longirostris

Blunt-jaw elephantnose
Campylomormyrus elephas

Discharges weak
electric currents

FACT FILE

Arapaima Also known as the pirarucu, this giant species is one of the largest freshwater fishes. Parental care by both males and females is well developed and begins with the making of nests for up to 50,000 eggs at a time. Once the eggs hatch, the parents guard the developing embryos and then protect the free-swimming fry.

⬥ Up to 15 ft (4.5 m)
⬥ Up to 440 lb (200 kg)
○ Oviparous
♀♂ Male & female
⚡ Data deficient

N. South America

Aba This species is in the same super-family as the elephantfishes, but differs from all other members of that group in that it has no caudal, anal, or pelvic fins. It does, however, have a very long dorsal fin that runs much of its body length and can be over 5 feet (1.5 m).

⬥ Up to 5½ ft (1.7 m)
⬥ Up to 40 lb (18 kg)
○ Oviparous, guarders
♀♂ Male & female
⚡ Common

N. & N.W. Central Africa

BIG BRAINS

enlarged "chin" used to probe
muddy substrates for food

Elephantfishes have a very large brain, proportionately as big as a human brain when body sizes are compared, and are said to have strong capabilities for learning. They have a particularly well-developed cerebellum. This processes the sensory information they gather by electroreception, their most important sense. They can detect and produce weak electric currents, and can create an electrical field around the body, which they use in murky river waters and for night navigation.

⚡ CONSERVATION WATCH

Of the 221 species contained in the Osteoglossiformes order (i.e. bony tongues and their allies), 4 are included on the IUCN Red List, as follows:

   1 Endangered
   2 Near threatened
   1 Data deficient

EELS AND ALLIES

CLASS	Actinopterygii
SUBDIVISION	Elopomorpha
ORDERS	5
FAMILIES	24
SPECIES	911

Many of the more than 900 species grouped together among these fishes, the elopomorphs, appear at first to have little to do with the elongated snake-like creatures most people know as eels. All, however, are linked because they invariably begin life as leptocephalus larvae. These translucent ribbon-shaped organisms drift in ocean currents for up to 3 years before metamorphosing into juvenile versions of their adult forms. This diverse assemblage of mainly marine and estuarine species includes three major groups: "true eels," such as moray, conger, and freshwater eels; the tarpons, ladyfishes, bonefishes, halosaurs, and spiny eels; and the bizarre deep-sea gulper eels.

Widespread distribution Most true eels are found in tropical and subtropical seas and oceans, although members of the family Anguillidae spend most of their lives in temperate fresh water. Tarpons and their closest relatives occur mainly in warm coastal and estuarine waters. Spiny eels are found throughout the world's oceans down to depths of 16,000 feet (4,900 m). All four families of deep-sea gulper eels also occur worldwide in deep ocean waters.

VARIED FORMS

With more than 700 species, the true eels are by far the largest group of elopomorphs. These have an elongated shape and almost all lack pelvic and pectoral fins. This streamlined body design suits a burrowing lifestyle and allows ease of movement into and out of holes around coral reefs.

Adult tarpons, bonefishes, and ladyfishes have large metallic scales, forked tails, and a typical fish shape. The former two groups are highly regarded as sportfish.

The spiny eels have deep-sea habits, are long, scaled, and feed on slow-moving, bottom-dwelling invertebrates. Deep-sea gulper eels have elongate, scaleless bodies. Enormous mouths and highly distensible stomachs take advantage of infrequent large meals.

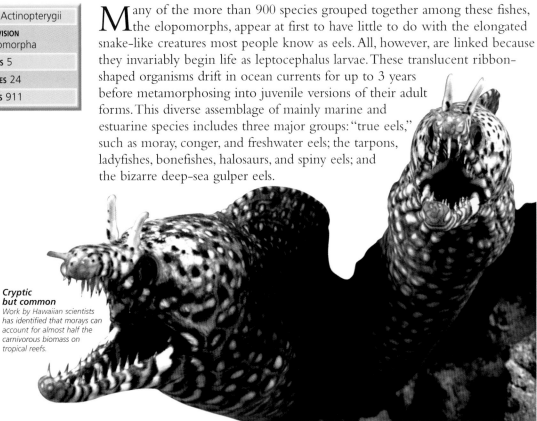

Cryptic but common Work by Hawaiian scientists has identified that morays can account for almost half the carnivorous biomass on tropical reefs.

Geometric moray
Gymnothorax griseus

Fierce faces Many morays, such as the dragon moray (*Enchelycore pardalis*) (above), have wide jaws filled with large sharp teeth. This species also has unusually prominent nose-tubes. Bold color patterns are also common among the morays; these help to camouflage them on the tropical reefs where they live. There are more than 200 species of this largely nocturnal family of carnivores. They have relatively small eyes and poor vision and locate prey using their keen sense of smell.

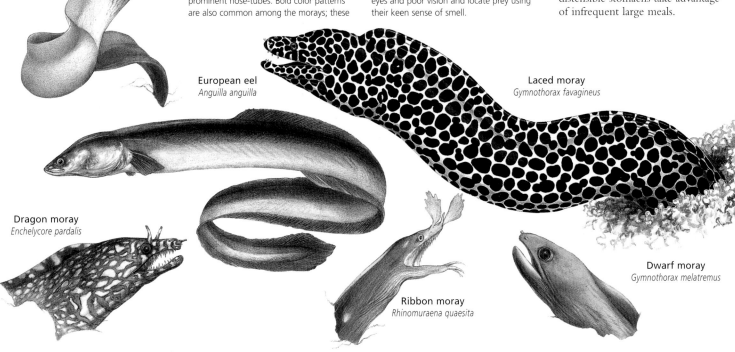

European eel
Anguilla anguilla

Laced moray
Gymnothorax favagineus

Dragon moray
Enchelycore pardalis

Ribbon moray
Rhinomuraena quaesita

Dwarf moray
Gymnothorax melatremus

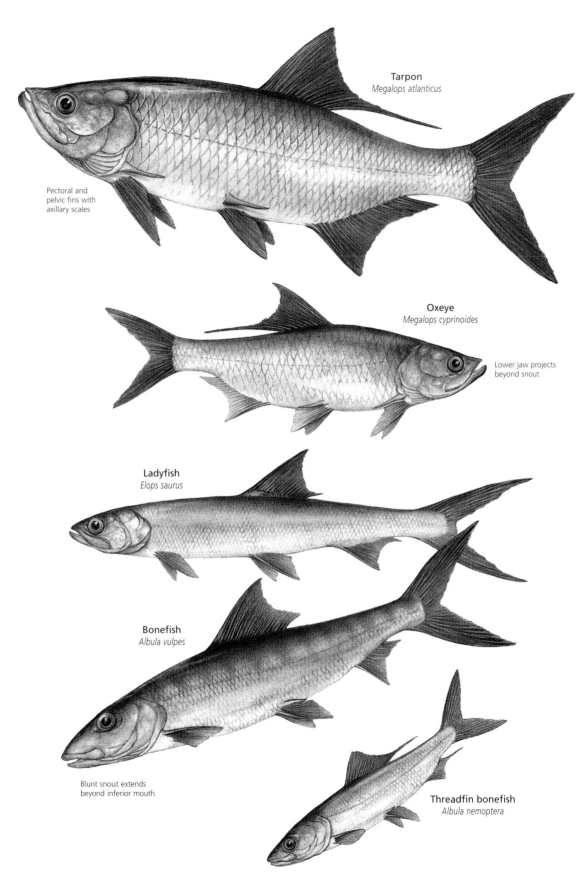

Tarpon
Megalops atlanticus

Pectoral and
pelvic fins with
axillary scales

Oxeye
Megalops cyprinoides

Lower jaw projects
beyond snout

Ladyfish
Elops saurus

Bonefish
Albula vulpes

Blunt snout extends
beyond inferior mouth

Threadfin bonefish
Albula nemoptera

FACT FILE

Tarpon Frequently the subject of sport-fishing legend, the Atlantic tarpon is sometimes called the "greatest gamefish in the world." They are hard to hook and land and will fight fiercely for many hours. During the struggle, they can leap explosively from the water.

- Up to 8 ft (2.4 m)
- Up to 300 lb (135 kg)
- ○ Oviparous
- ♀♂ Male & female
- ⚑ Common

Caribbean Sea, W. & E. Atlantic

Oxeye Like its close relative the tarpon, the oxeye is able to breathe atmospheric air in oxygen-poor environments and is often seen gulping air at the water's surface. These fishes are, however, far less tolerant of low temperatures and large kills can occur when waters suddenly become colder.

- Up to 5 ft (1.5 m)
- Up to 40 lb (18 kg)
- ○ Oviparous
- ♀♂ Male & female
- ⚑ Common

Indo-West Pacific

Bonefish A mainly coastal species that frequents river mouths and intertidal areas such as mangrove forests, the bonefish often occurs in highly coordinated schools of up to 100 individuals. They have a slender, typically torpedo-shaped body, and a protruding conical snout that they use to dig in substrates for prey, mostly invertebrate.

- Up to 3¼ ft (1 m)
- Up to 22 lb (10 kg)
- ○ Oviparous
- ♀♂ Male & female
- ⚑ Common

Worldwide in warm seas

TOOTHED TURKEY

Like other morays, the turkey moray (*Gymnothorax meleagris*) has large canine teeth and powerful jaws. Morays have been known to bite divers when scared or threatened but usually reserve their aggression for hunting prey.

morays have
small eyes and
poor vision

FACT FILE

European conger An elongated, scaleless, and nocturnal marine predator, the European conger can attain huge proportions. Juveniles occur around rocks and on sandy substrates near coastlines but adults live in deeper offshore waters. Like other congers, reproduction is a once-in-a-lifetime event, during which a female can produce up to 8 million eggs.

➴ Up to 9 ft (2.8 m)
⚖ Up to 143 lb (65 kg)
○ Oviparous, semelparous
♀♂ Male & female
🗡 Common

E. Atlantic, Mediterranean Sea, Black Sea

Pelican eel A typical gulper eel, this fish has an enormously enlarged mouth, distensible buccal cavity, and a stomach that can expand to two-thirds its total size: all adaptations to irregular feeding of large meals of invertebrates and other fishes. It dwells at depths of 1,650–24,500 feet (500–7,500 m) and bears a luminous tail organ.

➴ Up to 3¼ ft (1 m)
⚖ Up to 2 lb (900 g)
○ Oviparous
♀♂ Male & female
🗡 Common

Widespread in temperate and tropical seas

EEL GARDENS

At least 20 conger eel species spend their entire juvenile and adult lives in colonies—sometimes comprising thousands of individuals—permanently embedded in substrate. With tails down and heads waving, they feed on plankton. Known as garden eels, they occur at depths of up to 1,000 feet (300 m), retreating into their mucus-lined burrows when threatened.

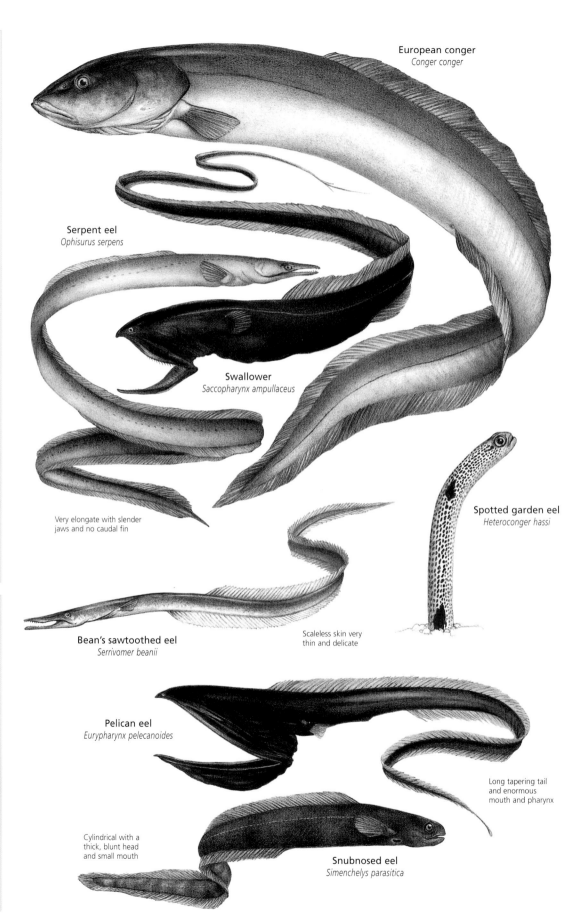

European conger
Conger conger

Serpent eel
Ophisurus serpens

Swallower
Saccopharynx ampullaceus

Very elongate with slender jaws and no caudal fin

Spotted garden eel
Heteroconger hassi

Scaleless skin very thin and delicate

Bean's sawtoothed eel
Serrivomer beanii

Pelican eel
Eurypharynx pelecanoides

Long tapering tail and enormous mouth and pharynx

Cylindrical with a thick, blunt head and small mouth

Snubnosed eel
Simenchelys parasitica

THE MYSTERIOUS LIVES OF FRESHWATER EELS

It was not until late in the 19th century that scientists began uncovering the truth about the life-cycles of the European eel (*Anguilla anguilla*) and its relative the American eel (*Anguilla rostrata*). Both species, it is now known, spawn in the Sargasso Sea, but precisely where remains a mystery. Their lives take decades and up to 7,000 miles (11,250 km) to complete. European eels head east, their American cousins west, both spending most of their lives in fresh water before returning to the Sargasso Sea to spawn once and die.

Small beginnings European eels are just 2–4 inches (5–10 cm) long when, after several years of oceanic drifting, they finally reach the estuaries that will take them to their freshwater adult habitats.

Parallel lives Adult European eels dwell in fresh water throughout Europe and parts of northern Africa. Their life history (below) mirrors that of the American species, in which adults occupy fresh water along North America's east coast. Research has shown that both species have suffered massive declines since the 1970s. Some investigations indicate adult European eel numbers may have fallen by 90 percent or more as a result of overfishing, pollution, disease, and habitat destruction.

Silver fishes Adult European eels occupy fresh water feeding the North Atlantic Ocean and Baltic and Mediterranean seas. Females mature between 9 and 20 years; males between 6 and 12 years.

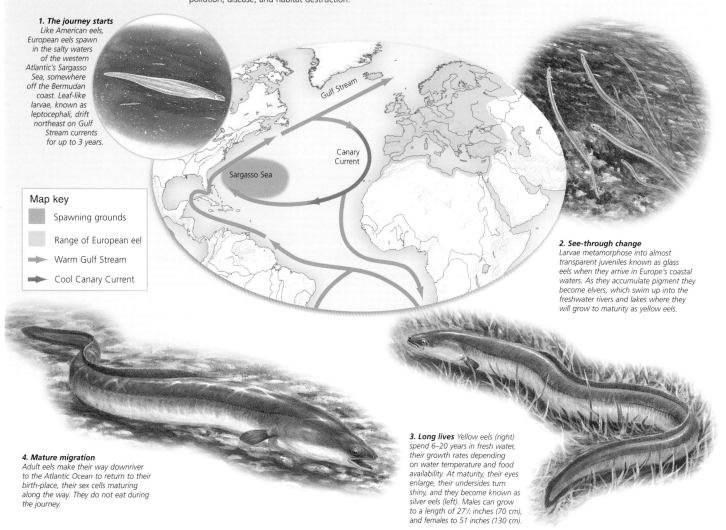

1. The journey starts
Like American eels, European eels spawn in the salty waters of the western Atlantic's Sargasso Sea, somewhere off the Bermudan coast. Leaf-like larvae, known as leptocephali, drift northeast on Gulf Stream currents for up to 3 years.

Gulf Stream

Canary Current

Sargasso Sea

Map key

- ▨ Spawning grounds
- ▨ Range of European eel
- ➡ Warm Gulf Stream
- ➡ Cool Canary Current

2. See-through change
Larvae metamorphose into almost transparent juveniles known as glass eels when they arrive in Europe's coastal waters. As they accumulate pigment they become elvers, which swim up into the freshwater rivers and lakes where they will grow to maturity as yellow eels.

4. Mature migration
Adult eels make their way downriver to the Atlantic Ocean to return to their birth-place, their sex cells maturing along the way. They do not eat during the journey.

3. Long lives Yellow eels (right) spend 6–20 years in fresh water, their growth rates depending on water temperature and food availability. At maturity, their eyes enlarge, their undersides turn shiny, and they become known as silver eels (left). Males can grow to a length of 27½ inches (70 cm), and females to 51 inches (130 cm).

SARDINES AND ALLIES

DIVISION Teleostei	
SUBDIVISION Clupeomorpha	
ORDER Clupeiformes	
FAMILIES 5	
SPECIES 378	

Known as the clupeoids, this group of 378 species comprises some of the world's most commercially important fishes, including herrings, sardines, anchovies, shads, and pilchards. Most are marine filter feeders of zooplankton that form large schools and lead pelagic lifestyles, although their distributions tend to be more coastal than open water. Spawning is frequently in near-shore surface waters and usually seasonal. Females mostly produce large numbers of small eggs which develop into larvae that can drift in surface currents for many months before metamorphosing into juveniles. Some species undertake extraordinary migrations as adults, covering thousands of miles (kilometres) and several years.

Coastal preferences The clupeoids are largely a Northern Hemisphere group. The adults of most species feed along the coasts of tropical and temperate seas, but more than 70 species inhabit freshwater rivers and lakes. Very few species occur far from shore in the open ocean and none are found in polar seas or at oceanic depths. Seasonal migrations to spawning grounds are common in the group.

Staying in school The gas bladder of all clupeoids extends into the inner ear. Known as an otophysic condition, this is thought to increase their abilities to detect low-frequency sounds such as those produced by beating tail fins, and may aid in schooling. Larvae old enough to have a swim bladder are thought to only inflate it, and rise in the water column, at night.

BOOM-BUST POPULATIONS

Renowned for natural boom–bust population cycles, clupeoids suffer from extremely high mortality rates—frequently up to 99 percent—partly because they fall victim to a wide range of predators at all stages of their life cycle. They are also very vulnerable as adults to variations in food availability, a result of increasingly fluctuating environmental conditions.

Most species, however, have very high reproductive rates and reach sexual maturity at early ages (rarely later than 3 years of age). As a consequence, populations tend to bounce back quickly when good conditions prevail.

The clupeoid group contributes enormously to the total biomass of the world's aquatic environments and is a crucial lower link in many aquatic food chains.

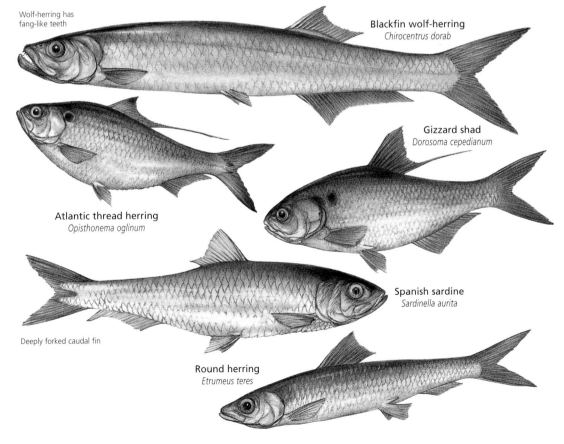

Wolf-herring has fang-like teeth

Blackfin wolf-herring
Chirocentrus dorab

Gizzard shad
Dorosoma cepedianum

Atlantic thread herring
Opisthonema oglinum

Spanish sardine
Sardinella aurita

Deeply forked caudal fin

Round herring
Etrumeus teres

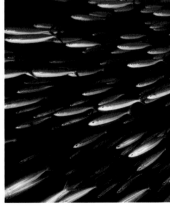

Archetypal fish Clupeoids, such as these herrings, are what most people visualize when they think of fishes: small—most attaining lengths less than 12 inches (30 cm)—streamlined, and schooling, with large silver scales and forked tails.

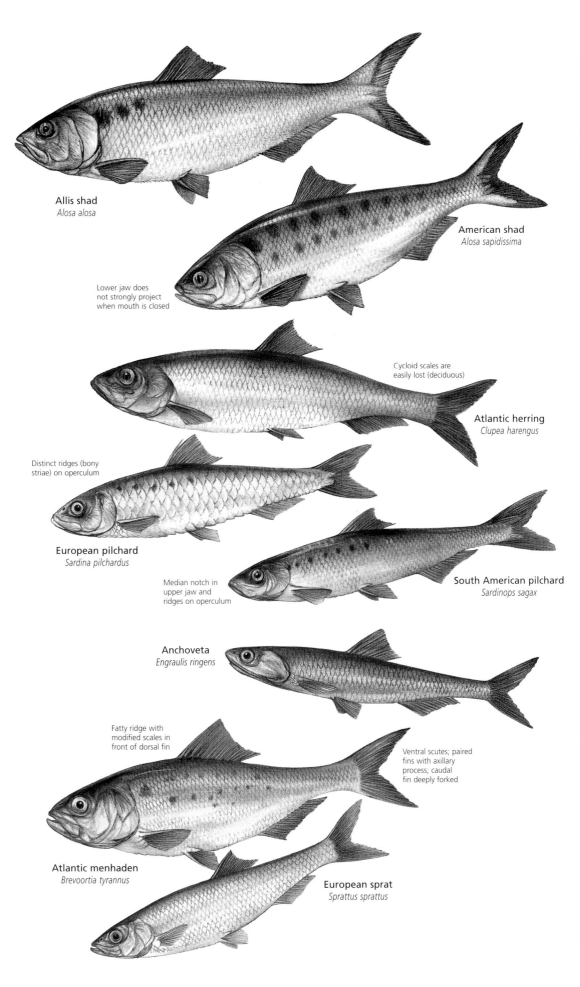

Allis shad
Alosa alosa

American shad
Alosa sapidissima

Lower jaw does
not strongly project
when mouth is closed

Cycloid scales are
easily lost (deciduous)

Atlantic herring
Clupea harengus

Distinct ridges (bony
striae) on operculum

European pilchard
Sardina pilchardus

Median notch in
upper jaw and
ridges on operculum

South American pilchard
Sardinops sagax

Anchoveta
Engraulis ringens

Fatty ridge with
modified scales in
front of dorsal fin

Ventral scutes; paired
fins with axillary
process; caudal
fin deeply forked

Atlantic menhaden
Brevoortia tyrannus

European sprat
Sprattus sprattus

FACT FILE

American shad Adult American
shad spend most of their lives in
coastal marine and brackish waters
but migrate without feeding to freshwater streams and rivers in eastern
North America to spawn, each female
releasing up to 600,000 eggs during
one evening. This species is the largest
of the shads and is considered to be
the best eating. Its roe is highly valued.

🛬 Up to 30 in (76 cm)
⚖ Up to 12 lb (5.4 kg)
○ Oviparous
♀♂ Male & female
🕯 Common

North America

Atlantic herring Sometimes listed in
the *Guinness Book of World Records* as
the planet's most numerous fish species,
the Atlantic herring has been observed
running in 17 mile (27 km) long schools
containing millions of individuals. It
is usual for schools of this species to
undertake daily vertical migrations.
During daylight hours they typically drop
to the ocean floor. By night they rise to
feed, mostly on zooplankton.

🛬 Up to 17 in (43 cm)
⚖ Up to 24 oz (680 g)
○ Oviparous
♀♂ Male & female
🕯 Common

North Atlantic

Anchoveta Perhaps the most
commercially exploited fish in human
history, the anchoveta was once taken
at an annual rate of over 11 million
tons (10 million tonnes). The natural
boom-bust population cycles of the
species appear to be exacerbated
periodically by overfishing.

🛬 Up to 8 in (20 cm)
⚖ Up to 2 oz (60 g)
○ Oviparous
♀♂ Male & female
🕯 Common

W. South American coast

⚡ CONSERVATION WATCH

Shad shortages The 12 clupeoids
on the IUCN Red List include the
endangered Alabama shad (*Alosa
alabamae*) and the Laotian shad
(*Tenualosa thibaudeaui*). The
former, a Gulf of Mexico species,
has declined dramatically due
to dam constructions that have
impeded spawning migrations into
the fresh waters of large rivers.
Declining numbers of Indochina's
freshwater Laotian shad may be
due to overfishing and dams.

CATFISHES AND ALLIES

CLASS	Actinopterygii
SUPERORDER	Ostariophysi
ORDERS	5
FAMILIES	62
SPECIES	7,023

These fishes, the ostariophysans, dominate the world's freshwater habitats. They form a massive group of over 7,000 species united by two key characteristics. A unique set of bones, the Weberian apparatus, connects the gas bladder and inner ear in an arrangement that enhances hearing. The second trait is an alarm response involving the release of distress chemicals from special skin cells. Many ostariophysans not only produce these substances but can also perceive and respond to them by fleeing. Some species that eat other ostariophysans, however, are not capable of the latter as most of their prey produce these alarm chemicals and reacting would thus inhibit feeding.

Freshwater dominance Most characins are tropical. Cyprinids are naturally found in North America and Africa, but most are found in Eurasia. Catfishes are found on all continents and the electric knifefishes are restricted to Central and South America. Many ostariophysans, however, are now found well beyond their original range, partly as a result of the aquarium trade.

ENORMOUS DIVERSITY

The most primitive ostariophysans are a tropical group with incomplete Weberian apparatus. The milk fish, an important protein source for people in Southeast Asia, is in this group.

The largest ostariophysan group, the cyprinids, includes the carps, minnows, and many of the world's most popular aquarium fishes, from goldfish to rasboras. These lack jaw teeth but grind food between a pharyngeal tooth-bearing bone and a hard keratinized pad at the base of the skull. This arrangement has been modified repeatedly to provide many different feeding strategies.

The mostly South American characins include species as diverse and well known as the piranhas and the aquarium favorites, the tetras.

There are well over 2,000 catfish species, all readily recognized by tentacle-like barbels around the mouth. All are substrate feeders.

The electric knifefishes make up a small, specialized, and poorly studied group that includes the legendary electric eel.

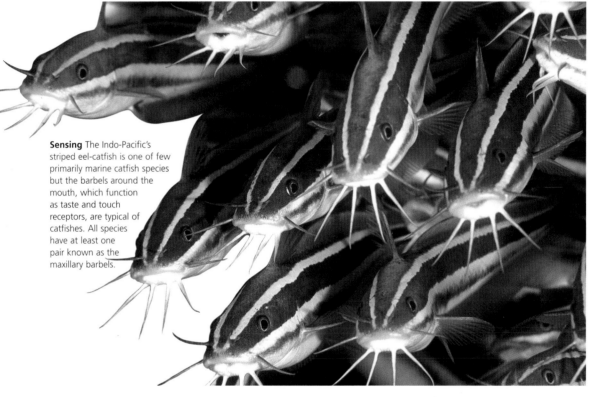

Sensing The Indo-Pacific's striped eel-catfish is one of few primarily marine catfish species but the barbels around the mouth, which function as taste and touch receptors, are typical of catfishes. All species have at least one pair known as the maxillary barbels.

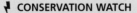

Barely there Once presumed extinct, the smoky madtom (*Noturus baileyi*) was rediscovered in a Tennessee stream in the eastern United States in 1980. Fewer than 1,000 adults of this tiny catfish, which grows to 2½ inches (6 cm), survive.

In Southern Africa's Twee River and tributaries, the Twee redfin (*Barbus erubescens*) leads a similarly precarious existence, facing threats such as agricultural pollution and excessive water extraction. Introduced fish species, too, have taken a toll. These are among 49 critically endangered ostariophysans worldwide. A further 45 are endangered and 18 have already gone extinct.

Efficient feeders In some but not all piranha species, short powerful jaws and sharp, interlocking teeth (above) enable flesh removal from prey in clean bites.

Popular fish Originally from the upper reaches of the Amazon River, neon tetras (left) have become a mainstay of the aquarium trade worldwide.

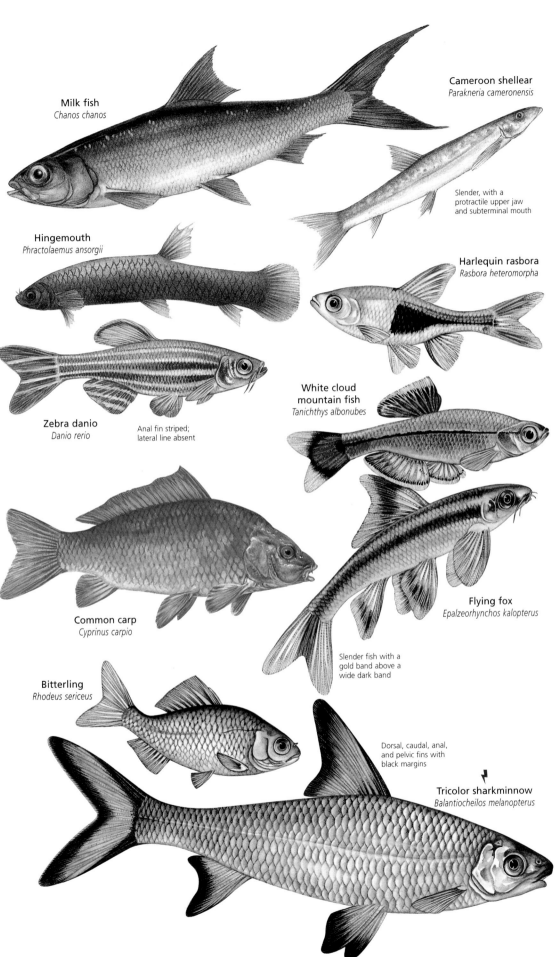

Milk fish
Chanos chanos

Cameroon shellear
Parakneria cameronensis

Slender, with a
protractile upper jaw
and subterminal mouth

Hingemouth
Phractolaemus ansorgii

Harlequin rasbora
Rasbora heteromorpha

Zebra danio
Danio rerio

Anal fin striped;
lateral line absent

White cloud
mountain fish
Tanichthys albonubes

Common carp
Cyprinus carpio

Flying fox
Epalzeorhynchos kalopterus

Slender fish with a
gold band above a
wide dark band

Bitterling
Rhodeus sericeus

Dorsal, caudal, anal,
and pelvic fins with
black margins

Tricolor sharkminnow
Balantiocheilos melanopterus

FACT FILE

Milk fish Larval milkfish develop
in brackish coastal wetlands and do
occasionally enter fresh water but
return to the sea to breed. This species
is an important protein source for
human populations throughout the
Indo-Pacific, where it is sometimes
farmed in ponds.

- Up to 6 ft (1.8 m)
- Up to 30 lb (14.5 kg)
- ○ Oviparous
- ♀♂ Male & female
- Common

E. Africa, S.E. Asia, Oceania & E. Pacific

Common carp Originally native to
parts of Europe and Asia, the common
carp has been introduced throughout
the world. It is a highly adaptable
species with an omnivorous diet and in
many places outside its original range
is now considered a pest because it
often outcompetes local native species.

- Up to 4 ft (1.2 m)
- Up to 82 lb (37 kg)
- ○ Oviparous
- ♀♂ Male & female
- Data deficient for wild carp

E. Europe to China; introd. worldwide
- Introduced range

BITTERLING

To breed, the female bitterling develops
a long ovipositor down which she
passes eggs into the gill chamber
of a freshwater mussel. Eggs
are fertilized when sperm,
released by a male near
the bivalve's inhalant
aperture, is drawn
in and mixes with
the eggs. Bitterling larvae
develop inside the mussel
for up to a month.

Fish pregnancy tests
*Female bitterlings were
once used to test for pregnancy
in women. Urine carrying traces of
pregnancy hormones encourages ovipositor
development when injected into these fish.*

FACT FILE

Spanner barb Many of the barbs are popular within the aquarium trade, although the spanner barb tends to grow a little too large to be a very highly sought-after species.

  ➙ Up to 7 in (18 cm)
  🗖 Up to 8 oz (225 g)
  ◯ Oviparous
  ♀♂ Male & female
  🕯 Common

Malay Peninsula to Borneo

Chinese sucker Native to the Yangtze River basin, this species' common name stems from its thickened lips, which are used to suck up food such as invertebrates and algae from sediments and around river rocks and plants. In some specimens, the dorsal fin can be almost as tall as the body is long.

  ➙ Up to 23½ in (60 cm)
  🗖 Up to 8 lb (3.6 kg)
  ◯ Oviparous
  ♀♂ Male & female
  🕯 Locally common

China (Yangtze River basin)

Reticulate loach A tropical, Asian freshwater species, this fish is also commonly known as the Pakistani loach. It is a nocturnal feeder with an omnivorous diet of mainly bottom-dwelling invertebrates, such as snails and worms, as well as algae. When threatened, it will rapidly bury the front end of its body in sediment leaving only its tail exposed.

  ➙ Up to 4¾ in (12 cm)
  🗖 Up to 4 oz (115 g)
  ◯ Oviparous, hiders
  ♀♂ Male & female
  🕯 Common

Pakistan, India, Bangladesh, Nepal

Stone loach A nocturnal bottom-dweller with three pairs of barbels, the stone loach feeds on benthic invertebrates. Because of its sensitivity to pollution and low levels of dissolved oxygen, it is regarded as a good biological indicator of water quality.

  ➙ Up to 8½ in (21 cm)
  🗖 Up to 7 oz (200 g)
  ◯ Oviparous
  ♀♂ Male & female
  🕯 Common

Eurasia

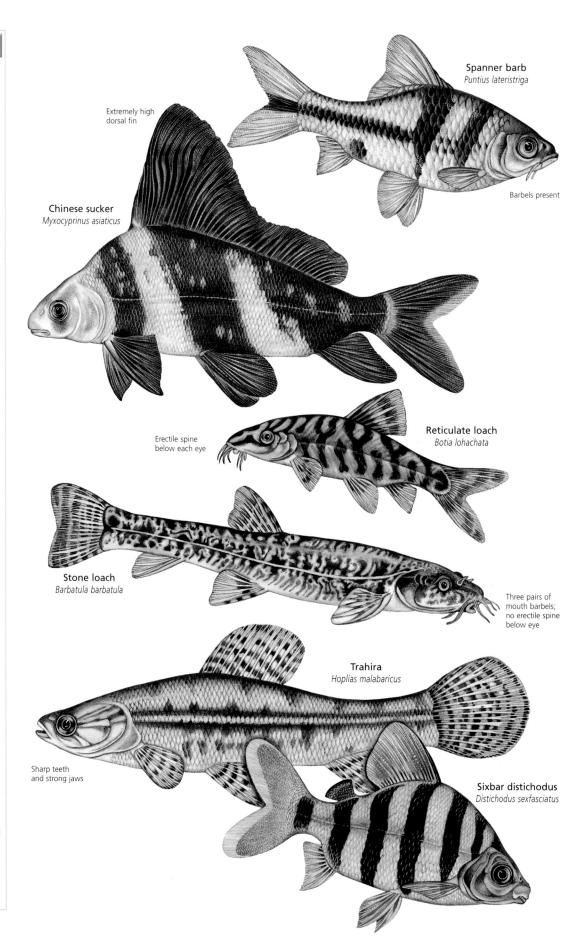

Spanner barb
Puntius lateristriga

Barbels present

Extremely high dorsal fin

Chinese sucker
Myxocyprinus asiaticus

Erectile spine below each eye

Reticulate loach
Botia lohachata

Stone loach
Barbatula barbatula

Three pairs of mouth barbels; no erectile spine below eye

Trahira
Hoplias malabaricus

Sharp teeth and strong jaws

Sixbar distichodus
Distichodus sexfasciatus

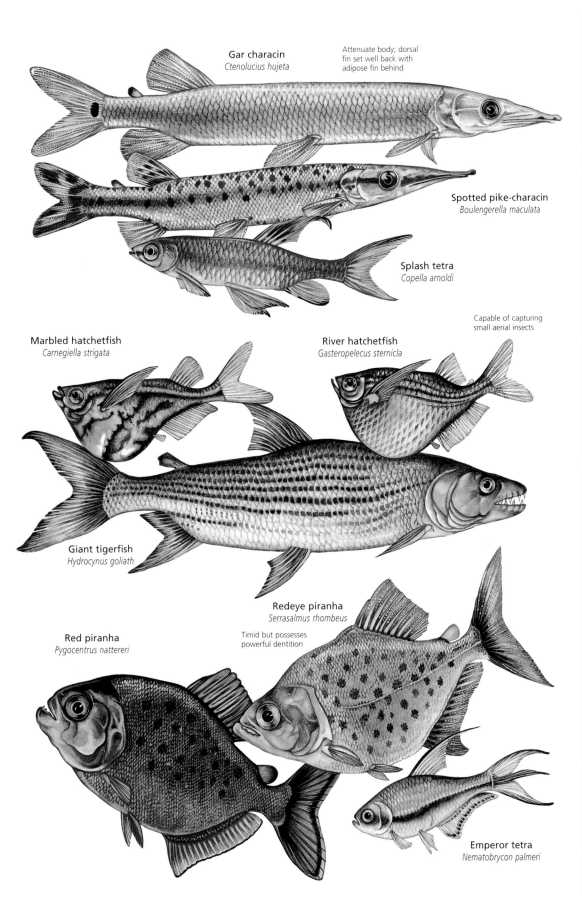

Gar characin
Ctenolucius hujeta

Attenuate body; dorsal fin set well back with adipose fin behind

Spotted pike-characin
Boulengerella maculata

Splash tetra
Copella arnoldi

Capable of capturing small aerial insects

Marbled hatchetfish
Carnegiella strigata

River hatchetfish
Gasteropelecus sternicla

Giant tigerfish
Hydrocynus goliath

Redeye piranha
Serrasalmus rhombeus

Timid but possesses powerful dentition

Red piranha
Pygocentrus nattereri

Emperor tetra
Nematobrycon palmeri

SPLASH TETRA

Splash tetras avoid aquatic egg-eaters by spawning out of water. Females leap into the air and, as their wet bodies stick to overhanging leaves, deposit up to eight eggs at a time before falling. Males follow quickly and the pair interlock fins to hang from the leaves as the male deposits his sperm. The procedure is repeated until several hundred fertilized eggs are laid. Tail flicks from males keep the eggs moist until fry hatch and fall into the water.

PARASITIC BROODERS

Africa's cuckoo catfish is the only fish species known to parasitize the reproductive brooding of another fish. Pairs of these catfish spawn near pairs of mouth-brooding cichlids doing the same, and consume most of the cichlid eggs. As the female cichlid collects what eggs she can into her mouth, she also gathers up the catfish eggs. The catfish fry hatch first, grow strong on their yolk sacs, and often consume the young cichlids as they hatch.

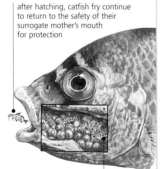

after hatching, catfish fry continue to return to the safety of their surrogate mother's mouth for protection

cichlid eggs are protected by their tough skin, but larval cichlids have no such protection

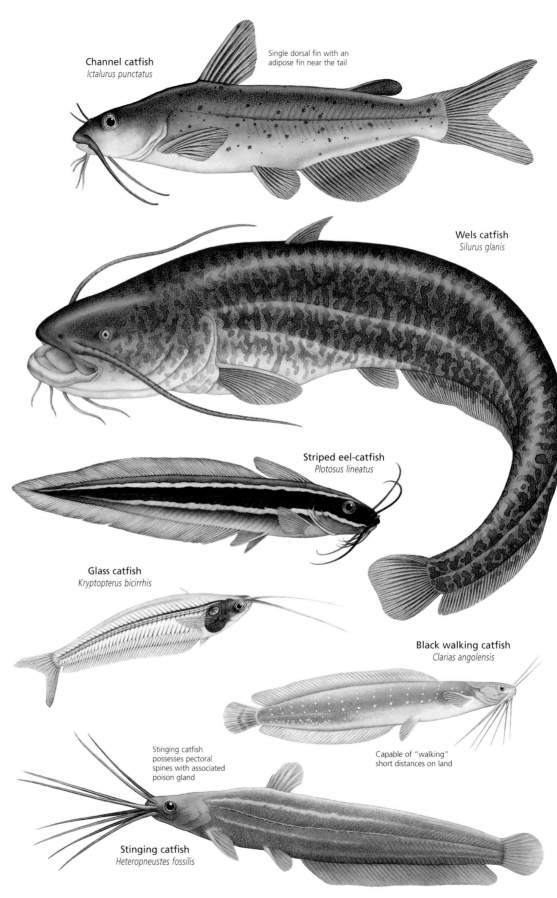

Channel catfish
Ictalurus punctatus

Single dorsal fin with an adipose fin near the tail

Wels catfish
Silurus glanis

Striped eel-catfish
Plotosus lineatus

Glass catfish
Kryptopterus bicirrhis

Black walking catfish
Clarias angolensis

Capable of "walking" short distances on land

Stinging catfish possesses pectoral spines with associated poison gland

Stinging catfish
Heteropneustes fossilis

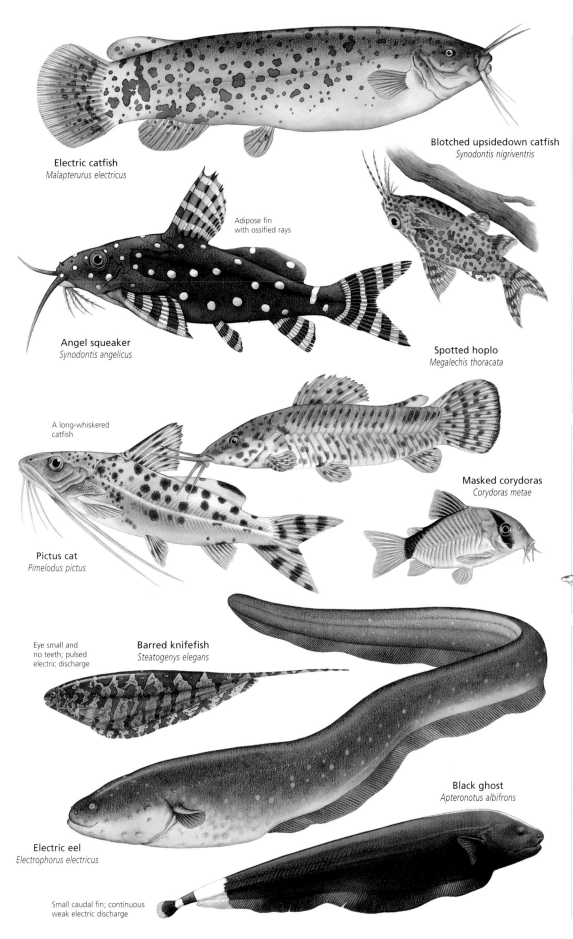

Electric catfish
Malapterurus electricus

Blotched upsidedown catfish
Synodontis nigriventris

Adipose fin
with ossified rays

Angel squeaker
Synodontis angelicus

Spotted hoplo
Megalechis thoracata

A long-whiskered
catfish

Masked corydoras
Corydoras metae

Pictus cat
Pimelodus pictus

Eye small and
no teeth; pulsed
electric discharge

Barred knifefish
Steatogenys elegans

Black ghost
Apteronotus albifrons

Electric eel
Electrophorus electricus

Small caudal fin; continuous
weak electric discharge

FACT FILE

Electric catfish With modified muscle around its body capable of discharging about 400 volts, this tropical African catfish generates electric shocks to stun prey and deter predators.

⇀ Up to 4 ft (1.2 m)
⚖ Up to 44 lb (20 kg)
○ Oviparous, guarders
♀♂ Male & female
🗡 Common

Nile River & Central Africa

Electric eel A South American native, the electric eel is unrelated to true eels but is a relative of piranhas and tetras. It is unable to absorb sufficient oxygen through gills and gulps air to breathe by using its mouth, which is richly lined with blood vessels, as a lung.

⇀ Up to 8 ft (2.4 m)
⚖ Up to 44 lb (20 kg)
○ Oviparous, guarders
♀♂ Male & female
🗡 Common

N. South America

CANDIRU

The candiru, a catfish that grows to just 1 inch (2.5 cm) in length, leads a uniquely parasitic lifestyle. Found in the fresh water of the Amazon, it swims into gill cavities of larger fish to feed on blood. It has been known to enter and lodge in the urethras of urinating human bathers, causing hemorrhaging and sometimes even death.

ELECTRIC EELS

Most of an electric eel's body is a tail endowed with many thousands of electricity-producing cells derived from modified muscle tissue. This fish navigates by generating and detecting weak electrical discharges. But it also stuns prey with intense electricity bolts of up to 550 volts in large specimens, strong enough to knock over a person.

electricity-generating organs are
arranged along each flank

anal fin extends almost entire body
length and generates swimming motion

SALMONS AND ALLIES

CLASS Actinopterygii	
SUPERORDER Protacanthopterygii	
ORDERS 3	
FAMILIES 15	
SPECIES 502	

Some of the world's most sought-after angling and table fishes are among this group, the protacanthopterygians, which has ancient origins stemming back to the cretaceous (144–65 million years ago). It is usually divided into 15 families comprising more than 500 species arranged into three groups: esociforms, osmeriforms, and salmoniforms. These fishes are mostly carnivores equipped with large mouths and sharp teeth. Many are powerful and agile swimmers with elongate, streamlined bodies and well-developed tail fins. This swimming prowess is particularly evident in the salmons and trouts, many of which make extraordinarily long and arduous spawning migrations, journeys that have perplexed and fascinated scientists for centuries.

Expanding distribution Esociforms are found only in temperate North America, Europe, and Asia. Osmeriforms are also temperate fishes, with representatives in both hemispheres. The salmoniforms are Northern Hemisphere natives but many species have been introduced into suitable temperate-water ponds, streams, and rivers worldwide for the purposes of recreational angling and aquaculture.

Trout farming Several trout species, particularly rainbow trout (*Oncorhynchus mykiss*), are farmed in temperate areas in both purpose-built ponds and netted enclosures within natural water courses. These are stocked with fry obtained by artificial fertilization from hatcheries and grown to marketable size on a diet of protein pellets derived as a byproduct from the meat-packing industry.

Voracious carnivores Ambush predators with large mouths and sharp teeth, pikes lurk motionlessly in the water column or among vegetation before striking suddenly at prey with great speed. Quarry usually includes aquatic invertebrates and other fishes but some larger species will also take small mammals and birds.

DIVERSE PREDATORS

The esociform group includes the pikes, muskellunge, pickerels, and mudminnows, all of which lead exclusively freshwater lives. Most are formidable ambush predators and the location of the median fins back toward the tail is an adaptation to this lifestyle.

There are more than 230 species of osmeriforms. Among the most bizarre are the barreleyes, which have upward-directed tubular eyes and other specialist adaptations for life in the high-pressure darkness of oceanic depths. Commercially important osmeriforms include the Northern Hemisphere's smelts—small, streamlined fishes that are particularly abundant in temperate coastal waters. In the Southern Hemisphere, the best-known osmeriforms are the galaxiids, a scale-less family with complex lifecycles in which juvenile stages move between fresh and saltwater.

The salmoniforms include the whitefishes, ciscoes, graylings, chars, salmons, and trouts. Most are commercially important but it is the homing and migratory capabilities of the salmons that attract particular attention.

The six species of Pacific salmons spend most of their lives at sea but all attempt to return to the fresh-water streams in which they were spawned when they reach sexual maturity. The sense of smell is believed to be critical to these fishes in locating their birthplaces.

Migratory marvels Each stream has a distinct odor created by the soil and vegetation in its drainage. It is believed that this may become imprinted on young salmon and help them to find their way home during spawning migrations.

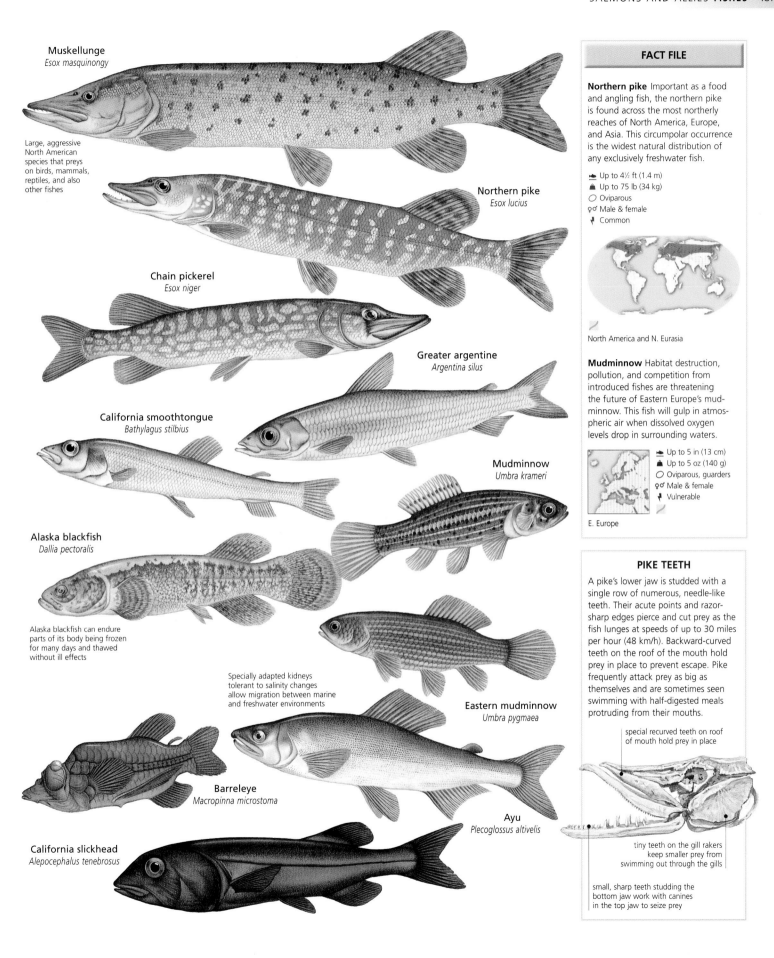

Muskellunge
Esox masquinongy

Large, aggressive
North American
species that preys
on birds, mammals,
reptiles, and also
other fishes

Northern pike
Esox lucius

Chain pickerel
Esox niger

Greater argentine
Argentina silus

California smoothtongue
Bathylagus stilbius

Mudminnow
Umbra krameri

Alaska blackfish
Dallia pectoralis

Alaska blackfish can endure
parts of its body being frozen
for many days and thawed
without ill effects

Specially adapted kidneys
tolerant to salinity changes
allow migration between marine
and freshwater environments

Eastern mudminnow
Umbra pygmaea

Barreleye
Macropinna microstoma

Ayu
Plecoglossus altivelis

California slickhead
Alepocephalus tenebrosus

FACT FILE

Northern pike Important as a food
and angling fish, the northern pike
is found across the most northerly
reaches of North America, Europe,
and Asia. This circumpolar occurrence
is the widest natural distribution of
any exclusively freshwater fish.

- Up to 4½ ft (1.4 m)
- Up to 75 lb (34 kg)
- ○ Oviparous
- ♀♂ Male & female
- ⚑ Common

North America and N. Eurasia

Mudminnow Habitat destruction,
pollution, and competition from
introduced fishes are threatening
the future of Eastern Europe's mudminnow. This fish will gulp in atmospheric air when dissolved oxygen
levels drop in surrounding waters.

- Up to 5 in (13 cm)
- Up to 5 oz (140 g)
- ○ Oviparous, guarders
- ♀♂ Male & female
- ⚑ Vulnerable

E. Europe

PIKE TEETH

A pike's lower jaw is studded with a
single row of numerous, needle-like
teeth. Their acute points and razor-
sharp edges pierce and cut prey as the
fish lunges at speeds of up to 30 miles
per hour (48 km/h). Backward-curved
teeth on the roof of the mouth hold
prey in place to prevent escape. Pike
frequently attack prey as big as
themselves and are sometimes seen
swimming with half-digested meals
protruding from their mouths.

special recurved teeth on roof
of mouth hold prey in place

tiny teeth on the gill rakers
keep smaller prey from
swimming out through the gills

small, sharp teeth studding the
bottom jaw work with canines
in the top jaw to seize prey

Atlantic salmon Most young Atlantic salmon remain in fresh water for up to 4 years before migrating to the North Atlantic Ocean. Between 1 and 4 years later, they return to their birthplaces to spawn, often surviving to repeat the migration. Some landlocked populations migrate between deep-water feeding grounds and shallow shorelines.

⊷ Up to 5 ft (1.5 m)
⬤ Up to 79 lb (36 kg)
○ Oviparous, hiders
♀♂ Male & female
↟ Common

N. Atlantic, N.W. Europe, N.E. North America

Sea trout This species is also born in fresh water, migrates to live at sea, and then returns to fresh water to spawn. It is long-lived and thought capable of surviving for up to 20 years, during which it makes repeated annual spawning migrations.

⊷ Up to 4½ ft (1.4 m)
⬤ Up to 33 lb (15 kg)
○ Oviparous, hiders
♀♂ Male & female
↟ Common

N.W. Europe; introd. widely

Brown trout A freshwater non-migratory form of the sea trout, this fish is prized by anglers and gourmet alike. Originally from Europe, it has been introduced to many other areas worldwide. As in other salmonids, females create nests, known as "redds," with their tails in clean gravel sediments.

⊷ Up to 4½ ft (1.4 m)
⬤ Up to 33 lb (15 kg)
○ Oviparous
♀♂ Male & female
↟ Common

Europe, W. Asia, N.W. Africa; introd. worldwide

European smelt Like their close relatives the salmons and trouts, smelts have an adipose fin. This small fleshy lump on the dorsal side near the tail fin is considered a primitive feature.

⊷ Up to 12 in (30 cm)
⬤ Up to 7 oz (200 g)
○ Oviparous
♀♂ Male & female
↟ Data deficient

N.W. Europe

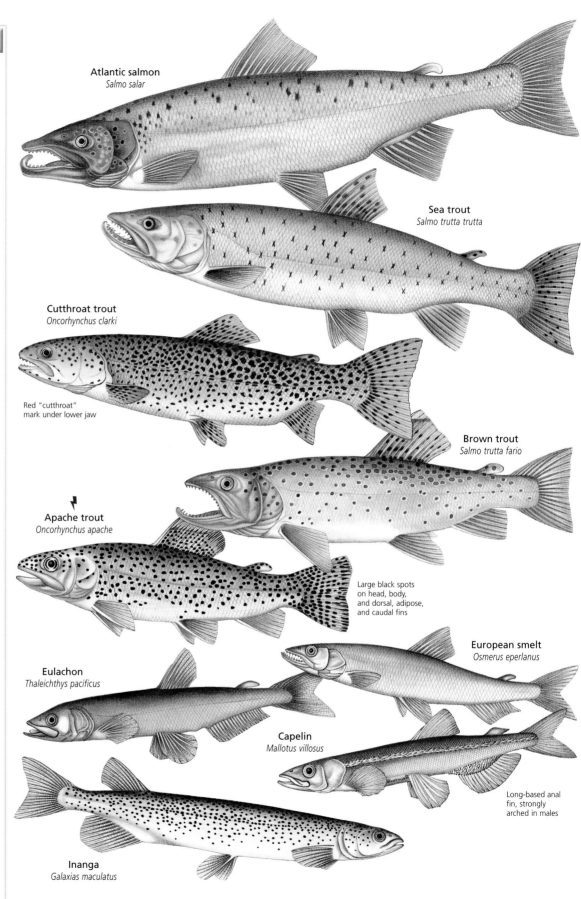

Atlantic salmon
Salmo salar

Sea trout
Salmo trutta trutta

Cutthroat trout
Oncorhynchus clarki

Red "cutthroat" mark under lower jaw

Brown trout
Salmo trutta fario

Apache trout
Oncorhynchus apache

Large black spots on head, body, and dorsal, adipose, and caudal fins

European smelt
Osmerus eperlanus

Eulachon
Thaleichthys pacificus

Capelin
Mallotus villosus

Long-based anal fin, strongly arched in males

Inanga
Galaxias maculatus

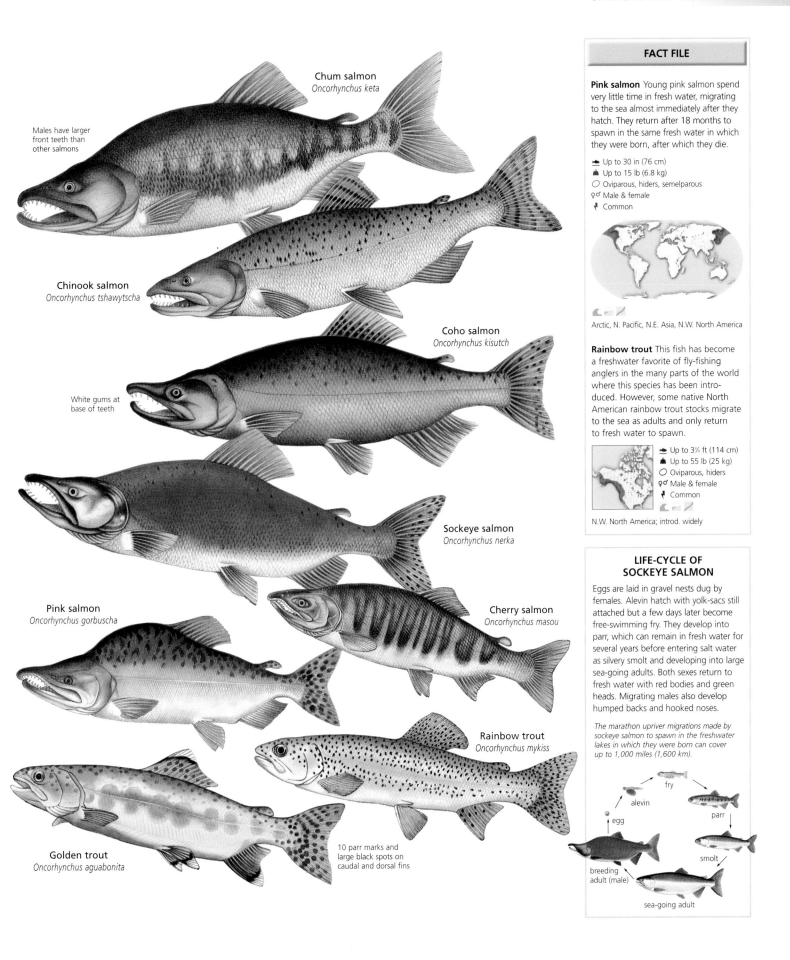

Chum salmon
Oncorhynchus keta

Males have larger
front teeth than
other salmons

Chinook salmon
Oncorhynchus tshawytscha

Coho salmon
Oncorhynchus kisutch

White gums at
base of teeth

Sockeye salmon
Oncorhynchus nerka

Pink salmon
Oncorhynchus gorbuscha

Cherry salmon
Oncorhynchus masou

Rainbow trout
Oncorhynchus mykiss

Golden trout
Oncorhynchus aguabonita

10 parr marks and
large black spots on
caudal and dorsal fins

FACT FILE

Pink salmon Young pink salmon spend
very little time in fresh water, migrating
to the sea almost immediately after they
hatch. They return after 18 months to
spawn in the same fresh water in which
they were born, after which they die.

- Up to 30 in (76 cm)
- Up to 15 lb (6.8 kg)
- Oviparous, hiders, semelparous
- Male & female
- Common

Arctic, N. Pacific, N.E. Asia, N.W. North America

Rainbow trout This fish has become
a freshwater favorite of fly-fishing
anglers in the many parts of the world
where this species has been introduced. However, some native North
American rainbow trout stocks migrate
to the sea as adults and only return
to fresh water to spawn.

- Up to 3¾ ft (114 cm)
- Up to 55 lb (25 kg)
- Oviparous, hiders
- Male & female
- Common

N.W. North America; introd. widely

LIFE-CYCLE OF SOCKEYE SALMON

Eggs are laid in gravel nests dug by
females. Alevin hatch with yolk-sacs still
attached but a few days later become
free-swimming fry. They develop into
parr, which can remain in fresh water for
several years before entering salt water
as silvery smolt and developing into large
sea-going adults. Both sexes return to
fresh water with red bodies and green
heads. Migrating males also develop
humped backs and hooked noses.

*The marathon upriver migrations made by
sockeye salmon to spawn in the freshwater
lakes in which they were born can cover
up to 1,000 miles (1,600 km).*

fry

alevin

parr

egg

smolt

breeding
adult (male)

sea-going adult

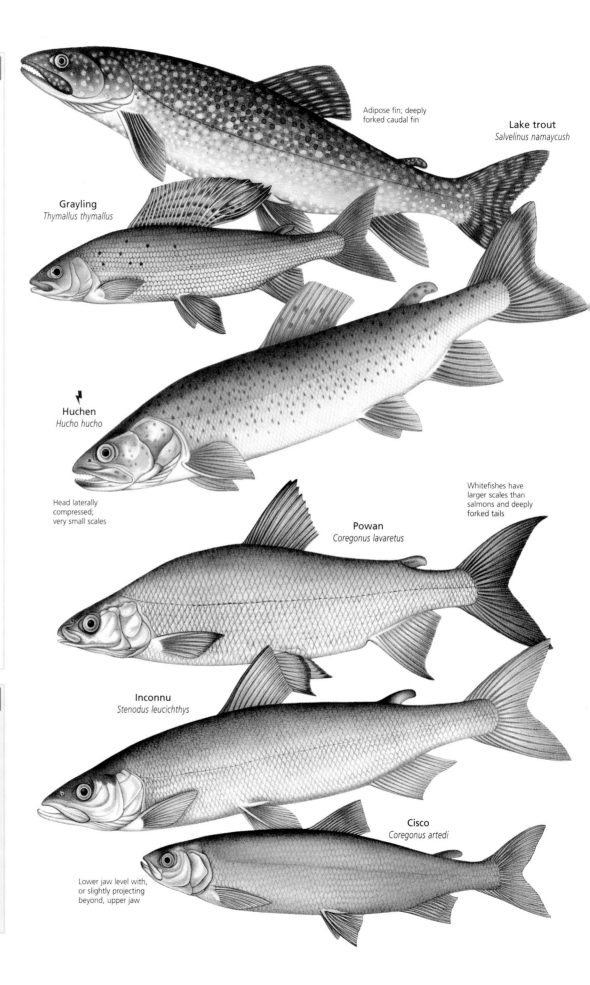

Adipose fin; deeply forked caudal fin

Lake trout
Salvelinus namaycush

Grayling
Thymallus thymallus

Huchen
Hucho hucho

Head laterally compressed; very small scales

Whitefishes have larger scales than salmons and deeply forked tails

Powan
Coregonus lavaretus

Inconnu
Stenodus leucichthys

Cisco
Coregonus artedi

Lower jaw level with, or slightly projecting beyond, upper jaw

DRAGONFISHES AND ALLIES

CLASS Actinopterygii	
SUPERORDER Stenopterygii	
ORDERS 2	
FAMILIES 5	
SPECIES 415	

Although diverse and widespread, the stomiiforms are rarely encountered by people because of their mid- to deep-water lifestyle. All are predators that are adapted to habitats defined by high pressure, limited light, and low productivity. Most have long teeth and big mouths that can cope with large, infrequently encountered meals. And all but one species are equipped with light organs—photophores. Dotted along their sides and underbellies, these serve as camouflage from predators below by masking their appearance against weak light from above. Most also use them for hunting; suspended from a chin barbel or fin ray they lure prey like moths to a flame.

Deep-sea distribution The stomiiforms are found in the deep, open waters of all major temperate and subtropical oceans. Some extend into polar waters. They are also found over a large depth range, migrating upward at night and returning down during the day.

Fierce predator This deep-sea dragonfish (*Idiacanthus* sp.) has features typical of the family Stomiidae, including photophores along its belly and a long barbel on the lower jaw. These fishes remain in the complete darkness of the ocean's extreme depths by day, but migrate vertically at night to feed at more moderate depths. Males of this genus are much smaller than the females and lack teeth and pelvic fins.

DEEP-SEA LIFE

Stomiiforms typically have big heads, long bodies, and dark coloration, although some are translucent or silvery. Common names such as lightfishes, bristlemouths, loosejaws, snaggletooths, and viperfishes suggest their often bizarre appearance.

Hermaphroditism is common, an adaptation to a habitat where members of the same species may be infrequently encountered.

Eggs and larvae are buoyant, floating within plankton in surface currents, but young descend as juveniles to pursue deep-sea life. As adults, many stomiiforms rest by day in deep water but migrate at dusk to shallower depths, where food in the form of small fishes and invertebrates is more plentiful.

⚡ CONSERVATION WATCH

Deep questions Because deep-sea habitats can be as inaccessible as outer space, very little is known about the population structure, size, or breeding behavior of the stomiiforms. It is possible that one genus, *Cyclothone*, includes more individuals than any other vertebrate genus: billions of these tiny, mesopelagic fishes live in the oceans. It is also possible that some are disappearing as a consequence of oceanic pollution before their existence has even been recorded.

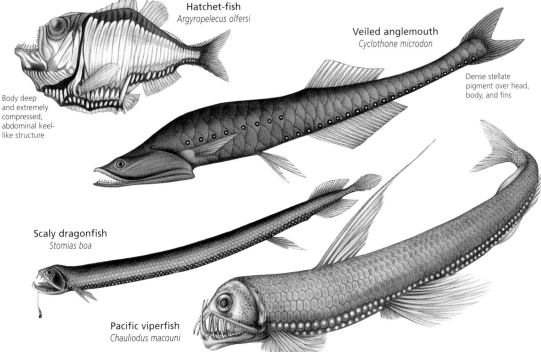

Hatchet-fish
Argyropelecus olfersi

Veiled anglemouth
Cyclothone microdon

Dense stellate pigment over head, body, and fins

Body deep and extremely compressed; abdominal keel-like structure

Scaly dragonfish
Stomias boa

Pacific viperfish
Chauliodus macouni

LIZARDFISHES AND ALLIES

CLASS	Actinopterygii
SUPERORDER	Cyclosquamata
ORDER	Aulopiformes
FAMILIES	13
SPECIES	229

This marine group, the aulopiforms, is represented in both coastal habitats and great depths. It includes the lizardfishes, Bombay ducks, greeneyes, ipnopids, lancetfishes, daggertooths, and telescopefishes. They exhibit an unusual mix of primitive and advanced features and are of particular interest to scientists for various reasons, including their range of peculiar eye modifications. Many deep-sea species are notable for their mode of reproduction: they are bisexual synchronous hermaphrodites, which means that they function as both sexes at the same time and may even be capable of self-fertilization. The aulopiforms are also known for the often extreme metamorphoses they pass through as they change from larvae into juveniles.

Wide distribution Lizardfishes occur in all warm seas. Greeneyes are found worldwide in tropical to warm-temperate waters. Most Bombay ducks have an Indo-Pacific distribution. The lancetfishes range widely in Atlantic and Pacific mid-waters.

ASSORTED ASSEMBLAGE

Lizardfishes, Bombay ducks, and greeneyes are inshore bottom-dwelling ambush predators. Typically well-camouflaged, they often sit propped up at the front end by their pectoral fins.

The ipnopids dwell at great depths and usually have flat heads, pencil-like bodies, and poorly developed eyes.

Attaining lengths of nearly 7 feet (2.1 m), lancetfishes are among the largest of all deep-sea predatory fishes. They have huge mouths and large, dagger-like teeth, and have a reputation for destroying undersea cables. They are sometimes seen floundering in the surf.

Cryptic coloration The patterns and colors of most lizardfishes, including this Indo-Pacific species, the variegated lizardfish (*Synodus variegatus*), blends in well with bottom sediments and the lizardfishes' benthic habitat.

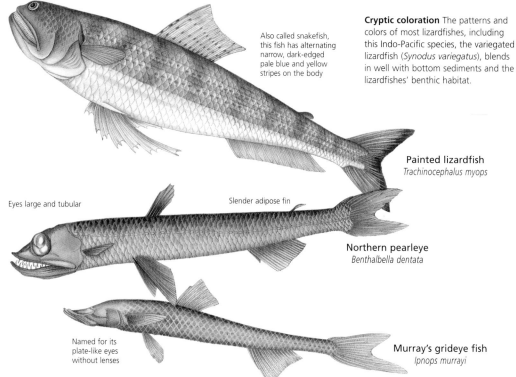

Also called snakefish, this fish has alternating narrow, dark-edged pale blue and yellow stripes on the body

Painted lizardfish
Trachinocephalus myops

Eyes large and tubular

Slender adipose fin

Northern pearleye
Benthalbella dentata

Named for its plate-like eyes without lenses

Murray's grideye fish
Ipnops murrayi

Say cheese Propped up on hugely elongated pelvic and tail fins, tripod fishes hold themselves clear of the substrate at great ocean depths awaiting prey. These ipnopid fishes grow to a maximum length of about 14 inches (36 cm).

LANTERNFISHES

CLASS Actinopterygii	
SUPERORDER Scopelomorpha	
ORDER Myctophiformes	
FAMILIES 2	
SPECIES 251	

Of all the deep, open-water marine fishes, the lanternfishes are the most widely distributed, diverse, and abundant. Because these small plankton-eaters form large dense schools preyed upon by sea birds (particularly penguins), marine mammals, and a huge range of carnivorous fishes, they play a crucial role in virtually all marine ecosystems. They are fished to a relatively small extent for fish meal and oil but, with a global mass estimated at a staggering 660 million tons (600 million tonnes), they are otherwise regarded as a largely underused commercial resource. Together with their less abundant allies the neoscopelids, lanternfishes make up the group known as the myctophiforms.

Global resource Lanternfishes can be found in all open oceanic waters. Species distributions are related to both ocean currents and the physical and biological characteristics of the water. Many exist at great depths by day but will migrate upward to feed in surface waters by night.

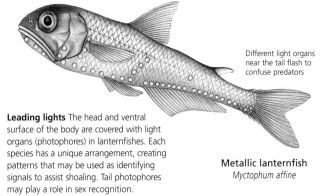

Different light organs near the tail flash to confuse predators

Leading lights The head and ventral surface of the body are covered with light organs (photophores) in lanternfishes. Each species has a unique arrangement, creating patterns that may be used as identifying signals to assist shoaling. Tail photophores may play a role in sex recognition.

Metallic lanternfish
Myctophum affine

BEARDFISHES

CLASS Actinopterygii	
SUPERORDER Polymixiomorpha	
ORDER Polymixiiformes	
FAMILY Polymixiidae	
SPECIES 10	

Just 10 species in one genus make up this small but puzzling group of deep-water marine fishes, which derives its common name from the pair of sensory barbels they all have hanging from the chin. Like the more advanced teleosts (the spiny-rayed fishes), they have true fin spines (as opposed to soft rays): between four and six in the dorsal fin and four in the anal fin. However, along with this modern feature they also exhibit several primitive and unique characteristics, creating a perplexing array of attributes that has fuelled much debate about which other fish groups are their closest relatives.

Bottom dwellers Beardfishes are known from tropical and subtropical oceanic waters. They live on the outer continental shelf and upper slope at various depths between 65 and 2,500 feet (20 and 760 m), usually on the bottom.

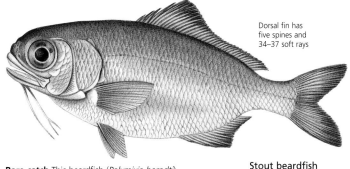

Dorsal fin has five spines and 34–37 soft rays

Rare catch This beardfish (*Polymixia berndti*) is mostly encountered as bycatch taken by commercial trawlers working the deep tropical and subtropical waters of the Indo-Pacific.

Stout beardfish
Polymixia nobilis

OPAHS AND ALLIES

| SUBDIVISION Euteleostei |
| SUPERORDER Lampridiomorpha |
| ORDER Lampridiformes |
| FAMILIES 7 |
| SPECIES 23 |

Twenty-three species of deep-sea fishes, known as the lampridiforms, make up this group. Outwardly, their external appearance is extremely variable although most are either large and moon-shaped or elongated and snake-like. They all, however, share four unique and characteristic features, all but one of which relates to an unusual and specialized jaw arrangement that gives them highly protrusible mouths. They usually lack scales, tend to have dorsal fins that run the length of the body, and most have pelvic fins that are well forward on the body. Many have bodies adorned in radiant hues with bright red fins. This group probably arose some 65 million years ago.

Widespread enigmas Lampridiforms are usually found at depths of between 330 and 3,300 feet (100 and 1,000 m). Opahs, ribbonfishes, and tube-eyes have worldwide distributions in warm waters. Velifers are found only in the Indian and Pacific Oceans. The open-ocean, deep-water lifestyles of lampridiforms mean that they are rarely encountered by people.

Mythical serpent Accounts of sea monsters probably stem from sightings of the oarfish, the world's longest fish. There has been at least one report of a specimen measuring 56 feet (17 m) long but most probably attain lengths of around 26 feet (8 m). These fish usually inhabit depths up to 660 feet (200 m) but have been seen at the surface and washed up on beaches.

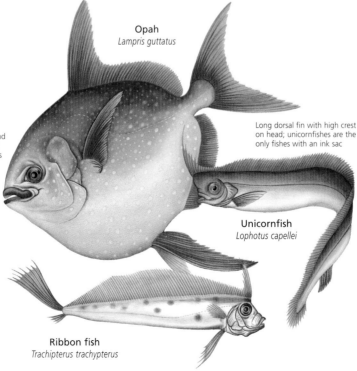

Body oval and compressed; vermilion lips and fins

Opah
Lampris guttatus

Long dorsal fin with high crest on head; unicornfishes are the only fishes with an ink sac

Unicornfish
Lophotus capellei

Ribbon fish
Trachipterus trachypterus

CONSERVATION WATCH

Enigmatic rarities No species in this group are listed as endangered, but most are considered rare. Our lack of understanding about their true conservation status is just one of many areas in which our knowledge of these fishes is wanting. Most of what is known comes from tales of ancient mariners and the brief observations of captured and stressed individuals.

STRANGE LIFESTYLES

The lampridiforms include the crestfish, opahs, ribbonfishes, tube-eyes, and velifers, some of which exhibit unusual feeding behaviors. The tube-eye *Stylephorus chordatus*, for example, rises hundreds of feet (meters) daily from depths of about 2,600 feet (800 m) to feed in a head-up, tail-down position on tiny crustaceans.

The ribbon fish *Trachipterus trachypterus* is thought to adopt a similarly vertical position to feed on other fishes and squid.

All lampridiforms produce large eggs, up to ¼ inch (6 mm) in diameter. Most are bright shades of red, which may protect against ultraviolet rays penetrating the surface waters in which they float for up to a month before hatching. Unlike the eggs of most bony fishes, which produce feeble larvae requiring a rich yolk sac for food, lampridiform embryos develop early and are vigorous swimmers.

CODS, ANGLERFISHES, AND ALLIES

Several small but significant features of the skeleton indicate relatedness between species in this group. There are also similarities in their habitat preferences. Most, for example, are bottom dwellers although some species (particularly those that are commercially important, such as haddock, hake, and the cods) form large pelagic schools. All but about 20 species comprising the group are marine. They tend to be active at night or live in dark habitats such as underwater caves or the deep sea. An unusual feature of some is that, by using special muscles located on the swim bladder, they can produce sounds. These may be important in courtship and communicating distress.

Distribution Cods and anglerfishes are found in all oceans. The cods are exclusively marine with the exception of one species—the burbot—which is found in fresh water. Troutperches and allies only live in fresh water and are restricted to North America. Toadfishes live along tropical coastlines.

SUBDIVISION	Euteleostei
SUPERORDER	Paracanthopterygii
ORDERS	5
FAMILIES	37
SPECIES	1,382

FACT FILE

Starry handfish The expanded head of this dorsoventrally flattened bottom-dweller forms a flat rounded disc. It uses its hand-like pectoral fins, which project sideways from the back of this disc, to move across the substrate. An anglerfish, it carries a fleshy lure in a bony recess on the snout.

- ➡ Up to 12 in (30 cm)
- ⬛ Up to 2 lb (900 g)
- ○ Oviparous
- ♀♂ Male & female
- ⚘ Common

Indo-West Pacific

Trout perch A native to freshwater streams and lakes in the eastern United States, the trout perch seeks rocky cover by day but ventures out by night to make short feeding migrations to shallower waters.

- ➡ Up to 8 in (20 cm)
- ⬛ Up to 6 oz (170 g)
- ○ Oviparous
- ♀♂ Male & female
- ⚘ Locally common

N. North America

⚡ CONSERVATION WATCH

Threatened futures The critically endangered Alabama cavefish (*Speoplatyrhinus poulsoni*) is entirely white and blind. It occurs only in Key Cave, Alabama, in the United States. Interference with bat populations has indirectly affected the cavefish's food chain and the contamination of groundwater is another major threat. An introduced sea star and the increased silt in river sediments due to land clearing have almost certainly contributed to the critically endangered status of the spotted handfish (*Brachionichthys hirsutus*), a marine fish endemic to the Australian island of Tasmania.

CRITICAL FOOD FISHES

Some of the most significant fisheries worldwide exploit species within the family Gadidae, which includes the Alaska pollock and Atlantic cod and annually accounts for well over 10 percent of the world's total fish catch.

These fishes produce enormous numbers of eggs, among the most of any fish. A large Alaskan pollock, for example, may spawn 15 million eggs in a year. Because, however, gadiform fishes are also long-lived and late to mature, they have proved to be highly susceptible to overexploitation and some fisheries are now in trouble.

Anglerfish Many cod species are strong swimmers that actively search for prey along ocean substrates. Anglerfishes (left), in contrast, are mostly slow-moving and sluggish predators that wait for food to come to them. They can, however, move with lightning speed as they capture unsuspecting prey.

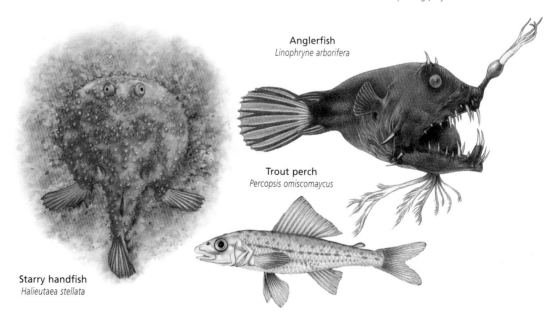

Anglerfish
Linophryne arborifera

Trout perch
Percopsis omiscomaycus

Starry handfish
Halieutaea stellata

FACT FILE

Burbot A freshwater lover of the dark with a circumarctic distribution, the burbot dwells mainly in the depths of lakes and slow-moving rivers. It shelters among vegetation, in suitable crevices, and down deep holes.

- Up to 5 ft (1.5 m)
- Up to 75 lb (34 kg)
- ○ Oviparous
- ♀♂ Male & female
- Common

N. North America & N. Eurasia

European hake After decades of commercial exploitation, there is wide-spread concern the European hake has been overfished. Although the maximum weight recorded for the species is about 25 pounds (11.5 kg), individuals over 11 pounds (5 kg) are now rarely caught.

- Up to 4½ ft (1.4 m)
- Up to 25 lb (11.5 kg)
- ○ Oviparous
- ♀♂ Male & female
- Common

E. North Atlantic, Mediterranean & Black seas

SEXUAL PARASITES

Ceratoid anglerfishes dwell in low-density populations in the high-pressure darkness of oceanic depths. To reduce the hit-or-miss probability of the sexes meeting, males of some species have evolved into little more than parasitic testes. As adults, they live permanently attached to and derive nourishment from their considerably larger female counterparts. In the triplewart seadevil (*Cryptopsaras couesii*) below, females have been found with up to four males attached.

"fishing pole" (illicium) and "lure" (esca)

large median and two small lateral oval caruncles

dwarf parasitic male

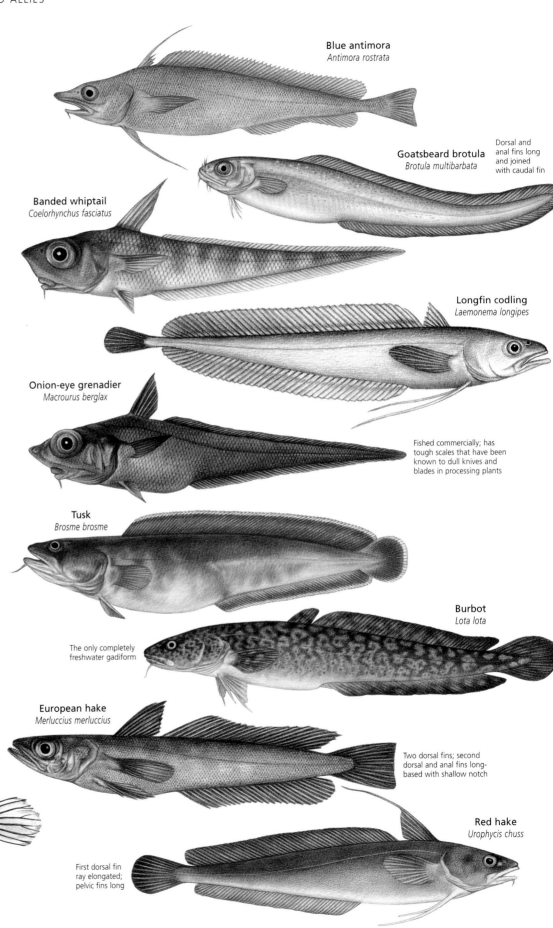

Blue antimora
Antimora rostrata

Goatsbeard brotula
Brotula multibarbata

Dorsal and anal fins long and joined with caudal fin

Banded whiptail
Coelorhynchus fasciatus

Longfin codling
Laemonema longipes

Onion-eye grenadier
Macrourus berglax

Fished commercially; has tough scales that have been known to dull knives and blades in processing plants

Tusk
Brosme brosme

Burbot
Lota lota

The only completely freshwater gadiform

European hake
Merluccius merluccius

Two dorsal fins; second dorsal and anal fins long-based with shallow notch

Red hake
Urophycis chuss

First dorsal fin ray elongated; pelvic fins long

Haddock
Melanogrammus aeglefinus

Plainfin midshipman
Porichthys notatus

Tadpole fish
Raniceps raninus

Stout-bodied with a broad
depressed head; minute first
dorsal fin and small chin barbel

American angler
Lophius americanus

Three dorsal fins

Bib
Trisopterus luscus

Angler
(monkfish)
Lophius piscatorius

Atlantic cod
Gadus morhua

Two anal fins

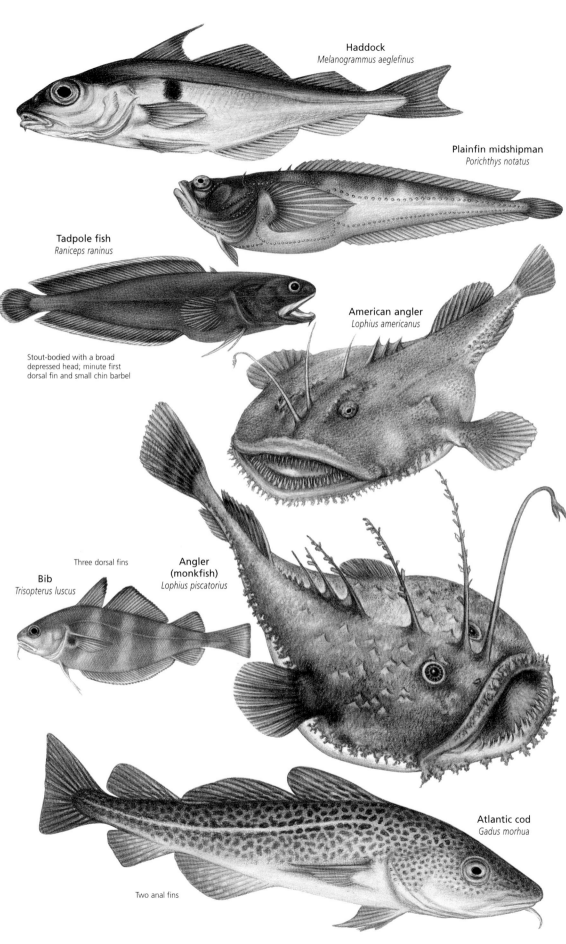

EXQUISITE CAMOUFLAGE

Anglerfishes have no need for stream-
lining. Instead, these sit-and-wait
predators are perhaps better served
by being shaped and colored like their
surroundings. The sargassumfish
(*Histrio histrio*), for example, blends
in with sargassum weed, and the
roughbar frogfish (*Antennarius
avalonis*) looks like a rock.

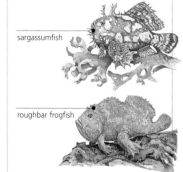

sargassumfish

roughbar frogfish

SPINY–RAYED FISHES

CLASS Actinopterygii	
SUPERORDER Acanthopterygii	
ORDERS 15	
FAMILIES 269	
SPECIES 13,262	

This, the largest single group of fishes, comprises the most advanced and recently evolved teleosts. Known also as acanthopterygians, the spiny-rayed fishes include more than 13,000 species in over 250 families. Common characteristics include a highly mobile and protrusible upper jaw, which has facilitated a wide range of feeding strategies, ctenoid scales that in some cases have become lost or further developed into hardened plates, and hard spines in the fins. As a group, these fishes have a ubiquitous distribution but are particularly abundant in coastal marine waters. They display an enormous breadth of specialized reproductive, behavioral, and anatomical adaptations that has seen them exploit niches not available to other fishes.

Extensive distribution The most wide-spread of all fish groups, the spiny-rays are found in almost every available aquatic habitat throughout the world from fresh water, through brackish to salt water, and from coastal shallows to the ocean's depths. Specialized morphological and behavioral adaptations allow them to inhabit a wide range of environments, from frozen seas to ponds that dry up.

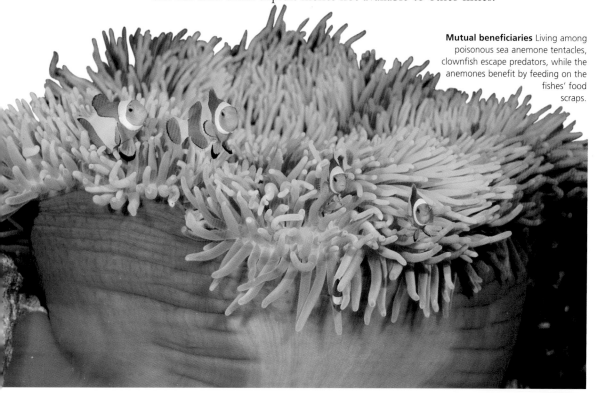

Mutual beneficiaries Living among poisonous sea anemone tentacles, clownfish escape predators, while the anemones benefit by feeding on the fishes' food scraps.

ADVANCED ADAPTATIONS

Almost every variation on the basic fish body plan can be found among the spiny-rayed fishes. The flatfishes, for example, depart dramatically from the normal symmetry seen in most fishes. The wrasses and parrot-fishes have a modified pharyngeal apparatus that acts as a second set of jaws in the throat, facilitating specialized feeding.

The cyprinodontiforms include guppies, swordtails, and other highly resilient freshwater aquarium favorites. Silversides can form large marine schools and many are commercially important baitfish. The flyingfishes are remarkably adapted for above-water gliding.

Gobies are mostly small fishes, such as the amphibious mudskippers, in which the pelvic fins are fused to form a cup-shaped disk.

In many of the triggerfishes and their relatives, scales have developed into protective body armor.

Damselfishes, including the well-known clownfishes, often display highly developed territorial behavior. The cichlids take parental care to its extreme. The groupers are among the most robust of all marine predators, and billfishes such as marlin and sailfish are among the fastest swimming.

The scorpionfishes include some of the deadliest fishes to humans. The drums and croakers make deep distinctive noises when threatened and the butterflyfishes and angelfishes are among the most exquisitely colored aquatic creatures.

Sexual variety The spiny-rayed fishes display a wide range of sexual strategies but monogamy and male pregnancies are particularly unusual features seen in sea-horse reproduction. Few other fishes adopt such an approach.

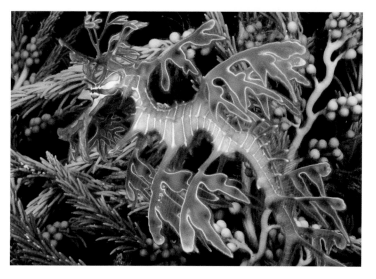

Masterful masquerade With frilled leaf-like fins, the leafy seadragon is so well-camouflaged that neither prey nor predators can detect it among the seagrass and seaweed beds where it dwells.

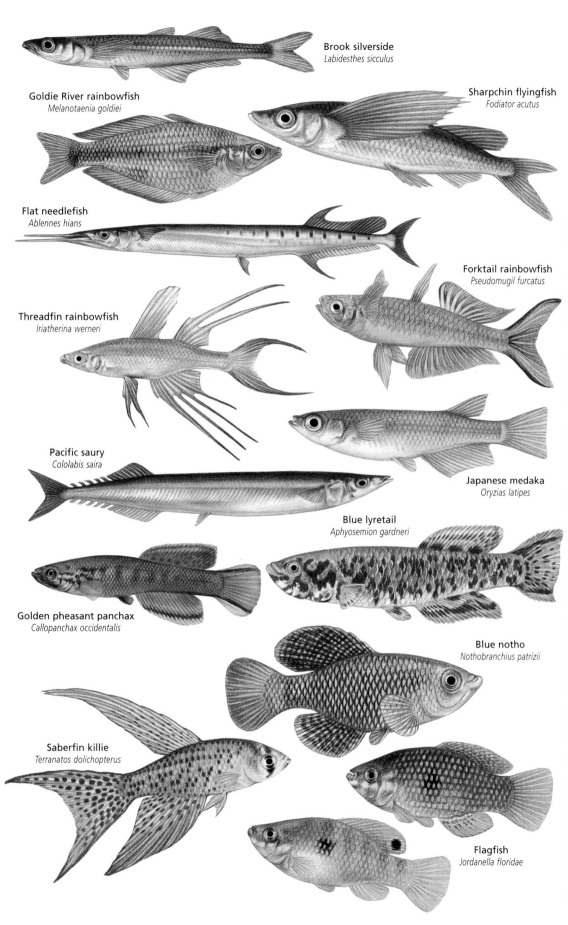

Brook silverside
Labidesthes sicculus

Goldie River rainbowfish
Melanotaenia goldiei

Sharpchin flyingfish
Fodiator acutus

Flat needlefish
Ablennes hians

Threadfin rainbowfish
Iriatherina werneri

Forktail rainbowfish
Pseudomugil furcatus

Pacific saury
Cololabis saira

Japanese medaka
Oryzias latipes

Blue lyretail
Aphyosemion gardneri

Golden pheasant panchax
Callopanchax occidentalis

Blue notho
Nothobranchius patrizii

Saberfin killie
Terranatos dolichopterus

Flagfish
Jordanella floridae

FACT FILE

Brook silverside A native of sub-tropical North American freshwater lakes and rivers, this small translucent fish is usually found in schools near the water's surface, feeding on tiny crustaceans, insect larvae, and small flying adult insects. Spawning takes place in clear water, usually around vegetation, and each egg becomes anchored to the substrate by its own sticky filament.

- Up to 5 in (13 cm)
- Up to 4 oz (115 g)
- ○ Oviparous
- ♀♂ Male & female
- Common

S.E. North America

Sharpchin flyingfish Members of the genus *Fodiator* are among the weakest "flyers" of the flyingfish. Nevertheless, this species is capable of gliding over water on enlarged wing-like pectoral fins for 165 feet (50 m) or more. They live close to the surface of the open ocean, feed on plankton, and "fly" to evade attacks from predators below.

- Up to 9½ in (24 cm)
- Up to 8 oz (225 g)
- ○ Oviparous
- ♀♂ Male & female
- Common

E. Pacific & E. Atlantic

Golden pheasant panchax This shortlived freshwater species occurs in short-term pools and puddles in West African rainforest and humid forest. Adults die after spawning, at the onset of the dry season, leaving fertilized eggs to survive in mud for as long as 3 months. They hatch with the onset of the rainy season.

- Up to 3 in (7.5 cm)
- Up to 1 oz (30 g)
- ○ Oviparous, hiders
- ♀♂ Male & female
- Common

W. Africa

Saberfin killie This is another short-lived species in which adults die not long after spawning. Left in drying mud, fertilized eggs enter a resting phase until they are stimulated to hatch by the first raindrops of the rainy season. There is virtually no larval period and the young are sexually mature in just over a month after hatching.

- Up to 1½ in (4 cm)
- Up to ½ oz (15 g)
- ○ Oviparous, hiders
- ♀♂ Male & female
- Common

Venezuela

FACT FILE

Guppy The guppy belongs to one of the few bony fish families (Poeciliidae) with members that give birth to live young. Originally from South America, this popular aquarium fish has been introduced to warm freshwater lakes throughout the world. This is an adaptable species that tolerates a wide range of habitats and water quality.

- Up to 2 in (5 cm)
- Up to ¾ oz (20 g)
- Viviparous; live bearers
- ♀♂ Male & female
- Common

N.E. South America, Barbados, Trinidad; introd. widely

Crown squirrelfish Most adult squirrelfishes live in shallow water around tropical reefs but their larvae have a long pelagic life and are often found way out to sea. This species has a short venomous cheek spine. Squirrelfishes are active at night, sheltering in caves and under ledges during the day. They feed on small fishes and invertebrates.

- Up to 6½ in (17 cm)
- Up to 8 oz (225 g)
- ○ Oviparous
- ♀♂ Male & female
- Common

Indo-Pacific, Red Sea, Oceania

Velvet whalefish Named because of their whale-like appearance not their size, whalefishes are deep-sea ocean dwellers. They have reduced eyes but a highly developed lateral line system to provide information about their environment. They also have big mouths and highly distensible stomachs to cope with the large, infrequently encountered meals typical of life in the deep sea.

- Up to 14 in (36 cm)
- Up to 16 oz (450 g)
- Not known
- ♀♂ Not known
- Common

Worldwide in tropical and temperate seas

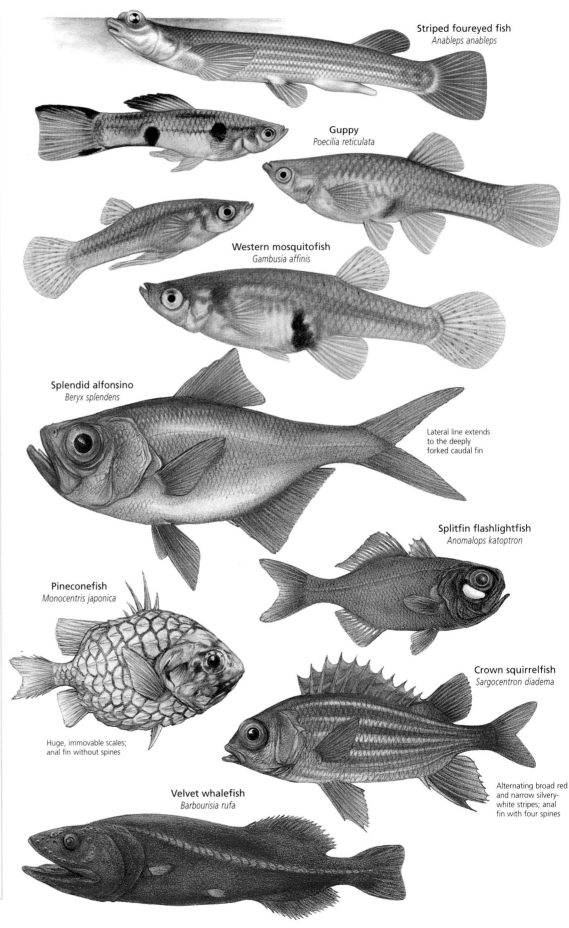

Striped foureyed fish
Anableps anableps

Guppy
Poecilia reticulata

Western mosquitofish
Gambusia affinis

Splendid alfonsino
Beryx splendens

Lateral line extends to the deeply forked caudal fin

Splitfin flashlightfish
Anomalops katoptron

Pineconefish
Monocentris japonica

Crown squirrelfish
Sargocentron diadema

Huge, immovable scales; anal fin without spines

Velvet whalefish
Barbourisia rufa

Alternating broad red and narrow silvery-white stripes; anal fin with four spines

Three-spined stickleback
Gasterosteus aculeatus

Ninespine stickleback
Pungitius pungitius

7–12 free spines
in front of dorsal fin

Longspine snipefish
Macrorhamphosus scolopax

Compressed scaleless
body; mouth at end
of long tubular snout

Indo-Pacific boarfish
Antigonia rubescens

3 anal spines with
24–48 soft rays; body
highly compressed

Shrimpfish
Aeoliscus strigatus

Massive head
with large, highly
protrusible jaws

John Dory
Zeus faber

FACT FILE

Shrimpfish With a long-snouted, semi-transparent, shrimp-like body, this fish congregates in synchronized schools in a head-down, tail-up vertical position. It lives among sea urchins and staghorn coral branches and feeds on zooplankton. It has been found cast up on Indo-Pacific beaches after storms.

➤ Up to 6 in (15 cm)
⬛ Up to 4 oz (115 g)
○ Oviparous
♀♂ Male & female
⬦ Common

Indo-West Pacific

John Dory The John Dory is a weak swimmer that leads a mostly solitary life near the ocean floor. A highly compressed body makes its head-on profile extremely narrow, helping it to remain hidden from prey as it stalks from behind. Its highly protrusible jaws are very efficient at capturing unsuspecting small fishes and crustaceans.

➤ Up to 26 in (66 cm)
⬛ Up to 13 lb (6 kg)
○ Oviparous
♀♂ Male & female
⬦ Common

E. Atlantic, Mediterranean Sea, W. Pacific & W. Indian Ocean

DEVOTED DADS

By cementing plants together with sticky secretions from their kidneys, male sticklebacks build elaborate nests for the females to lay their eggs in. After fertilizing the eggs, the males usually drive the females away but continue to maintain and defend the nests while attending to the eggs, fanning oxygenated water across them with their pectoral fins.

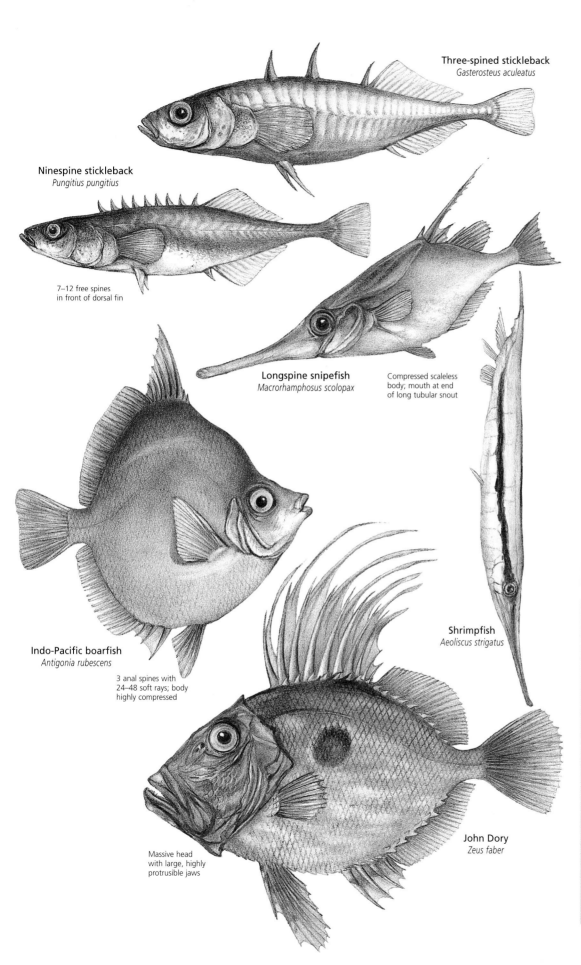

FACT FILE

Harlequin ghost pipefish Pipefishes and seahorses have a series of bony plates beneath their skin. As a result, they cannot swim by flexing the body like most other fishes but use rapid fanning movements of the fins instead.

↤ Up to 4¾ in (12 cm)
⛁ Up to 1 oz (30 g)
○ Oviparous, brooder
♀♂ Male & female
⌁ Uncommon

Indo-West Pacific

Leafy seadragon Like their close relatives the seahorses, male leafy seadragons incubate eggs. Females lay up to 250 onto a patch of spongy tissue on the tail's underside, where they develop for about 6 weeks before hatching.

↤ Up to 16 in (40 cm)
⛁ Up to 8 oz (225 g)
○ Oviparous, brooder
♀♂ Male & female
⌁ Data deficient

S. Australia

Swamp eel This highly adaptable, air-breathing, freshwater carnivore that resembles an eel can survive out of water for long periods. It has been labeled a potential "ecological night-mare" in some areas to which it has been introduced.

↤ Up to 18 in (46 cm)
⛁ Up to 24½ oz (700 g)
○ Oviparous, guarders
♀♂ Hermaphrodite
⌁ Common

S.E. Asia, Australia

SEAHORSE FEEDING

Seahorses are ambush predators that pluck mostly small, planktonic animals from the waters around them. Like their closest relatives, including the seadragons, trumpetfishes, and cornetfishes (families Syngnathidae, Aulostomidae, and Fistulariidae), they have an elongate, tube-shaped mouth. It functions like a straw, generating a powerful vacuum action to draw in prey.

Toothless predators Seahorses and their relatives do not have any teeth and so swallow their prey whole.

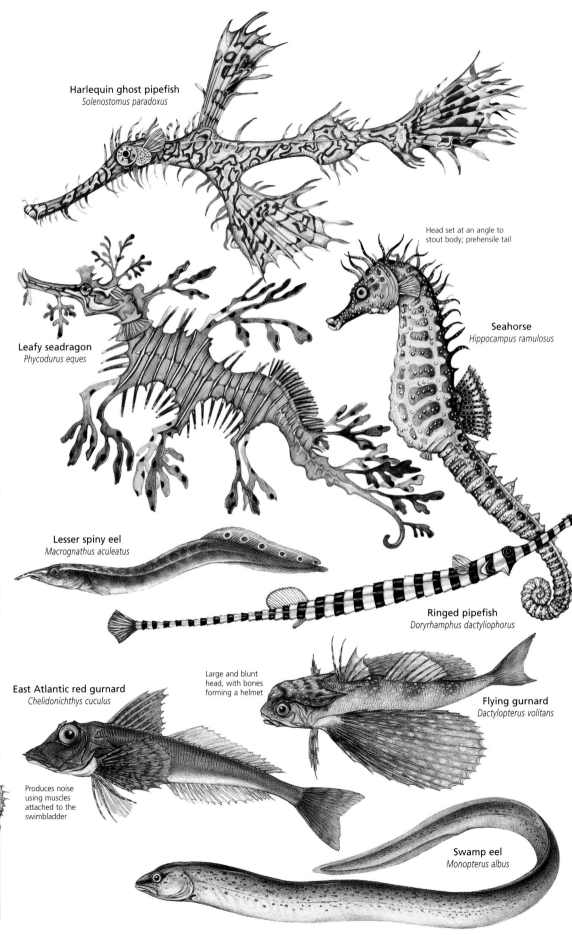

Harlequin ghost pipefish
Solenostomus paradoxus

Head set at an angle to stout body; prehensile tail

Leafy seadragon
Phycodurus eques

Seahorse
Hippocampus ramulosus

Lesser spiny eel
Macrognathus aculeatus

Ringed pipefish
Doryrhamphus dactyliophorus

East Atlantic red gurnard
Chelidonichthys cuculus

Large and blunt head, with bones forming a helmet

Flying gurnard
Dactylopterus volitans

Produces noise using muscles attached to the swimbladder

Swamp eel
Monopterus albus

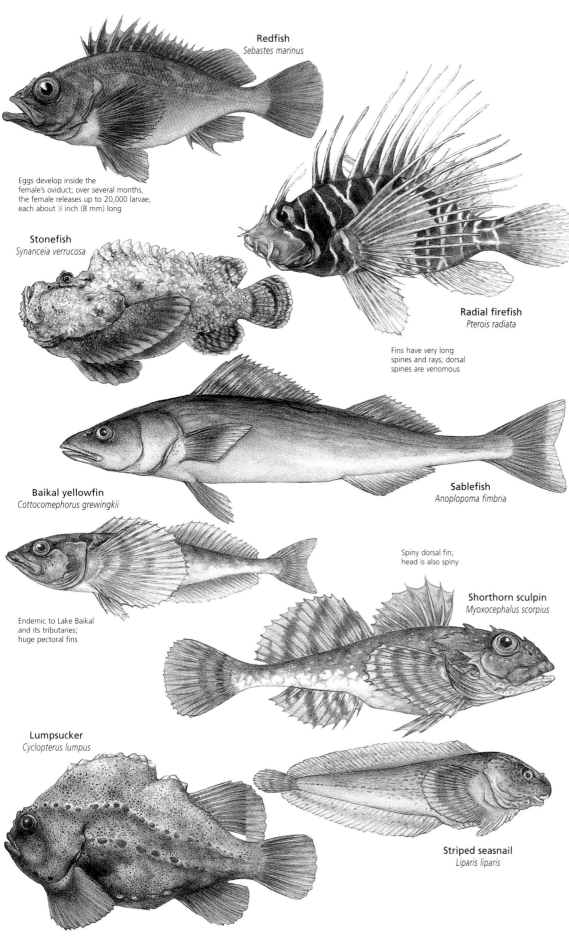

Redfish
Sebastes marinus

Eggs develop inside the
female's oviduct; over several months,
the female releases up to 20,000 larvae,
each about ⅓ inch (8 mm) long

Stonefish
Synanceia verrucosa

Radial firefish
Pterois radiata

Fins have very long
spines and rays; dorsal
spines are venomous

Baikal yellowfin
Cottocomephorus grewingkii

Sablefish
Anoplopoma fimbria

Spiny dorsal fin;
head is also spiny

Endemic to Lake Baikal
and its tributaries;
huge pectoral fins

Shorthorn sculpin
Myoxocephalus scorpius

Lumpsucker
Cyclopterus lumpus

Striped seasnail
Liparis liparis

LUMPSUCKER ANATOMY

The female lumpsucker
deposits her large egg
mass on the shore's
edge near the low
tide mark. The
male guards
it aggressively,
maintaining his
position in the
turbulent intertidal
zone by attaching
to rocks or seaweed
with a ventrally
located sucking
disk formed from
modified pelvic fins.

Cheap treat
*Lumpsucker eggs are
sold commercially as
inexpensive caviar.*

FACT FILE

Barramundi perch Most barramundi are protandrous hermaphrodites: they begin life as males, reach sexual maturity after about 3 years, and become females after about 5 years. As a result, larger individuals are inevitably female, while smaller fishes are mostly males.

🐟 Up to 6 ft (1.8 m)
⚖ Up to 132 lb (60 kg)
○ Oviparous
♀♂ Hermaphrodite
🗝 Common

Indo-West Pacific

Sea goldie The sexes in this fish look like different species. Females are colored orange to yellow while males are mainly purple. Males also have an enormously elongated dorsal fin spine and longer tail-fin lobes.

🐟 Up to 6 in (15 cm)
⚖ Up to 8 oz (225 g)
○ Oviparous
♀♂ Sequential hermaphrodite
🗝 Common

Indo-West Pacific & Red Sea

INGENIOUS MIMICRY

The comet deters predators by hiding in holes exposing only its tail, which mimics the fierce-looking head of the identically patterned turkey moray eel (*Gymnothorax meleagris*). An ocellus at the base of the dorsal fin looks like an eye. A gap between the anal and tail fins forms a "mouth."

Butterfly effect *This type of so-called "Batesian mimicry" is similar to that adopted by many butterflies.*

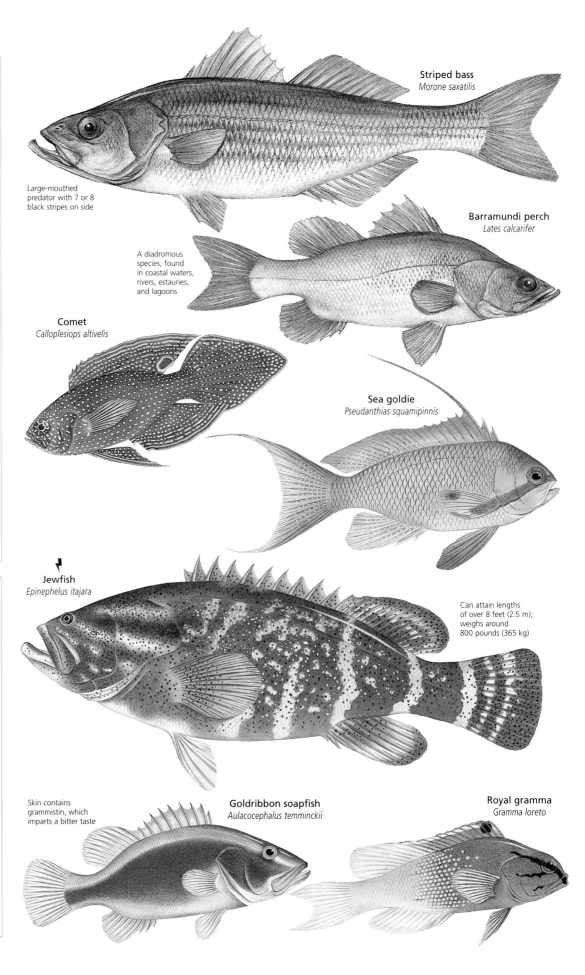

Striped bass
Morone saxatilis

Large-mouthed predator with 7 or 8 black stripes on side

Barramundi perch
Lates calcarifer

A diadromous species, found in coastal waters, rivers, estauries, and lagoons

Comet
Calloplesiops altivelis

Sea goldie
Pseudanthias squamipinnis

Jewfish
Epinephelus itajara

Can attain lengths of over 8 feet (2.5 m); weighs around 800 pounds (365 kg)

Skin contains grammistin, which imparts a bitter taste

Goldribbon soapfish
Aulacocephalus temminckii

Royal gramma
Gramma loreto

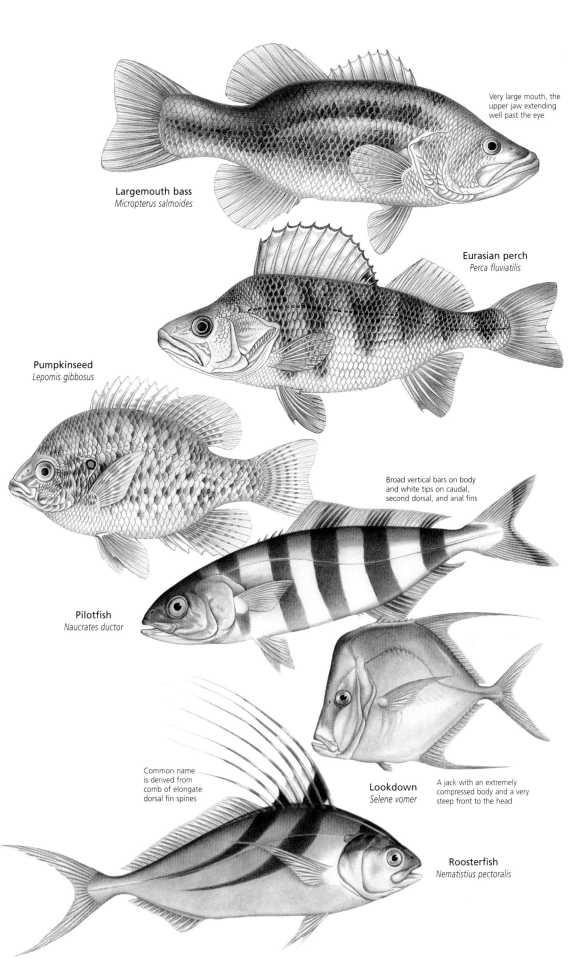

Largemouth bass
Micropterus salmoides

Very large mouth, the upper jaw extending well past the eye

Eurasian perch
Perca fluviatilis

Pumpkinseed
Lepomis gibbosus

Broad vertical bars on body and white tips on caudal, second dorsal, and anal fins

Pilotfish
Naucrates ductor

Lookdown
Selene vomer

A jack with an extremely compressed body and a very steep front to the head

Common name is derived from comb of elongate dorsal fin spines

Roosterfish
Nematistius pectoralis

FACT FILE

Eurasian perch The females of this species lay their eggs, several tens of thousands at a time, connected in long, white sticky mucous ribbons. These can be up to a meter in length and are left draped over submerged rocks and vegetation.

↦ Up to 20 in (51 cm)
⚖ Up to 10½ lb (4.7 kg)
○ Oviparous
♀♂ Male & female
⚑ Common

N. Eurasia

Pumpkinseed As occurs within most species of the freshwater North American sunfish family, the male pumpkinseed is a nest-builder. Eggs and sperm are deposited as the male and female swim around the nest in a circular pattern. The male guards the eggs and larvae for about 11 days after hatching.

↦ Up to 16 in (40 cm)
⚖ Up to 22 oz (630 g)
○ Oviparous, guarders
♀♂ Male & female
⚑ Common

E. North America; introd. widely

Pilotfish The common name of this species stems from its habit of swimming with sharks and other large marine creatures, feeding on food scraps and parasites. Sailors once thought they guided their "hosts" to food. Juveniles swim with jellyfish.

↦ Up to 27½ in (70 cm)
⚖ Up to 15 lb (6.8 kg)
○ Oviparous
♀♂ Male & female
⚑ Common

Tropical seas worldwide

Roosterfish The roosterfish (Nematistiidae) is a jack-like fish and, like many species in the jack family (Carangidae), is a popular gamefish that fights strongly when hooked. Its distinctive, elongate, backward-curving dorsal spines look like a cockscomb.

↦ Up to 4 ft (1.2 m)
⚖ Up to 100 lb (45 kg)
○ Oviparous
♀♂ Male & female
⚑ Common

E. Pacific

FACT FILE

Oriental sweetlips The loose rubbery lips of sweetlips are adapted for "vacuuming" invertebrates from sandy bottoms. They feed so vigorously that silt sweeps back through their gill arches to emerge as a cloud in the water behind. Being a member of the circumtropical grunt family (Haemulidae), this species is also able to produce sounds by grinding its pharyngeal teeth.

- ➤ Up to 20 in (51 cm)
- ⚖ Up to 4 lb (1.8 kg)
- ○ Oviparous
- ♀♂ Male & female
- ♦ Common

Indo-West Pacific

Red mullet This species is not a true mullet but a goatfish (family Mullidae). It has the long sensory chin barbels, characteristic of its family, and uses these to probe bottom sediments for small invertebrates. Goatfishes, like their mammalian namesakes, have a reputation for an omnivorous diet.

- ➤ Up to 16 in (40 cm)
- ⚖ Up to 35 oz (1 kg)
- ○ Oviparous
- ♀♂ Male & female
- ♦ Common

E. North Atlantic & Mediterranean Sea

Emperor snapper Juvenile emperor snapper live in shallow tropical waters, often in close association with sea urchins. Large adults, which dwell at greater depths, are sometimes implicated in cases of ciguatera poisoning. The vivid red stripes that characterize young fish of this species fade with age. Mature adults are colored red-pink all over.

- ➤ Up to 3¼ ft (1 m)
- ⚖ Up to 35½ lb (16 kg)
- ○ Oviparous
- ♀♂ Male & female
- ♦ Common

Indo-West Pacific, Red Sea

Jack-knifefish A greatly elongated dorsal fin gives this Caribbean endemic the most recognizable appearance of all the drums and croakers, a family of fishes (Sciaenidae) known for their ability to produce sounds by resonating their highly specialized swim bladder with special muscles. This species is nocturnal, hiding in coral caves by day.

- ➤ Up to 10 in (25 cm)
- ⚖ Up to 10 oz (285 g)
- ○ Oviparous
- ♀♂ Male & female
- ♦ Common

W. Atlantic

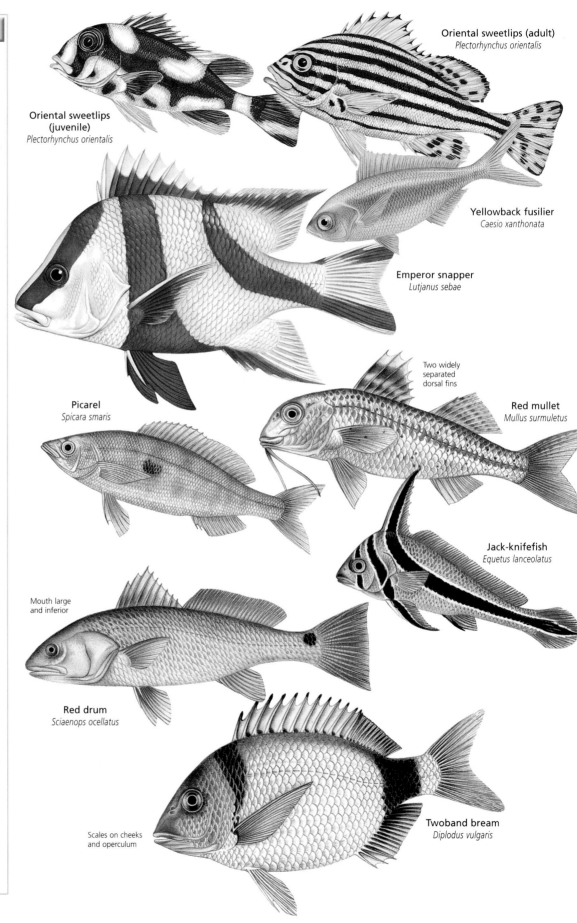

Oriental sweetlips (adult)
Plectorhynchus orientalis

Oriental sweetlips (juvenile)
Plectorhynchus orientalis

Yellowback fusilier
Caesio xanthonata

Emperor snapper
Lutjanus sebae

Picarel
Spicara smaris

Two widely separated dorsal fins

Red mullet
Mullus surmuletus

Jack-knifefish
Equetus lanceolatus

Mouth large and inferior

Red drum
Sciaenops ocellatus

Scales on cheeks and operculum

Twoband bream
Diplodus vulgaris

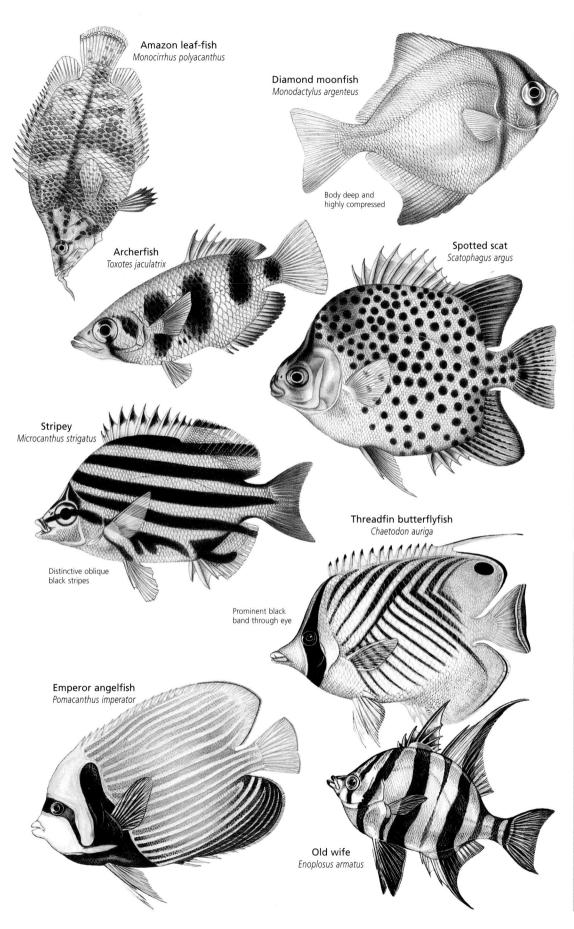

Amazon leaf-fish
Monocirrhus polyacanthus

Diamond moonfish
Monodactylus argenteus

Body deep and
highly compressed

Archerfish
Toxotes jaculatrix

Spotted scat
Scatophagus argus

Stripey
Microcanthus strigatus

Distinctive oblique
black stripes

Threadfin butterflyfish
Chaetodon auriga

Prominent black
band through eye

Emperor angelfish
Pomacanthus imperator

Old wife
Enoplosus armatus

FACT FILE

Amazon leaf-fish This species from
South America looks and behaves like
a floating dead leaf. Colored shades
of brown with a twig-like chin barbel,
it moves by means of transparent fins,
lunging at unsuspecting prey with its
enormous mouth.

⬌	Up to 3 in (7.5 cm)
⚖	Up to 1 oz (30 g)
○	Oviparous, guarder
♀♂	Male & female
❂	Common

N. South America

Emperor angelfish As occurs in
most angelfishes, juvenile coloration
in this species differs strikingly from
that of adults. Young fish have
incomplete rings of white and blue
overlaying a blue-black background
but develop light blue and yellow
stripes as they mature.

⬌	Up to 16 in (40 cm)
⚖	Up to 3 lb (1.4 kg)
○	Oviparous
♀♂	Male & female
❂	Common

Indo-Pacific & Red Sea

TAKING AIM

Archerfishes fire water jets from the
mouth to knock insects and other prey
from overhanging branches. Rapid oral
cavity compressions force water through
a tube formed by the tongue pressing
against the specially grooved palate.
The archerfish *Toxotes jaculatrix* has a
shooting range of about 5 feet (1.5 m).
They will also shoot at swarms of flying
insects hovering above the surface.

FACT FILE

Bluestreak cleaner Pairs or small groups of bluestreak cleaners set up permanent "cleaning stations" near cave entrances and beneath rocky overhangs in tropical reefs throughout the Indo-Pacific. Here they service "clients"—larger fishes looking to be picked free of parasites.

- Up to 4½ in (11.5 cm)
- Up to 3 oz (85 g)
- Oviparous
- Male & female
- Common

Indo-Pacific & Red Sea

Freshwater angelfish Eggs laid onto the submerged leaf of an aquatic plant are guarded and fanned constantly by freshwater angelfish parents for the 3 days until hatching. Fry are gathered into a dense pack each evening for ease of protection.

- Up to 3 in (7.5 cm)
- Up to 8 oz (225 g)
- Oviparous, guarder
- Male & female
- Common

N.E. South America

ORAL NURSERIES

All of the cichlids show some form of parental care, the most advanced being continuous mouth-brooding. Usually the female scoops her eggs up into her mouth quickly after laying them, nuzzles the male near his genital opening, and then draws his ejaculated sperm into her mouth where she may brood embryos and young for more than 3 weeks. Toward the end of that period, the young are released at intervals to forage.

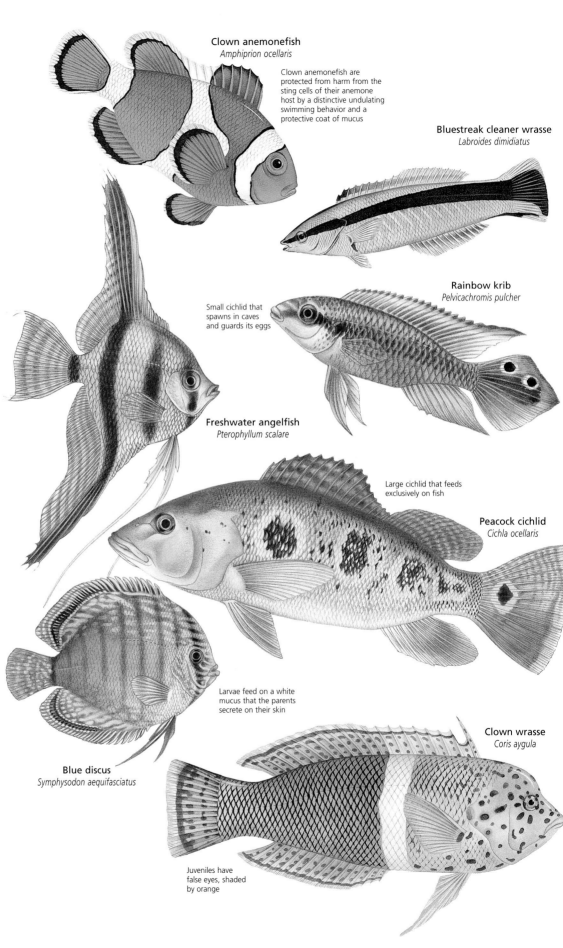

Clown anemonefish
Amphiprion ocellaris

Clown anemonefish are protected from harm from the sting cells of their anemone host by a distinctive undulating swimming behavior and a protective coat of mucus

Bluestreak cleaner wrasse
Labroides dimidiatus

Small cichlid that spawns in caves and guards its eggs

Rainbow krib
Pelvicachromis pulcher

Freshwater angelfish
Pterophyllum scalare

Large cichlid that feeds exclusively on fish

Peacock cichlid
Cichla ocellaris

Larvae feed on a white mucus that the parents secrete on their skin

Blue discus
Symphysodon aequifasciatus

Clown wrasse
Coris aygula

Juveniles have false eyes, shaded by orange

Yellowhead jawfish
Opistognathus aurifrons

Constructs elaborate
burrows; male broods
the eggs orally

Viviparous blenny
Zoarces viviparus

Greater weever
Trachinus draco

Lies buried in sand; spines
of the first dorsal fin and
gill cover have venom

Atlantic mudskipper
Periophthalmus barbarus

Atlantic stargazer
Uranoscopus scaber

False cleanerfish
Aspidontus taeniatus

Lesser sandeel
Ammodytes tobianus

Burrows in sand but also forms
huge schools in mid-water

Redtail surgeonfish
Acanthurus achilles

Named for the
sharp spine or
spines they
possess on the
caudal peduncle

Moorish idol
Zanclus cornutus

FACT FILE

Atlantic stargazer A camouflaged bottom-dweller, the Atlantic stargazer has a small, worm-like appendage on its bottom lip that it wriggles to attract prey. Morphological adaptations to a life spent almost completely buried in sandy sediments include having a high-set mouth, nostrils, and eyes. Defenses against predators include a venomous spine behind the operculum and electric organs behind the eyes.

↗ Up to 16 in (40 cm)
⚖ Up to 33 oz (940 g)
○ Oviparous
♀♂ Male & female
⚡ Locally common

E. Atlantic; Mediterranean & Black seas

False cleanerfish This blenny species mimics the bluestreak cleaner wrasse *Labroides dimidiatus*, which large fishes seek out for the removal of parasites. The two look so alike in their natural habitat that few other fishes seem capable of recognizing the interloping master of disguise. When the false cleaner is approached, however, it nips off pieces of flesh instead.

↗ Up to 4½ in (11.5 cm)
⚖ Up to 2½ oz (75 g)
○ Oviparous
♀♂ Male & female
⚡ Locally common

Indo-Pacific

AMPHIBIOUS FISHES

Mudskippers use their muscular tail and pectoral fins to "skip" over mud at low tide and even climb trees with the aid of a pelvic fin "suction cup." There are more than 30 species of mudskipper, found mainly in the muddy, intertidal mangrove forests of Southeast Asia and Africa. They take in oxygen through their moist skin, which is richly endowed with blood vessels.

Land lovers *The males of some mudskipper species display their masculine prowess and define territories by leaping about in the mud.*

FACT FILE

Swordfish Adult swordfish lack scales, teeth, and a lateral line. They mostly feed on pelagic fish—taken anywhere from surface waters to depths of about 2,100 feet (650 m)—using their sword to slash prey. This species is the only member of the family Xiphiidae.

- Up to 16 ft (4.9 m)
- Up to 1,435 lb (650 kg)
- ○ Oviparous
- ♀♂ Male & female
- Data deficient

Worldwide in tropical and temperate seas

Indo-Pacific sailfish No other fish is known to swim faster than the sailfish, which has been clocked in excess of 68 mph (110 km/h). It uses its rapier-like beak to stun and maim prey, which it then scoops up with toothless jaws.

- Up to 11½ ft (3.5 m)
- Up to 220 lb (100 kg)
- ○ Oviparous
- ♀♂ Male & female
- Common

Tropical and temperate Indian & Pacific oceans

CONSERVATION WATCH

Big decline The blue marlin is one of the most sought-after gamefishes, but sportfishing is not its biggest threat. Huge numbers are caught and killed accidentally on long-lines set for tuna and swordfish by commercial fishing operations. Being predators at the top of the food chain, they occur at naturally low levels of abundance, making their populations vulnerable to overexploitation of any sort.

The results of a 10-year study by German and Canadian scientists, published in 2003 in the international science journal *Nature*, indicate that blue marlin numbers, along with those of other large fishes in the world's oceans, declined dramatically during the last half of the 20th century.

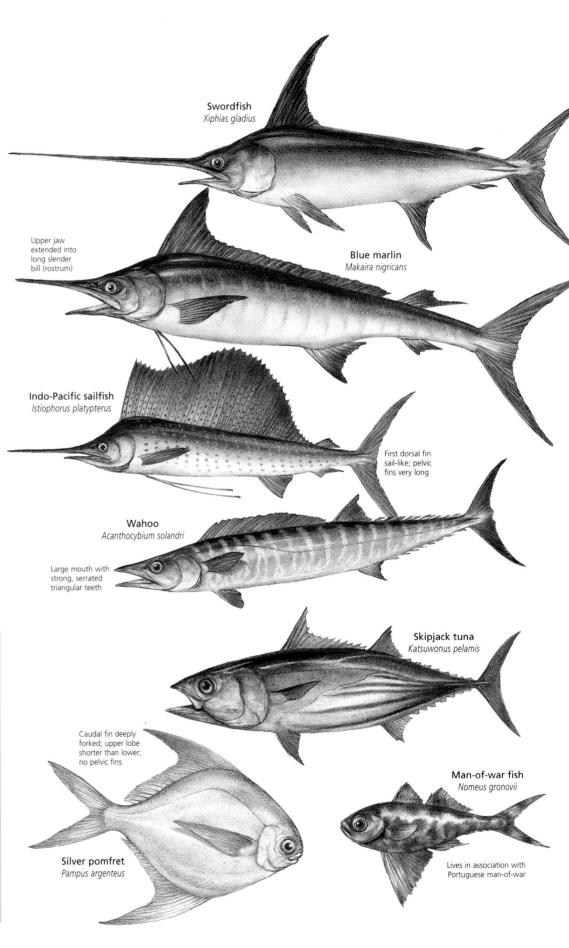

Swordfish
Xiphias gladius

Blue marlin
Makaira nigricans

Upper jaw extended into long slender bill (rostrum)

Indo-Pacific sailfish
Istiophorus platypterus

First dorsal fin sail-like; pelvic fins very long

Wahoo
Acanthocybium solandri

Large mouth with strong, serrated triangular teeth

Skipjack tuna
Katsuwonus pelamis

Caudal fin deeply forked; upper lobe shorter than lower; no pelvic fins

Man-of-war fish
Nomeus gronovii

Silver pomfret
Pampus argenteus

Lives in association with Portuguese man-of-war

Ornate ctenopoma
Microctenopoma ansorgii

Siamese fighting fish
Betta splendens

Pearl gourami
Trichogaster leerii

Gourami
Osphronemus goramy

Can breathe moist
air and survive long
periods out of water

Lips have horny
teeth; filter feeder
and grazer on
benthic algae

Kissing gourami
Helostoma temminckii

Sucking lips used
for kissing other
fishes, plants and
other objects

Brill
Scophthalmus rhombus

Climbing perch
Anabas testudineus

Zebra sole
Zebrias zebra

Very large flatfish that
is an active predator
of other fishes

Distinctive stripes
give this sole its
common name

Atlantic halibut
Hippoglossus hippoglossus

Summer flounder
Paralichthys dentatus

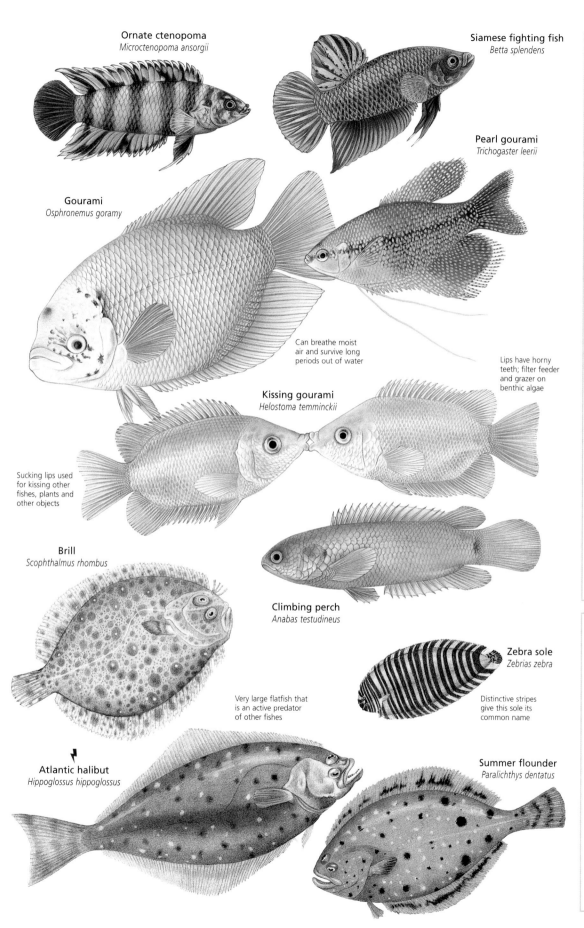

MIGRATING EYES

Larval flatfishes look like other young
fishes to begin with but soon begin
leaning sideways as one eye (left in
some species, right in others) migrates
across to take up position beside the
other. Simultaneously, the front of the
skull twists to bring the jaws into an
oblique sideways position.

larva with
normal eye position

left eye migrates
to top of head

adult eyes both on the right side

FACT FILE

Harlequin filefish The presence of this tropical coral reef specialist is a sign of environmental health. It lives almost exclusively on coral polyps and is one of the first species to disappear from reefs affected by pollution or global warming.

- Up to 4¾ in (12 cm)
- Up to 3 oz (85 g)
- ○ Oviparous
- ♀♂ Male & female
- Common

Indo-West Pacific

Guineafowl puffer This species is typical of pufferfishes in that it is poisonous to eat and has its teeth incorporated into the jawbone to form a parrot-like beak. This enables it to feed mainly on the soft-bodied polyps in the tips of branching coral species.

- Up to 19½ in (50 cm)
- Up to 4 lb (1.8 kg)
- ○ Oviparous
- ♀♂ Male & female
- Common

Indo-Pacific: E. Africa to Americas

Thornback cowfish Like their close relatives the boxfishes, cowfishes are encased in a protective outer armor. They also have a unique form of locomotion—octraciform swimming—which involves a sculling motion of the tail fin and no body flexing. This does not provide speed, but allows almost motionless hovering.

- Up to 9 in (23 cm)
- Up to 1 lb (460 g)
- ○ Oviparous
- ♀♂ Male & female
- Common

Indo-West Pacific: E. Africa to Hawaii

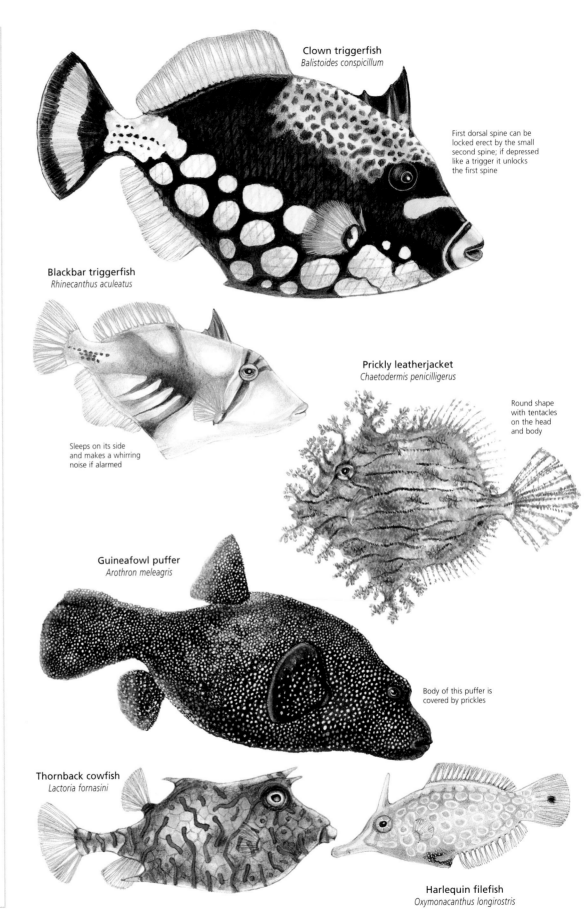

Clown triggerfish
Balistoides conspicillum

First dorsal spine can be locked erect by the small second spine; if depressed like a trigger it unlocks the first spine

Blackbar triggerfish
Rhinecanthus aculeatus

Sleeps on its side and makes a whirring noise if alarmed

Prickly leatherjacket
Chaetodermis penicilligerus

Round shape with tentacles on the head and body

Guineafowl puffer
Arothron meleagris

Body of this puffer is covered by prickles

Thornback cowfish
Lactoria fornasini

Harlequin filefish
Oxymonacanthus longirostris

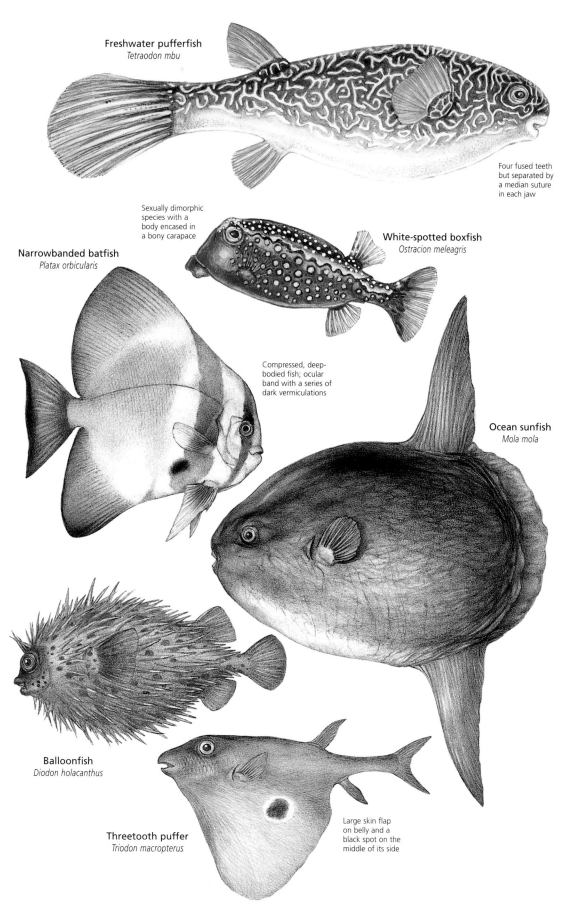

Freshwater pufferfish
Tetraodon mbu

Four fused teeth but separated by a median suture in each jaw

Sexually dimorphic species with a body encased in a bony carapace

White-spotted boxfish
Ostracion meleagris

Narrowbanded batfish
Platax orbicularis

Compressed, deep-bodied fish; ocular band with a series of dark vermiculations

Ocean sunfish
Mola mola

Balloonfish
Diodon holacanthus

Threetooth puffer
Triodon macropterus

Large skin flap on belly and a black spot on the middle of its side

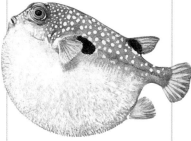

INVERTEBRATES

INVERTEBRATES

INVERTEBRATES	
PHYLA	> 30
CLASSES	> 90
ORDERS	> 370
SPECIES	> 1.3 million

Constituting more than 95 percent of all known animal species, invertebrates are not distinguished by a single positive characteristic. Instead the group is defined by what its members lack: they have no backbone, no bones, and no cartilage. As a term of classification, *invertebrate* is commonly used but has little scientific validity. Unlike vertebrates, which belong to a single phylum, invertebrates are a collection of more than 30 phyla, some of which are more closely related to vertebrates than to each other. Invertebrates encompass such diverse forms as porous sponges, floating jellyfishes, parasitic flatworms, jet-propelled squids, hard-cased crabs, venomous spiders, and fluttering butterflies.

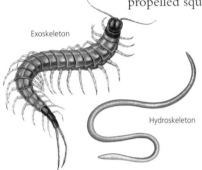

Exoskeleton

Hydroskeleton

Inside and outside An internal fluid-filled hydroskeleton supports the body of worms and many other invertebrates, but is only viable in moist environments or water. An external exoskeleton helped arthropods such as centipedes to colonize land.

SOFT-BODIED CREATURES
The first animals to evolve were invertebrates, but their soft bodies left no traces. Although tracks and burrows appear in the fossil record about 1 billion years ago, the oldest fossilized animal remains date back to about 600 million years ago, near the end of the Precambrian period. Known as Ediacaran fauna, these include forms that resemble sponges, jellyfishes, soft corals, segmented worms, and echinoderms. Beginning roughly 60 million years later, the Cambrian period is associated with the explosion of invertebrate life. By its conclusion, about 500 million years ago, all of today's invertebrate phyla seem to have appeared.

Although their greatest diversity occurs in the sea, invertebrates are now found in virtually all land and water habitats. Most species are small, and some are microscopic: many rotifers are less than 1/25,000 inch (0.001 mm) long. However, a few invertebrates reach staggering sizes: the elusive giant squid can be up to 59 feet (18 m) long and weigh up to 1,980 pounds (900 kg).

There are two basic invertebrate body plans. Species with radial symmetry, such as jellyfish and sea anemones, have a circular body plan and a central mouth. Others, such as worms and insects, have bilateral symmetry, with a distinct head and right and left sides.

Invertebrates lack bones, but are supported by some sort of skeleton. While many appear soft, they are held together by protein fibers, a feature of all animals. Many worms have a hydroskeleton, with fluid held under pressure inside the body cavity. Sponges and echinoderms have endoskeletons, with hard elements inside the tissues. Most mollusks and all arthropods have an exoskeleton, a hard, external casing. A mollusk's exoskeleton is a hard shell, while an arthropod's tough exoskeleton is jointed and flexible.

Most invertebrates reproduce sexually, laying vast numbers of fertilized eggs that are usually left to hatch on their own. Some species, however, develop from unfertilized eggs, and others reproduce by fragmenting or budding, with parts of their own body becoming the offspring. Invertebrate young are often quite unlike their parents and must go through metamorphosis to transform into adults.

External fertilization Staghorn coral (*Acropora* sp.) can reproduce asexually by fragmentation, or sexually by releasing eggs and sperm into the water. Each of the polyps is a hermaphrodite, able to release both eggs and sperm bundles. Spawning is synchronized with phases of the moon.

Modes of travel Sea anemones such as *Heteractis crispa* (opposite) are sessile as adults, remaining fixed in one spot and waving their tentacles to catch food. Many other invertebrates, however, are highly mobile and include swimming, burrowing, creeping, running, and flying forms.

Hunting spider
Tarantulas and other spiders are predacious invertebrates. They use venom to paralyze prey.

INVERTEBRATE CHORDATES

PHYLUM	Chordata
SUBPHYLA	2
CLASSES	4
ORDERS	9
FAMILIES	47
SPECIES	> 2,000

The phylum Chordata is made up of three subphyla. The largest of these is Vertebrata, which includes all the vertebrates (mammals, birds, reptiles, amphibians, and fishes). The other two subphyla are marine invertebrate groups: Urochordata, which contains about 2,000 species of sea squirts and their relatives, and Cephalochordata, which has about 30 species of lancelets. These invertebrate chordates do not have a backbone made of vertebrae, but they do have a flexible skeletal rod known as a notochord. The notochord is present in vertebrate embryos, but is resorbed and replaced by the backbone. Vertebrates appear to have evolved from invertebrate chordates.

From the deep The predatory tunicate (*Megalodicopia hians*) lives on the floor of deep oceans. After a small animal such as a krill floats into its mouth-like hood, the hood quickly closes to capture the prey.

Colonies This photo shows several colonies of magnificent ascidians (*Botrylloides magnicoecum*). The dark spots represent individual inhalant openings; the sieve-like plates are communal exhalant openings.

FIXED AND FREE

Also known as tunicates, sea squirts hatch from eggs as a tadpole-shaped larva with a notochord in its tail. After dispersing to a new location, the tadpole usually attaches itself to the seafloor, resorbs the tail and notochord, and moves its mouth to its free end. This sessile sea squirt is a bag-like creature. Water enters through an inhalant opening and leaves via an exhalant opening. On the way, the water passes through a perforated pharynx (throat), which uses mucus to filter out particles of food. Sea squirts can be solitary or colonial. Most are fixed as adults, but a small number remain free-living throughout their lives. They are almost all hermaphrodites that release eggs and sperm into the water to produce their young.

Superficially resembling little eels, lancelets (or amphioxus) can swim well, but usually stay partly buried in sand or gravel in shallow waters, with only the head protruding. They filter-feed using the same method as sea squirts, with water entering one opening, passing through the pharnx, and exiting via a second opening. The sexes are separate and the eggs are fertilized externally.

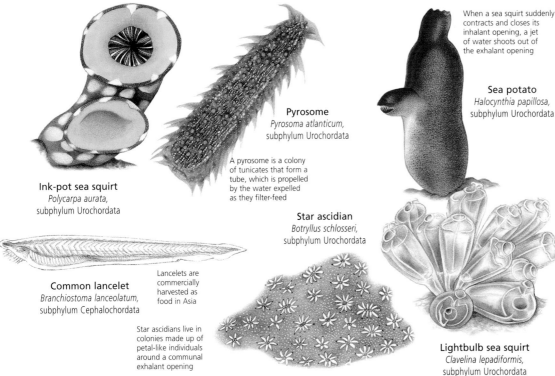

Ink-pot sea squirt
Polycarpa aurata,
subphylum Urochordata

Pyrosome
Pyrosoma atlanticum,
subphylum Urochordata

A pyrosome is a colony of tunicates that form a tube, which is propelled by the water expelled as they filter-feed

When a sea squirt suddenly contracts and closes its inhalant opening, a jet of water shoots out of the exhalant opening

Sea potato
Halocynthia papillosa,
subphylum Urochordata

Star ascidian
Botryllus schlosseri,
subphylum Urochordata

Common lancelet
Branchiostoma lanceolatum,
subphylum Cephalochordata

Lancelets are commercially harvested as food in Asia

Star ascidians live in colonies made up of petal-like individuals around a communal exhalant opening

Lightbulb sea squirt
Clavelina lepadiformis,
subphylum Urochordata

SPONGES

PHYLUM	Porifera
CLASSES	3
ORDERS	18
FAMILIES	80
SPECIES	about 9,000

More than 2,000 years ago, Aristotle considered sponges to be animals, but his claim remained without proof until 1765. In the interim, most scientists believed sponges, with their lack of movement and often branching form, to be plants. Sponges are unique in the animal kingdom. They do not possess a nervous system, muscles, or stomach. Their cells do not form tissues or organs, but are specialized for particular functions, such as food collection, digestion, defense, or skeleton formation. Able to migrate throughout a sponge, the cells can transform from one type to another, which allows the sponge to completely regenerate from a fragment or even from individual cells.

Electric sponge Although they lack nerves, glass sponges (class Hexactinellida) react to disturbances by sending electrical impulses through their body, which prompt their food-filtering system to shut down.

Supportive spicules A sponge's skeleton is made of units known as spicules, which can be scattered throughout the sponge or combined into fibers. This magnified view shows the needle-like and star-like spicules of a sponge at 100 times their actual size.

Budding young While most sponges reproduce sexually, many also use asexual reproduction. Some species fragment, while others grow buds (far right), which break away to become new sponges. Sponges may also expel gemmules, collections of cells and food granules. The gemmules can remain in a dormant state, growing into a sponge only when conditions are favorable.

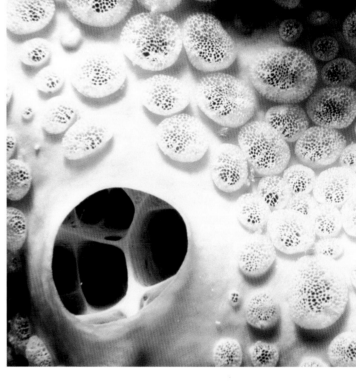

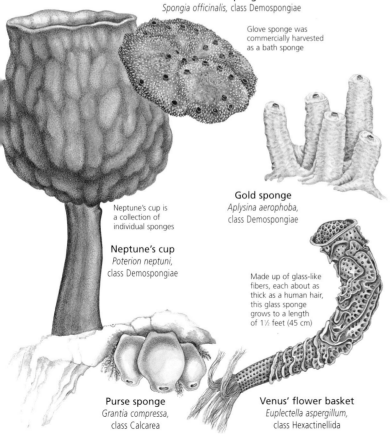

Glove sponge
Spongia officinalis, class Demospongiae

Glove sponge was commercially harvested as a bath sponge

Neptune's cup is a collection of individual sponges

Neptune's cup
Poterion neptuni,
class Demospongiae

Gold sponge
Aplysina aerophoba,
class Demospongiae

Made up of glass-like fibers, each about as thick as a human hair, this glass sponge grows to a length of 1½ feet (45 cm)

Purse sponge
Grantia compressa,
class Calcarea

Venus' flower basket
Euplectella aspergillum,
class Hexactinellida

POROUS FILTER-FEEDERS

Ranging in length from less than ½ inch to 6½ feet (1 cm to 2 m), sponges may be shaped like trees, bushes, vases, barrels, balls, cushions, or carpets, or they may simply form a shapeless mass. They are found in every marine habitat, from the shallows to the depths, and a small number of species have colonized freshwater lakes and rivers.

The skeletons of sponges are made of minerals, protein, or both. Species in the class Calcarea have skeletons of calcium carbonate and tend to be small and drab. Those in the class Hexactinellida are known as glass sponges and have skeletons made of silica. More than 90 percent of sponge species belong to the class Demospongiae. They have skeletons of silica and/or protein.

Sponges usually feed by filtering microorganisms from the water. After entering via tiny pores called ostia, the water travels through a system of canals, and is expelled via a large opening called the osculum. The collar cells that line the sponge's interior beat their whip-like flagella to maintain a constant current. A few carnivorous sponges use hook-like filaments to capture crustaceans.

Most sponges are hermaphrodites. Sperm are released into the water and carried to other sponges to fertilize their eggs. The larvae are free-swimming for a short period before attaching to a surface and developing into an adult sponge.

CNIDARIANS

PHYLUM	Cnidaria
CLASSES	4
ORDERS	27
FAMILIES	236
SPECIES	about 9,000

The phylum Cnidaria is a diverse assortment of mostly marine invertebrates, including sea anemones, corals, jellyfishes, and hydroids. All members are carnivores and use cells containing stinging nematocysts to subdue prey and deter predators. The cnidarian body is organized around a gastrovascular cavity that digests food and acts as a hydroskeleton. Food enters and waste leaves via a single opening, the mouth, which is often surrounded by tentacles. Cnidarians occur in two forms: polyp and medusa. Polyps are cylindrical and attached to a surface, with the mouth and tentacles at the free end. Medusae are free-swimming and umbrella-shaped, with the mouth and tentacles hanging down.

Fixed in position The jewel anemone (*Corynactis viridis*) has up to 100 tentacles arranged in three rings around its mouth. Attached to the substrate by a sucker-like disk, sea anemones rarely move but some are able to glide very slowly.

Floating home At up to 1½ feet (45 cm) wide, the purple jellyfish (*Pelagia panopyra*) offers a temporary haven to fish and young crabs. Like other jellyfishes, it has only weak jet propulsion and relies mainly on ocean currents for transport.

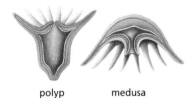

polyp   medusa

Jelly sandwich Between a cnidarian's two cell layers, the outer ectoderm and the inner endoderm, there is a jelly-like layer known as the mesoglea (shown as orange). In polyps, the mesoglea is thin, but in medusae, it forms the bulk of the animal.

POLYPS AND MEDUSAE

A small number of cnidarians live in fresh water, but most are marine, occupying all latitudes and levels of the ocean and reaching their greatest numbers in shallow tropical waters. They feed primarily on the fishes and crustaceans that swim past. Found mainly on the tentacles, a cnidarian's nematocysts (stingers) hold a coiled, barbed thread that can be ejected to spear and paralyze the prey. The tentacles then move the food to the mouth.

While corals and sea anemones exist only as polyps, many other cnidarians alternate between polyp and medusa during their life-cycle. Usually, the polyps asexually produce medusae, and the medusae sexually produce larvae that become polyps. Polyps and some medusae may grow buds or divide to create more of their own kind asexually. If the offspring detach from the parent they become clones, but if they remain attached they form colonies, such as the vast colonies that make up coral reefs. In some colonies, each member has a particular role and form, and may be specialized for feeding, defense, reproduction, or locomotion, for example.

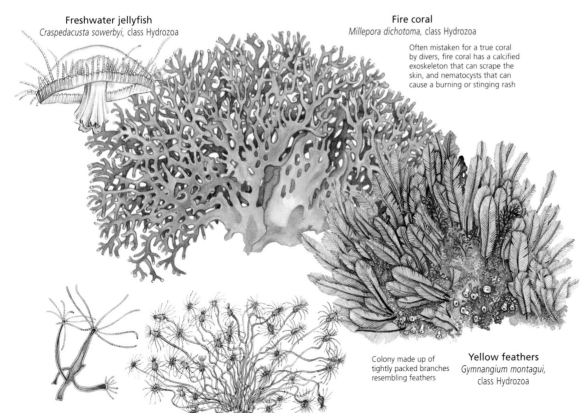

Freshwater jellyfish
Craspedacusta sowerbyi, class Hydrozoa

Fire coral
Millepora dichotoma, class Hydrozoa

Often mistaken for a true coral by divers, fire coral has a calcified exoskeleton that can scrape the skin, and nematocysts that can cause a burning or stinging rash

Colony made up of tightly packed branches resembling feathers

Yellow feathers
Gymnangium montagui,
class Hydrozoa

Green hydra
Chlorohydra viridis, class Hydrozoa

Oaten pipes hydroid
Tubularia indivisa, class Hydrozoa

Upside-down jellyfish
Cassiopeia andromeda,
class Scyphozoa

**Portuguese
man-of-war**
Physalia physalis,
class Hydrozoa

Attaches to sandy
ocean floor

The Portuguese man-of-war is a free-
floating colony, with a modified medusa
forming the gas-filled float, and some
polyps delivering a potent sting, some
digesting food, and others reproducing

Hula skirt siphonophore
Physophora hydrostatica,
class Hydrozoa

Lion's mane jellyfish
Cyanea arctica,
class Scyphozoa

Dead man's fingers
Alcyonium digitatum, class Anthozoa

Sea wasp
Chironex fleckeri,
class Cubozoa

A sea wasp
contains enough
venom to kill
60 people

By-the-wind sailor
Velella velella, class Hydrozoa

The stalked jellyfish
exists as a polyp
attached to algae
and does not occur
as a medusa

Stalked jellyfish
Haliclystus auricula, class Scyphozoa

Phosphorescent sea pen
Pennatula phosphorea,
class Anthozoa

Formosan soft coral
Sarcophyton glaucum, class Anthozoa

Organ-pipe coral
Tubipora musica, class Anthozoa

FACT FILE

Class Scyphozoa True jellyfishes belong
to the class Scyphozoa. Generally, they
spend most of their lives as medusae,
but these medusae produce larvae that
settle on the seabed as minute polyps.
The polyps then divide horizontally
and break off to become medusae.
Most jellyfishes are free-swimming. By
pulsating their bell, they squirt out jets
of water that provide weak propulsion.

Species 200

Worldwide; marine

Common jelly
*Found worldwide
in coastal waters,
the moon jellyfish
(Aurelia aurita) is
often seen washed
up onto beaches.*

Class Hydrozoa This class contains
species that have both polyps and
medusae in the life-cycle, but the
medusae are often reduced to buds
on the polyp's surface. Many species
are colonial, often with the individuals
specialized for different functions.

Species 3,300

Worldwide

Freshwater hydra
*The brown hydra
(Pelmatohydra oligactis)
is found in lakes and
ponds. Its tentacles
can reach lengths of
10 inches (25 cm).*

Class Cubozoa Viewed from above,
the box jellies in this class have a square
shape, which distinguishes them from
true jellyfishes. They are also faster and
more agile swimmers, and include
some of the ocean's deadliest species.

Species 36

Worldwide; tropical & temperate seas

YOUNG CNIDARIANS

When cnidarians sexually reproduce,
they produce a microscopic planula
larva, which either swims using its
beating cilia (tiny hairs) or crawls along
the bottom. After a time, the planula
transforms into a polyp by attaching its
front end to a surface and developing
tentacles at its free end. For corals and
sea anemones, which do not become
medusae, the larval stage is the only
chance for the species to disperse.

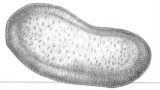

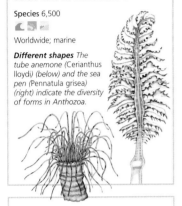
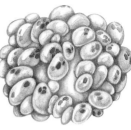

GREAT BARRIER REEF

Stretching more than 1,400 miles (2,240 km) along the northeastern coast of Australia, the Great Barrier Reef is a collection of coral reefs that together form the largest natural feature on Earth. It was created over millions of years from the calcium carbonate exoskeletons of true corals. Symbiotic algae live inside the coral polyps, supplying most of the corals' energy. The corals are restricted to clear, shallow waters where the algae can photosynthesize.

Teeming with life *With 400 species of coral, 1,500 species of fish, and 4,000 species of mollusk, the Great Barrier Reef's biodiversity rivals that of tropical rain forests.*

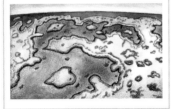

⚡ CONSERVATION WATCH

Dying reefs Corals are extremely sensitive to environmental stress and can die from even small changes in the temperature, salinity, or nitrogen levels of the water. Consequently, they are particularly vulnerable to ocean pollution and global warming. Mass tourism, over-collecting, and introduced species have also taken their toll, and half of all the world's reefs could die in the next 50 years.

Red brain coral
Lobophyllia hemprichii, class Anthozoa

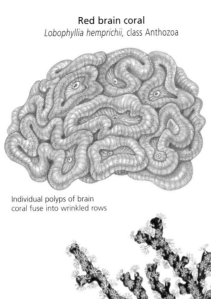

Individual polyps of brain coral fuse into wrinkled rows

Clubbed finger coral
Porites porites, class Anthozoa

Bubble coral
Plerogyra sinuosa, class Anthozoa

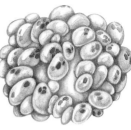

Black coral
Antipathes furcata, class Anthozoa

Red coral forms tree-like colonies

Red coral
Corallium rubrum, class Anthozoa

The tentacles of beadlet anemones retract when the tide goes out

Beadlet anemone
Actinia equina, class Anthozoa

Lophelia pertusa, class Anthozoa

Lophelia pertusa is a deep-water coral that forms reefs in the cold North Atlantic

West Indian sea fan
Gorgonia flabellum, class Anthozoa

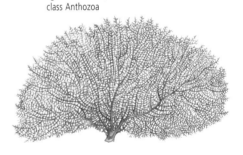

Stalk is inserted into deep crevices or buried in mud or sand

Daisy anemone
Cereus pedunculatus, class Anthozoa

Mushroom coral
Fungia fungites, class Anthozoa

Caryophyllia smithi, class Anthozoa

FLATWORMS

PHYLUM	Platyhelminthes
CLASSES	4
ORDERS	35
FAMILIES	360
SPECIES	13,000

The flatworms that make up the phylum Platyhelminthes range from microscopic free-living species, to tapeworms up to 100 feet (30 m) long that live inside humans and other vertebrates. The simplest animals possessing bilateral symmetry, flatworms have no body cavity and lack respiratory and circulatory systems. A few parasitic species also do without a digestive system. In most species, the gut has a single opening to take in food and expel waste. The indistinct head contains a brain and many of the sense organs, including ocelli (simple eyes) that can perceive light and dark, and receptors that can detect chemicals, balance, gravity, and water movement.

A long colony A parasite of rats, the tapeworm *Hymenolepis diminuta* is shown here at 50 times its actual size and with its color enhanced, revealing its individual proglottids and the suckers at its tip.

A PARASITIC LIFE

While the flukes of Monogenea have a single host in their lifetime, most parasitic flatworms make use of different hosts at different stages in their life cycle.

egg released in human feces and eaten by snail

final-stage larva emerges from snail

juvenile attaches to fish

adult matures inside a human's liver

Eggs of the Chinese liver fluke Opisthorchis sinensis (class Trematoda) eggs are eaten by aquatic snails. The final-stage larvae emerge to attach to fish, which are eaten by humans. Once the flukes mature into adults, their eggs are passed out in the human host's feces.

Crawling leaf Although some small species are cylindrical, most flatworms are flattened. Many marine species, such as *Pseudoceros dimidiatus*, are leaf-like.

DIFFERENT LIFESTYLES

Flatworms are divided into four classes. The first of these, Turbelleria, is predominantly free-living. Most species are marine, but many are found in lakes, ponds, and rivers, and a few can even tolerate both fresh and salt water. There are also a number of species that live on land in moist habitats. Turbellarians usually prey on invertebrates. They move along slime trails produced by special glands, beating cilia (tiny hairs) to propel themselves.

The other three flatworm classes are all parasitic during at least part of their often complex life-cycle. Members of Monogenea are small flukes. At the rear end, they have an opisthaptor, a bulb bearing suckers and/or hooks that attach to a fish's gills or a frog's bladder. Most flukes, however, belong to Trematoda. The adults are parasites of vertebrates, using suckers to attach to the gut and other organs. The class Cestoda contains the tapeworms, internal parasites that are long, flat colonies made up of individuals called proglottids. Almost all flatworms are hermaphrodites, and tapeworms have a set of male and female sex organs in each proglottid.

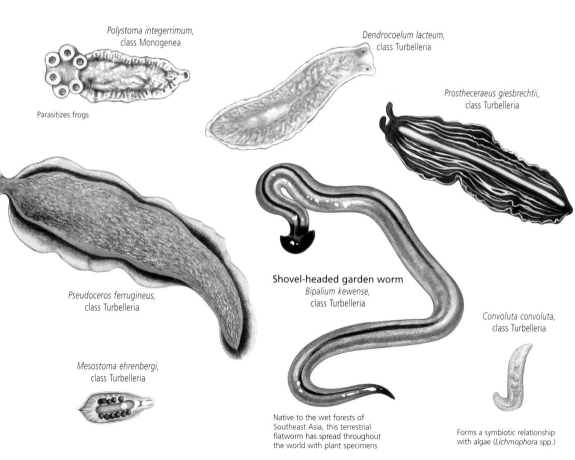

Polystoma integerrimum,
class Monogenea

Parasitizes frogs

Dendrocoelum lacteum,
class Turbelleria

Prosthecereaus giesbrechtii,
class Turbelleria

Pseudoceros ferrugineus,
class Turbelleria

Shovel-headed garden worm
Bipalium kewense,
class Turbelleria

Native to the wet forests of Southeast Asia, this terrestrial flatworm has spread throughout the world with plant specimens

Mesostoma ehrenbergi,
class Turbelleria

Convoluta convoluta,
class Turbelleria

Forms a symbiotic relationship with algae (*Lichmophora* spp.)

ROUNDWORMS

PHYLUM	Nematoda
CLASSES	4
ORDERS	20
FAMILIES	185
SPECIES	> 20,000

Although a species of roundworm that parasitizes sperm whales can grow to a length of 43 feet (13 m), most members of the phylum Nematoda are microscopic. Also known as nematodes, roundworms are among the most abundant of all animals—one rotting apple was found to contain about 90,000 individual roundworms. Free-living roundworms occur in virtually every aquatic and terrestrial habitat, and those living in soil play a crucial role in recycling detritus. Parasitic roundworms are found in most groups of plants and animals. Common roundworms, hookworms, pinworms, threadworms, and other roundworms infest more than half the world's human population.

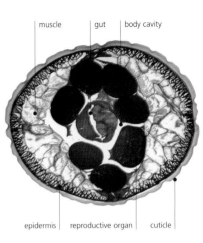

muscle | gut | body cavity

epidermis | reproductive organ | cuticle

Uncomplicated design The simple body plan of a roundworm, shown here in a magnified cross section, is not necessarily a sign of primitiveness, as it may have been derived from a more complex ancestor.

Growing up After hatching, this juvenile parasitic roundworm possesses all the features of an adult except for a mature reproductive system. It will molt four times before reaching adulthood.

Adaptable worms Free-living roundworms may be marine (right), freshwater, or terrestrial. Species occur in ice, hot springs, and even in acids, such as vinegar.

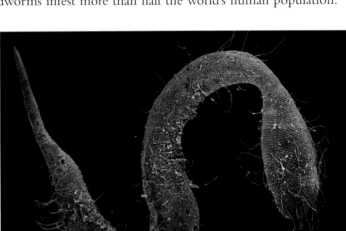

SIMPLE AND RESILIENT

Roundworms have long, slender, cylindrical bodies and often resemble tiny threads, a shape that allows them to live in tiny spaces between grains of soil, for example. They are bilaterally symmetrical and often tapered at both ends. The epidermis (skin) secretes a tough but flexible outer cuticle. As in arthropods such as insects, roundworms must molt as they grow. Most shed the cuticle four times before reaching maturity.

When a roundworm moves, it may appear to be thrashing aimlessly back and forth. Between the gut and the body wall, a body cavity known as a pseudocoel contains fluid under pressure. As the worm contracts its muscles, which run only lengthwise down its body, this pressure causes the body to flex from side to side—a method of locomotion that works well against soil particles or in water films.

After food enters a roundworm's mouth, it is pumped through the simple gut by a muscular pharynx (throat). Waste is released through an anus at the rear of the worm.

If they encounter unfavorable conditions, such as extreme heat or cold or drought, roundworms can enter a death-like state, known cryptobiosis, for months or even years. When conditions improve, they resume their life processes.

Although some roundworms are hermaphrodites, the sexes are more often separate. During copulation, the male holds onto the female with his hooked rear end. The young resemble miniature adults.

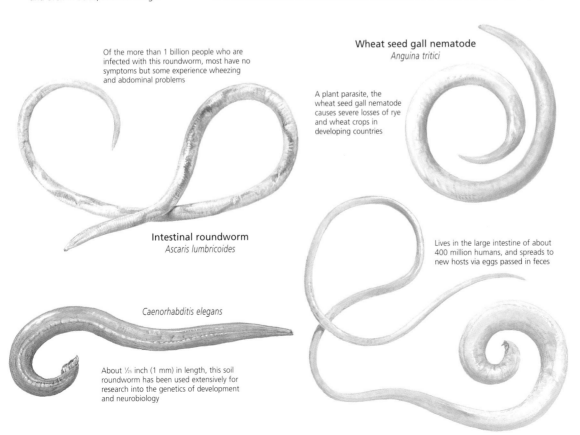

Of the more than 1 billion people who are infected with this roundworm, most have no symptoms but some experience wheezing and abdominal problems

Wheat seed gall nematode
Anguina tritici

A plant parasite, the wheat seed gall nematode causes severe losses of rye and wheat crops in developing countries

Intestinal roundworm
Ascaris lumbricoides

Lives in the large intestine of about 400 million humans, and spreads to new hosts via eggs passed in feces

Caenorhabditis elegans

About ⅟₂₅ inch (1 mm) in length, this soil roundworm has been used extensively for research into the genetics of development and neurobiology

MOLLUSKS

PHYLUM	Mollusca
CLASSES	7
ORDERS	35
FAMILIES	232
SPECIES	75,000

Abundant and adaptable, mollusks have diversified to fill most ecological niches. They are predominantly marine and are found in every level of the ocean, but have also colonized fresh water and land throughout the world. The diversity of their habitats is reflected in an immense variety of forms, from jet-propelled squid to creeping snails and fixed clams. Features shared by most mollusks include a well-developed head; a body cavity containing the internal organs; a specialized skin called the mantle that covers the body and secretes a shell of calcium carbonate; an intucking of the mantle to form a mantle cavity that houses the gills; and a muscular, often mucus-secreting foot.

Giants of the sea While many mollusks are small or even minute, some species reach great sizes. The largest bivalve, the giant clam (below), can measure almost 5 feet (1.5 m) across. It is dwarfed, however, by the giant squid, which grows to the astounding length of 59 feet (18 m).

Bottom-dweller Also known as sea slugs, nudibranchs are shell-less gastropods with external, feathery gills. *Chromodoris bullocki* is shown here feeding on coral.

Changing display Many cephalopods can change their colors and patterns, often for camouflage but also for communication. In a ritualized competition, male Caribbean reef squids (*Sepioteuthis sepioidea*) both try a zebra display to win the right to mate.

VARIED DESIGN

Mollusks are split into eight classes. The Aplacophora is a small class of worm-like mollusks that lack a shell. About 20 species that possess a low, rounded shell constitute the Monoplacophora. The chitons of Polyplacophora are protected by eight plates and creep on a sucker-like foot. Scaphopoda contains the burrowing tusk shells. The clams, oysters, mussels, and other bivalves of the class Bivalvia have a hinged, two-part shell and a tiny head. Gastropods such as snails and slugs (Gastropoda) sometimes have a spiral shell, but many have no shell at all. Octopuses, squids, and other cephalopods (Cephalopoda) have mobile arms and often lack a shell. They include the largest and most intelligent of all invertebrates.

Feeding habits among mollusks are varied. Species may feed on detritus, scrape algae from rocks, eat leaves, filter tiny organisms from the water, or actively pursue prey such as crustaceans and fishes. Food is taken in through the mouth, often with the help of a rasping, toothed organ called a radula, then passes through a complex digestive system terminating in an anus.

Most mollusks have separate sexes; some release eggs and sperm into the sea, while others copulate. Larvae tend to be free-swimming.

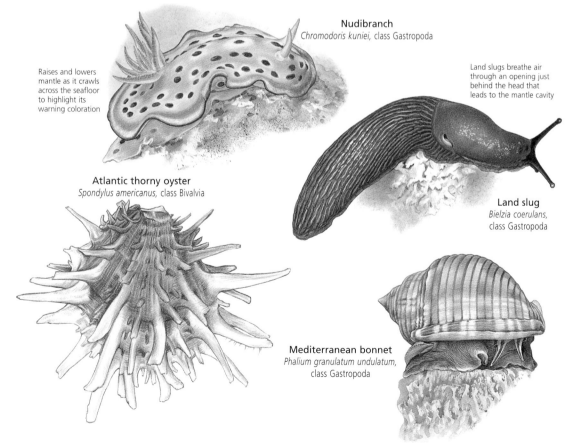

Raises and lowers mantle as it crawls across the seafloor to highlight its warning coloration

Nudibranch
Chromodoris kuniei, class Gastropoda

Land slugs breathe air through an opening just behind the head that leads to the mantle cavity

Atlantic thorny oyster
Spondylus americanus, class Bivalvia

Land slug
Bielzia coerulans,
class Gastropoda

Mediterranean bonnet
Phalium granulatum undulatum,
class Gastropoda

FACT FILE

Class Monoplacophora Known only from fossil shells until 1952, this rare class is made up of small, limpet-like mollusks living at great depths, from roughly 660 feet (200 m) to ocean trenches at 23,000 feet (7,000 m). They are distinguished by the repetition of body parts: there are five or six pairs of gills, six pairs of kidneys, and eight pairs of retractor muscles.

Species 20

Atlantic, Pacific, Indian Ocean; on seafloor

Deep crawler
Monoplacophorans, such as Vema hyalina, *are about 1 inch (2.5 cm) long and have a cap-like shell. They use a large flat foot to crawl across the seafloor.*

Class Polyplacophora Ranging from ¹⁄₁₀ inch to 16 inches (3 mm to 40 cm) in length, the chitons in this class are long and flat and have a shell of eight overlapping plates. They creep slowly on their broad, flat foot, often using their long radula to scrape algae from rocks or shells. If disturbed, they will clamp down to create a vacuum that makes them very difficult to dislodge.

Species 500

Worldwide; intertidal zone to deeper seafloor

Shell clinger
As in other chitons, the shell of the common eastern chiton Chaetopleura apiculata *of North America is surrounded by a girdle formed from its mantle. It uses this girdle and its foot to cling to rocks or shells.*

Class Scaphopoda This class contains the tusk shells, burrowing mollusks with long, tubular shells that are open at both ends. A muscular foot and a small head bearing sticky, thread-like tentacles protrude from the large end of the shell. Tusk shells feed on tiny organisms that are collected by the tentacles and broken up by the radula.

Species 500

Worldwide; sandy or muddy seafloor

Burrowing foot
Scaphopods, such as this Entalina *sp., use their well-developed foot to burrow into sand or mud on the seafloor.*

Glistenworm
Chaetoderma canadense,
class Aplacophora

Mottled red chiton
Tonicella marmorea,
class Polyplacophora

Browses on sedentary invertebrates such as sponges and bryozoans

Stenochiton longicymba,
class Polyplacophora

Limpets living higher on a rocky shore tend to have taller shells than those living lower down

Common limpet
Patella vulgata, class Gastropoda

Chiton olivaceus,
class Polyplacophora

Formosan tusk
Pictodentalium formosum,
class Scaphopoda

Neopilina galatheae,
class Monoplacophora

West Indian green chiton
Chiton tuberculatus,
class Polyplacophora

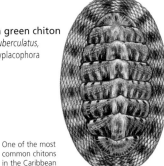

One of the most common chitons in the Caribbean

Elephant tusk
Dentalium elephantinum,
class Scaphopoda

Unlike the jade-green *Dentalium elephantinum,* most tusk shells are white or yelllowish

Antalis tarentinum,
class Scaphopoda

Tusk shells range from ³⁄₁₆ inch to 6 inches (4 mm to 15 cm) in length

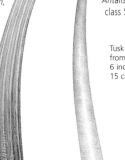

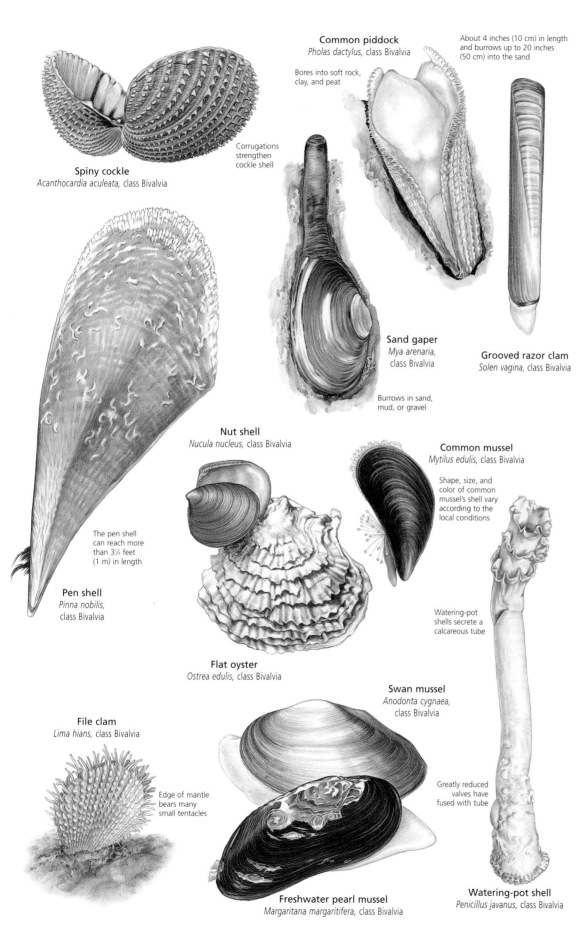

Spiny cockle
Acanthocardia aculeata, class Bivalvia

Corrugations strengthen cockle shell

Common piddock
Pholas dactylus, class Bivalvia

Bores into soft rock, clay, and peat

About 4 inches (10 cm) in length and burrows up to 20 inches (50 cm) into the sand

Sand gaper
Mya arenaria, class Bivalvia

Burrows in sand, mud, or gravel

Grooved razor clam
Solen vagina, class Bivalvia

The pen shell can reach more than 3¼ feet (1 m) in length

Pen shell
Pinna nobilis, class Bivalvia

Nut shell
Nucula nucleus, class Bivalvia

Common mussel
Mytilus edulis, class Bivalvia

Shape, size, and color of common mussel's shell vary according to the local conditions

Watering-pot shells secrete a calcareous tube

Flat oyster
Ostrea edulis, class Bivalvia

File clam
Lima hians, class Bivalvia

Edge of mantle bears many small tentacles

Swan mussel
Anodonta cygnaea, class Bivalvia

Greatly reduced valves have fused with tube

Freshwater pearl mussel
Margaritana margaritifera, class Bivalvia

Watering-pot shell
Penicillus javanus, class Bivalvia

FACT FILE

Class Bivalvia Clams, oysters, mussels, and other bivalves live inside a hinged two-valve shell. Most are burrowers, but some cement one valve to a firm surface or attach themselves with a byssus, sticky fibers secreted by the foot. Other bivalves live free on the sediment or in the water, or parasitize other aquatic animals.

Species 10,000

Worldwide

Brooder *The greater European pea clam (*Pisidium amnicum*) broods its eggs inside its shell and releases young that resemble miniature adults.*

GIANT CLAM

The largest bivalve of all lives on the tropical coral reefs of the Indian and Pacific oceans. Weighing up to 700 pounds (320 kg), giant clams (*Tridacna gigas*) remain embedded in the sediment, filtering plankton from the water. The bulk of their nutrition, however, comes from the algae that they host in the exposed thick lips of the mantle.

INSIDE A BIVALVE

A bivalve's mantle forms a sheet of tissue lining the two valves. The adductor muscles are used to close the shell, and the blade-like foot is often used for burrowing. Bivalves use their gills to filter particles from the water and their palps to sort them by size.

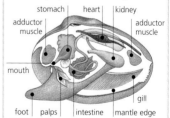

stomach | heart | kidney
adductor muscle | | adductor muscle
mouth | | gill
foot | palps | intestine | mantle edge

⚡ CONSERVATION WATCH

Unwelcome mussels First noticed in the Great Lakes in the 1980s, the European zebra mussels (*Dreissena* sp.) have now invaded many of the United States' major rivers and lakes. By filtering vast amounts of phytoplankton from the water, they are having a dramatic effect on aquatic food webs and endangering native mollusks and other species.

FACT FILE

Class Gastropoda This is the largest class of mollusks and includes snails, slugs, limpets, and nudibranchs. Some gastropods have lost their shells, but most have a spiral shell that contains a twisted body mass. Gastropods usually have a distinct head with eyes and tentacles. The mantle cavity lies over the head, allowing the head to retract into the shell. The muscular foot is used for creeping, swimming, or burrowing.

Species 60,000

Worldwide

Gastropod cannibal *The banded tulip (*Fasciolaria hunteria*) can be very aggressive and will eat other members of its species.*

RADULAS

A mollusk's mouth usually includes a tongue-like structure known as a radula. In some species, such as deep-sea limpets (*Neomphalus* sp.), a covering of hardened teeth makes the radula a rasping tool, able to scrape algae from rocks. In others, such as the slit shell mollusk (*Scissurella* sp.), the surface of the radula is feathery and is used to sweep detritus from sand. In a few carnivorous species, such as cone shells (*Conus* sp.), the radula has become a sharp, venom-injecting tooth.

THE GASTROPOD SHELL

While gastropod shells display immense variety in their details, the typical shell is a conical spire. The mantle adds new material to the inner and outer lips of the shell at slightly different rates, resulting in the spiral form.

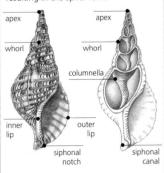

apex

whorl

apex

whorl

columnella

inner lip

outer lip

siphonal notch

siphonal canal

Plain marginella
Marginella cornea,
class Gastropoda

Purple bubble-raft snail
Janthina janthina, class Gastropoda

Floats at the ocean's surface, buoyed by mucus-covered bubbles

Phyllidia ocellata,
class Gastropoda

Color of *Phyllidia ocellata* is variable, but it can be identified by its knobby tubercles and white-ringed black spots

Crown conch
Melongena corona, class Gastropoda

European edible abalone
Haliotis tuberculata,
class Gastropoda

The mucus of *Murex brandarius* was used by the ancient Greeks to produce a dye called Tyrian purple, which was so rare that it was used only for the clothes of royalty

Trumpet triton
Charonia tritonis,
class Gastropoda

Purple-dye murex
Murex brandarius, class Gastropoda

Ruffled flaps of tissue (parapodia) resemble lettuce leaves

Lettuce sea slug
Elysia crispata,
class Gastropoda

Floats upside down and feeds on Portuguese man-of-war, ingesting the host's nematocysts (stinging cells) to use in its own defense

Tiger cowry
Cypraea tigris, class Gastropoda

Queen conch
Strombus gigas, class Gastropoda

Blue sea slug
Glaucus atlanticus, class Gastropoda

Common egg cowrie
Ovula ovum, class Gastropoda

Chocolate arion
Arion rufus, class Gastropoda

Gray garden slug
Deroceras reticulatus,
class Gastropoda

Tree slug
Lehmania marginata,
class Gastropoda

Wormslug
Boettgerilla pallens, class Gastropoda

Eared pond snail
Lymnaea auricularia, class Gastropoda

The eared pond snail
lives in overgrown,
stagnant waters

Daudebardia rufa,
class Gastropoda

River limpet
Ancylus fluviatilis, class Gastropoda

Manus Island tree snail
Papustyla pulcherrima,
class Gastropoda

White-lipped garden snail
Cepaea hortensis, class Gastropoda

The giant tiger snail's shell
can grow to 1 foot (30 cm)
in length, making it one
of the largest snails

Escargot
Helix pomatia,
class Gastropoda

Giant tiger snail
Achatina achatina,
class Gastropoda

The escargot has been
eaten by humans since
prehistoric times

LIFE ON LAND

Several times during the course of evolution, gastropods have adapted to life on land. There are now roughly 20,000 land snail species. In most, gills have been lost and replaced by a lung. Some species, however, have both a gill chamber and a lung. Out of water, the shell has become lightweight and now protects against desiccation as well as against predators. The mantle and the mucus that cover the body also help to keep the snail moist. If conditions become too dry, a snail will attach itself to a plant or other surface and become dormant. Land slugs, which have lost the shell, live largely in crevices.

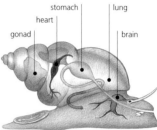

stomach   lung

heart

gonad   brain

Air breather *Most land snails lack gills. Instead, the mantle wall has become filled with blood vessels and acts as a lung.*

MATING HERMAPHRODITES

Land snails are hermaphrodites but do not self-fertilize. Before mating, two snails will usually circle each other, touch tentacles, intertwine bodies, and bite each other. In the copulation that follows, sperm are exchanged so that the eggs of both snails can be fertilized.

Love darts *The courtship of some land snails, such as Helix sp., is bizarre. When the snails are intertwined, one will shoot a chalky dart from its body into its mate. The dart appears to carry a chemical that helps the sperm reach the storage area in the female reproductive system.*

⚡ CONSERVATION WATCH

Mollusk losses Non-marine mollusks appear to be the most threatened of all animal groups. In the last 300 years, 284 species of non-marine mollusk are known to have become extinct—far more than in mammals (74 extinctions) or birds (129 extinctions). Although the oceans are by no means pristine, terrestrial and freshwater habitats have been drastically modified by human activities and their mollusk populations have suffered as a result.

FACT FILE

Class Cephalopoda The cephalopod foot lies close to the head and has been modified to form arms, tentacles, and a funnel. Nautiluses retain a large external shell; in squids and cuttlefishes the shell has been reduced and covered by tissue; and octopuses have lost their shell altogether. With the most developed brains of any invertebrates, cephalopods display complex behavior and are able to learn.

Species 600

Worldwide; marine

Predator Like all cephalopods, the European common squid (Loligo subulata) is carnivorous. It captures prey with its tentacles and arms and breaks it up with its beak and radula.

THE CHAMBERED NAUTILUS

The most primitive of the cephalopods, chambered nautiluses live in the last chamber of their large shell. The other chambers are filled with gas, and the nautilus controls its buoyancy by adding fluid to or removing fluid from them.

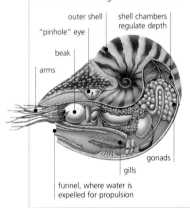

outer shell
shell chambers regulate depth
"pinhole" eye
beak
arms
gonads
gills
funnel, where water is expelled for propulsion

CEPHALOPOD SIGHT

Most cephalopods have excellent vision. Like the eyes of vertebrates, their eyes have a cornea, iris, lens, and retina. While humans bring objects into focus by changing the shape of the lens, cephalopods move the entire lens closer to or farther from the retina. Guided by the cephalopod's statocysts (balance organs), the slit-shaped pupils maintain a horizontal position no matter what the angle of the head.

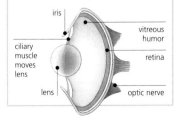

iris
vitreous humor
ciliary muscle moves lens
retina
lens
optic nerve

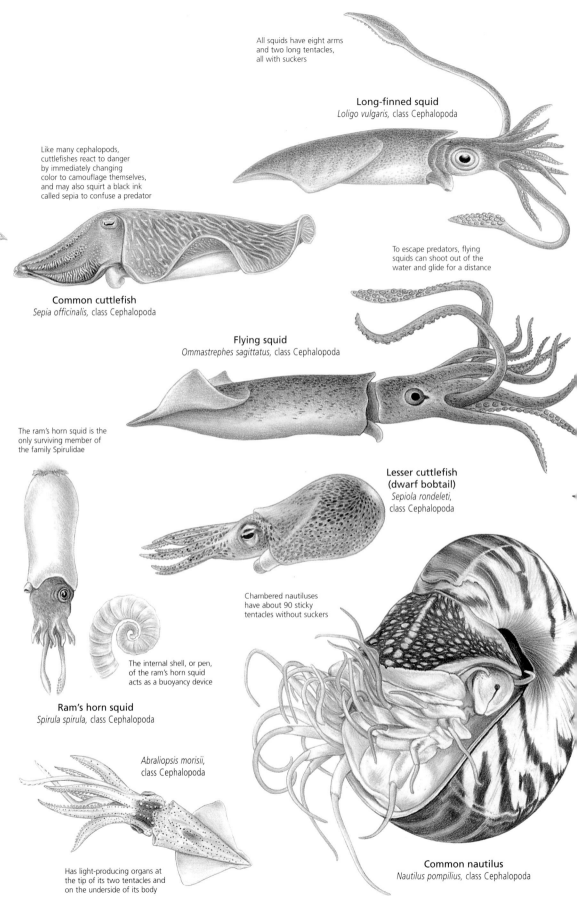

All squids have eight arms and two long tentacles, all with suckers

Long-finned squid
Loligo vulgaris, class Cephalopoda

Like many cephalopods, cuttlefishes react to danger by immediately changing color to camouflage themselves, and may also squirt a black ink called sepia to confuse a predator

To escape predators, flying squids can shoot out of the water and glide for a distance

Common cuttlefish
Sepia officinalis, class Cephalopoda

Flying squid
Ommastrephes sagittatus, class Cephalopoda

The ram's horn squid is the only surviving member of the family Spirulidae

Lesser cuttlefish (dwarf bobtail)
Sepiola rondeleti, class Cephalopoda

Chambered nautiluses have about 90 sticky tentacles without suckers

The internal shell, or pen, of the ram's horn squid acts as a buoyancy device

Ram's horn squid
Spirula spirula, class Cephalopoda

Abraliopsis morisii, class Cephalopoda

Has light-producing organs at the tip of its two tentacles and on the underside of its body

Common nautilus
Nautilus pompilius, class Cephalopoda

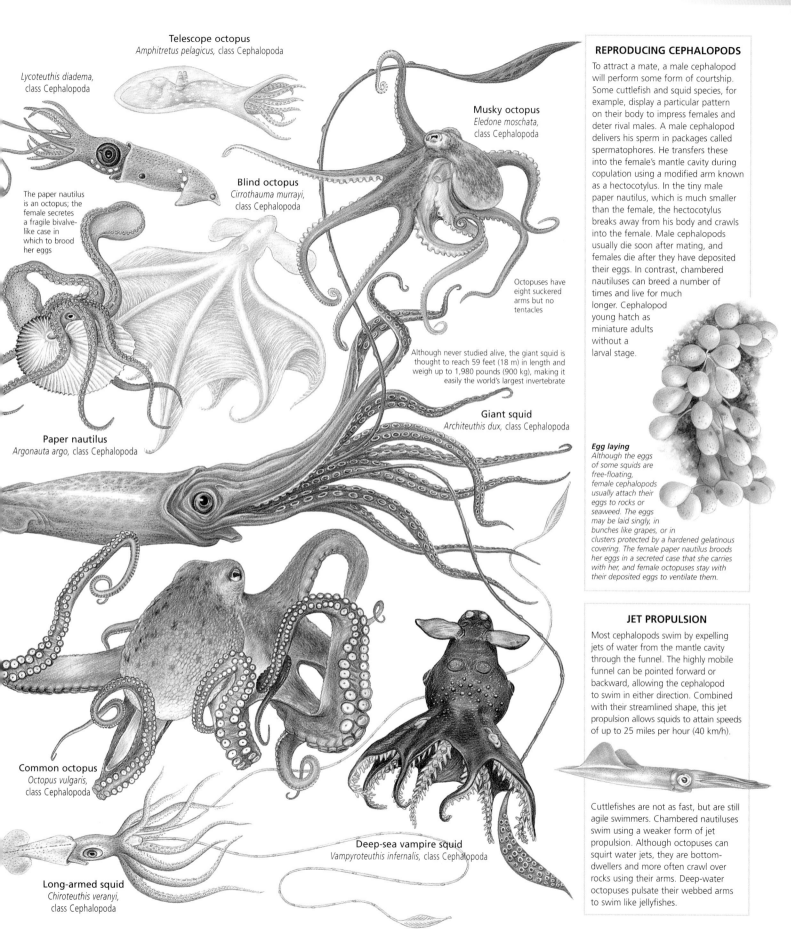

Lycoteuthis diadema,
class Cephalopoda

Telescope octopus
Amphitretus pelagicus, class Cephalopoda

Musky octopus
Eledone moschata,
class Cephalopoda

Blind octopus
Cirrothauma murrayi,
class Cephalopoda

The paper nautilus
is an octopus; the
female secretes
a fragile bivalve-
like case in
which to brood
her eggs

Octopuses have
eight suckered
arms but no
tentacles

Although never studied alive, the giant squid is
thought to reach 59 feet (18 m) in length and
weigh up to 1,980 pounds (900 kg), making it
easily the world's largest invertebrate

Giant squid
Architeuthis dux, class Cephalopoda

Paper nautilus
Argonauta argo, class Cephalopoda

Common octopus
Octopus vulgaris,
class Cephalopoda

Deep-sea vampire squid
Vampyroteuthis infernalis, class Cephalopoda

Long-armed squid
Chiroteuthis veranyi,
class Cephalopoda

REPRODUCING CEPHALOPODS

To attract a mate, a male cephalopod
will perform some form of courtship.
Some cuttlefish and squid species, for
example, display a particular pattern
on their body to impress females and
deter rival males. A male cephalopod
delivers his sperm in packages called
spermatophores. He transfers these
into the female's mantle cavity during
copulation using a modified arm known
as a hectocotylus. In the tiny male
paper nautilus, which is much smaller
than the female, the hectocotylus
breaks away from his body and crawls
into the female. Male cephalopods
usually die soon after mating, and
females die after they have deposited
their eggs. In contrast, chambered
nautiluses can breed a number of
times and live for much
longer. Cephalopod
young hatch as
miniature adults
without a
larval stage.

Egg laying
Although the eggs
of some squids are
free-floating,
female cephalopods
usually attach their
eggs to rocks or
seaweed. The eggs
may be laid singly, in
bunches like grapes, or in
clusters protected by a hardened gelatinous
covering. The female paper nautilus broods
her eggs in a secreted case that she carries
with her, and female octopuses stay with
their deposited eggs to ventilate them.

JET PROPULSION

Most cephalopods swim by expelling
jets of water from the mantle cavity
through the funnel. The highly mobile
funnel can be pointed forward or
backward, allowing the cephalopod
to swim in either direction. Combined
with their streamlined shape, this jet
propulsion allows squids to attain speeds
of up to 25 miles per hour (40 km/h).

Cuttlefishes are not as fast, but are still
agile swimmers. Chambered nautiluses
swim using a weaker form of jet
propulsion. Although octopuses can
squirt water jets, they are bottom-
dwellers and more often crawl over
rocks using their arms. Deep-water
octopuses pulsate their webbed arms
to swim like jellyfishes.

SEGMENTED WORMS

PHYLUM	Annelida
CLASSES	2
ORDERS	21
FAMILIES	130
SPECIES	12,000

Also known as annelids, segmented worms have a head and a long body made up of segments that look like rings from the outside. Each segment has its own fluid-filled body cavity that acts as a hydroskeleton and contains a separate set of excretory, locomotory, and respiratory organs. The segments are united, however, by common digestive, circulatory, and nervous systems. Segmented worms crawl or swim by wriggling from side to side, or burrow by waves of contractions that pass down the length of the body. Stiff bristles known as chetae protrude from each segment and provide traction. While many species are highly active, others live in burrows or tubes.

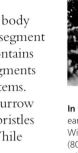

In the soil Long and thin with a small head, earthworms are shaped for burrowing. With up to 650 individuals per square yard (800 per sq. m), earthworms play a crucial role in aerating and fertilizing the soil.

Food-filtering fans
Fanworms are annelids that live attached to the seafloor. Most of their body is hidden inside a tube that they have secreted or built from sediment. A crown of fine tentacles, each covered in tiny beating hairs called cilia, filters particles of food from the water.

IN WATER AND ON LAND
Segmented worms are found in all levels of the ocean, in freshwater lakes and rivers, and on land. They include filter-feeders, predators, and blood-suckers, as well as deposit-feeders such as earthworms, which eat sediment to extract nutrients.

Commonly called bristleworms, the species in the class Polychaeta are almost entirely marine and usually swim or crawl using paddle-like appendages known as parapodia. The parapodia are often lost in tube-living forms, which have a modified head region for food-gathering. Typically, the sexes are separate and release eggs and sperm into the water for fertilization.

The class Clitellata includes many terrestrial and freshwater species such as earthworms and leeches, as well as some marine species. Most are hermaphrodites that exchange sperm by copulation. They all bear a clitellum, a ring of glandular skin that secretes a cocoon for the eggs.

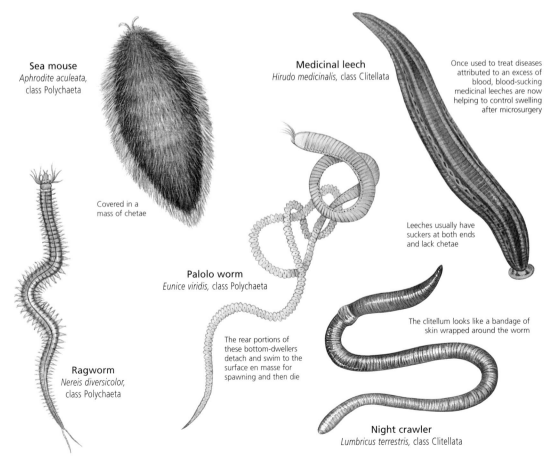

Sea mouse
Aphrodite aculeata,
class Polychaeta

Covered in a
mass of chetae

Ragworm
Nereis diversicolor,
class Polychaeta

Palolo worm
Eunice viridis, class Polychaeta

The rear portions of these bottom-dwellers detach and swim to the surface en masse for spawning and then die

Medicinal leech
Hirudo medicinalis, class Clitellata

Once used to treat diseases attributed to an excess of blood, blood-sucking medicinal leeches are now helping to control swelling after microsurgery

Leeches usually have suckers at both ends and lack chetae

The clitellum looks like a bandage of skin wrapped around the worm

Night crawler
Lumbricus terrestris, class Clitellata

Instantly withdraws its feathery crown if disturbed

Red feather duster
Spirographis spallanzanii,
class Polychaeta

METAMORPHOSIS

While some newly hatched invertebrates resemble little adults, most look quite distinct from their parents and have a very different lifestyle. To become adults, these young must go through a transformation known as metamorphosis. Corals, clams, many crustaceans, and most other invertebrates metamorphose into adults after a brief larval stage. For most insects, the larval stage is prolonged. In some insect groups, such as crickets and bugs, metamorphosis is incomplete. The hatched form, called a nymph, is structured like the adult, but lacks wings and full sex organs until it emerges from the final molt. Other insects, such as butterflies and bees, have complete metamorphosis. The young, called larvae, differ so much from their parents that they must go through a pupal stage, in which their bodies break down and are built up again as adults.

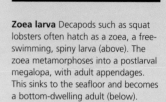

Veliger larva Marine gastropods such as sea snails have a free-swimming veliger larva, which feeds and propels itself using tiny hairs known as cilia. The larva gradually develops a shell, mantle cavity, and foot (above), then metamorphoses into an adult (below).

Zoea larva Decapods such as squat lobsters often hatch as a zoea, a free-swimming, spiny larva (above). The zoea metamorphoses into a postlarval megalopa, with adult appendages. This sinks to the seafloor and becomes a bottom-dwelling adult (below).

Egg to nymph to adult A dragonfly develops through incomplete metamorphosis. **1.** Male and female mate. **2.** Female lays eggs in water or in stem of water plant. **3.** Aquatic nymph hatches. **4.** Nymph feeds on tadpoles and worms and grows through a series of molts. **5.** Nymph emerges from the water for its final molt. **6.** Adult bursts from the nymphal skin. **7.** Adult rests while its wings dry out. **8.** Feeding on flying insects, adult flies to a new location to find a mate.

ARTHROPODS

PHYLUM	Arthropoda
CLASSES	22
ORDERS	110
FAMILIES	2,120
SPECIES	> 1.1 million

The insects, spiders, crustaceans, centipedes, and other invertebrates in the phylum Arthropoda account for three-quarters of all known animal species, and millions more remain to be discovered. Arthropods have adapted to fill virtually every ecological niche on land, in fresh water, and in the oceans. Consequently, their anatomy and lifestyles are extraordinarily diverse, but they share a number of defining features. The name arthropod means "jointed feet," and arthropods do have jointed appendages and segmented bodies. They are distinguished from other invertebrates by their tough but flexible exoskeleton, which provides both protection and support.

Fossilized marine class Arthropods called trilobites dominate marine fossil beds from about 500 million years ago, and include more than 15,000 known species. They disappeared about 250 million years ago.

Insects everywhere Insects account for about 90 percent of all arthropod species. Ubiquitous on land and in fresh water, but rare in the ocean, insects include predators such as the praying mantis, and nectar-feeders such as the swallowtail butterfly.

EARLY ARRIVALS

The first arthropods appeared in the oceans about 530 million years ago and included crustaceans; ancestors of horseshoe crabs; and the now-extinct trilobites. These early species tended to have numerous body segments, each bearing a similar pair of appendages. Over time, the appendages became specialized for particular tasks, such as locomotion, food collection, sensory perception, and copulation. Furthermore, the segments became arranged into distinct regions of the body called tagma, such as the head, thorax, and abdomen of insects.

Arthropods were the first animals to leave the sea and the first to take to the sky. Scorpions had ventured onto land by 350 million years ago and were soon followed by the first terrestrial insects. Before long, winged insects had emerged. The success of arthropods on land rests in part on their waxy exoskeleton, which stops them from drying out. Jointed legs also make arthropods highly mobile, thus enhancing their ability to find food and mates, evade predators, and colonize new places. In insects, the development of wings increases the advantage.

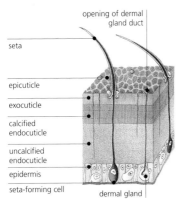

Arthropod armor Secreted by epidermal cells, the arthropod exoskeleton has thin layers of wax and protein, known as the epicuticle, overlying various layers of chitin and protein. Tanning of the protein in the exocuticle hardens the exoskeleton. Like all arthropods, the red carnation coral crab (*Hoplophrys oatesii*) (right) must replace its exoskeleton by molting as it grows.

Legs and bodies Tarantulas (left) and other spiders have eight legs and a body divided into two parts: the cephalothorax (fused head and thorax) and the abdomen. Insects have six legs and a three-part body made up of a head, thorax, and abdomen.

Deadly bites While some arthropods are venomous, few deliver a bite that is fatal to humans. Far more deadly are the diseases transmitted by blood-sucking arthropods such as mosquitoes (below), which transfer malaria and other illnesses from one vertebrate host to another.

WAYS OF BREATHING

The arthropod exoskeleton offers great protection, but it is also usually too impermeable to allow gas exchange. While some tiny arthropods can breathe directly through the body wall, most have developed specialized structures. Aquatic arthropods tend to have gills. Derived from gills are book lungs, found in many arachnids. Terrestrial arthropods often rely on minute air ducts known as tracheae, which deliver air to all parts of the body.

Gills Gills collect oxygen from water and help to maintain the arthropod's salt balance. In crustaceans, the gills are external and often positioned on the legs, but are protected by the exoskeleton. Horseshoe crabs have unique structures called book gills, with flaps like the pages of a book.

Limbs in motion By modifying their jointed limbs for walking, running, leaping, pushing, burrowing, and swimming, arthropods have been able to occupy diverse habitats. In some, such as grasshoppers (right), the legs also hold their hearing organs.

While many arthropods are themselves predators, almost all are prey to vertebrates or to other invertebrates. The compound eyes found on many arthropod species are made up of multiple lenses and are extremely effective at detecting motion. Simple eyes, with a single lens, detect light intensity. Hairs (known as setae), projections, and slits on the antennae, mouthparts, and legs detect subtle vibrations. A cerebral ganglion (brain) connects via a nerve cord to ganglia (clusters of nerve cells) in each body segment. In arthropods' open circulatory system, a heart pumps hemolymph (blood) to bathe the organs.

The sexes are separate in most arthropod species, and reproduction usually involves the male's sperm fertilizing the female's eggs. Often, the sperm is transferred in packages known as spermatophores, which the female can use when she is ready. The life stages of arthropods vary greatly. Some species are like small adults when they hatch, while others add segments as they grow. Many insect larvae look nothing like their parents and must go through a pupal stage to transform into a winged adult.

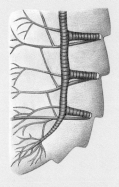

Book lungs Most likely derived from the book gills found on horseshoe crabs, book lungs are found in many arachnids. While blood circulates within the hollow, flat plates of a book lung, air circulates through the space between the plates. Arachnids often combine book lungs with tracheae.

Tracheae Found in insects, arachnids, and centipedes and millipedes, tracheae are tubes that collect air through minute openings in the exoskeleton called spiracles, and transport it to the tissues or blood. The spiracles can be closed to minimize moisture loss.

ARACHNIDS

PHYLUM	Arthropoda
SUBPHYLUM	Chelicerata
CLASS	Arachnida
ORDERS	17
FAMILIES	450
SPECIES	80,000

Including some of the most feared and fascinating of all invertebrates, the class Arachnida encompasses spiders, scorpions, harvestmen (or daddy-longlegs), and mites and ticks, along with several lesser known groups. Apart from some families of aquatic mites and a few species of water spiders, all arachnids are terrestrial, and the majority are predators of other invertebrates. Many spiders use silk webs to snare prey, and both scorpions and spiders inject their prey with venom to paralyze or kill it. Most arachnids are unable to swallow solid food and must squirt digestive enzymes into the prey and then suck the liquefied meal into their mouth.

Diurnal vision Most arachnid species are nocturnal and their simple eyes detect only variations in light. In daytime hunters, such as jumping spiders (family Salticidae), the two primary simple eyes provide very sharp vision at close range.

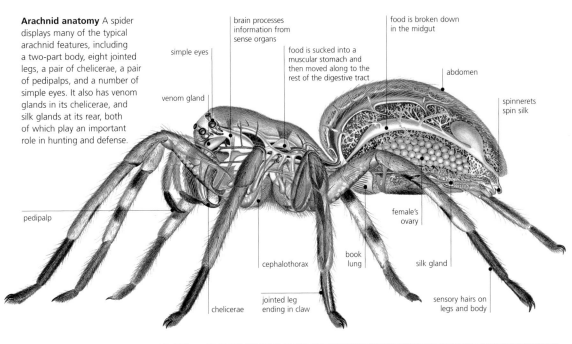

Arachnid anatomy A spider displays many of the typical arachnid features, including a two-part body, eight jointed legs, a pair of chelicerae, a pair of pedipalps, and a number of simple eyes. It also has venom glands in its chelicerae, and silk glands at its rear, both of which play an important role in hunting and defense.

brain processes information from sense organs

simple eyes

food is sucked into a muscular stomach and then moved along to the rest of the digestive tract

venom gland

food is broken down in the midgut

abdomen

spinnerets spin silk

pedipalp

female's ovary

cephalothorax

book lung

silk gland

chelicerae

jointed leg ending in claw

sensory hairs on legs and body

EIGHT-LEGGED PREDATORS

Like other arthropods, arachnids have a segmented body, a tough but flexible exoskeleton, and jointed limbs. Mites and ticks have rounded bodies made up of a single region, but other arachnids have two-part bodies, with a cephalothorax holding the eyes, mouthparts, and limbs, and an abdomen containing many of the internal organs.

Arachnids have eight legs, unlike insects, which have six. They also have two pairs of appendages near the mouth. The first pair, known as chelicerae, may be pincer-like or can form fangs that inject venom. They are used for subduing prey. The second pair, the pedipalps, are often used like antennae to sense the surroundings. Scorpions and some other arachnids use pedipalps to seize prey, while male spiders use them to transfer sperm to females.

To find prey and avoid danger, arachnids rely on the fine sensory hairs on their body and legs. They also have a number of simple eyes. Fine slits in the cuticle may detect odors, gravity, or vibrations.

Arachnids breathe using book lungs or a tracheal system, or both. Book lungs probably developed from gills and are made up of stacked leaves of tissue. The tracheal system takes air in through tiny pores called spiracles and distributes it to tissues via a network of tubes.

Development of arachnids is direct, with the young hatching as small versions of their parents and growing through a series of molts. Most arachnids are solitary.

Blood-sucking parasite While most arachnids hunt live prey, ticks and some mites are parasites. A tick uses its hooked mouthparts to pierce the skin of a mammal, then feed on its blood. Once engorged, the tick drops off the host and molts.

Stinging defense While arachnids such as spiders and scorpions are widely feared, only a tiny minority of species deliver a bite or sting that can be fatal to humans. The giant desert hairy scorpion (*Hadrurus arizonensis*) has a painful sting, but the effect of its venom on humans is mild.

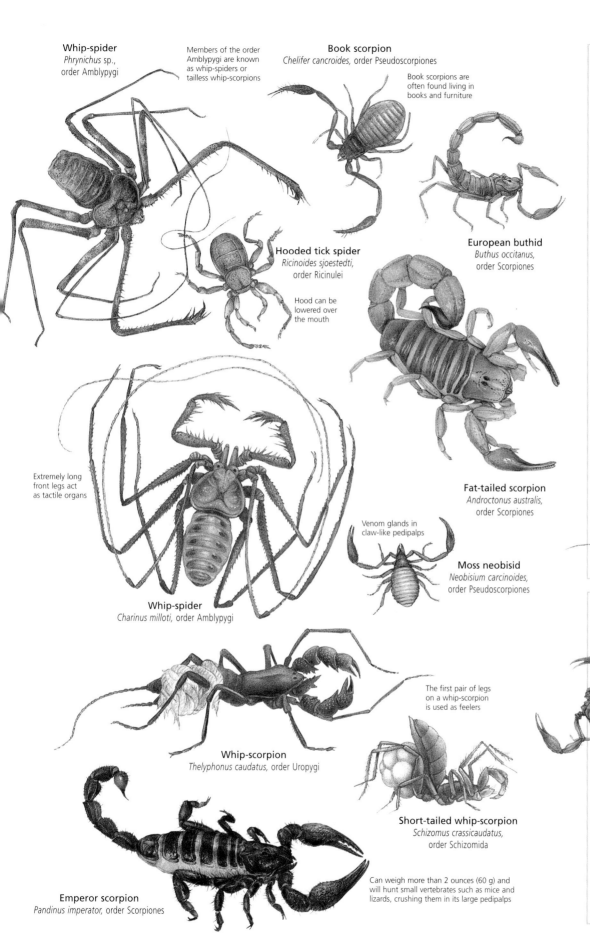

Whip-spider
Phrynichus sp.,
order Amblypygi

Members of the order
Amblypygi are known
as whip-spiders or
tailless whip-scorpions

Book scorpion
Chelifer cancroides, order Pseudoscorpiones

Book scorpions are
often found living in
books and furniture

Hooded tick spider
Ricinoides sjoestedti,
order Ricinulei

Hood can be
lowered over
the mouth

European buthid
Buthus occitanus,
order Scorpiones

Extremely long
front legs act
as tactile organs

Whip-spider
Charinus milloti, order Amblypygi

Fat-tailed scorpion
Androctonus australis,
order Scorpiones

Venom glands in
claw-like pedipalps

Moss neobisid
Neobisium carcinoides,
order Pseudoscorpiones

The first pair of legs
on a whip-scorpion
is used as feelers

Whip-scorpion
Thelyphonus caudatus, order Uropygi

Short-tailed whip-scorpion
Schizomus crassicaudatus,
order Schizomida

Emperor scorpion
Pandinus imperator, order Scorpiones

Can weigh more than 2 ounces (60 g) and
will hunt small vertebrates such as mice and
lizards, crushing them in its large pedipalps

FACT FILE

Order Scorpiones Scorpions are
distinguished from other arachnids
by two large clawed pedipalps and
a segmented abdomen. The tip of
the tail holds a prominent stinger,
which is sometimes used to subdue
prey but more often plays a defensive
role. Scorpions shelter in crevices or
burrows by day, and hunt by night.

Species 1,400

Warmer regions;
under rocks & bark

Houseguest
The slenderbrown scorpion
(*Centruroides gracilis*) usually lives in
tropical forests, but may dwell in houses
in regions where it has been introduced.

Order Uropygi Known as whip-
scorpions or vinegaroons, the arachnids
in this order resemble scorpions but
their pedipalps are stouter, and their
tails are long and thin. If threatened,
a whip-scorpion will spray acid from
glands at the base of its whip-like tail.

Species 100

Mainly tropical regions; under rocks

Takeaway meals The giant whip-scorpion
(*Mastigoproctus giganteus*) seizes and crushes
its invertebrate prey with its
pedipalps, then carries the
meal back to its burrow.

PARENTAL CARE

Most female arachnids lay their eggs in
soil or another safe location and leave
them to hatch on their own. In
Euscorpius carpathicus (left)
and other scorpions,
however, the
fertilized eggs
develop
inside the
mother and
she produces
live young. The
mother catches
them in her first or
first and second pairs of
legs. At this stage, the larval young are
rather helpless as they lack pedipalps
and cannot sting. They crawl onto the
mother's back and are carried around
until after their first molt, about 3 to
14 days later. Now equipped with their
pedipalps and a functioning stinger,
they are able to hunt. They quickly
scatter to establish their own territories
before their mother eats them.

FACT FILE

Order Araneae This order contains the spiders, arachnids that have silk glands with which they produce webs and protective egg cases. All spiders are carnivorous, feeding mainly on other invertebrates, including other spiders. Web spiders build webs to snare their prey, while ground spiders are active hunters. Most spiders are venomous, injecting poison into prey or enemies through their fang-like chelicerae. Despite their fearsome reputation, they rarely bite humans unless they feel threatened, and only about 30 species will cause illness.

Species 40,000

Worldwide; in all terrestrial habitats

Night hunter
Wolf spiders such as Trochosa terricola are active nocturnal hunters and do not spin webs to catch their invertebrate prey.

EYES OF THE SPIDER

Most spiders are nocturnal and tend to rely more on touch than on vision. Day-active species, however, generally have excellent eyesight at close range. Spiders usually have four pairs of simple eyes grouped in a characteristic pattern according to the family.

Wide field
The huntsman spider (family Heteropodidae) is an active hunter. Its wideset eyes provide all-round vision.

Night hunter *The woodlouse-eating spider (family Dysderidae) has six tiny eyes instead of the usual eight and relies on touch to uncover its prey.*

Close range *The diurnal crab spider (family Thomisidae) depends on its keen close-up vision as it ambushes insect prey.*

Seeing in the dark *The two huge eyes of the ogre-faced spider (family Deinopidae) enable it to see prey in near-total darkness.*

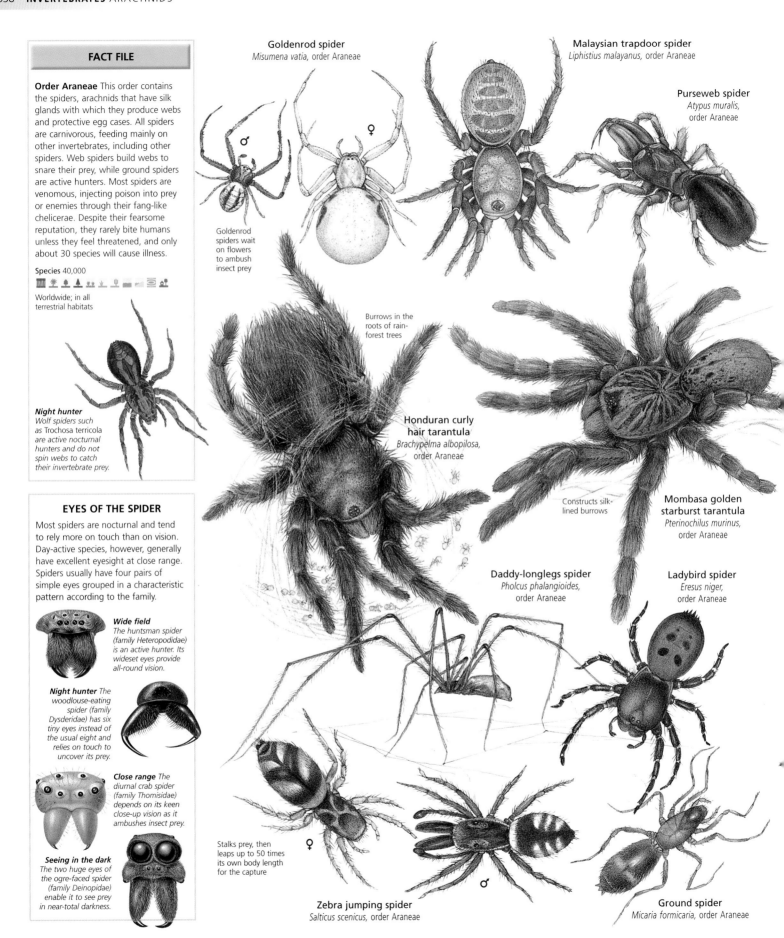

Goldenrod spider
Misumena vatia, order Araneae

Goldenrod spiders wait on flowers to ambush insect prey

Malaysian trapdoor spider
Liphistius malayanus, order Araneae

Purseweb spider
Atypus muralis, order Araneae

Burrows in the roots of rain-forest trees

Honduran curly hair tarantula
Brachypelma albopilosa, order Araneae

Constructs silk-lined burrows

Mombasa golden starburst tarantula
Pterinochilus murinus, order Araneae

Daddy-longlegs spider
Pholcus phalangioides, order Araneae

Ladybird spider
Eresus niger, order Araneae

Stalks prey, then leaps up to 50 times its own body length for the capture

Zebra jumping spider
Salticus scenicus, order Araneae

Ground spider
Micaria formicaria, order Araneae

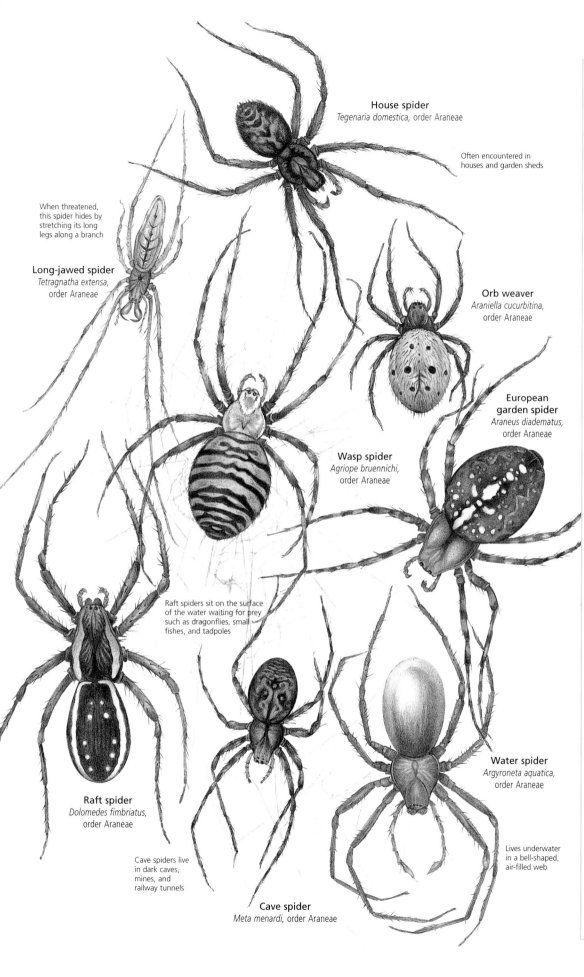

House spider
Tegenaria domestica, order Araneae

Often encountered in
houses and garden sheds

When threatened,
this spider hides by
stretching its long
legs along a branch

Long-jawed spider
Tetragnatha extensa,
order Araneae

Orb weaver
Araniella cucurbitina,
order Araneae

**European
garden spider**
Araneus diadematus,
order Araneae

Wasp spider
Agriope bruennichi,
order Araneae

Raft spiders sit on the surface
of the water waiting for prey
such as dragonflies, small
fishes, and tadpoles

Raft spider
Dolomedes fimbriatus,
order Araneae

Cave spiders live
in dark caves,
mines, and
railway tunnels

Water spider
Argyroneta aquatica,
order Araneae

Lives underwater
in a bell-shaped,
air-filled web

Cave spider
Meta menardi, order Araneae

CAPTURE METHODS

Not all spiders sit waiting for prey to
fall into their webs. The spitting spider
(*Scytodes thoracica*) has an enlarged
cephalothorax that houses linked
venom and silk glands. It hunts its prey
by night, stealthily approaching until it
is within spitting distance. The spider
then ejects two poisonous silk threads,
covering the prey in a zig-zag pattern.
The prey is glued to the ground by the
silk and paralyzed by the venom.

Spitting spider *With only six
small eyes, the spitting spider has poor
vision and uses the sensory hairs on its front
legs to find prey such as flies and moths.*

Bolas spiders belong to the orb-weaver
family, Araneidae, but they do not
build orb webs. Instead, they "fish"
for their prey, suspending a single
strand of silk tipped with a sticky
globule known as a bolas. The bolas
seems to be laced with pheromones
that mimic those of particular female
noctuid moths. When male moths are
attracted to the bolas and become
stuck, the spider reels them in.

Bolas spider
*When it senses a
moth approaching,
a bolas spider will
start swinging its
fishing line in a circle.*

Spiders in various families have
adapted their anatomy to convincingly
mimic ants. They are even thought to
produce pheromones that allow them
to infiltrate ant colonies. Some ant-
mimics use their disguise to prey on
the unsuspecting ants. Others may
merely be protecting themselves from
predators such as wasps and birds that
avoid ants because of the formic acid
ants produce when threatened.

Brazilian ant-mimicking spider *Like other
ant-mimics,* Aphantochilus rogersi *has a
long waist and a long cephalothorax that is
partially divided into two segments to give the
impression of a three-part insect body. The
first pair of legs is held in front like antennae.*

SILK AND WEBS

While only some species build webs, almost all spiders produce silk. Made of the protein fibroin, spider silk is as strong as nylon thread but has greater elasticity. A spider may have up to eight silk glands in its abdomen, each yielding a different kind of silk. One kind of silk is used for the dragline a spider usually spins after itself like a mountain climber's safety line. Other types are used for cocooning fertilized eggs or wrapping up captured prey. Male spiders may use silk to hold their packages of sperm. Some spiderlings use threads of silk like balloons to carry them aloft so they can disperse. The best known use of silk is among the webspinning spiders, which build a remarkable variety of webs to snare their prey.

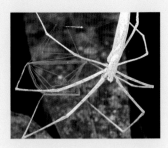

Captured prey A net-casting spider (*Deinopus* sp.) prepares a little snare net each night and holds it in its legs as it awaits prey. When a potential victim passes by, the spider flings the net over it, quickly wraps the catch in extra silk, and then bites it.

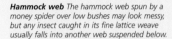

Scaffold web A scaffold web has stretched traplines with sticky ends attached to the ground. If a crawling insect walks into a trapline, the line snaps back, with the victim dangling in the air.

Hammock web The hammock web spun by a money spider over low bushes may look messy, but any insect caught in its fine lattice weave usually falls into another web suspended below.

Triangle web A spider holds a triangular web in its front legs while anchored by a thread of silk to the twig behind it. When an insect hits the web, the spider releases its grip and the web collapses, entangling the prey.

Orb web The most elaborate of all webs, the orb web covers a large area using a minimum of silk. An outer framework supports a continuous spiral and a series of spokes. The web is suspended between tree branches to capture flying insects.

Lace-sheet web The trap spun by a lace-web weaver is made of fine, woolly silk. It may not be sticky, but insects soon get tangled up in its many fibers, and remain trapped until the spider is ready for food.

No sticky feet Spiders such as this orb weaver (*Argiope* sp.) (right) avoid getting trapped in their own webs because only some of the threads are sticky and they know where to tread.

Shared nest While most spiders are solitary, some share communal nests. Some species are truly colonial, and hunt and feed together. In others, such as *Cryptophora* sp., each spider guards its own orb web and feeds alone, but the webs share frame lines, creating a vast network of silk (below).

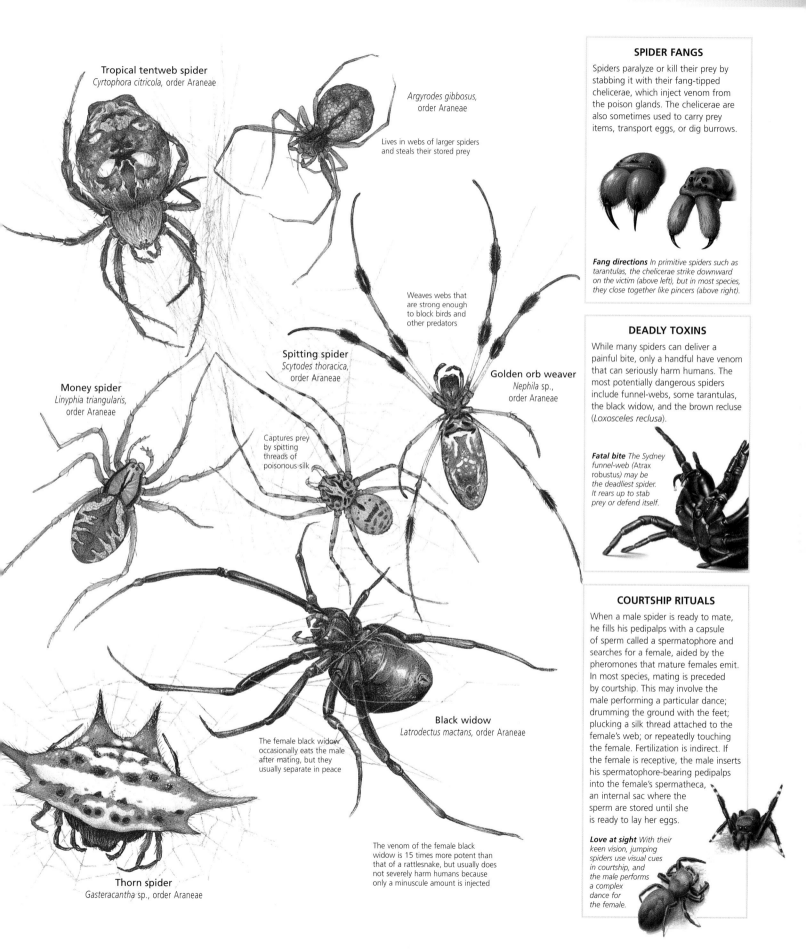

Tropical tentweb spider
Cyrtophora citricola, order Araneae

Argyrodes gibbosus, order Araneae

Lives in webs of larger spiders and steals their stored prey

Weaves webs that are strong enough to block birds and other predators

Spitting spider
Scytodes thoracica, order Araneae

Golden orb weaver
Nephila sp., order Araneae

Money spider
Linyphia triangularis, order Araneae

Captures prey by spitting threads of poisonous silk

Black widow
Latrodectus mactans, order Araneae

The female black widow occasionally eats the male after mating, but they usually separate in peace

The venom of the female black widow is 15 times more potent than that of a rattlesnake, but usually does not severely harm humans because only a minuscule amount is injected

Thorn spider
Gasteracantha sp., order Araneae

SPIDER FANGS

Spiders paralyze or kill their prey by stabbing it with their fang-tipped chelicerae, which inject venom from the poison glands. The chelicerae are also sometimes used to carry prey items, transport eggs, or dig burrows.

Fang directions *In primitive spiders such as tarantulas, the chelicerae strike downward on the victim (above left), but in most species, they close together like pincers (above right).*

DEADLY TOXINS

While many spiders can deliver a painful bite, only a handful have venom that can seriously harm humans. The most potentially dangerous spiders include funnel-webs, some tarantulas, the black widow, and the brown recluse (*Loxosceles reclusa*).

Fatal bite *The Sydney funnel-web (*Atrax robustus*) may be the deadliest spider. It rears up to stab prey or defend itself.*

COURTSHIP RITUALS

When a male spider is ready to mate, he fills his pedipalps with a capsule of sperm called a spermatophore and searches for a female, aided by the pheromones that mature females emit. In most species, mating is preceded by courtship. This may involve the male performing a particular dance; drumming the ground with the feet; plucking a silk thread attached to the female's web; or repeatedly touching the female. Fertilization is indirect. If the female is receptive, the male inserts his spermatophore-bearing pedipalps into the female's spermatheca, an internal sac where the sperm are stored until she is ready to lay her eggs.

Love at sight *With their keen vision, jumping spiders use visual cues in courtship, and the male performs a complex dance for the female.*

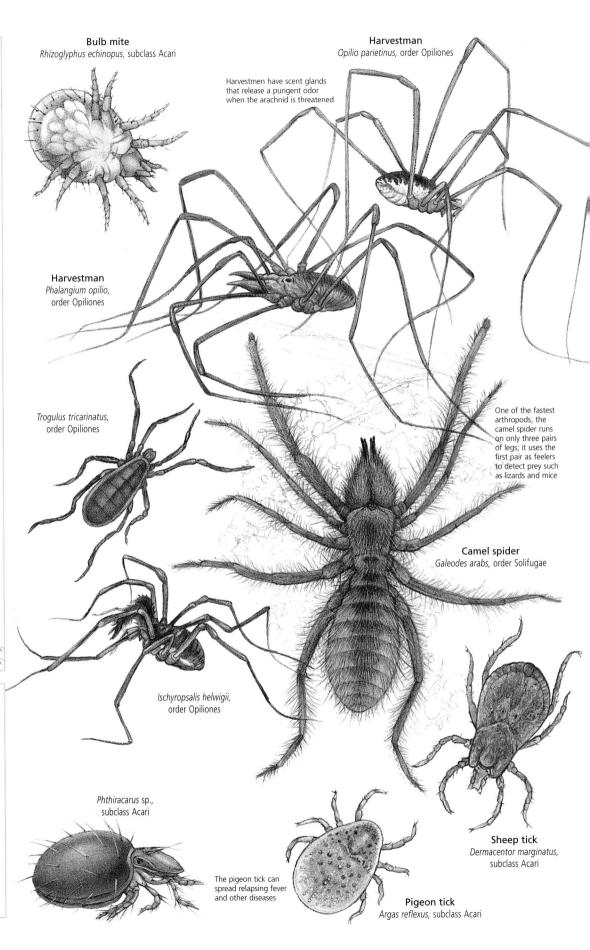

FACT FILE

Order Opiliones The harvestmen, or daddy-longlegs, in this order differ from spiders in that they lack venom glands, silk glands, and a waist. Most species have very long, fine legs. Unlike most other arthropods, fertilization is direct, with the male using a penis to deposit sperm in the female.

Species 5,000

Worldwide; in leaf litter & under stones

Bringing in the harvest The harvestman Phalangium opilio *looks for small arthropods in the leaf litter. It is also common in crops, where it feeds on pests such as aphids.*

Subclass Acari Classified in seven orders in the subclass Acari, mites and ticks are the most varied and abundant arachnids. They flourish in almost every kind of habitat, from polar caps to deserts, thermal springs to ocean trenches. With some species small enough to live inside a human hair follicle, however, mites are so tiny that they are rarely noticed. Ticks tend to be larger, but rarely exceed ½ inch (1 cm) in length. Mites and ticks have only six legs when they hatch as larvae and acquire an extra pair as nymphs before becoming adults.

Species 30,000

Worldwide

Dust to dust Flour mites (Acarus siro) *are found living in the dust of grain crops, in animal cages, and in food stores.*

DISEASE SPREADERS

While most members of Acari are free-living inhabitants of soil, leaf litter, and water, some are parasites of other animals or plants. Those mites and ticks that feed on vertebrates often transmit bacteria that cause potentially fatal diseases in humans. Several species of ticks also commonly trigger acute paralysis in humans and domestic animals.

Sick tick The European castor bean tick (Ixodes ricinus) *infests livestock, dogs, and humans, and can spread Lyme disease and other illnesses.*

Bulb mite
Rhizoglyphus echinopus, subclass Acari

Harvestman
Opilio parietinus, order Opiliones

Harvestmen have scent glands that release a pungent odor when the arachnid is threatened

Harvestman
Phalangium opilio,
order Opiliones

Trogulus tricarinatus,
order Opiliones

One of the fastest arthropods, the camel spider runs on only three pairs of legs; it uses the first pair as feelers to detect prey such as lizards and mice

Camel spider
Galeodes arabs, order Solifugae

Ischyropsalis helwigii,
order Opiliones

Sheep tick
Dermacentor marginatus,
subclass Acari

Phthiracarus sp.,
subclass Acari

The pigeon tick can spread relapsing fever and other diseases

Pigeon tick
Argas reflexus, subclass Acari

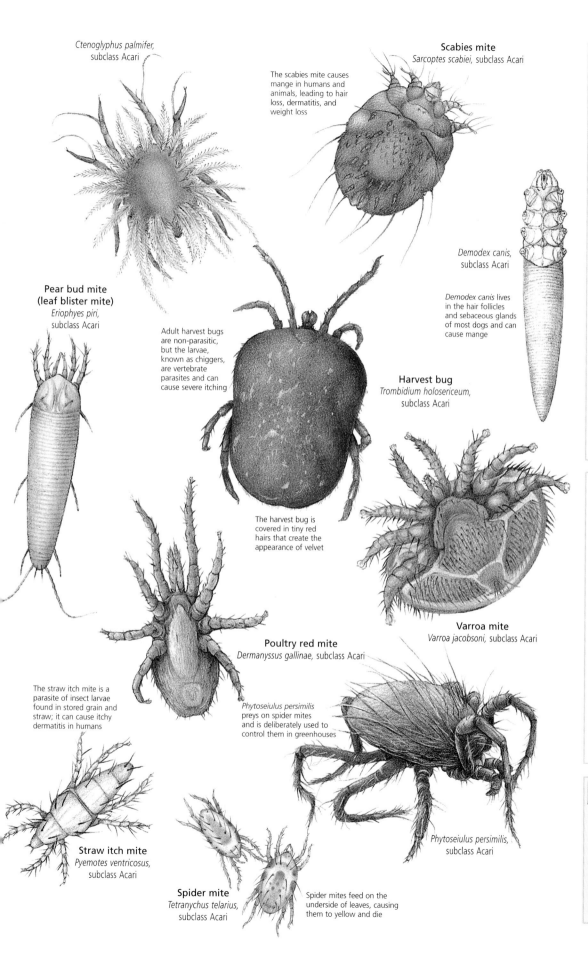

Ctenoglyphus palmifer, subclass Acari

Scabies mite
Sarcoptes scabiei, subclass Acari

The scabies mite causes mange in humans and animals, leading to hair loss, dermatitis, and weight loss

Demodex canis, subclass Acari

Demodex canis lives in the hair follicles and sebaceous glands of most dogs and can cause mange

Pear bud mite (leaf blister mite)
Eriophyes piri, subclass Acari

Adult harvest bugs are non-parasitic, but the larvae, known as chiggers, are vertebrate parasites and can cause severe itching

Harvest bug
Trombidium holosericeum, subclass Acari

The harvest bug is covered in tiny red hairs that create the appearance of velvet

Varroa mite
Varroa jacobsoni, subclass Acari

The straw itch mite is a parasite of insect larvae found in stored grain and straw; it can cause itchy dermatitis in humans

Poultry red mite
Dermanyssus gallinae, subclass Acari

Phytoseiulus persimilis preys on spider mites and is deliberately used to control them in greenhouses

Phytoseiulus persimilis, subclass Acari

Straw itch mite
Pyemotes ventricosus, subclass Acari

Spider mite
Tetranychus telarius, subclass Acari

Spider mites feed on the underside of leaves, causing them to yellow and die

AQUATIC MITES

Some mites have adapted to a life in water. They are found in every kind of freshwater habitat, from puddles to deep lakes, hot springs to raging rivers. Some have also colonized the ocean, where they live on mudflats, coral reefs, and the seafloor, absent only from open waters. While semiaquatic mites are found in many groups within Acari, more than 40 families of fully aquatic mites are gathered in the subcohort Hydracarina. These mites lay their eggs underwater on stones or plants or in sponges or mussels. The larvae are parasitic and quickly attach themselves to an invertebrate host. As the host is often a flying insect such as a dragonfly, it not only provides the mite larva with food, but also disperses it to a new body of water. As nymphs and adults, water mites are predators and feed on other mites, aquatic insects, and crustaceans.

Bright warning *Water mites such as Brachypoda versicolor often display bright colors, possibly to warn fishes and other predators that they are distasteful.*

HONEYBEE THREAT

A honeybee colony can be severely harmed or even killed by the varroa mite (*Varroa jacobsoni*). An adult female mite moves into a honeybee's brood cell to lay her eggs. Both the mite and her hatched offspring then feed on the bee larva as it matures. The mite offspring mate while still in the brood cell. The males then die, but the females emerge attached to the young adult bee. The female mites then find another brood cell where they can deposit their eggs. Parasitized honeybee pupae tend to emerge as weak, deformed adults.

Snug fit *The varroa mite usually inserts itself between the honeybee's body segments, making it difficult to see.*

⚡ CONSERVATION WATCH

Arachnid alert The conservation movement has paid little attention to arachnids. Only 18 species have been assessed by the IUCN, but all of these are on the Red List. Further research is needed, but arachnid diversity is no doubt threatened by habitat loss, pollution, pesticide use, and invasive exotic species.

HORSESHOE CRABS

PHYLUM	Arthropoda
SUBPHYLUM	Chelicerata
CLASS	Merostomata
ORDER	Xiphosura
FAMILY	Limulidae
SPECIES	4

More closely related to arachnids than to crustaceans, the horseshoe crabs of the class Merostomata have changed little in the past 200 million years. The four living species, found off the east coasts of North America and Asia, are marine. Their body is made up of a cephalothorax and abdomen, each protected by a hard shell. The horseshoe-shaped cephalothorax bears the pincer-like chelicerae, which are used for seizing worms and other prey from the muddy seafloor, and five pairs of legs, used for both walking and handling food. The formidable tail spine, or telson, does not act as a weapon but helps the horseshoe crab to right itself and to plow through mud.

Can grow 2 feet
(60 cm) long

Horseshoe crab
Limulus polyphemus

Time to breed
Breeding horseshoe crabs gather on beaches in spring. The eggs are buried in the sand just below the high-tide mark, where they stay moist but are warmed by the sun.

SEA AND SHORE

Horseshoe crabs spend most of their time on the seafloor at depths of about 100 feet (30 m), breathing through book gills on the abdomen. They push through mud, and can also swim on their back and walk.

In spring, horseshoe crabs come ashore to breed. A male will cling onto a female's back as she makes her way over the tidal zone. The female scoops out holes in the sand and lays up to 300 eggs in each, and the male then covers the eggs with sperm to fertilize them. The larvae reach adulthood after about 16 molts over 9 to 12 years.

SEA SPIDERS

PHYLUM	Arthropoda
SUBPHYLUM	Chelicerata
CLASS	Pycnogonida
ORDER	Pantopoda
FAMILIES	9
SPECIES	1,000

With their long legs, sea spiders bear a superficial resemblance to spiders, but they evolved separately and have many unique features. A sea spider's body is greatly reduced. The small head region, known as the cephalon, holds a long proboscis with the mouth; four simple eyes on a stalk; and, usually, two chelicerae (to grasp prey) and two pedipalps (used as sensors). It also bears the first pair of walking legs and a pair of appendages called ovigers, used for grooming and egg-carrying. A segmented trunk holds another three pairs of legs, and leads to a tiny abdomen at the rear. Because the abdomen is so small, the digestive and reproductive organs extend into the legs.

Antarctic forms Both the red sea spider (*Pycnogonum decalopoda*) and the smaller spider (*P. nymphon*) live under the sea ice.

BOTTOM-DWELLERS

Sea spiders live in every ocean, from warm, shallow waters to icy depths of 23,000 feet (7,000 m). Most are small, but some deep-sea species have a leg-span of up to 28 inches (70 cm). Although some sea spiders can swim, most are bottom-dwellers and feed on soft-bodied invertebrates such as sponges and corals.

To breed, a female sea spider releases her eggs into the water and the male covers them with sperm. The male then carries the fertilized eggs on his ovigers until they hatch.

Gracile sea spider
Nymphon gracile

Male can carry up to four egg masses from four different females on his ovigers

Littoral sea spider
Pycnogonum littorale

MYRIAPODS

PHYLUM	Arthropoda
SUBPHYLUM	Myriapoda
CLASSES	4
ORDERS	20
FAMILIES	140
SPECIES	13,500

Centipedes (class Chilopoda), millipedes (class Diplopoda), symphylids (class Symphyla), and pauropods (class Pauropoda) are all myriapods with long, segmented bodies; simple eyes; a pair of jointed antennae; and numerous pairs of legs. Most centipedes are carnivores and hunt through leaf litter for other small invertebrates. They paralyze prey with a pair of venomous fangs on the underside of the head that can deliver a nasty bite even to humans. Millipedes are plant-feeders and do not bite, but many react to danger by curling into a coil and emitting a toxic substance. Resembling tiny centipedes, symphylids and pauropods live in leaf litter and soil and feed on decaying plant matter.

Tropical species Centipedes are most diverse in tropical forests. Their many legs can be short and hooked, or long and slender. The final pair may be antenna-like (as in the *Scutigera* sp. above) or pincer-like.

MANY LEGS

Centipedes have flattened, worm-like bodies, with one pair of legs attached to each body segment except the last. They have at least 15 and as many as 191 pairs of legs. The largest centipede, *Scolopendra gigantea* of the American tropics, can grow to 11 inches (28 cm) long. It is strong enough to prey on mice, frogs, and other small vertebrates.

A millipede's body is rounded and made up of doubled segments known as diplosomites, most of which bear double pairs of legs. Millipedes range from 1/12 inch to 11 inches (2 mm to 28 cm) long, and have up to 200 pairs of legs.

Measuring no more than 1/2 inch (1 cm) in length, symphylids have 12 pairs of legs. Pauropods are even smaller, at less than 1/12 inch (2 mm) long, and bear nine pairs of legs.

Most myriapods are nocturnal, live in moist forests, and tend to be hidden in leaf litter or soil or under rocks or logs. Some species, however, are found in grasslands or deserts. Millipedes usually lay their eggs in a soil nest, while some centipedes curl around their eggs and young to protect them.

Coiled defense A forest log millipede (*Narceus americanus*) has curled into the typical millipede defensive posture that leaves only the tough, calcareous plates of its exoskeleton exposed. It may also squirt out a pungent, irritating liquid.

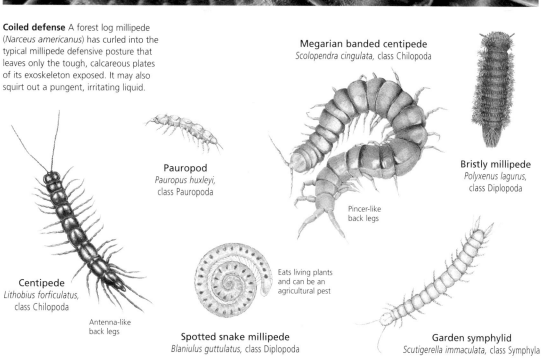

Centipede
Lithobius forficulatus,
class Chilopoda

Antenna-like
back legs

Pauropod
Pauropus huxleyi,
class Pauropoda

Megarian banded centipede
Scolopendra cingulata, class Chilopoda

Pincer-like
back legs

Eats living plants
and can be an
agricultural pest

Spotted snake millipede
Blaniulus guttulatus, class Diplopoda

Bristly millipede
Polyxenus lagurus,
class Diplopoda

Garden symphylid
Scutigerella immaculata, class Symphyla

CRUSTACEANS

PHYLUM	Arthropoda
SUBPHYLUM	Crustacea
CLASSES	11
ORDERS	37
FAMILIES	540
SPECIES	42,000

From water fleas less than 1/100 inch (0.25 mm) long, to giant spider crabs with legs spanning 12 feet (3.7 m), crustaceans are an extraordinarily diverse group. Although some species have adopted a terrestrial lifestyle, it is in aquatic environments that crustaceans have flourished, gradually evolving to exploit every marine and freshwater niche. Free-swimming planktonic species, such as krill, form the basis of vast aquatic food webs. Crabs and other bottom-dwellers may burrow into or crawl over the sediment. Barnacles remain cemented in place, filtering food from the water. There are also parasitic crustaceans, some of which exist in their adult form as a collection of cells inside the host.

LAND INVADERS

While crustaceans are abundant in water, only a minority have adapted to life on land. These tend to live in damp places and some return to water to breed. Sometimes known as pill bugs or sow bugs, the wood lice of family Oniscoidea are the most terrestrial crustaceans and even include a few desert species. Some wood lice curl into a tight ball when threatened (below). The limited exploitation by crustaceans of land habitats can be attributed to the lack of a waxy, water-tight cuticle and the reliance on gills for breathing.

Attractive claw In many crustaceans, the first pair of limbs has been modified to form claws known as chelipeds. As a male fiddler crab (*Uca* sp.) (above) matures, the right cheliped grows disproportionately, until it accounts for 65 percent of the crab's total weight. The crab waves this huge claw to attract mates and intimidate rival males. It may also use it to make sounds that entice females into its burrow.

CRUSTACEAN ANATOMY

Like other arthropods, crustaceans have a segmented body, jointed legs, and an exoskeleton, and grow by molting. The exoskeleton can be thin and flexible, as in water fleas, or rigid and calcified, as in crabs. The body usually has three regions—head, thorax, and abdomen—but in many of the larger species, the head and thorax form a cephalothorax, which is protected by a shield-like carapace. The abdomen often has a tail-like extension called a telson.

A typical crustacean's head holds two pairs of antennae; a pair of compound eyes, often on stalks; and three pairs of biting mouthparts. The thorax and sometimes the abdomen carry the limbs, each of which usually has two branches. In many species, particular limbs have become specialized for walking, swimming, food collection, or defense. Crabs, for example, have modified their first pair into claws called chelipeds.

While some female crustaceans lay eggs in water, many brood eggs on their body. The eggs of some species hatch into free-swimming nauplius larvae, with two pairs of antennae, a pair of mandibles, and a single simple eye. In most species, the eggs hatch at a more advanced stage or even as miniature adults.

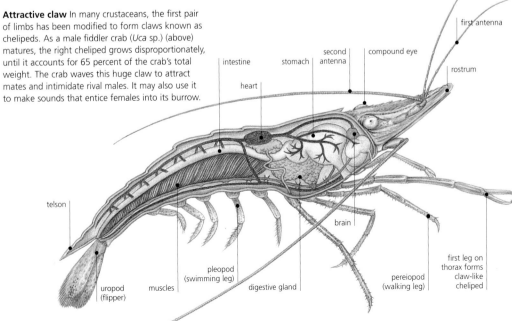

Shrimp design Like most crustaceans, shrimps have two pairs of antennae, the second of which is highly mobile; a pair of compound eyes, each with as many as 30,000 lenses; and specialized limbs. A shrimp's blood is pumped by a heart, and it breathes through gills attached to its legs.

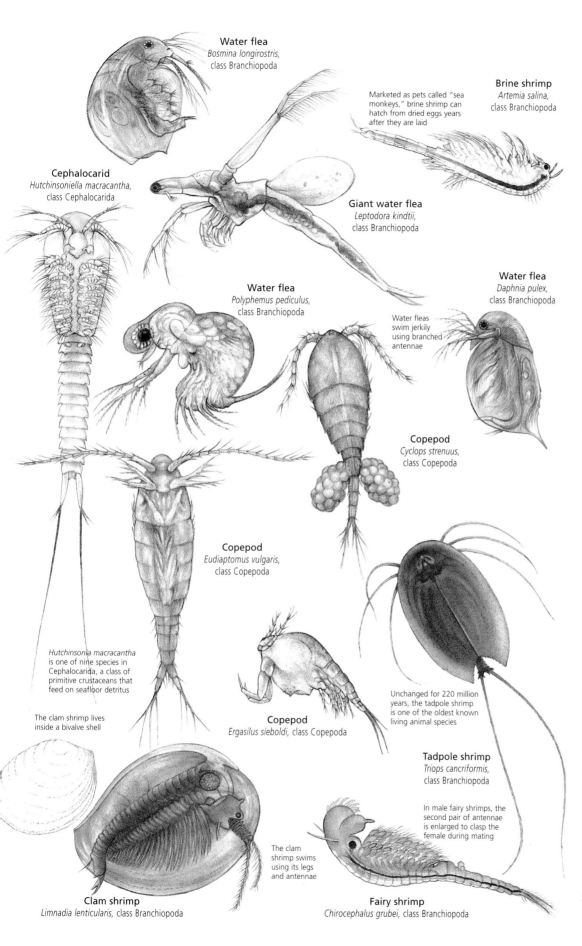

Water flea
Bosmina longirostris,
class Branchiopoda

Cephalocarid
Hutchinsoniella macracantha,
class Cephalocarida

Marketed as pets called "sea
monkeys," brine shrimp can
hatch from dried eggs years
after they are laid

Brine shrimp
Artemia salina,
class Branchiopoda

Giant water flea
Leptodora kindtii,
class Branchiopoda

Water flea
Polyphemus pediculus,
class Branchiopoda

Water flea
Daphnia pulex,
class Branchiopoda

Water fleas
swim jerkily
using branched
antennae

Copepod
Cyclops strenuus,
class Copepoda

Copepod
Eudiaptomus vulgaris,
class Copepoda

Hutchinsonia macracantha
is one of nine species in
Cephalocarida, a class of
primitive crustaceans that
feed on seafloor detritus

The clam shrimp lives
inside a bivalve shell

Copepod
Ergasilus sieboldi, class Copepoda

Unchanged for 220 million
years, the tadpole shrimp
is one of the oldest known
living animal species

Tadpole shrimp
Triops cancriformis,
class Branchiopoda

In male fairy shrimps, the
second pair of antennae
is enlarged to clasp the
female during mating

The clam
shrimp swims
using its legs
and antennae

Clam shrimp
Limnadia lenticularis, class Branchiopoda

Fairy shrimp
Chirocephalus grubei, class Branchiopoda

FACT FILE

Class Branchiopoda The tadpole
shrimps, water fleas, clam shrimps,
and fairy or brine shrimps in this class
are all small, ranging from 1/100 inch to
4 inches (0.25 mm to 10 cm) in length.
They have leaf-like legs, which they use
to gather food, swim, and breathe.
Found in almost every freshwater body
in the world, branchiopods include
species that live in temporary pools and
survive dry spells as dormant eggs.

Species 800

Worldwide; mainly
fresh water

Backstroker
Fairy shrimps such as Branchipus stagnalis
*swim on their back, using their legs to filter
organic particles from the water.*

Class Copepoda Found in vast
numbers in the oceans, copepods are
a crucial link in the marine food web.
The body tends to be cylindrical. They
lack compound eyes,
but have a single
simple eye retained
from the nauplius larval
stage. Some species are
parasitic on fish and
other aquatic animals.

Species 8,500

Worldwide; mainly marine

Forked tail
Canthocamptus staphylinus
*displays the branched
telson (tail-like projection)
typical of copepods.*

RECENT DISCOVERY

When *Speleonectes lucayensis* was
discovered in a sea cave in the Bahamas
in 1981, it prompted the description of
a new class of crustacean, Remipedia.
Since then, several more species, all
found in Caribbean or Australian caves
connected to the sea, have been added
to the class. These primitive forms
have long, worm-like bodies with up
to 32 segments, each bearing a pair
of legs. They are carnivorous.

Swimming blind Like other members
of Remipedia, Speleonectes lucayensis
*lacks eyes. It swims on its back using
its many legs as oars.*

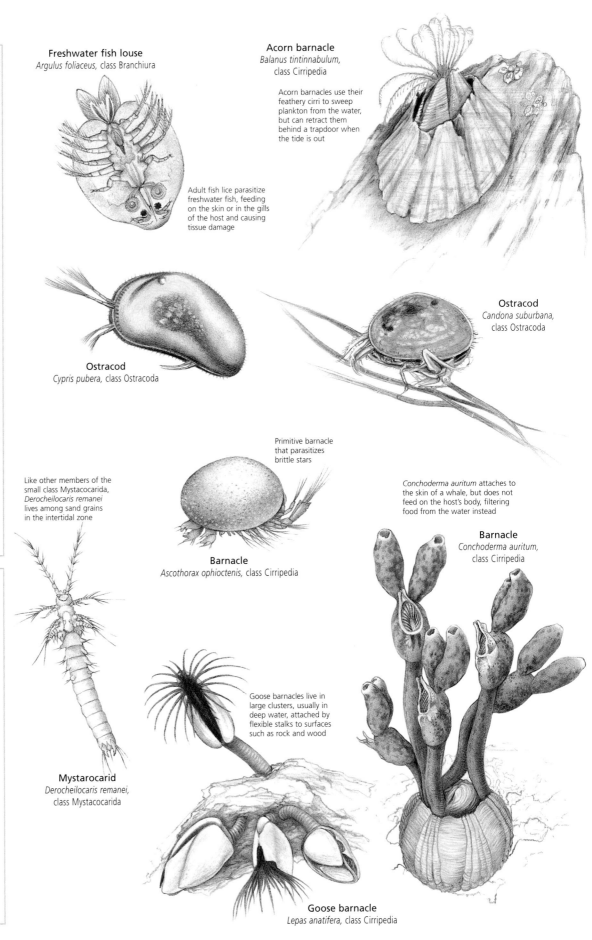

FACT FILE

Class Cirripedia The only sessile group of crustaceans, the barnacles that make up the class Cirripedia are entirely marine. Barnacles were believed to be mollusks until 1830, when it was discovered that they hatch from eggs as free-swimming nauplius larvae and therefore are crustaceans. The larvae mature and cement themselves head-down to rocks or boats or to hosts such as fish, turtles, and whales. Most adult barnacles are protected by calcareous plates and collect food particles from the water with their long, feathery legs (cirri). Barnacles are hermaphrodites. They live in tightly packed communities and can internally fertilize one another.

Species 900

Worldwide; in marine waters

Class Ostracoda Known as mussel shrimps or seed shrimps, the members of this class are distinguished by their carapace, which takes the form of a hinged, bivalve shell. Only the antennae and end bristles of the legs emerge from the shell.

Species 6,000

Worldwide; in all types of water bodies

Self-reproducing
The ostracod Ilyocypris gibba *can reproduce sexually but most often uses parthogenesis, with larvae developing from unfertilized eggs.*

TOTAL INVASION

The parasite *Sacculina carcini* can only be recognized as a barnacle by its free-swimming nauplius larvae. When a female larva has matured, it invades a host, usually the green crab (*Carcinus maenas*). The larva metamorphoses into a needle-like form that injects cells into the host. The cells form a root-like system, called an interna, throughout the crab's body. The interna eventually develops a reproductive body, an externa, outside the crab. Male larvae attach to the externa, into which they release cells that develop into sperm. Fertilization results in nauplius larvae.

Freshwater fish louse
Argulus foliaceus, class Branchiura

Adult fish lice parasitize freshwater fish, feeding on the skin or in the gills of the host and causing tissue damage

Acorn barnacle
Balanus tintinnabulum, class Cirripedia

Acorn barnacles use their feathery cirri to sweep plankton from the water, but can retract them behind a trapdoor when the tide is out

Ostracod
Cypris pubera, class Ostracoda

Ostracod
Candona suburbana, class Ostracoda

Like other members of the small class Mystacocarida, *Derocheilocaris remanei* lives among sand grains in the intertidal zone

Primitive barnacle that parasitizes brittle stars

Conchoderma auritum attaches to the skin of a whale, but does not feed on the host's body, filtering food from the water instead

Barnacle
Ascothorax ophioctenis, class Cirripedia

Barnacle
Conchoderma auritum, class Cirripedia

Goose barnacles live in large clusters, usually in deep water, attached by flexible stalks to surfaces such as rock and wood

Mystarocarid
Derocheilocaris remanei, class Mystacocarida

Goose barnacle
Lepas anatifera, class Cirripedia

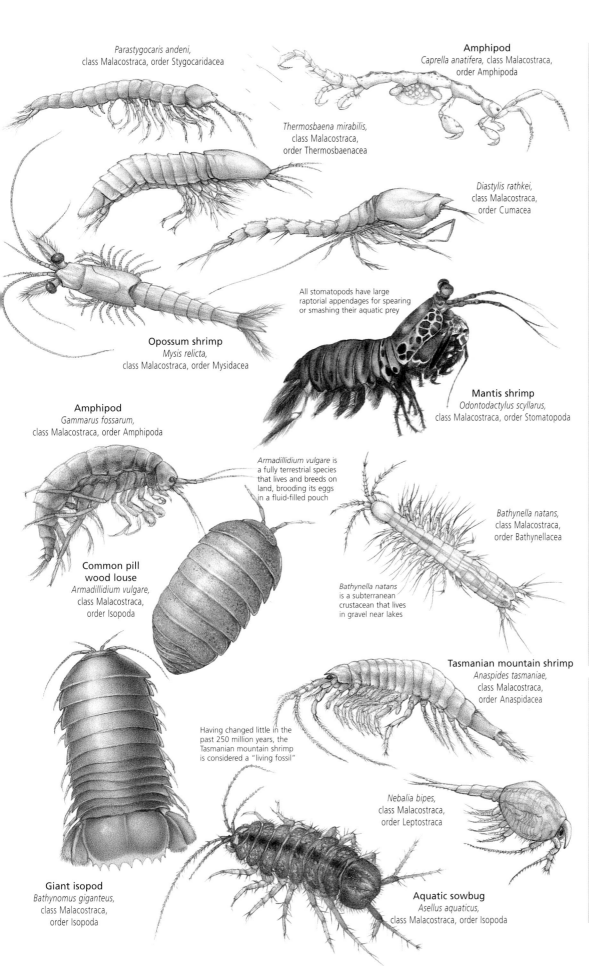

Parastygocaris andeni,
class Malacostraca, order Stygocaridacea

Thermosbaena mirabilis,
class Malacostraca,
order Thermosbaenacea

Amphipod
Caprella anatifera, class Malacostraca,
order Amphipoda

Diastylis rathkei,
class Malacostraca,
order Cumacea

Opossum shrimp
Mysis relicta,
class Malacostraca, order Mysidacea

All stomatopods have large
raptorial appendages for spearing
or smashing their aquatic prey

Mantis shrimp
Odontodactylus scyllarus,
class Malacostraca, order Stomatopoda

Amphipod
Gammarus fossarum,
class Malacostraca, order Amphipoda

Armadillidium vulgare is
a fully terrestrial species
that lives and breeds on
land, brooding its eggs
in a fluid-filled pouch

Bathynella natans,
class Malacostraca,
order Bathynellacea

Common pill
wood louse
Armadillidium vulgare,
class Malacostraca,
order Isopoda

Bathynella natans
is a subterranean
crustacean that lives
in gravel near lakes

Tasmanian mountain shrimp
Anaspides tasmaniae,
class Malacostraca,
order Anaspidacea

Having changed little in the
past 250 million years, the
Tasmanian mountain shrimp
is considered a "living fossil"

Nebalia bipes,
class Malacostraca,
order Leptostraca

Giant isopod
Bathynomus giganteus,
class Malacostraca,
order Isopoda

Aquatic sowbug
Asellus aquaticus,
class Malacostraca, order Isopoda

FACT FILE

Class Malacostraca By far the largest
class of crustaceans, Malacostraca
contains 13 orders. Almost all species
have six head segments, eight thoracic
segments, and six abdominal segments.
Except for the first head segment, each
segment usually bears two appendages.

Species 25,000

Worldwide; mainly aquatic

*Familiar crustaceans The
class Malacostraca includes
Daum's reef lobster (Enoplometopus
daumi), which joins other lobsters,
crabs, and shrimps in the order Decapoda.*

Order Amphipoda These small,
widespread members of the class
Malacostraca are found in marine,
freshwater, and moist terrestrial
habitats. An amphipod's
body is usually flattened
laterally, often creating
a shrimp-like appearance.

Species 6,000

Worldwide; mainly aquatic

*Varied salinities The amphipod Corophium
volutator lives in U-shaped burrows in
saltwater, brackish, and freshwater habitats.*

Order Isopoda Like amphipods,
isopods belong to Malacostraca. Their
bodies are flattened dorsoventrally
(from the back to the belly). They
include the truly terrestrial wood lice,
but most species crawl along the
bottom of aquatic habitats.

Species 4,000

Worldwide; mainly marine

*Aquatic isopod Like amphipods,
isopods such as Astacilla pusilla
lack a carapace. Most are
aquatic and can crawl and
swim. They tend to be omnivorous scavengers.*

PREDACIOUS EYES

The Stomatopoda is a highly specialized
order in the class Malacostraca. Known
as mantis shrimp, its members actively
prey on fish, crabs, and mollusks. Their
success as predators depends not only
on their raptorial appendages, but also
on their complex compound eyes.

*Stomatopod vision Each compound eye is
divided by a band that provides color vision
and also sees polarized light (contrast). The
rest of the eye provides monochromatic
vision and depth perception (perspective).*

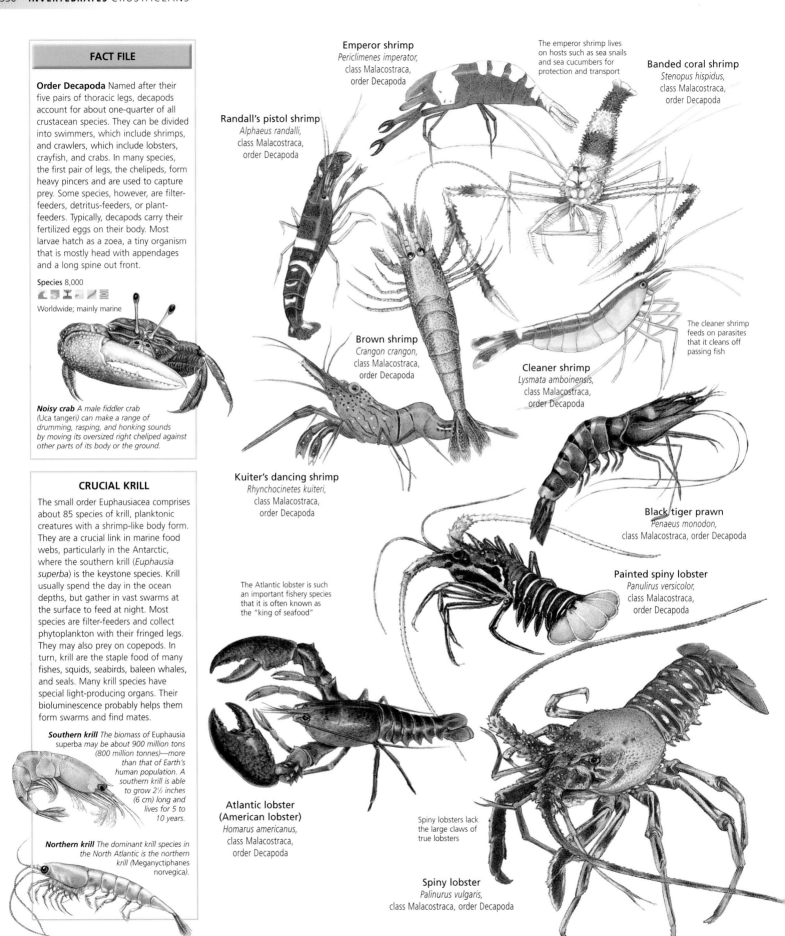

FACT FILE

Order Decapoda Named after their five pairs of thoracic legs, decapods account for about one-quarter of all crustacean species. They can be divided into swimmers, which include shrimps, and crawlers, which include lobsters, crayfish, and crabs. In many species, the first pair of legs, the chelipeds, form heavy pincers and are used to capture prey. Some species, however, are filter-feeders, detritus-feeders, or plant-feeders. Typically, decapods carry their fertilized eggs on their body. Most larvae hatch as a zoea, a tiny organism that is mostly head with appendages and a long spine out front.

Species 8,000

Worldwide; mainly marine

Noisy crab A male fiddler crab (Uca tangeri) can make a range of drumming, rasping, and honking sounds by moving its oversized right cheliped against other parts of its body or the ground.

CRUCIAL KRILL

The small order Euphausiacea comprises about 85 species of krill, planktonic creatures with a shrimp-like body form. They are a crucial link in marine food webs, particularly in the Antarctic, where the southern krill (Euphausia superba) is the keystone species. Krill usually spend the day in the ocean depths, but gather in vast swarms at the surface to feed at night. Most species are filter-feeders and collect phytoplankton with their fringed legs. They may also prey on copepods. In turn, krill are the staple food of many fishes, squids, seabirds, baleen whales, and seals. Many krill species have special light-producing organs. Their bioluminescence probably helps them form swarms and find mates.

Southern krill The biomass of Euphausia superba may be about 900 million tons (800 million tonnes)—more than that of Earth's human population. A southern krill is able to grow 2½ inches (6 cm) long and lives for 5 to 10 years.

Northern krill The dominant krill species in the North Atlantic is the northern krill (Meganyctiphanes norvegica).

Emperor shrimp
Periclimenes imperator,
class Malacostraca,
order Decapoda

The emperor shrimp lives on hosts such as sea snails and sea cucumbers for protection and transport

Banded coral shrimp
Stenopus hispidus,
class Malacostraca,
order Decapoda

Randall's pistol shrimp
Alphaeus randalli,
class Malacostraca,
order Decapoda

Brown shrimp
Crangon crangon,
class Malacostraca,
order Decapoda

The cleaner shrimp feeds on parasites that it cleans off passing fish

Cleaner shrimp
Lysmata amboinensis,
class Malacostraca,
order Decapoda

Kuiter's dancing shrimp
Rhynchocinetes kuiteri,
class Malacostraca,
order Decapoda

Black tiger prawn
Penaeus monodon,
class Malacostraca, order Decapoda

The Atlantic lobster is such an important fishery species that it is often known as the "king of seafood"

Painted spiny lobster
Panulirus versicolor,
class Malacostraca,
order Decapoda

Atlantic lobster (American lobster)
Homarus americanus,
class Malacostraca,
order Decapoda

Spiny lobsters lack the large claws of true lobsters

Spiny lobster
Palinurus vulgaris,
class Malacostraca, order Decapoda

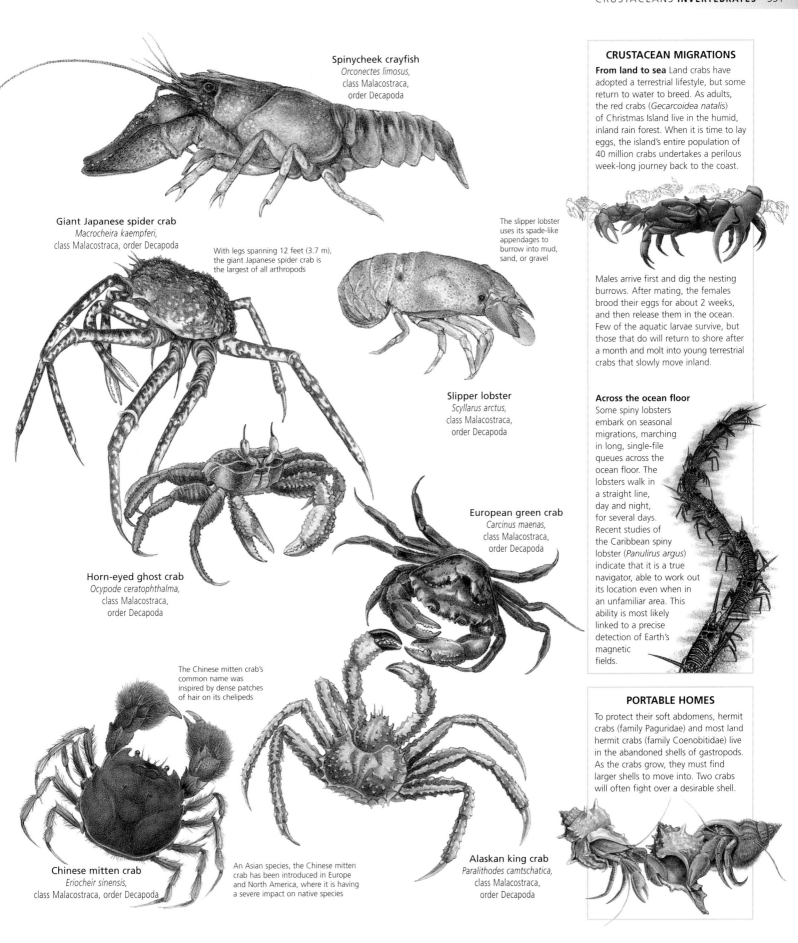

Spinycheek crayfish
Orconectes limosus,
class Malacostraca,
order Decapoda

Giant Japanese spider crab
Macrocheira kaempferi,
class Malacostraca, order Decapoda

With legs spanning 12 feet (3.7 m),
the giant Japanese spider crab is
the largest of all arthropods

The slipper lobster
uses its spade-like
appendages to
burrow into mud,
sand, or gravel

Slipper lobster
Scyllarus arctus,
class Malacostraca,
order Decapoda

Horn-eyed ghost crab
Ocypode ceratophthalma,
class Malacostraca,
order Decapoda

European green crab
Carcinus maenas,
class Malacostraca,
order Decapoda

The Chinese mitten crab's
common name was
inspired by dense patches
of hair on its chelipeds

Chinese mitten crab
Eriocheir sinensis,
class Malacostraca, order Decapoda

An Asian species, the Chinese mitten
crab has been introduced in Europe
and North America, where it is having
a severe impact on native species

Alaskan king crab
Paralithodes camtschatica,
class Malacostraca,
order Decapoda

CRUSTACEAN MIGRATIONS

From land to sea Land crabs have adopted a terrestrial lifestyle, but some return to water to breed. As adults, the red crabs (*Gecarcoidea natalis*) of Christmas Island live in the humid, inland rain forest. When it is time to lay eggs, the island's entire population of 40 million crabs undertakes a perilous week-long journey back to the coast.

Males arrive first and dig the nesting burrows. After mating, the females brood their eggs for about 2 weeks, and then release them in the ocean. Few of the aquatic larvae survive, but those that do will return to shore after a month and molt into young terrestrial crabs that slowly move inland.

Across the ocean floor
Some spiny lobsters embark on seasonal migrations, marching in long, single-file queues across the ocean floor. The lobsters walk in a straight line, day and night, for several days. Recent studies of the Caribbean spiny lobster (*Panulirus argus*) indicate that it is a true navigator, able to work out its location even when in an unfamiliar area. This ability is most likely linked to a precise detection of Earth's magnetic fields.

PORTABLE HOMES

To protect their soft abdomens, hermit crabs (family Paguridae) and most land hermit crabs (family Coenobitidae) live in the abandoned shells of gastropods. As the crabs grow, they must find larger shells to move into. Two crabs will often fight over a desirable shell.

INSECTS

PHYLUM	Arthropoda
SUBPHYLUM	Hexapoda
CLASS	Insecta
ORDERS	29
FAMILIES	949
SPECIES	>1 million

By most measures, insects are the most successful animals ever to have lived on Earth. With about a million described species, they account for more than half of all known animal species, and many more insects are still to be discovered, with estimates of the total number of insect species ranging from 2 million to 30 million. In terms of sheer numbers, insects overwhelm all other animal forms: some scientists believe that ants and termites alone make up 20 percent of the world's animal biomass. Some form of insect has managed to colonize the hottest deserts and the coldest polar zones, as well as virtually every land and freshwater habitat in between, and a handful live in the sea.

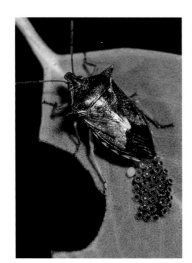

Parental guard Most insects display little parental care, depositing large numbers of eggs and then leaving them to hatch by themselves. Some shield bugs, however, will guard their eggs and then stay with the nymphs until at least their first molt.

Pollinators Pollen sticks to a honeybee as it drinks nectar from a flower. When the bee visits the next flower, some pollen will be dislodged. Most flowering plants rely on insects for pollination and have evolved flowers with different shapes, colors, and odors to attract particular species and force them to collect pollen as they feed.

Extreme habitats Brine flies (family Ephydridae) are found along the shores of California's salty Mono Lake, where their larvae feed on algae. Other insects have been found living in pools of crude oil, in hot springs, and in the ice of Antarctica.

SUCCESS STORY

Insects owe their incredible success to various aspects of their biology. Their tough, flexible exoskeleton provides protection without overly restricting movement. It is covered in a waxy coating that minimizes moisture loss, enabling insects to survive in dry conditions.

As the first creatures and the only invertebrates ever to develop powered flight, insects were able to find both food and mates, evade their predators, and colonize new areas with great efficiency. In most species, the wings can be folded over the body when at rest, so that the insect can exploit confined spaces inside bark, dung, leaf litter, or soil, for example.

Insects breathe through spiracles, small openings in the sides of their body that can be closed to prevent moisture loss. Rather than being transported around the body in the blood, oxygen is distributed directly into the body tissues by a series of tiny pipes known as tracheae. This method of gas diffusion works well only over short distances, which may be the main reason insects have remained small. Most insects are no more than an inch or two (a few centimeters) long. Their small size allowed insects to take advantage of a wide range of microhabitats, a major factor in their great diversity.

Sensory organs are scattered over an insect's body. The head usually has two compound eyes as well as three simple eyes, or ocelli. The two antennae can detect scent, taste, touch, and sound. Hearing organs

Spontaneous young Scarab beetles (family Scarabaeidae) use their sturdy front legs to roll dung into a large ball that may be taken underground as food, or used by the female as a nest for her eggs. Possibly because the young seemed to suddenly erupt from the dung, scarab beetles were revered by the ancient Egyptians.

INSECT FLIGHT

While some insects are wingless, the vast majority are winged as adults. Beetles and grasshoppers fly with a slow wing beat of 4–20 beats per second (bps) and limited maneuverability. Bees and true flies, with wing beats of about 190 bps, can hover and dart. Some midges have wing beats of 1,000 bps. The fastest sustained fliers are dragonflies, which can maintain speeds of 31 miles per hour (50 km/h).

Folding wings
The hindwings of a mantis fold up like fans when not in use. This protects the wings and allows the mantis to squeeze into tight spaces.

Purpose-built mouthparts The highly varied feeding habits of insects are reflected by the shape of their mouthparts. A moth (above) has a long, coiled tube, or proboscis, that can be unfurled to collect nectar. Mosquitoes have beaks that can pierce the skin of prey and suck up the body fluids, while aphids have piercing mouthparts for gathering plant juices. The mandibles, or jaws, of carnivorous insects such as ground beetles are sharp for cutting, while those of herbivores such as grasshoppers are adapted for grinding.

Insect anatomy While insects display a staggering diversity of shapes and forms, they follow the same basic body plan. Like other arthropods, insects have segmented bodies made up of a head, thorax, and abdomen; segmented limbs; and a tough exoskeleton. They are distinguished by having three pairs of legs and, usually, two sets of wings, a single pair of antennae, and a pair of compound eyes.

three ocelli (simple eyes) detect light variations

one pair of compound eyes, each made up of 100–30,000 lenses

two pairs of wings supported by thickened veins

head carries eyes, a pair of antennae, and mouthparts

Stabilizers
True flies appear to have only one pair of wings. In fact, the second pair has been reduced to knob-like structures called halteres that provide stability in flight.

abdomen contains digestive and reproductive organs

Plumed wings
In thrips and plume moths, the wings look like tiny feathers and are made up of fine hairs supported by a midrib.

antennae help insect smell, taste, feel, and hear

thorax carries wings and legs

three pairs of segmented legs

Hooked wings
Connected by tiny hooks, a wasp's forewings and hindwings move up and down in unison.

can also be on the body or legs. Other sensors detect changes in air pressure, humidity, and temperature.

The life-cycles of many insects allow for very rapid reproduction, enabling populations to respond quickly to favorable conditions and recover from catastrophes. Most insects use internal fertilization, but the female is able to store the male's sperm and use it over time. Some insects, such as aphids, can reproduce without fertilization to rapidly increase their numbers, but also use sexual reproduction to ensure genetic diversity. Insect eggs are protected by a shell-like membrane called the chorion that prevents them from drying out, another factor in the successful

colonization of dry habitats. While some insects resemble small versions of their parents when they hatch, most do not and must go through some form of metamorphosis. The juveniles and adults often occupy different ecological niches and eat different foods, which helps to avoid competition between them.

Because some species can cause human discomfort, spread disease, and destroy food stores and crops, insects are often regarded as pests. This ignores the fact that the vast majority of insects cause little harm and play a crucial role in the world's environments. About three-quarters of all flowering plants rely on insects for pollination, and many animals depend on insects for food.

Separate wings
A dragonfly's forewings and hindwings can beat in unison or independently. As in early flying insects, the wings do not fold back over the body.

Protective wings
In lady beetles and other beetles, the forewings have been modified to form protective cases known as elytra. The hindwings are used for flying.

Giant insect About the size of a small rat, New Zealand's giant weta (*Deinacrida heteracantha*) can weigh up to 2½ ounces (70 g), making it among the heaviest of all insect species. The smallest insect species is probably the parasitic wasp *Dicopomorpha echmepterygis*, at 1/200 inch (0.14 mm) long.

DRAGONFLIES AND DAMSELFLIES

PHYLUM	Arthropoda
SUBPHYLUM	Hexapoda
CLASS	Insecta
ORDER	Odonata
FAMILIES	30
SPECIES	5,500

The first members of the order Odonata emerged more than 300 million years ago—100 million years before dinosaurs appeared—and included the largest insect that has ever lived, a dragonfly with a wingspan of 28 inches (70 cm). Today's dragonflies and damselflies are much smaller, with wingspans from ¾ inch to 7½ inches (18 mm to 19 cm). Often seen flying near water, these voracious aerial predators are most abundant in the tropics, but can be found worldwide except for the polar zones. Damselflies have a fluttery flight and usually rest with the wings near the body, while dragonflies are strong, agile flyers and always rest with the wings held out from the body.

Sharp sight A dragonfly's large compound eyes may be made up of 30,000 lenses, providing superb vision and a wide field of view. Strong, biting mouthparts indicate the insect's carnivorous diet.

Odonate aerobatics With two pairs of large, veined wings that can beat together or independently, a dragonfly has great maneuverability and can hover and fly backward. Some species reach speeds of 20 miles per hour (30 km/h).

Mating position To mate, a male odonate grasps a female's head with claspers at the tip of his abdomen (below). The female will then bend her abdomen under the male to collect sperm from his sperm pouch.

AQUATIC BEGINNINGS

A dragonfly or damselfly spends most of its life as a wingless aquatic nymph. The nymph breathes via gills and feeds on other insect larvae, tadpoles, and small fishes. It has a specialized lower mouthpart known as a mask that usually rests under its head but can shoot out to snatch prey in its pincers. Over a period of a few weeks to 8 years, depending on the species, the nymph may molt up to 17 times. For the final molt, it climbs out of the water and sheds its nymphal skin to reveal an adult with prominent eyes, sharp mouthparts, two pairs of transparent wings, a sloping thorax, and a long, slender abdomen.

A newly emerged adult dragonfly or damselfly flies away from water and starts to feed, snatching flying insects on the wing. To find mates, dragonflies will gather along a body of water, and males may engage in aerial contests. After mating, the male usually guards the female as she lays her eggs in the water. Most adults live for only several weeks.

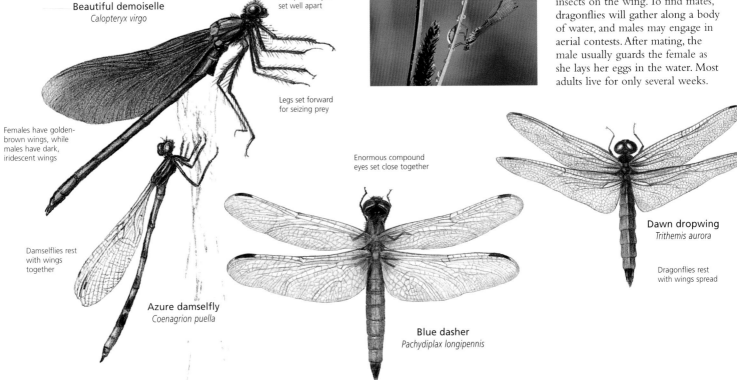

Beautiful demoiselle
Calopteryx virgo

Compound eyes set well apart

Legs set forward for seizing prey

Females have golden-brown wings, while males have dark, iridescent wings

Damselflies rest with wings together

Azure damselfly
Coenagrion puella

Enormous compound eyes set close together

Blue dasher
Pachydiplax longipennis

Dawn dropwing
Trithemis aurora

Dragonflies rest with wings spread

MANTIDS

PHYLUM	Arthropoda
SUBPHYLUM	Hexapoda
CLASS	Insecta
ORDER	Mantodea
FAMILIES	8
SPECIES	2,000

Waiting to ambush prey, a mantid will sit perfectly still, holding its large forelegs folded up before it—a posture that inspired the common name of praying mantis for some species. With lightning-quick reflexes, the raptorial forelegs will shoot out to snatch insect prey. Most mantids are medium-sized, measuring about 2 inches (5 cm) long, but some tropical species reach lengths of 10 inches (25 cm). These giant mantids may add small birds and reptiles to their primarily insectivorous diet. While the majority of mantids live in the tropics or subtropics, some are found in the warm temperate areas of southern Europe, North America, South Africa, and Australia.

Precision hunting With its forward-facing eyes supplying binocular vision, a praying mantis has used its spiky, hooked forelimbs to impale a passing wasp. The mantis's strong mandibles quickly devour the meal.

Foaming protection A female lays her eggs in a gummy fluid that she whips into a foam. When the foam hardens, it forms a case, or ootheca, that protects the eggs.

Cannibalistic mating In some mantid species, the transfer of sperm is more rapid if the female bites off the male's head. The male's self-sacrifice makes ecological sense as he is providing nourishment that will help his offspring survive.

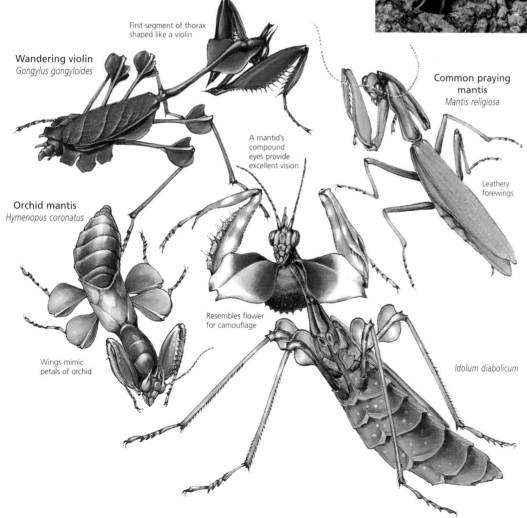

First segment of thorax shaped like a violin

Wandering violin
Gongylus gongyloides

A mantid's compound eyes provide excellent vision

Common praying mantis
Mantis religiosa

Leathery forewings

Orchid mantis
Hymenopus coronatus

Resembles flower for camouflage

Wings mimic petals of orchid

Idolum diabolicum

SILENT HUNTERS

Mantids are masters at hiding from both predators and prey. They are the only insects that can turn their head without moving other parts of the body, allowing them to silently observe potential victims. Most species also have cryptic coloration that helps them blend in with the grass, leaves, twigs, or flowers of their environment. An alarmed mantid will adopt a threat posture, raising and rustling its wings and rearing up to show off its vivid warning coloration. Mantids that are targeted by bats at night possess a single "ear" on their thorax that can detect a bat's ultrasonic signals.

In some species, and especially in captivity, the female mantid will devour the male during copulation. Females mate only once, but from that single mating, they can produce up to 20 ootheca, or egg cases, each containing from 30 to 300 eggs. The young emerge from the ootheca as highly active nymphs, looking like miniature, wingless versions of their parents. Ready to hunt, they may even eat each other. The survivors disperse. After a series of molts, the nymphs mature and develop wings and adult coloration.

COCKROACHES

PHYLUM	Arthropoda
SUBPHYLUM	Hexapoda
CLASS	Insecta
ORDER	Blattodea
FAMILIES	6
SPECIES	4,000

As scavenging insects willing to feed on almost any plant or animal product, including stored food, trash, paper, and clothing, some cockroaches thrive in human environments and are widely regarded as repulsive pests. In fact, less than 1 percent of all cockroach species are pests—the remainder perform an important ecological role by recycling leaf litter and animal excrement in forests and other habitats. While most cockroaches are found in warm tropical zones, about 25 species have spread worldwide, accidentally transported on ships. Cockroaches are among the most primitive of all living insects, with an anatomy that has changed little in more than 300 million years.

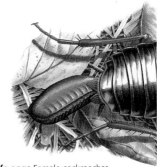

Safe eggs Female cockroaches deposit their eggs in an ootheca, or egg case. The nymphs emerge soft and white, but soon harden and turn brown. They eat the ootheca as their first food.

Maternal care Weighing up to 1¼ ounces (50 g), the giant burrowing cockroach (*Macropanesthia rhinoceros*) of Australia is the heaviest of all cockroaches. As many as 30 young are born alive and are fed by the mother in her burrow for up to 9 months.

SENSITIVE INSECTS

A range of sensors helps cockroaches to detect minute changes in their surroundings. Their long antennae can find minuscule amounts of food and moisture. Sensors on the legs and abdomen can pick up tiny air movements, prompting the insect to flee danger in a split second. With its oval, flattened shape, a cockroach can scuttle into tiny crevices. Not all cockroaches have wings, but in those that do the forewings are usually hardened and opaque, and the hindwings are translucent.

Mating between cockroaches may be initiated when a female emits a pheromone to attract males. Females usually lay 14–32 eggs in a hardened case known as an ootheca. The ootheca may then be left to hatch alone; carried at the base of the mother's abdomen; or incubated inside the mother's body so that she produces live young. Nymphs molt up to 13 times before emerging as adults. A cockroach usually lives for 2–4 years.

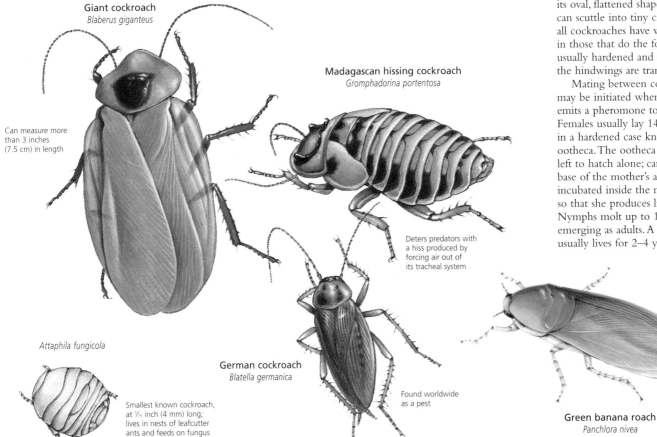

Giant cockroach
Blaberus giganteus

Can measure more than 3 inches (7.5 cm) in length

Madagascan hissing cockroach
Gromphadorina portentosa

Deters predators with a hiss produced by forcing air out of its tracheal system

Attaphila fungicola

Smallest known cockroach, at ³⁄₁₆ inch (4 mm) long; lives in nests of leafcutter ants and feeds on fungus

German cockroach
Blatella germanica

Found worldwide as a pest

Green banana roach
Panchlora nivea

TERMITES

PHYLUM Arthropoda	
SUBPHYLUM Hexapoda	
CLASS Insecta	
ORDER Isoptera	
FAMILIES 7	
SPECIES 2,750	

Termites may be the world's most monogamous animals. A king and queen mate for life, producing thousands or millions of offspring that operate together as a highly structured colony. These small insects are often referred to as white ants, but although their social system and anatomy resemble those of ants, they evolved independently and are most closely related to cockroaches. Found throughout much of the world, termites are most abundant in tropical rain forests, where there can be as many as 25,000 individuals per square mile (10,000 per sq km). By feeding on dead wood, they recycle nutrients in their natural habitats but can severely damage buildings in urban environments.

Wood-eaters With saw-toothed jaws that can shear through wood, termites cause costly damage to buildings. Infestations often involve introduced species that never found a niche in the natural environment.

Egg-laying machine A termite queen, shown here attended by other colony members, can grow 4½ inches (11 cm) long and produces 36,000 eggs a day from her enormous abdomen.

CASTE SOCIETY

A mature termite colony has three castes: reproductives, workers, and soldiers. The main reproductives are the king and queen, who produce all the colony's other members and can live for 25 years. There may also be secondary reproductives ready to take over if the king or queen dies.

Workers and soldiers can be male or female. They are wingless, usually lack eyes and mature reproductive organs, and may live for 5 years. The pale and soft-bodied workers feed all other colony members and maintain the nest. Soldiers defend the colony, usually using powerful mandibles. Termite nymphs can develop into any caste, depending on the needs of the colony.

At a particular time each year, a swarm of winged males and females will emerge from a termite nest and disperse. They then shed their wings and form pairs, becoming the kings and queens of new colonies.

Male reproductive of *Macrotermes natalensis*

Nasutitermes triodiae soldier shoots a sticky liquid out of its long snout to entangle enemies

Spinifex termite
Nasutitermes triodiae

Macrotermes natalensis soldier has an enlarged head with hooked mandibles for attacking invaders

Large fungus-growing termite
Macrotermes natalensis

INSIDE A NEST

Termite nests create a stable, humid microclimate sealed from the elements. While some termites nest inside wood, most construct subterranean nests that may rise partly above ground as mounds. Worker termites use saliva or feces to glue together particles of soil or wood, building a hard outer wall to protect the softer internal network of chambers.

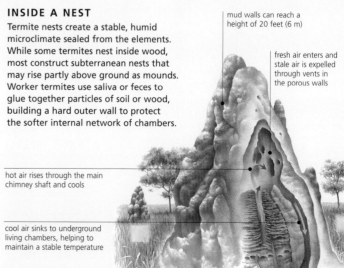

mud walls can reach a height of 20 feet (6 m)

fresh air enters and stale air is expelled through vents in the porous walls

hot air rises through the main chimney shaft and cools

cool air sinks to underground living chambers, helping to maintain a stable temperature

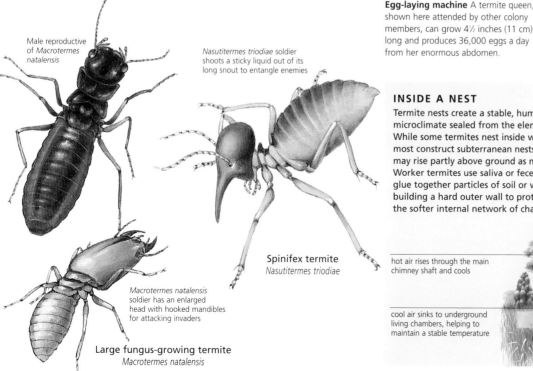

CRICKETS AND GRASSHOPPERS

PHYLUM	Arthropoda
SUBPHYLUM	Hexapoda
CLASS	Insecta
ORDER	Orthoptera
FAMILIES	28
SPECIES	> 20,000

Renowned for their songs and leaping ability, the members of the order Orthoptera include grasshoppers, locusts, crickets, katydids (or bush-crickets), and their kin. They are distinguished by elongated hindlegs that allow them to jump. Most species are winged, with slender, toughened forewings protecting the membranous, fan-shaped hindwings. While most orthopterans are ground-dwellers in grasslands and forests, some are arboreal, burrowing, or semiaquatic, and species can be found in deserts, caves, bogs and marshes, and seashores. Grasshoppers, groundhoppers, and a few katydids are herbivores, while most other orthopterans are omnivorous.

Threat display Colored to match the vegetation of its Amazon rain-forest home, a spiky-headed katydid (*Copiphora* sp.) is covered in sharp spines and will rear up in a threat display to deter predators.

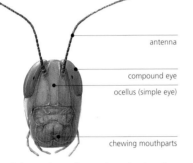

antenna
compound eye
ocellus (simple eye)
chewing mouthparts

Information collector An orthopteran's head features two fine, sensitive antennae, two large compound eyes, and mouthparts adapted for chewing vegetation.

Warning colors While many crickets and grasshoppers have cryptic coloration that helps them blend in with the background, others, such as this rain-forest grasshopper, are vividly colored as a warning to potential predators that they are unpalatable.

SINGING INSECTS

Most male orthopterans can make sounds by rubbing two parts of the body together, a technique known as stridulation. Crickets and long-horned grasshoppers move a scraper on one forewing along a row of teeth on the other forewing, while short-horned grasshoppers rub a ridged surface on the hindleg against a forewing. The sounds can be heard by tympanal organs located on the legs or abdomen.

There are three kinds of song: a calling song, to attract females from a distance; a courtship song, to entice a nearby female to mate; and a battle call, to deter rival males. The songs, some of which are too high-pitched to be heard by human ears, are unique to each species. Cricket songs tend to be affected by the weather, with the rate of chirps increasing with the temperature.

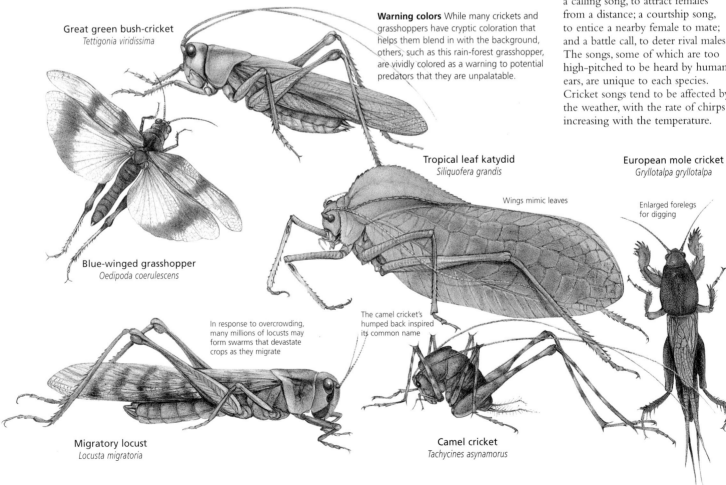

Great green bush-cricket
Tettigonia viridissima

Blue-winged grasshopper
Oedipoda coerulescens

Tropical leaf katydid
Siliquofera grandis

Wings mimic leaves

European mole cricket
Gryllotalpa gryllotalpa

Enlarged forelegs for digging

In response to overcrowding, many millions of locusts may form swarms that devastate crops as they migrate

The camel cricket's humped back inspired its common name

Migratory locust
Locusta migratoria

Camel cricket
Tachycines asynamorus

BUGS

PHYLUM	Arthropoda
SUBPHYLUM	Hexapoda
CLASS	Insecta
ORDER	Hemiptera
FAMILIES	134
SPECIES	> 80,000

Ranging from minute wingless aphids to giant frog-catching water bugs, the members of the order Hemiptera are extremely diverse. The name *Hemiptera* means "half wing," a reference to the fact that many, but not all, species have forewings that are leathery at the base and membranous at the tip. The one feature that all bugs have in common is their piercing and sucking mouthparts. With these, they pierce the surface of a plant or animal and inject saliva to start digesting the food, which they then suck into their mouth. Most bug species feed on plant sap, and some of these are significant agricultural pests. Other bugs suck the blood of vertebrates, or prey on other insects.

Malodorous defense The bright colors of many shield bugs serve to warn predators of the noxious odor produced when the insect is disturbed. Shield bugs are usually sap suckers, but a few are predatory.

Mimicry In thornbugs, or treehoppers, the first plate of the thorax forms a thorn-shaped spine. This camouflages the insect as it sucks sap, and, even if an enemy does spot the insect, the spine deters an attack.

Predator bugs While many bugs use their sucking mouthparts to feed on plant juices, others employ them to snare live prey. This assassin bug lies in wait, ready to ambush an unwitting insect victim.

DIFFERENT BUGS

Distributed throughout the world, bugs can be found in almost all terrestrial habitats. Some species have specialized for an aquatic life—these include sea skaters (*Halobates* sp.), the only insects on the open ocean. Bugs range from ⅟₂₅ inch to 4½ inches (1 mm to 11 cm) in length. They undergo incomplete metamorphosis, with nymphs that usually resemble small, wingless versions of the adults. In cicadas and some other species, however, the burrowing nymphs and the adults look quite different.

The order Hemiptera is divided into three suborders. In the true bugs (suborder Heteroptera), the forewings are toughened at the base and sit flat, concealing membranous hindwings. True bugs can flex their head and mouthparts forward, and many have stink glands. Cicadas and hoppers (Auchenorrhyncha) hold their uniform forewings over the abdomen like a tent. The head and mouthparts point down and back. Aphids, scale insects, mealy bugs, whiteflies, and their kin all belong to the suborder Sternorrhyncha. Most have soft bodies kept moist by a covering of wax or froth. Wings are often absent or reduced in adults.

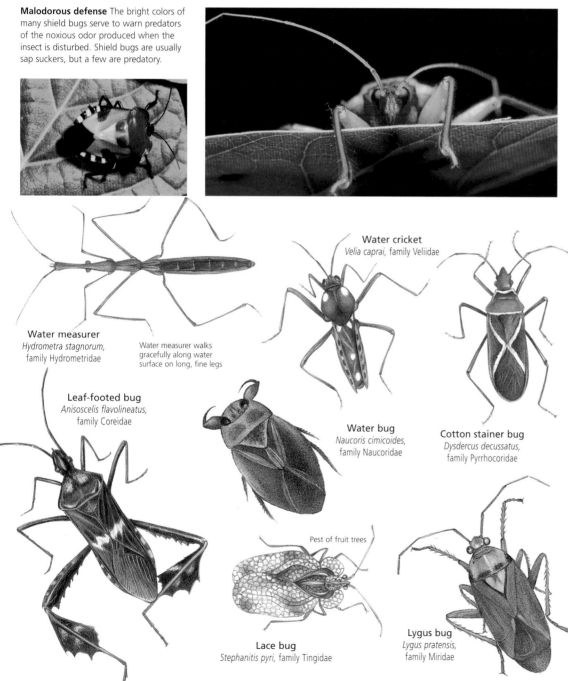

Water measurer
Hydrometra stagnorum,
family Hydrometridae

Water measurer walks gracefully along water surface on long, fine legs

Water cricket
Velia caprai, family Veliidae

Leaf-footed bug
Anisoscelis flavolineatus,
family Coreidae

Water bug
Naucoris cimicoides,
family Naucoridae

Cotton stainer bug
Dysdercus decussatus,
family Pyrrhocoridae

Pest of fruit trees

Lace bug
Stephanitis pyri, family Tingidae

Lygus bug
Lygus pratensis,
family Miridae

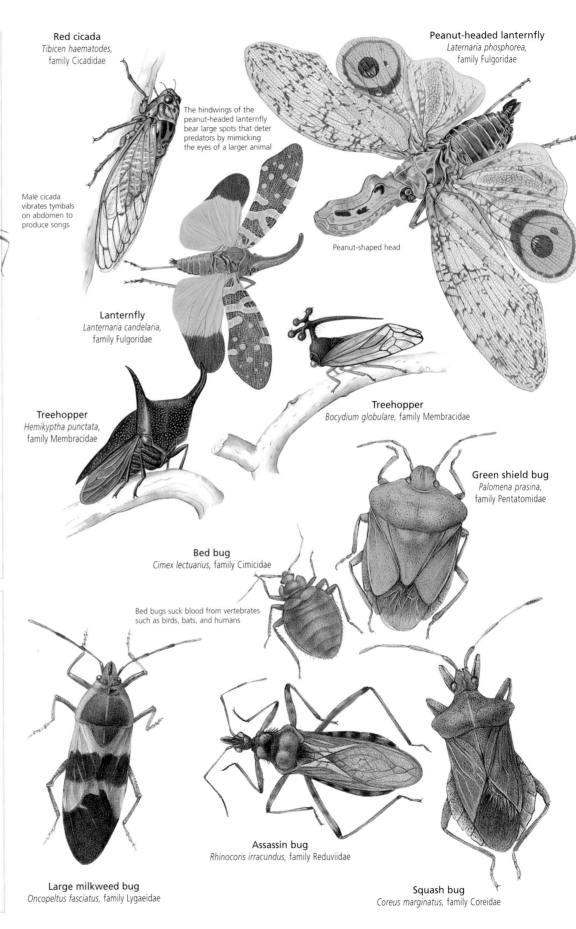

Red cicada
Tibicen haematodes,
family Cicadidae

The hindwings of the peanut-headed lanternfly bear large spots that deter predators by mimicking the eyes of a larger animal

Peanut-headed lanternfly
Laternaria phosphorea,
family Fulgoridae

Male cicada vibrates tymbals on abdomen to produce songs

Peanut-shaped head

Lanternfly
Laternaria candelaria,
family Fulgoridae

Treehopper
Bocydium globulare, family Membracidae

Treehopper
Hemikyptha punctata,
family Membracidae

Green shield bug
Palomena prasina,
family Pentatomidae

Bed bug
Cimex lectuarius, family Cimicidae

Bed bugs suck blood from vertebrates such as birds, bats, and humans

Assassin bug
Rhinocoris irracundus, family Reduviidae

Large milkweed bug
Oncopeltus fasciatus, family Lygaeidae

Squash bug
Coreus marginatus, family Coreidae

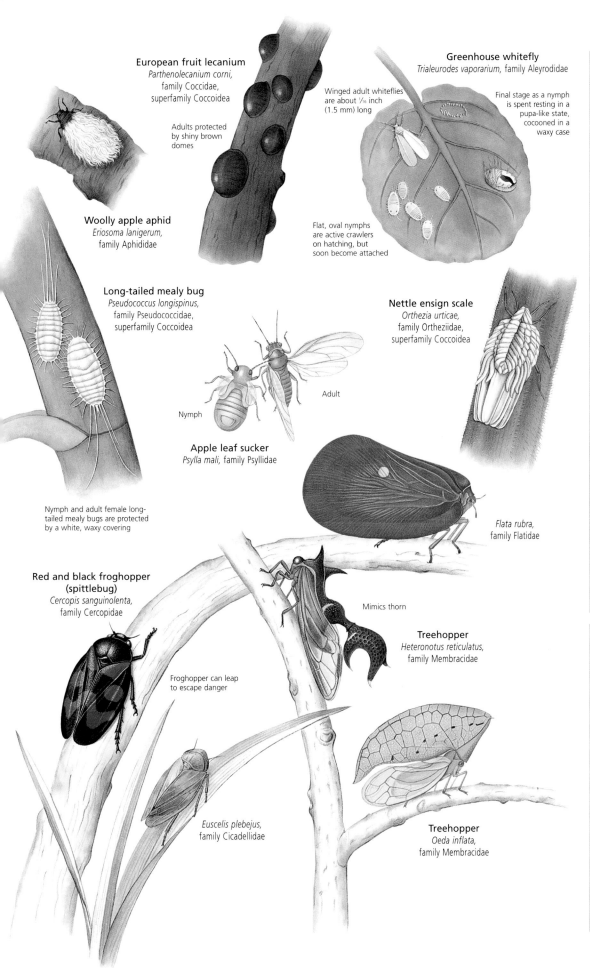

European fruit lecanium
Parthenolecanium corni,
family Coccidae,
superfamily Coccoidea

Adults protected
by shiny brown
domes

Woolly apple aphid
Eriosoma lanigerum,
family Aphididae

Greenhouse whitefly
Trialeurodes vaporarium, family Aleyrodidae

Winged adult whiteflies
are about 1/16 inch
(1.5 mm) long

Final stage as a nymph
is spent resting in a
pupa-like state,
cocooned in a
waxy case

Flat, oval nymphs
are active crawlers
on hatching, but
soon become attached

Long-tailed mealy bug
Pseudococcus longispinus,
family Pseudococcidae,
superfamily Coccoidea

Nettle ensign scale
Orthezia urticae,
family Ortheziidae,
superfamily Coccoidea

Nymph

Adult

Apple leaf sucker
Psylla mali, family Psyllidae

Nymph and adult female long-
tailed mealy bugs are protected
by a white, waxy covering

Flata rubra,
family Flatidae

**Red and black froghopper
(spittlebug)**
Cercopis sanguinolenta,
family Cercopidae

Mimics thorn

Treehopper
Heteronotus reticulatus,
family Membracidae

Froghopper can leap
to escape danger

Euscelis plebejus,
family Cicadellidae

Treehopper
Oeda inflata,
family Membracidae

FACT FILE

Family Aphididae Aphids are tiny,
soft-bodied plant-suckers that cause
great damage to crops. They are able
to rapidly increase their numbers by
producing wingless or winged females
from unfertilized eggs. In autumn, the
winged females produce males and
females that mate. The fertilized eggs
lay dormant through winter and hatch
in spring as females to begin
the cycle again.

Species 2,250

Worldwide; on plants

Cabbage lover
Native to Europe,
the cabbage aphid
(Bravicoryne brassicae)
has spread to many
parts of the world. It
feeds on both wild and
cultivated plants, causing
particular damage to cabbage crops.

Superfamily Coccoidea Spending
most of their life protected by a waxy
secretion, the scale insects collected in
this superfamily barely resemble insects
at all. Adult females have sack-like,
wingless bodies and often lack
legs and eyes. Adult males
usually have wings and a
more defined head, thorax,
and abdomen, but, lacking
mouthparts, they cannot
feed and are short-lived.

Species 7,300

Worldwide; on plants

Well protected
The oyster-shell scale
(Lepidosaphes ulmi) usually
overwinters as eggs.

APHIDS AS FOOD

Aphids provide food for many other
insects. They are prey for many lady
beetles (family Coccinellidae), hover
flies (family Syrphidae), and lacewings
(order Neuroptera). They also have a
happier relationship with some species
of ants, which protect them from
predators and the elements. In return,
the ants stroke, or "milk,"
the aphids to feed on
honeydew, the sweet
waste product
of the aphids'
plant-sap diet.

Aphid predator
In its lifespan of
1–2 months, a lady
beetle can consume
more than 2,000 aphids.

WATER INSECTS

Only about 3 percent, or 30,000, insect species are truly aquatic for at least part of their life-cycle. Water insects have had to adapt to the low oxygen content of water, as well as to the difficulty of moving about in still water or staying put in flowing water. Some insects, such as water striders, live only on the surface. Many, such as backswimmers, breathe air but carry it with them when they swim underwater. Others, such as dragonfly nymphs, spend all their time submerged, using gills to absorb oxygen from the water. The vast majority of aquatic insects live in fresh water, and only 300 or so species live in saltwater habitats, possibly because crustaceans evolved in the sea first and offer too much competition.

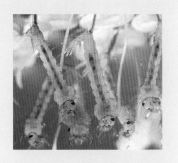

Spinning swimmers Whirligig beetles (family Gyrinidae) swim in small circles on the surface, using their flattened hindlegs like oars. They carry bubbles of air when they dive. Their eyes are split into two parts so they can see both above and below the water.

Bubble breather Predaceous diving beetles (family Dytiscidae) spend virtually their entire lives in water. The adults come to the surface to collect air, which they store in a bubble under their elytra (forewings). They swim using their hindlegs, which are fringed in thick hairs.

Aquatic locomotion Aquatic insects have evolved various strategies for moving around. Water scorpions (family Nepidae) can swim but often crawl along the bottom of a pond and spend much of their time hanging from pond weeds, waiting to ambush prey. They breathe air through a tube that sticks out above the surface. Water boatmen (family Corixidae) have long, hairy hindlegs that can "row" powerfully through the water. Water striders (family Gerridae) have fine hairs on their feet that allow them to make small jumps along the surface.

Snorkelers Mosquito larvae hang from the surface and filter plant matter from the water. While most aquatic larvae breathe through gills, mosquito larvae breathe air through a snorkel-like tube. This allows them to survive in pools of stagnant, oxygen-poor water.

Aquatic predators While many water insects are omnivores or plant-feeders, some, such as the predaceous diving beetles, are highly carnivorous. Even as larvae, diving beetles will tackle tadpoles, small fishes, and other prey larger than themselves. The adults use their powerful mandibles on substantial prey such as salamanders.

water scorpion hangs from pond weeds as it attacks a tadpole

water strider uses surface tension to "walk" on water

water boatman swims by using its legs as oars

BEETLES

<table>
<tr><td>PHYLUM</td><td>Arthropoda</td></tr>
<tr><td>SUBPHYLUM</td><td>Hexapoda</td></tr>
<tr><td>CLASS</td><td>Insecta</td></tr>
<tr><td>ORDER</td><td>Coleoptera</td></tr>
<tr><td>FAMILIES</td><td>166</td></tr>
<tr><td>SPECIES</td><td>> 370,000</td></tr>
</table>

When asked what his study of the natural world had revealed about its Creator, the scientist J. B. S. Haldane replied, "an inordinate fondness for beetles." Of all known animal species, about one in four is a beetle. Members of the order Coleoptera have colonized almost all of Earth's habitats, from Arctic tundra and exposed mountaintops to deserts, grasslands, woodlands, and lakes. They reach their greatest diversity, however, in the lush rain forests of the tropics. Most beetles are distinguished by their hardened, leathery forewings, known as elytra, which protect the membranous, flying hindwings and allow the insect to live in tight spaces under bark or in leaf litter.

Pollinating beetle Although considered pests by many gardeners, spotted cucumber beetles (*Diabrotica undecimpunctata*) and many other beetles that eat pollen and foliage play an important role in helping to pollinate the plants on which they feed.

Competing males The huge, branched mandibles of male stag beetles resemble a stag's antlers and perform a similar function, being used in competitions with rival males to win the right to mate. In some species, the mandibles are as long as the beetle.

DIVERSE FORMS

From lady beetles to diving beetles, scarab beetles to fireflies, the order Coleoptera encompasses enormous diversity. The feather-winged beetle (*Nanosella fungi*) is just ⅟₁₀₀ inch (0.25 mm) long, while the South American longhorn beetle (*Titanus giganteus*) can exceed 6½ inches (16 cm) in length. Adult beetles can be oval and flattened, long and slender, or squat and domed. While a few beetles are strong fliers, most are clumsy in flight, and some lack wings and cannot fly at all.

The mouthparts of beetles are all adapted for biting, but they have been put to many purposes. Plant-eating beetles may eat roots, stems, leaves, flowers, fruit, seeds, or wood. Carnivorous beetles usually attack invertebrate prey. A key ecological role is filled by scavenging beetles, which recycle dead animals and plants, excrement, and other waste.

Most beetles live on the ground, but some have specialized to live in trees, in water, and underground— a few even make their home in the nests of ants and termites.

Beetles develop by complete metamorphosis, with larvae looking markedly different from the parents and going through a non-feeding, pupal stage before becoming adults.

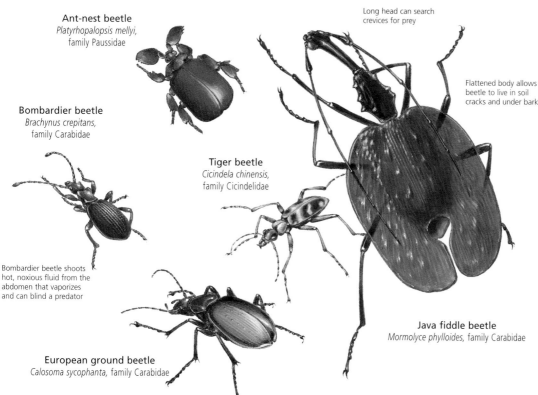

Ant-nest beetle *Platyrhopalopsis mellyi*, family Paussidae

Bombardier beetle *Brachynus crepitans*, family Carabidae

Long head can search crevices for prey

Flattened body allows beetle to live in soil cracks and under bark

Tiger beetle *Cicindela chinensis*, family Cicindelidae

Bombardier beetle shoots hot, noxious fluid from the abdomen that vaporizes and can blind a predator

Java fiddle beetle *Mormolyce phylloides*, family Carabidae

European ground beetle *Calosoma sycophanta*, family Carabidae

FACT FILE

Family Buprestidae With spectacular colors and a metallic sheen, the jewel beetles in this family include some of the world's most attractive insects. They have a long oval body and are good fliers. Many adults feed on nectar, while larvae tend to be wood-borers.

Species 15,000

Worldwide; on plants

Decorative wings
The iridescent wings of South America's metallic wood-boring beetle (Euchroma gigantea) are used in jewelry by local people.

Family Coccinellidae Also called ladybirds or ladybugs, the lady beetles of this family are among the best known of all insects. They are usually brightly colored and spotted, with a compact, rounded shape and short, clubbed antennae. Lady beetles are desired guests in gardens and farms. While a minority of species are plant-feeders, most lady beetles prey on aphids and other pests. A larva may eat 25 aphids a day, while an adult may eat as many as 50 aphids a day.

Species 5,200

Worldwide; on foliage

Lines and spots The convergent lady beetle (Hippodamia convergens) is named after the converging white lines on the first segment of its thorax. It usually has 13 spots.

Family Lampyridae This family is made up of fireflies. All firefly larvae glow, producing light via a chemical reaction while giving off little heat. Their luminescence may be a warning to predators of their unpleasant taste. The larvae themselves are predaceous, following slime trails to slugs and snails. Using organs on the abdomen, most adult fireflies also produce light, which they flash in a species-specific pattern to attract mates.

Species 2,000

Worldwide; on plants, in soil, and under rocks

Aggressive mimic Some female fireflies mimic the flash patterns of other species in order to lure males as prey.

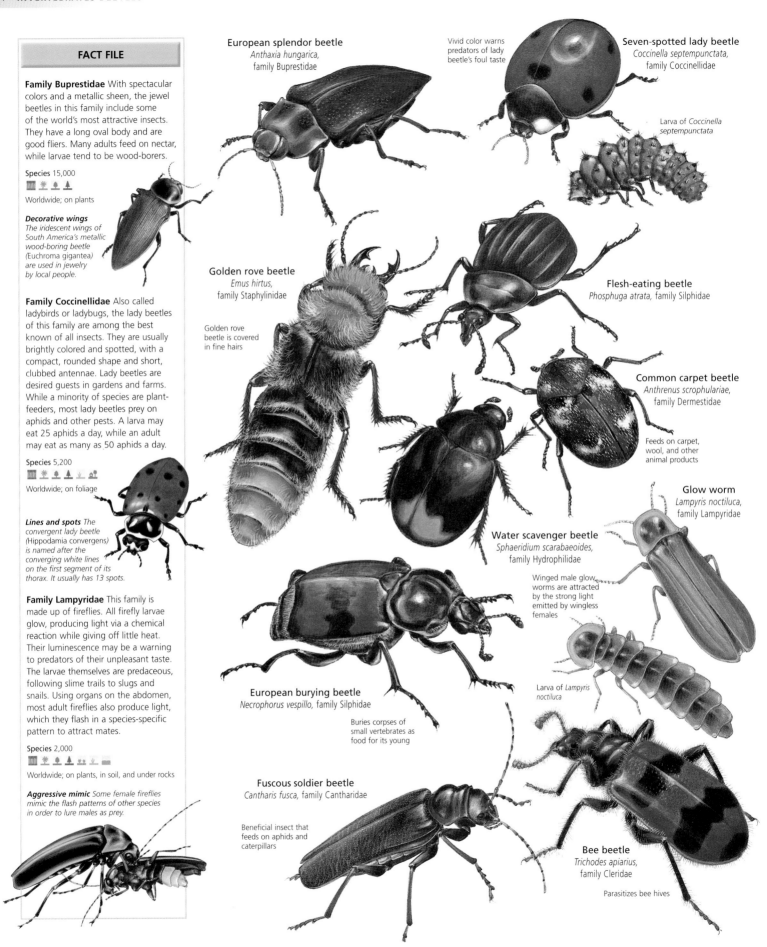

European splendor beetle
Anthaxia hungarica,
family Buprestidae

Vivid color warns predators of lady beetle's foul taste

Seven-spotted lady beetle
Coccinella septempunctata,
family Coccinellidae

Larva of *Coccinella septempunctata*

Golden rove beetle
Emus hirtus,
family Staphylinidae

Golden rove beetle is covered in fine hairs

Flesh-eating beetle
Phosphuga atrata, family Silphidae

Common carpet beetle
Anthrenus scrophulariae,
family Dermestidae

Feeds on carpet, wool, and other animal products

Glow worm
Lampyris noctiluca,
family Lampyridae

Water scavenger beetle
Sphaeridium scarabaeoides,
family Hydrophilidae

Winged male glow worms are attracted by the strong light emitted by wingless females

Larva of *Lampyris noctiluca*

European burying beetle
Necrophorus vespillo, family Silphidae

Buries corpses of small vertebrates as food for its young

Fuscous soldier beetle
Cantharis fusca, family Cantharidae

Beneficial insect that feeds on aphids and caterpillars

Bee beetle
Trichodes apiarius,
family Cleridae

Parasitizes bee hives

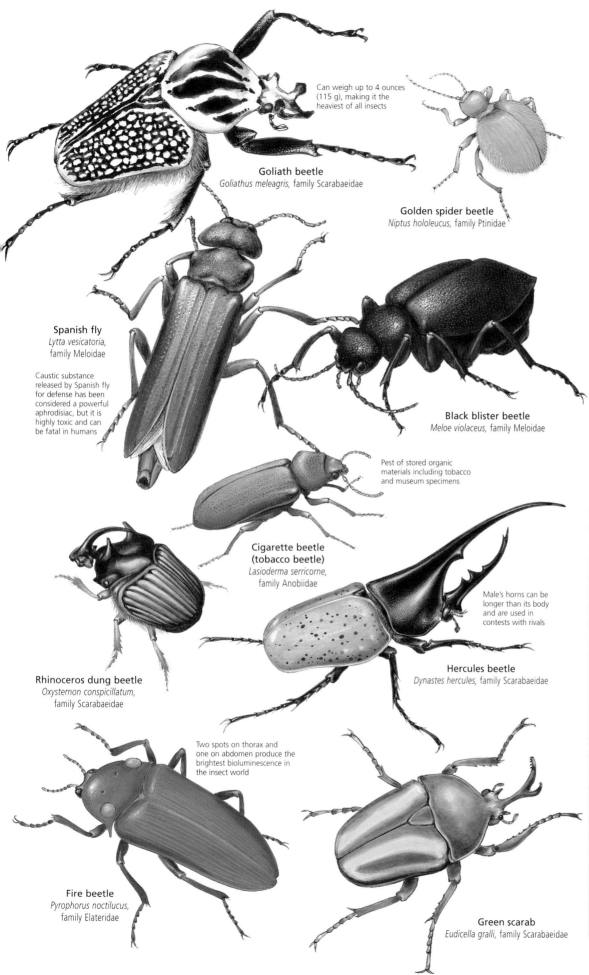

Can weigh up to 4 ounces (115 g), making it the heaviest of all insects

Goliath beetle
Goliathus meleagris, family Scarabaeidae

Golden spider beetle
Niptus hololeucus, family Ptinidae

Spanish fly
Lytta vesicatoria,
family Meloidae

Caustic substance released by Spanish fly for defense has been considered a powerful aphrodisiac, but it is highly toxic and can be fatal in humans

Black blister beetle
Meloe violaceus, family Meloidae

Pest of stored organic materials including tobacco and museum specimens

**Cigarette beetle
(tobacco beetle)**
Lasioderma serricorne,
family Anobiidae

Male's horns can be longer than its body and are used in contests with rivals

Rhinoceros dung beetle
Oxysternon conspicillatum,
family Scarabaeidae

Hercules beetle
Dynastes hercules, family Scarabaeidae

Two spots on thorax and one on abdomen produce the brightest bioluminescence in the insect world

Fire beetle
Pyrophorus noctilucus,
family Elateridae

Green scarab
Eudicella gralli, family Scarabaeidae

FACT FILE

Family Meloidae When disturbed, the blister beetles in this family release a caustic, toxic fluid from their joints that can cause blistering in human skin. Most adults feed on flowers and foliage. The larvae, however, tend to be specialized predators, with some species living in bee hives and eating the eggs, larvae, and stored food of the bees.

Species 2,500

Worldwide; on plants

Distinctive form
Blister beetles such as
Cerocoma muehlfeldi *have
a soft, slender body with long legs.*

Family Elateridae This family comprises the click beetles. If a click beetle ends up lying upside down, it will straighten its back and propel itself into the air to land right way up. As it does so, a hinge-like structure on the elytra makes a loud click. Adults feed on foliage, while the yellow or brown larvae live in soil and eat roots and bulbs.

Species 9,000

Worldwide; near plants and in soil

Groovy wings *Like other
click beetles,* Semiotus
distinctus *has a long,
slender body with
grooved elytra.*

BEETLE LIFECYCLE

All beetles develop through complete metamorphosis. In most species, eggs are fertilized sexually, although a few beetles lay unfertilized eggs without mating. The eggs hatch into a larva that has chewing mouthparts but otherwise bears little resemblance to the adult it will become. The larva molts several times, growing larger until it is ready to pupate. Enclosed in a cocoon, the larva transforms into a pupa with adult features. It is soft and pale when it emerges but soon hardens and colors, now an adult that is ready to reproduce.

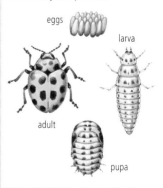

eggs

larva

adult

pupa

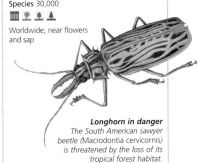
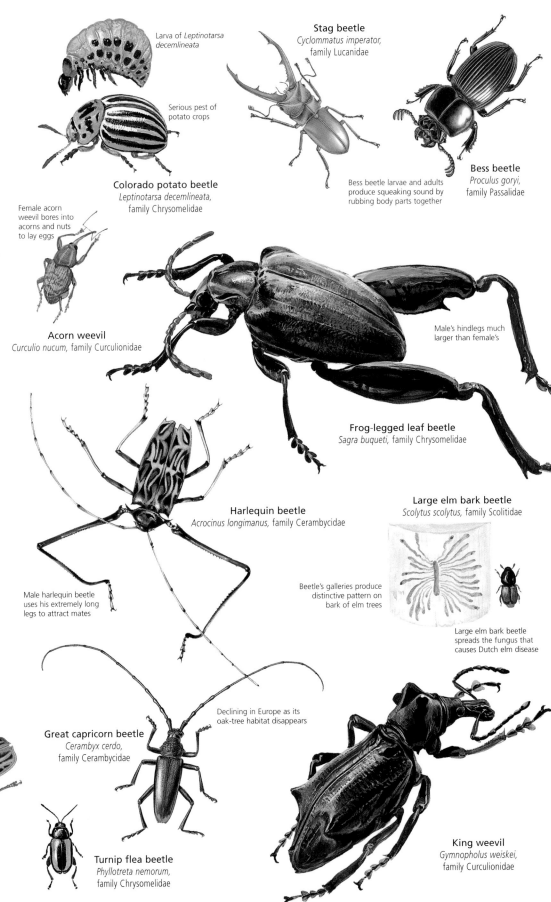

Larva of *Leptinotarsa decemlineata*

Serious pest of potato crops

Colorado potato beetle
Leptinotarsa decemlineata,
family Chrysomelidae

Female acorn weevil bores into acorns and nuts to lay eggs

Acorn weevil
Curculio nucum, family Curculionidae

Stag beetle
Cyclommatus imperator,
family Lucanidae

Bess beetle larvae and adults produce squeaking sound by rubbing body parts together

Bess beetle
Proculus goryi,
family Passalidae

Male's hindlegs much larger than female's

Frog-legged leaf beetle
Sagra buqueti, family Chrysomelidae

Male harlequin beetle uses his extremely long legs to attract mates

Harlequin beetle
Acrocinus longimanus, family Cerambycidae

Large elm bark beetle
Scolytus scolytus, family Scolitidae

Beetle's galleries produce distinctive pattern on bark of elm trees

Large elm bark beetle spreads the fungus that causes Dutch elm disease

Great capricorn beetle
Cerambyx cerdo,
family Cerambycidae

Declining in Europe as its oak-tree habitat disappears

Turnip flea beetle
Phyllotreta nemorum,
family Chrysomelidae

King weevil
Gymnopholus weiskei,
family Curculionidae

FLIES

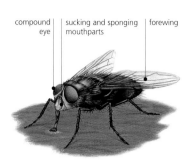

compound eye | sucking and sponging mouthparts | forewing

Eyes and wings Mostly active by day, flies rely on their large compound eyes, each with up to 4,000 lenses. The first segment of the thorax is enlarged to hold the huge muscles that propel the forewings.

PHYLUM	Arthropoda
SUBPHYLUM	Hexapoda
CLASS	Insecta
ORDER	Diptera
FAMILIES	130
SPECIES	120,000

Houseflies and mosquitoes are the most ubiquitous members of the order Diptera, which also includes gnats, midges, blowflies, fruit flies, crane flies, horseflies, hover flies, and other true flies. While most insects fly with four wings, dipterans are generally distinguished by their single pair of functional wings. The hindwings are reduced to halteres, tiny clubbed stalks that vibrate up and down in time with the forewings, helping the fly to balance during flight. Some species have lost their wings altogether and are flightless. Although often associated with moist environments and decaying matter, flies live almost everywhere in the world except Antarctica.

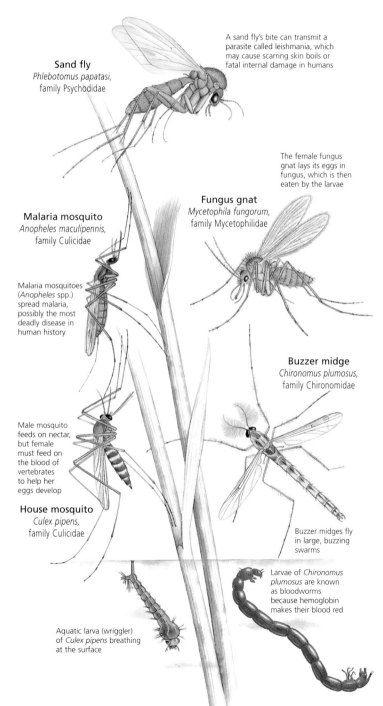

Sand fly
Phlebotomus papatasi, family Psychodidae

A sand fly's bite can transmit a parasite called leishmania, which may cause scarring skin boils or fatal internal damage in humans

The female fungus gnat lays its eggs in fungus, which is then eaten by the larvae

Fungus gnat
Mycetophila fungorum, family Mycetophilidae

Malaria mosquito
Anopheles maculipennis, family Culicidae

Malaria mosquitoes (*Anopheles* spp.) spread malaria, possibly the most deadly disease in human history

Buzzer midge
Chironomus plumosus, family Chironomidae

Male mosquito feeds on nectar, but female must feed on the blood of vertebrates to help her eggs develop

House mosquito
Culex pipens, family Culicidae

Buzzer midges fly in large, buzzing swarms

Larvae of *Chironomus plumosus* are known as bloodworms because hemoglobin makes their blood red

Aquatic larva (wriggler) of *Culex pipens* breathing at the surface

Daddy long legs Crane flies make up Tipulidae, the largest family of flies. Adults have mosquito-like bodies with very long legs. They often feed on nectar in humid forests. Larvae live in moist soil or water.

STICKY SUCKERS

Flies are small insects, ranging from midges ⅟₂₅ inch (1 mm) in length, to robber flies about 3 inches (7 cm) long. Their feet have sticky pads with tiny claws, allowing the insects to walk on smooth surfaces, even upside down along a ceiling.

With mouthparts designed for sucking, flies can consume only liquid food. Houseflies have fleshy pads on their mouth that can sponge up a meal. Mosquitoes have piercing mouthparts for feeding on nectar and blood. Robber flies use their mouthparts to stab insect prey.

Flies begin life as eggs, from which larvae hatch. Often known as maggots, the larvae usually have pale, soft bodies and lack true legs. After several molts, they pupate into the adult form.

Their feeding habits, abundance, and wide distribution have made flies responsible for the spread of deadly diseases such as malaria and sleeping sickness. They also play key ecological roles—as pollinators, as decomposers of organic matter, and as links in the food chain.

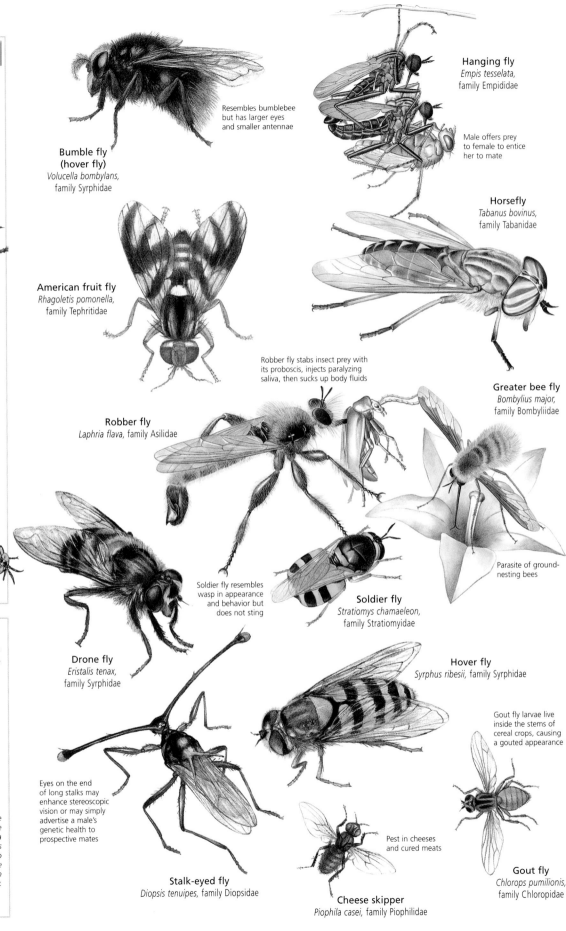

Resembles bumblebee but has larger eyes and smaller antennae

Bumble fly (hover fly)
Volucella bombylans,
family Syrphidae

Hanging fly
Empis tesselata,
family Empididae

Male offers prey to female to entice her to mate

Horsefly
Tabanus bovinus,
family Tabanidae

American fruit fly
Rhagoletis pomonella,
family Tephritidae

Robber fly stabs insect prey with its proboscis, injects paralyzing saliva, then sucks up body fluids

Greater bee fly
Bombylius major,
family Bombyliidae

Robber fly
Laphria flava, family Asilidae

Parasite of ground-nesting bees

Soldier fly resembles wasp in appearance and behavior but does not sting

Soldier fly
Stratiomys chamaeleon,
family Stratiomyidae

Drone fly
Eristalis tenax,
family Syrphidae

Hover fly
Syrphus ribesii, family Syrphidae

Gout fly larvae live inside the stems of cereal crops, causing a gouted appearance

Eyes on the end of long stalks may enhance stereoscopic vision or may simply advertise a male's genetic health to prospective mates

Pest in cheeses and cured meats

Stalk-eyed fly
Diopsis tenuipes, family Diopsidae

Cheese skipper
Piophila casei, family Piophilidae

Gout fly
Chlorops pumilionis,
family Chloropidae

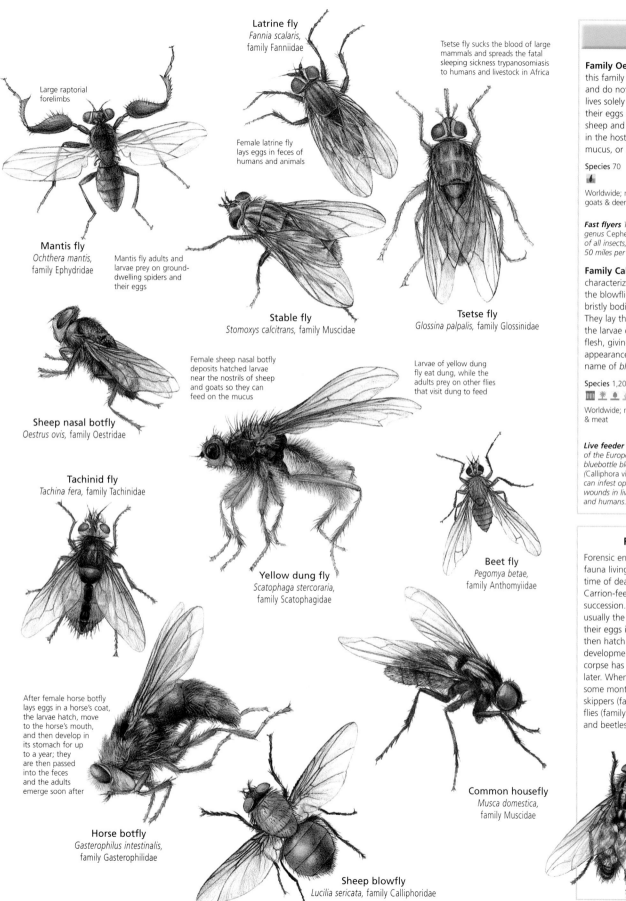

Mantis fly
Ochthera mantis,
family Ephydridae

Large raptorial
forelimbs

Mantis fly adults and
larvae prey on ground-
dwelling spiders and
their eggs

Latrine fly
Fannia scalaris,
family Fanniidae

Female latrine fly
lays eggs in feces of
humans and animals

Tsetse fly sucks the blood of large
mammals and spreads the fatal
sleeping sickness trypanosomiasis
to humans and livestock in Africa

Stable fly
Stomoxys calcitrans, family Muscidae

Tsetse fly
Glossina palpalis, family Glossinidae

Sheep nasal botfly
Oestrus ovis, family Oestridae

Female sheep nasal botfly
deposits hatched larvae
near the nostrils of sheep
and goats so they can
feed on the mucus

Larvae of yellow dung
fly eat dung, while the
adults prey on other flies
that visit dung to feed

Tachinid fly
Tachina fera, family Tachinidae

Yellow dung fly
Scatophaga stercoraria,
family Scatophagidae

Beet fly
Pegomya betae,
family Anthomyiidae

After female horse botfly
lays eggs in a horse's coat,
the larvae hatch, move
to the horse's mouth,
and then develop in
its stomach for up
to a year; they
are then passed
into the feces
and the adults
emerge soon after

Horse botfly
Gasterophilus intestinalis,
family Gasterophilidae

Sheep blowfly
Lucilia sericata, family Calliphoridae

Common housefly
Musca domestica,
family Muscidae

FACT FILE

Family Oestridae The adult botflies in
this family lack functional mouthparts
and do not feed, spending their short
lives solely on reproduction. They lay
their eggs on live mammals such as
sheep and deer. The larvae may live
in the host's nostrils and feed on
mucus, or burrow into its flesh.

Species 70

Worldwide; near sheep,
goats & deer

*Fast flyers The deer nose botflies in the
genus Cephenemyia are among the swiftest
of all insects, flying at speeds of up to
50 miles per hour (80 km/h).*

Family Calliphoridae Usually
characterized by a loud, buzzing flight,
the blowflies in this family have stout,
bristly bodies with a metallic sheen.
They lay their eggs in carrion so that
the larvae can feed on the decaying
flesh, giving the meat a blown-up
appearance that inspired the common
name of *blowfly.*

Species 1,200

Worldwide; near carrion
& meat

*Live feeder The larvae
of the European
bluebottle blowfly
(Calliphora vicina)
can infest open
wounds in livestock
and humans.*

FLIES AS CLUES

Forensic entomologists can analyze the
fauna living on a corpse to determine
time of death and other information.
Carrion-feeders arrive in a predictable
succession. Houseflies and blowflies are
usually the first to turn up. They lay
their eggs in the flesh, and the larvae
then hatch and develop. Their stage of
development can reveal how long the
corpse has been dead. Flesh flies arrive
later. When the corpse has dried out,
some months after death, cheese
skippers (family Piophilidae) and coffin
flies (family Phoridae), along with mites
and beetles, pick clean the skeleton.

*Flesh-feeder
Flesh flies (family
Sarcophagidae) will
colonize a corpse later
than houseflies and
blowflies, but catch
up by depositing
live larvae rather
than eggs.*

BUTTERFLIES AND MOTHS

PHYLUM	Arthropoda
SUBPHYLUM	Hexapoda
CLASS	Insecta
ORDER	Lepidoptera
FAMILIES	131
SPECIES	165,000

With their delicate fluttering flight and often intricate wing patterns, butterflies and moths are the most studied and admired of all insects. They belong to the order Lepidoptera, which means "scale wing" in Greek. Their four broad wings are covered by tiny, overlapping scales—hollow, flattened hairs that create the bright colors and iridescence of diurnal species. Almost all butterflies and moths are plant-feeders. Their larvae, known as caterpillars, have biting mouthparts for chewing plants, which can make them pests. Most adults have a long proboscis for sucking flower nectar and are important pollinators, though a few lack mouthparts and do not feed at all.

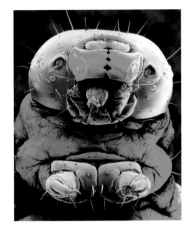

Plant muncher A caterpillar is an eating machine, with biting mouthparts for munching through foliage or other plant food. Many species are brightly colored to warn predators that they are unpalatable.

Nectar-feeder To drink nectar from flowers, butterflies and moths extend a long proboscis, which sits coiled like a watchspring when not in use. Many flowering plants evolved with colors and shapes to attract these pollinating insects.

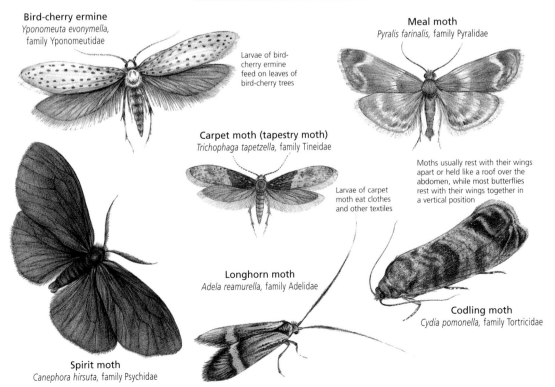

European pine shoot moth
Rhyacionia buoliana,
family Tortricidae

Larvae of European pine shoot moth feed on pine shoots, causing considerable damage in pine forests

Bird-cherry ermine
Yponomeuta evonymella,
family Yponomeutidae

Larvae of bird-cherry ermine feed on leaves of bird-cherry trees

Meal moth
Pyralis farinalis, family Pyralidae

Carpet moth (tapestry moth)
Trichophaga tapetzella, family Tineidae

Larvae of carpet moth eat clothes and other textiles

Moths usually rest with their wings apart or held like a roof over the abdomen, while most butterflies rest with their wings together in a vertical position

Longhorn moth
Adela reamurella, family Adelidae

Codling moth
Cydia pomonella, family Tortricidae

Spirit moth
Canephora hirsuta, family Psychidae

DAY AND NIGHT FLYERS
Absent only from the polar ice caps and the oceans, butterflies and moths are most abundant in the tropics but are found virtually anywhere that land plants grow. Most are highly specialized for feeding on particular flowering species.

As adults, butterflies and moths have slim bodies, broad wings, long antennae, and two large compound eyes. Their wingspans range from about 3/16 inch to 12 inches (4 mm to 30 cm). More than 85 percent of species in Lepidoptera are moths, most of which fly at night and have dull wings that are coupled by a spine-like structure known as a frenulum. Butterflies fly by day, display vibrant colors, have clubbed antennae, and lack the frenulum. Some moth species, however, are also diurnal and brightly colored.

Caterpillars are covered in hairs, and walk using three pairs of true legs and several false legs. When ready to pupate, they usually enclose themselves in a cocoon or chrysalis, using silk from modified salivary glands. The entire life-cycle takes from a few weeks to a few years.

⚡ CONSERVATION WATCH

Deadly trade Only 303 lepidopterids have been assessed by the IUCN, but of these, 94 percent are on the Red List and 37 are extinct. Habitat loss is the major cause of the decline, but over-collecting has also had a major impact on attractive species such as the birdwing butterflies.

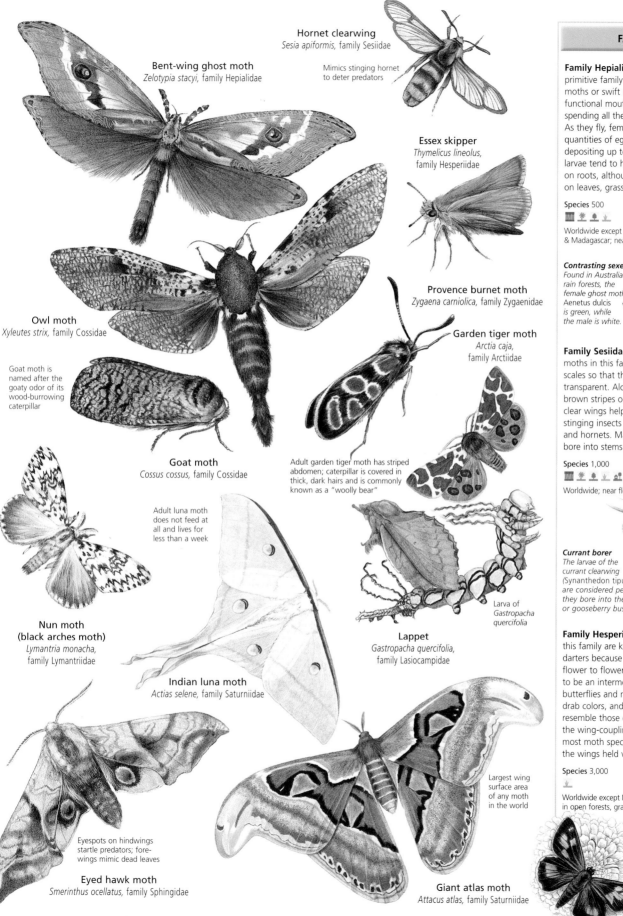

Bent-wing ghost moth
Zelotypia stacyi, family Hepialidae

Hornet clearwing
Sesia apiformis, family Sesiidae

Mimics stinging hornet to deter predators

Essex skipper
Thymelicus lineolus,
family Hesperiidae

Owl moth
Xyleutes strix, family Cossidae

Goat moth is named after the goaty odor of its wood-burrowing caterpillar

Provence burnet moth
Zygaena carniolica, family Zygaenidae

Garden tiger moth
Arctia caja,
family Arctiidae

Goat moth
Cossus cossus, family Cossidae

Adult garden tiger moth has striped abdomen; caterpillar is covered in thick, dark hairs and is commonly known as a "woolly bear"

Adult luna moth does not feed at all and lives for less than a week

Larva of *Gastropacha quercifolia*

**Nun moth
(black arches moth)**
Lymantria monacha,
family Lymantriidae

Lappet
Gastropacha quercifolia,
family Lasiocampidae

Indian luna moth
Actias selene, family Saturniidae

Eyespots on hindwings startle predators; forewings mimic dead leaves

Largest wing surface area of any moth in the world

Eyed hawk moth
Smerinthus ocellatus, family Sphingidae

Giant atlas moth
Attacus atlas, family Saturniidae

FACT FILE

Family Hepialidae The moths in this primitive family are known as ghost moths or swift moths. The adults lack functional mouthparts and do not feed, spending all their time on reproduction. As they fly, females release enormous quantities of eggs, with some species depositing up to 30,000 eggs. The larvae tend to hatch in soil and feed on roots, although some species feed on leaves, grasses, moss, and wood.

Species 500

Worldwide except C.W. Africa & Madagascar; near plants

Contrasting sexes
Found in Australian rain forests, the female ghost moth Aenetus dulcis is green, while the male is white.

Family Sesiidae The clear-winged moths in this family lose most of their scales so that their wings are largely transparent. Along with yellow and brown stripes on the abdomen, the clear wings help many species mimic stinging insects such as bees, wasps, and hornets. Many larvae bore into stems and roots.

Species 1,000

Worldwide; near flowers

Currant borer
The larvae of the currant clearwing (Synanthedon tipuliformis) are considered pests because they bore into the stems of currant or gooseberry bushes to feed.

Family Hesperiidae The members of this family are known as skippers or darters because they rapidly flit from flower to flower. They are considered to be an intermediate form between butterflies and moths. Their small size, drab colors, and solid, hairy body do resemble those of moths, but they lack the wing-coupling structure found in most moth species and often rest with the wings held vertically.

Species 3,000

Worldwide except New Zealand; in open forests, grasslands, and fields

Hooked antennae
Like other skippers, the Eliena skipper (Trapezites eliena) of Australia has clubbed antennae that end in fine hooks.

FACT FILE

Family Papilionidae Often large and brightly colored, the swallowtails of this family are named after the tail-like extensions on the hindwings of most species. All swallowtail caterpillars have an osmetrium, a fork-shaped organ on their head that can release a foul odor to deter a predator. In some species, the caterpillars feed on poisonous plants, absorbing enough toxins to make both themselves and their adult form unpalatable.

Species 600

Worldwide; near flowers

Vivid warning
The bright wing markings of the crimson rose (Pachliopta hector) and many other swallow-tails warn predators that they have an unpleasant taste.

Family Noctuidae Noctuid or owlet moths make up the largest family in the order Lepidoptera. These night flyers have hearing organs on their thorax that can pick up echolocation signals emitted by bats, their main predators. Adults feed on nectar, sap, rotting fruit, tears, or dung, while the larvae may eat foliage or seeds, or bore into stems and fruit, with some being serious crop pests.

Species 35,000

Worldwide; on plants

Background colors
Like many noctuid moths, Moma alpium has cryptic coloration, the mottled wings blending in with its woodland habitat.

Family Geometridae Known as geometer, looper, or inchworm moths, the members of this family tend to be small with slender bodies. The larvae, which often resemble twigs when resting, lack at least one middle pair of false legs. They move by stretching out the front part of the body, then bringing the tail forward to meet it, creating a "looping" or "inching" gait.

Species 20,000

Worldwide; on foliage

Leaf mimic The wings of the Southeast Asian geometer moth (Sarcinodes restitutaria) resemble leaves.

Moth butterfly
Liphyra brassolis, family Lycaenidae

Caterpillars of moth butterfly feed on larvae of citrus ants and have tough skin that protects them from the ants' bites

African giant swallowtail
Papilio antimachus, family Papiolinidae

Cabbage butterfly
Pieris brassicae, family Pieridae

Orange-barred sulphur
Phoebis philea, family Pieridae

African giant swallowtail is the largest of all African butterflies, with a wingspan of 10 inches (25 cm) or more

Scarce copper
Lycaena virgaureae, family Lycaenidae

Giant agrippa
Thysania agrippina, family Noctuidae

Large blue
Maculinea arion, family Lycaenidae

Union Jack butterfly
Delias mysis, family Pieridae

Giant agrippa has wingspan of up to 12 inches (30 cm), the largest of any butterfly or moth

Bhutan glory butterfly
Bhutanitis lidderdalii, family Papilionidae

Bogong moth
Agrotis infusa, family Noctuidae

Newly emerged adult bogong moths avoid the summer heat by migrating to mountain caves and becoming dormant until autumn, when they return to the plains to breed

Apollo butterfly
Parnassius apollo, family Papilionidae

Red underwing moth
Catocala nupta, family Noctuidae

Only male mottled umbers fly; females are wingless

Mottled umber
Erannis defoliaria, family Geometridae

Large eyespot on hindwing deters predators

The postman
Heliconius melpomene,
family Nymphalidae

Adult owl butterfly feeds on rotting fruit

Owl butterfly
Caligo idomeneus,
family Nymphalidae

Helena butterfly
Morpho rhetenor, family Morphidae

♀

All Helena butterflies have yellow-brown wings, but, on the males, air spaces between the scales scatter light and produce blue iridescence

♂

Silver-washed fritillary
Argynnis paphia,
family Nymphalidae

Patches on underside of wings mimic holes and other blemishes found on dead leaves

Dead leaf butterfly
Kallima inachus,
family Nymphalidae

Kallima inachus in flight

Northern jungle queen butterfly
Stichophthalma camadeva,
family Nymphalidae

Female mimic is palatable but imitates the unpalatable African monarch to discourage predators

African monarch (lesser wanderer)
Danaus chrysippus,
family Nymphalidae

Mimic
Hypolimnas misippus,
family Nymphalidae

♂

♀

Male mimic waits on ground for females and is colored to blend with the background

---

FACT FILE

Family Nymphalidae Known as brush-footed butterflies, the species in this family have very small and hairy forelegs, and walk using only the middle and back legs. The underside of the wings tends to be drab, while the upperside usually features strongly contrasting colors.

Species 5,200

Worldwide; near flowers

Seasonal variation The map butterflies (Araschnia levana) that emerge in spring have orange-marked wings, while those that emerge in summer are black and white.

COCOONS AND CHRYSALISES

When they are ready to transform into adults, caterpillars stop eating and find a safe spot. Some pupate underground without any further protection, but most moths spin a cocoon of silk, while almost all butterflies create a chrysalis, a hard shell formed from the caterpillar's skin and secured by silk to a branch.

Sticky cocoon *The larvae of bagworm moths (Thyridopteryx ephemeraeformis) live in cases of silk and plant debris, which harden into cocoons when they are ready to pupate.*

Transforming chrysalis *The pale green chrysalis of the monarch butterfly becomes transparent when the pupa is ready to emerge.*

BUTTERFLY MIGRATION

A monarch butterfly (*Danaus plexippus*) may travel more than 1,800 miles (2,900 km) in the 9 months of its life. Every autumn, millions of monarchs migrate south to their wintering sites in California and Mexico. Here, they shelter in great clusters on trees. In autumn, they begin the return journey north, during which they lay eggs and die. The new generation completes the trip to the summer feeding grounds.

BEES, WASPS, ANTS, AND SAWFLIES

PHYLUM	Arthropoda
SUBPHYLUM	Hexapoda
CLASS	Insecta
ORDER	Hymenoptera
FAMILIES	91
SPECIES	198,000

The order Hymenoptera is named after the Greek words for "membrane wing," and most of its species have two pairs of transparent wings. Although many bees, wasps, and sawflies are solitary creatures, some species of bees and all species of ants live in highly structured societies, whose thousands or millions of members belong to different castes and perform specific tasks. Many of the world's most beneficial insects are hymenopterids. Bees and some wasps are the chief pollinators of both crops and wild plants, while many parasitic wasps play a crucial role in controlling the populations of other insects. Honeybees have been domesticated for their honey and wax.

Bite and sting Despite their formidable jaws, adult bull ants mainly eat nectar and plant juices, but they do capture insect prey to feed their young. They use an abdominal stinger to subdue prey and deter predators.

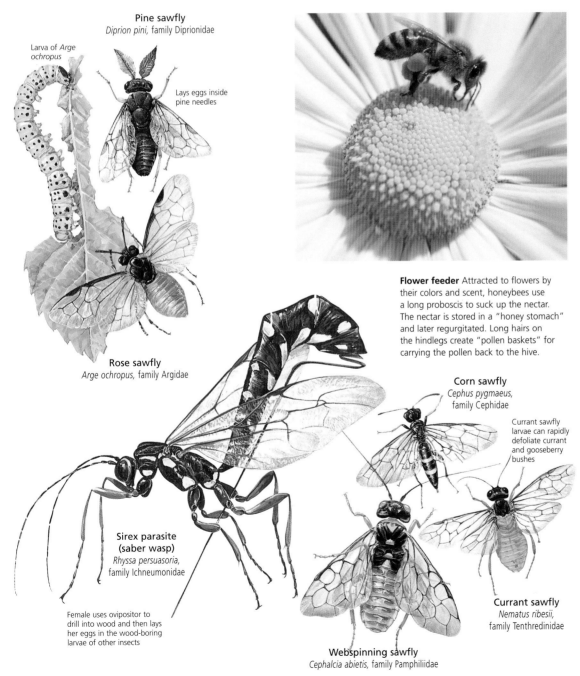

Pine sawfly
Diprion pini, family Diprionidae

Larva of *Arge ochropus*

Lays eggs inside pine needles

Rose sawfly
Arge ochropus, family Argidae

Flower feeder Attracted to flowers by their colors and scent, honeybees use a long proboscis to suck up the nectar. The nectar is stored in a "honey stomach" and later regurgitated. Long hairs on the hindlegs create "pollen baskets" for carrying the pollen back to the hive.

Corn sawfly
Cephus pygmaeus,
family Cephidae

Currant sawfly larvae can rapidly defoliate currant and gooseberry bushes

Sirex parasite (saber wasp)
Rhyssa persuasoria,
family Ichneumonidae

Female uses ovipositor to drill into wood and then lays her eggs in the wood-boring larvae of other insects

Currant sawfly
Nematus ribesii,
family Tenthredinidae

Webspinning sawfly
Cephalcia abietis, family Pamphiliidae

ANATOMY AND LIFE-CYCLE

Generally small to medium-sized, hymenopterids include the parasitic *Megaphragma caribea,* which, at just 3/500 inch (0.17 mm) long, is one of the smallest insects. Hymenopterids' hindwings are attached to the larger forewings by tiny hooks so that they beat together. The more primitive sawflies and their kin lack the slim waist that separates the thorax and abdomen of bees, wasps, and ants.

Hymenopterids have mouthparts designed for biting or for biting and sucking. Sawflies, gall wasps, and some ants and bees are herbivorous, while most other species in the order are predatory or parasitic. In many species, the female's ovipositor (egg-laying organ) has become modified—sawflies use it for sawing into plant stems, where they lay their eggs; while bees, wasps, and ants often use it for piercing or stinging predators or prey.

Female hymenopterids may lay their eggs in soil, in plants, in nests or hives, or in living insect hosts. In many species, the female determines whether the eggs are fertilized, with unfertilized eggs producing male larvae, and fertilized eggs producing females. In solitary species, the larvae hatch near a food source and develop independently, but in social species they rely on constant care from the adults. Development is by complete metamorphosis, and larvae usually pupate in a cocoon.

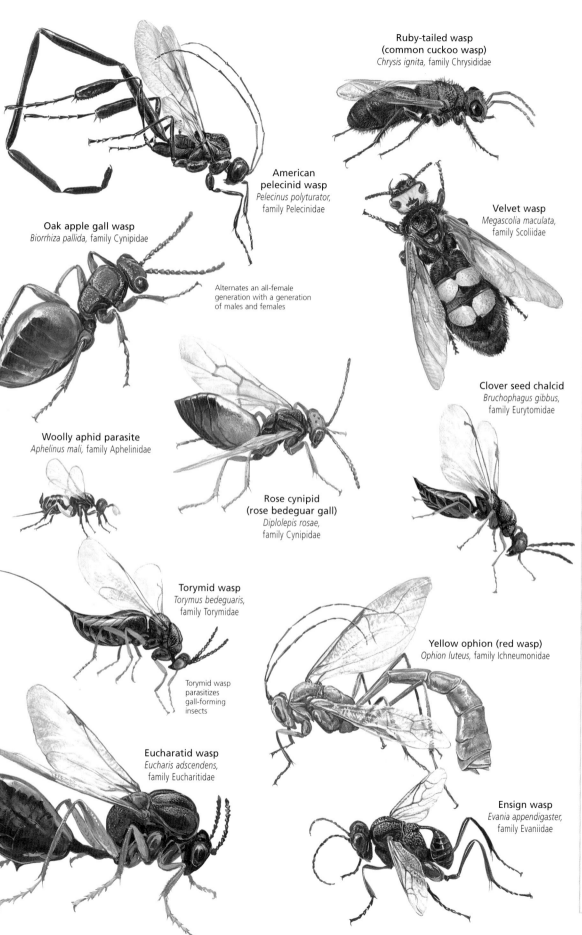

Ruby-tailed wasp
(common cuckoo wasp)
Chrysis ignita, family Chrysididae

American
pelecinid wasp
Pelecinus polyturator,
family Pelecinidae

Oak apple gall wasp
Biorrhiza pallida, family Cynipidae

Alternates an all-female
generation with a generation
of males and females

Velvet wasp
Megascolia maculata,
family Scoliidae

Clover seed chalcid
Bruchophagus gibbus,
family Eurytomidae

Woolly aphid parasite
Aphelinus mali, family Aphelinidae

Rose cynipid
(rose bedeguar gall)
Diplolepis rosae,
family Cynipidae

Torymid wasp
Torymus bedeguaris,
family Torymidae

Torymid wasp
parasitizes
gall-forming
insects

Yellow ophion (red wasp)
Ophion luteus, family Ichneumonidae

Eucharitid wasp
Eucharis adscendens,
family Eucharitidae

Ensign wasp
Evania appendigaster,
family Evaniidae

HIVE OF ACTIVITY

The social life of the honeybee (genus *Apis*) is among the most complex in the animal world. A queen bee can lay up to 1,500 eggs a day. Fertilized eggs develop into workers or queens, and unfertilized eggs become drones. The workers feed the hatched larvae before they are capped inside cells to pupate. When a hive reaches its optimum size, the old queen flies off to found a new colony, followed by thousands of workers. The first new queen to emerge will rule the old nest.

Close view This electron micrograph shows a honeybee's large compound eyes, each made up of more than 4,000 lenses. Two sensitive antennae emerge between the eyes.

Inside a beehive All the drones and workers in a hive (below) are the offspring of the queen, whose only task is to lay eggs. Drones are male bees, and their only job is to mate with new queens. Worker bees gather food, maintain the hive, and feed the queen and her young. Glands in their head produce royal jelly for the larvae. Workers also make honey by regurgitating nectar and spreading it in cells to dehydrate, then capping it with wax.

Being queen *A queen bee is the largest bee in the hive. She can live for up to 5 years.*

Hard workers *Workers are females that cannot breed. They live for only a few weeks.*

Bee young Worker bees bring food to the larvae as they mature inside water-resistant waxen cells. These larvae (above) are transforming into pupae, and will emerge as adults.

Honeybee swarm A swarm of bees gathers when an old queen, along with up to 70,000 workers, leaves an established hive to establish a new colony. The swarm waits while scouts find a good location for the nest.

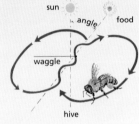

sun

angle

food

waggle

hive

Dance of the bees A worker bee uses a figure-eight dance to convey information about a new food source. The rate of tail-waggling indicates the distance to the food source, while the direction is signaled by the angle of the dance to the sun.

Larva cells *Most larvae become workers. They are first fed royal jelly, and later pollen and honey. Queen larvae are fed only royal jelly. Drones develop from unfertilized eggs.*

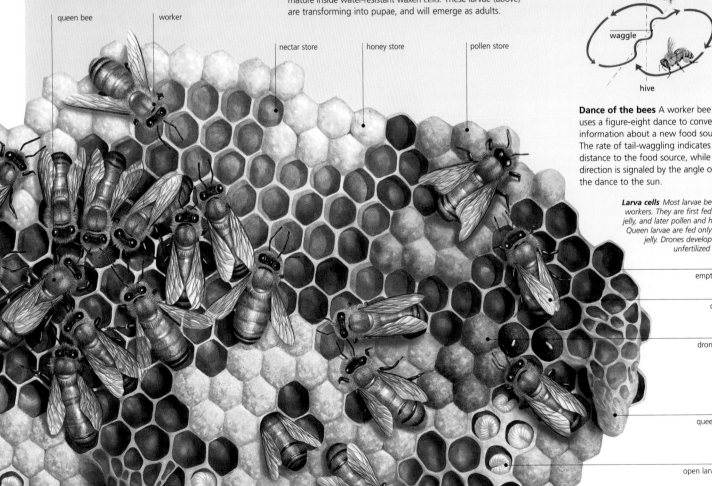

queen bee

worker

nectar store

honey store

pollen store

empty cell

drone

drone cell

queen cell

open larva cell

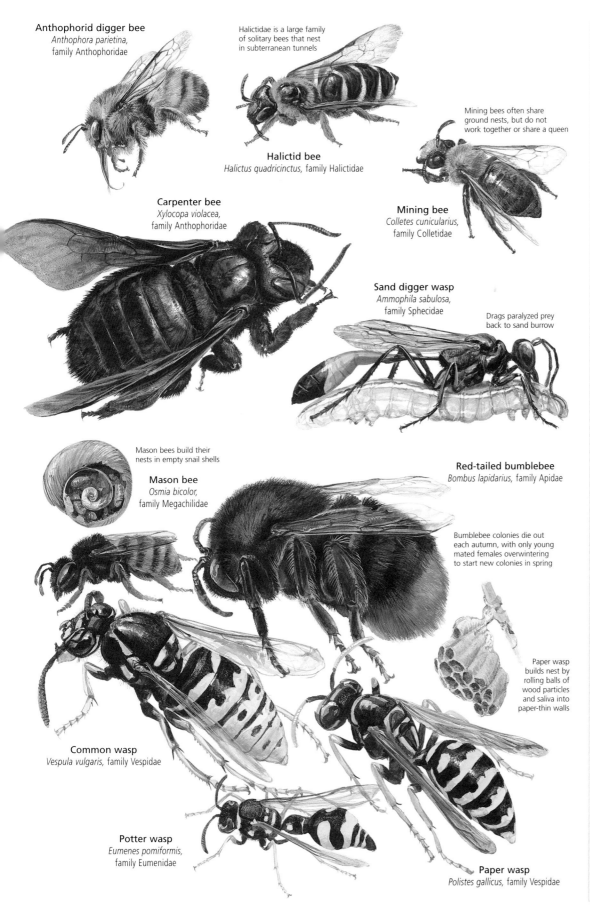

Anthophorid digger bee
Anthophora parietina,
family Anthophoridae

Halictidae is a large family
of solitary bees that nest
in subterranean tunnels

Halictid bee
Halictus quadricinctus, family Halictidae

Mining bees often share
ground nests, but do not
work together or share a queen

Mining bee
Colletes cunicularius,
family Colletidae

Carpenter bee
Xylocopa violacea,
family Anthophoridae

Sand digger wasp
Ammophila sabulosa,
family Sphecidae

Drags paralyzed prey
back to sand burrow

Mason bees build their
nests in empty snail shells

Mason bee
Osmia bicolor,
family Megachilidae

Red-tailed bumblebee
Bombus lapidarius, family Apidae

Bumblebee colonies die out
each autumn, with only young
mated females overwintering
to start new colonies in spring

Paper wasp
builds nest by
rolling balls of
wood particles
and saliva into
paper-thin walls

Common wasp
Vespula vulgaris, family Vespidae

Potter wasp
Eumenes pomiformis,
family Eumenidae

Paper wasp
Polistes gallicus, family Vespidae

FACT FILE

Family Sphecidae This family includes species known as digger wasps, sand wasps, mud-dauber wasps, and thread-waisted wasps. A female sphecid wasp paralyzes arthropod prey with her sting and deposits it in her nest, where her larvae later feed on its body. Most species have their nests underground in burrows, but some use plant stems or rotten wood, and mud daubers build rows of cells from pellets of mud.

Species 8,000

Worldwide; often in sandy soil

Solitary hunter The female bee wolf (Philanthus triangulum) will carry paralyzed honeybees back to its sandy burrow.

Family Apidae The honeybees, bumblebees, and stingless bees in this family live in communal nests centered on a queen. Worker bees forage for nectar and pollen, which they bring back to the nest to feed the queen and her larvae. While honeybees often attach their hives to trees, bumblebees usually create their nests in the soil.

Species 1,000

Worldwide; near flowers

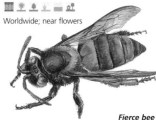

Fierce bee When provoked, a colony of Southeast Asia's giant honeybees (Apis dorsata) is capable of stinging a person to death.

Family Vespidae This family contains all the social wasps, as well as some solitary ones. Hornets, yellowjackets, and paper wasps all use chewed wood and saliva to build their papery nests. In most species, the colony dies out in late autumn, leaving only a few mated queens to overwinter and establish new colonies in spring.

Species 4,000

Worldwide

Self-sacrifice To defend its colony, a worker of the social wasp species Brachygastra lecheguana will sting an attacker and may well die when the sting detaches from its abdomen.

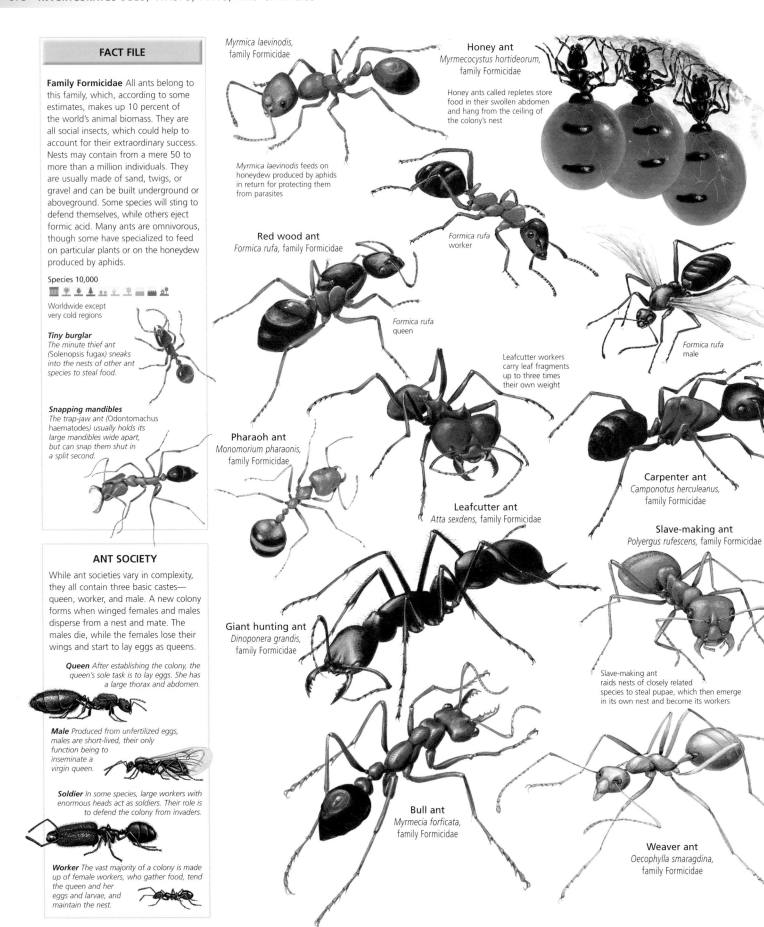

FACT FILE

Family Formicidae All ants belong to this family, which, according to some estimates, makes up 10 percent of the world's animal biomass. They are all social insects, which could help to account for their extraordinary success. Nests may contain from a mere 50 to more than a million individuals. They are usually made of sand, twigs, or gravel and can be built underground or aboveground. Some species will sting to defend themselves, while others eject formic acid. Many ants are omnivorous, though some have specialized to feed on particular plants or on the honeydew produced by aphids.

Species 10,000

Worldwide except very cold regions

Tiny burglar
The minute thief ant (*Solenopsis fugax*) sneaks into the nests of other ant species to steal food.

Snapping mandibles
The trap-jaw ant (*Odontomachus haematodes*) usually holds its large mandibles wide apart, but can snap them shut in a split second.

ANT SOCIETY

While ant societies vary in complexity, they all contain three basic castes—queen, worker, and male. A new colony forms when winged females and males disperse from a nest and mate. The males die, while the females lose their wings and start to lay eggs as queens.

Queen After establishing the colony, the queen's sole task is to lay eggs. She has a large thorax and abdomen.

Male Produced from unfertilized eggs, males are short-lived, their only function being to inseminate a virgin queen.

Soldier In some species, large workers with enormous heads act as soldiers. Their role is to defend the colony from invaders.

Worker The vast majority of a colony is made up of female workers, who gather food, tend the queen and her eggs and larvae, and maintain the nest.

Myrmica laevinodis, family Formicidae

Myrmica laevinodis feeds on honeydew produced by aphids in return for protecting them from parasites

Honey ant
Myrmecocystus hortideorum, family Formicidae

Honey ants called repletes store food in their swollen abdomen and hang from the ceiling of the colony's nest

Red wood ant
Formica rufa, family Formicidae

Formica rufa worker

Formica rufa queen

Formica rufa male

Pharaoh ant
Monomorium pharaonis, family Formicidae

Leafcutter workers carry leaf fragments up to three times their own weight

Leafcutter ant
Atta sexdens, family Formicidae

Carpenter ant
Camponotus herculeanus, family Formicidae

Slave-making ant
Polyergus rufescens, family Formicidae

Giant hunting ant
Dinoponera grandis, family Formicidae

Slave-making ant raids nests of closely related species to steal pupae, which then emerge in its own nest and become its workers

Bull ant
Myrmecia forficata, family Formicidae

Weaver ant
Oecophylla smaragdina, family Formicidae

OTHER INSECTS

PHYLUM	Arthropoda
SUBPHYLUM	Hexapoda
CLASSES	Insecta
ORDERS	19*
FAMILIES	218*
SPECIES	> 3,000*

* Totals refer only to the 19 orders that make up "Other insects."

The remaining 19 orders of Insecta are a diverse assortment. They include primitive, wingless forms such as silverfish and bristletails; mammal parasites such as fleas and lice; insect parasites such as strepsipterans (order Strepsiptera); crop pests such as thrips; and insects that spend most of their lives in water, such as mayflies, caddisflies, dobsonflies, and stoneflies. Stick insects and leaf insects are plant-feeders; angel insects (order Zorotypidae), snakeflies, antlions, lacewings, and hanging flies prey on other invertebrates; rock crawlers (order Grylloblattodea), webspinners, earwigs, and scorpionflies recycle detritus; and some booklice species will feed on paper or stored cereals.

Webspinners The only insects other than danceflies with silk-producing glands in the legs, webspinners (order Embioptera) use their silk to create a shared network of nests and tunnels. They develop by incomplete metamorphosis, with nymphs resembling their parents.

Earwigs The order Dermaptera contains the earwigs, small flattened insects that feed on detritus. The large "forceps" extending from the abdomen are used in defense and courtship. Females guard their eggs and newly hatched larvae.

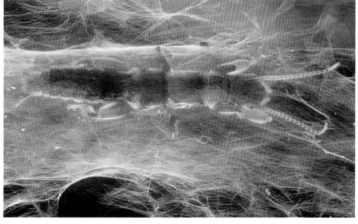

Elongated thorax allows head to be lifted in a snake-like fashion

Snakefly
Raphidia notata,
order Raphidioptera

Spines on hindlegs used for defense and battles with rivals

Spiny devil walking stick
Eurycantha horrida, order Phasmatodea

Mimics foliage to avoid predation

Leaf insect
Phyllium siccifolium,
order Phasmatodea

Silverfish
Lepisma saccharina,
order Thysanura

Bristletails cannot fly but can run and jump

Bristletail
Machilis helleri,
order Archaeognatha

FACT FILE

Order Thysanura This order contains the silverfish, flightless insects that may closely resemble the earliest insects. They have long antennae on their head, and three "tails" extending from the abdomen. Silverfish hatch from eggs laid in crevices near food sources.

Species 370

Worldwide; in trees, caves & buildings

House guest *This silverfish (Lepismodes inquilinus) is often found in houses, where it feeds on paper, glue in book bindings, starched clothing, and dry foods.*

Order Phasmatodea The supreme mimics in this order are known as stick insects and leaf insects. They feed on foliage. Most species are wingless, and, when wings are present, they usually are fully developed and functional only in the males. Many females can produce offspring from unfertilized eggs.

Species 3,000

Worldwide, esp. in the tropics; near plants

Stick mimic
When disturbed, stick insects such as Bacillus rossii may freeze for hours or they may sway as if blowing in the wind.

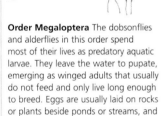

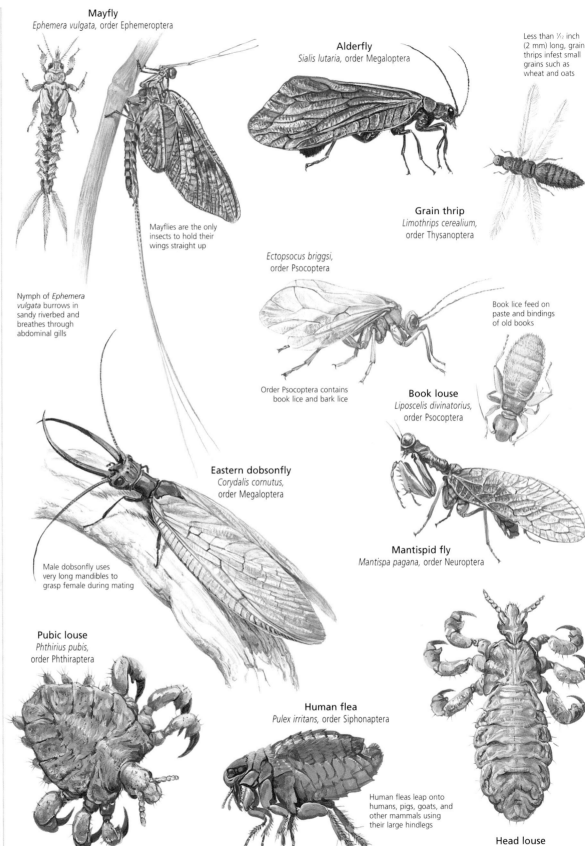

Mayfly
Ephemera vulgata, order Ephemeroptera

Mayflies are the only insects to hold their wings straight up

Nymph of *Ephemera vulgata* burrows in sandy riverbed and breathes through abdominal gills

Alderfly
Sialis lutaria, order Megaloptera

Less than ½ inch (2 mm) long, grain thrips infest small grains such as wheat and oats

Grain thrip
Limothrips cerealium, order Thysanoptera

Ectopsocus briggsi, order Psocoptera

Order Psocoptera contains book lice and bark lice

Book lice feed on paste and bindings of old books

Book louse
Liposcelis divinatorius, order Psocoptera

Eastern dobsonfly
Corydalis cornutus, order Megaloptera

Male dobsonfly uses very long mandibles to grasp female during mating

Mantispid fly
Mantispa pagana, order Neuroptera

Pubic louse
Phthirius pubis, order Phthiraptera

Pubic lice are parasites of humans and gorillas

Human flea
Pulex irritans, order Siphonaptera

Human fleas leap onto humans, pigs, goats, and other mammals using their large hindlegs

Head louse
Pediculus capitis, order Phthiraptera

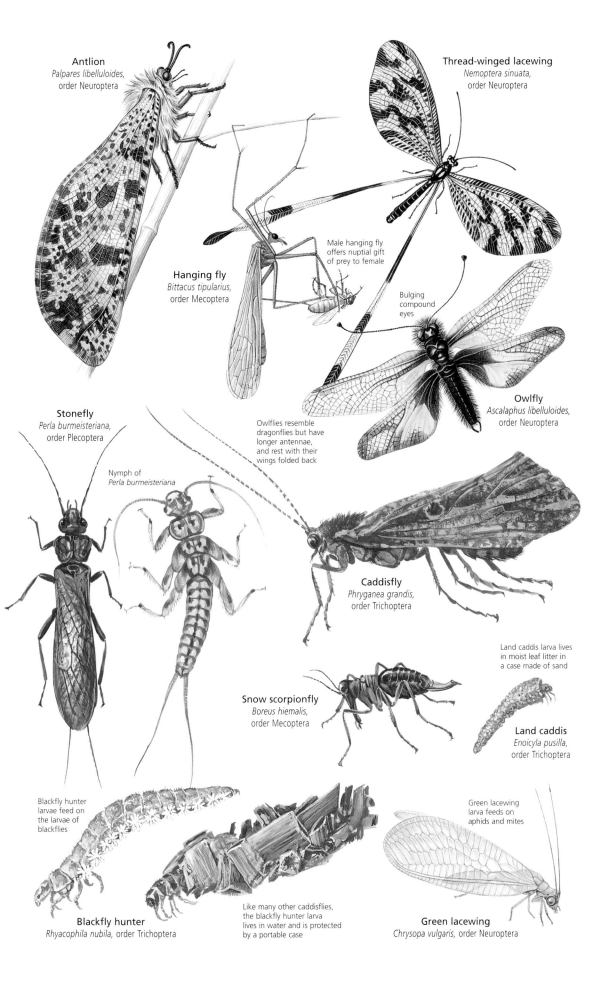

Antlion
Palpares libelluloides,
order Neuroptera

Thread-winged lacewing
Nemoptera sinuata,
order Neuroptera

Hanging fly
Bittacus tipularius,
order Mecoptera

Male hanging fly
offers nuptial gift
of prey to female

Bulging
compound
eyes

Owlfly
Ascalaphus libelluloides,
order Neuroptera

Owlflies resemble
dragonflies but have
longer antennae,
and rest with their
wings folded back

Stonefly
Perla burmeisteriana,
order Plecoptera

Nymph of
Perla burmeisteriana

Caddisfly
Phryganea grandis,
order Trichoptera

Land caddis larva lives
in moist leaf litter in
a case made of sand

Snow scorpionfly
Boreus hiemalis,
order Mecoptera

Land caddis
Enoicyla pusilla,
order Trichoptera

Blackfly hunter
larvae feed on
the larvae of
blackflies

Green lacewing
larva feeds on
aphids and mites

Like many other caddisflies,
the blackfly hunter larva
lives in water and is protected
by a portable case

Blackfly hunter
Rhyacophila nubila, order Trichoptera

Green lacewing
Chrysopa vulgaris, order Neuroptera

FACT FILE

Order Neuroptera The members of this order all have complex branching veins on their wings. Most are predators as both larvae and adults. Lacewing larvae are welcome in gardens because they devour aphids, mites, and scale insects. Antlion larvae live in sand and feed mainly on ants. Owlfly larvae live in leaf litter near stones, and prey on invertebrates. Spongefly larvae live in fresh water and eat only sponges.

Species 4,000

Worldwide; often near vegetation

Pit trap Antlion larvae, known as doodlebugs in North America, dig pits in loose sand to trap ants and other prey.

Order Mecoptera This order includes scorpionflies, named after the large tail the male uses to hold onto the female during mating, and hanging flies, which hang from leaves to hunt and mate. Males often offer their mates a nuptial gift of a prey item or a pellet of saliva that the female eats during copulation. Most species feed on detritus, insects, and plants.

Species 500

Worldwide; near
damp vegetation

Stingless scorpion
Like other males in its genus, the male common scorpionfly (Panorpa communis) is harmless but has an enlarged "tail" that resembles the stinger of a scorpion.

Order Plecoptera The nymphs of the stoneflies in this order can indicate the environmental health of streams. They are entirely aquatic and can survive only in cool, clean, flowing water with high concentrations of oxygen. After going through up to 30 molts, nymphs crawl from the water and emerge as adults, which are usually winged.

Species 3,000

Worldwide except Australia;
near running water

Feeding strategies
While adult stoneflies of the genus Perla may not feed at all, the aquatic nymphs are carnivorous and feed on other larvae.

INSECT ALLIES

PHYLUM	Arthropoda
SUBPHYLUM	Hexapoda
CLASSES	3
ORDERS	5
FAMILIES	31
SPECIES	8,300

Hexapods are arthropods with six legs and a body divided into three parts—head, thorax, and abdomen. The vast majority of hexapods are insects, but there are also three classes of non-insect hexapods—the springtails (class Collembola), proturans (Protura), and diplurans (Diplura). Found in soil and leaf litter, these tiny creatures range from ⅟₅₀ inch to 1¼ inches (0.5 mm to 3 cm) in length. They are distinguished from insects by their mouthparts, which are fully tucked into the head. As in the most primitive of insects, the silverfish and the bristletails, these tiny soil-dwellers evolved from a wingless ancestor and they continue to molt throughout their life.

Ready to leap Springtails are ubiquitous, occurring in freshwater, coastal marine, and most terrestrial habitats, including deserts and polar zones. They feed mainly on microorganisms. Surface-dwellers tend to be covered in hairs (above) or scales.

PRIMITIVE HEXAPODS

The springtails that make up the class Collembola are named after a forked organ on the abdomen that can propel them up to 100 times their own body length—a quick way to escape danger. They are among the most abundant animals, with populations of 300 million individuals per acre (750 million per ha) in some grassland habitats.

With a length of less than ⅟₁₂ inch (2 mm), proturans are rarely seen and were not even discovered until 1907. They live in soil and leaf litter, where they feed on fungi and decaying matter. They are the most primitive of all hexapods and lack eyes and antennae. Instead, the forelegs perform a sensory function and are held out in front of the animal rather than being used for walking. With each molt, proturans gain another segment of abdomen.

Diplurans lack eyes but feel their way around in soil or leaf litter with their long antennae and two tail-like structures known as cerci. In some carnivorous species, the cerci have become powerful pincers that can capture other soil-dwellers. Other species are herbivorous and feed on soil fungi and detritus. Diplurans are able to regenerate some of their body parts.

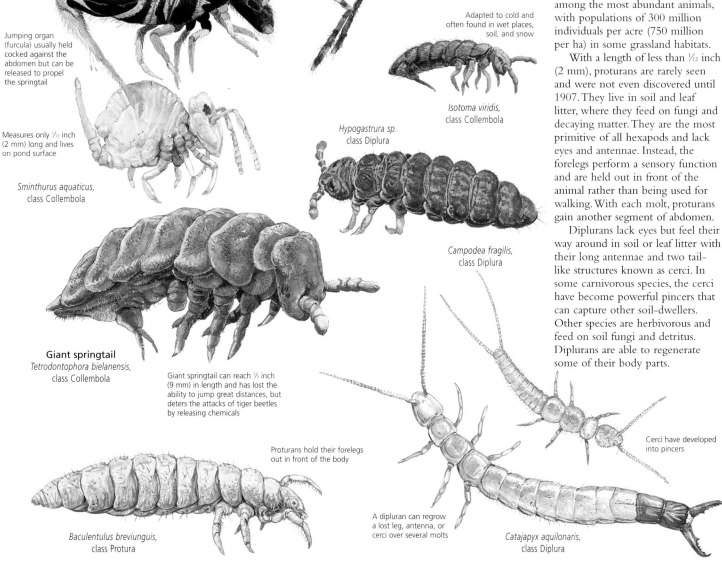

Entomobrya superba, class Collembola

Long legs often found on surface-dwelling springtails

Jumping organ (furcula) usually held cocked against the abdomen but can be released to propel the springtail

Measures only ⅟₁₂ inch (2 mm) long and lives on pond surface

Sminthurus aquaticus, class Collembola

Giant springtail
Tetrodontophora bielanensis, class Collembola

Giant springtail can reach ⅓ inch (9 mm) in length and has lost the ability to jump great distances, but deters the attacks of tiger beetles by releasing chemicals

Eosentomon ribagai, class Protura

Adapted to cold and often found in wet places, soil, and snow

Isotoma viridis, class Collembola

Hypogastrura sp. class Diplura

Campodea fragilis, class Diplura

Cerci have developed into pincers

Proturans hold their forelegs out in front of the body

A dipluran can regrow a lost leg, antenna, or cerci over several molts

Baculentulus breviunguis, class Protura

Catajapyx aquilonaris, class Diplura

ECHINODERMS

PHYLUM	Echinodermata
CLASSES	5
ORDERS	36
FAMILIES	145
SPECIES	6,000

The sea stars, sea urchins, brittle stars, feather stars, and sea cucumbers that belong to the phylum Echinodermata display diverse body forms, but share some defining characteristics. Adults usually have pentamerous symmetry, with the body arranged into a five-part radial pattern around a central axis. Most internal organs are also arranged in this pattern. An internal skeleton of calcareous plates provides protection and support. The plates often bear spines or small lumps—the name echinoderm means "spiny skin." The body cavity includes a water-vascular system, a network of water vessels that controls the tube feet used for locomotion, feeding, respiration, and sensory functions.

Sticky defense In sea cucumbers, the five-part symmetry is internal and the skeleton is reduced. This leopard sea cucumber (*Bohadschia argus*) has ejected a mass of sticky white threads, known as cuvierian tubules, to confuse or entangle a predator.

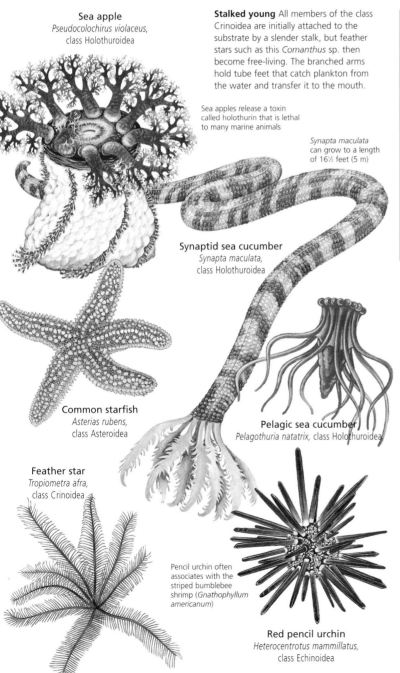

Sea apple
Pseudocolochirus violaceus,
class Holothuroidea

Stalked young All members of the class Crinoidea are initially attached to the substrate by a slender stalk, but feather stars such as this *Comanthus* sp. then become free-living. The branched arms hold tube feet that catch plankton from the water and transfer it to the mouth.

Sea apples release a toxin called holothurin that is lethal to many marine animals

Synapta maculata can grow to a length of 16½ feet (5 m)

Synaptid sea cucumber
Synapta maculata,
class Holothuroidea

Common starfish
Asterias rubens,
class Asteroidea

Feather star
Tropiometra afra,
class Crinoidea

Pelagic sea cucumber
Pelagothuria natatrix, class Holothuroidea

Pencil urchin often associates with the striped bumblebee shrimp (*Gnathophyllum americanum*)

Red pencil urchin
Heterocentrotus mammillatus,
class Echinoidea

SPINY BOTTOM-DWELLERS

All echinoderms are marine and most are mobile bottom-dwellers, although sea lilies are fixed to the seafloor by a long stalk, and a few sea cucumbers float in the open ocean. Echinoderms are found throughout the world at all depths. Sea urchins and sea stars are very commonly seen along the seashore.

The phylum Echinodermata is divided into five classes. Members of the class Crinoidea include the sessile sea lilies as well as the mobile feather stars. They are filter-feeders and, unlike other echinoderms, have their mouth facing away from the substrate. The other echinoderm classes include predators, grazers of algae, and deposit-feeders. The sea stars of class Asteroidea and brittle stars of class Ophiuroidea have arms radiating from their body. Their skeletal plates are held together by muscles, which allows flexibility. In the class Echinoidea, sea urchins and sand dollars have a rigid skeleton of fused plates that supports a globular or flattened form, usually featuring

Spiky gathering Many echinoderms, such as these sea urchins, gather in large groups, influenced by the availability of food. The strategy may also provide some protection from predators and enhance reproduction.

prominent spines. Sea cucumbers (class Holothuroidea) are generally soft-bodied, with their skeletons reduced to minute spicules.

Although a few echinoderms can reproduce asexually, most have separate sexes that release eggs and sperm into the water for fertilization. The larvae are often free-swimming and metamorphose into bottom-dwelling adults. In some species, the larval stage is omitted and the newly hatched young resemble small adults.

FACT FILE

Class Asteroidea Also known as starfish, the sea stars in this class move and feed using the suckered tube feet along their arms. Most are predators that evert their stomach over sessile or slow-moving prey to begin digesting it. Some use their tube feet to prise open the shells of bivalves such as clams. Sea stars are usually vibrant colors such as red, orange, or purple and may have a smooth, spiny, or ridged surface.

Species 1,500

Worldwide; on the seafloor

Many arms The common sun star (Crossaster papposus) can have as many as 40 arms and grow to about 1 foot (30 cm) in diameter.

Class Ophiuroidea In contrast to sea stars, in which the arms grade into the central disk, the brittle stars and basket stars of Ophiuroidea have sharply demarcated arms. Rather than using their tube feet to crawl, they move by using any two of their flexible, jointed arms in a breast-stroke action. The class includes predators, scavengers, and filter-feeders.

Surface star The brittle star (Ophiura ophiura) usually lives on the surface of sandy or muddy sediment but may also burrow shallowly.

Species 2,000

Worldwide; on the seafloor

Class Echinoidea The sea urchins in this class have a spherical body protected by a skeletal case called a test, and long, movable spines. Sand dollars are flattened and covered in smaller spines. Many echinoids have a system of plates, muscles, and teeth known as the Aristotle's lantern, with which they scrape algae from rocks.

Species 950

Worldwide; on or in the seafloor

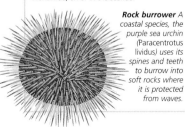

Rock burrower A coastal species, the purple sea urchin (Paracentrotus lividus) uses its spines and teeth to burrow into soft rocks where it is protected from waves.

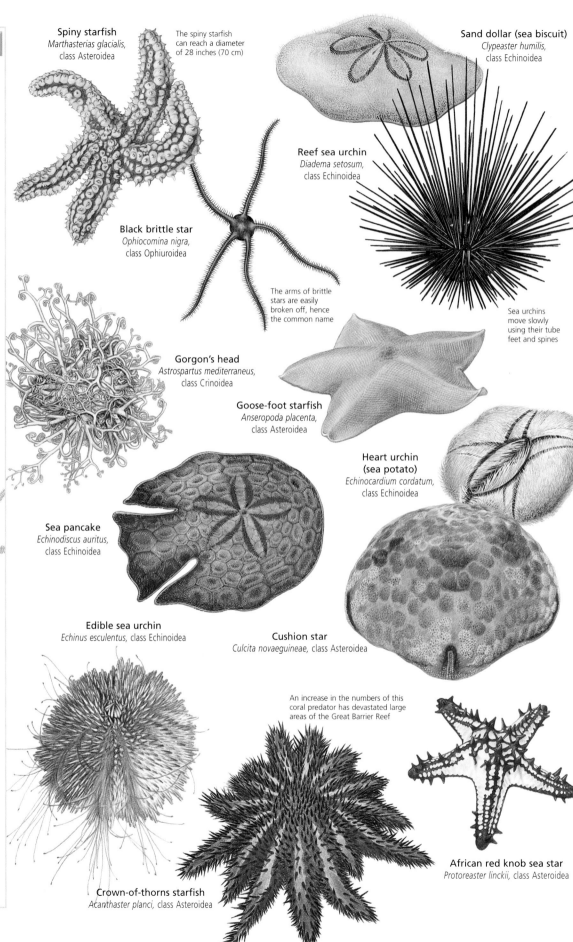

Spiny starfish
Marthasterias glacialis, class Asteroidea

The spiny starfish can reach a diameter of 28 inches (70 cm)

Sand dollar (sea biscuit)
Clypeaster humilis, class Echinoidea

Reef sea urchin
Diadema setosum, class Echinoidea

Black brittle star
Ophiocomina nigra, class Ophiuroidea

The arms of brittle stars are easily broken off, hence the common name

Sea urchins move slowly using their tube feet and spines

Gorgon's head
Astrospartus mediterraneus, class Crinoidea

Goose-foot starfish
Anseropoda placenta, class Asteroidea

Heart urchin (sea potato)
Echinocardium cordatum, class Echinoidea

Sea pancake
Echinodiscus auritus, class Echinoidea

Edible sea urchin
Echinus esculentus, class Echinoidea

Cushion star
Culcita novaeguineae, class Asteroidea

An increase in the numbers of this coral predator has devastated large areas of the Great Barrier Reef

African red knob sea star
Protoreaster linckii, class Asteroidea

Crown-of-thorns starfish
Acanthaster planci, class Asteroidea

OTHER INVERTEBRATES

The preceding pages have covered eight major invertebrate phyla, as well as two subphyla of invertebrate chordates. There are, however, another 25 phyla of invertebrates, many of which are profiled in this chapter. Although the phylum Bryozoa contains 5,000 species, most of these minor groups are small—Phoronida (horseshoe worms), for example, contains a mere 20 species. The members of these phyla also tend to be physically small and are often microscopic. Most are marine, although many are found in fresh water and some occur on land. Often overlooked, the minor invertebrate phyla all display fascinating solutions to the challenges presented by their environment.

MINOR INVERTEBRATE PHYLA	
PHYLA	25
CLASSES	> 40
ORDERS	> 60
SPECIES	> 12,000

Ribbon worms Roughly 1,000 species of long ribbon worms make up the phylum Nemertea. The majority are marine, but some live in fresh water or moist soil. They trap or stab their invertebrate prey using a unique projectile proboscis.

Invading comb jelly Native to America's Atlantic coast, the sea walnut (*Mnemiopsis leidyi*) was accidentally introduced into the Black Sea in the early 1980s. Explosive population growth followed, and soon this predator of zooplankton and fish eggs and larvae had devastated the ecosystem and triggered fishery crashes.

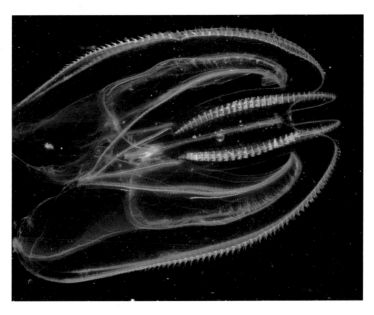

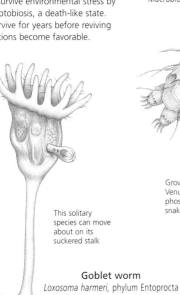

Extreme survivors Water bears such as *Echiniscus testudo* (shown here magnified) are able to survive environmental stress by entering cryptobiosis, a death-like state. They can survive for years before reviving when conditions become favorable.

FACT FILE

Phylum Entoprocta Attached by a stalk to the substrate, goblet worms use a ring of mucus-secreting tentacles to filter particles from the water. Many species form colonies by budding.

Species 150

Worldwide; mostly on the seafloor

Phylum Tardigrada Unknown until 1773, following the invention of the microscope, water bears crawl over ocean sediment, soil, or plants with a slow, bear-like gait. These minute creatures occur in almost all aquatic and moist terrestrial environments.

Species 600

Worldwide; aquatic & moist land habitats

Phylum Ctenophora The marine comb jellies that make up this phylum swim by beating the fused cilia (tiny hairs) arranged in eight comb rows along their body. They engulf their prey directly or capture it with a pair of sticky, retractable tentacles.

Species 100

Worldwide; mostly planktonic

Water bear
Macrobiotus hufelandi, phylum Tardigrada

This solitary species can move about on its suckered stalk

Goblet worm
Loxosoma harmeri, phylum Entoprocta

Growing up to 5 feet (1.5 m) long, Venus' girdle glows with a greenish phosphorescence and swims with a snake-like movement

Venus' girdle
Cestus veneris, phylum Ctenophora

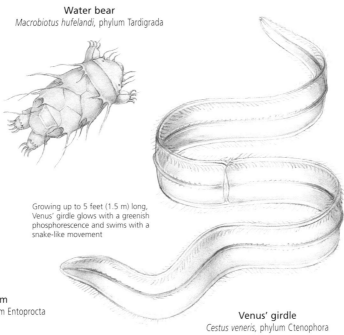

Melon jellyfish
Beroe cucumis, phylum Ctenophora

Can engulf prey (such as other comb jellies) that are as large as itself

FACT FILE

Phylum Rotifera Sometimes known as wheel animals, rotifers are microscopic aquatic creatures. A crown of cilia (hair-like structures) around the mouth beat rapidly to collect food and propel the rotifer, creating the impression of a whirling wheel. The trunk is often protected by a stiffened epidermis and sometimes bears defensive spines. Most rotifer species are found in fresh water, where they make up much of the zooplankton.

Species 1,800

Worldwide; in aquatic vegetation & moist terrestrial habitats

Swimming rotifer
Many rotifer species, including Brachionus plicatilis, *are free-swimming. Others use their cement-producing toes to attach to a surface.*

Phylum Hemichordata Known as acorn worms or hemichordates, the members of this phylum have a three-part body made up of a proboscis at the head; a collar in the middle, which bears tentacles in some species; and a trunk containing the digestive and reproductive organs. They feed on small particles, either filtered from the water or consumed with sediment.

Species 90

Worldwide; on the seafloor

Close relatives *Hemichordates, such as* Saccoglossus cambrensis, *are closely related to the chordates but lack a notochord.*

Phylum Chaetognatha The arrow worms of this phylum are voracious predators. Their head has up to 14 large grasping spines for seizing prey such as copepods or small fishes. When the spines are not in use, a hood formed from the body wall covers the head to protect them. Most species are plank-tonic and many migrate from deeper waters to feed at the surface at night.

Species 90

Worldwide; in plankton & on the seafloor

Torpedo worm
Like other arrow worms, Sagitta setosa *has a torpedo-shaped body with dorsal and ventral fins for balance and a tail fin for thrust.*

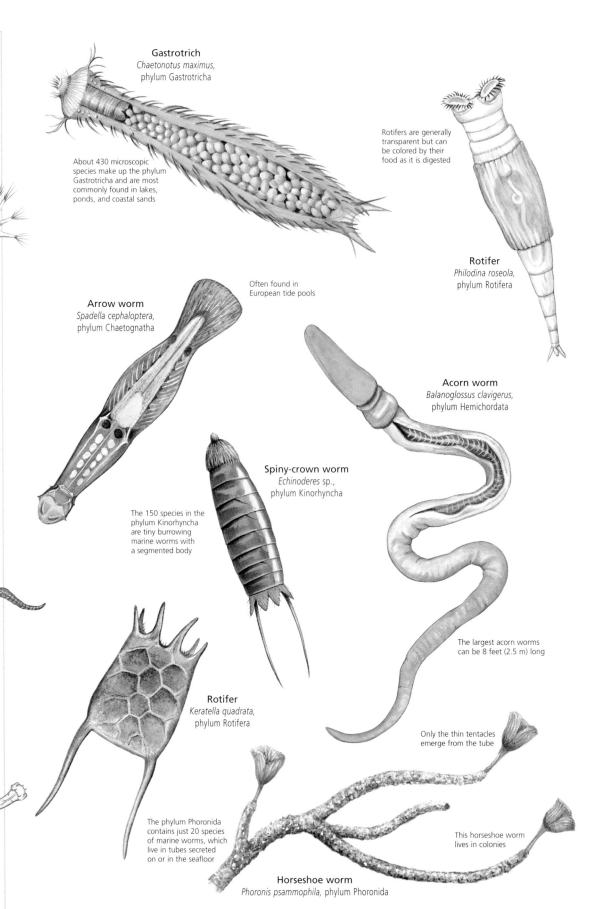

Gastrotrich
Chaetonotus maximus,
phylum Gastrotricha

About 430 microscopic species make up the phylum Gastrotricha and are most commonly found in lakes, ponds, and coastal sands

Rotifers are generally transparent but can be colored by their food as it is digested

Rotifer
Philodina roseola,
phylum Rotifera

Often found in European tide pools

Arrow worm
Spadella cephaloptera,
phylum Chaetognatha

Acorn worm
Balanoglossus clavigerus,
phylum Hemichordata

Spiny-crown worm
Echinoderes sp.,
phylum Kinorhyncha

The 150 species in the phylum Kinorhyncha are tiny burrowing marine worms with a segmented body

The largest acorn worms can be 8 feet (2.5 m) long

Rotifer
Keratella quadrata,
phylum Rotifera

Only the thin tentacles emerge from the tube

The phylum Phoronida contains just 20 species of marine worms, which live in tubes secreted on or in the seafloor

This horseshoe worm lives in colonies

Horseshoe worm
Phoronis psammophila, phylum Phoronida

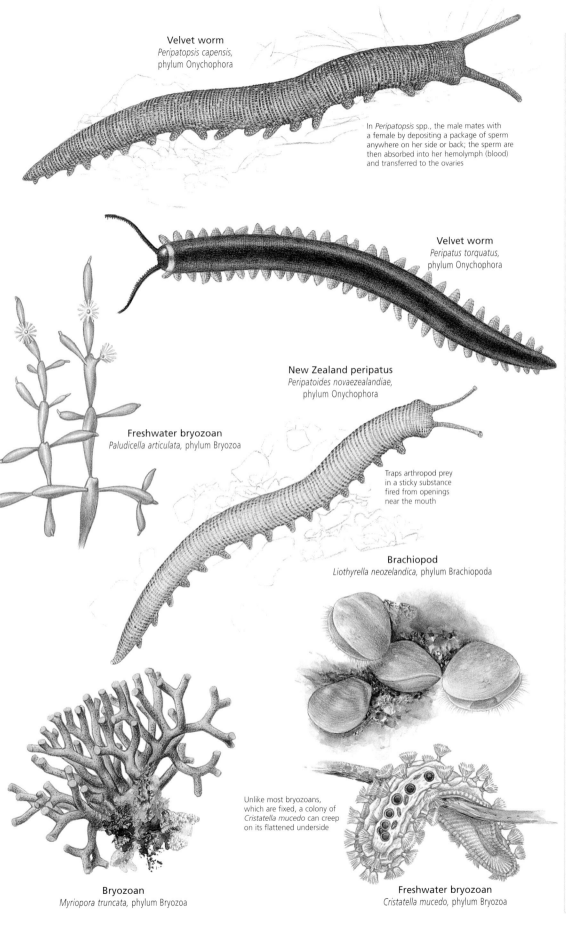

Velvet worm
Peripatopsis capensis,
phylum Onychophora

In *Peripatopsis* spp., the male mates with a female by depositing a package of sperm anywhere on her side or back; the sperm are then absorbed into her hemolymph (blood) and transferred to the ovaries

Velvet worm
Peripatus torquatus,
phylum Onychophora

New Zealand peripatus
Peripatoides novaezealandiae,
phylum Onychophora

Freshwater bryozoan
Paludicella articulata, phylum Bryozoa

Traps arthropod prey in a sticky substance fired from openings near the mouth

Brachiopod
Liothyrella neozelandica, phylum Brachiopoda

Bryozoan
Myriopora truncata, phylum Bryozoa

Unlike most bryozoans, which are fixed, a colony of *Cristatella mucedo* can creep on its flattened underside

Freshwater bryozoan
Cristatella mucedo, phylum Bryozoa

FACT FILE

Phylum Onychophora Since their thin, non-waxy cuticle does little to prevent water loss, the velvet worms of this phylum are restricted to moist land habitats. They are found in leaf litter and soil and under stones and rotting logs. To entangle insects and other prey, velvet worms squirt out sticky threads from slime glands near the mouth, a technique that can also be used in defense.

Species 70

Tropical & southern temperate zones

Like velvet *Like other velvet worms,* Peripatus trinitatis *is covered in tiny projections that give it a velvety appearance. The cylindrical body can have 13 to 43 pairs of stumpy legs.*

Phylum Brachiopoda With their bivalved shells, brachiopods look like clams or mussels, but are in fact much more closely related to bryozoans. Both brachiopods and bryozoans filter food particles from the water with a lophophore, a crown of hollow tentacles around the mouth.

Species 350

Worldwide; on the seafloor at all levels of the ocean

Burrowing stalk
The brachiopod Lingula anatina *uses it fleshy stalk, or pedicle, to burrow into sediment. Most brachiopods, however, attach themselves to rocks.*

Phylum Bryozoa Sometimes called moss animals, bryozoans are tiny, aquatic, and colonial. Most secrete a protective covering around the body with an opening for the lophophore (feeding tentacles). Created by budding, the colonies usually grow as encrustations on rocks or seaweed, or have a branching form, but a few are free-living.

Species 5,000

Worldwide; on the bottom of aquatic habitats

Fixed in place *The bryozoan* Bowerbankia imbricata *has an erect branching form rising from an attached stolon.*

GLOSSARY

abdomen The part of the body containing the digestive system and the reproductive organs. In insects and spiders, the abdomen makes up the rear of the body.

adaptation A change in an animal's behavior or body that allows it to survive and breed in new conditions.

adaptive radiation A situation in which animals descended from a common ancestor evolve to exploit different ecological niches that are not being filled by other animals, as in the finches of the Galápagos Islands and the lemurs of Madagascar.

algae The simplest forms of plant life.

altricial Helpless at birth. Describes newly hatched chicks of some birds; cf precocial.

amphibious Able to live on land and in water. Amphibians (frogs, toads, salamanders, caecilians, and newts) are vertebrates that are similar to reptiles, but have moist skin and lay their eggs in water.

amplexus A breeding position of frogs and toads, in which the male holds the female with his front legs.

anadromous Living in the ocean but returning to fresh water to spawn, as in salmon; cf catadromous.

anal fin An unpaired fin on the lower surface of a fish's abdomen. It plays an important role in swimming.

annulus (pl. annuli) A marking resembling a ring, as in caecilians; or a growth ring, as on a scale of a fish, that is used in estimating age.

antenna (pl. antennae) A slender organ on an animal's head, used to sense smells and vibrations around it. Insects have two antennae.

antlers Bony, often branched growths from the head of deer and moose. In most cases, they are grown and shed each year. They are used as weapons and for display; cf horns.

aquaculture Cultivation of the resources of the sea or inland waters, as opposed to their exploitation.

aquatic Living all or most of the time in water; cf amphibious, terrestrial.

arachnid An arthropod with four pairs of walking legs; includes spiders, scorpions, ticks, and mites.

arboreal Living all or most of the time in trees.

arthropod An animal with jointed legs and a hard exoskeleton. Arthropods make up the largest group of animals on Earth and include insects, spiders, crustaceans, centipedes, and millipedes.

avian Of or about birds. Birds form the class Aves in the animal kingdom.

baleen The comb-like, fibrous plates found in some whales; often referred to as whalebone. The plates hang from the upper jaw and are used to sieve food from sea water.

barb A part of a bird's feather. Barbs emerge from the central shaft of a feather in a parallel arrangement, like the teeth on a comb.

barbel A slender, fleshy outgrowth on an animal's lips or near its mouth. It is equipped with sensory and chemical receptors and is used by such animals as bottom-dwelling fishes to find food.

benthic Relating to or occurring at the bottom of a body of water or the depths of the ocean; bottom dwelling (as in fishes).

biodiversity The total number of species of plants and animals in a particular location.

birds of prey Flesh-eating land birds that hunt and kill their prey. Hawks, eagles, kites, falcons, buzzards, and vultures are diurnal birds of prey; owls are nocturnal birds of prey.

blubber A thick layer of insulating fat in whales, seals, and other large marine mammals.

bony fish A fish with a bony skeleton. Other characteristics are a covering over the gills, and a swim bladder. Most bony fishes also have scales on their skin.

book lungs Lungs found in primitive insects and arachnids, consisting of "pages" of tissue through which blood and oxygen circulate.

brachiation A form of movement in which an animal (such as an ape) swings from hold to hold by its arms.

brood parasite An animal, often a bird, that tricks another species into raising its young, as in some cuckoos. A young brood parasite often kills all its nestmates so that they do not compete with it for food or care.

browser A plant-eating mammal that uses its hands or lips to pick leaves from trees and bushes (as in koalas and giraffes) or low-growing plants (as in the black rhino).

camouflage The colors and patterns of an animal that enable it to blend in with the background. Camouflage conceals the animal from predators and helps it to ambush prey.

canine teeth The teeth between the incisors and molars of mammals.

carnassials Special cheek teeth with sharp, scissor-like edges used by carnivores to tear up food.

carnivore An animal that eats mainly meat. Most carnivorous mammals are predators; others are both predators and scavengers. Most carnivores eat some plant material as well as meat.

carrion The rotting flesh and other remains of dead animals.

cartilaginous fish A fish with a skeleton made of cartilage, such as a shark, ray, or chimaera.

catadromous Living in fresh water but returning to the sea to spawn, as in freshwater eels; cf anadromous.

caudal Relating to an animal's tail. In fishes, the caudal fin is the tail fin.

cephalic Of, relating to, or situated on or near the head.

cephalothorax In arachnids and crustaceans, a region of the body that combines the head and thorax. It is covered by a hard body case.

chelicera (pl. chelicerae) A pincer-like, biting mouthpart, such as a fang, as in spiders, ticks, scorpions, and mites.

chitin A hard, plastic-like substance that gives an exoskeleton its strength.

chrysalis The form taken by a butterfly in the pupal stage of its metamorphosis, or the case in which it undergoes this stage.

claspers The modified inner edges of the pelvic fins of male sharks, rays, and chimaeras, used for the transferring of sperm to the female.

cloaca An internal chamber in fishes, amphibians, reptiles, birds, and monotremes, into which the contents of the reproductive ducts and the waste ducts empty before being passed from the body.

clutch The full and completed set of eggs laid by a female bird or reptile in a single nesting attempt.

cocoon In insects and spiders, a case made of silk used to protect themselves or their eggs. In amphibians, a protective case made of mud, mucus, or a similar material, in which the animal rests during estivation.

cold-blooded See ectothermic.

complete metamorphosis A form of development in which a young insect changes shape from an egg to a larva, to a pupa, to an adult. Insects such as beetles and butterflies develop by complete metamorphosis.

compound eye An eye that is made up from many smaller eyes, each with its own lens; found in most insects, but not in spiders.

conspecific Of the same species.

convergent evolution The situation in which totally unrelated groups develop similar structures to cope with similar evolutionary pressures.

coral reef A structure made from the skeletons of coral animals, or polyps, found in warm waters.

crest In lizards, a line of large, scaly spines on the neck and back. In birds, a line of feathers on the top of the head. Lizards and many birds can raise or lower the crest as a means of communication with others.

crop A thin-walled, saclike pocket of the gullet, used by birds to store food before digestion or to feed chicks by regurgitation.

crocodilians Crocodiles, caimans, alligators, and gavials. Members of the order Crocodilia.

crustacean A mostly aquatic animal, such as a lobster, crab, or prawn, that has a hard external skeleton.

deforestation The cutting down of forest trees for timber, or to clear land for farming or building.

dewlap In a lizard, a flap of skin, sometimes brightly colored, on the throat. A dewlap usually lies close to the neck and can be extended to communicate with other lizards.

dimorphic Having two distinct forms within a species. Sexual dimorphism is the situation in which the male and female of a species differ in size and/or appearance.

dinosaurs A group of reptiles that dominated Earth from the Triassic to the Cretaceous Period (245–65 million years ago). The largest land animals that ever lived, dinosaurs are more closely related to today's birds and crocodilians than they are to other living reptiles.

dioecious Born either male or female and remaining that way through life; cf hermaphroditic.

display Behavior used by an animal to communicate with its own species, or with other animals. Displays can include postures, actions, or showing brightly colored parts of the body, and may signal threat, defense, or readiness to mate.

diurnal Active during the day. Most reptiles are diurnal because they rely on the Sun's heat to provide energy for hunting and other activities.

divergent evolution The situation in which two or more similar species become more and more dissimilar due to environmental adaptations.

DNA A molecule, found in chromosomes of a cell nucleus, that contains genes. DNA stands for deoxyribonucleic acid.

domestication The process of taming and breeding an animal for human use. Domesticated animals include pets, as well as animals used for sport, food, or work, such as sheep, horses, and dairy cattle.

dorsal fin The large fin on the back of some fishes and aquatic mammals, which helps the animal to keep its balance as it moves through the water. Some fishes have two or three dorsal fins.

dragline A strand of silk that spiders leave behind them as they move about.

drone A male honeybee. Drones mate with young queens, but unlike worker bees, they do not help in collecting food or maintaining the hive.

echolocation A system of navigation that relies on sound rather than sight or touch. Dolphins, porpoises, many bats, and some birds use echolocation to tell them where they are, where their prey is, and if anything is in their way.

ecosystem A community of plants and animals and the environment to which they are adapted.

ectothermic Unable to internally regulate the body temperature, instead relying on external means such as the Sun, as in cold-blooded animals such as reptiles; cf endothermic.

egg sac A silk bag that some spiders spin around their eggs. Some egg sacs are portable, so the eggs can be carried from place to place. Others are more like a blanket, and are attached to leaves or suspended from twigs.

egg tooth A special scale on the tip of the upper lip of a hatchling lizard or snake. It is used to break a hole in the egg so that the newborn animal can escape. The egg tooth falls off within a few days of hatching.

electroreceptors Specialized organs found in some fishes (such as sharks) and mammals (such as platypuses) that detect electrical activity from the bodies of other animals. They also help the animal to navigate by detecting distortions in the electrical field of its surroundings—for example, those caused by a reef.

elytra The thickened front wings of beetles that cover and protect the back wings.

embryo An unborn animal in the earliest stages of development. An embryo may grow inside its mother's body, or in an egg outside her body.

endothermic Able to regulate the body temperature internally, as in warm-blooded animals; cf exothermic.

estivate To spend a period of time in a state of inactivity to avoid unfavorable conditions. During times of drought, many frogs greatly reduce their overall metabolic rate and estivate underground until rain falls.

evolution Gradual change in plants and animals, over many generations, in response to their environment.

exoskeleton A hard external skeleton or body case. All arthropods are protected by an exoskeleton.

exotic A foreign or non-native species of animal or plant, often introduced into a habitat by humans.

feral A wild animal or plant, or a species that was once domesticated but has returned to its wild state.

fledgling A young bird that has recently grown its first true feathers and has just left its nest.

flippers The broad front (and often, back) legs of some aquatic animals. Flippers are composed mainly of the bones of the fingers and hand, and act like paddles to row the animal through the water.

food chain A system in which one organism forms food for another, which in turn is eaten by another, and so on. The first organism is usually an alga or other single-celled life form in an aquatic food chain, and a plant in a terrestrial food chain.

fossil A remnant, impression, or trace of a plant or animal from a past geological age, usually found in rock.

fossorial Adapted to digging or burrowing underground.

frill In a lizard, a collar around the neck. Like a crest or dewlap, a frill can be raised to signal to other lizards or to surprise a predator.

fry Young or small fishes.

gastroliths Stones swallowed by such animals as crocodilians, that stay in the stomach to help crush food.

gestation period The period of time during which a female animal is pregnant with her young.

gills Organs that collect oxygen from water and are used for breathing. Gills are found in many aquatic animals, including fishes, some insects, and the larvae of amphibians.

gizzard In birds, the equivalent of the stomach in mammals. Grit and stones inside the gizzard help to grind up food. Food passes from the gizzard to the intestines.

global warming The increase in the temperature of Earth and its lower atmosphere due to human activity such as deforestation, land degradation, intensive farming, and the burning of fossil fuels. This causes the absorption of heat by water vapor and "greenhouse" gases, including carbon dioxide and methane. The trapped heat, which would otherwise be radiated out to space, may cause the polar icecaps to melt and ocean levels to rise. Global warming is also known as "the greenhouse effect."

Gondwana Ancient southern supercontinent, comprising the present-day continents of Australia, India, Africa, South America, and Antarctica; see also Laurasia, Pangea.

gravid Full of eggs; pregnant.

grazer An animal that eats grasses and plants that grow on the ground.

greenhouse effect See global warming.

groom Of an animal, to clean, repair, and arrange the fur.

grub An insect larva, usually that of a beetle.

gullet Found in birds, the gullet is the equivalent of the esophagus in mammals. This tube passes food from the bill to the gizzard.

habitat The area in which an animal naturally lives. Many different kinds of animals live in the same environment (for example, a rain forest), but each kind lives in a different habitat within that environment. For example, some animals in a rain forest live in the trees, while others live on the ground.

harem A group of female animals that mate and live with one male.

hatchling A young animal, such as a bird or reptile, that has recently hatched from its egg.

heat-sensitive pit Sense organs in some snakes that detect tiny changes in temperature.

herbivore An animal that eats only plant material, such as leaves, bark, roots, and seeds; cf carnivore, omnivore.

hermaphrodite A plant or animal either possessing both male and female sex organs at the same time, and thus able to fertilize itself (known as simultaneous herma-phroditism), or one that starts life as one sex and later changes to the

other (sequential hermaphroditism); see also protandrous, protogynous.

hibernate To remain completely inactive during the cold winter months. Some animals eat as much as they can before the winter, then curl up in a sheltered spot and fall into a very deep sleep. They live off stored fat, and slow their breathing and heartbeat to help conserve energy until spring. Insects may hibernate as eggs, larvae, pupae, or adults.

hoof The toe of a horse, antelope, deer, or related animal that is covered in thick, hard skin with sharp edges.

horns In mammals such as ruminants, pointed, hollow, bony outgrowths on the head. Unlike antlers, horns are not shed.

hybrid The offspring of parents of two different species.

incisors The front teeth of an animal, located between the canines, used for cutting food.

incomplete metamorphosis A system of development in which a young insect gradually changes shape from an egg, to a nymph, to an adult.

incubate To keep eggs in an environment, outside the female's body, in which they can develop and hatch. Most birds incubate their eggs by warming them with their body heat. The eggs of other birds and reptiles may incubate in soil, leaf litter, or a similar covering, with no further input from the parents.

insectivore An animal that eats only or mainly insects or invertebrates. Some insectivores also eat small vertebrates, such as frogs, lizards, and mice. In mammals, a member of the order Insectivora.

introduced An animal or plant species imported from another place by humans and deliberately or accidentally released into a habitat.

invertebrate An animal with no backbone. Many invertebrates are soft-bodied animals, such as worms, leeches, or octopuses, but most have an exoskeleton, or hard external skeleton, such as insects.

ivory The tusks or enlarged teeth of elephants and walrus, or objects manufactured from them.

Jacobson's organ Two small sensory pits on the top part of the mouth of snakes, lizards, and mammals. They use this organ to analyze small molecules picked up from the air or ground with their tongue.

keratin A protein found in horns, hair, scales, feathers, and fingernails.

krill Small, shrimplike crustaceans that live in huge numbers in Arctic and Antarctic waters.

lateral line A sensory canal system along the sides of a fish. It detects moving objects by registering disturbances (or pressure changes) in the water.

larva (pl. larvae) A young animal that looks completely different from its parents. An insect larva, sometimes called a grub, maggot, or caterpillar, changes into an adult by either complete or incomplete metamorphosis. In amphibians, the larval stage is the stage before metamorphosis that breathes with gills rather than lungs (for example, tadpoles).

Laurasia Ancient northern supercontinent, comprising the present-day Asia, North America, and Europe; see also Gondwana, Pangea.

live-bearing Giving birth to young that are fully formed.

luminous Reflecting or radiating light. Some deepsea fishes can produce their own light source by using luminous bacteria.

maggot The larva of a fly.

mammal A warm-blooded vertebrate that suckles its young with milk and has a single bone in its lower jaw. Although most mammals have hair and give birth to live young, some, such as whales and dolphins, have very little or no hair; others, the monotremes, lay eggs.

mandibles Biting jaws of an insect.

marsupial A mammal that gives birth to young that are not fully developed. These young are usually protected in a pouch (where they feed on milk) before they can move around independently.

mesopelagic Of or associated with the deep sea, from about 600 to 3,000 feet (200 to 1,000 m).

metamorphosis A way of development in which an animal's body changes shape. Many invertebrates, including insects, as well as some vertebrates, such as amphibians, metamorphose as they mature.

migration A usually seasonal journey from one habitat to another. Many animals migrate vast distances to another location to find food, or to mate and lay eggs or give birth. Some deep-sea fishes migrate vertically at night to feed.

molars A mammal's side cheek teeth, used for crushing and grinding.

mollusk An animal, such as a snail or squid, with no backbone and a soft body that is often partly or fully enclosed by a shell.

molt To shed an outer layer of the body, such as hair, skin, scales, feathers, or the exoskeleton.

monotreme A primitive mammal with many features in common with reptiles. Monotremes lay eggs and have a cloaca. They are the only mammals that lack teats, although they feed their young milk released through ducts on their belly.

morph A color or other physical variation within, or a local population of, a species.

musth In elephants, a time of high testosterone levels when the musth gland between the eye and ear secretes fluid; associated with mating and characterized by increased aggression and searching for females.

mutualism An alliance between two species that is beneficial to both, as in ox-peckers and grazing herbivores.

neoteny The retention of some immature or larval characteristics into adulthood. Some salamander species are usually or often neotenic.

niche The ecological role played by a species within an animal community.

nocturnal Active at night. Nocturnal animals have special adaptations, such as large, sensitive eyes or ears, to help them find their way in the dark. All nocturnal animals rest during the day.

nomadic Lacking a fixed territory, instead wandering from place to place in search of food and water.

nymph The young stage of an insect that develops by incomplete metamorphosis. Nymphs are often similar to adults, but do not have fully developed wings.

ocellus A kind of simple eye with a single lens. Insects usually have three ocelli on the top of their head.

omnivore An animal that eats both plant and animal food. Omnivores have teeth and a digestive system designed to process almost any kind of food.

opposable Describing a thumb that can reach around and touch all of the other fingers on the same hand, or a toe that can similarly touch all of the other toes on the same foot.

order A major group used in taxonomic classification. An order forms part of a class, and is further divided into one or more families.

osteoderm In reptiles, a lump of bone in the skin that provides protection against predators. Most crocodilians and some lizards are protected by osteoderms, as well as by thick, strong skin.

oviparous Reproducing by laying eggs. Little or no development occurs within the mother's body; instead, the embryos develop inside the egg; cf ovoviviparous, viviparous.

ovipositor A tubelike organ through which female insects lay their eggs. The stinger of bees and wasps is an modified ovipositor.

ovoviviparous Reproducing by giving birth to live young that have developed from eggs within the mother's body. The eggs may hatch as they are laid or soon after; cf oviparous, viviparous.

pair bond A partnership maintained between a male and a female animal, particularly birds, through one or several breeding attempts. Some species maintain a pair bond for life.

paleontology The scientific study of life in past geological periods.

Pangea Ancient supercontinent in which all the present-day continents were once joined; see also Gondwana, Laurasia.

parallel evolution The situation in which related groups living in isolation develop similar structures to cope with similar evolutionary pressures.

parasitism The situation in which an animal or plant lives and/or feeds on another living animal or plant, sometimes with harmful effects.

passerine Any species of bird belonging to the order Passeriformes. A passerine is often described as a songbird or a perching bird.

pectoral fins In fishes and aquatic mammals, the paired fins attached to each side of the animal and used for lift and control of movement. In fishes, they are usually located just behind the gills.

pedicel The narrow "waist" that connects an insect's head to its thorax, or a spider's cephalothorax to its abdomen.

pedipalp One of a pair of small, leglike organs on the head of insects and the cephalothorax of spiders and scorpions, used for feeling or for handling food. In spiders, they are also used for mating.

pelagic Swimming freely in the open ocean; not associated with the bottom; cf benthic.

pelvic fins Paired fins, located on the lower part of a fish's body.

photophores Luminous organs on some deep-sea fishes.

pheromone A chemical released by an animal that sends a signal and affects the behavior of others of the same species. Many animals use pheromones to attract mates, or to signal danger.

phytoplankton Tiny, single-celled algae that float on or near the surface of the sea.

placental mammal A mammal that does not lay eggs (as monotremes do), or give birth to underdeveloped young (as marsupials do). Instead, it nourishes its developing young inside its body with a blood-rich organ called a placenta.

plankton The plant (phytoplankton) or animal (zooplankton) organisms that float or drift in the open sea. Plankton forms an important link in the food chain.

pollen A dustlike substance produced by male flowers, or by the male organs in a flower, and used in reproduction.

pollination The process by which the pollen produced by the male organs of a flower come into contact with the female parts of the flower, thus fertilizing the flower and enabling seeds to form.

pore A minute opening, as in the skin of an animal or the leaf surface of a plant.

precocial Active and self-reliant at birth. Describes newly hatched chicks of some birds, such as ducks and chickens; cf altricial.

predator An animal that lives mainly by killing and eating other animals.

preen Of a bird, to clean, repair, and arrange plumage.

prehensile Grasping or gripping. Some tree-dwelling mammals and reptiles have prehensile feet or a tail that can be used as an extra limb to help them stay safely in a tree. Elephants have a prehensile "finger" on the end of their trunk so they can pick up small pieces of food. Browsers, such as giraffes, have prehensile lips to help them grip leaves. The prehensile tongue of parrots enables them to extract kernels from shells.

proboscis In insects, a long, tubular mouthpart used for feeding. In some mammals, a proboscis is an elongated nose, snout, or trunk. That of an elephant has many functions, such as smelling, touching, and lifting.

protandrous Of sequential hermaphrodites (as in some fishes), starting life as a male and later becoming female.

protogynous Of sequential hermaphrodites (as in some fishes), starting life as a female and later becoming male.

pupa (pl. pupae) The stage during which an insect transforms from a larva to an adult.

queen A female insect that begins a social insect colony. The queen is normally the only member of the colony that lays eggs.

quill Long, sharp hair of an echidna, porcupine, anteater, and a few other mammals; used for defense.

rain forest A tropical forest that receives at least 100 inches (250 cm) of rain each year. Rain forests are home to a vast number of plant and animal species.

raptor A diurnal bird of prey, such as a hawk or falcon. The term is not used to describe owls.

regurgitate To bring food back up from the stomach to the mouth. Many hoofed mammals use this process to break down their food into a more liquid form. This is called chewing the cud. Birds also regurgitate partially digested food to feed to their chicks.

roost A place or site used by some animals, such as birds and bats, for sleeping. Also, the act of traveling to or gathering at such a place.

refection A practise of some rodents, in which they eat their own fecal matter to gain maximum nutrition from the food, before passing feces as dry pellets.

retractile claws The claws of cats and similar animals that are usually protected in sheaths. Such claws spring out when the animal needs them to capture prey or to fight.

rostrum A tubular, beak-like feeding organ on the head of some insects.

rudimentary Describes a simple, undeveloped, or underdeveloped part of an animal, such as an organ or wing. The rudimentary parts of some modern-day animals are the traces of the functional parts of an early ancestor, but now serve no purpose.

ruminants Hoofed animals—cattle, buffalo, bison, antelopes, gazelles, sheep, goats, and other members of the family Bovidae—with a four-chambered stomach. One of these chambers is the rumen, in which food is fermented by microorganisms before being regurgitated and chewed a second time. This efficient digestive system allows bovids to make the most of low-nutrient foods such as grasses, and to colonize a wide array of habitats.

savanna Open grassland with scattered trees. Most savannas are found in subtropical areas that have a distinct summer wet season.

scales In reptiles, distinct thickened areas of skin that vary in size, from very small to large; in fishes, small plates that form part of the external body covering.

scavenger An animal that eats carrion—often the remains of animals killed by predators.

scutes In a turtle or tortoise, the horny plates that cover the bony shell.

sedentary Having a lifestyle that involves little movement; also used to describe animals that do not migrate.

semi-aquatic Living some of the time in water, and some of the time on land.

silk A strong but elastic substance made by many insects and spiders. Silk is liquid until it leaves the animal's body.

social Living in groups. Social animals can live in breeding pairs, sometimes together with their young, or in a colony or herd of up to thousands of animals.

spawn To release eggs and sperm together directly into the water.

species A group of animals with very similar features that are able to breed together and produce fertile young.

spermatophore A container or package of sperm that is passed from male to female during mating.

spicule A minute, needle-like, siliceous or calcareous body, found in invertebrates.

spinnerets The fingerlike appendages of spiders that are connected to silk glands, found near the tip of the abdomen.

spiracle A small opening that leads into the trachea, or breathing tube, of an insect or spider. In cartilaginous fishes, spiracles are located behind the eyes. They take in water for breathing when the fish is at rest on the bottom, or when the mouth is being used for feeding.

squalene A substance found in the liver oil of deep-sea sharks. It is refined and used as a high-grade machine oil in high-technology industries, as a human health and dietary supplement, and in cosmetics.

stinger A hollow structure on the tail of insects and scorpions that pierces flesh and injects venom.

streamlined Having a smooth body shape to reduce drag, as in seals.

stereoscopic vision Vision in which both eyes face forward, giving an animal two overlapping fields of view and thus allowing it to judge depth.

stridulate To make a sound by scraping objects together. Many insects communicate in this way, some by scraping their legs against their body.

stylet A sharp mouthpart used for piercing plants or animals.

subantarctic Of the oceans and islands just north of Antarctica.

swim bladder A gas-filled, baglike organ in the abdomen of bony fishes. It enables the fishes to remain at a particular depth in the water.

symbiosis An alliance between two species that is usually (but not always) beneficial to both. Animals form symbiotic relationships with plants, microorganisms, and other animals; cf mutualism, parasitism.

sympatric Of two or more species, occurring in the same area.

syndactylic Having fused toes on the hindfoot, as in bandicoots, kangaroos, and wombats.

tadpole The larva of a frog or toad. Tadpoles are aquatic and take in oxygen from the water through gills.

taxonomy The system of classifying living things into various groups and subgroups according to similarities in features and adaptations.

teleosts Bony fishes.

terrestrial Living all or most of the time on land; cf amphibious, aquatic.

temperate Describes an environ-ment or region that has a warm (but not very hot) summer and a cool (but not very cold) winter. Most of the world's temperate regions are located between the tropics and the polar regions.

tentacles On marine invertebrates, long, thin structures used to feel and grasp, or to inject venom. On caecilians, sensory organs located on the sides of the head, possibly used for tasting and smelling.

territory An area of land inhabited by an animal and defended against intruders. The area often contains all the living resources required by the animal, such as food and a nesting or roosting site.

thermal A column of rising air, used by birds to gain height, and on which some birds soar to save energy.

thorax The middle part of an animal's body. In insects, the thorax is divided from the head with a narrow "waist," or pedicel. In spiders, the thorax and head make up a single unit.

torpid In a sleep-like state in which bodily processes are greatly slowed. Torpor helps animals to survive difficult conditions such as cold or lack of food. Estivation and hiber-nation are types of torpor.

trachea A breathing tube in an animal's body. In vertebrates, there is one trachea (or windpipe), through which air passes to the lungs. Insects and some spiders have many small tracheae that spread throughout the body.

tropical Describes an environment or region near the Equator that is warm to hot all year round.

tropical forests Forests growing in tropical regions, such as central Africa, northern South America, and southeast Asia, that experience little difference in temperature throughout the year. See rain forest.

tundra A cold, barren area where much of the soil is frozen and the vegetation consists mainly of mosses, lichens, and other small plants adapted to withstand intense cold. Tundra is found near the Arctic Circle and on mountain tops.

tusks The very long teeth of such mammals as elephants, pigs, hippos, musk deer, walruses, and narwhals; used in fights and for self-defense.

ungulate A large, plant-eating mammal with hoofs. Ungulates include elephants, rhinoceroses, horses, deer, antelope, and wild cattle.

venom Poison injected by animals into a predator or prey through fangs, stingers, spines, or similar structures.

venomous Describes an animal that is able to inject venom into another. Venomous animals usually attack by biting or stinging.

vertebrate An animal with a backbone. All vertebrates have an internal skeleton of cartilage or bone. Fishes, reptiles, birds, amphibians, and mammals are vertebrates.

vertebral column The series of vertebrae running from head to tail along the back of vertebrates, and which enclsoes the spinal cord.

vestigial Relating to an organ that is non-functional or atrophied.

vibrissae Specialized hairs, or whiskers, that are extremely sensitive to touch.

viviparous Reproducing by means of young that develop inside the mother's body and are born live; sometimes called placental viviparity. Most mammals and some fishes (such as sharks) are viviparous.

warm-blooded See endothermic.

worker A social insect that collects food and tends a colony's young, but which usually cannot reproduce.

zooplankton The tiny animals that, together with phytoplankton, form the plankton that drifts on or near the sea's surface. Zooplankton are eaten by some whales, fishes, and seabirds.

zygodactylous Of birds, having two of the four toes pointing forward, and the other two pointing backward.

INDEX

t=top; l=left; r=right; tl=top left; tcl=top center left; tc=top center; tcr=top center right; tr=top right; cl=center left; c=center; cr=center right; b=bottom; bl=bottom left; bcl=bottom center left; bc=bottom center; bcr=bottom center right; br=bottom right

AAP = Australian Associated Press; AFP = Agence France-Presse; APL = Australian Picture Library; APL/CBT = Australian Picture Library/ Corbis ; APL/MP = Australian Picture Library/Minden Pictures; ARL = Ardea London; AUS = Auscape International; BCC = Bruce Coleman Collection; COR = Corel Corp.; GI = Getty Images; IQ3D = imagequestmarine.com; NGS = National Geographic Society; NHPA = Natural History Photographic Agency; NPL=Nature Picture Library; NV = naturalvisions.co.uk; OSF = Oxford Scientific Films; PL = photolibrary.com; SP=Seapics.com; WA = Wildlife Art Ltd.

PHOTOGRAPHS
Front cover Rick Stevens/The Sydney Morning Herald
1 GI; **2-3** GI; **4**c GI; **6**tc APL/ CBT; tl, tr GI; **7**tcl, tcr, tl, tr GI; **8**c GI; **12–13**c GI; **14**bl APL/MP; br, tr APL/CBT; **15**cr PL; tr GI; **16**cr Artville; tr GI; **17**bl, br, tl, tr APL/CBT; cr GI; **22**bl AUS/Reg Morrison; c APL/CBT; tr Queensland Museum; **23**bc GI; c PL; cr NPL/Rachel Hingley; **25**br GI; tl, tr APL/CBT; **26**bl, br, tr APL/CBT; **27**cr APL/CBT; **28**bl, tr APL/ CBT; tl AAP/103; **29**bl APL/CBT; br, c, cl, tc GI; **30**cr, tc APL/ CBT; **31**br PL; cr, tl APL/CBT; tr APL/ MP; **32**bl, t APL/CBT; **33**cl APL/CBT; tl GI; **34**bl, br, c, t APL/ CBT; **35**br GI; t APL/ CBT; **36**l GI; **37**bl, br, tl, tr APL/CBT; **38**t APL/CBT; **39**bl AUS/Martyn Colbeck/OSF; br PL; c AUS/Kitchen & Hurst; tr AUS/OSF; **40**bc GI; br PL; **41**br, tl APL/CBT; **42**bl PL; cl, tr APL/CBT; **43**bl, br, cl, tr APL/CBT; **44**b GI; c Esther Beaton; cl, tr PL; **45**bl, c, cr APL/CBT; tr PL; **46**br, cl, tc APL/CBT; c APL/MP; **47**bl GI; br, cl, tr APL/CBT; **48**bl, cr, tr APL/CBT; br, cl GI; **49**bc, br, tr APL/CBT; **50**b APL/CBT; t GI; **51**bl, cl, cr APL/ CBT; tr GI; **52**bl, br, tr APL/CBT; **53**bc, bl, br, cl, tr APL/CBT; **54**br, cl APL/CBT; tr Esther Beaton; **55**br AAP/AP; Photo/Dita Alangkara; **56**bl, tr AAP Image; c PL; **57**bl, br, tr AAP Image; c PL; **58**bcr, cl, cr, tr APL/CBT; br PL; **59**bl, tc GI; br, c APL/CBT; **60–61**c GI; **62**bc AUS/Michael Maconachie; bl APL/MP; **63**c PL; **64**bl, br, tcr, tl APL/CBT; **65**br, tl PL; c, cr APL/CBT; **66**bc PL; c APL/CBT; tr Kathie Atkinson; **68**bl, cl PL; **79**tr GI; **81**bl, tr APL/CBT; c PL; **84**bc APL/CBT; c PL; **89**br PL; **90**bl GI; cl National Geographic Image Collection; **91**cr PL; **92**cl APL/CBT; tl AUS/D Parker & E Parer-Cook; **98**bl AUS/Daniel Cox/OSF; c AUS/David Haring/OSF; cr AUS/Ferrero-Labat; **99**br, c, tl, tr PL; **100**br AUS/Rod Williams; cl AUS/T-Shivanandappa; r APL/CBT; **106**bl, cr PL; tr APL/CBT; **121**b, tr APL/MP; cl F W Frohawk; cr NHPA/ Mirko Stelzner; **122**bc, c GI; tr APL/MP; **123**c APL/CBT; cr GI; t APL/MP; **124**cl PL; **130**br PL; cl APL/MP; **133**c, tl GI; tr PL; **134**bl, br, c APL/CBT; **139**bl AUS/Daniel Cox/OSF; **142**bc PL; cl APL/MP; **146**tr APL/MP; **147**cr APL/CBT; **148**bl GI; **154**bl GI; **160**tr GI; **161**bc, cl APL/CBT; **162**bl PL; tr APL/MP; **163**bl APL/MP; cr APL/CBT; tl GI; **164**b APL/CBT; c PL; tr GI; **166**c, cl, tr APL/CBT; **168**bc, c APL/CBT; bl GI; **171**cr APL/CBT; **172**bl, c GI; tr PL; **174**c APL/MP; cl PL; **175**br PL; **176**bc APL/CBT; cl APL/MP; **189**br APL/CBT; tc Kirk Olson; tl APL/MP; **190**b PL; c APL/CBT; cl GI; **196**bl, tr APL/CBT; c PL; **197**b GI; **198**bl, br, cl APL/CBT; **200**c, tr GI; **202**c GI; cl APL/MP; **203**c GI; tr APL/MP; **204**bl, tr APL/MP; cl APL/CBT; **205**br Spectrogram Program by Richard Horne (Original from Cornell Laboratory of Orinthology); cr GI; t APL/CBT; **206**bl, c APL/CBT; tr GI; **208**bl COR; **211**br AUS; cr, tl PL; **213**br GI; cl PL; tr APL/MP; **216**b, cl GI; c PL; **217**c, tr PL; tl GI; **218**bl PL; cl GI; **222**c PL; tl APL/CBT; tr BCC; **225**bl APL/CBT; cl, tr PL; **232**bc APL/CBT; tr GI; **239**bc APL/ MP; c APL/CBT; cl GI; **243**cl, cr PL; **244–245**c GI; **246**bc GI; tl APL/CBT; **247**r GI; **248**br, c, cl, tl, tr GI; **249**br PL; cl APL/CBT; tl GI; **250**c PL; cl APL/CBT; **252**br GI; c APL/CBT; **256**c APL/MP; cl PL; **260**bc PL; bl, c APL/MP; **262**bl, c APL/CBT; cl PL; **264**c APL/MP; cl PL; **266**tl, tr APL/CBT; **267**c APL/MP; **268**c, cl GI; **271**br, c PL; **274**bc GI; c, cr PL; **281**cr, tr PL; **283**c, cl PL; **287**bc, br PL; cl GI; **294**c PL; cr GI; **296**bl PL; br APL/CBT; c GI; **301**br, cl APL/CBT; **303**bc APL/CBT; c GI; **306**c, cl APL/CBT; **308**bl ARL/Jean-Paul Ferrero; br APL/CBT; c APL/MP; **312**br APL/MP; **313**br GI; cl APL/ CBT; **314**bl PL; c APL/CBT; **318**bl, br, cr, tl APL/CBT; **319**bl, br, tl, tr APL/MP; **320**c PL; cl APL/MP; **324**br APL/MP; cl APL/CBT; tr GI; **325**bl, c, tr APL/MP; br APL/ CBT; tl PL; **329**tr PL; **336**br GI; tl APL/MP; **342**tr APL/MP; **343**tl, tr APL/MP; **352**tl VIREO/Peter La Tourette; **353**br APL/MP; cl ARL/Alan Greensmith; **354–355**c BCC; **356**bl APL/MP; tl PL; **357**c PL; **358**bc PL; cl APL/MP; cr SP/Doug Perrine; **366**bc, bl GI; br APL/CBT; tl PL; **367**c, cl APL/MP; **370**bc APL/CBT; cr APL/MP; **371**cr PL; **372**bc APL/CBT; cr GI; tl APL/MP; **378**bl, br, c, tr PL; **386**c GI; tr PL; **394**bc APL/MP; bl APL/CBT; **395**br, c PL; t APL/MP; **400**cl APL/MP; tl AUS/Jean-Paul Ferrero; **401**br AUS/Joe McDonald; c GI; tl PL; tr APL/CBT; **407**bc, cl GI; bl APL/MP; cr AUS/Paul de Oliveira/OSF; tl AUS/Glen Threlfo; tr APL/CBT; **416–417**c GI; **418**bl GI; cl PL; r BCC; **420**c, t APL/CBT; **426**cr PL; tl APL/CBT; **427**cr TPL; **428**cl AUS/Satoshi Kuribayashi/ OSF; cr AUS/Kitchin & Hurst; **429**c AUS/Michael Fogden/OSF; r AUS/ Stephen Dalton/OSF; tl GI; **430**bl, r MG; **435**b, c APL/CBT; cl AUS/Kathie Atkinson; tl AUS/ Jean-Paul Ferrero; **441**cl AUS/Michael Fogden/OSF; cr APL/CBT; tl APL/MP; **448–449**c GI;

450bc MG; cl GI; **451**c GI; **452**tl PL; **453**c NPL/Reijo Juurinen/ Naturbild; **454**cl PL; tr APL/CBT; **455**br SP/Doug Perrine; c SP/Phillip Colla; **456**c AFP; cl PL; **460**bc, bl, tl PL; c SP/Ben Cropp Productions; **462**bc APL/CBT; br PL; cl GI; **466**bc AUS/Kevin Deacon; cl AUS/Alby Ziebell; **467**cr SP/Richard Herrman; tl AUS/Kevin Deacon; tr APL/CBT; **468**cl PL; cr SP/Mark V Erdmann; **469**c AUS/Tom & Therisa Stack; cl PL; **471**cl APL/CBT; cr SP/Masa Ushioda; **472** PL; cl NV/Heather Angel; **474**c PL; **477**tl PL; tr NPL/Tim Martin; **478**br GI; c IQ3D/Masa Ushioda; **480**bc, br PL; cl APL/CBT; **486**bl PL; c SP/Mark Conlin; cl APL; **491**cl AUS/Paulo De Oliveira/OSF; **492**br Picture taken with the ROV Victor 6000 of Ifremer, at 2500 meter depth, copyright Ifremer/biozaire2-2001; cl Kevin Deacon; **493**bl John E. Randall; cl PL; **494**c Carlos Ivan Garces del Cid & Gerardo Garcia; **495**c GI; **498**bl, br PL; cl GI; **514–515**c APL/MP; **516**bc GI; l APL/MP; **517**c APL/MP; **518**tcl AUS/Jean-Paul Ferrero; tr APL/MP; **519**c, tr APL/CBT; **520**c APL/MP; tr GI; **523**l GI; tc PL; **524**c PL; tr GI; cl PL; **532**c APL/CBT; tr GI; **533**cl, cr, tr APL/CBT; tl PL; **534**bc APL/MP; c APL/CBT; tr PL; **535**c GI; c PL; tl APL/CBT; **536**bc GI; cl, tr APL/CBT; **540**bl AUS/ Jean-Paul Ferrero; c PL; **544**c, cr APL/CBT; **545**c PL; tr APL/CBT; **546**c, cr APL/CBT; **552**c PL; tl AUS/Andrew Henely; tr PL; **553**bc APL/MP; tc GI; tl APL/CBT; tr APL/MP; **554**c GI; cl PL; cl PL; tr APL/MP; **555**cl APL/CBT; cr, tr PL; **556**c AUS/Kathie Atkinson; **557**cl PL; tr GI; **558**c GI; tr APL/MP; **559**c PL; cl GI; tr AUS/John Cancalosi; **562**c Austral; tl PL; tr APL/CBT; **563**cl PL; tr APL/CBT; **567**cr PL; **570**c PL; tr GI; **574**c APL/MP; tr GI; **576**tl PL; tr PL; **579**c NHPA; tr AUS/Pascal Goetgheluck; **582**tr APL/CBT; **583**cr GI; cr, APL/MP; **585**c GI; cl PL; tr AUS/Karen Gowlett-Holmes/OSF.

ILLUSTRATIONS
All illustrations © MagicGroup s.r.o. (Czech Republic) – www.magicgroup.cz:
Pavel Dvorský, Eva Göndörová, Petr Hloušek, Pavla Hochmanová, Jan Hošek, Jaromír a Libuše Knotkovi, Milada Kudrnová, Petr Liška, Jan Maget, Vlasta Matoušová, Jiří Moravec, Pavel Procházka, Petr Rob, Přemysl Vranovský, Lenka Vybíralová; except for the following:
Susanna Addario 576bl; **Alistair Barnard** 28br, 67br; **Alistair Barnard/Frank Knight/John Bull** 426bl; **Andre Boos** 150cl, 151br; **Anne Bowman** 20br, 205bl, 367bc, 539br, 558tl; **Peter Bull Art Studio** 62tl; **Martin Camm** 106tr, 159bl br, 211c, 455tr, 458bl, 460br; **Andrew Davies/Creative Communication** 15tl, 16br, 24c, 55tr, 188tl; **Kevin Deacon** 455tl, 562b; **Simone End** 19cr, 21bl c; **Christer Eriksson** 80cl, 215br, 365br, 368bl, 373br, 376cl; **Alan Ewart** 541cr; **John Francis/Bernard Thornton Artists UK** 380bl, 395c cl, 338bl; **Giuliano Fornari** 540cr, 454b, 457br, 567cr, 573cr; **Jon Gittoes** 33r, 110bl, 118b; **Mike Golding** 530cl; **Ray Grinaway** 21tr, 552r; **Gino Hasler** 209br, 249c cr tr, 456bl br, 459br, 460tr; **Robert Hynes** 371c, 557br; **Ian Jackson** 556tr; **Frits Jan Maas** 273br; **David Kirshner** 18bl, c, 19bl br, 20bl cl, 35cr, 55cr, 68bc, 102cl, 109br, 132bl, 204c, 246cc, 257br, 295br, 300bl, 303cl, 327cr, 330bl, 334bl, 335br cr, 344cl, 345tr, 346cl, 347tr, 349cr, 351br, 356c, 358bl, 360bl, 363br, 369br, 370bl br, 371bl, 373cr, 375cr tr, 377br, 383cr, 384tl, 393br, 394c, 395tl, 399cr, 401cr, 409br, 413cr, 415cr, 426tr, 427cl, 436bl, 437br, 441tr, 450tl, 452tr, 467b, 474cr, 489br, 511br, 553; **David Kirshner/John Bull** 443bl; **Frank Knight** 125cr, 131br, 153br, 162br, 163br, 359br, 398cl; **Alex Lavroff** 211tr, 526tcr; **John Mac/FOLIO** 97br, 194bl tl, 248bl, 250br, 287bl; **Robert Mancini** 21cl, 337tr, 258bl, 540cr, c, 541tr; **Map Illustrations** 40cl; **Map Illustrations & Andrew Davies/Creative Communication** 50cl, 356bl, 366c; **James McKinnon** 222b, 235br, 370c; **Karel Mauer** 290bl, 344tl; **Erik van Ommen** 302bl, 307br, 340bl; **Tony Pyrzakowski** 274bl, 551cr; **John Richards** 576cr; **Edwina Riddell** 576c, 231br, 242bl, 359br, 364bl, 372cr tr, 374cl, 382bl, 393tr, 396bl cl; **Barbara Rodanska** 20cr, 79bl cl tl, 298bl, 300bl; **Trevor Ruth** 188b, 412bl; **Peter Schouten** 385br; **Kevin Stead** 66bl, 120cl, 204cl, 533c, 562cl, 578bl; **Roger Swainston** 461br cr, 462bl; **Guy Troughton** 15br, 30bl, 79bc c cr, 92b, 114b cr, 128bl, 133b, 152t, 160b, 161t, 175t, 182t, 266bl, 281c, 292bl, 336c, 342c, 352c, 386l, 400br, 452b, 468c, 477b; **Trevor Weekes** 315br; **Rod Westblade** 458bl; **WA/Priscilla Barret** 32br, 275br **WA/B. Croucher** 312l; **WA/Sandra Doyle** 21tl, 296bc, 297cr, 536c, 538bcl bl, 541br; **WA/Phil Hood** 27br; **WA/Ken Oliver** 380tl, 420bl, 428bl; **WA/Mick Posen** 22br, 23tr, 275br; **WA/Peter Scott** 19cl; **WA/Chris Shields** 539tr, cr; **WA/Chris Turnbull** 20c, 465cr, 561bl; **WA** 207br, 212bl, 464cl, 465br, 476bl, 481cl, 501br, 502bl, 513br.

MAPS/GRAPHICS
All maps by **Andrew Davies/Creative Communication**, except for those appearing on 450–499 by **Brian Johnston**.

INDEX
Tonia Johansen/Johansen Indexing Services.

The publishers wish to thank Brendan Cotter, Helen Flint, Frankfurt Zoological Society, Paul McNally, Kathryn Morgan, and Tanzania National Parks for their assistance in the preparation of this volume.